Methods in Enzymology

Volume 185
GENE EXPRESSION TECHNOLOGY

METHODS IN ENZYMOLOGY

EDITORS-IN-CHIEF

John N. Abelson   Melvin I. Simon

DIVISION OF BIOLOGY
CALIFORNIA INSTITUTE OF TECHNOLOGY
PASADENA, CALIFORNIA

FOUNDING EDITORS

Sidney P. Colowick and Nathan O. Kaplan

Methods in Enzymology

Volume 185

Gene Expression Technology

EDITED BY

David V. Goeddel

MOLECULAR BIOLOGY DEPARTMENT
GENENTECH INC.
SOUTH SAN FRANCISCO, CALIFORNIA

ACADEMIC PRESS, INC.
Harcourt Brace Jovanovich, Publishers
San Diego   New York   Boston
London   Sydney   Tokyo   Toronto

ACADEMIC PRESS, INC.
San Diego, California 92101

United Kingdom Edition published by
ACADEMIC PRESS LIMITED
24-28 Oval Road, London NW1 7DX

LIBRARY OF CONGRESS CATALOG CARD NUMBER: 54-9110

ISBN   0-12-182086-6   (alk. paper)

PRINTED IN THE UNITED STATES OF AMERICA
90 91 92 93   9 8 7 6 5 4 3 2 1

Table of Contents

Section I. Introduction

Section II. Expression in *Escherichia coli*

A. Vectors

B. Promoters

C. Translation

Section IV. Expression in Yeast

E. Proteases

F. Protein Secretion and Modification

Section V. Expression in Mammalian Cells

A. Vectors

B. Transfection Methods

C. Markers for Selection and Amplification

D. Growth of Cell Lines

E. Posttranslation Processing, Modification, and Secretion

Section VI. Mutagenesis

Contributors to Volume 185

Article numbers are in parentheses following the names of contributors.
Affiliations listed are current.

LARS ABRAHMSÉN (13), *Department of Biomolecular Chemistry, Genentech Inc., South San Francisco, California 94080*

PAULINA BALBAS (3), *Departamento de Biologia Molecular, Centro de Investigacion sobre Ingenieria Genética y Biotechnologia, Universidad Nacional Autonoma de Mexico, Cuernavaca, Morelos, Mexico*

CLINTON E. BALLOU (36), *Department of Biochemistry and Molecular Biology, University of California at Berkeley, Berkeley, California 94720*

RAMA M. BELAGAJE (8), *Division of Molecular and Cell Biology, Lilly Research Laboratories, Indianapolis, Indiana 46285*

ANNE BELL (29), *Zymo Genetics Inc., Seattle, Washington 98105*

FRANCISCO BOLIVAR (3), *Departamento de Biologia Molecular, Centro de Investigacion sobre Ingenieria Genética y Biotechnologia, Universidad Nacional Autonoma de Mexico, Cuernavaca, Morelos, Mexico*

DAVID BOTSTEIN (23), *Genentech Inc., South San Francisco, California 94080*

RALPH A. BRADSHAW (33), *Department of Biological Chemistry, College of Medicine, University of California at Irvine, Irvine, California 92717*

ANTHONY J. BRAKE (34), *Chiron Corporation, Emeryville, California 94608*

JAMES R. BROACH (22), *Department of Biology, Princeton University, Princeton, New Jersey 08544*

COLIN R. CAMPBELL (41), *Department of Genetics, University of Illinois College of Medicine, Chicago, Illinois 60612*

DAVID F. CARMICHAEL (16), *Synergen Inc., Boulder, Colorado 80303*

A. CHAMBERS (27), *Department of Biochemistry, University of Oxford, Oxford OX1 3QU, England*

CHUNG NAN CHANG (35), *Protein Design Laboratories, Palo Alto, California 94304*

CHRISTINA Y. CHEN (35, 37), *Department of Cell Genetics, Genentech Inc., South San Francisco, California 94080*

VANESSA CHISHOLM (35, 37), *Department of Cell Genetics, Genentech Inc., South San Francisco, California 94080*

A. MARK CIGAN (30), *Laboratory of Molecular Genetics, National Institutes of Health, Bethesda, Maryland 20892*

WILLIAM CLEVENGER (25), *Department of Molecular Biology, Immunex Corporation, Seattle, Washington 98101*

DORIS COIT (28), *Chiron Corporation, Emeryville, California 94608*

D. COUSENS (27), *Wellcome, Beckenham, Kent, England*

HERMAN A. DE BOER (9), *Gorlaeus Laboratories, University of Leiden, 2300 RA Leiden, The Netherlands*

DANIEL DIMAIO (45), *Department of Human Genetics, Yale University School of Medicine, New Haven, Connecticut 06510*

THOMAS F. DONAHUE (30), *Department of Biology, Indiana University, Bloomington, Indiana 47405*

ANDREW J. DORNER (44), *Department of Molecular and Cellular Genetics, Genetics Institute Inc., Cambridge, Massachusetts 02140*

DONALD J. DOWBENKO (35), *Department of Molecular Immunology, Genentech Inc., South San Francisco, California 94080*

JOHN W. DUBENDORFF (6), *Department of Biology, Brookhaven National Laboratory, Upton, New York 11973*

JOHN J. DUNN (6), *Department of Biology, Brookhaven National Laboratory, Upton, New York 11973*

SCOTT D. EMR (21), *Division of Biology, California Institute of Technology, Pasadena, California 91125*

TINA ETCHEVERRY (26), *Department of Fermentation Research, Genentech Inc., South San Francisco, California 94080*

DAVID V. GOEDDEL (1), *Molecular Biology Department, Genentech Inc., South San Francisco, California 94080*

LARRY GOLD (2, 7), *Department of Molecular, Cellular and Developmental Biology, University of Colorado at Boulder, Boulder, Colorado 80309*

SUSAN GOTTESMAN (11), *Laboratory of Molecular Biology, National Cancer Institute, National Institutes of Health, Bethesda, Maryland 20892*

JAY D. GRALLA (4), *Department of Chemistry and Biochemistry and the Molecular Biology Institute, University of California at Los Angeles, West Los Angeles, California 90024*

FRANCIS J. GRANT (29), *Zymo Genetics Inc., Seattle, Washington 98105*

ROBERT HAMILTON (35), *Department of Fermentation Research and Process Development, Genentech Inc., South San Francisco, California 94080*

DENNIS J. HENNER (5, 17, 20), *Department of Genetics, Genentech Inc., South San Francisco, California 94080*

RONALD A. HITZEMAN (35, 37), *Department of Cell Genetics, Genentech Inc., South San Francisco, California 94080*

KATHRYN J. HOFMANN (24), *Department of Virus and Cell Biology, Merck Sharp and Dohme Research Laboratories, West Point, Pennsylvania 19486*

JAMES E. HOPPER (24), *Department of Biological Chemistry, The Milton S. Hershey Medical Center, Pennsylvania State University, Hershey, Pennsylvania 17033*

BRUCE H. HORWITZ (45), *Department of Human Genetics, Yale University School of Medicine, New Haven, Connecticut 06510*

ANNA S. HUI (9), *Department of Pharmacology and Cell Biophysics, University of Cincinnati, Cincinnati, Ohio 45267*

ELIZABETH W. JONES (31), *Department of Biological Sciences, Carnegie Mellon University, Pittsburgh, Pennsylvania 15213*

RANDAL J. KAUFMAN (39, 42, 44), *Department of Molecular and Cellular Genetics, Genetics Institute Inc., Cambridge, Massachusetts 02140*

KIMBERLY KELSAY (29), *Department of Biochemistry, University of British Columbia, Vancouver, British Columbia V6K 157, Canada*

RICHARD L. KENDALL (33), *Department of Biological Chemistry, College of Medicine, University of California at Irvine, Irvine, California 92717*

WAYNE A. KEOWN (41), *Department of Genetics, University of Illinois College of Medicine, Chicago, Illinois 60612*

A. J. KINGSMAN (27), *Department of Biochemistry, University of Oxford, Oxford OX1 3QU, England*

S. M. KINGSMAN (27), *Department of Biochemistry, University of Oxford, Oxford OX1 3QU, England*

TADAHIKO KOHNO (16), *Synergen Inc., Boulder, Colorado 80303*

WILLIAM J. KOHR (35), *Department of Protein Chemistry, Genentech Inc., South San Francisco, California 94080*

MICHAEL KRIEGLER (40), *Cetus Corporation, Emeryville, California 94608*

RAJU S. KUCHERLAPATI (41), *Department of Genetics, University of Illinois College of Medicine, Chicago, Illinois 60612*

STUART F. J. LE GRICE (18), *Department of Infectious Diseases, Case Western Reserve University School of Medicine, Cleveland, Ohio 44106*

ARTHUR D. LEVINSON (38), *Department of Cell Genetics, Genentech Inc., South San Francisco, California 94080*

CHUNG LIU (35), *Department of Protein Chemistry, Genentech Inc., South San Francisco, California 94080*

HONG MA (23), *Division of Biology, California Institute of Technology, Pasadena, California 91125*

VIVIAN L. MACKAY (29), *Zymo Genetics Inc., Seattle, Washington 98105*

JENNIE P. MATHER (43), *Department of Cell Culture Research and Development, Genentech Inc., South San Francisco, California 94080*

TOMAS MOKS (12), *Department of Biochemistry, The Royal Institute of Technology, S-100 44 Stockholm, Sweden*

LAWRENCE M. MYLIN (24), *Department of Biological Chemistry, The Milton S. Hershey Medical Center, Pennsylvania State University, Hershey, Pennsylvania 17033*

VASANTHA NAGARAJAN (19), *Central Research and Development Department, E. I. du Pont de Nemours and Co., Wilmington, Delaware 19880*

BJÖRN NILSSON (13), *Department of Biochemistry and Biotechnology, Royal Institute of Technology, S-100 44 Stockholm, Sweden*

PETER O. OLINS (10), *Biological Sciences Department, Monsanto Company, St. Louis, Missouri 63198*

ROGER PAI (35), *Department of Product Recovery, Genentech Inc., South San Francisco, California 94080*

VIRGINIA L. PRICE (25), *Department of Molecular Biology, Immunex Corporation, Seattle, Washington 98101*

SHAUKAT H. RANGWALA (10), *Biological Sciences Department, Monsanto Company, St. Louis, Missouri 63198*

MARK E. RENZ (35), *Department of Developmental Biology, Genentech Inc., South San Francisco, California 94080*

ALAN B. ROSE (22), *Department of Biology, Princeton University, Princeton, New Jersey 08544*

ALAN H. ROSENBERG (6), *Department of Biology, Brookhaven National Laboratory, Upton, New York 11973*

STEVEN ROSENBERG (28), *Protos Corporation, Emeryville, California 94608*

BRIGITTE E. SCHONER (8), *Division of Molecular and Cell Biology, Lilly Research Laboratories, Indianapolis, Indiana 46285*

RONALD G. SCHONER (8), *Division of Molecular and Cell Biology, Lilly Research Laboratories, Indianapolis, Indiana 46285*

LOREN D. SCHULTZ (24), *Department of Virus and Cell Biology, Merck Sharp & Dohme Research Laboratories, West Point, Pennsylvania 19486*

ANDRZEJ Z. SLEDZIEWSKI (29), *Zymo Genetics Inc., Seattle, Washington 98105*

THOMAS J. SILHAVY (15), *Department of Biology, Princeton University, Princeton, New Jersey 08544*

NANCY J. SIMPSON (35, 37), *Department of Cell Genetics, Genentech Inc., South San Francisco, California 94080*

ARJUN SINGH (35), *Department of Cell Genetics, Genentech Inc., South San Francisco, California 94080*

ANDREAS SOMMER (16), *Biogrowth Inc., Richmond, California 94806*

JOAN A. STADER (15), *School of Basic Life Sciences, University of Missouri, Kansas City, Missouri 64110*

C. A. STANWAY (27), *Institute of Molecular Medicine, Headington, Oxford, England*

TIM STEARNS (23), *Department of Biochemistry and Biophysics, University of California, San Francisco, San Francisco, California 94143*

GARY D. STORMO (7), *Department of Molecular, Cellular and Developmental Biology, University of Colorado at Boulder, Boulder, Colorado 80309*

F. WILLIAM STUDIER (6), *Department of Biology, Brookhaven National Laboratory, Upton, New York 11973*

WAYNE E. TAYLOR (25), *Department of Chemistry and Biochemistry, California State University, Fullerton, California 92634*

PATRICIA TEKAMP-OLSON (28), *Chiron Corporation, Emeryville, California 94608*

ROBERT C. THOMPSON (16), *Synergen Inc., Boulder, Colorado 80301*

MATHIAS UHLÉN (12), *Department of Biochemistry, The Royal Institute of Technology, S-100 44 Stockholm, Sweden*

KEITH D. WILKINSON (32), *Department of Biochemistry, Emory University School of Medicine, Atlanta, Georgia 30322*

M. WILSON (27), *Delta Biotechnology Nottingham, England*

MARLIS WORTHINGTON (25), *Department of Biochemistry, University of Washington, Seattle, Washington 98195*

RYO YAMADA (33), *Department of Biological Chemistry, College of Medicine, University of California at Irvine, Irvine, California 92717*

DANIEL G. YANSURA (5, 14), *Department of Cell Genetics, Genentech Inc., South San Francisco, California 94080*

CARLI YIP (29), *Zymo Genetics Inc., Seattle, Washington 98105*

ELTON T. YOUNG (25), *Department of Biochemistry, University of Washington, Seattle, Washington 98195*

Preface

The articles in this volume were assembled to enable the reader to design effective strategies for the expression of cloned genes and cDNAs. More than a compilation of papers describing the multitude of techniques now available for expressing cloned genes, this volume provides a manual that should prove useful for solving the majority of expression problems one is likely to encounter.

The four major expression systems commonly available to most investigators are stressed: *Escherichia coli, Bacillus subtilis,* yeast, and mammalian cells. Each of these systems has its advantages and disadvantages, details of which are found in Chapter [1] and the strategic overviews for the four major sections of the volume. The papers in each of these sections provide many suggestions on how to proceed if initial expression levels are not sufficient.

DAVID V. GOEDDEL

METHODS IN ENZYMOLOGY

VOLUME 80. Proteolytic Enzymes (Part C)
Edited by LASZLO LORAND

VOLUME 81. Biomembranes (Part H: Visual Pigments and Purple Membranes, I)
Edited by LESTER PACKER

VOLUME 82. Structural and Contractile Proteins (Part A: Extracellular Matrix)
Edited by LEON W. CUNNINGHAM AND DIXIE W. FREDERIKSEN

VOLUME 83. Complex Carbohydrates (Part D)
Edited by VICTOR GINSBURG

VOLUME 84. Immunochemical Techniques (Part D: Selected Immunoassays)
Edited by JOHN J. LANGONE AND HELEN VAN VUNAKIS

VOLUME 85. Structural and Contractile Proteins (Part B: The Contractile Apparatus and the Cytoskeleton)
Edited by DIXIE W. FREDERIKSEN AND LEON W. CUNNINGHAM

VOLUME 86. Prostaglandins and Arachidonate Metabolites
Edited by WILLIAM E. M. LANDS AND WILLIAM L. SMITH

VOLUME 87. Enzyme Kinetics and Mechanism (Part C: Intermediates, Stereochemistry, and Rate Studies)
Edited by DANIEL L. PURICH

VOLUME 88. Biomembranes (Part I: Visual Pigments and Purple Membranes, II)
Edited by LESTER PACKER

VOLUME 89. Carbohydrate Metabolism (Part D)
Edited by WILLIS A. WOOD

VOLUME 90. Carbohydrate Metabolism (Part E)
Edited by WILLIS A. WOOD

VOLUME 91. Enzyme Structure (Part I)
Edited by C. H. W. HIRS AND SERGE N. TIMASHEFF

Section I

Introduction

[1] Systems for Heterologous Gene Expression

By DAVID V. GOEDDEL

This volume concentrates on the four major expression systems that are commonly available to most investigators: *Escherichia coli, Bacillus subtilis,* yeast, and mammalian cells. Each of these systems has its advantages and disadvantages, some of which I briefly outline in this introductory article.

One expression system which is not covered in this volume is the baculovirus system in insect cells since it has been primarily developed by one group, and a detailed manual describing the system is available from that laboratory.[1] There are also examples in the literature of heterologous gene expression in a variety of other organisms, including *Streptomyces,* a number of fungi, pseudomonads, and others. Although these systems might become relatively more important in the future, we feel their utility has not at present been demonstrated broadly enough to recommend their routine use for heterologous protein production.

In general, the expression of each cDNA or gene presents its own peculiar set of problems that must be overcome to achieve high-level expression. The synthesis of foreign proteins is still largely empirical. There is no set of hard-and-fast rules to follow. In fact, a particular protein is almost as likely to be the exception as it is to follow any set of rules. Keeping this caveat in mind, I will make some generalizations which I hope will aid the reader in selecting an initial expression system.

Types of Proteins to Be Expressed

For the purpose of gene expression in heterologous cells, proteins can be arbitrarily grouped into four broad classes. The first class covers small (less than ~80 amino acids) peptides. These are most easily expressed as fusion proteins, usually in *E. coli.* The second class are polypeptides that are normally secreted proteins (e.g., enzymes, cytokines, hormones) and range in size from about 80 to 500 amino acids. This class of proteins is often the most straightforward to express (in all four systems) and secretion or direct expression should be considered the method of choice. In particular, direct expression in *E. coli* has proved extremely effective for the subset of proteins in the 100–200 amino acid size range. A third class consists of very large (greater than about 500 amino acids) secreted proteins and cell

---

[1] V. A. Lukow and M. D. Summers, *Bio/Technology* **6,** 47 (1988).

surface receptor proteins. Unless the protein of interest is of microbial origin, one generally has the most success with this class using a mammalian cell expression system. The fourth group encompasses all nonsecreted proteins larger than ~80 amino acids. Many proteins fall into this class; however, relative to secreted proteins, much less work has been directed toward overexpression of proteins in this category. Therefore, the selection of an appropriate expression system should be based on the intended use for the protein (see below).

For What Purpose Is the Expressed Protein Needed?

Probably the most common reason for expressing a new gene or cDNA has been to verify that the correct sequence has, in fact, been isolated. If verification is all that is required, and if the protein being expressed is from a higher eukaryote, then transient expression in mammalian cells is the preferred expression route. Transient expression is not only easy to perform, but also gives answers quickly and has a very high probability of yielding a biologically active protein. Historically, *E. coli* has also been used with great success for identification of mammalian proteins, often yielding biologically active protein, and, nearly always, immunoreactive material. For verification of microbial proteins, the generalization "make yeast proteins in yeast and bacterial proteins in bacteria" can be made.

A relatively large amount of substantially pure protein is desirable for *in vivo* biological experiments or structural determination. In this case, it is worth taking the time and effort to evaluate more than one expression system to find the optimal one for that protein. Bacteria, yeast, and mammalian cells have all been used to produce clinical grade proteins in large amounts.

Escherichia coli can be considered a good bet for the preparation of proteins, or portions thereof, to be used for the generation of antibodies. It is usually possible to express large quantities of antigen rapidly in *E. coli,* even though the overexpressed protein may not be correctly folded. In Section II of this volume there are articles on *E. coli* that outline, in addition to the old and still reliable *trp* vector systems, various novel methods for the direct and fusion expression of polypeptides suitable for antibody preparation.

An increasingly common need for the expression of cloned genes is to obtain mutagenized protein for structure–function experiments (see [45] in this volume). The seiection of the appropriate expression system for such studies is usually based on the ease of generating many defined mutants in an assayable form. Therefore, *E. coli* is an obvious first choice.

However, if *E. coli* does not yield protein suitable for analysis, other systems must be examined. Obviously, if posttranslational modification contributes to the properties being studied, or if one is looking for phenotypic changes in the host cells, the selection of an expression system should mimic the natural source of the protein as closely as possible.

Escherichia coli Expression Systems

Considerable lore has accumulated over the years concerning *E. coli* expression. It is often assumed that any protein can be produced in *E. coli* as long as it is not too small, too large, too hydrophobic, and does not contain too many cysteines. To some extent, these generalizations are correct; if one wants to express Factor VIII or a complex mammalian cell surface receptor containing 40 disulfides, *E. coli* is not the proper host. However, *E. coli* is a suitable, and often desirable host, for a large number of other expression needs. *Escherichia coli* systems for direct expression, fusion protein expression, and secretion, each of which has its particular advantages (and disadvantages), have been developed and refined. While it might be difficult to predict *a priori* the relative difficulty of purification for proteins produced by these three methods, there are usually other reasons for selecting one approach over the others.

Fusion protein expression strategies ensure good translation initiation and often permit one to overcome the instability problems that can be encountered with small peptides. Furthermore, the high degree of certainty of expression by the fusion approach makes this a preferred method for generating immunogens. The fusion methods described in this volume also have purification advantages built into them, including insolubility of the fusion protein and purification protocols tailored to the fusion partner. Direct expression is usually the ideal method for *E. coli* production of a heterologous protein of 100–300 amino acids, assuming there is not an inordinate number of cysteines.

The reducing environment present in *E. coli* does not permit cysteine-rich proteins to form the disulfide bonds required for proper conformation. This problem can sometimes be overcome by secretion into a more oxidized environment. However, secretion from *E. coli* is still largely a hit-or-miss proposition and is most likely to work with a naturally secreted protein (as opposed to trying to engineer the secretion of an intracellular protein). There are many beneficial features to secretion: Problems with unwanted N-terminal methionines are avoided, proteins tend to fold better than during intracellular expression, and protein purification can often be simplified.

Bacillus subtilis and Yeast Expression Systems

Although *B. subtilis* and yeast are not usually thought of as first choices for heterologous gene expression, there are advantages of each that make them good candidates for specific expression needs. These advantages are discussed by Henner [17] and Emr [21] elsewhere in this volume. In particular, *B. subtilis* is a good choice for obtaining the secretion of a prokaryotic protein. Yeast offer many important advantages of both prokaryotic and eukaryotic systems. Yeast grow rapidly, achieve high cell densities, and are able to make posttranslational modifications that bacterial systems are unable to perform. Many advances in yeast expression have helped place yeast on a more equal footing with *E. coli* as an initial choice for an expression system.

Mammalian Cell Expression Systems

Progress in mammalian cell expression systems has been nothing short of astonishing in the last few years. Transient expression systems, which initially permitted the verification that a desired gene had been cloned, have become increasingly popular in a wide range of cDNA cloning strategies. Furthermore, transient expression assays allow one to evaluate rapidly the relative merits of different vectors for subsequent use in stable expression systems. Stable mammalian cell expression is more time-consuming than any of the other systems described in this volume; however, stable systems are essential for large-scale protein production in mammalian cells. The time and effort required for selection, amplification, and growth of stable cell lines are more than offset by the great versatility of mammalian cells in producing all of the different classes of protein described above; there are virtually no examples of proteins that cannot be made in at least detectable levels in mammalian cells.

Synthetic DNA

It is important to emphasize the benefits that can be gained by taking full advantage of the power of synthetic DNA. In making expression constructions, one should not be constrained by the DNA sequence or restriction sites present in the DNA to be expressed. Codon usage, potential RNA secondary structure, and ribosome-binding-site sequences can be changed quite easily by the use of synthetic DNA, and the potential value of such alterations at improving expression should not be underestimated. It is worth considering as many variables as possible in designing an optimal expression vector for the gene of interest. If the existing sequences cannot be easily manipulated, replace and/or alter with synthetic DNA as much sequence as is necessary to achieve the "idealized" construction.

Future Developments

It is interesting to try to predict what surprises the next 5–10 years might hold for the heterologous expression of cloned genes. It is likely that there will be continued improvements made in the four major expression systems covered by this volume. Perhaps microbial hosts will be engineered so that they are able to perform specific mammalian posttranslational modifications such as the addition of complex sugars. It is even conceivable that each of the four host systems described here might be sufficient for any expression need, thereby eliminating the necessity of having access to more than one system. And finally, the recent demonstrations that heterologous proteins can be produced in the milk of transgenic animals makes it intriguing to speculate that even this technology might become an important component of the repertoire of a typical gene expression laboratory.

Section II

Expression in *Escherichia coli*
Article 2

A. Vectors
Article 3

B. Promoters
Articles 4 through 6

C. Translation
Articles 7 through 10

D. Proteases
Article 11

E. Gene Fusions
Articles 12 through 14

F. Secretion
Article 15

G. Protein Refolding
Article 16

[2] Expression of Heterologous Proteins in *Escherichia coli*

By LARRY GOLD

Heterologous genes must be expressed at a very high level for basic research and for production of useful biologicals. The heterologous proteins then are purified for either purpose, and thus the most important parameter is the fractional abundance of the protein at the time the culture is harvested. Different fermentation schemes can achieve identical fractional abundances. One might induce the heterologous gene fully at some late stage of the fermentation, obtain a high fractional synthetic rate for that protein, and harvest when the fractional abundance of the protein is maximal. Conversely, one might arrange an expression module that yields a reasonable fractional synthetic rate throughout the growth period, and harvest the culture at the highest possible cell concentration; if the heterologous protein is stable, the fractional synthetic rate during the entire growth period equals the fractional abundance of the desired protein at harvest. The second method depends strongly on the impact on cell growth of any protein whose fractional synthetic rate is high, as well as any specific inhibitory activities of the heterologous protein. *Escherichia coli* can be grown to very high cell densities; thus for fractional abundances of around 10%, corresponding to a concentration twice that of the most abundant bacterial protein in *E. coli* (for cells grown aerobically in rich media), one should expect a yield of more than 1 g of heterologous protein per liter of culture.

One can obtain a high fractional synthetic rate for any heterologous protein in *E. coli*. The investigator can, in a cookbook fashion, arrange plasmid constructs that give high instantaneous synthetic rates for the protein of interest. The difficult problems arise when a high fractional synthetic rate does not lead directly to high fractional abundance (that is, concentration) of the desired protein at the time the fermentation is harvested. The heterologous expression problem should be divided into a small number of subproblems. Initially, the investigator decides if the heterologous protein is to be secreted or not, and if the protein will be fused to an amino-terminal fragment; fusions relate to the issues of secretion, the redox state of the cysteines during culture growth, and subsequent protein purification. Next, one decides on the plasmid, promoter, and translation initiation region, usually choosing a stable, high-copy-number plasmid, a strong and regulated promotor, and a strong translation initiation region. These decisions relate to the desired form of the fermentation. The most ferocious expression systems (including those driven by bacteriophage T7 promoters) kill cells, and thus require induction for a time that is designed to maximize the final concentration of heterologous protein in an extract. Less ferocious expression systems, usually driven by weaker promoters and translation initiation domains, can be used to express the heterologous protein continuously. In either case, most genes (from which the heterologous protein is neither degraded nor toxic to the cell) should yield 10 to 20% of the heterologous protein as a final concentration at harvest time.

Many stable, high-copy-number plasmids have been described. Similarly, many strong promoters for *E. coli* RNA polymerase have been described; the bacteriophage T7 promoters, driven by T7 RNA polymerase, yield induced cells containing only mRNAs corresponding to genes placed downstream from the T7 promoter. High gene dosage and rampant transcription after induction are portable to any heterologous gene. However, translation initiation domains are not portable in a strict sense; the problem of intramolecular secondary structures caused by restricted nucleotides immediately 3' to the initiation codon (for a heterologous protein of fixed amino-terminal protein sequence) is not solved by a single sequence for translation initiation. Nevertheless, translation initiation can always be arranged so as to yield high fractional expression of the heterologous protein. For a gene cloned downstream from a T7 promoter and containing a good translation initiation region that is not occluded in a stable RNA secondary structure, the fractional expression of the heterologous protein after induction of the T7 RNA polymerase approaches one!

Proteolysis of heterologous proteins is a problem, although frequently proteolysis is slow enough to allow the accumulation of a high fraction of undegraded protein. Proteolysis of heterologous proteins in bacteria or even lower eukaryotes is conceptually like a primitive immune system, and as such deserves far more experimental work; the mechanisms by which *E. coli* recognizes "self" are completely unknown. No generic method exists for blocking all host proteases, although some work has been done with a bacteriophage T4 gene (called *pin*) that may be useful in blocking some proteases. Since bacteriophages must confront the primitive immune system of the host, and since many bacteriophages have a wide host range limited only by the availability of a specific receptor on the surface of the organism, the protease problem intrinsic to production of heterologous proteins may be solved eventually by insights from work in phage systems. For the moment, investigators survey the available protease mutations in *E. coli,* although many proteins are reported to be unstable in every strain. In that case, secretion may be used as a means to remove the protein of interest from the proteases, at a substantial cost related to the low volumetric productivity in the fermentor. The alternative, production in another organism, is tried last.

Most aggravatingly, some heterologous proteins are extremely toxic to *E. coli*. In such cases, the induction of the heterologous gene is followed shortly thereafter by cessation of protein synthesis. In an extreme case, imagine a heterologous protein whose enzymatic activity inactivated ribosomes catalytically and rapidly. The induced cell would make a few molecules of the protein and inactivate the 10,000 to 20,000 ribosomes in the cell, leading to a low fractional abundance. This problem is not solved by a high fractional synthetic rate; efficient accumulation of a toxic protein does not improve the final yield. Imagine another protein that is inserted into the inner membrane of *E. coli* so as to destroy the membrane potential; in such a case, one molecule of heterologous protein per cell might be the production limit. These proteins are not likely to be encountered often, but they present problems that cannot be solved easily. Secretion of such proteins might be of limited help. The most sensible strategy is to try alternative hosts or to select bacterial mutants resistant to the toxic protein if such mutants can be found. Some toxic proteins will have a target that is organism-specific, and those targets probably can be avoided or altered.

Kohno *et al.* ([16] in this volume) suggest that disulfides can be formed *in vitro,* both accurately and with high yields. Proteins with many disulfides are no more difficult to renature than proteins with a single disulfide. The success of renaturation and oxidation must reflect the normal folding pathway, which is an intrinsic property of the primary amino acid sequence and the redox state of the *in vivo* environment in which the newly synthesized protein is placed. Intracellular production of heterologous proteins in *E. coli* is possible for even those proteins that must be oxidized for activity.

Two problems remain: (1) Many proteins of interest do not contain an amino-terminal methionine or formylmethionine. Heterologous proteins expressed at high levels in *E. coli* can contain either one, both, or neither. Although high level expression and sophisticated purification schemes tend to make the final product homogeneous, one cannot be certain *a priori* what the amino terminus of the protein will contain. In a pharmaceutical setting a gram of purified, active protein with an extra amino acid leads to many useful experiments, e.g., human growth hormone with an extra amino acid continues to be a useful pharmaceutical. (2) What does the prudent investigator do if the protein of interest is found (in nature) glycosylated, phosphorylated, ADP-ribosylated, bound to a lipid, or otherwise posttranslationally modified? No eukaryotic posttranslational modification can be done properly in *E. coli,* although one can imagine complex cloning strategies in which a second gene expressed in *E. coli* provides a simple enzymatic activity. Such a strategy will probably fail to yield baroque posttranslational modifications and adducts. By analogy with the problem of the amino terminus, one probably should ignore posttranslational modifications in the early stages of research. Experimentation with an abundant, inexpensive, slightly unusual derivative of a rare and fascinating protein makes a lot of sense.

Escherichia coli thus remains a valuable host for expression of heterologous proteins, and for many purposes is the best host. The articles in this volume provide methods that will yield large quantities of active protein. Proper attention to the gene dosage, promoters, and translation initiation domains leads to high fractional rates of synthesis and, usually, high concentrations of heterologous proteins in extracts or even in the extracellular media. These lead in turn to simple purification schemes. Thus, investigators should expect to achieve quickly a high fractional synthetic rate for any protein, and hope that neither proteolysis nor toxicity to the host are seen. Although the documentation is weak, perhaps 90 to 95% of the heterologous genes expressed correctly in *E. coli* yield high concentrations of the heterologous proteins, as though rampant proteolysis and extreme toxicity are slightly unusual. In both industrial and university settings, many scientists who want to synthesize a lot of protein choose first to explore tissue culture with cells harboring the heterologous gene. This choice is both expensive and time-consuming, but it is made because the high success rate for production in *E. coli* is not fully appreciated. Powerful reasons may exist in specific cases for using other expression systems, especially when one is preparing human or animal pharmaceuticals. Subsequent sections in this volume address several other expression systems, whose simplicity and usefulness are improving rapidly.

[3] Design and Construction of Expression Plasmid Vectors in *Escherichia coli*

By Paulina Balbas and Francisco Bolivar

Escherichia coli plasmids are widely used in recombinant DNA-based biotechnologies as vectors for the overproduction of proteins with theoretical and practical values. Even though standard cloning procedures have become routine, and a large variety of host/vector systems for the expression of genes are available, difficulties are usually encountered when the theoretical strategies for overproduction of a protein are put into practice. Issues such as the manipulation of the regulatory mechanisms for gene expression and the host/vector interactions usually require extensive research if a high-level expression of a gene is to be achieved. Furthermore, technical factors such as availability of oligonucleotide synthesis for DNA

METHODS IN ENZYMOLOGY, VOL. 185

edition, cell growth facilities, and suitable purification procedures are indispensable for the establishment of an efficient process to obtain recombinant proteins.

The advanced knowledge concerning the genetics and physiology of *E. coli* has accounted for the preferential use of *E. coli* as a host for gene expression. Additional advantages of *E. coli*-based expression systems include rapid generation of biomass due to high rates of cell growth and availability of low-cost culture conditions. The most important drawbacks in the use of this gram-negative bacterium as a host for the overproduction of proteins are its limited capacity to secrete proteins and its inability to exert certain posttranslational modifications of proteins (such as disulfide bond formation, glycosylation, and acetylation), which usually leads to wrong protein folding. Thus, certain proteins may be obtained in high quantities, but their purification and renaturalization, in order to obtain biological activity, may be difficult and expensive. Nevertheless, despite the aforementioned disadvantages, *E. coli* continues to be the preferred host for gene expression.

In plasmid-based systems, expression of a gene can be achieved in two principal ways: by cloning a DNA cartridge containing the necessary regions for efficient transcription and translation upstream of the gene in question, or by cloning such a gene in a specially designed expression vector. Although both approaches have been successfully utilized, the second approach is generally preferred for higher levels of protein synthesis.[1]

The minimal elements that an expression plasmid vector should supply are a well-characterized origin of replication and a selection marker for plasmid propagation and maintenance; a strong promoter (usually regulatable); a ribosome binding site, and in some cases a translation initiation ATG codon, to ensure transcription initiation and efficient translation. Occasionally, the region coding for the amino- or carboxy-terminal domain of a protein is also included to act as a carrier or protein stabilizer. The product of the last type of manipulation yields a hybrid protein (also known as a fusion protein) that requires a cleavage process for the liberation of the protein or peptide of interest from its precursor or carrier.

As the knowledge about the rate-limiting steps in the synthesis and degradation of mRNA and proteins increases, sophisticated expression vectors have emerged. These vectors contain DNA fragments that help reduce problems such as premature termination of transcription, mRNA instability, inefficient translation initiation, and plasmid instability. In

[1] P. Balbas, X. Soberon, E. Merino, M. Zurita, H. Lomeli, F. Valle, N. Flores, and F. Bolivar, *Gene* **50**, 3 (1986).

some instances, several strain features may be utilized in conjunction with the molecular elements incorporated in the expression plasmid. Table I shows some of the most important factors known to influence the high-rate synthesis of heterologous proteins in *E. coli*.

Once the gene is inserted in and enzymatically tailored to the expression vector, the gene product can be obtained in a suitable host strain that contains the genotypic features needed for promoter regulation and cell growth. Expression may be monitored and quantitated by gel electrophoresis, high-performance liquid chromatography (HPLC), immunological schemes, or functional assays.[2] Finally, the expressed protein can be purified and characterized in terms of its physicochemical and biological properties.

This article addresses some aspects of the current knowledge about the factors that affect the overall product yield of a plasmid-based expression system in *E. coli*, as well as the strategies currently in use to circumvent some of them for maximal protein production. The molecular elements that may be included in the design of an efficient host/plasmid expression system are discussed according to the organization presented in Table I. Sections on some special considerations about the selection of host strains, fermentation conditions, and improvement of an expression system are also included.

Design of Expression Plasmid Vectors

Plasmid-Related Elements

Replication is the primary function exhibited by a plasmid. This mechanism is not universal for all plasmids; in fact, a variety of replication mechanisms may be found among the large number of naturally occurring plasmids: uni- and bidirectional origins, single or multiorigin plasmids, and plasmids dependent on or independent of the initiation proteins of the host. Interestingly, all plasmids studied so far regulate their own replication by negative control, and inhibition of replication may be mediated mainly by protein, RNA, or a series of direct DNA repeats.

It is a well-established phenomenon that the amount of plasmid-encoded proteins synthesized in *E. coli* usually increases as the gene dosage is elevated. High-copy-number plasmids are useful for the production of proteins whose expression at high levels does not perturb the cell. In such instances, this type of plasmid with an elevated copy number is preferred as

---

[2] P. Dobner and L. Villa-Komaroff, *in* "Maximizing Gene Expression" (W. Reznikoff and L. Gold, eds.), p. 35, Butterworth, Stoneham, Massachusetts, 1986.

TABLE I

FACTORS AFFECTING OVERALL PRODUCT YIELD OF A PLASMID-DIRECTED EXPRESSION SYSTEM IN
Escherichia coli[a]

Factors influencing final protein concentration	Molecular aspects of plasmid vector that may be manipulated to increase product yield	Strain elements used in conjunction with expression vector
Plasmid-related elements		
Plasmid/gene copy number	Replicon	Elements for copy-number regulation
	Genes for selection	Selection-sensitive genotype
	Genes for retention	Elements for cell suicide
	Genes for partition	Not identified
	Instability-inducing DNA sequences	RecA⁻ genotype
mRNA-related elements		
Initiation rate of mRNA synthesis	Promoter and regulatory sequences	trans-Acting regulatory proteins (repressors and activators)
Elongation and termination of mRNA synthesis	Transcriptional terminators and attenuators	trans-Acting termination factors (rho, NusA, NusB)
	Antiterminators	trans-Acting antitermination factors (λN and λQ, NusA)
mRNA degradation rate	RNase recognition sequences	RNase⁻ or RNase *ts* genotype
	REP sequences	Not identified
	Other sequences	Not identified
Protein-related elements		
Translation initiation rate	RBS and AUG sequences	Ribosomes with altered specificity
	Distance and sequence between RBS and translation initiation	Ribosomes with altered specificity
	Specialized ribosomes	Not identified
	Other sequences	Not identified
Elongation and termination of protein synthesis	Gene codon usage	tRNA availability
	Termination codon usage	Availability of termination factors
Protein degradation rates	Fusion proteins	Altered proteolytic activities
	Secretion proteins	Not identified

[a] Under the assumption that the substrates (nucleotides and amino acids), catalytic components (RNA polymerase, DNA polymerase, tRNAs, and cofactors), and energy requirements are not rate limiting. Compiled from text Refs. 14–86.

the basis of a high-level expression system in order to increase the productivity of a fermentation.

The copy number and stability of the plasmid are determined primarily by the replicon region. Table II offers selected examples of natural and

TABLE II
FUNCTIONAL FEATURES OF COMMONLY USED PLASMID REPLICONS

Plasmid[a]	Incompatibility group	Copy number	Mobilization	Direction of replication	General replication control	Host functions required for replication	Host range	Refs.[c]
ColEI-type (pBR322)	ColEI	~20	Mobilizable by conjugative plasmids	Unidirectional	RNA, Rop protein	polA, dnaA, dnaZ, dnaB, dnaC, dnaG, RNA polymerase	E. coli and few others	b–g
F (pDF41)	FI	1–2	Self-transmissible, capable of mobilizing nonconjugative plasmids	Bidirectional	Direct DNA repeats, E protein	dnaE, dnaB, dnaC	E. coli and few others	b, h
R1, R100, R6, NR1 (pRK356, pBEU50)	FII	~2	Self-transmissible, capable of mobilizing nonconjugative plasmids	Unidirectional	RNA, RepA1 protein	dnaB, dnaC, dnaE, dnaF, dnaG	E. coli and few others	b, i
RK2 (pRK2501)	P	1–4	Capable of conjugal transfer	Unidirectional	trfA protein product, oriV	Not determined in refs. cited	Broad host range for gram-negative	h, j–l
RSF1010	Q	>10	Capable of conjugal transfer	Bidirectional	oriV, repA, repB, repC	dnaB, dnaC, dnaG	Broad host range for gram-negative	m–o
R6K (pRK353)	X	13–40	Capable of conjugal transfer	Bidirectional	ori αβγ, DNA repeats, pi (π) protein	RNA polymerase	E. coli and few others	h, p

P1	Y	1–2	Capable of conjugal transfer	Bidirectional	Direct DNA repeats	$dnaC$, $dnaE$, $dnaB$, $dnaG$	Broad host range for gram-negative	b

[a] A prototype plasmid is shown in parentheses.

[b] J. R. Scott, *Microbiol. Rev.* **48**, 1 (1984).

[c] B. Polisky, *in* "Maximizing Gene Expression" (W. Reznikoff, and L. Gold, eds.), p. 143. Butterworth, Stoneham, Massachusetts, 1986.

[d] M. Filutowicz, M. McEachern, A. Greener, P. Mukhopadhyay, E. Uhlenhopp, R. Durland, and D. Helinski, *in* "Plasmids in Bacteria" (D. R. Helinski, S. N. Cohen, D. B. Clewell, D. A. Jackson, and A. Hollaender, eds.), Basic Life Sciences, Vol. 30, p. 125. Plenum, New York, 1985.

[e] J. Tomizawa and T. Itoh, *Cell* **31**, 575 (1982).

[f] R. M. Lacatena, D. W. Banner, C. Castagniolli, and G. Cesareni, *Cell* **37**, 1009 (1984).

[g] H. Masukata and J. Tomizawa, *Cell* **44**, 125 (1986).

[h] J. J. Nijkamp, B. van Gemen, M. J. Hakaart, A. J. van Putten, and E. Veltkamp, *in* "Plasmids in Bacteria" (D. R. Helinski, S. N. Cohen, D. B. Clewell, D. A. Jackson, and A. Hollaender, eds.), Basic Life Sciences, Vol. 30, p. 283. Plenum, New York, 1985.

[i] R. H. Rownd, D. D. Womble, X. Dong, V. A. Luckow, and R. P. Wu, *in* "Plasmids in Bacteria" (D. R. Helinski, S. N. Cohen, D. B. Clewell, D. A. Jackson, and A. Hollaender, eds.), Basic Life Sciences, Vol. 30, p. 335. Plenum, New York, 1985.

[j] C. M. Thomas, *Plasmid* **5**, 10 (1981).

[k] C. M. Thomas, C. A. Smith, V. Shingler, M. A. Cross, A. K. Hussain, and M. Pinkey, *in* "Plasmids in Bacteria" (D. R. Helinski, S. N. Cohen, D. B. Clewell, D. A. Jackson, and A. Hollaender, eds.), Basic Life Sciences, Vol. 30, p. 261. Plenum, New York, 1985.

[l] M. Kahn, R. Kolter, C. Thomas, D. Figursky, R. Meyer, E. Remaut, and D. Helinski, this series, Vol. 68 p. 286.

[m] D. H. Figurski, C. Young, H. C. Schreiner, R. F. Pohlman, D. H. Bechhofer, A. S. Prince, and T. F. D'Amico, *in* "Plasmids in Bacteria" (D. R. Helinski, S. N. Cohen, D. B. Clewell, D. A. Jackson, and A. Hollaender, eds.), Basic Life Sciences, Vol. 30, p. 227. Plenum, New York, 1985.

[n] E. Scherzinger, M. M. Bagdasarian, P. Schloz, R. Lurz, B. Ruckert, and M. Bagdasarian, *Proc. Natl. Acad. Sci. U.S.A.* **81**, 654 (1984).

[o] P. Schloz, V. Haring, E. Scherzinger, R. Lurz, M. M. Bagdasarian, H. Schuster, and M. Bagdasarian, *in* "Plasmids in Bacteria" (D. R. Helinski, S. N. Cohen, D. B. Clewell, D. A. Jackson, and A. Hollaender, eds.), Basic Life Sciences, Vol. 30, p. 243. Plenum, New York, 1985.

[p] R. J. Meyer, L. S. Lin, K. Kim, and M. A. Brasch, *in* "Plasmids in Bacteria" (D. R. Helinski, S. N. Cohen, D. B. Clewell, D. A. Jacksn, and A. Hollaender, eds.), Basic Life Sciences, Vol. 30, p. 172. Plenum, New York, 1985.

prototype plasmids derived from natural replicons, and their salient features concerning copy number, mobilization capacity, replication features, and host range.

The genes contained in the plasmid, the growth conditions, and the physiology of the host strain employed will also influence plasmid stability and the overall product yield as well.[3] These considerations call for a careful analysis of the elements that may affect plasmid copy number and stability.

Plasmid copy number is maintained within defined limits by several factors that control initiation of DNA replication at the specific origin. Relaxed replicating plasmids, such as pBR322[4] and its derivatives,[1] have been preferentially used for expression systems. The main reasons for this preference are: pBR322 fulfills the major criteria for a suitable vector,[5] its nucleotide sequence has been determined, and the mechanisms concerning its replication, stability, mobilization, and amplification have been well studied (for reviews, see Refs. 1 and 6).

For relaxed replicating plasmids, the actual copy number in a growing culture is influenced at least by the growth rates.[7] In fact, this relationship is inversely proportional, so at high growth rates, the plasmid copy number per cell is lower, thus resulting in segregational instability.[8] The segregational instability arises due to defective cell partitioning of plasmid DNA between daughter cells during cell division.[9]

The incorporation of selection schemes into the vector has been a general practice for years. In general terms, selection refers to the use of growth environments in which only cells possessing certain genetic traits are able to grow. Four alternatives are available to diminish the effects of segregational instability: the incorporation into the plasmid of a gene for selection (usually an antibiotic resistance gene), the use of specially designed "suicidal" mechanisms for those cells that segregate the plasmid, the incorporation into the plasmid of a partition locus, and the immobilization of cells in organic supports.

In the first case, a single gene is involved in the selection scheme, so its

---

[3] D. W. Zabriskie and E. J. Acuri, *Enzyme Microbiol. Lett.* **22**, 239 (1984).
[4] F. Bolivar, R. L. Rodriguez, P. J. Greene, M. C. Betlach, H. L. Heyneker, H. W. Boyer, J. H. Crosa, and S. Falkow, *Gene* **2**, 95 (1977).
[5] R. L. Rodriguez and R. C. Tait, "Recombinant DNA Techniques: An Introduction." Addison-Wesley, Cambridge, Massachusetts, 1983.
[6] P. Balbas, X. Soberon, F. Bolivar, and R. L. Rodriguez *in* "Vectors: A Survey of Molecular Cloning Vectors and Their Uses" (R. L. Rodriguez and D. T. Denhardt, eds.), p. 5. Butterworth, Stoneham, Massachusetts, 1988.
[7] J. Seo and J. E. Bailey, *Biotechnol. Bioeng.* **27**, 1668 (1985).
[8] R. A. Klotsky and I. Schwartz, *Gene* **55**, 141 (1987).
[9] B. D. Ensley, *CRC Crit. Rev. Biotechnol.* **4**, 263 (1985).

presence accounts for the survival of the organism. This is the case of the antibiotic resistance genes present in almost every cloning vector. However, when considering genes in multicopy plasmids, the conceptual basis of selection becomes more complicated due to the probability of substantial variability in single cell plasmid content in growing populations. When the selective antibiotic is decomposed, deactivated, or tightly bound to the product of the selection gene, cells with a high content of the selection genes reduce the concentration of the selective agent, lowering its effects on the more sensitive cells.[10] Multiple additions of the antibiotic during fermentation eliminate this problem, but this is usually expensive and inefficient in high-volume fermentors. In addition, the final product may be contaminated with antibiotics, which may be a serious problem. Although this is an inherent limitation implicating practical difficulties for all selection strategies of this type, the antibiotic resistance markers are the most commonly used selection schemes. Table III summarizes the modes of action, modes of resistance, and working concentrations of the antibiotics commonly used for selection pressure in *E. coli*.[1,11-13]

The second alternative to avoid plasmid segregational instability, selective retention of the vector, forces the retention of the plasmid when used in conjunction with a strain that depends on a plasmid-encoded function for survival.[14-17] Two practical limitations of this scheme are (1) the expression of the plasmid-encoded survival function may interfere with cellular functions, and (2) if the survival function is a product that accumulates in the cytoplasm (i.e., an auxotrophic requirement), the daughter cells may inherit the product while losing the plasmid and still survive.

The third approach, the incorporation of a cis-acting partitioning element into the plasmid vector, promotes the accurate distribution of the plasmid between the daughter cells during cell division.[18-21] It has been shown, however, that the utility of this approach is restricted because

[10] K. Dennis, F. Srienc, and J. E. Bailey, *Biotechnol. Bioeng.* **27**, 1490 (1985).
[11] F. M. Ausubel, R. Brent, R. E. Kingston, D. D. Moore, J. G. Seiman, J. A. Smith and K. Strhul, "Current Protocols in Molecular Biology." Wiley, New York, 1987.
[12] T. J. Foster, *Microbiol. Rev.* **47**, 361 (1983).
[13] D. Moazed and H. F. Noller, *Nature (London)* **327**, 389 (1987).
[14] P. R. Rosteck, Jr. and C. L. Hershberger, *Gene* **25**, 29 (1983).
[15] G. Skogman and J. Nilsson, *Gene* **31**, 117 (1984).
[16] K. Gerdes, P. B. Rasmussen, and S. Molin, *Proc. Natl. Acad. Sci. U.S.A.* **83**, 3116 (1986).
[17] S. Molin, P. Klemm, K. Poulsen, H. Biehl, K. Gerdes, and P. Andersson, *Bio/Technology* **5**, 1315 (1987).
[18] G. Skogman, J. Nilsson, and P. Gustaffson, *Gene* **23**, 105 (1983).
[19] M. Zurita, F. Bolivar, and X. Soberon, *Gene* **28**, 119 (1984).
[20] T. Ogura and S. Hiraga, *Cell* **32**, 351 (1983).
[21] D. K. Summers and D. J. Sherrat, *Cell* **36**, 1097 (1984).

TABLE III

MODES OF ACTION, MODES OF BACTERIAL RESISTANCE, AND WORKING CONCENTRATIONS OF ANTIBIOTICS IN *Escherichia coli*[a]

Antibiotic[b]	Mode of action	Mode of resistance	Working concentration (μg/ml)	Refs.
Ampicillin	Bacteriocidal; inhibits cell wall synthesis by inhibiting the formation of peptidoglycan cross-link; kills only growing cells	β-Lactamase inactivates ampicillin by hydrolyzing β-lactam ring	50–100	1,11,12
Chloramphenicol (in methanol)	Bacteriostatic; inhibits protein synthesis by interacting with 50S ribosomal subunit and inhibiting peptidyltransferase reaction	Chloramphenicol acetyltransferase inactivates chloramphenicol by acetylation	20	1,11,12
Gentamicin	Bacteriocidal; inhibits protein synthesis by binding to L6 protein of 50S ribosomal subunit	Aminoglycoside acetyltransferase and aminoglycoside nucleotidyltransferase inactivate gentamicin by acetylation	15	11–13
Kanamycin	Bacteriocidal; inhibits protein synthesis by hampering translocation and causing misreading of mRNA	Aminoglycoside phosphotransferase, aminoglycoside acetyltransferase, and aminoglycoside nucleotidyltransferase inactivate kanamycin by chemical modification	30	11–13

Nalidixic acid (pH 11 with NaOH)	Bacteriostatic; inhibits DNA synthesis by inhibiting DNA topoisomerase	Mutations in host DNA topoisomerase prevent nalidixic acid from binding	15	11
Rifampicin (in methanol)	Bacteriostatic; inhibits RNA synthesis by binding to β subunit of RNA polymerase	Mutations in β subunit of RNA polymerase prevents rifampicin from complexing	150	11
Spectinomycin	Bacteriostatic; inhibits protein synthesis by inhibiting translocation of peptidyl-tRNA	Mutation in gene coding for S5 protein (rpsE) prevents spectinomycin from binding	100	11,13
Streptomycin	Bacteriocidal; inhibits protein synthesis by binding to S12 protein of 30S ribosomal subunit, which inhibits translation	Aminoglycoside phosphotransferase inactivates streptomycin. Mutations in gene coding for S12 protein (rpsL) prevent spectinomycin from binding	30	11,13
Tetracycline (in 70% ethanol)	Bacteriostatic; inhibits protein synthesis by preventing binding of aminoacyl-tRNA to ribosome A site	One polypeptide that prevents tetracycline from entering cell by decreasing permeability of cell wall	15	1,11,13

[a] From Ref. 11.
[b] All antibiotics should be dissolved in sterile water and sterilized by filtration, unless otherwise indicated.

23

cloning of a highly expressed gene into such plasmids usually renders them unstable despite the presence of the partition element.[22,23]

The last method, the immobilization of living cells, has proved useful for highly unstable plasmids. The advantage of using growing immobilized microorganisms is the possibility of obtaining higher cell concentrations within the organic matrix, with a consequent reduction in reactor size and process investment cost.[24] Its principal drawback is the restrained number of generations that may be obtained after immobilization.[24-26]

A different type of plasmid instability, structural instability, is the result of physical changes in the plasmid DNA, such as deletions, insertions, and rearrangements. Structural instabilities are more insidious than segregational instabilities because the changes selected usually result in the loss or alteration of the DNA of interest, while the remainder of the plasmid DNA, including the selectable markers and partition elements is retained.[9] DNA sequences such as inverted repeats or highly palindromic sequences within the vector and the passenger DNA should be avoided as they raise the probability of rearrangements. To date there is no way in which an accurate prediction can be made as to whether a particular gene or DNA construction will be stable. The design of alternative strategies for DNA cloning and the use of RecA⁻ strains are two ways to cope with a potential structural instability problem.

Controlled gene expression may be difficult with steady-state high-copy-number plasmids, and if the protein of interest is deleterious to the cell, undesired instability or even cell death may result in poor yields. The use of runaway replication plasmids and plasmids with an amplifiable copy number is an alternative strategy to overcome this problem partially. Runaway replicating plasmids are based in a plasmid R1 mutant that, on a temperature shift from 35 to 42°, exhibits uncontrolled plasmid replication, which eventually inhibits the host cell growth and results in a conditionally lethal phenotype.[27]

In plasmids with an amplifiable copy number, the initiation of the synthesis of the RNA molecule that primes the replication function is manipulated by replacing the native promoter with a foreign regulatable

[22] S. W. Lee and G. Edlin, *Gene* **39,** 173 (1985).

[23] S. A. Jones and J. Melling, *FEMS Microbiol. Lett.* **22,** 239 (1984).

[24] P. De Taxis du Poet, P. Dhuster, J. N. Barbotin, and D. Thomas, *J. Bacteriol.* **165,** 871 (1986).

[25] D. S. Inoles, W. J. Smith, D. P. Taylor, N. Cohens, A. S. Michaels, and C. R. Robertson, *Biotechnol. Bioeng.* **25,** 2653 (1983).

[26] M. Nasri, S. Sayadi, J. N. Barbotin, and D. Thomas, *J. Biotechnol.* **6,** 147 (1987).

[27] N. Panayotatos, *in* "Vectors: A Survey of Molecular Cloning Vectors and Their Uses " (R. L. Rodriguez and D. T. Denhardt, eds.), p. 227. Butterworth, Stoneham, Massachusetts, 1988.

promoter that may be induced at will.[1,27,28] However, three drawbacks limit the usefulness of runaway replicating and amplifiable-copy-number plasmids: (1) repression of the system is likely to be abolished by titration of repressors, (2) amplification of the plasmid-encoded accompanying genes may compromise cellular functions and complicate purification of the desired product, and (3) elevated transcription and extensive replication may interfere with each other. An additional limitation of all systems using heat induction is the simultaneous induction of heat-shock genes of the host, especially proteases.[29,30] These factors may render the high-copy-number property disadvantageous for product output.

In conclusion, the advantages and limitations for high-level gene expression reflect the properties of the plasmid vector. The genetic background of the host, in conjunction with the growth environment, will ultimately determine a good host/vector interaction to achieve stable maintenance of the expression plasmid.

mRNA-Related Elements

The first step in gene expression is the transcription initiation event, a process catalyzed by the RNA polymerase holoenzyme. Furthermore, this process is rate-limiting for mRNA synthesis.

In E. coli, the frequency of transcription initiation is programmed by a promoter sequence, which can be modulated by a variety of mechanisms, namely, the interaction of one or more regulatory proteins with specific sequences in the vicinity of the promoter. Consequently, every expression vector attempts to maximize the efficiency of transcription initiation with the introduction of a strong, regulatable promoter sequence. The most utilized native promoters used in E. coli expression systems are trp, lacUV5, lpp, λPL, and λPR. Hybrid and tandem promoters have been also constructed to combine specific features, thus increasing strength or regulation capacity.[1,31-33] Selection of a host cell for controllable gene expression is then based on the particularities of the regulatory region, because repressors and activators are usually supplied in trans by the host cell. Reviews of some of these, and other efficient promoters are included in this volume.

[28] R. N. Rao and S. G. Rogers, Gene 3, 247 (1978).
[29] F. C. Niedhardt, R. A. Van Bogelen, and V. Vaughn, Annu. Rev. Genet. 18, 295 (1984).
[30] S. A. Goff, L. P. Casson, and A. L. Goldberg, Proc. Natl. Acad. Sci. U.S.A. 81, 6647 (1984).
[31] J. Brosius, in "Vectors: A Survey of Molecular Cloning Vectors and Their Uses" (R. L. Rodriguez and D. T. Denhardt, eds.), p. 205. Butterworth, Stoneham, Massachusetts, 1988.
[32] D. T. Denhardt and J. Colasanti, in "Vectors: A Survey of Molecular Cloning Vectors and Their Uses" (R. L. Rodriguez and D. T. Denhardt, eds.), p. 179. Butterworth, Stoneham, Massachusetts, 1988.
[33] K. Nakamura and M. Inouye, EMBO J. 1, 771 (1982).

Ideally, specific transcription from the promoter of an expression vector should be tightly repressed during the early stages of the fermentation to avoid problems such as instability, lower growth rates, or lethality. This condition is not easily fulfilled in multicopy systems: some transcripts escape from regulation. So far, there has only been one efficient system designed to overcome this problem: the strong promoter is oriented away from the gene to be expressed before the culture reaches a desired optical density. This promoter is then inverted by addition of the λInt product and a short heat pulse in order to "turn on" transcription, so there is no possibility of basal transcription from the promoter into the gene of interest.[34,35]

Once transcription is initiated at the promoter, the mRNA is elongated and finally terminated at specific sites. However, two mechanisms modulate this process by inducing premature termination of the transcript: attenuation and nonspecific termination within the transcript. The attenuator site has the features of a simple terminator (as defined by Platt and Bear[36]) and is located between the promoter and the first structural gene of several biosynthetic operons.[37] Because attenuators are negative regulatory elements they are usually avoided in expression systems.[1]

Nonspecific transcription termination may occur within any transcript due to fortuitous terminator-like sequences in the mRNA. Antitermination elements have also been included in expression systems to ensure complete transcription of long messages. In general, antitermination events seem to require the presence of a trans-acting element supplied by the host (i.e., bacteriophage λ proteins Q and N, and *E. coli* NusA protein) and specific sites at which the polymerase becomes modified and "resistant" to subsequent termination sites that it encounters.[34,35,38]

Transcriptional terminators determine the points where the mRNA–RNA polymerase–DNA complex dissociates, thereby ending transcription. Numerous parameters collaborate on the termination response, including not only the nucleotide sequence, but also the presence of secondary structure in the RNA transcript and the participation of protein factors (i.e., ρ, NusA, NusB). These regulatory elements are not indispensable in an expression plasmid vector, but their presence at the end of a highly expressed gene is certainly advantageous: as transcription terminators act as barriers to elongation, they minimize sequestering of RNA

[34] A. J. Podhajska, N. Hasan, and W. Szybalski, *Gene* **40**, 163 (1985).
[35] N. Hasan and W. Szybalski, *Gene* **56**, 145 (1987).
[36] T. Platt and D. G. Bear, *in* "Gene Function in Prokaryotes" (J. Beckwith, J. Davies and J. A. Gallant, eds.), p. 123. Cold Spring Harbor Laboratory, Cold Spring Harbor, New York, 1983.
[37] C. Yanofsky, *Nature (London)* **289**, 751 (1981).
[38] A. R. Shatzman and M. Rosenberg this series, Vol. 152, p. 661.

polymerase that might be engaged in unnecessary transcription, and also restrict the mRNA length to the minimal, thus limiting energy expense. Furthermore, the transcription-directing promoter interacts dynamically with the plasmid vector, as strong transcription may interfere with the origin of replication. Therefore, the inclusion of transcriptional termination sequences to isolate the *ori* or *rep* region increases plasmid stability as a result of copy number maintenance.[38-40]

The degradation of mRNA is a general feature of biological systems that allows cells to adapt to changing environments. The variation in the half-life of individual mRNAs suggest that differential message stability has an important physiological role, mediated by the level of specific gene expression.[41,42]

Information on the enzymes and other factors involved in the mRNA decay process is still very limited, although two $3' \longrightarrow 5'$-exonucleases, RNase II and polynucleotide phosphorylase, seem to be primarily responsible for the major degradation of mRNA.[43-45] Some factors that may influence mRNA decay are (1) the presence or absence of ribosomes[46,47]; (2) the rate of translation[48]; (3) mRNA secondary structure at the 5' and 3' ends, including the stem loops of ρ-independent terminators and REP sequences[41,49-51]; (4) specific sequences in 5' or 3' ends of the DNA that respond to trans-acting effectors[52,53]; (5) stability determinants within specific genes[54]; (6) the cell growth rate[55]; and (7) the utilization of RNase-deficient or conditionally lethal mutants.[43]

In the literature, there are no examples of mRNA stabilization with high-level expression purposes. However, three general approaches have

[39] H. Bujard, C. Baldari, M. Brunner, U. Deuschle, R. Gentz, J. Hughes, W. Kammerer, and D. Stuber, *in* "Gene Amplification and Analysis" (Papas, M. Rosenberg and J. G. Chirikjian, eds.), Vol. 3, p. 65. Elsevier, Amsterdam, 1983.

[40] J.-D. Chen and D. A. Morrison, *Gene* **64**, 155 (1988).

[41] D. E. Kennell *in* "Maximizing Gene Expression" (W. Reznikoff and L. Gold, eds.), p. 101. Butterworth, Stoneham, Massachusetts, 1986.

[42] S. F. Newbury, N. H. Smith, and C. F. Higgins, *Cell* **51**, 1131 (1987).

[43] W. P. Donovan and S. R. Kushner, *Proc. Natl. Acad. Sci. U.S.A.* **83**, 120 (1986).

[44] M. P. Deutscher, *Cell* **40**, 731 (1985).

[45] G. Brawerman, *Cell* **48**, 5 (1987).

[46] S. R. Gupta and D. Schlessinger, *J. Bacteriol.* **125**, 84 (1976).

[47] R. Har-El, A. Siberstein, J. Kuhn, and M. Tal, *Mol. Gen. Genet.* **173**, 135 (1979).

[48] E. Schneider, M. Blundell, and D. E. Kennell, *Mol. Gen. Genet.* **160**, 121 (1978).

[49] V. J. Cannistraro and D. E. Kennell, *Nature (London)* **277**, 407 (1979).

[50] E. Merino, B. Becerril, F. Valle, and F. Bolivar, *Gene* **58**, 305 (1987).

[51] S. F. Newbury, J. Smith, E. C. Robinson, I. D. Hiles, and C. F. Higgins, *Cell* **48**, 297 (1987).

[52] K. Gorski, J.-M. Roch, P. Prentki, and H. M. Kirsch, *Cell* **43**, 461 (1985).

[53] M. N. Hayashi and M. Hayashi, *Nucleic Acids Res.* **13**, 5937 (1985).

[54] J. G. Belasco, G. Nilsson, A. Von Gabain, and S. N. Cohen, *Cell* **46**, 245 (1986).

[55] G. Nilsson, J. G. Belasco, S. N. Cohen, and A. von Gabain, *Nature (London)* **312**, 75 (1984).

been suggested for the manipulation of mRNA stability: the use of RNase mutants or conditionally lethal strains, the insertion of sequences that enhance the stability of the transcript, and the modulation of the growth conditions.

Mutants for the known major ribonucleases have been examined for altered mRNA stabilities, but the variety of findings suggests that no single RNase is solely responsible for the degradation of every RNA.[41]

The second approach is based on the fact that highly structured mRNA 3' ends act as a barrier for the 3' $\longrightarrow$ 5'-exonucleolytic degradation of the transcript. Therefore, the inclusion of a ρ-independent terminator or one or more REP sequences will theoretically increase the half-life of the transcript.[50,51] Some bacteriophage-derived stabilizing sequences, and the phage effector trans-acting function, may also be included in the expression system for this purpose.[52] However, none of these strategies has been successfully reported so far for high-level gene expression.

The last approach is the modulation of the growth rate. There is evidence indicating that some mRNAs exhibit differential stabilities during fast and slow growth. Since this is a characteristic of every transcript, the obvious consequence is that mRNA stability needs to be evaluated at different growth rates, and in various energy and nitrogen sources.[55]

In summary, gene expression is primarily limited by the rate of transcription initiation; much information on this process is available. In some instances, constitutive promoters may be used in expression systems,[56] although it is advantageous to work with an efficiently controlled system. Even when a constitutively expressed gene product is not toxic to the host cell, continuous transcription may result in segregational or structural instability of the plasmid DNA.[9]

Efficient elongation and termination of mRNA synthesis are two processes that usually remain overlooked in expression systems. For large transcripts, antitermination regions located between the promoter and the 5' end of the coding region may ensure complete mRNA elongation.[38] Transcriptional terminators placed at the 3' end of the expressed gene restrict the size of the mRNA and isolate the plasmid's replication functions from the effects of strong transcription going into the origin, further reducing the probability of plasmid instabilities.[40]

Finally, very little is known about those features of mRNA structure which determine mRNA stability. Although some alternatives have been proposed to tackle potential mRNA instability problems,[51,52] more work is

[56] E. Jay, J. Rommens, L. Pomeroy-Cloney, D. MacKnight, C. Lutze-Wallace, P. Wishart, D. Harrison, W.-Y. Lui, V. Asundi, M. Dawood, and F. Jay, *Proc. Natl. Acad. Sci. U.S.A.* **81**, 2290 (1984).

needed to establish the real impact that this issue will have on the final product yield of an *E. coli* plasmid-based expression system.

Protein-Related Elements

Translation is an mRNA-directed process by which amino acids are assembled into a polypeptide chain. The initiation of translation is a multicomponent process involving base-pairing among three different mRNA species: 16S rRNA, mRNA, and fMet-tRNA. Interactions occur also with the ribosomal protein S1 and three proteic initiation factors (IF1, IF2, and IF3) to facilitate the initiation step.[57]

The mRNA elements that condition the initiation of translation are an initiation codon, a Shine–Dalgarno or ribosome binding site (RBS), the spacing between these two regions, the nucleotide sequence of the spacing region, and the potential mRNA secondary structure. There is evidence that the translation initiation efficiency also depends on the sequence of the 5' untranslated region of the mRNA upstream of the Shine–Dalgarno sequence, and the 5' end of the protein-coding region.[58-61]

A set of general rules for maximal efficiency of translation initiation has been compiled by Stromo,[57] and they include (1) the preferential initiation codon is AUG, although GUG, UUG, AUU, and AUA are not uncommon; (2) the RBS sequence has at least four nucleotides taken from the sequence AGGAGG; (3) the spacing between the RBS and the AUG is 9 ± 3 nucleotides; (4) besides the RBS, the nucleotides located 5' of the AUG should be As and Us (A at position -3); (5) the sequence GCAU or AAAA located after the AUG enhances translation; (6) the region around the initiation site should be relatively unstructured.

A large variety of RBS from efficiently expressed genes have been adopted to suit gene expression. Furthermore, a specialized ribosome system for translation of a single mRNA has been constructed by changing the native anti-RBS in the 16S rRNA and its complementary sequence in the 5' end of the mRNA (actually the RBS). Consequently, the new pool of altered ribosomes will translate one exclusive species of mRNA[62] ([9] in this volume).

Although many RBS do not meet every requirement of sequence and distance, they work adequately.[58] This supports the observation that the most important parameter for translation initiation is the inhibitory effect

[57] G. D. Stormo, *in* "Maximizing Gene Expression" (W. Reznikoff and L. Gold, eds.), p. 195. Butterworth, Stoneham, Massachusetts, 1986.

[58] G. D. Stormo, T. D. Schneider, and L. M. Gold, *Nucleic Acids Res.* **10,** 2971 (1982).

[59] L. Gold, G. Stormo, and R. Saunders, *Proc. Natl. Acad. Sci. U.S.A.* **81,** 7061 (1984).

[60] P. Stranssens, E. Remaut, and W. Fiers, *Gene* **36,** 211 (1985).

[61] P. Stranssens, E. Remaut, and W. Fiers, *Cell* **44,** 711 (1986).

[62] A. Hui and H. A. de Boer, *Proc. Natl. Acad. Sci. U.S.A.* **84,** 4762 (1987).

of mRNA secondary structures.[63] The site-directed mutagenesis of the 5′ untranslated regions to destabilize stem–loop structures has been successfully utilized to enhance translation initiation.[64,65] An alternative strategy, easily incorporated into a general-use expression plasmid, is the inclusion of an upstream "cistron" with an extra RBS that will help produce a virtually unstructured mRNA 5′ end to facilitate the accessibility of the ribosome to the RBS.[66]

Elongation of the nascent polypeptide chain does not proceed at a constant rate. Such discontinuity of translation implies that the ribosome at various regions of the mRNA template acts as a regulatory mechanism for gene expression.[37] Moreover, it has been demonstrated that, in *E. coli,* the pattern of nonrandom codon usage varies with the levels of gene expression; therefore, weakly expressed genes are characterized by the occurrence of codons recognized by rare tRNA species.[67,68] The relevance of codon usage, and therefore of tRNA availability, in high-level expression systems has not been conclusively demonstrated.[69–71] Anyway, synthetic genes are usually designed incorporating those codons that are preferentially recognized in *E. coli,*[72,73] or avoiding those triplets that are very infrequently used.[74,75]

Peptide chain liberation in *E. coli* is mediated by two release factors which recognize the three termination codons: RF1 recognizes UAA and

[63] A. C. Looman, J. Bodlaender, M. de Gruyter, A. Vogelaar, and P. H. van Knippenberg, *Nucleic Acids Res.* **14,** 5481 (1986).

[64] L. H. Teissier, P. Sondermeyer, T. Faure, D. Dreyer, A. Benavente, D. Villeval, M. Courtney, and J. P. Lecocq, *Nucleic Acids Res.* **12,** 7663 (1984).

[65] C. R. Wood, M. A. Boss, T. P. Patel, and J. S. Emtage, *Nucleic Acids Res.* **12,** 3937 (1984).

[66] B. E. Schoner, R. M. Belagaje, and R. G. Schoner, *Proc. Natl. Acad. Sci. U.S.A.* **83,** 8506 (1986).

[67] H. A. de Boer and R. A. Kastelein, *in* "Maximizing Gene Expression" (W. Reznikoff and L. Gold, eds.), p. 225. Butterworth, Stoneham, Massachusetts, 1986.

[68] M. Bulmer, *Nature (London)* **325,** 728 (1987).

[69] M. Robinson, R. Lilley, S. Little, J. S. Emtage, G. Yarranton, P. Stephens, A. Millican, M. Eaton, and G. Hymphreys, *Nucleic Acids Res.* **12,** 6663 (1984).

[70] L. Holm, *Nucleic Acids Res.* **14,** 3075 (1986).

[71] J. F. Ernst and E. Kawashima, *J. Biotechnol.* **7,** 1 (1988).

[72] K. Itakura, T. Hirose, R. Crea, A. D. Riggs, H. L. Heyneker, F. Bolivar, and H. W. Boyer, *Science* **198,** 1056 (1977).

[73] D. V. Goeddel, D. G. Kleid, F. Bolivar, H. L. Heyneker, D. G. Yansura, R. Crea, T. Hirose, A. Krazewski, K. Itakura, and A. D. Riggs, *Proc. Natl. Acad. Sci. U.S.A.* **76,** 106 (1979).

[74] M. D. Edge, A. R. Greene, G. R. Heathcliffe, P. A. Meacock, W. Schuch, D. B. Scanlon, T. C. Atkinson, C. R. Newton, and A. F. Markham, *Nature (London)* **292,** 756 (1981).

[75] L. Ferretti, S. S. Harnik, H. G Khorana, M. Nassal, and D. D. Oprian, *Proc. Natl. Acad. Sci. U.S.A.* **83,** 599 (1986).

UAG, whereas RF2 recognizes UAA and UGA.[76] The three stop codons differ in their efficacy of translation termination, and there is a strong bias in favor of UAA in genes expressed at high levels.[77] In view of this observation, this triplet should be preferentially incorporated in synthetic genes to ensure termination. The common strategy has been the use of tandem termination codons.[72-75]

The last important parameter pertaining to protein yield is protein degradation. Proteolysis is a very selective and carefully regulated process that has a deep influence on the degree of protein accumulation in *E. coli*. Normally, protein breakdown serves as a cellular "sanitation" system that eliminated unnecessary polypeptides. In addition, this process can also provide a source for amino acids and energy in conditions of limited nutrient availability or starvation.

Protein degradation has important practical consequences since many cloned proteins are recognized as abnormal by the proteolytic system of the cell, and are therefore rapidly hydrolyzed.[72] This selective degradation implies that bacterial proteins must share certain conformational features that normally prevent their rapid turnover. If the conformation of a heterologous protein resembles that of a native product, the probability of degradation is lower. This fact accounts for the dramatic differences in stability of highly expressed products that have been reported so far.[78]

A preferred strategy to overcome selective degradation of cloned products has been the generation of gene fusions (translational fusions) that direct the synthesis of hybrid proteins or secretable proteins. In the first case, the carrier portion of the protein stabilizes against degradation by virtue of a conformational change. In the second case, degradation is avoided by exporting the protein product to the periplasmic space. These topics are discussed elsewhere in this volume.

Hybrid products require a cleavage process for the liberation of the protein of interest from the precursor molecule. One interesting aspect of rapidly synthesized abnormal proteins is that they accumulate in characteristic intracellular granules, known as inclusion bodies.[79] Also, these tight aggregates may be purified by a one-step procedure, which represents a clear advantage for the subsequent process.[80-82]

[76] C. T. Caskey, *Trends Biochem. Sci.* **5**, 234 (1980).

[77] P. M. Sharp and M. Bulmer, *Gene* **63**, 141 (1988).

[78] A. L. Goldberg and S. A. Goff, *in* "Maximizing Gene Expression" (M. Reznikoff and L. Gold, eds.), p. 287. Butterworth, Stoneham, Massachusetts, 1986.

[79] D. C. Williams, R. M. Van Frank, R. M. Muth, and J. P. Burnett, *Science* **215**, 687 (1982).

[80] R. G. Schoner, L. F. Ellis, and B. E. Schoner, *Bio/Technology* **3**, 151 (1985).

[81] J. M. Schoemaker, A. H. Brasnett, and F. A. O. Martson, *EMBO J.* **4**, 775 (1985).

[82] N. Flores, R. de Anda, L. Guereca, N. Cruz, S. Antonio, P. Balbas, F. Bolivar, and F. Valle, *Appl. Microbiol. Biotechnol.* **25**, 267 (1986).

The synthesis of secretable proteins is based on the fact that a variety of native proteins are exported into the periplasmic space of *E. coli* by virtue of a signal peptide placed at the amino-terminal end of the protein.[83] This is not a commonly used procedure because the presence of the leader peptide is not sufficient to accomplish transport of all proteins through the membranes.[84,85] Furthermore, saturation of the protein export channels (porins) will not only hamper secretion, but may also interfere with the normal cellular functions and lead to cell death (F. Valle, personal communication).

An additional strategy is available to reduce protein degradation: the utilization of temperature-sensitive, conditionally lethal mutant strains deficient in proteases. These specialized host strains have had a dramatic impact on the expression of some gene products in *E. coli*.[38,86] A drawback in the use of constitutive protease mutant strains is their abnormal growth and increased lability. The use of protease inhibitors is not recommended either because these chemicals also induce a number of heat-shock genes.[87]

In conclusion, gene expression is not limited solely at the level of transcription and mRNA stability. Translation initiation is mediated by the 5' untranslated mRNA region. There is now compelling evidence that not only the RBS and the AUG determine the efficiency of the process, but also both structure and sequence of this region affect the ribosome accessibility to the transcript. Therefore, this region may be manipulated to maximize translation initiation. In contrast to the limited number of inducible promoters that are used in expression plasmid vectors, a large variety of regions for translation initiation are in use.

Although codon usage has often been assigned a crucial role in determining the efficiency of protein synthesis, there is now growing evidence that this is not necessarily the case.[71]

The issue of protein degradation represents a particularly difficult problem that may be encountered when overproducing a protein in *E. coli*. Although some solutions to degradation problems are available (translational fusions, special strains), there is no way in which an accurate prediction can be made *a priori* on whether a protein will be stable. The design of alternative DNA constructions is, to date, the best way to cope with a potential stability problem.

[83] L. L. Randall and S. J. S. Hardy, *Microbiol Rev.* **48**, 290 (1984).

[84] P. J. Bassford, Jr., T. J. Silhavy, and J. R. Beckwith, *J. Bacteriol.* **139**, 19 (1979).

[85] K. Ito, P. J. Bassford, Jr., and J. Beckwith, *Cell* **24**, 707 (1981).

[86] R. A. Watt, A. R. Shatzman, and M. Rosenberg, *Mol. Cell. Biol.* **5**, 448 (1985).

[87] S. A. Goff and A. L. Goldberger, *Cell* **41**, 587 (1985).

General Considerations about Selection of Host Strain

Every expression system is composed of two elements: the DNA directing the protein synthesis and the host. In the previous sections, an overview of the first component has been presented, and some particular aspects of bacterial-encoded functions for expression have been addressed. However, some generalities about strain selection should not be overlooked when designing an efficient expression system.

The properties of any microbial cell are ultimately determined by its genome, which carries the information that enables the organism to maintain its functional and structural integrity. It also contains the potential to respond to changes in its environment, and hence, to adapt to different conditions.[88]

The host genetic background can play an important role in the final yield of accumulated gene product. The reasons for the rather dramatic differences seen in product yields from different strains are not well understood, although plasmid stability and product degradation often appear to be the determining factors.

In many cases, strains carrying recombinant high-level expression systems have been found to be unstable. The host which carries a plasmid inflicting a high metabolic load tends to be overgrown by cells without the extra energy requirement, thus leading to plasmid segregation.

Besides the genetic background of the host, there are two other important issues for the selection of a strain as a host: growth requirements and the mode of regulation of gene expression. Growth requirements include (1) nutrients (e.g., vitamins, amino acids, and metals), (2) temperature, and (3) oxygen, all of which directly affect the costs of the fermentation process. Therefore, a strain with fewer specific requirements is preferred over others with several auxotrophies or specific growth requirements. In addition to the genotype of the host strain, the mode of induction can also affect the overall accumulation of a gene product. Such modes of induction of an expression system will undoubtedly bias the selection of a host strain, since a particular genotype is usually needed for maximal induction. The most widely used induction methods include heat (λpL), chemical inducers (*trp, lac, λpL*), and nutrient starvation *(trp, phoA)*. The host response to any of these induction methods will lead to different cellular states that range from mild alterations (modulation of metabolic pathways) to dramatic changes (SOS or heat-shock response mechanisms).[38,89]

Moreover, it has been demonstrated that the same expression vector

[88] H. P. Meyer, O. Kappeli, and A. Fiechter, *Annu. Rev. Microbiol.* **39,** 299 (1985).
[89] S. Gottesman, *Annu. Rev. Genet.* **18,** 415 (1984).

exhibits different product yields in a variety of host strains.[90] The reasons for this observation remain obscure.

Based on the above considerations, it is always desirable to have a series of strains with the necessary genetic traits for the particular expression system, in order to scan for the highest productivity yields.

Important Parameters of Fermentation Conditions

Once a hypothetically good expression system has been constructed and an adequate host strain has been selected, establishment of the growth and induction conditions is necessary to obtain the maximal efficiency of the system.

The environmental parameters that commonly influence the growth of microbial cells can be divided into two categories: physical (temperature, oxygen flow rate) and chemical (media composition, pH). Usually, the alteration of one of these parameters will lead to a different behavior of the culture, and therefore to a change in the final productivity of the system.

A good productivity of a specific expression system is obtained when the result of a fermentation is a large biomass containing a large amount of the desired protein. Plasmid stability is then an essential issue in the fermentation of strains carrying recombinant DNA. The instability of a recombinant plasmid in a culture may reduce the overall levels of the desired product in the fermentation, and increase production costs, since growth substrates are consumed by nonproductive cells.

In general, the segregational instability of a plasmid may be affected not only by the genetic features of the host cells, as previously discussed, but also by the copy number of the plasmid, the culture conditions, and the genes contained in the plasmid.

Based on the observation that at high growth rates the copy number of plasmids per cell is lower,[8] the culture conditions must be settled to restrain the growth rates to avoid plasmid loss. Moreover, fast growth rates are usually achieved only at the expense of product output. Therefore, optimal culture conditions for product formation are usually incompatible with optimal growth conditions.[88] The obvious consequence is that product formation needs to be evaluated at different growth rates.

The growth rate may be modulated by nutrient input, aeration rate, or temperature. Nutrient limitation, however, may hamper the product yield, so low temperature is a better modulator of growth.[88,91]

---

[90] P. S. Kaytes, N. Y. Theriault, R. A. Poorman, K. Murakami, and C. C. Tomich, *J. Biotechnol.* **4**, 205 (1986).

[91] A. W. Emerick, B. L. Bertolani, A. Benbassart, and T. J. White, *Bio/Technology* **2**, 165 (1984).

The genes contained in the plasmid will also determine plasmid stability and the overall product yield as well. Cloning and expression vectors usually contain at least one gene for selection which, as discussed earlier, has to exert a beneficial effect to avoid segregation.

Manipulation of Plasmid DNA

Because plasmids are one of the central tools in molecular cloning, a number of efficient methods for their purification, storage, mutagenesis and transformation of hosts have been developed. There are protocol manuals and collections of methods available that give detailed procedures on plasmid manipulation, so they will not be described here.[5,11,92-94] However, two issues deserve careful consideration to ensure success in the application of techniques: physical manipulation and storage.

Low efficiencies in methodology performance are usually a result of coarse and careless manipulation of the DNA, resulting in the chemical modification of the DNA bases or in the alteration of the covalently closed circular nature of the plasmid. Furthermore, careless preparation of solutions, or the use of old stock solutions may induce problems due to pH variations, oxidation, biological contamination, precipitation of salts, or the presence of impurities. Excessive UV-induced damage to DNA is not uncommon if the wavelength cutoff of working lamps is lower than 300 nm.

A variety of plasmid and strain storage conditions have been reported. Purified plasmid may be stored dried, ethanol-precipitated, or in aqueous solution at different temperatures. DNA in solution, however, is usually more sensitive to temperature variations (nicking, breaking) and to biological contamination.[11,92-94]

Storage of transformed cells is a very important factor, especially in the case of expression systems. It is a common phenomenon to find substantial variation in the levels of expression obtained from cells stored under different conditions and for various periods of time. Segregational and structural instabilities of the expression plasmid are easily detected in stored bacteria. It is not known whether these variations are due to increased cell sensitivity and/or to specific selection against the population of cells with a higher metabolic load. The quality of the inoculum should be

[92] T. Maniatis, E. F. Fritsch, and J. Sambrook, "Molecular Cloning: A Laboratory Manual." Cold Spring Harbor Laboratory, Cold Spring Harbor, New York, 1982.

[93] J. M. Walker, "Methods in Molecular Biology, Vol. 2, Nucleic Acids." Humana Press, Clifton, New Jersey, 1984.

[94] L. G. Davis, M. D. Dibner, and J. F. Battey, "Basic Methods in Molecular Biology." Elsevier, New York, 1986.

standardized by transforming the host prior to every fermentation (P. Balbas and F. Bolivar, personal observations).

Improvement of Expression Systems

The existence of a multiplicity of elements that affect the overall efficiency of a plasmid-based expression system implies some difficulties in the selection and further improvement of a particular system. Selection of the molecular elements that direct transcription and translation may be based on the extensive body of information regarding each particular aspect, the availability of the system components, and personal preference. Improvements on a particular gene expression system, however, must be done on a case-by-case basis because every genetic construction, transcript, and protein has its own particular effects on the host, and, consequently, on the final product yield.

Maximizing gene expression is usually a costly and time-consuming enterprise based largely on experimentation. The data usually presented in the literature are concerned with the final protein output in relation to the total protein content, so the values reported do not reflect the efficiency of any particular aspect of the expression plasmid/host system.

There are a few quantitative data on the impact that some elements of the expression system exert on the final product yield. Although the numerical values may not be universally true, they indicate their relevance as key steps that may be preferentially improved. These selected data include (1) the induction range of the promoter (which is the level of maximal repression versus the level of maximal induction) may vary up to 1000-fold[95-97]; (2) The effect of the host genetic background may influence the final product yield of one DNA construction and induction scheme as much as 1000-fold[90]; (3) the influence of the mRNA 5'-end hairpin structures may affect translation initiation up to 100-fold[55,64,65]; (4) the growth rate of a culture may affect the half-life of some mRNA species up to 50-fold[55]; (5) the mRNA 5'-terminus sequence may affect translation initiation up to 3-fold.[98,99]

Almost inevitably, this information leads to transcription initiation and host strain development being the primary steps for improvement, followed by the analysis and modification of the structures in the mRNA 5' terminus. However, elements other than those mentioned above may be

[95] H. Bujard, *Trends Biochem. Sci.* **19**, 274 (1980).
[96] A. von Gabain and H. Bujard, *Proc. Natl. Acad. Sci. U.S.A.* **76**, 185 (1979).
[97] C. Yanofsky, R. L. Kelley, and V. Horn, *J. Bacteriol.* **158**, 1018 (1984).
[98] M. D. Matteucci and R. L. Heyneker, *Nucleic Acids Res.* **11**, 3113 (1983).
[99] H. M. Shepard, E. Yelverton, and D. V. Goeddel, *DNA* **1**, 125 (1982).

decisive for a particular protein; therefore, experimentation involving any of the aspects presented in Table I may produce a more efficient system.

Acknowledgments

We are grateful to Enrique Merino and Xavier Soberon for fruitful discussions during the preparation of this manuscript.

[4] Promoter Recognition and mRNA Initiation by Escherichia coli $E\sigma^{70}$

By Jay D. Gralla

The purpose of this article is to give an overview of considerations relevant to expression from promoters recognized by $E\sigma^{70}$, the major RNA polymerase in *Escherichia coli*. The article is organized into three main sections. The first section discusses the complex mechanism by which mRNA initiation occurs. The goal here is to describe the mechanistic steps involved, with special emphasis on how expression could be limited at each of these steps. The second section discusses the likely roles of the three core promoter DNA elements, the -10 region, the -35 region, and the spacer. The third section discusses how promoter-dependent expression is controlled by trans-acting factors. Rather than being a survey, this section emphasizes the mechanistic considerations important to modulating transcription. Almost all of this information is available, collectively, in various excellent reviews,[1-14] and these will be the primary reference sources for

[1] P. H. von Hippel, D. B. Bear, W. D. Morgan, and J. A. McSwiggen, *Annu. Rev. Biochem.* **53**, 389 (1984).

[2] M. J. Chamberlin, *Annu. Rev. Biochem.* **43**, 721 (1974).

[3] W. R. McClure, *Annu. Rev. Biochem.* **54**, 171 (1985).

[4] W. Gilbert, *in* "RNA Polymerase" (R. Losick and M. Chamberlin, eds.), p. 193. Cold Spring Harbor Laboratory, Cold Spring Harbor, New York, 1976.

[5] O. Raibaud and M. Schwartz, *Annu. Rev. Genet.* **18**, 173 (1984).

[6] M. Rosenberg and D. Court, *Annu. Rev. Genet.* **13**, 319 (1979).

[7] U. Siebenlist, R. B. Simpson, and W. Gilbert, *Cell* **20**, 269 (1980).

[8] T. D. Yager and P. H. von Hippel, *in* "*Escherichia coli* and *Salmonella typhimurium*" (F. C. Neidhardt, ed.), p. 1241. American Society for Microbiology, Washington, D.C., 1987.

[9] M. Ptashne, "A Genetic Switch." Blackwell, Cambridge, Massachusetts, 1985.

[10] C. Yanofsky, *J. Biol. Chem.* **263**, 609 (1988).

[11] T. Platt, *Ann. Rev. Biochem.* **55**, 339 (1986).

[12] W. Reznikoff, D. Siegel, D. Cowing, and C. Gross, *Annu. Rev. Genet.* **19**, 355 (1985).

[13] B. C. Hoopes and W. R. McClure, *in* "*Escherichia coli* and *Salmonella typhimurium*" (F. C. Neidhardt, ed.), p. 1231. American Society for Microbiology, Washington, D.C., 1987.

[14] F. C. Neidhardt, *in* "*Escherichia coli* and *Salmonella typhimurium*" (F. C. Neidhardt, ed.), p. 1313. American Society for Microbiology, Washington, D.C., 1987.

this article. Before discussion of details, an overview of the transcription pathway is presented. This pathway consists of many steps and there is probably the potential for limitation of expression at every step.

The first steps involve formation of the holoenzyme $E\sigma^{70}$ from core polymerase and σ^{70} protein, followed by initial DNA search and recognition by this enzyme. The search process begins with nonspecific DNA binding by the enzyme, and, by facilitated transfer mechanisms, the promoter is subsequently located. It is not yet known if there is ever physiological control of the search process, a phenomenon which would require preferential accessibility of genes targeted for expression, as appears to occur in the chromatin of eukaryotes.

The second steps involve a series of conformational changes which end when the RNA polymerase has succeeded in melting a stretch of DNA that exposes the "start point" of transcription. These changes occur slowly, and for a typical promoter it takes several minutes for the polymerase to locate the DNA and expose the transcription start point. Since these steps are so slow, they may limit expression for many promoters. The rate of these steps can be changed by altering promoter sequences or by regulators.

The third series of steps involve mRNA chain initiation by the bound RNA polymerase. These steps involve the RNA polymerase cycling reaction in which small abortively initiated RNA chains are produced. Eventually, an RNA primer of sufficient stability is produced and this is elongated into productive mRNA. σ factor is probably lost as the RNA polymerase breaks the strong contacts that bind it to the promoter and it escapes to transcribe the downstream DNA. This process probably takes many seconds, and thus has the potential to limit expression, especially from strong promoters.

In the last series of steps under consideration here, the enzyme elongates the full-length mRNA. Under normal circumstances the mRNA is produced at the rate of 100 nucleotides every 2 or 3 sec, and this is too fast to contribute significantly to limited expression. Instead, flow through the promoter, as described above, is normally slow, and this is followed by rapid elongation. However, there are clear examples of polymerase pausing during elongation. Some of these pauses can last for many seconds and thus can contribute to lessened expression. Such pauses are usually purposeful, as they are often sites of regulation.

Overall, then, the mRNA production pathway is extremely complex. There are many steps at which expression can be limited and these are also targets of a variety of regulatory circuits. The following sections present an overview of the mechanisms by which these steps proceed and are controlled.

Steps That May Limit Expression

Promoter Search and Initial Recognition

The first steps in promoter utilization lead to the loose, but specific, association of RNA polymerase with the promoter DNA.[1,2,15] This must be accomplished in the presence of a huge excess of competing nonpromoter DNA sites. However, the effect of nonpromoter DNA may be complex, since by participating in facilitated protein transfer, it could actually increase the rate at which the polymerase finds the promoter. Thus, one imagines that the enzyme is initially bound nonspecifically to a nonpromoter site, perhaps principally via electrostatic forces. It is then transferred to a second DNA site which may have some resemblance to a promoter, perhaps involving degenerate -10 or -35 core sequences. Such transfers may continue, with the rate of each step dependent on how closely each sequence resembles that of a promoter; the stronger the resemblance, the longer specific contacts may hold the polymerase there and inhibit its removal. Occasionally, the enzyme may be lost from very weak sites to free solution. These transfers may occur by the well-established DNA looping mechanism.[16,17] Also, as part of each transfer it is likely that the polymerase scans a short region of the genome as a DNA region slides by.[15,18] The end result of this process is that a large segment of the genome is scanned rapidly. The details of this explicit version of the facilitated transfer model are largely untested.

During this search process the polymerase will inevitably encounter authentic promoters. There are two ways in which one imagines the enzyme to be held there long enough for it to melt the DNA and form a stable, "open" complex. Both mechanisms require stabilizing contacts. First, the bases in the -35 or -10 region may form molecular contacts with the enzyme and hold it there. Second, an auxiliary factor, such as catabolite activator protein, may be bound nearby and contact the polymerase, holding it near the promoter. In either case, the polymerase is held near the promoter long enough for it to have a reasonable chance of making the extensive contacts to DNA that are required for steps leading to mRNA initiation.

The complex that has now formed is certainly one species of "closed" complex, in which the DNA is unmelted.[1-3] The stability of such com-

[15] P. H. von Hippel, D. B. Bear, R. B. Winter, and O. G. Berg, *in* "Promoters: Structure and Function" (R. Rodriguez and M. Chamberlin, eds.), p. 3. Praeger, New York, 1982.

[16] M. Ptashne, *Nature (London)* **322**, 697 (1986).

[17] R. Schleif, *Nature (London)* **327**, 369 (1987).

[18] M. Richetti, W. Metzger, and H. Heumann, *Proc. Natl. Acad. Sci. U.S.A.* **85**, 4610 (1988).

plexes varies significantly among promoters[3,13]; in many cases they are likely so unstable that they will dissociate and have to form again after the polymerase researches the nonpromoter DNA. It is not known whether any particular region of the promoter is of greater importance in stabilizing such initial closed complexes. Indeed, the region of paramount importance could vary among promoters, depending on which region provided the best initial contacts or on whether an activating factor was bound nearby.[19] Promoter search and initial recognition may be considered over when a complex is formed that is long-lived enough to initiate conformational changes that stabilize and specifically position the enzyme at the promoter.

Conformational Changes in Polymerase–DNA Complex

Early models for what happens after initial recognition emphasized simple DNA melting. Thus, the DNA was "opened" by RNA polymerase in a closed complex,[2] or an intermediate (A*) complex "converted" to an open complex,[20] or the initial complex "isomerized" by melting the DNA.[21] All of these models are now known to be oversimplifications, as a series of conformational changes precedes the stable positioning of the enzyme.[3] These conformational changes can be relatively slow, and for many promoters this is likely the slow process that limits expression. The rate of these steps can be changed only by changing the promoter DNA sequence or by auxiliary factors.

The purpose of these conformational changes is to melt an approximately 12-base-pair region and expose the start point of transcription so that it can be read.[7] The substantial energy barrier involved in melting DNA is likely the cause of the slow rate of these steps. A key consideration here is whether the contacts that hold the enzyme at the promoter are sufficiently stabilizing to prevent dissociation from occurring before melting is accomplished. This stabilization likely increases progressively as the series of conformational changes occur. In one version of this process,[19] a key step occurs when the enzyme finally firmly contacts the -10 and -35 promoter regions simultaneously, and thus is strongly stabilized. Alternatively, the polymerase may be stabilized by contacts with bound stimulatory proteins. In this manner, such proteins may promote the necessary conformational changes by holding the polymerase at the promoter long enough for the multistep melting process to occur. It is also possible that the bound stimulatory factors promote the conformational changes directly by inducing structural transitions in the DNA or the polymerase.

[19] J. E. Stefano and J. D. Gralla, *Proc. Natl. Acad. Sci. U.S.A.* **79,** 1069 (1982).
[20] J. E. Stefano and J. D. Gralla, *J. Biol. Chem.* **255,** 10423 (1980).
[21] W. R. McClure, *Proc. Natl. Acad. Sci. U.S.A.* **77,** 5634 (1980).

The detailed nature of these conformational changes is not yet known. Some current models suggest that their goal is to induce topological stress in the complex that is relieved when the DNA is melted. Minimally, the process should involve changing contacts between σ factor and the two DNA recognition regions near -10 and -35.[22] The fact that the complex is bonded principally at two sites separated by more than 50 Å, and probably not on the same face of the helix, may be helpful in promoting topological stress, which requires two points of contact. Alternatively, the two points of contact could be, for example, to the promoter -10 region and to a bound stimulatory factor. These changes are strongly affected by negative DNA supercoiling[23,24] which assists in driving topological changes that involve net DNA unwinding.

Since these steps may limit promoter expression there is a natural interest in how to change their rate. The most obvious possibility is to provide the maximum number of stabilizing contacts, either by changing sequences near -10 and -35, or by building a nearby site for a stimulatory protein. The main problem here is that, by increasing the stability of the polymerase–DNA complex, one may be increasing the difficulty of breaking the complex, which is required for mRNA chain initiation. Thus, in principle, by increasing the rate of open complex formation one might slow the subsequent RNA chain initiation steps and end up without a net gain in expression. These initiation steps are discussed in the next section.

Escape from Promoters

The factors controlling the rate at which polymerase initiates an RNA chain and escapes from the promoter are not well understood. This is unfortunate, since for very strong promoters this process could well limit expression. This section summarizes relevant aspects of what is known about promoter escape or clearance.[1,3] Some more detail will be given, since this subject has not been reviewed extensively.

An early hint that RNA chain initiation could limit expression came from study of the *lac*UV5 promoter. At this promoter, chain initiation was slow and under certain circumstances could be slower than the rate at which open complexes could form.[25] This behavior contrasted sharply with the T7 promoters, which were clearly "rapid start."[26] Eventually, it became clear that this rate could vary widely among promoters, between

[22] R. Losick and J. Pero, *Cell* **25**, 585 (1981).
[23] J. A. Borowiec and J. D. Gralla, *J. Mol. Biol.* **184**, 587 (1985).
[24] J. A. Borowiec and J. D. Gralla, *J. Mol. Biol.* **195**, 89 (1987).
[25] J. E. Stefano and J. D. Gralla, *Biochemistry* **18**, 1063 (1979).

"slow start" promoters like *lac*UV5 and Tn7 and rapid start promoters like T7 A$_3$ and A$_2$.[25-33]

Part of the basis for this effect is related to the ability of RNA polymerase to catalyze an "abortive initiation" reaction. That is, if bound RNA polymerase has available only the first two complementary nucleotide triphosphates, then it reiteratively produces a complementary dinucleotide.[34] This is an intrinsic property of the enzyme and does not occur only at slow start promoters. At many slow start promoters, however, this type of process probably occurs frequently even in the presence of all four nucleoside triphosphates. The enzyme repeatedly produces and releases short initiation products. It does this while it is held firmly at the promoter in a so-called "cycling" pathway.[27] Since this cycling to produce abortive initiation products takes time, the synthesis of a properly initiated mRNA chain can be quite slow.[31,32]

It is not yet possible to predict from DNA sequence data which promoters will be slow starters. One clue comes from study of a related set of mutant *lac* promoters, where the chain initiation rate and the rate of open complex formation were both studied. In that case, the rates of the two processes changed oppositely, as the rate of open complex increased the chain initiation rate slowed down. The reason for this was suggested to be that the same contacts that accelerated open complex formation by holding the enzyme at the promoter also held it there subsequently, thus preventing rapid escape.[32] This may be partly responsible for the inability of the enzyme to move rapidly away from the promoter and make a long RNA transcript. Thus, the nontranscribed region of the promoter clearly has a role in determining the rate of RNA polymerase escape and promoter clearance. In principle, these same considerations could apply to factor-dependent transcription, since factors may mimic DNA sequence in contacting the polymerase and holding it at the promoter.

It has also been suggested that the transcribed region, coding for the potential abortive products of cycling, has a role in setting the rate of polymerase escape and promoter clearance. In one case, a relatively strong promoter could be further strengthened by changing the DNA sequence

[26] W. Niermann and M. Chamberlin, *J. Biol. Chem.* **254**, 7921 (1979).

[27] A. J. Carpousis and J. D. Gralla, *Biochemistry* **19**, 3245 (1980).

[28] L. M. Munson and W. S. Reznikoff, *Biochemistry* **20**, 2081 (1981).

[29] M. A. Grachev and E. F. Zaychikov, *FEBS Lett.* **115**, 23 (1980).

[30] J. R. Levin, B. Krummel, and M. Chamberlin, *J. Mol. Biol.* **196**, 85 (1987).

[31] J. D. Gralla, A. J. Carpousis, and J. E. Stefano, *Biochemistry* **19**, 5864.

[32] A. J. Carpousis, J. E. Stefano, and J. D. Gralla, *J. Mol. Biol.* **157**, 619 (1982).

[33] A. J. Carpousis and J. D. Gralla, *J. Mol. Biol.* **183**, 165 (1985).

[34] D. E. Johnston and W. R. McClure, *in* "RNA Polymerase" (R. Losick and M. Chamberlin, eds.), p. 413. Cold Spring Harbor Laboratory, Cold Spring Harbor, New York, 1976.

within this region.[35] In another case, it was shown that a transcript starting with guanosine produced fewer cycling products than a nearly identical transcript initiated with adenosine.[32] Cycling reactions are likely ended when σ is released, contacts with the promoter are broken, and a downstream elongation complex is formed. The point of σ release may vary with DNA sequence and this too should have an effect on cycling and the rate of escape. Unfortunately, there have been no systematic studies addressing these issues.

The possibility that polymerase escape can be regulated has also been discussed.[33,36,37] In one case, the region coding for cycling products was included in the element controlling the responsiveness of a promoter to physiological regulation by DNA supercoiling.[37] In another case, a stimulatory transcription factor was proposed to work by allowing a tightly bound polymerase to escape and initiate mRNA synthesis.[38] The general possibility that transcription could be controlled at this step has been recognized for some time,[25] and more recently has been discussed principally in terms of limiting expression from potentially very strong promoters. Very recent evidence in fact suggests that the ribosomal P_1 promoter, probably the strongest in *E. coli,* may be rate-limited by the slow escape of tightly bound polymerase.[39] There is much yet to be learned concerning the rules governing RNA polymerase escape and promoter clearance.

Elongation of mRNA

Elongation of the mRNA is generally rapid, and in most circumstances probably does not contribute to the limitation of RNA synthesis.[8,40,41] However, if such a limitation occurs it could be difficult to eliminate genetically, since the controlling DNA sequence may be within the coding region. The main purpose of this section is to provide a brief overview of control of elongation. RNA 3′ end formation and processing are not considered.

Transcription elongation probably begins when the RNA polymerase cycling pathway ends. At this point, the enzyme has essentially synthesized a short, stable RNA primer and is irreversibly committed to long RNA

[35] U. Deuschle, W. Kammerer, R. Gentz, and H. Bujard, *EMBO J.* **5,** 2987 (1986).

[36] D. Straney and D. Crothers, *Cell* **43,** 449 (1985).

[37] R. Menzel and M. Gellert, *Proc. Natl. Acad. Sci. U.S.A.* **84,** 4185 (1987).

[38] M. Menendez, A. Kolb, and H. Buc, *EMBO J.* **6,** 4227 (1987).

[39] K. Olsen and J. Gralla, unpublished results (1988).

[40] M. Chamberlin, *in* "RNA Polymerase" (R. Losick and M. Chamberlin, eds.), p. 17. Cold Spring Harbor Laboratory, Cold Spring Harbor, New York, 1976.

[41] J. Krakow, G. Rhodes, and T. Jovin, *in* "RNA Polymerase" (R. Losick and M. Chamberlin, eds.), p. 127. Cold Spring Harbor Laboratory, Cold Spring Harbor, New York, 1976.

chain elongation.[33] This primed enzyme may have lost σ factor and its contacts to the promoter may have been broken. Elongation will in most cases be done by the core enzyme, which has quite different properties from the holoenzyme that was involved in open complex formation and mRNA initiation. The properties of the elongating core polymerase, however, can be changed by interactions occurring in transit, as will be discussed subsequently.

It has been known for some time that the elongating RNA polymerase pauses occasionally during transcription.[8,42] These pauses can last anywhere from a fraction of a second to many seconds. Since the longer pauses are comparable in length to the time required for rapid promoter recognition and escape by RNA polymerase, they can contribute to lowered levels of expression. Pausing has been best studied in λ phage gene transcription, but is probably widespread in cellular genes as well.

In the cases that have been studied, the long pauses occur at specific sequences and have specific functions. The λtR_1 pause occurs at a site in which *E. coli* host factor ρ causes regulated termination by releasing the RNA from the paused complex.[43] A conceptually related mechanism may occur at the *E. coli trp* attenuator, where controlled translation determines whether the polymerase can proceed past a potential DNA termination site and transcribe a regulated operon.[10] Examples of defined intragenic pause sites are the phage λ Nut and Qut sequences.[8] These are cis-acting control elements that determine whether downstream genes will be expressed. They are not, however, transcription termination sites. Instead, they are sites at which polymerase pauses to allow the elongation complex to be modified.[44,45] The modified polymerase is then relatively insensitive to downstream termination signals. In addition to these long pauses, shorter pauses frequently occur, as for example in the *lacZ* leader region downstream from the *lac*UV5 promoter.[42] The function of such pauses is not known, but they could also be involved in promoting the interaction of core RNA polymerase with other factors.

It is now known that many factors can associate with elongating core RNA polymerase and change its properties. Many of these are *E. coli* proteins known as *"nus"* factors.[8,44,45] NusA protein binds to core RNA polymerase, but not holoenzyme, and may be commonly associated with elongating RNA polymerases.[44] *In vitro,* elongating polymerase with *nusA* has a somewhat altered pattern of pausing and termination. When RNA polymerase transits through a λ phage Nut sequence it binds at least four other proteins; these are *E. coli* proteins *nusB, nusC,* and *nusE* (ribosomal

[42] N. Maizels, *Proc. Natl. Acad. Sci. U.S.A.* **70,** 3585 (1973).

[43] L. Lau, J. Roberts, and R. Wu, *J. Biol. Chem.* **258,** 9391 (1983).

[44] R. J. Horwitz, J. Li, and J. Greenblatt, *Cell* **51,** 631 (1987).

[45] S. Barik, B. Ghosh, W. Whalen, D. Lazinski, and A. Das, *Cell* **50,** 885 (1987).

protein S10) and the λ-encoded N protein.[44,45] This presumably occurs while the elongating polymerase is paused at the Nut site. It is likely that a modification of the elongation complex also occurs at the λQut site, involving λQ protein and perhaps host factors. The point of these events is to create an elongating complex with altered termination properties; enzymes that have been modified in this way are now resistant to termination at downstream sites. It has been suggested that analogous mechanisms exist for *E. coli* genes, especially downstream from the strong ribosomal RNA promoters. This view is supported by the involvement of three *E. coli*-encoded proteins in the λ phage mechanism.

Substantial information is available concerning the DNA sequences involved in pausing and regulation of elongation.[4,8,11] Many pause sites are one or two helical turns downstream from GC-rich regions. If such regions contain appropriate symmetry, their RNA transcripts have the potential to form particularly stable RNA hairpin loops. These loops probably form within microseconds of being transcribed and appear then to cause the downstream polymerase to pause. This may happen either by loops touching a hypothetical receptor site on the enzyme, or simply because they disrupt the DNA–RNA hybrid that is within the elongation complex. Not all such potential loops cause pausing, however, On the other hand, GC-rich regions without such symmetry may also cause pausing, perhaps by forming DNA–RNA hybrids of exceptional stability; these may be difficult to break and cause the enzyme to stall during the time it takes to dissociate them. Pause sites involved in antitermination control, such as the λNut sites, may have so-called "box A, B, or C" DNA sequences, that may have specialized roles in causing factors to bind to a paused elongation complex. Such considerations can be quite important to expression, not only because long pauses limit the flow of polymerase, but because pausing is believed to be a prerequisite for actual termination. The existence of a uridine-rich region downstream from a strong pause site, or a ρ factor site upstream, is probably sufficient to release the RNA and cause termination.

Obviously, the existence of strong pause sites within a cloned gene can have serious consequences for expression. Such sites can be identified by analysis of the distribution of transcripts along a gene. Since they may be within the coding region, there may be some difficulty in removing them by directed mutagenesis, if higher expression is desired. Existing knowledge about the sequences involved in causing pausing should be useful in this regard.

Role of Core Promoter DNA Sequences in Expression

The core DNA promoter elements for $E\sigma^{70}$ lie in an approximately 29-base-pair region between positions -7 and -36 with respect to the start point of transcription.[1,3,4] The hexanucleotide sequences at the two borders of this region constitute the two primary DNA sequence elements. Originally, the region near -10 was designated the Pribnow box and the one near -35 the recognition sequence.[4,46,47] This latter designation originated because it was thought possible that the polymerase entered the promoter at the upstream element before moving downstream. As discussed in the previous section, this is not the case, and the two elements are now thought of as simply two parts of the core promoter. The -10 and -35 elements are separated by the third core element, the spacer.[19,48,49] In general, the core promoter is that DNA region that is critically important for complex formation with σ^{70} RNA polymerase, in the absence of other factors. Polymerases containing other σ factors generally have different core DNA sequences.

These core elements were identified by both functional and statistical analyses.[1,3,4,6,7] Most importantly, they have been extensively mutagenized and the damaged elements were shown to be associated with defective transcription. Statistical analysis of compilations of promoter sequences confirmed that the elements were indeed concentrated within promoters. Probing of open promoter complexes showed that all three elements lay underneath the bound polymerase. Thus, the case for the importance of the three elements is overwhelming.

An important clue pertaining to the function of these three elements was the derivation of the "consensus rule." This rule states that the core promoter specifies more rapid transcription as its elements approach the consensus of the compilation of all promoters. For the -10 and -35 regions this refers to specific sequences, while for the spacer it refers to a particular length of DNA. This rule was derived from two general kinds of analyses. First, in a series of closely related mutant promoters, changes away from consensus invariably weakened the promoter.[48,49] Second, statistical analysis showed that promoters with rapid rates of open complex formation were closer to consensus, and a semiquantitative correlation could be established between rate and elements of the consensus.[50,51] It is important, however, to realize that this correlation applies to measurements of the rate of open complex formation measured *in vitro*. It is quite possible that certain classes of promoters will not obey the consensus rule when *in vivo* strength measurements are taken into account.

[46] D. Pribnow, *Proc. Natl. Acad. Sci. U.S.A.* **72**, 784 (1975).
[47] R. Dickson, J. Abelson, W. Barnes, and W. Reznikoff, *Science* **187**, 27 (1975).
[48] J. E. Stefano and J. D. Gralla, *J. Biol. Chem.* **257**, 13924 (1982).
[49] J. W. Ackerson and J. D. Gralla, *Cold Spring Harbor Symp. Quant. Biol.* **37**, 473 (1983).
[50] D. K. Hawley and W. R. McClure, *Nucleic Acids Res.* **11**, 2232 (1983).
[51] G. A. Mulligan, D. E. Hawley, R. Entriken, and W. R. McClure, *Nucleic Acids Res.* **12**, 789 (1984).

The existence of the consensus rule *in vitro* provides important constraints on models for how the elements are used by RNA polymerase as it forms an open complex. Since changing the spacer length altered the rate of this process, it was inferred that the -10 and -35 regions, which the spacer separates, had to be properly aligned along the helix circumference. By changing the spacer length one changes the spatial relationship between these regions. Therefore, this property must be important to the RNA polymerase during recognition of the promoter. It was concluded that the ability of the enzyme to touch these two sequence elements simultaneously was an important determinant of promoter function.[19] As discussed in the previous section, the two sequence elements likely provide the stabilizing contacts that allow the enzyme to stay associated with the promoter while it attempts to melt the DNA. The conformational changes that occur as the polymerase attempts to bind simultaneously to the two remote regions could help trigger DNA melting. The upstream border of the melted region is within one of these promoter elements near position -9, and extends well beyond it to $+3$. In the final open complex the polymerase extends from nearly -45 to nearly $+20$.

DNA sequences both upstream and downstream from the core can have an important influence on expression. The most obvious examples are upstream activator regions which, in most cases, are binding sites for regulatory proteins. How these factors work is not yet known in detail, and the mechanisms involved may be quite diverse. Currently, it is convenient to think that they make protein–protein contact with polymerase and thereby change its properties. In the simplest cases, the interactions are thought to provide additional stabilizing contacts, complementing those made between the holoenzyme and core DNA sequences.

It has also been suggested that noncore DNA sequences can alter promoter expression without the required intervention of accessory protein factors. For example, sequences with strong tendencies to bend or melt could play a role in promoter function in some instances. *Escherichia coli* σ^{70} promoters are in general quite rich in AT pairs and the sequences near -45 show some tendency toward being conserved.[6,7,50,51] Overall, the ability of the promoter region to assume a structure that allows it to closely contact, and perhaps wrap, polymerase is likely a significant determinant of open complex formation.[52,53] Unfortunately, little is known about these potential functions of noncore DNA sequences.

Sequences downstream from the core can also be important for function. Clearly, the transcription start site is a significant site. Most *E. coli* promoters begin with purines and first bond formation is probably slow

[52] J. A. Borowiec and J. D. Gralla, *Biochemistry* **25**, 5051 (1986).
[53] G. Kuhnke, H. J. Fritz, and R. Ehring, *EMBO J.* **6**, 507 (1987).

when the mRNA begins with a pyrimidine. As discussed above, the initial coding region may be important in controlling the RNA polymerase cycling reaction and thus the rate of escape from the promoter. In one study, as the core elements approached consensus, the chain initiation rate *in vitro* slowed progressively, although not enough to lead to lowered expression *in vivo*.[32] In another case, a synthetic promoter with consensus core elements functioned much less well than expected, probably because of defective RNA chain initiation.[54] Since promoters with close to consensus core elements may be rate limited at escape, the engineering of downstream sequences may assume some future importance.

Control of Expression

The control of expression is complex because gene-specific mechanisms must coexist with a bewildering array of apparently competing global control mechanisms.[14,55,56] Moreover, many gene-specific repression mechanisms are constructed so as to allow significant basal levels of expression. In some cases, this is because there is a physiological need for a minimal level of operon proteins, for example, the need for *gal* operon enzymes in biosynthetic as well as degradative pathways.[57] Alternatively, the sensing mechanism for gene activation may require a low level of operon protein; for example, in the *lac* operon[58] β-galactosidase cleaves newly introduced lactose to produce the inducer that directs the *lac* promoter to produce more *lac* enzymes. Since most products of operon expression are nontoxic in low amounts, there is little physiological need for supertight control. In this section, selected important strategic aspects of control are discussed.

Gene-Specific Mechanisms

Two of the tightest known repression systems are in the *lac* operon and in λ phage. In each case, in the natural context repression is 99.7–99.9% effective (it can be leakier out of context, see below). Both the operators and repressors in these systems have certain common features that are

[54] J. Rossi, K. Saboron, Y. Muramoto, J. McMahon, and K. Itakura, *Proc. Natl. Acad. Sci. U.S.A.* **80**, 3203 (1983).

[55] B. Magasanik and F. C. Neidhardt, *in* "*Escherichia coli* and *Salmonella typhimurium*" (F. C. Neidhardt, ed.), p. 1318. American Society for Microbiology, Washington, D.C., 1987.

[56] M. Cashel and K. E. Rudd, *in* "*Escherichia coli* and *Salmonella typhimurium*" (F. C. Neidhardt, ed.), p. 1410. American Society for Microbiology, Washington, D.C., 1987.

[57] S. Adhya, *in* "*Escherichia coli* and *Salmonella typhimurium*" (F. C. Neidhardt, ed.), p. 1503. American Society for Microbiology, Washington, D.C., 1987.

[58] J. Beckwith, *in* "*Escherichia coli* and *Salmonella typhimurium*" (F. C. Neidhardt, ed.), p. 1444. American Society for Microbiology, Washington, D.C., 1987.

probably important for optimal repression and responsiveness. An important aspect of these features is that they allow for a cooperative repression mechanism.

In the case of λ,[9] repression is mediated by three tandem, nonidentical operator sites. These sites are particularly effective in part because the repressors that bind them can form a cooperative complex using protein–protein interactions. This requires that two of the sites be arranged with a spacing that allows two DNA-bound repressors to bind each other. The extra energy gained presumably increases the lifetime of the complex, increasing the probability that the promoter will be prevented from functioning. Thus, protein–protein interactions involving multiple operators increase the effectiveness of repression. In principle, a similar increase would be associated with a single, tighter binding repressor. However, such a hypothetical repressor might be difficult to remove during induction, leading to lowered maximal levels of expression.

A somewhat analogous process accounts for the tight repression in the *lac* operon. In this case, the multiple operators are separated by 400 base pairs, but still act cooperatively to strengthen repression.[59,60] One of these sites overlaps the promoter and the other is actually within the coding region for the gene being repressed. So far, all known repression systems contain at least one element overlapping the promoter, although the overlap is not necessarily with the core DNA elements. When repressor occupies both *lac* operator elements *in vivo,* a repression complex is established which is tighter than one involving either operator alone. The effect is not extremely large, probably because the internal site is intrinsically weak at binding repressor on its own. The two sites must be properly spaced to optimize interaction between them,[61,62] probably to allow facile formation of a DNA loop.

The *lac* repression complex is effective in part due to a dual mechanism of repression.[60] First, the loop involves increased affinity of repressor with the operator site overlapping the promoter, blocking initiation more effectively. Second, the loop can block expression within the *lacZ* gene, presumably by inhibiting elongation past the intragenic operator element. This internal blockage mechanism is a secondary effect, the primary control being over initiation of transcription. Blockage mechanisms in general may depend on context, since investigations of blockage by the *lac* operator in artificial systems give mixed results. Plasmids containing genes cloned downstream from the *lac* promoter–operator often lack an effec-

[59] E. Eismann, B. von Wilcken-Bergman, and G. Müller-Hill, *J. Mol. Biol.* **195,** 949 (1987).
[60] Y. Flashner and J. Gralla, *Proc. Natl. Acad. Sci. U.S.A.* **85,** 8968 (1988).
[61] M. Mossing and T. Record, *Science* **233,** 889 (1986).
[62] H. Kramer, M. Amouyal, A. Nordheim, and B. Müller-Hill, *EMBO J.* **7,** 547 (1988).

tive internal operator and thus are probably not fully repressible. Therefore, for tightest repression, one probably wants a tight DNA-binding protein that is capable of forming tight higher order structures when interacting with a suitably spaced operator element of optimal sequence.

The particular location of the operator within the promoter could be a significant determinant of repression. There are two potential effects related to this. First, the operator must be located in a position that prevents polymerase function. It is now clear that this inhibition need not be at the level of excluding repressor binding. *In vitro, lac* repressor and RNA polymerase can associate simultaneously with the DNA; the polymerase is not, however, in a functional open complex.[63] The detailed position rules concerning this blocking function of repressors are not known. A second consideration is also related to this effect. The average lifetime of repressors on operators is probably no more than a few minutes. In the absence of loosely bound polymerase, this should be sufficient for tight repression since a dissociated repressor would rebind rapidly, preventing expression. Repressor normally wins this competition because rebinding is much more rapid than open complex formation by RNA polymerase, which takes longer due to the requirement for conformational changes leading to DNA opening. However, if a polymerase is already loosely bound, it may occlude the promoter after repressor leaves, leading to a round of expression, under repression conditions. There may be certain operator locations that are more effective in preventing these loose associations,[64] for example, within the spacer, that could lead to tighter repression.

Thus, at the level of DNA, the location of the operator within the promoter, and the presence of a suitably spaced remote operator could have important consequences for repression. Little is known concerning how to optimize the structure of the repressor with respect to DNA binding and cooperativity. For repression of plasmid-based expression it is obviously important to have repressor present at unusually high levels. Natural repression systems take advantage of these variables to maintain appropriate, rather than optimal, levels of repression. Engineered systems have not yet been optimized with respect to these variables.

Another means of achieving tight control is by the use of protein activators of transcription.[1,3,5,13] The activity of these proteins is generally regulated in response to the metabolic state of the cell. $E\sigma^{70}$ activation is principally achieved by binding of such proteins to specific target DNA sequences in close proximity to the core promoter elements. Obviously, if plasmid-based expression is involved, there needs to be an artificially high level of activator present. It is not yet known how these activators work in

[63] S. Straney and D. Crothers, *Cell* **51,** 699 (1987).
[64] M. Lanzer and H. Bujard, *Proc. Natl. Acad. Sci. U.S.A.* **85,** 8973 (1988).

detail. The current evidence suggests that they likely contact RNA polymerase as part of activation. This means that the spacing between the activation sequence and the core elements should have a significant influence on expression.

Even if polymerase contact is eventually shown to underlie activation, it is already clear that there is not a single mechanism of activation. In fact, probably all the steps in the transcription pathway are potential targets for activator proteins. *Escherichia coli* catabolite activator protein probably stimulates *lac* transcription by providing stabilizing contacts that assist in promoter location.[65-67] The nitrogen regulator NR-1 appears to cause already bound $E\sigma^{54}$ to melt the promoter DNA.[68,69] Catabolite activation of the maltose promoter may be via a polymerase that has already formed an open complex escaping from the promoter to begin productive RNA chain initiation.[38] Thus, it appears that any step that may limit expression may be accelerated by the intervention of an appropriate activator protein.

The use of activator proteins in conjunction with operators offers advantages in tightness and flexibility of control. For example, the *lac* operon is subject to lactose-specific control of up to 1000-fold and catabolite control of up to 20-fold. Thus, by growing in the absence of inducer and on rich media, very tight repression can be achieved in the natural *lac* operon. Exceptionally tight repression in a plasmid system would probably require strains that overproduce repressor and catabolite activator protein (CRP) and the presence of an appropriately spaced *lac* operator pair. The use of other gene-specific regulators with compatible core elements could possibly allow even tighter control.

Global Control

Global control mechanisms[14] usually supersede gene-specific control mechanisms, or at least modulate them strongly. Such controls usually sense the overall availability of groups of nutrients or the presence of various types of stresses. The global pattern of transcription is then adjusted accordingly. Although there are undoubtedly some hierarchical arrangements involved, these mechanisms likely work independently and can exert reinforcing or opposing effects on particular genes. There is probably no escape from the influence of these controls since one or another is likely operative under any given growth condition.

[65] T. Malan, A. Kolb, H. Buc, and W. McClure, *J. Mol. Biol.* **180**, 881 (1984).
[66] A. L. Meiklejohn and J. D. Gralla, *Cell* **43**, 769 (1985).
[67] Y. Ren, S. Garges, S. Adhya, and J. Krakow, *Proc. Natl. Acad. Sci. U.S.A.* **85**, 4138 (1988).
[68] A. Ninfa, L. Reitzer, and B. Magasanik, *Cell* **50**, 1039 (1987).
[69] S. Sasse-Dwight and J. D. Gralla, *Proc. Natl. Acad. Sci. U.S.A.* **85**, 8934 (1988).

Consider first several mechanisms that are operative when cells are growing exponentially, without external stress. If the cells are growing rapidly on glucose, glucose repression will be in effect. This involves the global repression of many genes involved in the utilization of alternative energy sources;[55] thus, even if lactose is present the *lac* operon is repressed, although not fully. This is related to the phenomenon of catabolite repression in which the expression of catabolic enzymes is repressed roughly in concert with the increasing richness of the media. In part, this repression is mediated by the lowering of cyclic AMP levels, which in turn deactivates those promoters that depend on cyclic AMP receptor protein for expression. There are probably other mechanisms involved as well.[55,70]

Superimposed on these "catabolite control" mechanisms is the phenomenon of "growth control." [56,71,72] Many *E. coli* genes have their expression level linked directly to the cell growth rate. That is, for many genes there is a strict correlation between transcription and doubling time; this holds even as the composition of the media changes drastically. The genes most strongly affected are those involved in macromolecular synthesis, these being needed in amounts appropriate to the rate at which new cells are being constructed. However, other genes are undoubtedly affected by growth control, both positively and negatively. A primary effector of growth control is believed to be the small "magic spot" nucleotide, ppGpp. This molecule accumulates with decreasing growth and inhibits the transcription machinery in an as yet uncharacterized manner. Taken together with the effects of catabolite control, it is easy to see how the same promoter can be affected simultaneously by different global regulatory pathways; there is no reason why cAMP, ppGpp, and other effectors cannot influence the same transcription machinery at the same time.

These considerations also highlight why global control mechanisms are likely to influence expression whatever growth medium is used. Since many of these global pathways have been defined genetically, it should be experimentally straightforward to decide if the particular promoter under study is subject to these controls. However, attempts to use strains deficient in these pathways to minimize global effects on expression of cloned genes may be difficult, since such strains may be generally expression-deficient, due to suboptimal levels of critical proteins.

[70] A. Ullman, *Biochimie* **67**, 29 (1985).
[71] S. Jinks-Robertson and M. Nomura, *in* "*Escherichia coli* and *Salmonella typhimurium*" (F. C. Neidhardt, ed.), p. 1358. American Society for Microbiology, Washington, D.C., 1987.
[72] H. Bremer and P. P. Dennis, *in* "*Escherichia coli* and *Salmonella typhimurium*" (F. C. Neidhardt, ed.), p. 1527. American Society for Microbiology, Washington, D.C., 1987.

Another set of global pathways involve starvation responses in which the cell senses that the supply of an important nutrient is dangerously low. Under these circumstances, the pattern of gene expression can change drastically. The "stringent response" to amino acid deprivation is the best characterized of these,[56] but there are also strong responses to restriction of essential elements such as carbon, oxygen, nitrogen, and phosphorus.[14,55,73] Transient responses of this type probably occur any time the bacteria adjust to a poorer growth medium. Such responses are mediated by small molecule effectors such as ppGpp, by changes in DNA supercoiling that influence supercoiling-sensitive promoters,[74,75] as well as by control proteins. Although the detailed mechanisms are still emerging, the effects on expression are clear.

Among the most dramatic global control mechanisms are the stress responses that sense a potentially toxic environmental challenge. These range from the well-characterized heat-shock[76] and SOS responses[77] to important but less well understood responses to toxins such as heavy metals and oxidizing agents. Such responses are always accompanied by the induction of specific protecting genes and sometimes involve global repression as well. The repression pathways here are not understood in terms of mechanism, although hypotheses are emerging.

From these considerations, it should be clear that the transcription machinery is targeted for control by many effectors. The expression from a given promoter is set by its intrinsic potential strength, by the state of gene-specific effectors, and by the powerful influence of global regulatory networks. Only in a few cases, such as repression via CRP–cAMP, do we know enough to inspect DNA sequence and then hazard a guess as to whether the promoter will be influenced by a particular global control pathway. In most cases, the importance of global control in limiting expression of a particular gene can be tested by assaying transcription in varying environments and strains. In some cases, expression may also be limited by "position" or "context" effects. These range from inhibition by

[73] B. L. Wanner, *in "Escherichia coli* and *Salmonella typhimurium"* (F. C. Neidhardt, ed.), p. 1326. American Society for Microbiology, Washington, D.C., 1987.

[74] V. L. Balke and J. D. Gralla, *J. Bacteriol.* **169**, 4499 (1987).

[75] C. Higgins, C. Dorman, D. Stirling, L. Waddell, I. Booth, G. May, and E. Breman, *Cell* **52**, 569 (1988).

[76] F. C. Neidhardt and R. A. VanBogelen, *in "Escherichia coli* and *Salmonella typhimurium"* (F. C. Neidhardt, ed.), p. 1334. American Society for Microbiology, Washington, D.C., 1987.

[77] G. C. Walker, *in "Escherichia coli* and *Salmonella typhimurium"* (F. C. Neidhardt, ed.), p. 1346. American Society for Microbiology, Washington, D.C., 1987.

opposing replication forks or transcription units to changes in DNA super-coiling that depend on the location of the promoter. Overall, then, expression is set by core and accessory sequences, modulated by genome position, and strongly influenced by gene-specific effector proteins and global effectors.

Acknowledgment

Research and preparation of this article were supported by USPHS grant GM35754.

[5] Use of *Escherichia coli trp* Promoter for Direct Expression of Proteins

By DANIEL G. YANSURA and DENNIS J. HENNER

Introduction

The promoter for the *Escherichia coli* tryptophan operon *(trp)* has been used for the expression of heterologous genes for more than a decade now, and in our hands is still the promoter of choice for the routine intracellular expression of heterologous proteins. A cursory survey of the literature shows that at least 100 proteins have been successfully synthesized with this promoter, testifying to its widespread utility. This promoter has several features which make it useful. First, it is very well understood,[1] making its behavior reasonably predictable. Second, it is strong, and heterologous genes transcribed by this promoter often result in protein accumulation to levels of 2–30% of the total cell protein (see Fig. 1). Third, its basal expression level is reasonably low. This usually avoids problems that can occur due to high-level constitutive expression of foreign proteins, such as slow growth of the cells, the accumulation of plasmid deletions, and the accumulation of cells that have lost the plasmid. Fourth, expression vectors using this promoter can be transferred to almost any *E. coli* strain. This makes it simple to survey a large number of *E. coli* strains for desirable properties, without having to either first move other control elements for the promoter into the desired strains or place them on the expression vector. The last feature worth mentioning is that the induction of the promoter is very simple, both in shake flasks and in fermenters.

There are general problems that may be encountered in attempting to express a protein in *E. coli* which are not specific to the *trp* promoter.

[1] C. Yanofsky, T. Platt, I. P. Crawford, B. P. Nichols, G. E. Christie, H. Horowitz, M. VanCleemput, and A. M. Wu, *Nucleic Acids Res.* **9**, 6647 (1981).

A   B

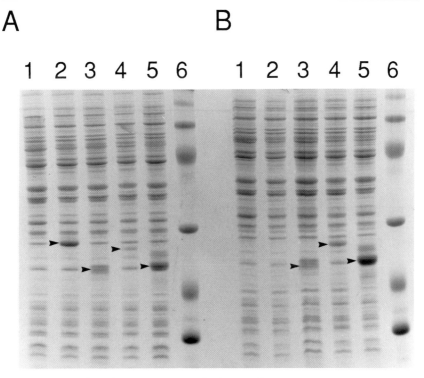

FIG. 1. Expression of heterologous proteins in *E. coli*. An expression vector similar to pHGH207-1 [H. A. de Boer, L. J. Comstock, and M. Vasser, *Proc. Natl. Acad. Sci. U.S.A.* **80,** 21 (1983)] was used to produce human prolactin (lane 2), human interleukin 6 (lane 3), murine prolactin (lane 4), and human growth hormone (lane 5). Lane 1 shows a culture transformed with pBR322. The cultures were prepared and analyzed as described in the text. (A) Five-hour samples; (B) 18-hr samples. Arrowheads point to the induced band in each culture. In each case the identity of the band was proved by further analysis (not shown). The protein molecular weight standards (lane 6) from bottom to top are 14K, 18K, 25K, 43K, 60K, and 97K.

These include lack of adequate translation initiation, effects of the 3' untranslated region on expression, and protein degradation. As these are dealt with in some detail in other articles in this volume, they will only be touched on here. In certain cases, one might prefer a promoter that is more tightly controlled than the *trp* promoter. Its induction ratio when on a pBR322-based plasmid is about 50-fold.[2,3] Certain proteins are toxic to the cell even at rather low levels. DNA encoding such proteins can be difficult to place downstream of the *trp* promoter and cells containing the resultant

[2] R. A. Hallewell and S. Emtage, *Gene* **9,** 27 (1980).
[3] G. Bogosian and R. L. Somerville, *Mol. Gen. Genet.* **193,** 110 (1984).

plasmids usually grow poorly. In such cases, a choice from one of the more tightly regulated promoters described in this volume might be appropriate. Finally, higher copy number plasmids using the *trp* promoter will usually not be regulated due to lack of sufficient *trp* repressor, and a strategy to increase the *trp* repressor level will be necessary to regulate expression from such plasmids.[4] However, in our experience, higher copy number plasmids do not usually have any advantage in foreign protein expression and we prefer the standard pBR322-based vectors.

Vector Construction

The design and construction of a particular direct expression plasmid using the *trp* promoter basically involves engineering a coding sequence downstream of the *trp* promoter and Shine–Dalgarno sequence (SD) of the leader peptide. The coding sequence should start with a methionine translational initiation codon, and this should be situated at a reasonable distance from the SD, usually in the range of 7–13 base pairs (bp). The natural distance in the *trp* operon is 7 bp,[1] and as this distance gets shorter than 7 or longer than about 13 bp expression of a heterologous protein can drop dramatically.[5,6]

There are a number of available plasmids containing the *trp* promoter that can be used in vector constructions.[7] The natural promoter sequence contains a *Taq*I restriction endonuclease site between the SD and the start codon for the leader peptide which can be used to prepare a promoter fragment. However, another *Taq*I site lies in the middle of the promoter requiring the use of a partial digestion. It is much easier to use one of the generally available *trp* expression vectors which have convenient restriction sites engineered either just before or just after the translational initiation codon. The promoter sequences of several such vectors are shown in Fig. 2. All of these vectors contain convenient restriction sites after the SD, while the vector pTrS3 contains an additional *Sph*I site after the methionine initiation codon.[8] Treatment of pTrS3 with *Sph*I, followed by treatment with DNA polymerase I (Klenow fragment) in the presence of deoxyribonucleotide triphosphates results in a blunt end immediately after the initiation codon.

[4] S. R. Warne, C. M. Thomas, M. E. Nugent, and W. C. A. Tacon, *Gene* **46,** 103 (1986).
[5] S. Itoh, T. Mizukami, T. Matsumoto, T. Nishi, A. Saito, T. Oka, A. Furuya, C. Takaoka, and T. Tanaguchi, *DNA* **3,** 157 (1984).
[6] H. M. Shepard, E. Yelverton, and D. V. Goeddel, *DNA* **1,** 125 (1982).
[7] B. P. Nichols and C. Yanofsky, this series, Vol. 101, p. 155.
[8] T. Nishi, M. Sato, A. Saito, S. Itoh, C. Takaoka, and T. Taniguchi, *DNA* **2,** 265 (1983).

We usually use short synthetic DNA oligonucleotides to link the promoter to an available restriction site within the coding sequence of the gene of interest. This DNA sequence will encode the initiation codon and first several codons of the gene. Since this part of the coding sequence is also part of the ribosome binding site, its nucleotide sequence can have significant effects on translation initiation. This is a major concern, especially for heterologous proteins, and one should refer to the article in this volume on ribosome binding sites for proper design of this region.[9] Fortunately, problems of poor translation initiation can almost always be overcome by simply altering the codons at the beginning of the gene.[10]

All the vectors described in Fig. 2 have transcription from the *trp* promoter proceeding in the direction of the tetracycline resistance gene *(tet)* of the plasmid pBR322. Any of the sites within *tet* can be used to anchor the 3' end of the gene. Sequences downstream of the translational termination codon can affect expression, and it is generally a good idea to remove large tracts of 3' untranslated sequences from expression vectors.[11] Also, the exact placement of the 3' end of the gene in the expression vector can sometimes affect the final expression levels. For example, we placed the 3' end of one gene at five different sites within the *tet* gene and observed 3- to 4-fold variations in the expression levels.[12] Such small differences in the expression levels due to subtle difference in the placement of the gene in the plasmid are not usually of concern for most routine expression projects.

trp Induction

Induction of the *trp* promoter is accomplished by starvation for tryptophan of *E. coli* cultures containing an expression plasmid. The procedure for induction of a directly expressed protein is essentially the same as that used for a *trpE* fusion.[13] Depending on the protein being expressed, some *E. coli* strains may produce more of the specific protein than others, and therefore one might wish to try several strains. Usually, a culture is grown overnight at 37° in a rich medium such as L broth[14] containing the appropriate antibiotic. This culture is then diluted 25- to 50-fold into a minimal medium such as M9[14] containing the same antibiotic and incu-

[9] L. Gold and G. D. Stormo, this volume [7].

[10] C. R. Wood, M. A. Boss, T. P. Patel, and J. S. Emtage, *Nucleic Acids Res.* **12,** 3937 (1984).

[11] T. Sato, H. Matsui, S. Shibahara, T. Kobayashi, Y. Morinaga, N. Kashima, S. Yamasaki, J. Hamuro, and T. Taniguchi, *J. Biochem.* **101,** 525 (1987).

[12] D. Yansura, unpublished (1988).

[13] D. Yansura, this volume [14].

[14] J. H. Miller, *in* "Experiments in Molecular Genetics," p. 431. Cold Spring Harbor Laboratory, Cold Spring Harbor, New York, 1972.

FIG. 2. Expression plasmid sequences. Sequences of the *trp* promoter (−35 and −10 regions), Shine–Dalgarno region (SD), and initiation codon of the *trpL* are shown. Restriction sites introduced to allow convenient construction of expression plasmids are shown below the natural sequence. Initiation codons are underlined. Sequence: *trpL* [C. Yanofsky, T. Platt, I. P. Crawford, B. P. Nichols, G. E. Christie, H. Horowitz, M. VanCleemput, and A. M. Wu, *Nucleic Acids Res.* 9, 6647 (1981)]; pHGH207-1 [H. A. de Boer, L. J. Comstock, and M. Vasser, *Proc. Natl. Acad. Sci. U.S.A.* 80, 21 (1983)]; pWT551 [W. C. A. Tacon, W. A. Bonass, B. Jenkins, and J. S. Emtage, *Gene* 23, 255 (1983)]; pSTP1 [J. D. Windass, C. R. Newton, J. DeMaeyer-Guignard, V. E. Moore, A. F. Markham, and M. D. Edge, *Nucleic Acids Res.* 21, 6639 (1982)]; pTrS3 [T. Nishi, M. Sato, A. Saito, S. Itoh, C. Takaoka, and T. Taniguchi, *DNA* 2, 265 (1983)].

bated with aeration at 37°. After 1–2 hr, indole-3-acrylic acid is added to a final concentration of approximately 25 μg/ml.[15] Incubation of the culture should be continued for an additional 3–5 hr, at which time samples are removed for analysis. Allowing the culture to continue incubating overnight sometimes results in higher levels of accumulation, depending on the protein of interest. The temperature of the culture can also affect the accumulation of certain proteins, and temperatures from 27 to 37° might be tried. It is wise to include controls of cultures with an expression vector without an expressed gene and, if available, with an expression vector for a protein which is known to accumulate well.

There are many possible ways to analyze the culture for expression of the specific protein. If the protein of interest has biological or enzymatic activity, the cell pellets from the induced culture can often be sonicated, treated with lysozyme and SDS, or dissolved in guanidine hydrochloride to

[15] W. F. Doolittle and C. Yanofsky, *J. Bacteriol.* 95, 1283 (1968).

release the proteins from the cell.[16,17] As the protein level in the cell is usually high enough to be visualized on a Coomassie brilliant blue-stained SDS–polyacrylamide gel, this method can also be used to follow expression of the protein of interest. Often the expressed protein accumulates in intracellular inclusion bodies. This allows a simple preliminary purification of the expressed protein based on differential centrifugation.[18]

Troubleshooting

For those proteins that are initially not successfully expressed, several options are open. The first would be to determine whether a fusion protein might not be adequate for the purposes of the experiment. This might be especially useful for small proteins which tend to not accumulate well due to proteolysis. As discussed in other articles in this volume, fusion proteins can almost always be successfully expressed, and in many cases techniques exist to process the fusion protein and release the desired protein. The next option would be to redesign the first portion of the coding sequence of the gene, while retaining the desired amino acid sequence. We usually use an empirical approach of screening a large number of N termini for the gene of interest. A mixture of synthetic oligonucleotides utilizing all the potential codons for the correct amino acid sequence is used at the N terminus. A variety of methods can then be used to select or screen for one which gives high expression levels. Another approach would be the rational design outlined by Gold and Stormo.[9] The third option is to screen a number of strains with mutations in various proteases, as described elsewhere in this volume.[19] The final option would be to attempt to use one of the various vectors designed for secretion.[20] We have had experience with several proteins which did not accumulate to high levels on our initial attempts to express them intracellularly, but which were expressed at high levels when secreted into the periplasm. All of these strategies can be carried out rapidly, usually within a month or two. The reward is often great, as one can purify several hundred milligrams to a gram of the desired protein from only a few liters of culture in a very short time.

[16] D. V. Goeddel, E. Yelverton, A. Ullrich, H. L. Heyneker, G. Miozzari, W. Holmes, P. H. Seeburg, T. Dull, L. May, N. Stebbing, R. Crea, S. Maeda, R. McCandliss, A. Sloma, J. M. Tabor, M. Gross, P. S. Familletti, and S. Pestka, *Nature (London)* **287,** 411 (1980).
[17] P. W. Gray, D. W. Leung, D. Pennica, E. Yelverton, R. Najarian, C. C. Simonsen, R. Derynck, P. J. Sherwood, D. M. Wallace, S. L. Berger, A. D. Levinson, and D. V. Goeddel, *Nature (London)* **295,** 503 (1982).
[18] S. J. Shire, L. Bock, J. Ogez, S. Builder, D. Kleid, and D. M. Moore, *Biochemistry* **23,** 6474 (1984).
[19] S. Gottesman, this volume [11].
[20] J. A. Stader and T. J. Silhavy, this volume [15].

Conclusion

With the strategy outlined above, we find that we can successfully express most heterologous proteins directly in *E. coli*. Our definition of success is that an obvious band is visible on a Coomassie brilliant blue stain of an SDS–polyacrylamide gel of the total cell lysate following induction.

[6] Use of T7 RNA Polymerase to Direct Expression of Cloned Genes

By F. WILLIAM STUDIER, ALAN H. ROSENBERG, JOHN J. DUNN, and JOHN W. DUBENDORFF

The RNA polymerase of bacteriophage T7 is very selective for specific promoters that are rarely encountered in DNA unrelated to T7 DNA.[1,2] Efficient termination signals are also rare, so that T7 RNA polymerase is able to make complete transcripts of almost any DNA that is placed under control of a T7 promoter. A very active enzyme, T7 RNA polymerase elongates chains about five times faster than does *Escherichia coli* RNA polymerase.[3,4] These properties, together with the availability of the cloned gene,[5,6] make T7 RNA polymerase attractive as the basis for expression systems in *E. coli*[6,7] and, potentially, in a variety of cell types.[8-10]

In principle, T7 expression systems can be completely selective, if the host cell RNA polymerase and T7 RNA polymerase recognize completely different promoters and if the host cell DNA contains no T7 promoters. In this situation, any DNA that can be cloned in the cell can be placed under control of a T7 promoter, since addition of the T7 promoter will not increase expression by the host RNA polymerase. T7 RNA polymerase introduced into such a cell will transcribe actively and selectively only the

[1] M. Chamberlin, J. McGrath, and L. Waskell, *Nature (London)* **228,** 227 (1970).
[2] J. J. Dunn and F. W. Studier, *J. Mol. Biol.* **166,** 477 (1983).
[3] M. Chamberlin and J. Ring, *J. Biol. Chem.* **248,** 2235 (1973).
[4] M. Golomb and M. Chamberlin, *J. Biol. Chem.* **249,** 2858 (1974).
[5] P. Davanloo, A. H. Rosenberg, J. J. Dunn, and F. W. Studier, *Proc. Natl. Acad. Sci. U.S.A.* **81,** 2035 (1984).
[6] S. Tabor and C. C. Richardson, *Proc. Natl. Acad. Sci. U.S.A.* **82,** 1074 (1985).
[7] F. W. Studier and B. A. Moffatt, *J. Mol. Biol.* **189,** 113 (1986).
[8] T. R. Fuerst, E. G. Niles, F. W. Studier, and B. Moss, *Proc. Natl. Acad. Sci. U.S.A.* **83,** 8122 (1986).
[9] W. Chen, S. Tabor, and K. Struhl, *Cell* **50,** 1047 (1987).
[10] J. J. Dunn, B. Krippl, K. E. Bernstein, H. Westphal, and F. W. Studier, *Gene* **68,** 259 (1988).

DNA under control of the T7 promoter. In at least some cases,[7] transcription by T7 RNA polymerase is so active that transcription by the host RNA polymerase apparently cannot compete, and almost all transcription in the cell rapidly becomes due to T7 RNA polymerase.

This article describes materials and techniques for using T7 RNA polymerase to express cloned DNA in *E. coli*.[7,11] Other configurations have been described[6] and further improvements are being explored, but we concentrate here on the current state of the system we have developed. In this system, T7 RNA polymerase is delivered to the cell from the cloned gene by induction or by infection. Vectors for cloning and expressing target DNA derive from the multicopy plasmid pBR322; all carry a T7 promoter followed by a unique cloning site, and some also carry translation initiation signals, a transcription terminator, or an RNase III cleavage site. This system is capable of expressing a wide variety of DNAs from prokaryotic and eukaryotic sources. Under favorable circumstances, the resources of the cell are devoted almost entirely to the production of target RNAs and proteins: within 1–3 hr, target RNA can accumulate to amounts comparable to ribosomal RNA, and target proteins can constitute the majority of total cell protein (Fig. 1).

Hosts and Vectors for Expression

HMS174 and BL21

The bacterial hosts for cloning and expression are the *E. coli* K12 strain HMS174 (F^- *recA* $r^-_{K12}m^+_{K12}$Rif^R)[12] and the B strain BL21 (F^- *ompT* $r^-_B m^-_B$).[7,13] HMS174 is used as the host for initial cloning of target DNA into pET vectors and for maintaining plasmids. As an expression strain, BL21 has the potential advantage that, as a B strain, it should be deficient in the *lon* protease, and it also lacks the *ompT* outer membrane protease that can degrade proteins during purification.[13] Thus, at least some target proteins might be expected to be more stable in BL21 than in host strains that contain these proteases. HMS174 has the potential disadvantage that rifampicin cannot be used to inhibit transcription by the host RNA polymerase in cases where a reduction of background synthesis of host RNA and proteins may be desirable.

[11] A. H. Rosenberg, B. N. Lade, D. Chui, S. Lin, J. J. Dunn, and F. W. Studier, *Gene* **56**, 125 (1987).

[12] J. L. Campbell, C. C. Richardson, and F. W. Studier, *Proc. Natl. Acad. Sci. U.S.A.* **75**, 2276 (1978).

[13] J. Grodberg and J. J. Dunn, *J. Bacteriol.* **170**, 1245 (1988).

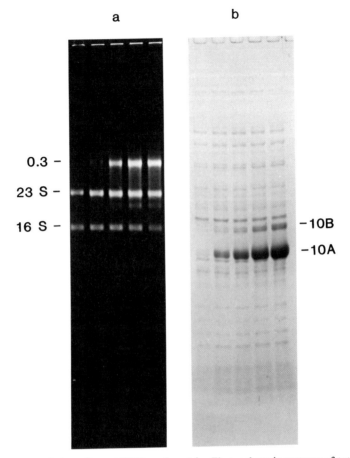

a   b

0.3 –

23 S –

16 S –

–10B

–10A

FIG. 1. Accumulation of target RNA and protein. Electrophoresis patterns of total cell RNA [visualized by ethidium bromide fluorescence (a)], and protein [visualized by Coomassie brilliant blue staining (b)] are shown for samples collected immediately before and 0.5, 1, 2, and 3 hr after induction of expression of target DNA (left to right within each set). Cultures of BL21(DE3) containing target plasmid were grown in M9ZB + ampicillin at 37° and expression was induced by addition of 0.4 mM IPTG when the OD_{600} reached about 0.7. For both cultures, the plating assay showed that more than 99% of the cells were capable of expressing target DNA at the time of induction. (a) The target plasmid was pAR946, which carries the T7 R0.3 RNase III cleavage site and gene 0.3 under control of the $\phi 10$ promoter (in a configuration equivalent to cloning in pET-1). The position of plasmid-length RNA produced by RNase III cleavage of T7 transcripts that have gone completely around the plasmid is indicated, as are the positions of the 16 and 23S ribosomal RNAs. After samples of culture were centrifuged, the cell pellet was resuspended in 5 volumes of sample buffer, heated for 2 min in a boiling water bath, and 5 μl was subjected to electrophoresis through a 1.4% agarose gel in 40 mM Tris-acetate (pH 8.0), 2 mM Na$_3$EDTA. (b) BL21(DE3) contained both pLysS and the target plasmid pAR3625 and was therefore grown in the presence of chloramphenicol as well as ampicillin. This target plasmid contains T7 gene 10 and its flanking $\phi 10$ and Tϕ signals (the equivalent of gene 10 cloned in a pET-3 translation vector). The positions of the major gene 10 protein, 10A, and a minor protein, 10B, which arises by

BL21(DE3) and HMS174(DE3) Lysogens

Bacteriophage DE3 is a λ derivative that has the immunity region of phage 21 and carries a DNA fragment containing the *lacI* gene, the *lac*UV5 promoter, the beginning of the *lacZ* gene, and the gene for T7 RNA polymerase.[7] This fragment is inserted into the *int* gene, and, because the *int* gene is inactivated, DE3 needs a helper for either integration into or excision from the chromosome. Once a DE3 lysogen is formed, the only promoter known to direct transcription of the T7 RNA polymerase gene is the *lac*UV5 promoter, which is inducible by isopropyl-β-D-thiogalactopyranoside (IPTG). (T7 RNA polymerase is produced from its own translation start and not as a fusion to the beginning of the *lacZ* protein.) Addition of 0.4 mM IPTG to a growing culture of either the BL21(DE3) or HMS174(DE3) lysogen induces T7 RNA polymerase, which in turn transcribes the target DNA in the plasmid.

pLysS and pLysE

Target genes whose products are sufficiently toxic cannot be established in BL21(DE3) or HMS174(DE3) because the basal level of T7 RNA polymerase activity will promote some transcription of the target gene in the uninduced cell. One way to reduce this basal activity (and thereby increase the range of target genes that can be maintained and expressed in these cells) is through the use of T7 lysozyme, a natural inhibitor of T7 RNA polymerase.[14,15]

T7 lysozyme is a bifunctional protein: it cuts a specific bond in the peptidoglycan layer of the *E. coli* cell wall,[16] and it binds to T7 RNA polymerase and inhibits transcription.[14] When produced from the cloned gene, relatively high levels of T7 lysozyme can be tolerated by *E. coli*, apparently because the protein is unable to pass through the inner membrane to reach the peptidoglycan layer. Treatments that disrupt the inner membrane but do not normally cause lysis, such as addition of chloroform or mild detergents, induce rapid lysis of cells that contain even small amounts of T7 lysozyme.

[14] B. A. Moffatt and F. W. Studier, *Cell* **49**, 221 (1987).
[15] F. W. Studier, unpublished work, 1986.
[16] M. Inouye, N. Arnheim, and R. Sternglanz, *J. Biol. Chem.* **248**, 7247 (1973).

---

translational frameshifting, are indicated. Samples of culture (20 μl) were mixed with 3× concentrated sample buffer (10 μl) and heated for 2 min in a boiling water bath, and the entire sample was subjected to electrophoresis through a 10–20% gradient polyacrylamide gel in the presence of sodium dodecyl sulfate essentially as described [F. W. Studier, *J. Mol. Biol.* **79**, 237 (1973)].

T7 lysozyme can be provided to the cell from a clone of the T7 lysozyme gene in the *Bam*HI site of pACYC184.[17] The cloned fragment we have used (bp 10,665–11,296 of T7 DNA[2]) also contains the φ3.8 promoter for T7 RNA polymerase immediately following the lysozyme gene. A plasmid having this fragment oriented so that the lysozyme gene is expressed from the *tet* promoter of pACYC184 is referred to as pLysE; cells carrying this plasmid accumulate substantial levels of lysozyme. A plasmid having the fragment in the opposite orientation is referred to as pLysS; cells carrying this plasmid accumulate much lower levels of lysozyme. These plasmids confer resistance to chloramphenicol and are compatible with the pET vectors for cloning target genes (described below). Neither lysozyme plasmid interferes with transformation of cells that contain it; pLysS has little effect on growth rate but pLysE causes a significant decrease in the growth rate of cells that carry it.

The presence of either pLysS or pLysE increases the tolerance of BL21(DE3) or HMS174(DE3) for toxic target plasmids: unstable plasmids become stable, and plasmids that would not otherwise be established can be maintained and expressed. Some target plasmids are too toxic to be established in the presence of pLysS but are able to be established in the presence of pLysE, and a few are too toxic to be established even in the presence of pLysE.

The low level of lysozyme provided by pLysS usually has little effect on expression of target genes on induction of T7 RNA polymerase, except for a short lag in the appearance of target gene products. Apparently, more T7 RNA polymerase is induced than can be inhibited by the small amount of lysozyme. (The level of lysozyme might be expected to increase somewhat on induction, since T7 RNA polymerase should be able to transcribe completely around the pLysS plasmid from the φ3.8 promoter to make lysozyme mRNA; however, the φ3.8 promoter is relatively weak,[18] and most transcription should be from the much stronger φ10 promoter used in the target plasmids.)

The higher level of lysozyme provided by pLysE can substantially increase the lag and substantially reduce the maximum level of expression of target genes on induction of T7 RNA polymerase. This damping of expression is sufficient that cells containing a target gene whose product is relatively innocuous can continue to grow indefinitely in the presence of IPTG, a property that may be useful in some circumstances. (In contrast, the high level of expression in the absence of lysozyme or in the presence of pLysS almost always prevents continued growth of the cell.) Because of this damping of expression, most target genes will be expressed to higher levels

[17] A. C. Y. Chang and S. N. Cohen, *J. Bacteriol.* **134**, 1141 (1978).
[18] W. T. McAllister, C. Morris, A. H. Rosenberg, and F. W. Studier, *J. Mol. Biol.* **153**, 527 (1981).

by CE6 infection (described below) than by induction in the presence of pLysE.

The presence of pLysS (or pLysE) has the further advantage of facilitating the preparation of cell extracts. After the target protein has accumulated, the cells are collected and suspended in a buffer such as 50 mM Tris-Cl, 2 mM Na$_3$EDTA, pH 8.0. Simply freezing and thawing, or adding 0.1% Triton X-100, will allow the resident lysozyme to lyse the cells efficiently. This property can make it advantageous to carry pLysS in the cell even when it is not required for stabilizing the target plasmid.

Bacteriophage CE6

Target plasmids that are too toxic to be established in DE3 lysogens (even in the presence of pLysE) can be expressed by infecting with a bacteriophage that provides T7 RNA polymerase to the cell. No T7 RNA polymerase will be present in the cell before infection, so any target DNA that can be cloned under control of a T7 promoter should be expressible in this way.

A convenient bacteriophage for delivering T7 RNA polymerase to the cell is CE6, a λ derivative that carries the gene for T7 RNA polymerase under control of the phage p$_L$ and p$_I$ promoters and also has the $cI857$ thermolabile repressor and the $Sam7$ lysis mutations.[7] When CE6 infects HMS174, the newly made T7 RNA polymerase transcribes target DNA so actively that normal phage development cannot proceed. Comparable levels of target RNAs and proteins are produced whether T7 RNA polymerase is delivered to the cell by induction or by infection.

pET Vectors

The plasmid vectors we have developed for cloning and expressing target DNAs under control of a T7 promoter are designated pET vectors (*p*lasmid for *e*xpression by *T*7 RNA polymerase).[11] They contain a T7 promoter inserted into the *Bam*HI site of the multicopy plasmid pBR322 in the orientation that transcription is directed counterclockwise, opposite to that from the *tet* promoter (Fig. 2). In the absence of T7 RNA polymerase, transcription of target DNAs by *E. coli* RNA polymerase is low enough that very toxic genes can be cloned in these vectors. However, some expression can be detected, so it is possible that an occasional gene may be too toxic to be cloned in them.

Most of the pET vectors described here confer resistance to ampicillin. In such vectors, the *bla* gene is oriented so that it will be expressed from the T7 promoter along with the target gene (Fig. 2). However, in the pET-9 series of vectors, the *bla* gene has been replaced by the *kan* gene in the opposite orientation (Tables I and II). In these vectors, the only coding

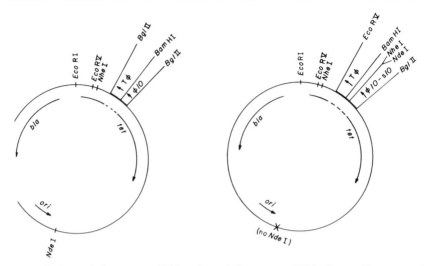

FIG. 2. Transcription vector pET-3 and translation vector pET-3a, b, or c. Fragments of T7 DNA containing the $\phi10$ promoter (Fig. 3), the $\phi10$ promoter plus the $s10$ translation initiation region for the gene *10* protein (Fig. 7), and the Tϕ transcription termination signal (Fig. 4) were inserted into the *Bam*HI site of pBR322 in the indicated orientations. Relative locations of selected restriction sites, the *bla* and interrupted *tet* genes, and the origin of replication are shown.

sequence transcribed from the T7 promoter will be that of the target gene.

The T7 promoter in the pET vectors is derived from the $\phi10$ promoter, one of six strong promoters in T7 DNA that have the identical nucleotide sequence from positions -17 to $+6$, where $+1$ is the position of the first nucleotide of the RNA transcribed from the promoter.[2] The $\phi10$ promoter fragments carried by the vectors all begin at bp -23 and continue to bp $+2$, $+3$, $+26$, and $+96$ or beyond. Some of the vectors also contain a transcription termination signal or an RNase III cleavage site downstream of the cloning site for the target DNA.

Details of the construction of most of the vectors described here are given in Ref. 11. The nucleotide sequence of each vector can be assembled from the known nucleotide sequences of pBR322,[19,20] the kan^{21} and $lacI^{22,23}$ genes, and the inserted fragment(s) from T7 DNA,[2] together with the sequences of linkers and any changes introduced in the construction of the vector (given in Tables I and II and Figs. 3–7). The linker sequences in most of the vectors contain restriction sites that allow easy removal of control elements for use in constructing other vectors, or that may be

[19] J. G. Sutcliffe, *Cold Spring Harbor Symp. Quant. Biol.* **43**, 77 (1979).
[20] K. W. C. Peden, *Gene* **22**, 277 (1983).
[21] A. Oka, H. Sugisaki, and M. Takanami, *J. Mol. Biol.* **147**, 217 (1981).
[22] M. P. Calos, *Nature (London)* **274**, 762 (1978).
[23] P. J. Farabaugh, *Nature (London)* **274**, 765 (1978).

TABLE I
TRANSCRIPTION VECTORS[a]

Vector	Selective marker[b]	Upstream site	Promoter	Leader RNA (nt)[c]	Cloning sites	Downstream elements	
pET-1	*bla*	ClaI[d]	φ10	29	BamHI		
pET-2	*bla*	BglII	φ10	29	BamHI		
pET-3	*bla*	BglII	φ10	29	BamHI	Tφ	BglII[e]
pET-4	*bla*	BglII	φ10	29	BamHI	R1.1	BglII
pET-5	*bla*	ClaI[d]	φ10	29	BamHI–EcoRI[f]	*tet* deletion	
pET-6	*bla*	BamHI	φ10	2	StuI–BamHI[g]		
pET-7	*bla*	BglII	φ10	2	StuI–BamHI[g]		
pET-9	*kan*	BglII	φ10	29	BamHI	Tφ	BglII[e]
pET-10	*bla*	BglII	T7lac[h]	31	BamHI		
pET-11	*bla*	BglII	T7lac[h]	31	BamHI	Tφ	BglII[e]

[a] Transcription and RNA processing signals from T7 DNA are directed counterclockwise in the *Bam*HI site of pBR322 (Fig. 2). The nucleotide sequences of these signals and the linkers used in cloning them are given in Figs. 3–6 and in footnote *h* below.

[b] The selective antibiotic resistance marker is the *bla* gene from pBR322 (Fig. 2) or the *kan* gene of Tn*903*, from pUC4KISS [F. Barany, *Gene* 37, 111 (1985)]. To convert from *bla* to *kan*, a *Bsp*HI–*Eco*RI fragment of pBR322, bp 3196–4361, was replaced by an 867-bp fragment that starts at a *Bsp*HI site 50 bp ahead of the *kan* initiation codon and ends at a newly introduced *Eco*RI site immediately following the termination codon. This places the *kan* coding sequence in the orientation opposite to that of the *bla* coding sequence. Plasmids of the pET-9 series confer resistance to kanamycin concentrations of at least 250 μg/ml.

[c] The length of RNA from the start point for T7 RNA polymerase to the first point of cleavage in the first cloning site is given in nucleotides (nt).

[d] The *Cla*I site will be methylated by the *dam* methylase of *E. coli* and will not be cut unless the plasmid DNA is isolated from a *dam⁻* host. The *Cla*I site of pET-5 is unique; pET-1 has a second *Cla*I site at bp 23 of the pBR322 DNA.

[e] We have just discovered that an *Ava*I cleavage site (CPyCGPuG) is also present downstream of Tφ in pET-3, pET-9, and pET-11, presumably because more than one *Bgl*II linker (GAGATCTC) was incorporated when the *Bgl*II site was introduced into the common precursor of these vectors.

[f] In pET-5, the *Bam*HI cloning site is followed immediately by a unique *Eco*RI site in the sequence GGATCCGAATTC; the DNA between the *Bam*HI and *Eco*RI sites of pBR322 (including the *tet* promoter) has been deleted.

[g] When cloning into the *Stu*I site, pET-6 and pET-7 should be isolated from a *dcm⁻* host such as BL21 (see text).

[h] The T7*lac* promoter was made by inserting a 25-bp *lac* operator sequence into the *Stu*I site 2 bp downstream of the start site of the T7 promoter in pET-7 (Fig. 6). The nucleotide sequence from the RNA start site to the *Bst*NI–*Bam*HI site immediately beyond the operator sequence is GGGGAATTGTGAGCGGATAACAATTCCCCTGGATCC, where the inserted *lac* operator sequence is underlined. The pET-10 and pET-11 series of vectors also contain a *lacI* fragment (bp −50 to +1152 relative to the first nucleotide of the *lacI* mRNA) cloned into the *Sal*I site so that the *lacI* and T7*lac* promoters diverge. The *lacI* fragment was cloned initially into the *Bam*HI site of pBR322 by use of linkers (CCGGATCCGG upstream and CGGGATCCCG downstream), and was subsequently moved into the *Sal*I site by blunt-end ligation between filled-in *Bam*HI and *Sal*I ends. pET-11 contains the downstream elements of pET-3.

TABLE II
TRANSLATION VECTORS[a]

Vector	Selective marker[b]	Upstream site	Expression signals[c]	ATG cloning site[d]	Fusion cloning aa[e]	Fusion cloning Site	Fusion cloning rf[f]	Downstream elements	
pET-1a	bla	ClaI[g]	φ10–s10	NdeI	12	BamHI	GGA		
pET-1b	bla	ClaI	φ10–s10	NdeI	12	BamHI	GAT		
pET-1c	bla	ClaI	φ10–s10	NdeI	12	BamHI	ATC		
pET-2a	bla	BglII	φ10–s10	NdeI	12	BamHI	GGA		
pET-2b	bla	BglII	φ10–s10	NdeI	12	BamHI	GAT		
pET-2c	bla	BglII	φ10–s10	NdeI	12	BamHI	ATC		
pET-3a	bla	BglII	φ10–s10	NdeI	12	BamHI	GGA	Tφ	EcoRV[h]
pET-3b	bla	BglII	φ10–s10	NdeI	12	BamHI	GAT	Tφ	EcoRV
pET-3c	bla	BglII	φ10–s10	NdeI	12	BamHI	ATC	Tφ	EcoRV
pET-3d	bla	BglII	φ10–s10	NcoI	12	BamHI	ATC	Tφ	EcoRV
pET-3xa	bla	BglII	φ10–s10	NdeI	260[i]	BamHI	GGA	Tφ	EcoRV[h]
pET-3xb	bla	BglII	φ10–s10	NdeI	261	BamHI	GAT	Tφ	EcoRV
pET-3xc	bla	BglII	φ10–s10	NdeI	261	BamHI	ATC	Tφ	EcoRV
pET-3xd	bla	BglII	φ10–s10	NcoI	261	BamHI	ATC	Tφ	EcoRV
pET-4a	bla	BglII	φ10–s10	NdeI	12	BamHI	GGA	R1.1	BglII
pET-4b	bla	BglII	φ10–s10	NdeI	12	BamHI	GAT	R1.1	BglII
pET-4c	bla	BglII	φ10–s10	NdeI	12	BamHI	ATC	R1.1	BglII
pET-5a	bla	BglII	φ10–s10	NdeI	12	BamHI	GGA	EcoRI	GAA[j]
pET-5b	bla	BglII	φ10–s10	NdeI	12	BamHI	GAT	EcoRI	AAT
pET-5c	bla	BglII	φ10–s10	NdeI	12	BamHI	ATC	EcoRI	ATT
pET-8c	Previous designation for pET-3d								
pET-9a	kan	BglII	φ10–s10	NdeI	12	BamHI	GGA	Tφ	EcoRV[h]
pET-9b	kan	BglII	φ10–s10	NdeI	12	BamHI	GAT	Tφ	EcoRV
pET-9c	kan	BglII	φ10–s10	NdeI	12	BamHI	ATC	Tφ	EcoRV
pET-9d	kan	BglII	φ10–s10	NcoI	12	BamHI	ATC	Tφ	EcoRV
pET-10a	bla	BglII	T7lac–s10	NdeI	12	BamHI	GGA		
pET-10b	bla	BglII	T7lac–s10	NdeI	12	BamHI	GAT		
pET-10c	bla	BglII	T7lac–s10	NdeI	12	BamHI	ATC		
pET-11a	bla	BglII	T7lac–s10	NdeI	12	BamHI	GGA	Tφ	EcoRVΔ[k]
pET-11b	bla	BglII	T7lac–s10	NdeI	12	BamHI	GAT	Tφ	EcoRVΔ
pET-11c	bla	BglII	T7lac–s10	NdeI	12	BamHI	ATC	Tφ	EcoRVΔ
pET-11d	bla	BglII	T7lac–s10	NcoI	12	BamHI	ATC	Tφ	EcoRVΔ

[a] Transcription, translation, and RNA processing signals from T7 DNA are directed counterclockwise in the BamHI site of a derivative of pBR322 in which the NdeI site has been eliminated by opening, filling in, and religating (Fig. 2). The nucleotide sequences of these signals and the linkers used in cloning them are given in Figs. 3, 4, 5, and 7 and in the footnotes to the tables.

[b] See footnote b of Table I.

[c] The φ10–s10 segment contains the φ10 promoter and the translation initiation region for the gene 10 protein (Fig. 7). The T7lac promoter is described in footnote h in Table I. The junction between the T7lac promoter and the s10 translation initiation region was made by blunt-end ligation of the filled-in BstNI end downstream of the T7lac promoter and the filled-in XbaI site upstream of the

useful for moving cloned target DNA linked to the transcription, translation, or processing signals.

Transcription Vectors. Vectors containing a T7 promoter but no potential translation initiation site ahead of the cloning site are referred to as transcription vectors (Table I). Vectors pET-1 to pET-5 all carry the same $\phi 10$ promoter fragment (-23 to $+26$) followed by a unique *Bam*HI cloning site (Fig. 3). This fragment should contain all of the upstream and downstream sequences normally used by T7 RNA polymerase to initiate transcription at the $\phi 10$ promoter. RNA initiated at the promoter in these vectors will have 29 nucleotides ahead of the GATC of the *Bam*HI cloning site, the first 21 of which can form an 8-bp stem-and-loop structure (Fig. 3). Retention of this naturally occurring stem-and-loop structure in the RNA made from these vectors may confer some as yet undefined advantage in, for example, stability or translatability.

The pET-1 to pET-5 vectors differ from each other in the restriction sites found in the linkers immediately upstream of the promoter and in the signals found downstream of the *Bam*HI cloning site, as indicated in Table I. pET-1 and pET-2 have no additional signals downstream of the *Bam*HI cloning site, and T7 RNA polymerase is capable of transcribing completely around these plasmid DNAs multiple times.[7,18] pET-3 has a T7 DNA fragment containing the transcription terminator Tϕ (Fig. 4) downstream

translation start (Fig. 7). This ligation recreated a unique *Xba*I site, which is convenient for exchanging target genes between the $\phi 10$ and T7*lac* promoters in the pET translation vectors.

d Coding sequences can be joined to the gene *10* initiation codon, using a unique *Nde*I site (CA'TATG) or a unique *Nco*I site (C'CATGG) (Fig. 7).

e The number of codons before the first in-frame codon of the *Bam*HI fusion cloning site.

f The first codon within the *Bam*HI recognition sequence (GGATCC) that is in the same reading frame as the gene *10* initiation codon (see Fig. 7 and footnote *i*).

g See footnote *d* of Table I.

h Besides the *Eco*RV site in the downstream linker, a second *Eco*RV site is present at bp 185 of the pBR322 DNA.

i The pET-3x series is equivalent to the pET-3 series except that the *Bam*HI fusion site is after the two hundred and sixtieth rather than the eleventh codon of gene *10*. The same set of *Bam*HI linkers (Fig. 7) was attached after a filled-in *Asp*718 site (G'GTACC) at codons 259 and 260 (bp 23,740 of T7 DNA). The initial C of the linkers regenerated a unique *Asp*718 site in each vector. In pET-3xa the two hundred and sixty first codon is the GGA of the *Bam*HI site; in pET-3xb the two hundred and sixty first codon specifies alanine and is followed by the GAT of the *Bam*HI site; in pET-3xc and pET-3xd the two hundred and sixty first codon specifies glycine and is followed by the ATC of the *Bam*HI site.

j The pET-5 translation vectors have the same *Bam*HI–*Eco*RI sequence and *tet* deletion as in pET-5 itself (Table I). The *Eco*RI cloning site can be used instead of the *Bam*HI site to place coding sequences in the reading frame of the gene *10* initiation codon: the first in-frame codon within the *Eco*RI recognition sequence (GAATTC) is given.

k The *Nhe*I site at the second and third codons of gene *10* is unique in pET-11a, b, c, and d, because the *Nhe*I site at bp 229 of pBR322 has been removed from these vectors by deleting the DNA between the *Eco*RV sites immediately downstream of Tϕ and at bp 185 of pBR322.

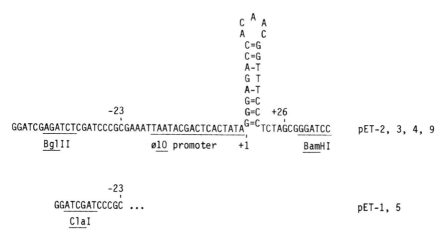

```
                                          C   A  A
                                          A     C
                                          C = G
                                          C = G
                                          A - T
                                          G   T
                                          A - T
                                          G = C
                  -23                      G = C  +26
                   '                       G = C   '
GGATCGAGATCTCGATCCCGCGAAATTAATACGACTCACTATA     TCTAGCGGGATCC        pET-2, 3, 4, 9
      BglII                      ø10 promoter  +1         BamHI


                  -23
                   '
         GGATCGATCCCGC ...                                           pET-1, 5
             ClaI
```

FIG. 3. Nucleotide sequence of the cloned $\phi10$ promoter in pET-1, 2, 3, 4, 5, and 9. The cloned $\phi10$ fragment (bp 22,880–22,928 of T7 DNA) extends from bp −23 to +26 relative to the RNA start at +1. Linker sequences through the original *Bam*HI site of pBR322 are shown on both sides. The upstream conserved T7 promoter sequence is underlined and the potential stem-and-loop structure at the beginning of the RNA is indicated.

of the *Bam*HI cloning site. The Tφ fragment causes efficient termination of transcription by T7 RNA polymerase both *in vivo* and *in vitro*.

pET-4 has a T7 DNA fragment containing the R*1.1* RNase III cleavage site (Fig. 5) downstream of the *Bam*HI cloning site. RNase III is a host nuclease that efficiently cuts RNA within the R*1.1* cleavage site to leave a potential stem-and-loop structure at the 3′ end of the upstream RNA,[2] a structure that appears to stabilize the RNA against degradation.[24] Cleavage at this site reduces to unit length the long RNAs made by transcribing multiple times around the plasmid DNA.[7] Such unit-length RNA can be detected by gel electrophoresis, a simple confirmation that full-length transcripts of target DNA have accumulated (Fig. 1).

In pET-5, the DNA between the unique *Bam*HI and *Eco*RI sites of pET-1 has been deleted. This removes the *tet* promoter of pBR322, so that transcription by *E. coli* RNA polymerase across these two cloning sites is low in both directions. Some target DNA fragments that have coding sequences in both directions can be cloned in pET-5 but not pET-1, apparently because expression of either coding sequence from the *tet* promoter is toxic to the cell.

pET-6 and pET-7 carry a shorter $\phi10$ promoter fragment (−23 to +2) ending in a unique *Stu*I site followed immediately by a *Bam*HI site (Fig. 6). The *Stu*I site (AGG′CCT) is placed so that target DNA can be inserted by

[24] N. Panayotatos and K. Truong, *Nucleic Acids Res.* **7**, 2227 (1985).

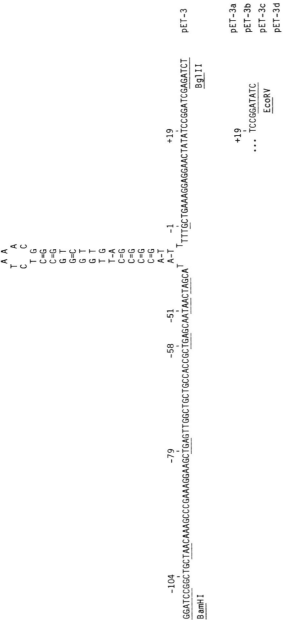

FIG. 4. Nucleotide sequence of the cloned Tφ transcription terminator. The cloned Tφ fragment (bp 24,106–24,228 of T7 DNA) extends from bp −104 to +19 relative to the last nucleotide of the terminated transcript at −1. Linker sequences from the upstream BamHI cloning site through the downstream BglII site of the pET-3, 9, and 11 transcription vectors or the downstream EcoRV site of the pET-3, 3x, 9, and 11 translation vectors are shown. We have just discovered that an AvaI cleavage site (CPyCGPuG) is also present downstream of Tφ in pET-3, 9, and 11, presumably because more than one BglII linker (GAGATCTC) was incorporated when the BglII site was introduced into the common precursor of these vectors. The DNA between the EcoRV site immediately downstream of Tφ and the one located at bp 185 of pBR322 has been eliminated from the pET-11 translation vectors. A potential stem-and-loop structure just before the last few nucleotides of the terminated transcript is indicated and all stop codons ahead of this structure are underlined: the TAA at nt −51 is the natural stop codon for the gene 10B protein of T7; the TAG at nt −47 is in a second reading frame; and the TAA at nt −100, the TGA at nt −79, and the TGA at nt −58 are all in the third reading frame.

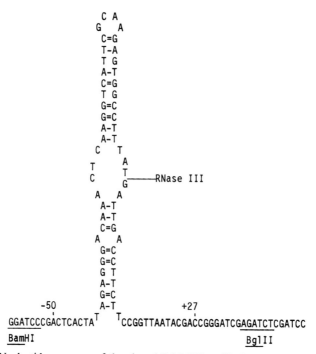

FIG. 5. Nucleotide sequence of the cloned R*1.1* RNase III cleavage site of pET-4. The cloned R*1.1* fragment (bp 5838–5914 of T7 DNA) extends from bp -50 to $+27$ relative to the cut site. Linker sequences from the upstream *Bam*HI cloning site through the original *Bam*HI site of pBR322 on the downstream side are shown. The potential stem-and-loop structure similar to that found in eight RNase III cleavage sites in T7 RNA is indicated, as is the known position of cleavage in this sequence.

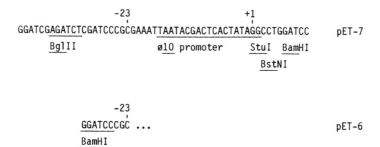

FIG. 6. Nucleotide sequence of the cloned $\phi 10$ promoter of pET-7 and pET-6. The cloned $\phi 10$ fragment (bp 22,880–22,904 of T7 DNA) extends from bp -23 to $+2$ relative to the RNA start at $+1$. Linker sequences through the original *Bam*HI site of pBR322 are shown on both sides. The conserved promoter sequence and the upstream *Bgl*II or *Bam*HI site are underlined, and the positions of the downstream overlapping *Stu*I (AGG′CCT), *Bst*NI (CC′TGG), and *Bam*HI (G′GATCC) sites are indicated.

blunt-end ligation after position +2 of the promoter, and T7 RNA polymerase will transcribe target DNA cloned at this site with only a GG ahead of the cloned sequence. The loss of a few nucleotides of conserved promoter sequence in the transcribed region may reduce promoter strength when certain sequences are cloned into the *Stu*I site, but we expect that almost any DNA cloned into this site will be actively transcribed if no competing T7 promoters are present. pET-6 contains a second *Bam*HI site upstream of the promoter whereas pET-7 is somewhat more versatile in having a unique *Bgl*II site at this position.

The *Stu*I–*Bam*HI sequence of both pET-6 and pET-7 contains the sequence CCTGG, which is both a cleavage site for *Bst*NI and a methylation site for the *dcm* methylase of *E. coli*. Methylation at this site prevents cleavage by *Stu*I, so plasmid DNA must be prepared from a *dcm⁻* strain if it is to be cut efficiently by *Stu*I. *E. coli* B strains are known to be *dcm⁻*, and BL21 is a good host for obtaining pET-6 or pET-7 that can be cut by *Stu*I.

pET-9 is equivalent to pET-3, except that the *bla* gene has been replaced by the *kan* gene in the opposite orientation (Table I, footnote *b*). pET-10 and pET-11 are equivalent to pET-2 and pET-3, except that they carry the T7*lac* promoter and the *lacI* gene (Table I, footnote *h*, and the section Vectors That Contain a *T7lac* Promoter, below).

Translation Vectors. Vectors for placing cloned DNA under control of both *φ10* and the efficient translation initiation signals for gene *10* protein (the major capsid protein of T7) are listed in Table II. Most of these translation vectors carry a fragment (−23 to +96 relative to the RNA start) that includes the first 11 codons for the gene *10* protein; they have been constructed in sets of three, such that cloning a DNA fragment into the unique *Bam*HI site places the cloned nucleotide sequence in a different reading frame relative to the gene *10* initiation codon in each of the three vectors (Fig. 7). The configuration of upstream restriction sites and downstream signals is indicated by the pET number; the three reading frames are identified by the suffixes a, b, and c to indicate whether the GGA, GAT, or ATC triplet of the *Bam*HI site (GGATCC) is in the open reading frame. In the pET-5 translation vectors, the *Bam*HI site is followed immediately by an *Eco*RI site, which can also be used to fuse coding sequences in each of the three reading frames.

The pET-3, pET-5, pET-9, and pET-11 translation vectors are not strictly equivalent to their respective transcription vectors: pET-3, 9, and 11 have a *Bgl*II linker downstream of Tφ, whereas the corresponding translation vectors have an *Eco*RV linker at this position (Figs. 2 and 4; Tables I and II); pET-5 itself has a *Cla*I linker upstream of *φ10* whereas the pET-5 translation vectors have the *Bgl*II linker of pET-2 at this position (Fig. 3; Tables I and II).

```
        C  A  A
        A     C
        C=G
        C=G
        A-T
        G-T
        A-T
        G=C
        G=C
   pppG=c
              +26
   C TCTAGAAATAATTTTGTTTAACTTTAAGAAGGAGATATACATATG
     XbaI                         SD        NdeI

                            1   2   3   4   5   6   7   8   9   10  11
              +50    +61   Met Ala Ser Met Thr Gly Gly Gln Gln Met Gly Arg Gly Ser
                           GCT AGC ATG ACT GGT GGA CAG CAA ATG GGT CGC GGA TCC      pET-3a
                           NheI                                              BamHI

                                                      ... Arg Asp
                                                      ... CGG GAT CC                 pET-3b
                                                                  BamHI

                                                      ... Arg Ile
                                                      ... CGG ATC C                  pET-3c
                                                                 BamHI

                            1   2   3
              +50    +61   Met Ala Ser ...
              ... AAGGAGATATACC ATG GCT AGC ...
                       SD      NcoI NheI

                                                      ... Arg Ile
                                                      ... CGG ATC C                  pET-3d
                                                                 BamHI
```

FIG. 7. Target mRNA and protein sequences specified by pET translation vectors. The cloned φ10–s10 fragment (bp 22,880–22,998 of T7 DNA) extends from bp −23 to +96 (relative to the RNA start at +1) for all of the translation vectors in Table II except the pET-3x series, which extends to bp +843 (bp 23,745 of T7 DNA), and the pET-10 and pET-11 series, in which RNA transcribed from the *lac* operator replaces the RNA ahead of the *XbaI* site (footnotes *h* of Table I and *c* of Table II). Nucleotide and amino acid sequences from the start of the mRNA through the *BamHI* fusion cloning site are shown for the pET-3 series. The potential stem-and-loop structure at the 5' end of the mRNA is indicated and the Shine–Dalgarno (SD) sequence is underlined. The indicated *XbaI*, *NdeI*, *NcoI*, and *BamHI* sites are unique in the vectors that contain them; the *NheI* site is unique in the pET-5 and pET-11 series, but the other vectors contain a second *NheI* site at bp 229 of the pBR322 DNA. The nucleotide sequence AT at bp +62 and +63 in pET-3c was replaced by C in pET-3d, thereby eliminating the *NdeI* site and creating an *NcoI* site; this is the only difference between these two vectors. In the pET-3x series, the gene *10* coding sequence extends through codon 260; the linker amino acids are given in footnote *i* of Table II.

A coding sequence in the proper reading frame in the *Bam*HI site of the pET translation vectors will be translated as a fusion protein linked to the first 11 amino acids of the gene *10* protein plus an arginine specified by the linker sequence (Fig. 7). However, coding sequences can also be joined directly to the gene *10* initiation codon at a unique *Nde*I site (CA′TATG) that includes the initiating ATG. Protein synthesis will be directed by the efficient upstream translation signals of gene *10* and the resulting protein will not contain any foreign amino acids. Potentially, any natural initiation codon (or any internal methionine codon) can be converted to an *Nde*I site and joined to the vector without altering the coding sequence.

ATG is also found in the cleavage sites for *Nco*I (C′CATGG), *Bsp*HI (T′CATGA), and *Afl*III (A′CPuPyGT). Initiation codons (or internal methionine codons) not already part of one of these sites can be converted to one with little or no change in coding sequence. To allow such coding sequences to be joined directly to the upstream gene *10* translation signals, the initiation codon in pET-3c was converted to an *Nco*I site to create pET-3d (formerly pET-8c) (Fig. 7). In our limited experience, we find that the gene *10* initiation site of pET-3d seems to direct protein synthesis about as well as that of pET-3c, indicating that the alteration in nucleotide sequence that created the *Nco*I site has little effect on translation efficiency.

Coding sequences can also be joined at the second and third codons of gene *10*, through the *Nhe*I site at this position (Fig. 7). Ends produced by *Nhe*I (G′CTAGC), *Xba*I (T′CTAGA), *Spe*I (A′CTAGT), and *Avr*II (C′CTAGG) can be joined directly to this site. Most of the pET vectors also contain a second *Nhe*I site at bp 229 of the pBR322 sequence, but this site has been deleted from the pET-5 and pET-11 series.

For some applications, such as making antibodies to relatively small peptides, fusions to a relatively large protein can be useful. Therefore, we have also created the pET-3x set of translation vectors, which is equivalent to the pET-3 set except that the *Bam*HI cloning site has been placed after the two hundred and sixtieth rather than the eleventh codon for the gene *10* protein. The 260-amino acid fragment of the gene *10* protein is itself insoluble, so the fusion proteins will likely be insoluble as well. Such insolubility can be an aid in the purification of protein for making antibodies and may also help to stabilize fused peptides that might otherwise be unstable. The pET-3x vectors are also convenient for placing coding sequences directly behind the gene *10* initiation codon: loss of the 780 bp of gene *10* coding sequence between the *Nde*I (or *Nco*I) and *Bam*HI sites reduces the size of the vector DNA enough that gel electrophoresis readily shows whether both cuts have been made.

Vectors That Reduce Colony Size. Some of the pET vectors, most notably pET-1, pET-1c, and pET-2c, cause a significant reduction in colony size of HMS174 and BL21. The effect is not understood in detail, but the growth rate in log phase cultures seems little affected. This effect does not impair the usefulness of the vectors; in fact, insertion of target sequences has always restored normal colony size, and this can be a useful phenotype for identifying plasmids that have inserts.

Vectors That Contain a T7lac Promoter. We have placed a T7 promoter under control of *lac* repressor by inserting a *lac* operator sequence just downstream of the start site of a T7 promoter.[25] In the presence of appropriate levels of *lac* repressor, transcription from this T7*lac* promoter is greatly reduced, but normal transcription can be restored by addition of IPTG. In order to provide sufficient *lac* repressor to occupy the operator sites on a multicopy plasmid vector, a DNA fragment containing the natural promoter and coding sequence for *lacI* was placed upstream of the T7*lac* promoter, oriented so that the *lacI* and T7*lac* promoters diverge. Vectors analogous to the pET-2 and pET-3 series of vectors but containing the T7*lac* promoter and *lacI* are designated the pET-10 and pET-11 series (Table I and II).

When the pET-10 or pET-11 vectors are used for expressing target genes in BL21(DE3) or HMS174(DE3), the *lac* repressor acts both to repress transcription of the T7 RNA polymerase gene by *E. coli* RNA polymerase and to block transcription of the target gene by any T7 RNA polymerase that is made. This double repression significantly reduces basal expression of the target gene in the uninduced cell but seems to have little effect on the level of expression on induction, relative to the analogous configuration using the *φ10* promoter rather than the T7*lac* promoter.

Only a few target genes have been encountered that are too toxic to be stable in the pET-10 or pET-11 vectors in BL21(DE3) or HMS174(DE3). However, addition of T7 lysozyme, as provided by pLysS, reduces the basal activity of T7 RNA polymerase sufficiently that each of these plasmids can be maintained. The slight delay in expression of target genes on induction in the presence of pLysS is increased when the gene is controlled by the T7*lac* promoter, but substantial levels of expression can nevertheless be attained.

Genes already cloned in any of the pET translation vectors can usually be placed under control of the T7*lac* promoter in the pET-10 or pET-11 translation vectors by using equivalent *Xba*I, *Nde*I, or *Nco*I sites to exchange the appropriate DNA fragments (Fig. 7).

Potential Problems with Other Vectors

Vectors for placing cloned DNA under control of a T7 promoter are widely available from commercial sources. In principle, many such vectors

[25] J. W. Dubendorff and F. W. Studier, unpublished work, 1988.

could be used for expressing target DNA in the expression system described here. However, some vectors are inappropriate, and the reason may not be immediately apparent. For example, many commercial vectors carry a *lac* operator, whose presence is not always obvious in descriptions of the vector. (Vectors that provide a complementing fragment of β-galactosidase for use in a blue–white color test for fragment insertion would carry a *lac* operator.) We have received several reports of plasmid instability in BL21(DE3) that turned out to be due to the use of such vectors. Apparently, the multicopy vector introduces more copies of the *lac* operator than can be titrated by the *lac* repressor present in the cell. As a result, T7 RNA polymerase (itself controlled by *lac* repressor) increases to a level where transcription of the target plasmid becomes too high for stability. Users should be alert to the possibility that this or other types of incompatibility may be encountered when trying to express DNA cloned in vectors that were not designed for use in this system.

Growth Media

BL21 and HMS174 grow on minimal or complex media, and presumably a wide range of growth media would be suitable for growth of these strains and expression of target DNAs. For routine growth of cultures we use ZB medium (10 g of N-Z-amine A and 5 g of NaCl in 1 liter of water). N-Z-amine A is obtained from Sheffield Products (P.O. Box 398, Memphis, TN 38101); Bacto Tryptone (Difco) in place of N-Z-amine A in any of the media described gives essentially equivalent results. Defined media are usually M9 medium, containing 1 g of NH_4Cl, 3 g of KH_2PO_4, 6 g of Na_2HPO_4, 4 g of glucose, and 1 ml of 1 M $MgSO_4$ in 1 liter of water, or B2 medium, which is essentially M9 medium in which all but 0.16 mM of the phosphate is replaced by salts and bis-Tris buffer.[26] M9mal and B2mal are the equivalent media in which glucose is replaced by maltose. Richer media include M9ZB, which contains the components of both M9 and ZB; ZY [10 g of N-Z-amine A, 5 g of Bacto yeast extract (Difco), and 5 g of NaCl in 1 liter of water]; or ZYG (ZY medium plus 4 g of glucose per liter). Sugars, $MgSO_4$, and phosphate solutions are autoclaved separately and added to the media after cooling.

The many small (1–3 ml) cultures generated in cloning DNA fragments and isolating recombinant plasmids are usually grown in M9ZB + antibiotic in standing 13 × 100 mm culture tubes in a 37° incubator. After overnight incubation of standing cultures, HMS174 remains largely dispersed throughout the culture whereas BL21 cells mostly settle to the bottom of the tube. Larger cultures are grown in shaking flasks at 37°. When cultures are grown overnight in shaking flasks, the growth medium

[26] F. W. Studier, *J. Bacteriol.* **124**, 307 (1975).

is usually ZB, because continued shaking after saturation in rich media containing glucose (such as M9ZB or ZYG) leads to some lysis of BL21.

Dilutions for titering bacteria or phage are made in ZB. Plating is done by mixing samples with 2.5 ml of melted top agar [ZB containing 0.7% (w/v) agar] and spreading on standard 100×15 mm plastic petri dishes containing 20 ml of hardened bottom agar (ZB containing 1% agar).

Where antibiotics are added to liquid growth media or bottom agar of petri dishes, ampicillin is typically used at 20 μg/ml and chloramphenicol or kanamycin at 25 μg/ml. It is not necessary to use preformed antibiotic plates: where bottom agar contains no antibiotic, selection can be accomplished by adding 200 μg of ampicillin or 250 μg of chloramphenicol or kanamycin per milliliter of top agar at the time of plating.

For induction of T7 RNA polymerase in BL21(DE3) or HMS174(DE3), a growing culture is made 0.4 mM in IPTG. For induction on plates, top agar is made 1 mM in IPTG at the time of plating.

Storage of Strains

For long-term storage, 1.5 ml of a growing or saturated culture is placed in a cryovial, mixed with one-tenth volume of 80% (v/v) glycerol, and the tube is stored directly in a $-75°$ freezer. We avoid higher concentrations of glycerol because they become increasingly toxic to cells at room temperature. Plasmid-bearing strains, particularly those having any tendency toward instability, are titered at the time of freezing to be sure that the vast majority of cells in the culture have the intended host–plasmid combination (see Toxic Genes and Plasmid Instability, below). To inoculate a culture from the frozen stock, a few microliters is scraped or melted from the surface, typically with a sterile pipet or plastic culture loop, and the remainder is returned to the freezer without thawing. Cells stored at $-75°$ in this way have remained viable for several years and presumably will remain viable for very long periods. (In our experience, cells survive for many months in a $-20°$ freezer, but survival for longer periods is variable.)

Cloning Target DNAs

Target DNAs are cloned into the pET vectors by standard techniques.[27] We use HMS174 as the host for initial cloning and analysis of plasmids, because plasmid DNAs remain monomers in the *recA* background and expression of the target DNA is minimal in the absence of T7 RNA

[27] T. Maniatis, E. F. Fritsch, and J. Sambrook, "Molecular Cloning, A Laboratory Manual." Cold Spring Harbor Laboratory, Cold Spring Harbor, New York, 1982.

polymerase. Any easily transformable, preferably *recA* strain should be suitable for this purpose. BL21 is not appropriate because it is *recA⁺* and it has a somewhat lower transformation efficiency. DE3 lysogens should not be used for initial cloning because of potential problems from expression of the target gene by the small amounts of T7 RNA polymerase present in the uninduced cell.

Once the desired plasmid is obtained, the target DNA can be expressed by infection with CE6, or, if the target plasmid is stable, by induction in BL21(DE3) or HMS174(DE3) or in one of these strains carrying pLysS. To test for stability, transformations with the target plasmid are attempted in a set of four strains: BL21, BL21(DE3), BL21(DE3)pLysS, and BL21(DE3)pLysE, or the equivalent set based on HMS174. Plasmids having no target gene, or whose target gene is relatively innocuous, will give transformants with about equal frequency in all four hosts; at some level of toxicity, plasmids will fail to transform the DE3 lysogen itself (where the basal activity of T7 RNA polymerase is highest); at somewhat higher toxicity the lysogen containing pLysS will also fail to be transformed; and a few target plasmids are so toxic that even the lysogen containing pLysE cannot be transformed, although the host that lacks the gene for T7 RNA polymerase will be transformed at normal frequency.

A few of the target genes we have worked with are more stable in HMS174(DE3) or its derivatives than in the equivalent derivative of BL21(DE3). A possible explanation for this difference (suggested by S. Shuman, personal communication) is that small amounts of some target gene products induce the SOS response of *E. coli.* This in turn induces the prophage and kills BL21(DE3), but the *recA* deficiency of HMS174 prevents induction of the prophage and killing of HMS174(DE3).

Expressing Target DNA by IPTG Induction of BL21(DE3) or HMS174(DE3)

If a target plasmid can be established in BL21(DE3), HMS174(DE3), or in one of these strains containing pLysS, induction of T7 RNA polymerase by IPTG is a convenient way to direct expression of the target DNA. We usually grow the cells in M9 or M9ZB containing the selective antibiotic (and also 25 μg of chloramphenicol/ml if the cells carry pLysS), and make the culture 0.4 mM in IPTG when the culture reaches an OD_{600} of 0.6–1.

Immediately before induction, the culture is titered to determine the fraction of cells that carry inducible plasmid. This involves plating on four plates, which differ in the composition of the top agar used in plating. Typically, the culture would be plated at a dilution of 10^5 on plates that have both IPTG and antibiotic or just IPTG added to the top agar, and at a dilution of 2×10^6 on plates that have just antibiotic or nothing added to the top agar. This test and its interpretation are described more fully in the next section. (We usually do not test for the relatively stable pLysS.)

If appropriate attention is paid to possible problems of plasmid instability, more than 98% of the cells in the culture will usually contain expressible target plasmids. Cells are usually harvested 2–3 hr after induction, enough time for substantial accumulation of target protein but before the culture can be overgrown with cells that have lost plasmid or are otherwise unproductive. However, some target proteins continue to accumulate for much longer times.

Occasional results have suggested that the basal activity of T7 RNA polymerase in uninduced cells may be somewhat lower when the growth medium contains glucose than when it does not. Induction of the *lac*UV5 promoter (which directs transcription of the T7 RNA polymerase gene in the DE3 lysogens) is not subject to catabolite repression,[28] but perhaps the repressed promoter retains some sensitivity. We have not analyzed this effect in detail, but we suspect that media containing glucose (such as M9 and M9ZB) may be more suitable than media without glucose (such as ZY) for growing DE3 lysogens for the induction of target genes.

Toxic Genes and Plasmid Instability

Plasmid pBR322 and many of its derivatives are relatively stable and are retained by a very high fraction of host cells even after growth for many generations in the absence of a selective antibiotic. However, problems of plasmid instability can arise when a gene whose product is toxic to the host cell is cloned in the plasmid. The level of expression may be such that the plasmid can be maintained but growth of the cell is impaired; segregation of cells lacking plasmid may also be increased because of decreased copy number or for other reasons. In such a situation, cells that lack the plasmid can rapidly overgrow the culture whenever selective antibiotic is lacking. If the plasmid is to be maintained in a significant fraction of the cells, the culture must not be allowed to grow in the absence of selection for the plasmid.

Use of ampicillin as a selective antibiotic requires special care, because β-lactamase is made in substantial amounts and is secreted into the medium, where it can destroy all of the ampicillin. This means that a culture whose cells carry an unstable plasmid will be growing under ampicillin selection only until enough β-lactamase has been secreted to destroy the

[28] A E. Silverstone, R. R. Arditti, and B. Magasanik, *Proc. Natl. Acad. Sci. U.S.A.* **66**, 773 (1970).

ampicillin in the medium; from that point on, cells that lack plasmid will not be killed and will begin to overgrow the culture. For a typical pBR322-based plasmid growing in a medium containing 20 μg of ampicillin per milliliter, this point is reached when the culture is barely becoming turbid, perhaps around 10^7 cells per milliliter. Growth in the presence of 200 μg of ampicillin per milliliter delays this point to a slightly higher cell density, but, given the catalytic activity of β-lactamase, it would not be feasible to add enough ampicillin to the medium to keep the cells under selection all the way to saturation.

A further complication is that certain toxic genes, while having little effect on cells that are growing logarithmically, kill cells at saturation. Almost all cells retain plasmid until saturation, but on continued incubation, fewer and fewer plasmid-containing cells survive and, because no ampicillin remains, cells that lack plasmid overgrow the culture.

A culture grown to saturation from selective conditions will have secreted a considerable amount of β-lactamase into the medium even if it becomes substantially overgrown by cells that lack plasmid. Subcultures might typically be grown from dilutions of 200- to 1000-fold into fresh ampicillin-containing medium. However, enough β-lactamase is typically present in the saturated culture that, even at these dilutions, enough remains to destroy all of the ampicillin before the cells that lack plasmid can be killed. Therefore, the subculture will grow completely in the absence of selection. The inoculum may already have had a substantial fraction of cells lacking plasmid, and by the time the subculture has grown to a density where expression of the target gene is induced, it is quite possible that only a minor fraction of the cells will contain the target plasmid. Failure to appreciate these potential problems can easily lead to the erroneous conclusion that certain target genes are poorly expressed, when in fact only a small fraction of cells in the cultures that were tested contained plasmid.

Most of the pET vectors described here have ampicillin as the selective antibiotic, and simple precautions are advisable to maximize retention of plasmid through the procedures for isolating, maintaining, and expressing target plasmids. We use the following isolation protocol, which usually produces the highest possible fraction of cells containing functional target plasmid. A colony from the transformation plate is inoculated into 2 ml of M9ZB + ampicillin and incubated for a few hours, until the culture becomes lightly turbid, when a sample is streaked on a plate containing ampicillin to obtain a single colony. As soon as the colony develops (usually overnight at 37°), it is inoculated into 2 ml of M9ZB + ampicillin and grown almost to saturation, when 1.5 ml of culture is mixed with 0.15 ml of 80% glycerol in a cryovial and stored in a −75 freezer. If there is any question about the possible stability of the plasmid, the culture is titered at the time of freezing to determine what fraction of the cells contain functional target plasmid.

For cells that carry a plasmid but no source of T7 RNA polymerase, titering in the presence and absence of ampicillin (200 μg/ml in the top agar) determines the fraction of cells that have plasmid. When the target plasmid is carried in BL21(DE3) or HMS174(DE3), the fraction of cells able to express the target gene can be tested by including 1 m*M* IPTG in the top agar, which will prevent colony formation by any cell that has both the inducible gene for T7 RNA polymerase and a functional target plasmid (but will not prevent growth of cells that lack plasmid or mutants that have lost the ability to express target DNA). In the presence of pLysS, IPTG also prevents colony formation (except in rare cases, including pET-3 itself). In the presence of pLysE, IPTG usually does not prevent colony formation unless the target gene product is toxic.

In practice, DE3 lysogens that carry a target plasmid that confers ampicillin resistance are titered on four plates, which have ampicillin, IPTG, both, or neither added to the top agar: all viable cells will grow on the plate with no additive; only cells that retain plasmid will grow in the presence of ampicillin; only cells that have lost plasmid or mutants that have lost the ability to express target DNA will grow in the presence of IPTG; and only mutants that retain plasmid but have lost the ability to express target DNA will grow in the presence of both ampicillin and IPTG. In a typical culture useful for producing target proteins, almost all cells will form colonies both on plates without additives and on plates containing only ampicillin, less than 2% of the cells will form a colony on plates containing only IPTG, and less than 0.01% will form a colony on plates containing both ampicillin and IPTG. With unstable target plasmids, the fraction of cells that have lost plasmid will be reflected by an increase in colonies on the IPTG plate and a decrease on the ampicillin plate. Mutants that retain plasmid but have lost the ability to express target DNA arise in some cases, but relatively infrequently.

If the plasmid is stable, cultures for expressing the target gene can be grown from the freezer stock without special precautions: even if the ampicillin in the fresh medium is destroyed or if the culture is incubated overnight at saturation, almost all of the cells will retain the target plasmid. However, if the target plasmid is unstable, cultures are grown from a dilution of 10^4 or higher from the freezer stock and grown directly to the density used for expression. Because of the potential for loss of plasmid, we always determine the composition of the cells in the culture by plating immediately before induction. This simple test can be invaluable in interpreting any unusual properties of an induction and in making sure that effort is not wasted on processing cells that had suboptimal levels of expression.

Some of the problems outlined here might be circumvented by using the pET-9 series of vectors, which provide resistance to kanamycin rather than ampicillin. However, the general principles and procedures should be useful in dealing with the problems of cloning toxic genes in any vector.

Expressing Target DNA by Infection with CE6

In principle, any target DNA that can be cloned in HMS174 can be transcribed by T7 RNA polymerase produced during infection with CE6. However, expression by CE6 infection is generally less convenient than expression by induction of DE3 lysogens because active phage stocks must be prepared and specialized growth conditions are needed to ensure good infection. Typically, we use CE6 infection only if a target plasmid is not able to be expressed in a DE3 lysogen, alone or in the presence of pLysS. (Lysogens of CE6 are not a good source of inducible T7 RNA polymerase because constitutive transcription from the p_I promoter in the lysogen provides basal levels that are too high to allow establishment of target plasmids.)

Preparation of CE6 Stocks

General procedures for working with λ have been described.[27,29] We grow CE6 stocks on host strain ED8739 ($r_K^- m_K^-$ metB supE supF),[30] which lacks the EcoK restriction and modification systems and provides supF for suppressing the Sam7 mutation of CE6. These phage effectively direct the expression of target genes in HMS174, and presumably in other nonrestricting hosts that adsorb λ. Although we have not examined the ability of unmodified CE6 to direct expression of target genes in a restricting host, one might expect such expression to be reduced. Where it may be desirable to express target genes in a host that has an active EcoK restriction system, CE6 stocks could be grown on a host that will provide EcoK modification, such as the ED8739 derivative ED8654 ($r_K^- m_K^+$ metB supE supF trpR)[30,31] or the equivalent LE392.[27] (HMS174 has no amber suppressor and therefore is not a host for growth of CE6.)

CE6 stock lysates are grown by adding a single plaque and 50 μl of a fresh overnight culture of ED8739 to 35 ml of ZY medium in a 125-ml flask and shaking at 37° until lysis. Larger volumes are grown by adding 10 μl of lysate and 1 ml of culture to 500 ml of ZY medium in a 1-liter flask. Lysates are made 0.5 M in NaCl and centrifuged for 10 min at

[29] R. W. Hendrix, J. W. Roberts, F. W. Stahl, and R. A. Weisberg, "Lambda II." Cold Spring Harbor Laboratory, Cold Spring Harbor, New York, 1983.
[30] K. Borck, J. D. Beggs, W. J. Brammar, A. S. Hopkins, and N. E. Murray, *Mol. Gen. Genet.* **146**, 199 (1976).
[31] N. E. Murray, W. J. Brammar, and K. Murray, *Mol. Gen. Genet.* **150**, 53 (1977).

10,000 g and 4°. The supernatant typically contains a few times 10^{10} infective phage particles per milliliter and is suitable either for long-term storage or for further purification.

CE6 phage particles are purified by precipitation with polyethylene glycol 8000 followed by rapid isopycnic banding in CsCl step gradients. All solutions used during purification, including CsCl solutions, contain 10 mM Tris-Cl (pH 8.0), 10 mM MgSO$_4$, and 100 μg of gelatin per milliliter to keep the phage intact. The purified phage are stored refrigerated in the CsCl solution, and dilutions are made in 0.1 M NaCl, 50 mM Tris-Cl (pH 8.0), 10 mM MgSO$_4$, 100 μg of gelatin/ml.

Lysates and purified phage stocks both lose titer on storage, so fresh stocks must be prepared from time to time. We find that either clarified lysates or purified phage in the above diluent can be mixed with one-tenth volume of 80% glycerol and stored frozen at $-75°$ and thawed without significant loss of titer. However, we don't as yet have sufficient experience to know whether this is a better means for long-term storage of phage stocks than simple refrigeration.

Expressing Target Genes

Efficient expression of target genes by CE6 infection requires that the host cell be appropriately receptive to infection: synthesis of the λ receptor is known to be stimulated by maltose but inhibited by glucose. An appropriate ratio of phage particles to receptive cells is also important, since too few phage particles will leave a substantial fraction of the cells uninfected and too many will completely inhibit protein synthesis.[7] The following protocol was developed to optimize expression of target genes in M9 medium, where proteins can be labeled by incorporation of radioactive amino acids. An equivalent protocol can be used in B2 medium, where RNA can also be labeled by ^{32}PO$_4$. Efficient expression of target genes can also be obtained by infecting cells that are growing in ZY medium supplemented with 0.04% (w/v) maltose.

HMS174 containing the target plasmid is grown in M9mal + antibiotic in a shaking flask at 37°. When a logarithmically growing culture reaches an OD$_{600}$ of 0.3, glucose is added to give a concentration of 4 mg/ml. After an additional 1–2 hr of growth, when the OD$_{600}$ reaches 0.6–1.0 (a cell concentration around 5×10^8/ml), MgSO$_4$ is added to give a final concentration of 10 mM, and purified CE6 phage is added to give a final concentration of 2×10^9 to 4×10^9 infective phage particles per milliliter (usually 1 μl of an OD$_{260}$ = 6 phage stock per milliliter of culture). The multiplicity of infection should be around 5–10 phage particles per cell. Immediately before infection, the culture is titered on plates with and without antibiotic to confirm that essentially all cells carry the target plasmid; the efficiency of infection, typically greater than 95%, is measured by titering surviving colony-forming units 5 min after infection. In this protocol, addition of glucose and $MgSO_4$ is not necessary but seems to give slightly better production of protein from the target genes. Cells are usually harvested 2–3 hr after infection, enough time for substantial accumulation of target protein but before the culture can be overgrown with uninfected cells.

Analyzing Target RNAs and Proteins

Target RNAs and proteins are analyzed by gel electrophoresis,[7] agarose gels being convenient for analyzing large RNAs, and gradient acrylamide gels for analyzing proteins or small RNAs (Fig. 1). The distribution of total RNA can be visualized by ethidium bromide fluorescence, total protein by staining with Coomassie brilliant blue, and radioactively labeled components by autoradiography. The amount of culture needed for analyzing RNAs and proteins depends on a number of factors, such as the host, culture medium, cell density, specific activity of label, and the gel system used. In a typical expression experiment using our standard protocols, 5–50 μl of culture is appropriate for visualizing proteins by Coomassie brilliant blue staining and 1–5 μl for visualizing RNAs by ethidium bromide fluorescence. For labeling proteins, 50 μl of culture labeled for 2–5 min with 1 μCi of [^{35}S]methionine usually gives enough label to visualize proteins after overnight exposure of the autoradiogram.

Total cell contents are analyzed by placing the cells in sample buffer [50 mM Tris-Cl (pH 6.8), 2 mM Na_3EDTA, 1% (w/v) sodium dodecyl sulfate, 1% (v/v) mercaptoethanol, 8% (v/v) glycerol, 0.025% (w/v) bromophenol blue], heating for 2–3 min in a boiling water bath, and applying directly to the gel. When analyzing proteins, samples of culture can be mixed directly with sample buffer; when analyzing RNAs, the gel patterns seem to be better if the cells are collected by centrifugation and suspended in sample buffer. Where samples are collected over a period of several hours, they can be left in sample buffer at room temperature and all heated together before loading onto a gel; where considerably more time will elapse before electrophoresis, they are kept in a $-20°$ freezer and heated before analysis. Placing sample buffer (itself or the appropriate mixture with growth medium) in the outside wells adjacent to the samples on the gel can help to keep patterns from spreading out in the gel during electrophoresis.

Where a target protein is made in relatively small amounts or is obscured by a host protein having a similar mobility, pulse labeling is a sensitive means of detecting its synthesis. Pulse–chase experiments can also be used to look for possible instability of target RNAs or proteins. Where a target protein is very difficult to detect, rifampicin can be used to inhibit transcription by host RNA polymerase and thereby reduce any background incorporation of label into host proteins. Typically, 200 μg of rifampicin per milliliter of culture is added 30 min after induction or infection, after which the host mRNAs decay while the target mRNA continues to be produced by the rifampicin-resistant T7 RNA polymerase. Within a few minutes, the only proteins still being made should be those under control of the T7 promoter. This test must of course be done in a host whose RNA polymerase is sensitive to rifampicin (such as BL21, but not HMS174).

A common problem encountered when proteins are expressed in *E. coli* is that the target protein is insoluble.[32] A quick test for this is to pellet a sample of cells (30 sec in an Eppendorf centrifuge at room temperature) and resuspend them in 50 mM Tris-Cl, 2 mM EDTA, pH 8.0, typically in a volume not much less than the original culture volume. If the cells contain T7 lysozyme, simply freezing and thawing this suspension will usually make a good lysate; otherwise, the suspended cells are first treated with commercial egg white lysozyme (100 μg/ml for 15 min at 30°). A further 2-min centrifugation pellets the insoluble protein. The protein contents of the total lysate, the supernatant, and the resuspended pellet are compared by gel electrophoresis.

Factors That Affect Production of Target Proteins

This T7 expression system has produced substantial amounts of target protein from a wide variety of genes, both prokaryotic and eukaryotic. However, some proteins are made in disappointingly small amounts, for reasons that are obvious in some cases and obscure in others. We here summarize briefly some of the known or likely reasons for obtaining low levels of expression.

The target protein itself may interfere with gene expression or with the integrity of the cell. Sometimes pulse labeling shows a gradual or rapid decrease in the rate of protein synthesis as target protein accumulates, or sometimes all protein synthesis stops before any target protein can be detected. Occasionally, considerable lysis of a culture is observed.

One might expect that instability of target mRNA might limit expression in some cases, although in each case we have examined, substantial amounts of target mRNA seem to accumulate. This apparent stability of target mRNA could be due to the stem-and-loop structures at both ends of RNAs that are initiated at the usual φ10 promoter and terminated at Tφ or

---

[32] F. A. O. Marston, *Biochem. J.* **240,** 1 (1986).

cut at R*1.1*; or the mRNA may be relatively inaccessible to exonucleases by being embedded in the long RNAs produced by T7 RNA polymerase in the absence of Tφ; or perhaps so much RNA is produced that the normal mRNA degradation system is overloaded.

Instability of certain target proteins might also be expected, although BL21 is apparently deficient in the *lon* and *ompT* proteases and many proteins produced in this strain are quite stable. Some relatively short proteins produced by out-of-frame fusions are also quite stable in this strain, whereas others are so rapidly degraded as to be undetected by pulse labeling.

Many target proteins seem to be made in equivalent amounts whether or not the Tφ transcription terminator is present in the vector. In some cases, however, having Tφ behind the target gene increases the production of target protein. In the cases we have encountered, the target mRNA is translated from its own translation initiation signals rather than from the strong T7 gene *10* signals. A possible interpretation is that some translation initiation signals do not compete well against the *bla* mRNA, which is made along with the target mRNA, and that Tφ, by reducing the amount of this competing mRNA, allows more target protein to be made. In the pET-9 vectors, where the *kan* gene and the target gene have opposite orientations, no competing mRNAs are known to be made along with the target mRNA.

Some target proteins are made in relatively small amounts even though both the mRNA and protein appear to be relatively stable and the coding sequence is joined to the efficient T7 translation initiation signals. The cause of the poor translation of these mRNAs is not well understood but could perhaps be due to factors such as unfavorable distributions of rare codons, relatively high levels of translational frameshifting, or interfering structures in the mRNA.

Potential Improvements

The inducible expression system is convenient to use and can produce high levels of target gene products. The most serious limitation is in its ability to maintain and express genes whose products are toxic to the cell. Addition of pLysS reduces basal T7 RNA polymerase activity enough that most target genes are stable yet still expressible at high level. The presence of T7 lysozyme in the cell has the additional advantage of facilitating the preparation of extracts. The repressible T7*lac* promoter reduces basal expression by a different mechanism, by blocking transcription from the target promoter. The T7*lac* promoter itself appears to stabilize more target genes than even the high level of lysozyme provided by pLysE, and seems to have little if any effect on the level of expression after induction. The combination of the T7*lac* promoter and pLysS appears to allow the maintenance and expression of almost any gene in BL21(DE3) or HMS174(DE3). Full expression from the T7*lac* promoter takes longer to induce in the presence of pLysS and may not reach as high a level as with the natural φ*10* promoter, but substantial levels of expression have been obtained in most cases tested so far.

Another way to reduce basal expression of target genes might be to place the gene for T7 RNA polymerase under control of inducible promoters that are more tightly repressed than the *lac*UV5 promoter. A high level of induction from the promoter may not be very important, since relatively small amounts of T7 RNA polymerase are capable of saturating the translation apparatus with mRNA. The appropriate way to test other inducible promoters would be to investigate whether they enable the system to tolerate and express ever more toxic genes. Even with more tightly regulated promoters, the system might well need the additional control provided by T7 lysozyme or the T7*lac* promoter.

Antisense promoters[33] may also be useful for reducing basal expression of toxic target genes. We find[34] that a promoter for *E. coli* RNA polymerase in the antisense direction can stabilize toxic target plasmids in BL21(DE3) or HMS174(DE3). However, the level of expression achieved on induction of T7 RNA polymerase seems to depend both on the strength of the antisense promoter and on the particular gene that is being expressed. The yield of target protein can sometimes be increased by adding rifampicin to shut off the synthesis of antisense RNA after T7 RNA polymerase has been induced. An antisense promoter has also permitted the cloning and expression of certain genes we had previously been unable to clone at all, apparently by antagonizing or neutralizing the effects of promoters for *E. coli* RNA polymerase in the target DNA fragment.

Other variations or improvements to the vectors for cloning and expressing target genes could also be considered. Additional antibiotic resistances could be substituted for ampicillin or kanamycin resistance as the selective marker. Equivalent vectors having compatible replicons might be useful for expressing different proteins in the same cell. Vectors with higher or lower copy numbers might also be useful in some applications. It may also be possible to reduce the already low basal expression of target genes in the absence of T7 RNA polymerase, perhaps by inserting transcription terminators ahead of the T7 promoter or by eliminating nonessential regions of the plasmid that may contain weak promoters for *E. coli* RNA polymerase.

The expression system can be expanded to other hosts that might be

---

[33] P. J. Green, O. Pines, and M. Inouye, *Annu. Rev. Biochem.* **55**, 569 (1986).
[34] L. Lin, A. H. Rosenberg, and F. W. Studier, unpublished work, 1988.

specialized for various purposes. A wide range of host strains can be easily lysogenized with DE3, using helper bacteriophages that both provide *int* function and select for growth of the appropriate lysogen. Derivatives of λ that have other host ranges or derivatives of other bacteriophages that carry no T7 promoters could also be used to deliver the gene for T7 RNA polymerase to the cell. Vehicles to deliver T7 RNA polymerase and vectors to carry target genes could in principle be developed for a wide variety of bacteria besides *E. coli.*

Acknowledgments

This work was supported by the Office of Health and Environmental Research of the United States Department of Energy and by Public Health Service grant GM21872 from the Institute of General Medical Sciences.

[7] High-Level Translation Initiation

By LARRY GOLD and GARY D. STORMO

Introduction

Promoters are cassettes; they work in a manner that is independent of the surrounding nucleotide sequences, with some perturbations allowed for the torsional strain or relaxed state of the DNA. Ribosome binding sites (RBS), the translation equivalent of promoters, are not known to be portable. However, we think translation initiation is simple, and portable RBS are easy to imagine.

Before we describe translation initiation in a simple manner, which leads directly to the design of portable RBS, we disclaim the extension of these ideas to explain the behavior of *all* mRNAs. We have studied the initiation activity of hundreds of different mRNAs; many of the RBS of those mRNAs are far more efficiently utilized than we would expect. Some mRNAs must have evolved nonstandard mechanisms for fast binding to ribosomes and the subsequent steps; examples are the RBS of the coat protein genes of the RNA phages of *Escherichia coli:*

```
                    * * *        1 1        * * *
QB coat RBS:    GUUGAAACUUUGGGUCAAUUUGAUCAUGGCAAAAUUAGAG
                     > > > > >      < < < < <
                     * * * *        8        * * *
MS2 coat RBS:   AGAGCCUCAACC GGAGUUUGAAGCAUGGCUUCUAACUUU
                          > > > > >      < < < < <
```

The QB coat protein RBS has a weak Shine–Dalgarno (SD) sequence, long spacing between the SD and the AUG, and a weak secondary structure ($> <$) that should diminish the rate of initiation.[1,2] The MS2 coat protein RBS has a better SD, more normal spacing, but a local secondary structure ($> <$) around the AUG that could slow the filling of the ribosomal P site.[1,2] These two RBS are nothing special by our criteria, yet they are used with high efficiency; in this article we ignore such RBS to deal with the straightforward methods by which any protein can be synthesized at a high level of translation.

Most mRNAs are initiated via a kinetic scheme that was first proposed by Gualerzi and collaborators.[3] A version of that scheme is as follows:

$$30S \rightleftharpoons 30S^*$$

$$30S^* + mRNA \underset{k_{-1}}{\overset{k_1}{\rightleftharpoons}} PC$$

$$PC \underset{k_{-2}}{\overset{k_2}{\rightleftharpoons}} IC$$

$$IC \overset{k_3}{\longrightarrow} TC$$

Here, 30S is the small ribosomal subunit; 30S*, that subunit complexed with initiation factors IF1, IF2, IF3, and fMet-tRNA$_f^{Met}$; PC, a preinitiation complex between 30S* and a specific messenger RNA in which the SD region of the mRNA is base-paired to the 3' end of the 16S RNA,[4] but no base pairing exists between the anticodon of fMet-tRNA$_f^{Met}$ and the initiation codon; IC, an initiation complex in which, in addition to the SD interaction, codon–anticodon pairing exists (i.e., resulting from a first-order rearrangement of PC); TC, translation complex, in which full 70S particles are actively engaged in peptide bond formation.

Our recent systematic experiments[5] and reviews of the literature suggest the domains of RBS that contribute to the various rate constants: (1) k_1 is made fast by mRNA sequences that have no intramolecular structures involving the SD; k_1 is slowed by any structures that include the SD, and can be very slow.[1,2] The forward rate constant could also be increased if the SD is long. (2) k_{-1} is fast if the SD is short or does not include the central

[1] L. Gold, *Annu. Rev. Biochem.* **57**, 199 (1988).

[2] G. D. Stormo, *in* "Maximizing Gene Expression" (W. Reznikoff and L. Gold, eds.), p. 195. Butterworth, Stoneham, Massachusetts, 1986.

[3] C. O. Gualerzi, C. L. Pon, R. T. Pawlik, M. A. Canonaco, M. Paci, and W. Wintermeyer, *in* "Structure, Function and Genetics of Ribosomes" (B. Hardesty and G. Kramer, eds.), p. 621. Springer-Verlag, Berlin and New York, 1986.

[4] J. Shine and L. Dalgarno, *Proc. Natl. Acad. Sci. U.S.A.* **71**, 1342 (1974).

[5] S. Shinedling, L. Green, D. Barrick, G. D. Stormo, and L. Gold, manuscript in preparation.

Cs of the SD;[6,7] k_{-1} is slow if the SD is long. (3) k_2 is a function of the spacing between the SD and the initiation codon; the optima are not sharp if the spacing is longer than the optimum, but can be extremely sharp if the spacing is too short;[5] k_2 will be faster if the initiation codon is not involved in an intramolecular structure, and slower if a structure is present. For the SD sequence UAAGGAGG the optimum spacing is seven to nine nucleotides.[5] The forward rate constant may be slightly faster if the initiation codon is AUG rather than GUG or UUG. (4) k_{-2} is made faster if the initiation codon is other than AUG, both because of the intrinsic weakness of the codon–anticodon pairing and (plausibly[8,9]) the action of IF3. (5) k_3 might be related to second codon choice, although this has not been tested directly. AAA and GCU are very abundant *E. coli* second codons, and could be used to hasten k_3. This is unimportant for most people, who will not use GCU or AAA if the price is an inexact protein sequence for the gene of interest. The effects are likely to be small compared to other parameters, especially when AUG is the initiation codon. (6) Other mRNA determinants present in statistical evaluations of *E. coli* RBS[6,7,10] can alter any rate constant, although impact on k_{-1} and/or k_{-2} is most easily imagined. For expression purposes this is not important: enough protein can be expressed routinely without worrying about other determinants. It may be these other determinants that allow the RNA phage coat genes to function efficiently.

Operating Procedure

The gene of interest should be cloned downstream from a strong, regulated promoter so that the RNA will have >25 nucleotides 5' to the initiation codon[11] and so that the RNA will include

* * * * * * * *   * * *

5' ppp (NNNNNNNNN)UAAGGAGGAAAAAAAAAUG-(codons)

We select nucleotides 5' to the SD to both minimize secondary structures within the RBS and to provide convenient restriction sites. We select nucleotides just 3' to the initiation codon to minimize secondary structures within the RBS. In practice, we use a vector in which a strong promoter fires toward a polylinker, and then we clone the gene of interest into the

[6] G. D. Stormo, T. D. Schneider, and L. M. Gold, *Nucleic Acids Res.* **10**, 2971 (1982).

[7] G. D. Stormo, T. D. Schneider, L. Gold, and A. Ehrenfeucht, *Nucleic Acids Res.* **10**, 2997 (1982).

[8] D. Hartz, D. S. McPheeters, and L. Gold, manuscript in preparation.

[9] B. Berkhout, C. J. Van der Laken, and P. H. Van Knippenberg, *Biochim. Biophys. Acta* **866**, 144 (1986).

[10] T. D. Schneider, G. D. Stormo, L. Gold, and A. Ehrenfeucht, *J. Mol. Biol.* **188**, 415 (1986).

[11] A. Bingham and S. Busby, *Mol. Microbiol.* **1**, 117 (1987).

polylinker using the closest sensible restriction site 3' to the AUG and another restriction site 3' to the chain-terminating codon. That construct will thus have no RBS (and no initiation codon) and will also be missing only a small number of codons. Synthetic DNA is then used to expand this construct and provide the appropriate RBS and flanking nucleotides. We would use codons that are found in highly expressed genes, although codon choice may not be a serious problem unless the protein of interest has adjacent arginines.[12]

Thus, we suggest a two-step procedure, in which the second step requires a pair of synthetic deoxyoligonucleotides. If the T7 gene 1 RNA polymerase system is chosen for the strong regulated promoter, the constructs built by Rosenberg *et al.*[13] around the T7 gene 10 RBS are suitable, since the RBS of gene 10 is

* * * * * *   * * *

UCUAGAAAUAAUUUUGUUUAACUUUAAGAAGGAGAUAUACAUAUGGCUAGC

The gene 10 RBS has no obvious structures to block initiation, and the SD is close to the optimum. However, as noted elsewhere in this volume,[13] the use of these vectors will almost always require synthetic DNA to obtain perfect protein sequence. In fact, although one can use our sequence as a portable RBS, no vector can be designed in which perfect protein sequence is obtained for any gene of interest unless one always anticipates the use of synthetic DNA for one cloning step. We have tested only A-rich constructs because of our previous statistical analyses, and because of the data of Dreyfus that show a strong bias toward As in the most active RBS.[14]

Finally, a number of genes have been driven by such constructs, although none is yet described in the literature. The RBS has only one chain-termination codon upstream of the RBS, and that UAA is not in the same frame as the initiating AUG. Other chain terminators should be included if the prospective transcript has an extensive 5' leader (through poor selection of vector, or because the synthetic RBS is being inserted into a preexisting clone via site-directed mutagenesis). Runs of As also are thought to be "slippery" to elongating ribosomes (and perhaps RNA polymerase as well),[15] and thus the problem of upstream initiation on a long 5' leader might be serious, regardless of the frame of the upstream initiation codon. Fortunately, the last step of the proposed cloning uses synthetic DNA, and so we suggest that the upstream nucleotides include chain

---

[12] R. A. Spanjaard and J. vanDuin, *Proc. Natl. Acad. Sci. U.S.A.* **85,** 7967 (1988).
[13] F. W. Studier, A. H. Rosenberg, and J. J. Dunn, this volume [6].
[14] M. Dreyfus, *J. Mol. Biol.* **204,** 79 (1988).
[15] R. Weiss, manuscript in preparation.

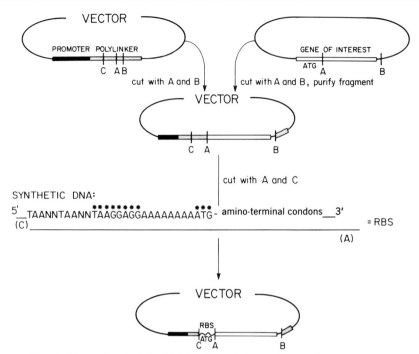

FIG. 1. Scheme for obtaining high-level translation and expression of any gene.

terminators when the 5' leader is long (the Ns can be chosen to avoid secondary structure between the RBS and codons in the gene):

* * * * * * * *   * * *

5' (UAANNUAANN)UAAGGAGGAAAAAAAAAAUG-codons

∧ ∧ ∧   ∧ ∧ ∧   ∧ ∧ ∧

The use of translation coupling or reinitiation is described by Schoner *et al.,*[16] in general, coupling will not be required for high-level translation, although one might be able to use a shorter piece of synthetic DNA in the final step of such constructions. In summary, we propose the scheme shown in Fig. 1 for obtaining high-level translation and expression of any gene.

Acknowledgments

This research was supported by Public Health Service Grants GM28685 (to L.G.) and GM28755 (to G.S.) from the National Institutes of Health.

[16] B. E. Schoner, R. M. Belagaje, and R. G. Schoner, *Proc. Natl. Acad. Sci. U.S.A.* **83,** 8506 (1986); B. E. Schoner, R. M. Belagaje, and R. G. Schoner, this volume [8].

[8] Enhanced Translational Efficiency with Two-Cistron Expression System

By BRIGITTE E. SCHONER, RAMA M. BELAGAJE, and
RONALD G. SCHONER

High-level expression of numerous prokaryotic and eukaryotic proteins in *Escherichia coli* has been achieved through recombinant DNA technology. Many of these proteins were previously unavailable for study because of their limited supply from natural sources. The ability to express foreign proteins in *E. coli* has greatly aided research in many disciplines and has led to the large-scale commercialization of some of these proteins.

High-level gene expression in *E. coli* typically requires the use of a multicopy expression plasmid with a strong promoter and an efficient ribosome binding site. There are some notable exceptions where, despite the use of such expression plasmids, very low levels of the desired protein are synthesized. Two common explanations are often proposed for this result: (1) the protein is unstable in *E. coli* and is rapidly degraded by proteases, or (2) the mRNA transcript is not efficiently translated. The former problem is most severe with low-molecular-weight proteins and can often be overcome by creating polyproteins,[1] fusion proteins,[2,3] or by the use of protease-deficient mutants of *E. coli*.[4] The latter problem is more elusive and is the topic of this article.

Features of mRNA Related to Translational Efficiency

Generally, a functional ribosome binding site containing a Shine–Dalgarno (SD) sequence properly positioned 5′ to an AUG initiation codon is essential for efficient translation (Fig. 1). Variations in the distance between the SD sequence and the AUG codon and the nucleotide sequence of the "window" region are known to affect mRNA translation, as do changes in the degree of complementarily between the SD sequence

---

[1] S.-H. Shen, *Proc. Natl. Acad. Sci. U.S.A.* **81**, 4627 (1984).

[2] D. V. Goeddel, D. G. Kleid, F. Bolivar, H. L. Heyneker, D. G. Yansura, R. Crea, T. Hirose, A. Kraszewski, K. Itakura, and A. D. Riggs, *Proc. Natl. Acad. Sci. U.S.A.* **76**, 106 (1979).

[3] K. Itakura, T. Hirose, R. Crea, A. D. Riggs, H. L. Heyneker, F. Bolivar, and H. W. Boyer, *Science* **198**, 1056 (1977).

[4] K. H. S. Swamy and A. L. Goldberg, *J. Bacteriol.* **149**, (1982).

One-Cistron mRNA

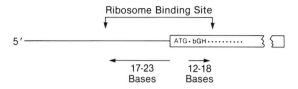

Two-Cistron mRNA

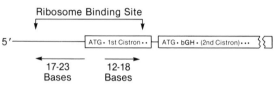

FIG. 1. Extent of the ribosome binding site. The ribosome binding site has been defined as that portion of mRNA protected from RNase digestion by ribosomes that have bound to a mRNA but have not initiated translation.[15-17] In typical one-cistron mRNAs the ribosome binding site overlaps with the 5′ coding sequence of the gene downstream from it. In two-cistron mRNAs it does not overlap with the coding sequence of the gene that comprises the second cistron.

and 16S rRNA.[5-8] Numerous studies, for example, those with *lamB*,[9] the *gal* operon,[10] the *trp* operon,[11] λ*cro*,[12] and the MS2 *lysis*[13] gene, suggest that mRNA translation is less efficient when the SD sequence or the AUG initiation codon is sequestered in a double-stranded region of the mRNA, thereby blocking the accessibility of these sequences to the ribosomes. Thus, sequences outside the ribosome binding site can alter translational efficiency if they base pair with the SD sequence or the initiation codon. Codon usage may also affect translational efficiency. Runs of very rare codons can decrease synthesis of a protein by causing ribosomes to stall on the tract of rare codons, thereby exposing the downstream portion of the

[5] L. Gold, D. Pribnow, T. Schneider, S. Shinedling, B. S. Singer, and G. Stormo, *Annu. Rev. Microbiol.* **35**, 365 (1981).

[6] M. G. Shepard, E. Yelverton, and D. V. Goeddel, *DNA* **1**, 125 (1982).

[7] H. A. deBoer, A. Hui, L. J. Comstock, E. Wong, and M. Vasser, *DNA* **2**, 231 (1983).

[8] A. Hui, J. Hayflick, K. Dinkelspiel, and H. A. deBoer, *EMBO J.* **3**, 623 (1984).

[9] M. N. Hall, J. Gabay, M. Debarbouille, and M. Schwartz, *Nature (London)* **295**, 616 (1982).

[10] C. Queen and M. Rosenberg, *Cell* **25**, 241 (1981).

[11] A. Das, J. Urbanowski, H. Weissbach, J. Nestor, and C. Yanofsky, *Proc. Natl. Acad. Sci. U.S.A.* **80**, 2879 (1983).

[12] D. Iserentant and W. Fiers, *Gene* **9**, 1 (1980).

[13] R. A. Kastelein, B. Berkhout, and J. van Duin, *Nature (London)* **305**, 741 (1983).

mRNA to attack by RNases.[14] Last, the sequence in the 5' coding region of most mRNAs is a part of the ribosome binding site and base changes in this region can dramatically alter gene expression, suggesting that these bases may be another factor in determining translational efficiency.[15-19]

Structural Elements in Two-Cistron mRNA

Most expression plasmids contain a single gene of interest inserted downstream from a promoter and a SD sequence. To eliminate some of the important problems described above, we have designed a two-cistron expression system (Fig. 2). We placed a short coding sequence (the first cistron) 5' to the gene encoding the product (the second cistron) such that the 5' end of this gene is no longer a part of the ribosome binding site. The role of the first cistron in this system is to provide an optimal sequence for translation initiation. Since the polypeptide encoded by the first cistron is not a product of interest, its DNA sequence can be manipulated to optimize mRNA translation without the need to preserve a particular amino acid sequence.

Our major assumption in designing the two-cistron expression system was that ribosome accessibility to mRNA would be enhanced if the SD sequence and the AUG initiation codon were located in an AU-rich sequence context free of local secondary structure (Fig. 2). Thus, the DNA encoding the first cistron is AT-rich (23/31 bases). Once bound to mRNA, ribosomes translating a coding region should be able to disrupt any secondary structures downstream from the ribosome binding site.[20,21] Further, since the first cistron serves to separate the 5' untranslated region from the second cistron, the opportunity for secondary structure formation between these regions should be minimized. The sequence of the first cistron was kept short to reduce the expenditure of cellular energy on the synthesis of the polypeptide that it encodes.

Based on the translation of polycistronic mRNAs from *E. coli* operons,

[14] M. Robinson, R. Lilley, S. Little, J. S. Emtage, G. Yarranton, P. Stephens, A. Milligan, M. Eaton, and G. Humphreys, *Nucleic Acids Res.* **12**, 6663 (1984).
[15] J. A. Steitz, *Nature (London)* **224**, 957 (1969).
[16] J. Hindley and D. H. Staples, *Nature (London)* **224**, 964 (1969).
[17] S. L. Gupta, J. Chen, L. Schaefer, P. Lengyel, and S. M. Weissman, *Biochem. Biophys. Res. Commun.* **39**, 883 (1970).
[18] B. E. Schoner, H. M. Hsiung, R. M. Belagaje, N. G. Mayne, and R. G. Schoner, *Proc. Natl. Acad. Sci. U.S.A.* **81**, 5403 (1984).
[19] H. J. George, J. J. L'Italien, W. P. Pilacinski, D. L. Glassman, and R. A. Krzyek, *DNA* **4**, 273 (1985).
[20] B. Berkhout and J. van Duin, *Nucleic Acids Res.* **13**, 6955 (1985).
[21] R. A. Kastelein, B. Berkhout, and J. van Duin, *Nature (London)* **305**, 741 (1983).

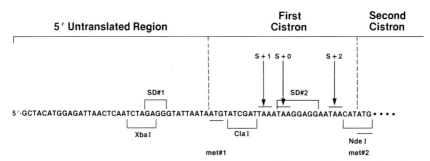

FIG. 2. DNA sequence corresponding to the 5′ end of the two-cistron mRNA. The 5′ untranslated region includes the SD 1 and extends to the ATG codon for Met 1. This sequence is identical to that of the *E. coli lpp* mRNA. The region 3′ to the ATG codon for Met 1 is the synthetic first-cistron sequence. Restriction sites and the SD sequences are indicated by brackets; the three stop codons are indicated by arrows.

we assumed that ribosomes would bind to the ribosome binding site of the first cistron, initiate translation at the AUG codon, and translate the first cistron. Following termination of translation at the end of the first cistron, these ribosomes may not dissociate, but rather reinitiate translation at the AUG codon of the second cistron and proceed with translation of the second cistron. This two-cistron system allows us to test some of these assumptions. The 5′ untranslated region is identical to the native *lpp* 5′ untranslated region (Fig. 2). Lipoprotein is one of the most abundant proteins in *E. coli,* suggesting that its mRNA is efficiently translated. The 5′ untranslated region is believed to be free of secondary structure and contains a convenient *Xba*I restriction site.[22] The first cistron contains two restriction sites, *Cla*I and *Nde*I, to facilitate the introduction of sequence changes and a SD sequence that is highly complementary to 16S rRNA. The first cistron also contains three translational stop codons (TAA), designated S + 0, S + 1, S + 2. Each is in a different reading frame relative to the AUG codon of the first cistron.

Methods

The series of two-cistron plasmids we constructed are identical except for changes in the first cistron. The plasmids were derived from pIMIA and contain a thermoinducible runaway replicon, a kanamycin resistance gene, the *lpp* promoter for transcription of the two-cistron mRNA and the *lpp* transcriptional terminator.[23] The methionyl bovine growth hormone

[22] K. Nakamura, R. M. Pirtle, I. L. Pirtle, K. Takeishi, and M. Inouye, *J. Biol. Chem.* **255**, 210 (1980).
[23] Y. Masui, J. Coleman, and M. Inouye, *in* "Experimental Manipulation of Gene Expression" (M. Inouye, ed.), p. 15. Academic Press, New York, 1983.

(Met-bGH) gene with its native codons was chosen for the second cistron. *Escherichia coli* K12 RV308 (su⁻, Δ*lacX74, gal*ISII::OP308, *strA*) was used in all of the experiments with identical conditions for growth and thermal induction of gene expression.[18] Cell pellets were dissolved directly in sample buffer and loaded onto polyacrylamide gels for analysis of Met-bGH accumulation. After staining with Coomassie blue, gels were scanned with a Shimadzu 930 TLC scanner that integrates the areas under the peaks.

Analysis of Alternations in First Cistron

We set out to determine how changes in the relationships between key elements in the first cistron would affect the translational efficiency of Met-bGH.[24] As shown in Fig. 3, no detectable accumulation of Met-bGH occurs in cells harboring pCZ140, a plasmid without a first cistron that contains the Met-bGH gene. Since we could detect high levels of Met-bGH-specific mRNA, we assumed that the translational efficiency of the mRNA was low. However, cultures harboring pCZ143, our first two-cistron plasmid in this series, did not produce measurable amounts of Met-bGH. This suggested that the first cistron present in pCZ143 was defective in terms of enhancing translational efficiency.

To investigate the lack of high-level expression of Met-bGH with this particular first-cistron sequence, we introduced the series of base changes shown in Fig. 3. In pCZ143, the ATG (for Met 1) is in frame with the S+0 stop codon. In pCZ144 and pCZ145, the reading frame was shifted either by inserting two bases (GC) at the *Cla*I restriction site or by deleting the base (T) from pCZ143. These changes placed the S+2 stop codon into frame with respect to the ATG codon (for Met 1). Measurements of Met-bGH synthesis in cells harboring plasmids pCZ144 or pCZ145 revealed a dramatic increase in the amount of Met-bGH produced compared with cells harboring pCZ143. Next, we shifted the reading frame to the S+1 stop codon by either inserting two bases (GC) at the *Cla*I site in plasmid pCZ145 or by adding one base (C) near the *Cla*I site in pCZ143. RV308 cells harboring these two plasmids (pCZ145 and pCZ147) produced undetectable amounts of Met-bGH. These results suggested that the level of expression of the second cistron might be very sensitive to changes in the first cistron that alter: (1) the length of the open reading frame in the first cistron, (2) the position of the stop codon that terminates translation of the first cistron, (3) the length of the intercistronic region, and (4) the phasing of the stop codon in the first cistron relative to the start codon for the second cistron. To determine the relative importance of these variables, we

---

[24] B. E. Schoner, R. M. Belagaje, and R. G. Schoner, *Proc. Natl. Acad. Sci. U.S.A.* **83**, 8506 (1986).

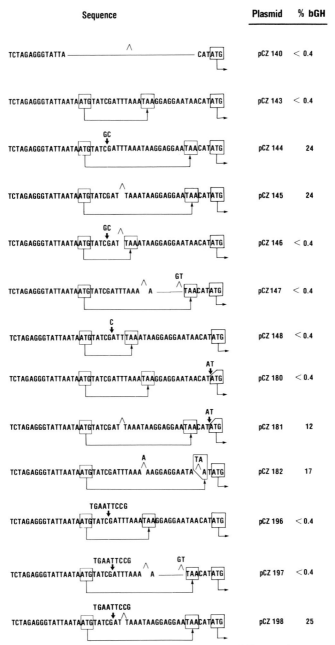

Sequence	Plasmid	% bGH

FIG. 3. First-cistron sequences for Met-bGH. The only differences between the various two-cistron expression plasmids are localized within the region shown. This region extends from the *Xba*I site in the 5' untranslated region of the two-cistron mRNA to the *Nde*I site at the beginning of the second cistron. Sequence changes are relative to the sequence in pCZ143. Some were introduced by "filling in" bases at restriction sites for *Cla*I (in pCZ144 and

analyzed the effect of changing the phasing of the stop codon in the first cistron relative to the restart codon for the second cistron. When the stop and restart codons in pCZ143 were shifted into frame by inserting two bases (AT) at the *Nde*I restriction site of pCZ143 (to create pCZ180), no detectable change in the expression of Met-bGH was observed. Shifting the stop and restart codons in pCZ145 out of frame and either two bases farther away from the restart codon (as in pCZ181) or two bases closer to the restart codon (as in pCZ182) decreases somewhat the expression of Met-bGH relative to the level obtained with plasmids pCZ144 and pCZ145.

The addition of two bases (AT) and the *Nde*I site in pCZ145 (to give pCZ181) not only changes the phasing of the stop and restart codons but also expands the "window" region between SD2 and ATG codon for Met 2. Thus, if a SD sequence in the first cistron is required for Met-bGH expression, a decrease in the amount of Met-bGH produced might be expected as a result of changing the distance between the SD sequence and the restart codon. To examine this point further, we deleted SD 2 as shown in pCZ148 and pCZ197. The stop codons in pCZ148 and pCZ197 are in a proper position relative to the restart codon, yet no detectable expression of Met-bGH was observed. From these experiments we concluded that the SD sequence is required for expression of the second cistron. Further, the observed reduction in Met-bGH expression with pCZ181 (where the window region was expanded by two bases) suggests that proper spacing between the SD sequence and the translational restart codon is important for optimal expression of the second cistron. The phasing of the stop codon and the restart codon for the second cistron appears to be relatively less important. These observations, however, do not adequately explain the large differences in the expression levels obtained between plasmids

---

pCZ146) and *Nde*I (in pCZ180 and pCZ181). This was accomplished by cutting the starting plasmids with *Cla*I or *Nde*I, treating the linearized plasmids with the Klenow fragment of DNA polymerase I in the presence of all four deoxyribonucleotides, and reclosing the plasmid with T4 DNA ligase. Other changes were made by replacing the region containing the sequence change by synthetic DNA fragments. Specifically, the *Cla*I and *Nde*I restriction sites were used to construct pCZ145, pCZ147, pCZ148, and pCZ182. pCZ196, pCZ197, and pCZ198 were constructed by inserting a synthetic linker, containing the bases shown, at the *Cla*I restriction site in pCZ143, pCZ148, and pCZ145, respectively. Arrows indicate positions at which bases have been added; carets indicate deletions of bases. A caret with associated bases indicates the deletion of bases found in pCZ143 and the addition of new bases at that position. The predicted reading frame of the first cistron in each plasmid begins at the boxed ATG codon and terminates at the boxed TAA codon. The amounts of Met-bGH measured in cells containing these plasmids are expressed as percent total cell protein as determined by scanning Coomassie blue-stained polyacrylamide gels.

pCZ143 and pCZ144 or pCZ145. The open reading frame in the first cistron in pCZ144 and pCZ145 contains nine and eight codons, respectively. Thus, we questioned whether the number of codons that are translated in the first cistron is important for the efficient expression of the second cistron. To test this possibility directly, we inserted three codons into the *Cla*I restriction site of pCZ143 to create an open reading frame with eight codons (pCZ196). Expression of Met-bGH from pCZ196 was undetectable. Therefore, we concluded that the position of the stop codon that terminates translation of the first cistron determines the efficiency of translation of the second cistron rather than the number of codons in the first cistron (i.e., five versus eight or nine). Insertion of the same three codons into plasmid pCZ145 (to give pCZ198) did not lower the level of expression of Met-bGH obtained with pCZ145.

To demonstrate that we were observing translational and not transcriptional effects, we measured the steady-state levels of the two-cistron mRNA in high and low producers of Met-bGH by dot-blot analysis (data not shown).[24] Intense hybridization with no more than 3-fold differences is seen with RNA transcribed from these plasmids. The highest and lowest amounts of hybridizing mRNA were found in cultures containing pCZ143 (a low producer of Met-bGH) and pCZ140 (the plasmid without the first cistron), respectively. Intermediate levels of hybridizing mRNA were found in cultures that overproduce Met-bGH. These measurements indicate that the two-cistron mRNAs are efficiently transcribed and confirmed our assumption that the large differences in Met-bGH expression are due to differences in the translational efficiency of these mRNAs. The 3-fold differences in the steady-state mRNA levels probably reflect differences in the mRNA turnover.

Expression of Human Proinsulin Gene with Two-Cistron System

From our studies, some rules have emerged (as outlined above) concerning the expression of Met-bGH with a synthetic two-cistron system. In the next series of experiments, we examined whether our findings are applicable to the expression of another mammalian gene. We also wanted to determine whether the presence of a first cistron affects the expression of a gene that can be expressed without the use of a first cistron.[25] An example of such a gene is the one that encodes human proinsulin (hPI).

The coding sequence for methionyl-hPI (Met-hPI) with its native codons was inserted into plasmids pCZ140, pCZ143, pCZ144, pCZ145, pCZ146, and pCZ148 by replacing exactly the coding sequences for Met-

[25] B. E. Schoner, R. M. Belagaje, and R. G. Schoner, this series, Vol. 153, p. 401.

bGH. The expression of Met-hPI from these plasmids (pCZ240, pCZ243, pCZ344, pCZ345, pCZ246, and pCZ248, respectively) was studied in RV308 cells (Fig. 4). Plasmid pCZ240 contains no first cistron, but still expresses Met-hPI at a level equal to 12% of cell protein. This suggests that the mRNA is efficiently translated. Significantly, however, the pattern of expression for the other plasmids mimics that found for the Met-bGH plasmids. For example, pCZ245, which contains a properly positioned stop codon in the first cistron, expresses Met-hPI at a high level. The two

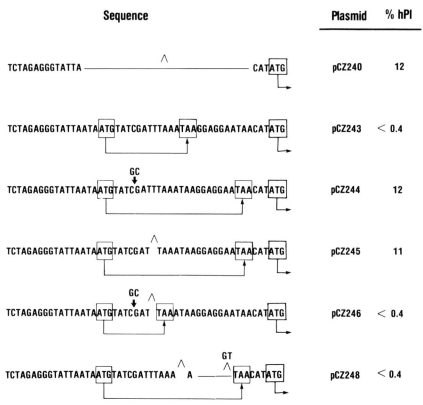

Sequence		Plasmid	% hPI

FIG. 4. First-cistron sequences for hPI. The sequences shown extend from the *Xba*I site in the 5′ untranslated region to the *Nde*I site at the beginning of the second cistron. The sequence changes are relative to the sequences in pCZ243. Arrows indicate positions at which bases have been inserted into the sequence; carets indicate deletions. A caret with associated bases indicates that the corresponding bases found in pCZ243 have been deleted and new bases inserted at that position. The predicted reading frame of the first cistron in each plasmid begins at the boxed ATG codon and terminates at the boxed TAA codon. The amounts of hPI measured in cells containing these plasmids are expressed as percent total cell protein as determined by scanning Coomassie blue-stained polyacrylamide gels.

plasmids with stop codons at inappropriate positions, pCZ243 and pCZ246, do not express Met-hPI at detectable levels. Thus, a mRNA with low translational efficiency can be created through the use of an inappropriate first cistron such as those found in pCZ243 and pCZ246. Such first cistrons appear to be able to suppress translational efficiency to very low levels. Last, the undetectable level of Met-hPI expression from pCZ248 again suggests that an appropriately positioned SD sequence is required for the second cistron. These results with the Met-hPI plasmids confirm and extend our results and conclusions from the studies with the Met-bGH plasmids.

Discussion

The two-cistron expression system described in this article was designed for the expression of proteins that are encoded by mRNAs with low translational efficiency. The role of the first cistron in this expression system is to allow for efficient ribosome binding and translation initiation. The sequence of the first cistron is AT-rich to reduce local secondary structure formation that might sequester the SD sequence or the initiation codon. Since this sequence also separates the 5' untranslated region of the mRNA from the coding sequence for the gene of interest, the potential for secondary structure formation between these two sequences should also be reduced. Further, the presence of a first cistron in the mRNA moves the ribosome binding site to a position upstream of the coding sequence for the gene of interest. Thus the entire first-cistron sequence can be optimized for translation without any constraints to maintain a particular amino acid sequence and, therefore, it can be modified by the addition or deletion of bases, as well as by base substitutions. While we have shown that such changes affect gene expression, it should be noted that this system may be quite suitable to carry out further experiments to gain a better understanding of how ribosomes interact with mRNA to initiate translation.

[9] Sequences within Ribosome Binding Site Affecting Messenger RNA Translatability and Method to Direct Ribosomes to Single Messenger RNA Species

By HERMAN A. DE BOER and ANNA S. HUI

Introduction

One of the most important elements in the success of biotechnology in serving science, industry, and health care is that genes from any source can be cloned and expressed in heterologous cells that can be cultured on a large scale in fermentors. Thus any protein can, in principle, be produced in large quantities.

Soon after it was shown that high expression levels of heterologous genes introduced into *Escherichia coli* could be obtained, it became clear that, in those cases where little or no expression was observed, the problem was at the translational rather than at the transcriptional level. In this article, we discuss the parameters affecting the translatability of introduced heterologous mRNAs and strategies that can be developed to correct for lack of translation of certain mRNAs. We also briefly discuss a system in which a special type of ribosomes is devoted to the translation of a heterologous mRNA species.

Ribosome Binding Site

In *E. coli,* initiation of protein synthesis begins at a start codon which in most cases is an AUG. This start codon is at the center of an RNA fragment which is 30 to 40 bases in length and which can be isolated from initiation complexes as an RNase-resistant oligonucleotide.[1] By definition, such regions are called ribosome binding sites (RBS). In most ribosome binding sites, the start codon is preceded by a purine-rich region at a distance of 5 to 12 bases. This so-called Shine–Dalgarno (SD) sequence[2] shows a variable degree of complementarity with a region close to the 3′ end of 16S rRNA; the anti-Shine–Dalgarno (ASD) sequence. The SD- and ASD-containing sequences can be coisolated from initiation complexes as an RNA duplex.[2] The SD region is thought to assist the 30S particle in positioning itself at the proper place with respect to the start codon on the mRNA.[3-5] Recently, we have shown that the presence of a SD sequence which is complementary to the ASD sequence enhances the efficiency of initiation greatly.[6-8] Lack of complementarity still allows for protein initiation to occur, but does so at reduced rates. These results were obtained

[1] J. A. Steitz, *in* "Biological Regulation and Development, Vol. 1: Gene Expression" (R. F. Goldberger, ed.), p. 349. Plenum, New York, 1979.

[2] J. Shine and L. Dalgarno, *Proc. Natl. Acad. Sci. U.S.A.* **71**, 1342 (1974).

[3] L. Gold, D. Pribnow, T. Scheider, S. Shinedling, B. W. Singer, and G. Stormo, *Annu. Rev. Microbiol.* **35**, 365 (1981).

[4] M. Kozak, *Microbiol. Rev.* **47**, 1 (1983).

[5] C. O. Gualerzi, R. A. Calogero, M. A. Canonaco, M. Brombach, and C. L. Pon, *in* "NATO–ASI Series, Vol. H14: Genetics of Translation" (M. F. Tuite *et al.,* eds.), p. 317. Springer-Verlag, Berlin and Heidelberg, 1988.

[6] H. A. de Boer, P. Ng, and A. S. Hui, *in* "Sequence Specificity in Transcription and Translation," p. 419. Alan R. Liss, New York, 1985.

[7] A. S. Hui and H. A. de Boer, *Proc. Natl. Acad. Sci. U.S.A.* **84**, 4762 (1987).

[8] H. A. de Boer, D. H. Eaton, and A. S. Hui, *in* "NATO–ASI Series, Vol. H14: Genetics of Translation" (M. F. Tuite *et al.,* eds.), p. 343. Springer-Verlag, Berlin and Heidelberg, 1988.

in experiments in which both the SD sequence and the ASD sequence were altered. A similar demonstration was reported by Dahlberg et al.[9]

In the early experiments[10,11] in which recombinant DNA techniques were for the first time applied to optimization of protein initiation, the distance between the AUG and the SD region was altered. This distance appeared to affect the efficiency of the translation initiation process quite dramatically (see below). Thus these three components of the initiation region: the AUG, the SD sequence, and their distance became the best known elements of the ribosome binding site, involved in determining the efficiency of translation at the mRNA level. However, later it became clear that sequences around these elements, including sequences downstream of the start codon, appeared to affect translational efficiency as well. Circumstantial evidence for this was provided by the observation that within the coding region of natural mRNAs, the sequence AUG could be found in and out of the natural reading frame, sometimes preceded at a proper distance by a SD-like sequence. Yet, such sequences are not recognized as functional initiation sites.[3,12] This has been confirmed experimentally by Dreyfus.[13] Apparently, in such pseudo-initiation sites, parameters other than the actual start codon, the SD sequence, and their distance are unfavorable for the translational process. Knowledge of these parameters is essential for those who wish to express a foreign gene in E. coli, since in most cases the ribosome binding sites used are hybrids of homologous and heterologous elements, sometimes joined together with synthetic DNA fragments (see, e.g., Ref. 11). In most cases, prokaryotic sequences are attached to the coding sequence of the heterologous genes to be expressed, and often sequences encoding convenient restriction sites are used, mostly for the purpose of easing plasmid assembly procedures.

Here, we discuss the parameters at the mRNA level that affect mRNA translatability and which must be taken into account when designing strategies for achieving maximal expression levels.

Start Codon

Although AUG is most abundantly used as a start codon in E. coli,[14] other start codons also function, but all have a lower efficiency than AUG and therefore should not be used for heterologous expression purposes.

[9] W. F. Jacob, M. Santer, and A. Dahlberg, Proc. Natl. Acad. Sci. U.S.A. **84,** 4757 (1987).
[10] H. M. Shepard, E. Yelverton, and D. Goeddel, DNA **1,** 125 (1982).
[11] H. A. de Boer and M. Shepard, in "Genes: Structure and Expression" (A. M. Kroon, ed.), p. 205. Wiley, New York, 1983.
[12] G. D. Stormo, T. D. Scheider, and L. M. Gold, Nucleic Acids Res. **10,** 2971 (1982).
[13] M. Dreyfus, in "NATO–ASI Series, Vol. H14: Genetics of Translation" (M. F. Tuite, et al., eds.), p. 295. Springer-Verlag, Berlin and Heidelberg, 1988.
[14] E. J. Gren, Biochimie **66,** 1 (1984).

SD Sequence

The endogenous mRNAs use a variety of SD sequences ranging from sequences like 5′GGAGG, with perfect 5-base complementarity, to barely recognizable SD sequences or no SD sequences at all.[14] Whether SD sequences classified as poor, i.e., with little complementarity with the ASD sequence, are still able to establish base pairing with the 3′ end of the 16S rRNA molecule, or whether other sequences of both RNA molecules are involved, is not known. An example of a mutation that maps within the SD sequence and which enhances SD–ASD complementarity has been reported by Chapon.[15] Here, it concerned a base change from 5′GAAG to 5′GGAG of the *malT* operon. This mutation increased *malT* mRNA translation more than 10-fold. In a different approach, we have varied the length of the double-stranded region that can be formed by increasing the length of the SD region from 4 to 13 bases. For that purpose a series of DNA segments was synthesized which were introduced in front of the start codon.[16] It appeared that the translatability of the mRNA did not increase as one might naively have expected, but decreased instead by about 50% with increasing SD length from 4 to 13 base pairs. This observation suggested that base interaction must be sufficient for mRNA–rRNA binding to occur but should not exceed a certain strength or else the ribosome may bind too tightly to the mRNA. As at the onset of elongation the SD–ASD interaction must be disrupted, a too-tight binding might impede the transition from initiation to the elongation phase of protein synthesis.

Distance between SD Sequence and Start Codon

Shepard *et al.*[10] have explored the effect of varying the distance between the SD sequence and the *trp* ribosome binding site and the AUG codon of the human fibroblast and leukocyte interferon genes. In both cases, a sharp optimal spacer length of 9 bases was found; one base more or one base less resulted in about a 50% decline of translation levels. In fact in one case, addition of one more base (i.e., a 10-base spacer) resulted in a decline of 84%. However, the sample size of these studies is too limited to conclude that 9 bases is the optimal distance for other mRNAs as well. The experiments of Shepard *et al.*[10] showed that the distance is a very important element and that it ought to be examined on a case-by-case basis when attempting heterologous gene expression in prokaryotic organisms.

[15] C. Chapon, *EMBO J.* **3**, 369 (1982).
[16] H. A. de Boer, L. J. Comstock, A. S. Hui, E. Wong, and M. Vasser, *Biochem. Soc. Symp.* **48**, 233 (1983).

Base Composition of Spacer between SD Sequence and Start Codon

From the computer analysis of Gold, Stormo, and co-workers,[3,12] it became clear that the composition of this spacer, and in fact of the entire ribosome binding site, is nonrandom. They found that an A residue was preferred in the spacer region, that C residues were random, that G residues were discriminated against in the spacer (and elsewhere in the ribosome binding site), and that U residues may be preferred at position -4 (the A of the AUG is position $+1$). Based on another computer analysis, Scherer *et al.*[17] proposed a sequence for an optimal ribosome binding site which consisted predominantly of A and U residues anywhere except in the SD region. These studies suggested that the sequences between the AUG and the SD sequence may play a role in determining the efficiency of translation initiation either directly or indirectly by preventing secondary structure formation of the ribosome binding site and thus keep the SD and the AUG accessible for the ribosome. Others (reviewed in Ref. 3) have isolated mutations that mapped in the spacer which stimulated, inhibited, or did not affect the translation efficiency. This issue was further investigated by changing, using synthetic DNA fragments and a portable SD sequence element, the four nucleotides that follow the SD sequence.[16] It was found that A and U residues stimulated translation over 2-fold compared to the translational efficiency of the same mRNA with a spacer containing a mixture of bases. C residues at this position had little effect, while G residues reduced translation 3-fold. Thus, these experiments showed a 6-fold range in translation levels caused by the four bases following the SD sequence alone. Recently, Bingham *et al.*[18] found also that A residues, following the SD sequence of the *galE* mRNA, enhance translatability.

Sequences Upstream of SD Sequence Affecting mRNA Translatability

Usually, there are several dozen bases between the 5′-terminal nucleotide of the mRNA and the beginning of the actual SD sequence. The function of the sequences in this area is poorly investigated. Only a few examples showing enhancement of translation of the *argE* mRNA by mutations mapping in this region have been reported.[19] It is not clear whether this area plays a role in protecting the mRNA from premature

[17] G. F. E. Scherer, M. D. Walkinshaw, S. Arnott, and D. J. Morre, *Nucleic Acids Res.* **8**, 3895 (1980).
[18] A. Bingham, F. Fulford, P. Murray, M. Dreyfus and S. Busby, *in* "NATO–ASI Series, Vol. H14: Genetics of Translation" (M. F. Tuite *et al.,* eds.), p. 307. Springer-Verlag, Berlin and Heidelberg, 1988.
[19] A. Boyen, J. Piette, R. Cunin, and N. Glansdorff, *J. Mol. Biol.* **162**, 715 (1982).

cleavage and whether it can confer some stability to the mRNA, as $5' \rightarrow 3'$-exonucleases have not yet been found. In several cases, the 5' untranslated part of the mRNA originates from the transcription of an operator, as exemplified by the *lac* operon. In a deletion experiment we eliminated the entire transcribed *lac* operator, thus creating a transcript which started immediately 5' adjacent to the SD sequence. Although formally all the elements are present for ribosome binding to occur, no translation took place on this mutated mRNA (our unpublished data). Either such a mRNA is unable to form initiation complexes with ribosomes, or the 5' end of such a mRNA is prone to degradation by nucleases. Alternatively, a critical element needed for efficient binding to ribosomes has been deleted. This may be the case, as Olins and Rangwala [10] have provided evidence that such a hitherto unknown element does indeed exist. Additional evidence for the existence of such an element also came from an analysis of deletions upstream of the SD sequence of the *galE* mRNA. This study showed that this region is needed for translation initiation to occur and that at least 10 bases must precede the SD sequence. On lengthening this untranslated region, a gradual increase of translatability was observed between 11 and 18 bases, at which point maximal translation rates were measured.[18] In conclusion, for the purpose of expressing a heterologous protein in *E. coli*, sufficient sequence information must be present at the 5' side of the SD sequence.

Sequences Immediately Upstream of Start Codon Profoundly Affect mRNA Translatability

Some of the mutations which were isolated via genetic means appeared to map in the region immediately 5' adjacent to the start codon. This observation, and the fact that bases immediately downstream of the SD sequence affect translatability, led us to analyze in more detail the three bases flanking the AUG codon on the 5' side of the *lacZ* mRNA. A procedure, which later became known as cassette mutagenesis, involving mutagenesis of a short synthetically derived DNA segment, was used.[20] This approach allowed us to compare 39 mutants of the same mRNA which only differed at any of these three bases. Thus, any difference in expression must have been caused solely by the nature of this sequence. We found a 20-fold range in expression levels, depending on the nature of the bases in this -1 triplet. It was found that the most favorable combinations of bases in the -1 triplet are UAU and CUU. Twenty-fold lower expression levels were observed with UUC, UCA, or AGG as the -1 triplet. In general, a U residue immediately preceding the start codon is more favorable for translation than any other base. Furthermore, an A residue at the

[20] A. S. Hui, J. Hayflick, K. Dinkelspiel, and H. A. de Boer, *EMBO J.* **3**, 623 (1984).

−2 position enhanced translation efficiency in most instances. In both of these cases, however, the degree of enhancement appeared to depend on its context. Although no rules could be derived from these studies, it became clear that G residues and, in particular, AGG-like sequences must be avoided, as they might mislead the ribosome by providing incorrect SD–ASD-like interactions. For a full account of all the data obtained from this study, see Ref. 20. As in many cases restriction sites are engineered in this area, the choice of the sequence should not solely be dictated by the sequence of the restriction site but should meet the requirements of the translation initiation process as well.

It should be noted that our study[20] was done using the *lacZ* mRNA and that the data thus obtained cannot blindly be used for expression of any other mRNA. For such purposes, a similar randomized cassette mutagenesis could be done combined with a selection or screening strategy (see [14] in this volume).

Influence of Codon Composition Following AUG Codon on mRNA Translatability

Several years ago, while attempting to express the human γ-interferon gene in *E. coli,* it was found that a certain mutation in the third position of the fourth codon resulted in a more than 30-fold increased expression level (D. Leung, Genentech, Inc. unpublished observation). This finding coincided with the publication of computer analyses mentioned earlier[3,12] which showed that the sequences around nucleotide +10 were in many cases AT-rich. Thus, the first surprising examples were provided showing that the actual coding sequence can have a great effect on mRNA translatability. The effects found by Leung could not be explained in terms of mRNA secondary structure effects, although mRNA structures within the coding sequence can dramatically influence translation.[21]

It seemed therefore, that translation initiation might be affected by the context of the start codon, as was first suggested by the work of Taniguchi and Weissmann.[22] Context sensitivity of suppression of stop codons is well established[23-25] and possibly may apply to sense codons as well[26] (reviewed in Ref. 27). These studies showed that context sensitivity is especially important where it concerns the 3′ adjacent nucleotides of stop codons.

[21] M. Schwarz, M. Roa, and M. Debarbouille, *Proc. Natl. Acad. Sci. U.S.A.* **78,** 2937 (1981).
[22] T. Taniguchi and C. Weissmann, *J. Mol. Biol.* **118,** 533 (1978).
[23] J. Miller and A. M. Albertini, *J. Mol. Biol.* **164,** 59 (1983).
[24] L. Bossi, *J. Mol. Biol.* **164,** 73 (1983).
[25] L. Bossi and J. R. Roth, *Nature (London)* **286,** 123 (1980).
[26] H. Engelberg-Kulka, *Nucleic Acids Res.* **9,** 983 (1981).
[27] H. A. de Boer and R. A. Kastelein, *in* "From Gene to Protein: Steps Dictating the Maximal Level of Gene Expression" (J. Davis, Man. ed.; W. S. Reznikoff and L. Gold, eds.), p. 225. Butterworth, Stoneham, Massachusetts, 1986.

Because of these considerations, we examined whether context sensitivity 3' adjacent to the start codon does play a role in determining the efficiency of translation.[28] To examine this issue, we constructed and analyzed, in collaboration with the group of Dr. van Knippenberg at the Leiden University, a series of mutants that differed only in the triplet at the 3' side of the AUG, again using cassette mutagenesis. The expression of these so-called second-codon variants was measured by assaying for the plasmid-coded protein product, in this case β-galactosidase. A 15-fold difference in expression was found among these variants. The best-performing codon following AUG appeared to be the lysine codon AAA (AAG performed 70% less well). Other relatively well-performing codons were UUU(Phe), AUC/A(Ile), GUA(Val), GCU(Ala), UAU(Tyr), CAU(His), AAU(Asn), CGU(Arg), and AGA(Arg). For a detailed account of all the data thus obtained, see Ref. 28.

There appeared to be no correlation between the efficiency of a given second-codon variant and the availability of its corresponding tRNA in the cell. This is apparent from the observation that AAG performs 4-fold worse than AAA and that UUC performs 9-fold worse than UUU, although they are recognized by the same tRNA^{Lys} and tRNA^{Phe}, respectively. Also, the Ile codon AUA performed quite efficiently, although there is only one minor species of tRNA corresponding to this codon.[27,29]

As far as the usage of the second codon in natural mRNAs is concerned, there is a strong bias which is quite different from the general bias in the codon usage.[28] The most frequently used codons at the second position of the 288 mRNA sequences[14] examined are AAA(Lys) and GCU(Ala); both performed well in our system. Another frequently used codon at the second position, ACA(Ala), performed poorly in this system. This codon is among the minor codons when the entire coding sequence of many *E. coli* genes is considered.[27-29] These data show that *E. coli* is able to regulate, with the choice of the second codon, the level of translation, presumably in accordance with the physiological needs for the translation product. For that reason, the choice of the second codon is under different constraints than the overall codon choice. We do not discuss the nature of the biased codon usage here, as we have reviewed this issue elsewhere.[27]

The observations discussed in this section must also be taken into account when attempting to express a foreign gene in *E. coli*. A similar strategy, as referred to in the previous section, is recommended (see [5] in this volume).

[28] A. C. Looman, J. Bodlaender, L. J. Comstock, D. Eaton, P. Jhurani, H. A. de Boer, and P. H. van Knippenberg, *EMBO J.* **6**, 2489 (1987).
[29] T. Ikemura, *J. Mol. Biol.* **151**, 389 (1981).

Directing Ribosomes to Single mRNA; Specialized
Ribosome System

Concept and Purpose

In this section, we discuss a substitution within the ribosomal RNA which affects the relative specificity with which ribosomes recognize various mRNAs.

In an attempt to improve yields of protein products from specific mRNAs, we designed the so-called specialized ribosome system.[7,8,30] This system was also designed in order to answer the question whether the SD–ASD interaction itself is an absolute requirement for initiation to occur and whether the actual sequences involved, i.e., 5'GGAGG as SD and 5'CCUCC as ASD sequence, are crucial or whether these complementary sequences can be replaced by any other pair of nucleotide sequences provided that they are complementary.

Construction

To build the specialized ribosome system, the natural SD and ASD sequences were replaced by their complements on both the mRNA and the rRNA using oligonucleotide-directed mutagenesis techniques. Thus the SD sequence was changed from 5'GGAGG to 5'CCUCC and the ASD sequence from 5'CCUCC to 5'GGAGG. This is system IX. Since we were concerned that such radical changes might impair ribosomal functions, a system that differed less radically from the wild-type system was made as well. In this system X, the SD sequence was changed from 5'GGAGG to 5'GUGUG, and the ASD sequence was changed accordingly from 5'CCUCC to 5'CACAC. In a control system, the wild-type SD and ASD sequences were present. In all three of these systems, the gene encoding the specialized mRNA [in this case, the human growth hormone gene, hGH (see Ref. 31), was used] is transcribed constitutively by a promoter derived from the *trp* promoter, whereas the wild-type and the mutated ribosomal RNA operons *(rrnB)* are under the control of the λP_L promoter. A detailed construction of the specialized ribosome system is given elsewhere.[30] The essential features of the plasmid harboring any one of the systems (VIII, IX, and X) are shown in Fig. 1.

[30] A. S. Hui, P. Jhurani, and H. A. de Boer, this series, Vol. 153, p. 432.
[31] D. V. Goeddel, H. L. Heyneker, T. Hozumi, R. Arentzen, K. Itakura, D. G. Yansura, M. J. Ross, G. Miozzari, R. Crea, and P. Seeburg, *Nature (London)* **281**, 544 (1979).

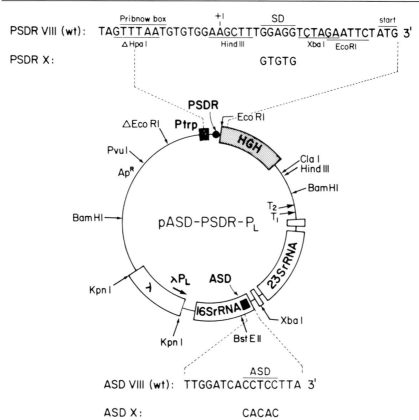

FIG. 1. Structure of the plasmids encoding the specialized ribosome systems IX and X and the control system VIII. Details of plasmid construction have been described elsewhere.[30] Note that in all systems, the hGH gene is constitutively transcribed by a modified *trp* promoter and that the specialized *rrnB* operon is under temperature control of the λP_L promoter. (From Ref. 30.)

Performance

On induction of the P_L promoter by a temperature shift, a rapid accumulation of the rRNA originating from the plasmid-borne mutated *rrnB* operon occurs.[30] We estimate that in 3 hr close to 50% of all the ribosomes are of the specialized type.[32] In Fig. 2, the accumulation of the protein made by specialized ribosomes is given. Figure 2 shows that very little hGH is made prior to induction of the specialized *rrn* operons, although the hGH mRNA is made constitutively. hGH accumulation clearly depends on the formation of specialized ribosomes in both systems. After 3 hr of

[32] R. L. Gourse, Y. Takebe, R. A. Sharrock and M. Nomura, *Proc. Natl. Acad. Sci. U.S.A.* **82,** 1069 (1985).

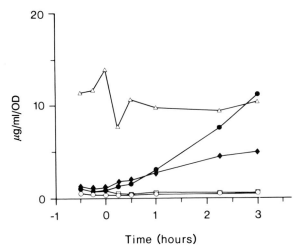

Time (hours)

Fig. 2. hGH synthesis in *E. coli* cells containing the systems shown in Fig. 1. △, Wild-type system VIII; ◆, system IX; ●, system X; □, control for system IX; ○, control for system X. The latter two controls reflect data obtained from cells that express a specialized hGH mRNA constitutively but lack a specialized *rrnB* operon. At $t = 0$, the temperature of the culture was raised from 30 to 42° for 15 min, after which incubation was continued at 37° for the indicated times. (Data from Ref. 30.)

induction of system X, the hGH level is the same as that in the control system VIII. In system IX induction of hGH synthesis also occurs but at a lower rate. Apparently, the dramatic alteration of the SD and the ASD sequences in system IX affects a step of protein initiation or ribosome assembly negatively. System X performs equally well or perhaps better than the wild-type system. Further optimization studies must be done to be able to utilize fully the potential of the specialized ribosome system. The fact that in this, and in other experiments, the hGH level is not higher than that in the wild-type system suggests that a rate-limiting factor other than the number and synthetic capacity of the specialized ribosomes operates. We suspect that the amount of hGH mRNA itself is limiting. As the amount of hGH made in system VIII is 11% of total proteins made, the relative hGH mRNA concentration must be about 11% of the steady-state concentration of total mRNA. The fraction of (wild-type) ribosomes translating hGH mRNA must therefore also be about 11%, assuming an average initiation frequency on the hGH mRNA in system VIII. Since the hGH gene is, in all three cases, transcribed by the same constitutive *trp* promoter, the same amount of mRNA will be synthesized in all three systems. This means that the specialized ribosome fraction that is engaged in the translation of the specialized mRNA is at most 10% of *all* ribosomes. As we assumed that after 3 hr of induction the specialized ribosome pool is 50%

of the total ribosome pool, only 20% of the specialized ribosome pool is involved in hGH mRNA translation in system IX. Therefore, we believe that this system has great potential for the biotechnological production of proteins, provided that the amount of specialized mRNA produced can be increased. In this case, a 5-fold increase would be needed to utilize fully the capacity of all of the specialized ribosomes.

Limitations and Opportunities

It should be noted that a specialized ribosome system will not solve translational problems inherent to the mRNA itself. It is unlikely that mRNAs which are difficult to translate by wild-type ribosomes can be translated any better by specialized ribosomes. However, there may be special cases where specialized ribosomes might have benefits over wild-type ones. For example, specialized ribosomes have been made spectinomycin resistant (by mutating the C1192 residue of the 16S rRNA molecule).[7,30] On addition of spectinomycin, the wild-type ribosomes could be inactivated. Under these conditions, there will be no competition with other wild-type ribosomes for, e.g., charged minor tRNAs which might restrict translation rates.[33] It would be interesting to see whether such observed reduction of translation rates could be alleviated using system X in the presence of spectinomycin.

After about 3–4 hr of induction, the growth rate of the cells is reduced to about 50%. Since the growth rate is proportional to the cellular (wild-type) ribosome concentration, the reduction in the growth rate is likely due to the reduction of the level of wild-type ribosomes. Although this system has not yet been examined extensively under fermentation conditions, cessation of growth might be beneficial, as the flow of energy and precursors would cease to go toward cell mass expansion and instead could be directed to the synthesis of specific proteins on specialized mRNAs. Further work using fermentors is needed to exploit the full potential of *E. coli* cells harboring the specialized ribosome system for biotechnological production of biomedically and industrially useful proteins.

[33] S. Pedersen, *in* "Gene Expression: The Translational Step and Its Control" (B. F. G. Clark and H. U. Petersen, eds.), p. 101. (Proceedings of the Alfred Benzon Symposium 19). Munksgaard, Copenhagen, Denmark, 1984.

[10] Vector for Enhanced Translation of Foreign Genes in *Escherichia coli*

By PETER O. OLINS and SHAUKAT H. RANGWALA

Introduction

Although significant progress has been made in the design of vectors for expression of foreign genes in *Escherichia coli,* high-level expression is still not a routine accomplishment. The simple juxtaposition of prokaryotic transcriptional and translational signals upstream of a foreign coding region often gives unpredictable results. In order to compensate for poor expression, one major approach has been to increase mRNA levels, either by using strong promoters or by employing very high plasmid copy number. However, in our experience, the most common block to efficient expression of foreign genes is poor translation initiation. The *E. coli* ribosome often does not recognize the chimeric junction between a prokaryotic ribosome binding site (RBS) and a foreign coding region.

Here we describe the use of a generic RBS (the T7 bacteriophage gene *10* leader, or "*g10*-L"), which is highly effective for enhancing translation of a wide variety of foreign genes in *E. coli.* Plasmid pMON5743 was constructed as a versatile system for engineering and expression of foreign genes. It includes the inducible *recA* promoter, the *g10*-L RBS, and an origin of single-stranded DNA replication.

Materials and Methods

Materials

Reagents were purchased from Fisher Scientific (Springfield, NJ) or Sigma Chemical Co. (St. Louis, MO), and enzymes for DNA manipulation were from New England Biolabs (Beverly, MA) or Promega (Madison, WI).

Bacterial Strains

All *recA+* strains of *E. coli* that we have tested work well for induction of the *recA* promoter. Typically, strain JM101[1] was used as a convenient host both for cloning and expression.

---

[1] J. Messing, technical bulletin, "Recombinant DNA," NIH Publ. 79–99, Vol. 2, No. 2, p. 43. National Institutes of Health, Bethesda, Maryland.

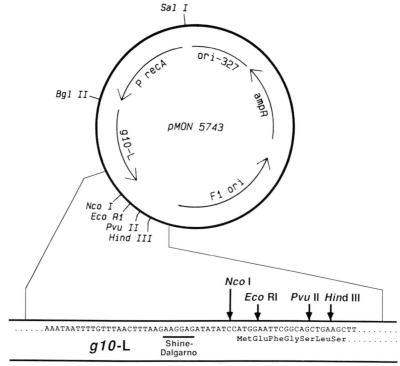

FIG. 1. Physical map of plasmid pMON5743. The *ampR* and *ori-327* regions are derived from the *Sal*I to *Eco*RI fragment of pBR327.[3] The *recA* promoter,[4] *g10*-L,[5] and multilinker segments were constructed as synthetic DNAs, and the F1 phage origin of single-stranded DNA replication is derived from the 512-bp *Rsa*I fragment of the phage.[6] All restriction sites shown are unique. The map is not to scale.

Plasmid Construction

Plasmid constructions were performed according to Maniatis *et al.*[2] For expression of foreign genes in the plasmid system, the heterologous coding region was cloned either into the unique *Eco*RI site or by blunt-end ligation into the *Pvu*II site of pMON5473 (shown in Fig. 1).[3-6] For precise expression of the coding region, unfused to any undesirable protein se-

[2] T. Maniatis, E. F. Fritsch, and J. Sambrook, "Molecular Cloning: A Laboratory Manual." Cold Spring Harbor Laboratory, Cold Spring Harbor, New York, 1982.

[3] X. Soberon, L. Covarrubias, and F. Bolivar, *Gene* **9**, 287 (1980).

[4] T. Horii, T. Ogawa, and H. Ogawa, *Proc. Natl. Acad. Sci. U.S.A.* **77**, 313 (1980).

[5] P. O. Olins, C. S. Devine, S. H. Rangwala, and K. S. Kavka, *Gene* **73**, 227 (1988).

[6] D. A. Meade, E. Szczesna-Skapura, and B. Kemper, *Nucleic Acids Res.* **14**, 1103 (1985).

quences, the following procedure was used. Single-stranded plasmid DNA was prepared according to Meade *et al.*[6] using the helper phage M13-KO7, and site-specific mutagenesis[7] was then used to create unique *Nco*I and *Hind*III sites at the N and C termini of the heterologous coding regions, respectively. Finally, all extraneous DNA was removed by deletion between the two *Nco*I sites, and then deletion between the two *Hind*III sites. This resulted in expression plasmids having a unique *Nco*I site correctly positioned at the junction between the *g10*-L RBS and the heterologous coding region.

Growth and Induction of Strains Carrying Expression Plasmids

Escherichia coli cells were transformed with the expression plasmids and selected at 37° on LB plates containing 200 μg/ml ampicillin. A single colony was picked into 2 ml of LB medium containing 200 μg/ml ampicillin and grown overnight at 37°. A 100-fold dilution was then made into 20 ml of the following medium: M9 minimal salts[2] supplemented with 0.8% (w/v) glucose, 1% (w/v) casamino acids (Difco), and 0.0005% (w/v) thiamin in distilled water. The culture was grown at 37° with vigorous aeration until it reached a density of approximately $OD_{550} = 0.5$ (or 150 Klett units, using a Klett–Summerson meter), and the *recA* promoter was induced by the addition of 50 μg/ml nalidixic acid (using a fresh solution of nalidixic acid at 10 mg/ml in 0.1 M NaOH). Growth was continued for a further 4 hr, and the cells harvested by centrifugation.

Results

Application of g10-L Ribosome Binding Site for Expression in Escherichia coli

As a consequence of a search for RBS sequences suitable for high-level expression of foreign genes, the RBS region from the highly expressed coat protein (gene *10*) of phage T7 "*g10*-L" was tested.[5,8] This RBS gave surprisingly efficient expression of several different genes, with the protein product often accumulating to a level corresponding to a large proportion of cellular protein. The stimulatory effect of the *g10*-L was the same even when various promoters were tested. The experiments described here employ the *recA* promoter of *E. coli*. Although this promoter has only been

[7] T. A. Kunkel, *Proc. Natl. Acad. Sci. U.S.A.* **82,** 488 (1985).
[8] P. O. Olins and S. H. Rangwala, *J. Biol. Chem.* **264,** 16973 (1989).

TABLE I
Expression Levels Obtained for Different Genes, Using
Either g*10*-L RBS or "Consensus" RBS[a]

Source of coding region	Coding region	Increase in expression (*g10*-L/consensus RBS)
E. coli	*lacZ*	100
Maize	GST1	50
Rat	Atriopeptigen	40
Agrobacterium	IPT1	340

[a] Parental plasmids were constructed carrying the *recA* promoter and either the *g10*-L or consensus RBS. Various coding regions were cloned into these vectors, and expression levels were measured after induction of the *recA* promoter.

used occasionally for expression of foreign genes,[5,8–12] it has a number of advantages which make it a convenient choice: it is well-regulated, even at high copy number; it can be induced simply by the addition of nalidixic acid; there are no restrictions on growth medium or temperature; and it is effective in most *E. coli* hosts except those with a *recA⁻* genotype.

Several genes have been expressed using the combination of the *recA* promoter and the *g10*-L, from sources as diverse as bacteria, plants, and mammals, and, in almost all cases, efficient expression was observed.[5,8,11–13] Table I illustrates the efficacy of the *g10*-L RBS, compared with similar plasmid constructs containing a designed "consensus" RBS sequence.[5] As can be seen, the presence of the *g10*-L RBS caused a remarkable stimulation in expression, often 100-fold or more. Further experiments[5,8] indicate that the effect of the *g10*-L is primarily at the level of translation initiation. Based on these results, a plasmid vector was designed which combined the *g10*-L RBS with a number of other desirable features, as described below.

[9] S. I. Feinstein, Y. Chernajovsky, L. Chen, L. Maroteaux, and Y. Mory, *Nucleic Acids Res.* **11**, 2927 (1983).

[10] G. G. Krivi, M. L. Bittner, E. Rowold, E. Y. Wong, K. C. Glenn, K. S. Rose, and D. C. Tiemeier, *J. Biol. Chem.* **260**, 10263 (1985).

[11] E. Y. Wong, R. Seetharam, C. E. Kotts, R. A. Heeren, B. K. Klein, S. B. Braford, K. J. Mathis, B. F. Bishop, N. R. Siegel, C. E. Smith, and W. C. Tacon, *Gene* in press (1988).

[12] J. K. Gierse, P. O. Olins, C. S. Devine, J. D. Marlay, M. G. Obukowicz, L. H. Mortensen, E. G. McMahon, E. H. Blaine, and R. Seetharam, *Biochim. Biophys. Acta* (submitted).

[13] R. J. Duronio, E. Jackson-Machelski, R. O. Heuckeroth, P. O. Olins, C. S. Devine, W. Yonemoto, L. W. Slice, S. S. Taylor, and J. I. Gordon, *Proc. Natl. Acad. Sci. U.S.A.* in press (1990).

Description of Plasmid pMON5743

Figure 1 is a schematic representation of pMON5743. This plasmid is based on pBR327.[3] In addition, the plasmid includes the following features: the *E. coli recA* promoter, which is inducible under conditions which induce the SOS response,[14] such as the addition of nalidixic acid[9]; the *g10*-L RBS,[5] followed by an *Nco*I site which flanks a methionine initiator codon; multiple cloning sites for ease of insertion of various coding regions; and the origin of single-stranded replication from phage F1,[6] which permits the rapid preparation of single-stranded DNA for mutagenesis or DNA sequencing.

Discussion

One outstanding question remains. How does the *g10*-L function to stimulate translation of such a variety of heterologous genes in *E. coli?* The RBS is derived from a highly expressed phage coat protein, and has therefore probably evolved to optimize expression of that gene. Our recent studies on this question[8] indicate that the efficacy of the *g10*-L is due to a combination of increased mRNA stability with a potent enhancement of translation initiation. While studies of the precise mechanism continue, plasmid vectors containing this element provide a simple and valuable addition to the systems available for expression in *E. coli.*

[14] J. W. Little and D. W. Mount, *Cell* **29,** 11 (1982).

[11] Minimizing Proteolysis in *Escherichia coli:* Genetic Solutions

By SUSAN GOTTESMAN

Many proteins cloned in *Escherichia coli* on high-copy-number vectors fail to accumulate. In many cases, the explanation for this failure to accumulate is the rapid turnover of the protein. Both foreign and native *E. coli* proteins seem to be subject to this degradation. Presumably, part of the difficulty may reflect the normal presence of such proteins in particular cellular compartments, or as part of multiprotein structures[1-3] which may not have counterparts in *E. coli*. Produced on their own, these proteins

[1] P. P. Dennis, *Mol. Gen. Genet.* **134,** 39 (1974).
[2] K. Nishi and J. Schneir, *Mol. Gen. Genet.* **212,** 177 (1988)
[3] H. A. Nash, C. A. Robertson, E. Flamm, R. A. Weisberg, and H. I. Miller, *J. Bacteriol.* **169,** 4124 (1987).

METHODS IN ENZYMOLOGY, VOL. 185

without partners presumably do not have conformations which favor intracellular stability. In other cases, fusion proteins seem to be unstable, sometimes being degraded entirely and sometimes being degraded to the reporter protein moiety.

A variety of evidence suggests that the initial steps in the proteolysis of many unstable proteins are energy-dependent.[4-7] In addition, the turnover of abnormal proteins, such as those containing amino acid analogs, is also dependent on energy.[4-7] Therefore, mutations which affect these energy-dependent proteases provide the best hope for stabilizing foreign, cloned proteins. In many cases, such mutants have proved useful for this purpose. It is clear, however, that the mechanism for specificity in the recognition of unstable proteins by proteases is not understood, and that no one protease mutant will suffice to stabilize all foreign proteins.

While the failure of a particular protein to accumulate in cells may be a hallmark of an unstable protein, it is clearly not the only explanation. Before investing much time in trying to solve proteolysis problems, the investigator should test directly for protein turnover with a pulse–chase experiment.

Protease Mutants

lon Mutants

Lon is a major ATP-dependent protease in *E. coli,* and seems to be responsible for the degradation of a number of naturally unstable proteins and many abnormal proteins.[6-8] The overall degradation of canavanine-containing proteins containing the arginine analog canavanine is reduced about 2-fold in *lon* mutants,[4] and, when individual canavanine-containing proteins are displayed on acrylamide gels, the degradation of essentially all unstable bands is slowed in *lon* mutants.[9] In addition, at least some temperature-sensitive mutations will be suppressed by *lon* mutations,[10] and in at least one case, this suppression is reflected in increased stability of the

---

[4] M. R. Maurizi, P. Trisler, and S. Gottesman, *J. Bacteriol.* **164,** 1124 (1985).

[5] A. L. Goldberg and A. C. St. John, *Annu. Rev. Biochem.* **45,** 747 (1976).

[6] S. Gottesman, *in "Escherichia coli* and *Salmonella typhimurium"* (F. C. Neidhardt, ed.), p. 1308. American Society for Microbiology, Washington, D.C., 1987.

[7] S. Gottesman, *Annu. Rev. Genet.* **23,** 163 (1989).

[8] C. G. Miller, *in "Escherichia coli* and *Salmonella typhimurium"* (F. C. Neidhardt, ed.), p. 680. American Society for Microbiology, Washington, D.C., 1987.

[9] M. R. Maurizi, Y. Katayama, and S. Gottesman, *in* "The Ubiquitin System: Current Communications in Molecular Biology" (M. J. Schlesinger, ed.), p. 147. Cold Spring Harbor Laboratory, Cold Spring Harbor, New York, 1988

[10] S. Gottesman and D. Zipser, *J. Bacteriol.* **113,** 844 (1978).

temperature-sensitive protein in *lon* mutants.[11] All of these observations suggest that Lon may have relatively broad specificity for unfolded or misfolded proteins *in vivo*. In fact, *lon* mutations have proved useful in a large number of instances to stabilize cloned wild-type and mutant proteins (see, for instance, Refs. 12 and 13). A *lon* mutation is part of the recommended host strain for the λgt11 system for detecting eukaryotic proteins as fusions to β-galactosidase; it was observed to increase the yield of a number of fusions in this vector.[14]

Both deletions and insertion mutations in *lon* are available, and can be introduced readily into other strain backgrounds by P1 transduction. Because the *lon* mutations are pleiotropic, they have two major phenotypes which may cause complications in using them as hosts for foreign proteins: mucoidy and UV sensitivity.[7,10] Both complications can be bypassed by either growth conditions or appropriate secondary mutations. Table I lists some available strain backgrounds carrying tight *lon* mutations as well as mutations to suppress some of the secondary effects of the *lon* mutation. Since *lon* mutations may affect the efficiency of transformation or stability of some plasmids, these hosts may not be appropriate for the initial selection of transformants after ligation.

lon::ΔTn10 can be introduced directly into new hosts by selecting for tetracycline (tet) resistance (15 μg/ml).[4] The mini-tet insertion will not transpose. A Δ*lon*-510 mutation can be introduced without a final antibiotic resistance marker, by two transductions. First, *proC zaj*-403::Tn10 is introduced from SG1039 (Table II). Transduction from SG4144 and selection for Pro+ gives about 20% *lon* mutations; most will have lost the Tn10. The *lon*-510 deletion removes *hupB* activity; this is without phenotype in the presence of *hupA*+.[4,15]

Genetic Methods to Eliminate Secondary Phenotypes of lon Mutations:
  Mucoidy

Lon strains overproduce the capsular polysaccharide colanic acid, particularly at low temperatures (34° and below) and on minimal media.[16] This phenotype can cause difficulties in the collection of cells from liquid media, in phage infection of mucoid cells, and in the isolation or screening of single colonies on solid media.

[11] A. D. Grossman, R. R. Burgess, W. Walter, and C. A. Gross, *Cell* **32**, 151 (1983).
[12] C. Miyamoto, R. Chizzonite, R. Crowl, K. Rupprecht, R. Kramer, M. Schaber, G. Kumar, M. Poonian, and G. Ju, *Proc. Natl. Acad. Sci. U.S.A.* **82**, 7232 (1985).
[13] L. C. Sambucetti, M. Schaber, R. Kramer, R. Crowl, and T. Curran, *Gene* **43**, 69 (1986).
[14] R. A. Young and R. W. Davis, *Proc. Natl. Acad. Sci. U.S.A.* **80**, 1194 (1983).
[15] Y. Kano, M. Wada, T. Nagase, and F. Imamoto, *Gene* **45**, 37 (1986).
[16] A. Markovitz, *in* "Surface Carbohydrates of the Prokaryotic Cell" (I. Sutherland, ed.), p. 415. Academic Press, London, 1977.

TABLE I
BACTERIAL STRAINS WITH MUTATIONS IN MAJOR PROTEASE SYSTEMS

Strain	Relevant genotype	Parental strain	Comments
lon mutant hosts			
SG1117	Δ*gal* Δ*lac lon*-146::ΔTn*10 leu sup⁺rec⁺*	HB101[a]	Tetracycline resistant; good transformation recipient
SG12036	Δ*gal* Δ*lon*-510 *sulA*	C600	UV and complex medium resistant; good transformation recipient
SG12041	Δ*gal* Δ*lon*-510 *sulA recA*	C600	Nonfilamenting; Rec⁻
SG1611	F'i^Q *lacZ* ΔM15/Δ(*lac-pro*) Δ*gal* Δ*lon*-510 *supE*	JM101	Can be used for M13 or pUC vectors
SG13646	Δ*gal* ΔBAM cI857 ΔH1 *lon*-146::ΔTn*10 pro his ilv*	N5340[b]	Useful for plasmids with p_L promoters
lon htpR mutant hosts			
SG21163	Δ*lon*-510 *supFts htp*Ram165	MC4100[c]	Temperature sensitive; will not grow at 39°
lon clp mutant hosts			
SG12044	Δ*gal* Δ*lon*-510 *sulA clpA*319::Δ*kan*	C600	Kanamycin resistant
SG21165	Δ*gal* Δ*lac* Δ*lon*-510 Δ*clpA*500	MC4100	
lon clp htpR mutant host			
SG21173	Δ*lon*-510*supFts hpt*Ram165	MC4100	Kanamycin resistant; temperature sensitive
	Δ*clpA*319::Δ*kan* Δ*lac*		

[a]SA2692, the immediate parent of SG1117, was derived by S. Adhya from HB101.
[b]S. Gottesman, M. Gottesman, J. E. Shaw, and M. L. Pearson, *Cell* **24**, 225 (1981).
[c]M. J. Casadaban and S. N. Cohen, *Proc. Natl. Acad. Sci. U.S.A.* **76**, 4530 (1979).

Growth of cells on rich media or at temperatures above 37° may be sufficient to avoid serious problems with mucoidy. In addition, some standard *E. coli* hosts are naturally nonmucoid.

The genetic solution to avoiding capsule synthesis is the inactivation of the structural or regulatory genes necessary for capsule synthesis. Mutations in *gal* and *cps* have been used most frequently.[17] Either of these genes can be readily inactivated as shown below. In some conditions, such cells may accumulate intermediates in capsule synthesis which are detrimental to the cell. In such cases, the use of regulatory mutations such as *rcsA* or *rcsB* may be called for.[18,19]

[17] P. Trisler and S. Gottesman, *J. Bacteriol.* **160**, 184 (1984).
[18] S. Gottesman, P. Trisler and A. Torres-Cabassa, *J. Bacteriol.* **162**, 1111 (1985).
[19] J. A. Brill, C. Quinlan-Walshe, and S. Gottesman, *J. Bacteriol.* **170**, 2599 (1988).

TABLE II
BACTERIAL DONOR STRAINS FOR TRANSFERRING *lon* PROTEASE MUTATIONS BY P1 TRANSDUCTION

Strain	Genotype	Parental strain	Comments
SG1039	Δ*lac proCYA221 zaj*-403::Tn10	MC4100[a]	*proC*-Tn10 linkage is 80%[b]
SG4144[c]	Δ*lon*-510 *pro*⁺	N99	
SG12047	*lon*-146::ΔTn10	C600	Defective ΔTn10 should show 100% linkage to *lon*⁻
SG21135	Δ*gal nad*::Tn10	MC4100	
GC4540	*pyrD sulA*::Tn10	From O. Huisman and R. D'Ari (Paris)	Screen for linked *pyrD*⁻ to avoid transposition of Tn5 during P1 transduction

[a] M. J. Casadaban and S. N. Cohen, *Proc. Natl. Acad. Sci. U.S.A.* **76**, 4530 (1979).
[b] P. Trisler and S. Gottesman, *J. Bacteriol.* **160**, 184 (1984).
[c] M. R. Maurizi, P. Trisler, and S. Gottesman, *J. Bacteriol.* **164**, 1124 (1985).

galE. Deletion of the whole *gal* operon or inactivation of *galE* will render *E. coli lon* mutants nonmucoid. Strains in Table I with Δ*gal* carry deletions of the whole operon.

To introduce a gal deletion into a strain of choice, use P1 grown on SG21135 (Table II) to select for tetracycline-resistant transductants (15 μg/ml tetracycline), and screen the transductants for inability to ferment galactose [white colonies on MacConkey base agar plates containing 1% (w/v) galactose]. P1 transduction is described by Silhavy *et al.*[20] The linkage of *gal* to tetracycline resistance should be at least 30%. The cells can be rid of the Tn10 (tetracycline-resistant) transposon in *nad* by growing cells to stationary phase, concentrating them by centrifugation at least 10-fold, and plating the concentrated cells on minimal plates lacking nicotinamide. *nad*⁺ colonies will arise within a few days; these colonies are generally tetracycline sensitive. Introducing the *gal* deletion in this fashion will allow the subsequent introduction of *lon* with a linked or inserted tetracycline resistance marker.

cps. A cluster of genes near 43′ is necessary for capsule synthesis. Insertions or deletions in this region will make the cells nonmucoid.[17] *cps*::Tn10 hosts are available; this mutation can be moved into other hosts by P1 transduction. Deletions of the *cps* region can be generated by introduction of *his*::Tn10 into hosts by P1 transduction, and isolation of tetracycline-sensitive deletions from this strain.[17] A certain fraction of these will

[20] T. J. Silhavy, M. L. Berman, and L. W. Enquist, "Experiments with Gene Fusions." Cold Spring Harbor Laboratory, Cold Spring Harbor, New York, 1984.

be *cps⁻*; if the *lon* mutation is already in the strain, these can be recognized easily as nonmucoid colonies.

rcsB and rcsA. These positive regulators of capsule synthesis can be introduced into strains with insertions of kanamycin resistance (Δkan) or tetracycline resistance (ΔTn10).[18,19]

Filamentation

lon mutants are sensitive to DNA-damaging agents such as UV or methylmethane sulfonate (MMS). In response to DNA damage, they form long, nonsepted filaments, which are frequently not viable.[21] In addition, cells devoid of Lon tend to show at least some filamentation when grown on rich (LB, containing yeast extract) growth medium. Transfer of cells from minimal growth medium to rich medium induces lethal filamentation.[22] The filamentation is due to the stabilization of the SOS-inducible cell division inhibitor SulA.[23] Generally, this phenotype is not troublesome enough to require remedies.

Avoidance of growth of cells on media with yeast extract will be sufficient to minimize filamentation. Tryptone broth and tryptone plates, rather than LB, should be used if possible.

Inactivation of *sulA* by mutation has no deleterious effects on cell growth and should eliminate the filamentation phenotype. SG12036 and SG12041 in Table I carry *sulA* mutations. *sulA* can be introduced into other hosts as a *sulA*::Tn5 (kanamycin resistance), from GC4540 (Table II). Alternatively, resistant hosts can be selected directly by plating cells on 0.05% (w/v) MMS in LB agar plates (LB-MMS agar plates).[10] The surviving cells have usually accumulated a mutation either in *sulA* or in the target for SulA action, *sulB*. Either mutation will suffice.

To make LB-MMS agar plates: mix 5 g NaCl, 10 g Tryptone, 5 g yeast extract, and 15 g agar in 1 liter of H_2O, autoclave, and add directly before pouring 0.5 ml of MMS. Plates remain active for 3–4 days at room temperature, and for 1–2 weeks if stored at 4°.

Caution: Because any low-level SOS induction in *lon⁻* hosts can be lethal, the combination of *lon* with other mutations (such as *dam⁻*) can be lethal. Such lethality is frequently overcome by *sulA* mutations.

Confirmation of Lon Phenotype in Strains

Cells which carry the *lon* mutation can frequently be easily recognized by the overproduction of capsular polysaccharide, seen particularly on

[21] P. Howard-Flanders, E. Simson, and L. Theriot, *Genetics* **49**, 237 (1964).
[22] R. C. Gayda, L. T. Yamomoto, and A. Markovitz, *J. Bacteriol.* **127**, 1208 (1976).
[23] S. Mizusawa and S. Gottesman, *Proc. Natl. Acad. Sci. U.S.A.* **80**, 358 (1983).

glucose minimal media at low temperatures. However, since many strains used for stabilizing cloned proteins are particularly engineered to avoid capsule overproduction, it may be necessary to confirm the presence of the *lon* mutation in some other fashion. The UV sensitivity of *lon* mutants can be demonstrated by comparing the growth of cells carrying the suspected *lon* mutation to appropriate control strains on LB-MMS plates. Cells carrying both the *lon* mutation and a *sulA* mutation will grow on LB-MMS plates.

A direct test for the defective degradation phenotype (Deg⁻) of *lon* is provided by the ability of *lon* hosts to support the growth at high temperatures of λ phage carrying a temperature-sensitive mutation in the O replication gene.[10] No secondary suppressor of this phenotype has yet been identified.

Testing for lon Mutations with Ots Phage. Do serial dilutions of the λcI857 Ots phage and its isogenic λcI857 O⁺ parent on lawns of the *lon* host and appropriate control hosts (*lon*⁺, and, if possible, *lon*⁻), and incubate the plates at 32 and 39°. The wild-type phage should grow well at both temperatures; the Ots phage will form plaques on all hosts at 32°, but give about 10^2- to 10^4-fold fewer plaques on a *lon*⁺ host at 39°, compared to 32°.[10] The *lon*⁻ host, on the other hand, should have an eop (efficiency of plating) at 39°, compared to 32°, of about 1.0. In some cases, the wild-type host will give tiny but visible plaques at the high temperature. At temperatures above 39°, the plating efficiency on the *lon* strain begins to decline. In order to carry out this test more efficiently, the serial dilutions can all be applied to a single plate as 10-μl spots.

A final alternative is to grow P1 on the suspected *lon* host, and transduce the *lon* region into a recipient such as SG1039 (Table II), selecting Pro⁺ recombinants. A *lon* donor should give 20% mucoid colonies in SG1039 on glucose minimal agar plates.

Cells Defective in Heat-Shock Response: htpR Mutants

Cells defective in the σ factor necessary for heat-shock response, σ32, coded for by the *htpR* gene, are generally defective in proteolysis,[24,25] and have been used to increase the accumulation of cloned proteins.[26] Since *lon* is itself a heat-shock protein,[27] a decrease in *lon* synthesis in *htpR* mutants may be part of the explanation of this defect. However, *htpR* mutants seem

---

[24] T. A. Baker, A. D. Grossman, and C. A. Gross, *Proc. Natl. Acad Sci. U.S.A.* **81**, 6779 (1984).

[25] S. A. Goff, L. P. Casson and A. L. Goldberg, *Proc. Natl. Acad. Sci. U.S.A.* **81**, 6647 (1984).

[26] G. Buell, F. Schulz, G. Selzer, A. Chollet, N. R. Movva, D. Semon, S. Escanez, and E. Kawashima, *Nucleic Acids Res.* **13**, 1923 (1985).

[27] T. A. Phillips, R. A. VanBogelen, and F. C. Neidhardt, *J. Bacteriol.* **159**, 237 (1984).

to be more defective for proteolysis in some instances than *lon* null mutants, and the usual *htpR* alleles available still make detectable levels of *lon*. Therefore, it seems likely that other proteases are under heat-shock control. Some mutations in heat-shock genes other than *lon* have also been shown to have defects in proteolysis.[28] *dnaJ* mutations may be the most generally useful of these.

Cells carrying both the *lon* and *htpR* mutations are available, and in at least some cases seem to prevent degradation of foreign proteins better than either one alone.

The *htpR* product is essential for cell growth at normal growth temperatures, and is needed in higher amounts at higher temperatures.[24,25,29] The most commonly used *htpR* allele is an amber mutation in the gene, used in strain backgrounds which provide conditional suppression of the mutation. Most frequently, *sup temperature-sensitive* hosts have been used.[24,25,29] In these strains, the low level of suppression which may allow growth at low temperatures still makes the strains proteolysis deficient at the permissive temperature. Thus expression of foreign proteins can be carried out at the permissive temperature. Raising the cells to higher temperatures can lead to cell lysis within relatively short periods of time.

Table I lists SG21163, a host carrying both a *lon* deletion and the *htpR*am mutation. In this case, the mutation is suppressed by a *supFts* allele. Cells grow well at 32°, poorly at 37°, and do not grow at temperatures above 37°. Transfer of the *htpR* mutation to other hosts is complicated by the necessity for the unlinked *sup temperature-sensitive* mutation.

The presence of an *htpR* mutation in a host can be confirmed by the temperature-sensitive phenotype.

Other Protease Mutants: clpA

The recent isolation of mutations in a second ATP-dependent protease, *clp*, may allow an improvement in the ability of *lon* mutant hosts to stabilize foreign proteins. *clp* mutations by themselves seem to have relatively little effect on the degradation of abnormal, canavanine-containing proteins.[30] *clpA* mutations, in *lon⁻* cells, however, slow the residual degradation of canavanine-containing proteins seen in *lon* cells.[9]

---

[28] D. B. Straus, W. A. Walter, and C. A. Gross, *Genes Dev.* **2**, 1851 (1988).

[29] F. C. Neidhardt and R. A. VanBogelen, in *"Escherichia coli and Salmonella typhimurium"* (F. C. Neidhardt, ed.), p. 1334. American Society for Microbiology, Washington, D.C., 1987.

[30] Y. Katayama, S. Gottesman, J. Pumphrey, S. Rudikoff, W. P. Clark, and M. R. Maurizi, *J. Biol. Chem.* **263**, 15226 (1988).

SG21165 and SG12044 (Table I) carry the *lon* and *clpA* mutations. SG21173 carries *lon, clpA,* and *htpR.*

Other Possible Protease and Peptidase Mutants

Mutations in other proteases have not thus far proved useful for stabilization of cytoplasmic proteins. Mutations such as *hfl*, which stabilize λcII protein,[31] have not been shown to have any general effects on protein turnover.

Mutations in *ompT* inactivate an outer membrane-localized protease which cuts specifically at paired basic residues.[32] *ompT* is apparently very active in extracts from *E. coli* cells, and is probably responsible for a number of reported specific cleavages of both *E. coli* proteins and bacteriophage proteins.[33,34] It is not yet clear if this protease has any important role *in vivo.*

degP mutations, isolated by Strauch and Beckwith,[35] inactivate a periplasmic protease which can cleave *phoA* fusion proteins to release the phosphatase portion. Presumably, this function plays a role in degrading periplasmic proteins, and mutations in it may help to stabilize some exported proteins. Null mutations in this gene render cells temperature sensitive.[36,37]

While most peptidase mutations would not be expected to have significant effects in stabilizing large foreign proteins, those with the ability to degrade polypeptides from the ends may be important.

Inhibition of Proteolysis by Phage Functions

Phage T4 infection of *E. coli* causes stabilization of unstable protein fragments.[38] One of the genes responsible for this effect, the *pin* gene, has been cloned onto a λ vector.[39] Introduction of the *pin* clone into cells may help to stabilize proteins, but data available thus far suggest that the inhibition may be limited to inhibition of the Lon protease. Thus, in most cases, use of the *lon* mutation may be more straightforward than use of the *pin* clone.

[31] M. A. Hoyt, D. M. Knight, A. Das, H. I. Miller, and H. Echols, *Cell* 31, 565 (1982).
[32] K. Sugimura and T. Nishihara, *J. Bacteriol.* 170, 5625 (1988).
[33] B. Sedgwick, *J. Bacteriol.* 171, 2249 (1989).
[34] J. Grodberg and J. J. Dunn, *J. Bacteriol.* 170, 1245 (1988).
[35] K. Strauch and J. R. Beckwith, *Proc. Natl. Acad. Sci. U.S.A.* 85, 1576 (1988).
[36] B. Lipinska, O. Fayet, L. Baird, and C. Georgopoulos, *J. Bacteriol.* 171, 1574 (1989).
[37] K. L. Strauch, J. Johnson, and J. Beckwith, *J. Bacteriol.* 171, 2689 (1989).
[38] L. D. Simon, K. Tomczak, and A. C. St. John, *Nature (London)* 275, 424 (1978).
[39] K. Skorupski, J. Tomaschewski, W. Ruger, and L. Simon, *J. Bacteriol.* 170, 3016 (1988).

Alterations to Cloned Protein

Fusion Proteins

In some cases, the fusion of extra material to the C terminus or N terminus of a protein will help protect it from degradation. Small peptides, which will be very sensitive to degradation in *E. coli*, have been successfully synthesized by making tandem end-to-end repeats, using sites cleavable *in vitro* between the repeat units.[40] Fusion of proteins and peptides to the C terminus of ubiquitin, made from an *E. coli* promoter, is also successful, for reasons that are not yet clear.[41]

Recent work on the turnover of fragments of the λ repressor in *E. coli* has suggested that the C-terminal five amino acids may be particularly important in targeting some unfolded proteins for degradation.[42] Hydrophobic amino acids are particularly resistant to this particular degradation pathway.[42]

Secretion

Proteins which are sensitive to degradation intracellularly may be protected from degradation by being targeted for secretion.[43] However, this does not seem to be a universally useful approach. In some cases, fusion proteins which become unstable during secretion may be stabilized if they remain cytoplasmic.[44]

Overproduction and Formation of Insoluble Precipitates

Abnormal proteins which are overproduced in *E. coli* will begin to accumulate in insoluble precipitates.[45] High levels of overproduction lead to even more insolubility. These precipitates, however, do seem to be resistant to the normal proteolysis pathways.[46,47] Therefore, for the synthe-

[40] S.-H. Shen, *Proc. Natl. Acad. Sci. U.S.A.* **81**, 4627 (1984).
[41] T. R. Butt, S. Jonnalagadda, B. P. Monia, E. J. Sternberg, J. A. Marsh, J. M. Stadel, D. J. Ecker, and S. T. Crooke, *Proc. Natl. Acad. Sci. U.S.A.* **86**, 2540 (1989).
[42] J. U. Bowie and R. T. Sauer, *J. Biol. Chem.* **264**, 7596 (1989).
[43] K. Talmadge and W. Gilbert, *Proc. Natl. Acad. Sci. U.S.A.* **79**, 1830 (1982).
[44] R. Gentz, Y. Kuys, C. Zwieb, D. Taatjes, H. Taajes, W. Bunnwarth, D. Stueber, and I. Ibrahim, *J. Bacteriol.* **170**, 2212 (1988).
[45] A. L. Goldberg and A. C. St. John, *Annu. Rev. Biochem.* **45**, 747 (1976).
[46] D. G. Kleid, D. Yansura, B. Small, D. Dowbenko, D. M. Moore, M. J. Grubman, P. D. McKercher, D. O. Morgan, B. H. Robertson, and H. L. Bachrach, *Science* **214**, 1125 (1981).
[47] Y.-S. E. Cheng, D. Y. Kwoh, T. J. Kwoh, B. C. Saltvedt, and D. Zipser, *Gene* **14**, 121 (1981).

sis of a protein, not necessarily in active form, extreme overproduction may be one mechanism for avoiding proteolysis.

Approaches to Proteolysis in Other Systems

The identification of genes involved in proteolysis in other systems, and the isolation of mutations in those genes, can be undertaken using the general procedures developed in *E. coli* and *Salmonella.*

Since many temperature-sensitive mutations in *E. coli* and *Salmonella* are temperature sensitive by virtue of degradation at the nonpermissive temperature, the isolation of second-site suppressors of the temperature sensitivity may include mutations in the proteases responsible for degradation. Such mutations are likely to be characterized by suppression at some but not all temperatures, and possibly by pleiotropic effects on different mutations, or on growth. Obviously such selections for suppressing mutations would best be done using functions known to be unstable as the result of a temperature-sensitive mutation.

[12] Gene Fusions for Purpose of Expression: An Introduction

By Mathias Uhlén and Tomas Moks

Introduction

Circumstantial evidence supports the hypothesis that the assembly of relatively small domains into functional proteins is an important factor in evolution.[1] These natural gene fusion events include processes such as gene duplications and "exon shuffling." Numerous examples of proteins consisting of structurally and functionally discrete domains exist,[1] well-known examples of which are structures such as the "kringles" and the "fingers" of mammalian proteins. The possibility of deleting and adding such domains in evolution suggests that proteins are often tolerant to changes involving whole domains.

Recombinant DNA technology has allowed *in vitro* fusions of genes or gene fragments in a simple and predictable manner. Gene fusions were used in the first described systems for heterologous bacterial expression of

---

[1] A. J. Jeffreys, *in* "Genetic Engineering" (R. Williamson, ed.), Vol. 2, p. 1. Academic Press, New York, 1981.

small peptides, such as somatostatin[2] and insulin.[3] The β-galactosidase system used in these experiments has since been extensively utilized in gene fusions, not only for expression but also as a probe for translational and transcriptional activity.[4] In addition, gene fusions have been used to follow the cellular localization of gene products[4,5] and to construct bifunctional proteins having the activity of each of the two gene products.[6]

There are several reasons to use gene fusions for expression of recombinant proteins in heterologous hosts. First, foreign proteins are often rapidly degraded by host proteases, and this may sometimes be avoided by a gene fusion strategy, as has been demonstrated for a large number of small peptides (for a review, see Ref. 5). Second, general and efficient purification schemes may be obtained which allow rapid recovery of gene products.[6-11] Third, the proteins can be localized to different compartments of the host cell (e.g., periplasm, cell wall, culture medium) through specific peptides fused to the protein.[5,11] In addition, a more reliable and reproducible method to obtain a native protein might be to use *in vitro* cleavage of the fusion protein, as compared to *in vivo* removal of the formylmethionine or cleavage of a signal peptide, which in both cases may yield a heterogeneous N terminus.

In this overview, the use of gene fusions for the expression of recombinant proteins is discussed. The article primarily deals with gene fusion strategies in *Escherichia coli,* although the concepts discussed can be applied to the expression in other prokaryotes or in eukaryotic cells. Several alternative gene fusion strategies are presented, as well as examples of approaches to facilitate protein purification.

[2] K. Itakura, T. Hiroso, R. Crea, A. D. Riggs, H. L. Heyneker, F. Bolivar, and H. W. Boyer, *Science* **198,** 1056 (1977).

[3] D. V. Goeddel, D. G. Kleid, F. Bolivar, H. L. Heyneker, D. G. Yansura, R. Crea, T. Hirose, A. Kraszewski, K. Itakura, and A. D. Riggs, *Proc. Natl. Acad. Sci. U.S.A.* **76,** 106 (1979).

[4] M. J. Casadaban, A. Martinez-Arias, S. K. Shapira, and J. Chou, this series, Vol. 100, p. 293.

[5] F. A. O. Marston, *Biochem. J.* **240,** 1 (1986).

[6] M. Uhlén, B. Nilsson, B. Guss, M. Lindberg, S. Gatenbeck, and L. Philipson, *Gene* **23,** 369 (1983).

[7] A. Ullman, *Gene* **29,** 27 (1984).

[8] J. Germino and D. Bastia, *Proc. Natl. Acad. Sci. U.S.A.* **81,** 4692 (1984).

[9] A. D. Bennett, S. K. Rhind, P. A. Lowe, and C. C. G. Hentchel, Eur. Pat. Appl. 0131363 (1984).

[10] H. M. Meade and I. L. Garwin, Pat. Appl. PCT/US85/01901 (1986).

[11] B. Nilsson and L. Abrahmsén, this volume [13].

Solubility of Gene Fusion Product

A major consideration of relevance for expression of gene fusions is whether the fusion product should be produced in soluble or insoluble ("inclusion body") form. Both methods have advantages and disadvantages, depending on the nature and final use of the gene product.

Inclusion bodies are dense particles containing precipitated proteins formed in the recombinant host upon expression of some proteins. In particular, gene fusions involving $trpE^{12}$ and cII^{13} often yield such insoluble material. The formation of the precipitates depends on the solubility of the fusion protein, the protein synthesis rate, and the growth conditions.[5] The inclusion body strategy has the advantage that the protein, in most cases, becomes protected from proteolysis. In addition, large amounts of gene product are normally obtained and the inclusion bodies can easily be recovered by differential centrifugation of cell lysates.[5] However, the product is recovered in a nonactive form and must be dissolved and renatured to obtain a biologically active protein. For applications where the aim is the immunogenic properties of the gene product, a native three-dimensional structure is often not necessary and sometimes not even desired, while other applications require a correctly folded molecule. A large number of recombinant proteins produced as fusion protein in *E. coli* have been obtained as inclusion bodies or aggregates, e.g., somatostatin,[2] insulin A and B chain,[3] calcitonin,[9] β-endorphin,[14] urogastrone,[15] β-globin,[13] myoglobin,[16] human growth hormone,[17] and angiotensin.[18]

The alternative expression strategy is to produce the fusion protein in a soluble form. This approach has the great advantage that a product with full biological activity can be obtained directly without renaturation. In addition, this allows the introduction of soluble "affinity handles" as fusion partners, thus facilitating the recovery of the recombinant protein. However, the soluble recombinant protein must be proteolytically stable in the heterologous host. The success of the strategy is therefore more difficult to predict as compared to the inclusion body approach. A large number of

[12] D. G. Yansura, this volume [14].
[13] K. Nagai and H. C. Thorgersen, *Nature (London)* **309**, 810 (1984).
[14] J. Shine, I. Fettes, N. C. Y. Lan, J. L. Roberts, and J. D. Baxter, *Nature (London)* **285**, 456 (1980).
[15] H. M. Sassenfeld and S. J. Brewer, *Bio/Technology* **2**, 76 (1984).
[16] R. Varadarajan, A. Szabo, and S. G. Boxes, *Proc. Natl. Acad. Sci. U.S.A.* **82**, 5681 (1985).
[17] P. R. Szoka, A. B. Schreiber, H. Chan, and J. Murthy, *DNA* **5**, 11 (1986).
[18] S. P. Kunnapuli, G. L. Prasad, and A. Kumar, *J. Biol. Chem.* **262**, 7672 (1987).

soluble fusion proteins have been produced in *E. coli,* both intracellularly[19,20] and secreted.[11,21,22]

Localization of Gene Fusion Product

Another important consideration for expression of gene fusions is whether a secretion system should be used to direct the product to a specific compartment of the cell. Such systems might offer advantages over intracellular expression systems. In particular, disulfide bond formation is enhanced in the oxidative environment outside the cytoplasm. Several correctly folded eukaryotic proteins containing disulfide bridges have been produced, as demonstrated for human epidermal growth factor (EGF),[21] human insulin-like growth factor I (IGF-I),[22] human IGF-II,[23] and bovine pancreatic trypsin inhibitor (BPTI).[11] Secretion of proteins to the periplasm can also protect protein from degradation, e.g., the half-life of proinsulin located in the periplasm was 10-fold longer than when located in the cytoplasm.[24] Since the periplasm of *E. coli* only contains approximately 4% of the cellular proteins, it is obvious that a large degree of purification can be achieved by secretion of the recombinant gene product followed by selective release of the periplasmic content. In addition, if a recombinant protein is toxic to the cell, the secretion approach might offer the only expression alternative.

Secretion strategies have, however, some limitations. Many proteins are not secretion "competent" as they cannot be translocated through membranes. Earlier secretion systems have also given rather low yields, e.g., 0.1–0.2% of total cell protein.[21] As a comparison, proteins produced intracellularly involving inclusion bodies often give yields higher than 10% of total protein content.[5] Efforts to increase the production of secreted protein using strong promoters have led to cell lysis[25] or accumulation of the gene product intracellularly.[26] However, a better understanding of the secretion mechanism as well as an adaptation of the growth conditions for

[19] L. Monaco, H. M. Bond, K. E. Howell, and R. Cortese, *EMBO J.* **6,** 3253 (1987).

[20] B. Nilsson, L. Abrahmsén, and M. Uhlén, *EMBO J.* **4,** 1075 (1985).

[21] T. Oka, S. Sakamoto, K.-I. Miyoshi, T. Fuwa, K. Yoda, M. Yamasaki, G. Tamura, and T. Miyake, *Proc. Natl. Acad. Sci. U.S.A.* **82,** 7212 (1985).

[22] T. Moks, L. Abrahmsén, E. Holmgren, M. Billich, A. Olsson, G. Pohl, C. Sterky, H. Hultberg, S. Josephson, A. Holmgren, H. Jörnvall, M. Uhlén, and B. Nilsson, *Biochemistry* **26,** 5239 (1987).

[23] B. Hammarberg, T. Moks, M. Tally, A. Elmblad, E. Holmgren, B. Nilsson, S. Josephson, and M. Uhlén, *J. Biotechnol.* in press (1990).

[24] K. Talmadge and W. Gilbert, *Proc. Natl. Acad. Sci. U.S.A.* **79,** 1830 (1982).

[25] J. Brosius, *Gene* **27,** 161 (1984).

[26] H. M. Hsiung, N. G. Mayne, and G. W. Becker, *Bio/Technology* **4,** 991 (1986).

secretory expression systems have led to substantially higher expression levels. Human growth hormone has been secreted to the periplasm of *E. coli* to a level corresponding to 30% of total cell proteins[27] and, recently, the development of systems for extracellular production in *E. coli*[22] has allowed production levels of more than 1000 mg/liter of gene fusion products containing mammalian peptide hormones.[28]

Gene Fusion Strategies

In Fig. 1, some examples of gene fusion strategies are outlined, with emphasis on soluble gene fusion products. The most simple fusion involves splicing the recombinant gene X directly after a suitable signal sequence (Fig. 1,I). One potential advantage of this strategy is that if the signal peptide is correctly processed during transport, it is possible to produce recombinant protein with a native N terminus. This was achieved in the production of human growth hormone in *E. coli* using the alkaline phosphatase signal sequence.[29]

Another conceptionally simple fusion strategy is to express a gene product fused to itself (Fig. 1,II). Proinsulin was found to be more stable in *E. coli* using such an approach.[30] Similarly, the yield of human IGF-I expressed intracellularly in *E. coli* was increased by a factor of 200, accomplished by an increase of the half-life from approximately 1–2 min for monomeric IGF-I to more than 60 min for the fusion.[31] In both of these cases, the effect was probably caused by changes in the folding of the gene products, resulting in the formation of inclusion bodies. Thus, the self-polymerization strategy may be an alternative to other types of fusions to obtain insoluble aggregates.

Two of the most common fusion strategies are C-terminal (Fig. 1,III) or N-terminal (Fig. 1,IV) fusions, often where the fusion partner A encodes an affinity handle to facilitate the purification. An advantage with C-terminal fusion, where the recombinant product X is positioned at the C-terminal side of the fusion partner A, is that the promoter and the translation initiation signals are all integrated in the 5′ end of the gene and are thus not changed by different fusions in the 3′ end. The expression level is therefore relatively predictable and several different promoters can rapidly be tested

[27] M. Tahahara, H. Sagai, S. Inouye, and M. Inouye, *Bio/Technology* **6**, 195 (1986).

[28] S. Josephson and R. Bishop, *Trends Biotechnol.* **6**, 218 (1988).

[29] G. L. Gray, J. S. Baldridge, K. S. McKeown, H. L. Heyneker, and C. N. Chang, *Gene* **39**, 247 (1985).

[30] S.-H. Shen. *Proc. Natl. Acad. Sci. U.S.A.* **81**, 4627 (1984).

[31] M.-F. Schulz, G. Buell, E. Schmid, R. Movva, and G. J. Selzer, *J. Bacteriol.* **169**, 5385 (1987).

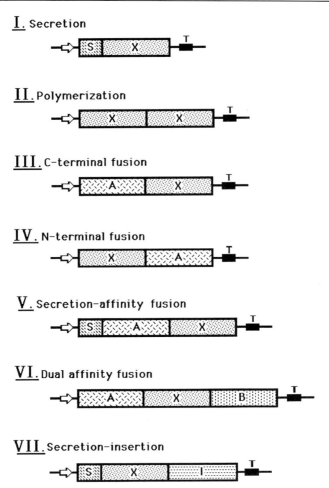

FIG. 1. Examples of gene fusions used for the expression of a recombinant protein (X). Promoters are symbolized by arrows and transcription terminator sequences by rectangles. The boxes represent signal sequence (S), different affinity "handles" (A and B), and a domain for insertion into membranes or cell wall (I).

for each gene product. N-Terminal fusion (Fig. 1,IV) has the disadvantage that a product-specific transcriptional and translational start must be engineered in the 5′ end of the gene X. In addition, when chemical methods are used to release X, a cleavage rest is usually obtained in the C terminus, thus giving a nonnative protein. Advantages with N-terminal fusions include the ease with which direct N-terminal sequencing of the gene fusion product can be performed and the possibility of designing various biological expression assays, such as HIV frameshift studies.[32]

Secretion – affinity fusion (Fig. 1,V) combines the advantages of secretion and affinity purification. Secretion to the culture medium provides ample opportunities for various procedures in which the product is continuously recovered from the medium during fermentation.[33,34] Such approaches might have great impact on large-scale purification (downstream processing) of recombinant proteins in the future.

A new concept recently described[35] is the dual-affinity fusion (Fig. 1, VI) in which the gene of interest, X, is fused between two heterologous domains, A and B, with specific affinity for two different ligands. The advantage is that a full-length protein can be selectively recovered by two subsequent affinity purification steps. This approach is suitable for expression of proteins which are highly susceptible to proteolysis, as even small amounts of full-length protein can be easily recovered. It is also attractive for proteins intended for structural and functional studies, where a heterogeneous population of products would lead to difficulties in the interpretation of the results. In the development of dual-affinity fusion systems, factors such as solubility, size, stability, binding constants, and subunit structure must be considered for both affinity handles. It is also important that the desired protein X is allowed to fold into a biologically active structure, without obstruction from the flanking heterologous domains. Recently, it was found that several eukaryotic proteins can fold into a biologically active form using the dual-affinity fusion approach.[35] However, the fact that two separate site-specific cleavages are necessary to cleave off the flanking domains makes this approach less attractive for applications where exact N- and C-terminal ends of the protein are desired.

Finally, an interesting fusion approach (Fig. 1,VII) is to combine a secretion signal S with a sequence I that inserts the gene product into the cell wall or one of the cellular membranes. This approach can be used to expose receptors or antigens on the outer surface of bacteria[36] or to assem-

[32] T. Jacks, M. D. Power, F. R. Masiarz, P. A. Luciw, P. J. Barr, and H. E. Varmus, *Nature (London)* **331,** 280 (1988).

[33] E. Pungor, N. B. Afeyan, N. F. Godon, and C. L. Cooney, *Bio/Technology* **5,** 604 (1987).

[34] T. Moks, L. Abrahmsén, B. Österlöf, S. Josephson, M. Östling, S.-O. Enfors, I.-L. Persson, B. Nilsson, and M. Uhlén, *Bio/Technology* **5,** 379 (1987).

[35] B. Hammarberg, P.-Å. Nygren, E. Holmgren, A. Elmblad, M. Tally, U. Hellman, T. Moks, and M. Uhlén, *Proc. Natl. Acad. Sci. U.S.A.* **86,** 4367 (1989).

[36] M. Agterberg, H. Adriaanse, and J. Tommassen, *Gene* **59,** 145 (1987).

ble the fusion protein into viruslike particles.[37] Such systems might prove beneficial for the development of vaccines and other systems to generate immunogenic complexes.

Affinity Purification of Fusion Proteins

The basic concepts for using gene fusion for affinity purification of recombinant proteins are shown in Fig. 2. A cell lysate or a culture medium containing the fusion protein, consisting of the desired protein X fused to an affinity handle A, is passed through an affinity column containing a ligand L that specifically interacts with the affinity handle. The fusion protein AX thus binds to the ligand while all other proteins can be washed out of the column using an appropriate buffer. After elution, a chemical or enzymatic method is used to cleave the purified fusion protein at the junction between the two protein moieties. The cleavage mixture is again passed through the column to allow the affinity handle to bind, while the desired product X is collected in the flow-through fraction. The column is regenerated simply by elution of bound material. Obviously, this simple purification scheme offers many advantages over traditional processes which have to be developed and optimized for each gene product.

As indicated in Fig. 2, production of recombinant proteins with an affinity handle could have three different goals. First, the affinity interaction can be used to immobilize enzymes and receptors on a solid support, such as a biosensor or an affinity column, without any prior purification of the protein.[20] Second, the immobilized fusion protein can be eluted and used directly for structural or functional studies or be used as an immunogen to generate antibodies.[11,38] Third, in a number of cases, the affinity-purified fusion protein can be processed by site-specific cleavage to release the desired product.[22]

Examples of Gene Fusion Systems

The choice of gene fusion system depends on the properties and final use of the gene product to be expressed. In Table I, examples of gene fusion systems are listed that have been used to facilitate purification of soluble

---

[37] J. R. Hynes, I. Cunningham, A. von Seefried, M. Lennich, R. T. Garvin, and S.-H. Shen, *Bio/Technology* **4**, 637 (1986).
[38] M. D. Winther, R. H. Bromford, G. Allen, and F. Brown, *in* "Vaccines 86," p. 79. Cold Spring Harbor Laboratory, Cold Spring Harbor, New York, 1986.

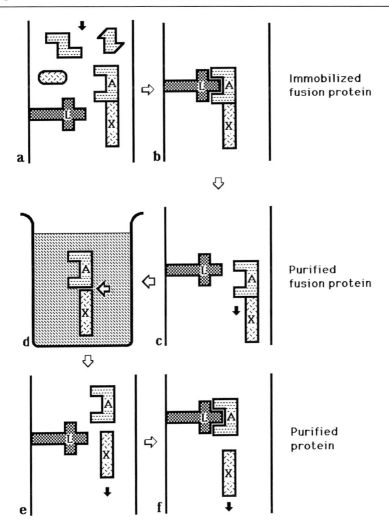

FIG. 2. Basic concepts for using fusions to purify a gene product (X) by the use of an affinity column. A crude extract from a culture expressing the fusion protein (AX) is passed through the column (a) containing a specific ligand (L) for the affinity handle (A). The fusion protein (AX) will bind to the column, and a pure and immobilized fusion protein is obtained (b). Elution is performed by an appropriate method (c), and the fusion protein is site-specifically cleaved (d). The cleavage mixture is again passed through the column (e), and pure product X can be collected as flow-through (f).

TABLE I
GENE FUSION SYSTEMS USED TO FACILITATE PROTEIN PURIFICATION[a]

Gene product	Origin	Molecular Weight ($\times 10^3$)	Sec.	Ligand	Refs.
β-Galactosidase	*Escherichia coli*	116	−	TPEG, APTG	7, 8, 39
Protein A	*Staphylococcus aureus*	31	+	IgG	11, 20, 40
CAT	*Escherichia coli*	24	+	Chloramphenicol	9
Poly(Arg)	Synthetic	1–3	−	Ion-exchange	15
Streptavidin	*Streptomyces*	13	+	Biotin	10
Poly(Glu)	Synthetic	1–2	−	Ion-exchange	41
Z	Synthetic	7	+	IgG	22, 23, 34
PhoS	*Escherichia coli*	36	+	Hydroxylapatite	42
Cysteine	Synthetic	<1	+	Thiol	43
Protein G	Streptococci	28	+	Albumin	44
MBP	*Escherichia coli*	40	+	Starch	45
GST	*Escherichia coli*	26	−	Glutathione	46
Flag peptide	Synthetic	2–5	+	Specific IgG	47
Poly(His)	Synthetic	1–7	+	Zn^{2+}, Cu^{2+}	48

[a] A few references for each gene fusion are listed. The molecular weight of the most common fusion part is indicated, as well as if the secretion (Sec.) of the fusion proteins has been demonstrated (+). CAT, Chloramphenicol acetyltransferase; Z, IgG-binding fragment based on staphylococcal protein A; PhoS, phosphate-binding protein; MBP maltose-binding protein; GST, glutathione *S*-transferase; TPEG, (*p*-aminophenyl-β-D-thiogalactosidase); APTG, *p*-aminophenyl-β-D-thiogalactoside.

recombinant proteins. Systems such as *trpE*[12] and cII,[13] which almost exclusively yield inclusion bodies are therefore not included in the list. Interestingly, C-terminal fusions to β-galactosidase normally yield insoluble inclusion bodies, while most fusion proteins obtained after N-terminal

[39] S. Scholtissek and F. Grosse, *Gene* **62**, 55 (1988).
[40] B. Nilsson, E. Holmgren, S. Josephson, S. Gatenbeck, L. Philipson, and M. Uhlén, *Nucleic Acids Res.* **13**, 1151 (1985).
[41] H. Dalbøge, H.-H. M. Dahl, J. Pedersen, J. W. Hansen, and T. Christensen, *Bio/Technology* **5**, 161 (1987).
[42] J. Anba, D. Baty, R. Lloubes, J.-M. Pages, E. Joseph-Liauzun, D. Shire, W. Roskam, and C. Lazdunski, *Gene* **53**, 219 (1987).
[43] P. Carter and J. A. Wells, *Science* **237**, 394 (1987).
[44] P.-Å. Nygren, M. Eliasson, E. Palmcrantz, L. Abrahamsén, and M. Uhlén, *J. Mol. Recognit.* **1**, 69 (1988).
[45] C. diGuan, P. Li, P. D. Riggs, and H. Inouye, *Gene* **67**, 21 (1988).
[46] D. B. Smith and K. S. Johnson, *Gene* **67**, 31 (1988).
[47] T. P. Hopp, K. S. Prickett, V. L. Price, D. P. Cerretti, C. J. March, and P. J. Conlon, *in* "ICSU Short Report: Advances in Gene Technology," p. 138. IRL Press, Oxford, 1988.
[48] M. C. Smith, T. C. Furman, T. D. Ingolia, and C. Pidgeon, *J. Biol. Chem.* **263**, 7211 (1988).

fusions are soluble and allow affinity purification with substrate analogs as ligands. This makes the β-galactosidase system versatile, as demonstrated by numerous authors. However, the large size of the enzyme reduces the attractiveness of the system, as even high expression levels give rather low yield of the desired product. In addition, the active enzyme consists of a tetramer, which might interfere with the folding of the fused gene products. Finally, it is not possible to obtain secretion using β-galatosidase as the fusion partner.

The protein A system[11] has several advantages, such as smaller size, higher solubility, and secretion "competence" of the protein A moiety. An IgG-binding domain, designated Z, was designed based on region B of protein A.[22] Domain Z was made resistant to some of the most common chemical methods used to cleave fusion proteins. A similar system is the albumin-binding domains of protein G[44] with high affinity for human serum albumin (HSA). These systems have the disadvantages that the elutions of the fusions have to be performed at low pH (3.4 for protein A and 2.9 for protein G) and that the affinities are based on protein ligands which might be degraded by bacterial proteases in the extracts. However, IgG affinity columns has been used over a hundred times without significant loss of binding capacity (T. Moks, unpublished data). In addition, the binding strength can be decreased by changing the origin of the IgG or serum albumin used as a ligand, thus giving milder elution conditions.

Systems based on streptavidin are promising, in particular for applications involving immobilization on solid supports. This protein, with an extremely high affinity for biotin, is normally secreted from *Streptomyces avidinii* and can be recovered on a biotin affinity column. However, the active protein is potentially lethal to *E. coli* and the structure is rather complex, consisting of four identical 13-kDa subunits which are processed at both ends from the 17-kDa precursor.[10]

Several systems based on small synthetic peptides have also been described. The advantages of these systems are small affinity handles and protease-resistant and cost-effective gel matrices. Polyarginine residues fused to the C-terminal part of human urogastrone were used to recover the recombinant protein by ion-exchange chromatography,[15] but the low solubility of the fusion protein made it necessary to use 5 *M* urea in all the purification steps, including the cell disruption. The strategy of using acidic residues, such as glutamic acid residues, might therefore be more attractive. This approach was used to facilitate the recovery of authentic human growth hormone using ion-exchange chromatography.[41] A third system, based on the metal-binding properties of histidine residues, has been used to recover a recombinant leuteinizing hormone-releasing hormone (LHRH) analog.[48] Similarly, proteins can be efficiently recovered by thiol-containing matrices if a cysteine is incorporated into the gene product by protein engineering.[43] In these two latter cases, the concentrations of metal ions and redox potentials must be carefully controlled to ensure reproducible purifications.

Recently, a number of alternative fusion systems have been described. The phosphate-binding protein (PhoS) of *E. coli* was used to recover human growth hormone-releasing factor by hydroxylapatite chromatography.[42] In addition, the maltose-binding protein of *E. coli* was used to express and purify alkaline phosphate on an amylose (starch) column.[45] Finally, glutathione *S*-transferase was used to purify malaria antigen by affinity chromatography on immobilized glutathione.[46]

Site-Specific Cleavages of Fusion Proteins

When a native gene product is desired, a site-specific cleavage of the fusion protein must be performed. There are two principal ways to obtain specific cleavages of proteins: chemical and enzymatic. A list of methods used to cleave recombinant fusion proteins is given in Table II.[49-51] The chemical methods cleave after a single amino acid (Met) or between residues in a dipeptide sequence (Asn-Gly or Asp-Pro). The low specificity limits their use because the recombinant product frequently contains the corresponding residue or peptide. Sequences recognized by enzymes can be more specific (e.g., enterokinase and factor Xa), although some proteases recognize single amino acid residues or dipeptide sequences (trypsin and clostripain). Proteases also often cleave at the C-terminal side of a recognition sequence, thus leaving no "linker residues" after cleavage on the newly created N terminus. Enzymatic cleavages are more affected by steric factors as compared to chemical methods. The cleavage site must therefore be carefully engineered to be structurally accessible to the enzyme.

Cyanogen bromide, which cleaves after methionine residues, has been used to cleave a large number of recombinant fusion proteins, produced both in soluble and insoluble form (Table II). The acid-labile Asp-Pro dipeptide cleavage has been used with variable success. With a *trpE*–EGF fusion, containing an Asp-Pro linker, no or little EGF could be site-specifically cleaved,[49] while a protein A–IGF-I fusion with an Asp-Pro linker was efficiently cleaved after 48 hr incubation in 70% (v/v) formic acid.[40] Both cyanogen bromide and Asp-Pro cleavages are performed under acidic conditions, which means that the protein of interest must be acid-stable. Hydroxylamine, however, can efficiently cleave an Asn-Gly peptide bond

---

[49] G. Allen, C. A. Paynter, and M. D. Winther, *J. Cell Sci. Suppl.* **3**, 29 (1985).

[50] T. Doi, T. Tokunaga, E. Ohtsuka, Y. Hiraki, F. Suzuki, and M. Ikehara, *Protein Eng.* **1**, 17 (1986).

[51] R. Varadarajan, A. Szabo, and S. G. Boxer, *Proc. Natl. Acad. Sci. U.S.A.* **82**, 5681 (1985).

TABLE II
CHEMICAL AND ENZYMATIC AGENTS THAT HAVE
BEEN USED TO CLEAVE FUSION PROTEINS SITE-
SPECIFICALLY[a]

Cleavage method	Recognition sequence	Refs.
Chemical		
Cyanogen bromide	-Met▾-	2, 3, 35
Formic acid	-Asp▾-Pro-	17, 40, 49
Hydroxylamine	-Asn▾-Gly-	22, 34
Enzymatic		
Collagenase	-Pro-Val▾-Gly-Pro-	8, 39
Enterokinase	-Asp-Asp-Asp-Lys▾-	47
Factor Xa	-Ile-Glu-Gly-Arg▾-	13, 46
Thrombin	-Gly-Pro-Arg▾-	46, 50
Trypsin	-Arg▾-	51
Clostripain	-Arg▾-	9
Ala^{64}-subtilisin	-Gly-Ala-His-Arg▾-	43

[a] A few references for each cleavage method are listed. The recognition sequences used in the references are shown and the cleavage site is indicated with an arrowhead.

at pH 9.0, leaving a glycine at the N terminus. Native IGF-I, which has an N-terminal glycine, has been produced on a large scale using hydroxylamine cleavage of a soluble fusion protein.

Several enzymatic methods have been used to cleave gene fusion products (Table II). Some of these have been performed under denaturing conditions, while others are confined to cleaving fusion proteins that are soluble at neutral pH. A proteolytic enzyme which is immobilized on a solid support facilitates the recovery of the recombinant material after cleavage and allows the protease to be reused. An elegant approach for obtaining a site-specific protease has been described[43] in which a subtilisin protease was designed using protein engineering and shown to cleave proteins with a histidine residue at the P2 position of a subtilisin recognition sequence.

Exopeptidases has also been used for processing the ends of fusion proteins. In these cases, the fusion only contains a few extra residues at either end, which are trimmed down to the native state. The positive charge of a poly(Arg) tail fused to the carboxy-terminal part of urogastrone made it possible to purify the protein on an ion-exchange column.[15] Carboxypeptidase B removed the arginines to yield free urogastrone. As the poly(Arg) fusion was insoluble, all the purification steps, as well as the carboxypeptidase reaction, were performed in buffers containing 5 M urea.

Another example of the use of exopeptidases is the production of human growth hormone (hGH), which was expressed as a precursor with a three-residue extension, including a glutamic acid at the -1 position.[41] The extension was removed after treatment with exopeptidase DAP 1, and the free authentic hGH could easily be purified from the remaining fusion protein by ion-exchange chromatography. This approach made it possible to produce native hGH intracellularly in *E. coli*.

In conclusion, the cleavage method must be determined separately for each product. For small proteins, chemical cleavages will probably, but not necessarily, be preferable. For larger proteins, the cleavage is usually restricted to specific proteases and the aim must be to design the cleavage site in such a way that the recognition sequence is accessible to the protease.

Concluding Remarks

Bacterial gene fusion systems have been used to facilitate protein purification, to overcome protease degradation problems, and to allow secretion of gene products through the cytoplasmic membrane. However, a large number of considerations have to be taken into account in order to choose an optimal expression strategy. Whether to use a soluble or an insoluble gene fusion product, and secretion versus intracellular production are primary important decisions that need to made when trying to recover the recombinant protein. When a fusion system is designed to facilitate protein purification, the structures of the affinity handle as well as the ligand must be designed to give an appropriate binding strength to allow elution of fusion protein without denaturing the recombinant protein. The choice of gene fusion system influences factors such as solubility, folding, proteolysis, and purification yield, which emphasizes the need for several alternative expression systems.

For immobilization of enzymes, receptors, and other proteins, ligand stability and the binding strength are important factors. Affinity systems based on streptavidin–biotin, histidine–Zn^{2+}, and cysteine–thiol are therefore suitable for such applications, due to the protease-resistant ligands. Streptavidin binds very strongly to biotin, which is attractive for many applications.

In the field of protein engineering, there is a great need for efficient expression purification schemes to allow experiments in which amino acid residues are changed systematically. The fact that protein engineering often changes the properties of the protein makes it attractive to use an affinity handle for the purification, thus avoiding the necessity of developing a new purification scheme for each mutant protein. When the recombinant protein and the affinity handle will fold independently, there is no need to go through site-specific cleavage before studying the properties of the engineered protein.

Similarly, proteins to be utilized for immunizations or in diagnostics can be used directly as intact fusion proteins. Both *trpE* and β-galactosidase fusions have been used to raise antibodies against the fusion protein purified by SDS–polyacrylamide gel electrophoresis.[52] Protein A fusions have also been used to obtain both monoclonal and polyclonal antibodies.[11] It is likely that such systems will be combined with fusions to mitogens, toxins, etc. to allow the assembly of molecules with improved immunological properties.

Novel expression systems have been designed to combine the advantages of two different gene fusion systems. An example of this is a dual-expression system, in which the protein A system is used for immunization, while the protein G system is used to produce the immunogen fused to an albumin-binding receptor.[53] The protein G-fused immunogen can thus conveniently be immobilized to HSA-columns for epitope-specific purification of antibodies obtained by immunization with the protein A fusion. The protein G fusion can also be directed to HSA-coated microtiter wells for analysis of the antibody response. Thus, the B-cell mitogen property of protein A is used to obtain a good immune response, while the albumin-binding system is used for screening and purification of the monospecific antibodies. The similarity of the two fusion partners with respect to size, solubility, and stability makes the dual-expression system general and relatively predictable.

In conclusion, the use of gene fusions for expression of proteins can be described as protein engineering, and is aimed at improving the yield and/or facilitating the recovery of the recombinant protein.

Acknowledgments

We acknowledge the valuable suggestions and advice from Birger Jansson, Björn Nilsson, and Staffan Bergh. We are also grateful for patient secretarial help from Gerd Benson.

[52] S. A. Tooze and K. Stanley, *J. Virol.* **60,** 928 (1986).
[53] S. Ståhl, P.-Å. Nygren, A. Sjölander, P. Perlmann, and M. Uhlén, *J. Immunol. Methods* **124,** 43 (1989).

[13] Fusions to Staphylococcal Protein A

By BJÖRN NILSSON and LARS ABRAHMSÉN

Introduction

Recombinant DNA techniques to construct hybrid genes for the expression of fusion proteins have greatly facilitated the recovery of heterologous gene products expressed in bacteria (for a review, see Uhlén[1]). This article presents the use and the development of a gene fusion expression system based on staphylococcal protein A (SPA). SPA binds to the Fc portion of IgG, permitting the fusions to be purified in a single step at >95% yields.[2] This fusion technique has been used successfully for the high-level expression of peptide hormones,[3-6] for the immobilization of enzymes,[2,7] and for the production of specific antibodies against gene products.[8-10] SPA fusions are expressed in bacteria. In *Escherichia coli*, vectors have been constructed to direct the fusion protein to the cytoplasm,[2,10,11] the periplasmic space,[2,5] or into the medium.[3-5] Alternatively, the fusions can be expressed in the Gram-positive *Staphylococcus aureus*, where the fusions will be secreted extracellularly.[2,6,12]

The SPA fusion systems have been developed sequentially, and here we present the most recent version by describing six new *spa* gene fusion plasmids. There are two separate sets of new plasmids, and each set comprises all three reading frames. One set includes the signal sequence of SPA

[1] M. Uhlén and T. Moks, this volume [12].
[2] B. Nilsson, L. Abrahmsén, and M. Uhlén, *EMBO J.* **4,** 1075 (1985).
[3] T. Moks, L. Abrahmsén, E. Holmgren, M. Bilich, A. Olsson, G. Pohl, C. Sterky, H. Hultberg, S. Josephson, A. Holmgren, H. Jörnvall, M. Uhlén, and B. Nilsson, *Biochemistry* **26,** 5239 (1987).
[4] T. Moks, L. Abrahmsén, B. Österlöf, S. Josephson, M. Östling, S.-O. Enfors, I. Persson, B. Nilsson, and M. Uhlén, *Bio/Technology* **5,** 379 (1987).
[5] L. Abrahmsén, T. Moks, B. Nilsson, and M. Uhlén, *Nucleic Acids Res.* **14,** 7487 (1986).
[6] B. Nilsson, E. Holmgren, S. Josephson, S. Gatenbeck, and M. Uhlén, *Nucleic Acids Res.* **13,** 1151 (1985).
[7] M. Uhlén, B. Nilsson, B. Guss, M. Lindberg, S. Gatenbeck, and L. Philipson, *Gene* **23,** 419 (1983).
[8] B. Löwenadler, B. Nilsson, L. Abrahmsén, T. Moks, L. Ljungqvist, E. Holmgren, S. Paleus, S. Josephson, L. Philipson, and M. Uhlén, *EMBO J.* **5,** 2393 (1986).
[9] B. Löwenadler, B. Jansson, S. Paleus, E. Holmgren, B. Nilsson, T. Moks, G. Palm, S. Josephson, L. Philipson, and M. Uhlén, *Gene* **58,** 87 (1987).
[10] K. Valerie, G. Fronko, W. Long, E. E. Henderson, B. Nilsson, M. Uhlén, and J. K. de Riel, *Gene* **58,** 99 (1987).
[11] L. Monaco, H. M. Bond, K. E. Howell, and R. Cortese, *EMBO J.* **6,** 3253 (1987).
[12] M. Uhlén, B. Guss, B. Nilsson, F. Götz, and M. Lindberg, *J. Bacteriol.* **159,** 713 (1984).

which directs the fusion protein out of the cytoplasm; the second set has the signal sequence deleted, and the fusion protein will thus remain in the cytoplasm. The use of the SPA fusion strategy is described for both types of plasmids. Expression and purification are specifically demonstrated for an SPA fusion to a mutant form of bovine pancreatic trypsin inhibitor (BPTI) lacking the disulfide bond between cysteine-14 and cysteine-38.

Staphylococcal Protein A as "Affinity Handle"

SPA is a protein bound to the cell wall of the pathogenic bacterium *S. aureus*. It binds to the Fc portion of most mammalian class G immunoglobulins.[13] SPA functions most likely by contributing to pathogenicity by its IgG binding activity, but the detailed mechanism for this contribution has not to date been clarified.[14] The nucleotide sequence of the cloned gene shows that the protein consists of three structurally and functionally different regions (Fig. 1A).[15] Starting at the N terminus, the first region is the signal sequence consisting of 36 amino acid residues which are cleaved off during or after translocation.[16] The signal sequence is functional and correctly processed not only in *S. aureus* but also in *E. coli*,[16] *Bacillus subtilis*,[17] and *Streptomyces lividans* (B.N., unpublished). The second region consists of the five highly homologous IgG binding domains, A–E, each with approximately 58 amino acid residues.[18] The third region is the C-terminal region X, part of which is anchored in the cytoplasmic membrane with the rest located in the cell wall.[19] In the gene fusion vectors described here, region X has been deleted.

Many of the characteristics of SPA, and its expression in different bacteria, make it useful as an "affinity handle" in gene fusion expression systems. The most important of those features are (1) SPA binds IgG tightly with a dissociation constant of 2×10^{-8} M for isolated domain B to human IgG.[13] This strong and specific affinity permits the purification of the fusion proteins in a single step by IgG affinity chromatography. (2) The

---

[13] R. Lindmark, K. Thorén-Tolling, and J. Sjöquist, *J. Immunol. Methods* **62**, 1 (1983).

[14] A. H. Patel, P. Nowlan, E. D. Weavers, and T. Foster, *Infect. Immun.* **55**, 3103 (1987).

[15] M. Uhlén, B. Guss, B. Nilsson, S. Gatenbeck, L. Philipson, and M. Lindberg, *J. Biol. Chem.* **259**, 1695 (1984).

[16] L. Abrahmsén, T. Moks, B. Nilsson, U. Hellman, and M. Uhlén, *EMBO J.* **4**, 3901 (1985).

[17] C. W. Saunders, C. D. B. Banner, S. R. Fahnestock, M. Lindberg, M. S. Mirot, C. S. Rhodes, C. F. Rudolph, B. I. Schmidt, L. D. Thomson, M. Uhlén, and M. Guyer, *in* "Protein Transport and Secretion" (D. L. Oxender, ed.), p. 329. Alan R. Liss, New York, 1984.

[18] T. Moks, L. Abrahmsén, B. Nilsson, U. Hellman, J. Sjöquist, and M. Uhlén, *Eur. J. Biochem.* **14**, 7487 (1986).

[19] B. Guss, M. Uhlén, B. Nilsson, M. Lindberg, J. Sjöquist, and J. Sjödahl, *Eur. J. Biochem.* **138**, 413 (1984).

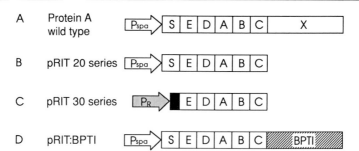

Fig. 1. Linear representation of different gene fragments based on staphylococcal protein A. (A) Protein A wild-type gene *(spa)* from *S. aureus;* (B) Gene fragment used in the pRIT20 series of gene fusion plasmids; (C) Gene fragment used in the pRIT30 series of gene fusion plasmids; (D) Fusion between SPA and BPTI made from pRIT23. The boxed letters E, D, A, B, and C refer to the five IgG binding domains, X is the cell wall binding domain, and S is the signal sequence. P_{spa} is the promoter region from the *spa* gene, and P_R refers to the right promoter from phage λ [M. Zabeau and K. Stanley, *EMBO J.* 1, 1217 (1982)]. The black box shows the 5' end of the structural gene in the pRIT30 vector series originating from the *cro* gene of phage λ, which encodes the 11 N-terminal amino acids of the CRO protein.[2] The figure is not to scale.

IgG binding region of SPA has a proposed extended tertiary structure consisting of small globular domains.[15,18,20] This characteristic facilitates independent folding of the product and SPA and minimizes the risk of steric interference of the product on IgG binding. (3) SPA is proteolytically stable not only in the homologous host *S. aureus* but also in the cytoplasmic and the periplasmic spaces of *E. coli*.[2,7] (4) High-level expression and secretion systems of SPA fusions have been developed for several bacteria.[2-7,10-12,16-18] To date, the most useful hosts are *E. coli* and *S. aureus.* (5) The half-life of gene products is often significantly increased *in vivo* when fused to SPA.[3,6] This protection from proteolysis is a general observation in all hosts and for most products where a comparative study has been undertaken. The effect is most striking for secreted basic peptide hormones.[3-6] A possible general explanation might be the increase in size of the product when fused to SPA. Specifically for the basic products, electrostatic interactions between the basic product and acidic SPA due to their opposite charges at neutral pH might cause the formation of protected soluble aggregates. Even though it has not been fully explained, the protection against proteolysis is an important advantage of the SPA fusion strategy for the expression of heterologous gene products in bacteria.

[20] J. Deisenhofer, T. A. Jones, R. Huber, J. Sjödahl, and J. Sjöquist, *Hoppe Seyler's Z. Physiol. Chem.* 359, 975 (1978).

Selection of Fusion Point in SPA

In the new fusion vectors, a fusion point was selected at the junction between domain C of the IgG binding region and region X of SPA (Fig. 1). This position was selected for three reasons. First, the IgG binding capacity will be maximized by including all five IgG binding domains and the cell wall binding region is deleted. Second, the gene fusions place the product gene 3' of *spa* gene and thus maintain the expression signals from the *spa* gene. Third, the point of fusion is situated in an area of the SPA protein molecule which is thought to be flexible. Flexibility is supported by three independent observations: (1) The position is located between independently functional domains.[15,19] (2) The only trypsin-sensitive site is situated in this region, which, due to the overall homologous structure, is found between each IgG binding domain.[15] (3) The three-dimensional structure, which was solved for domain B in a complex with IgG to 2.8 Å resolution by X-ray crystallography,[20] suggests flexibility. The residues which are implicated by the structure to participate in the binding are all located within, or adjacent to, the two α helices, while the residues outside the helical region show few potential side-chain interactions with the α helices. The fusion point is located 10 amino acid residues beyond the last C-terminal residue shown in the structure in Fig. 2. The missing residues were not determined in the structure as they did not diffract in the crystal,[20] which is most likely explained by a flexible three-dimensional structure.

Use of SPA Fusions

A major use of SPA fusions has been for the production of specific antibodies,[8-10] which are raised in rabbits, where the SPA moiety seems to act as an adjuvant. This was demonstrated first with an SPA fusion to insulinlike growth factor 1 (IGF-1),[8] which is a poor immunogen by itself. The enhanced immune response could simply be explained by the increase in protein size in the fusion, but alternative explanations are suggested by SPA's B-cell mitogenic activity[21] and its ability to stimulate polyclonal antibody response.[22] Recently, it was shown by Löwenadler *et al.* that antibodies against short peptide sequences could also be obtained by making SPA fusions.[9] In this experiment, an oligonucleotide linker encoding the amino acids 57 to 70 of IGF-1 was fused to the coding sequence of a two-domain SPA molecule. The titer of the polyclonal antisera raised in rabbits to the fusion was comparable to antisera raised against the peptide chemically conjugated to bovine serum albumin (BSA). These results

[21] G. E. Rodey, T. Davis, and P. G. Quie, *J. Immunol.* **108,** 178 (1972).
[22] A. Moraguchi, T. Kishimoto, T. Kuritani, T. Watanaki, and Y. Yamamura, *J. Immunol.* **125,** 564 (1980).

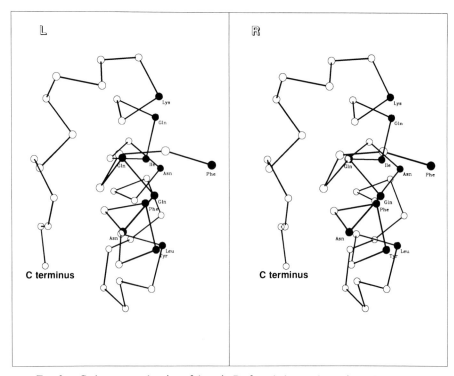

FIG. 2. α-Carbon stereo drawing of domain B of staphylococcal protein A. The α carbons are represented by circles and are connected following the polypeptide chain. The α carbons of the amino acid residues (three-letter code) suggested to be involved in the binding to IgG[20] are filled. The C terminus of the fragment is shown, and the gene fusions are made about 10 amino acid residues on the C-terminal side of this point. The IgG molecule, which is not shown, would bind at the right-hand side of the domain B molecule.

make SPA fusions to a synthetic DNA linker an attractive alternative to peptide synthesis for the production antibodies against short peptide structures. The scheme not only has advantages due to the recent improvements in oligonucleotide synthesis, making the synthesis of a small gene fragment faster than the synthesis of corresponding peptide, but, in addition, the supply of antigen will be unlimited and each preparation of antigen will be identical, in contrast to peptide coupled to BSA, which may show batch-to-batch variations.

The SPA fusion protein can often be used directly for structural or functional analysis if the fusion protein is biologically active.[2,7] The direct expression and purification of mutant forms of proteins are potentially time consuming and difficult. Single amino acid replacements can sometimes dramatically change the properties of the protein in terms of susceptibility to proteases, solubility, and behavior during purification. An SPA fusion could not only solve the purification and proteolysis problems, but could also facilitate the analysis of the mutations on biological activity if the fusion protein is fully active.

SPA fusions have been used extensively to express and purify heterologous gene products in bacteria.[2-7,10] To recover the product in a pure form, the strategy is to purify the SPA fusion protein and cleave the hybrid product at the junction. In principle, by a second passage through the IgG column, the product should be the only protein in the flow-through. This process is, however, dependent on the unique cleavage of the fusion protein. Currently, this is a serious bottleneck in a process based on any gene fusion strategy. Several sequence-specific proteases exist and have successfully been tried,[23-25] e.g., from the blood coagulation cascade, but purity, availability, and stability of these enzymes are limitations in their use. The vectors presented in this article all include the recognition sequences for specific proteases adjacent to the cloning linkers (Fig. 3). Currently, these are being investigated. The peptide hormones that have to date been expressed and purified using SPA fusions have all been cleaved by chemical methods such as cyanogen bromide,[29] weak acids,[5] or hydroxylamine.[2,3] A mutant two-domain SPA was recently presented in which some of the amino acid sequences for chemical cleavages were removed,[30] making a unique cleavage possible in the junction between this SPA molecule and the product. The chemical cleavages have obvious technical and economical advantages.[4] However, since chemical methods suffer from limited specificity, the general application of SPA fusions for the expression of isolated gene products will be dependent on future development of proteolytic cleavage methods for fusion proteins. Recently, promising results have been obtained for the cleavage of SPA fusions by an engineered subtilisin which was made specific.[31] Specifically and quantitatively, this enzyme cleaves a target amino acid sequence in a SPA fusion to alkaline phophatase, with no detectable cleavage at additional sites.[32]

[23] K. Nagai and H. C. Thøgersen, *Nature (London)* **309**, 810 (1983).

[24] J. Germino and D. Bastia, *Proc Natl. Acad. Sci. U.S.A.* **81**, 4692 (1984).

[25] S. Scholtissek and F. Grosse, *Gene* **62**, 55 (1988).

[26] S. Iordanescu, *J. Bacteriol.* **124**, 597 (1975).

[27] L. Dente, G. Cesarini, and R. Cortese *Nucleic Acids Res.* **11**, 1645 (1983).

[28] L. Abrahmsén, R. Hjort, L. Ljungqvist, M. Ulhén, and B. Nilsson, unpublished.

[29] B. Hammarberg, T. Moks, M. Tally, A. Elmblad, E. Holmgren, M. Murby, B. Nilsson, S. Josephson, and M. Uhlén, *Biotechnol.* in press (1990).

[30] B. Nilsson, T. Moks, B. Jansson, L. Abrahmsén, A. Elmblad, A. Holmgren, C. Henrichson, T. A. Jones, and M. Uhlén, *Protein Eng.* **1**, 107 (1987).

[31] P. Carter and J. A. Wells, *Science* **237**, 394 (1987).

[32] P. Carter, B. Nilsson, J. P. Burnier, and J. A. Wells, *Proteins* **6**, 240 (1990).

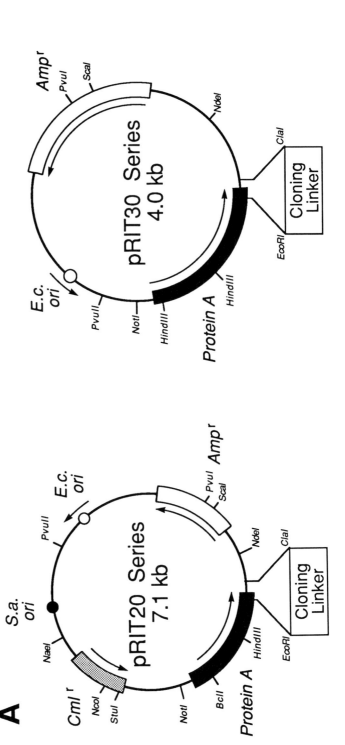

B

Cloning Linker 1 (in pRIT21 and pRIT31):

```
                                                                      SmaI
       EcoRI SacI KpnIAvaI BamHI XbaI  SalI PstI SphI                                                    EcoRV  ClaI
AAGCCTGATGACGATAAAGAATTCGAGCTCGGTACCCGGGATCCTCTAGAGTCGACCTGCAGGCATGCAAGCTAGCTTAAGTAAGTAAGTAAGCCGGCCAGTTCCGCTGGCGGCATTTTTTTGATATCATCGAT
LysProAspAspAspLysGluPheGluLeuGlyThrArgGlySerSerArgValAspLeuGlnAlaCysLysLysLeuAla***
```

→ Enteropeptidase

Cloning Linker 2 (in pRIT22 and pRIT32):

```
                                                                     SmaI
       EcoRI SacI KpnIAvaI BamHI XbaI  SalI PstI SphI                                                    EcoRV  ClaI
AAGCCTGGTGGACCAGGAATTCGAGCTCGGTACCCGGGATCCTCTAGAGTCGACCTGCAGGCATGCAAGCTAGCTTAAGTAAGTAAGTAAGCCCCGGCCAGTTCCGCTGGCGGCATTTTTTTGATATCATCGAT
LysProGlyGlyProGlyIleArgAlaArgTyrProGlyIleLeu***
```

→ Collagenase

Cloning Linker 3 (in pRIT23 and pRIT33):

```
                                                                     SmaI
       EcoRI SacI KpnIAvaI BamHI XbaI  SalI PstI SphI                                                    EcoRV  ClaI
AAGCCTATCGAAGGTAGGATTCGAGCTCGGTACCCGGGATCCTCTAGAGTCGACCTGCAGGCATGCAAGCTAGCTTAAGTAAGTAAGTAAGCCGGCCAGTTCCGCTGGCGGCATTTTTTTGATATCATCGAT
LysProIleGluGlyArgAsnSerSerValProGlyAspProLeuGluSerThrCysArgHisAlaSer***
```

→ FactorXa

FIG. 3. pRIT20 and pRIT30 gene fusion vectors based on staphylococcal protein A. (A) Plasmid maps of pRIT20 and pRIT30 drawn to scale, including some restriction enzyme sites. The pRIT20 type of plasmid directs secretion of the SPA fusion product, and the pRIT30 type of plasmid directs an intracellular expression of the fusion. A linear representation of each *spa* expression fragment used is shown in Fig. 1. *AMp^r* indicates the gene encoding β-lactamase, *Cml^r* indicates the gene encoding the chloramphenicol acyltransferase from the *S. aureus* plasmid pC194.[26] E.c. ori refers to the origin of replication for *E. coli* from pEMBL8+,[27] and S.a. ori refers to the origin of replication for *S. aureus* from pC194.[26] (B) Cloning linkers used in the SPA fusion vectors. Restriction cleavage sites, putative amino acid sequences, and the cleavage sites for the three different specific proteases are shown. Opposite arrows indicate the transcription terminator.[28]

Design and Construction of Gene Fusion Vectors pRIT21, 22, and 23 and pRIT31, 32, and 33

Two sets of improved gene fusion vectors were constructed. The first set, the pRIT20-series (pRIT21, 22, and 23), can replicate and be selected for in both *E. coli* and *S. aureus.* These plasmids include the coding sequence of the signal sequence of *spa.* Therefore, the fusion proteins are secreted through the cytoplasmic membrane of either host. In the gram-positive *S. aureus,* the protein will be located extracellularly. In *E. coli,* the expression of SPA fusions by these vectors will translocate the protein to the periplasmic space. These proteins can be recovered after an osmotic shock procedure.[2] For some gene fusion expression from these vectors, the expressed SPA fusion induces a "leaky" phenotype in the *E. coli* cell, and the periplasmic proteins, including the SPA fusion protein, will leak out to the growth medium.[3-5,9] Although the mechanism is not fully understood, it has technical advantages when it occurs, as the fusion protein can be recovered directly from the growth medium. Similar, perhaps identical, induction of a leaky phenotype has been observed by Suominen *et al.* when expressing α-amylase from *Bacillus stearothermophilus* in *E. coli.*[33]

The second set of vectors (pRIT31, 32, and 33) is based on the previously described pRIT2,[2] which has the promoter and signal sequence of *spa* deleted and the transcription is instead initiated from the strong P_R promoter from phage λ.[34] SPA fusions will consequently be located intracellularly in *E. coli* when expressed from this type of plasmid.[2]

Both vector series pRIT20 and pRIT30 are derived from the cloning linkers from the recently described pRIT11, 12, and 13.[8] The new plasmids are shown in Fig. 3A, and linear representations of each *spa* gene fragment used are shown in Fig. 1. The three different linkers represent the different reading frames. Additionally, each linker encodes the recognition sequence of a specific protease: enteropeptidase[35] by linker 1, collagenase[24,25] by linker 2, and factor X_a[23] by linker 3. These proteolytic sites potentially could be utilized to cleave the fused gene product from the SPA moiety.

The general improvements of the new vectors compared to the previously described pRIT2, pRIT5,[2] pRIT11, 12, and 13[8] are the following: (1) The overall expression levels of SPA fusion protein are higher in *E. coli* with the pRIT20 and pRIT30 series and in *S. aureus* with the pRIT20 series, compared to their respective parents (pRIT2 and pRIT11, 12, and 13). (2) The cloning linker is based on mp18 instead of mp8. This change increases the number of available unique cloning sites (Fig. 3). (3) A

[33] I. Suominen, M. Karp, M. Lähde, A. Kopio, T. Glumoff, P. Meyer, and P. Mäntsäl, *Gene* **61,** 165 (1987).

[34] M. Zabeau and K. Stanley, *EMBO J.* **1,** 1217 (1982).

[35] F. Hesford, B. Hadorn, K. Blaser, and C. H. Schneider, *FEBS Lett.* **71,** 279 (1976).

synthetic transcription terminator has been included 3' of the mp18 cloning linker. (4) Translation stop codons (TAA) in the three reading frames are included between the cloning linker and the transcription terminator. These ensure a translational stop when cloning and expressing open reading-frame fragments. (5) A *Not*I restriction site is included 5' of the promoter to facilitate the recloning of the expression fragment into other vectors. (6) A *lac*UV5 promoter[36] is situated 5' of each major promoter (= the promoter from the *spa* gene for the pRIT20 series and the λ P_R for the pRIT30 series). This *lac*UV5 promoter has only limited effect on the expression of SPA fusions from the pRIT20 series, but may be useful in the pRIT30 series to obtain a weaker induction of SPA fusion than from the P_R promoter to avoid the formation of refractor bodies.[28]

The pRIT20 series and pRIT30 series were constructed using the plasmids pEMBL8+,[27] pC194,[26] pRIT11, pRIT12, pRIT13,[8] pRIT2,[2] M13mp18, and a synthetic linker encoding a stop signal for translation and a transcription terminator (KabiGen AB, Stockholm, Sweden). The details of these constructions will be presented elsewhere.[28] The new plasmids are shown in Fig. 3A, and all the DNA fragments are composed of available nucleotide sequences. A list of characteristics and a comparison of the two sets of vectors are shown in Table I.

Choice of Vector and Host

The choice of fusion vector is dependent on the origin and properties of the gene product to which SPA is to be fused. Mammalian proteins that are secreted in their natural host are often proteolytically unstable and rapidly degraded or precipitate and form so-called refractor bodies in the intracellular matrix.[37] Possibly, these problems are due to protein-folding problems in the reducing environment found inside most bacteria, in which disulfide bonds are rarely formed. For the expression of secreted proteins the pRIT20 series is strongly recommended. Alternatively, proteins which are found in the cytoplasm or in a membrane compartment in their natural host may be incompatible with translocation through a bacterial membrane and should therefore be expressed in the pRIT30 series.

The pRIT30 type of plasmid is used exclusively in *E. coli,* while the pRIT20 vectors can be hosted in either *E. coli* or *S. aureus.* Potentially, expression in *S. aureus* has several advantages compared to *E. coli.* First, a number of well-characterized plasmids exist which are stably maintained without selection. Second, efficient secretion to the growth medium is accomplished in the Gram-positive host using an appropriate signal sequence, like the SPA signal sequence. Finally, the amount of extracellular

[36] U. Deuschele, W. Kamnerer, R. Gentz, and H. Bujard, *EMBO J.* **5,** 2987 (1986).
[37] J. M. Schoemaker, A. H. Brasnett, and F. A. O. Marston, *EMBO J.* **4,** 775 (1985).

TABLE I
COMPARISON OF pRIT20 AND pRIT30 SERIES OF GENE FUSION VECTORS
BASED ON PROTEIN A FROM *S. aureus*[a]

	Plasmids		
Characteristic	pRIT20 series in *E. coli*	PRIT20 series in *S. aureus*	pRIT30 series in *E. coli*
Selection	Ampicillin [250 mg/liter]	Chloramphenicol [10 mg/liter]	Ampicillin [250 mg/liter]
Localization of fusion	Periplasm/extracellular	Extracellular	Intracellular
Promoter	*spa* (*lac*UV5)	*spa* (*lac*UV5)	λP_R *lac*UV5
Expression levels in shake flask [mg/liter]	3–200	5–50	5–300

[a] The wide range in the expression values is due to their dependence on fusion product and the method of cell growth. The *lac*UV5 in parentheses shows that the promoter has only limited effect on expression in the pRIT20 series of plasmids. The designation *spa* means the promoter from the staphylococcal protein A gene (*spa*), λP_R means the right promoter from phage λ,[34] *lac*UV5 means the *lac*UV5 promoter.[36]

proteases is much lower in staphylococci cultures relative to most other gram-positive bacteria like *Bacillus* and *Streptomyces*. However, *S. aureus* is a pathogenic bacterium and produces several endotoxins. Therefore, use of other species such as the nonpathogenic *Staphylococcus xylosus*[12] and *Staphylococcus carnosus*[12] might be preferable in the future.

There are several examples where *S. aureus* instead of *E. coli* has been the preferred host for the expression of SPA fusion proteins. The expression of insulinlike growth factor 2 (IGF-2) was unsuccessful in *E. coli* due to a specific cleavage in the IGF-2 moiety. In contrast, when expressed in *S. aureus,* proteolytic degradation was minimal.[29] In addition, SPA fusions to IGF-1 are more stable when secreted from *S. aureus* compared to the same fusion expressed and secreted into the *E. coli* periplasm.[6] The use of *S. aureus* as a host is, however, hampered by the protoplast transformation technique needed.[12,38] DNA constructions are first made in *E. coli,* and the plasmids are subsequently used to transform *S. aureus*. In spite of the labor-intensive transformation procedure, the use of *S. aureus* should be considered as an important alternative to *E. coli* for the expression of SPA fusions.

[38] F. Götz, S. Ahrne, and M. Lindberg, *J. Bacteriol.* **145**, 74 (1981).

Procedures

Method 1: Construction of spa Gene Fusions

Construction of Gene Fusion between spa and BPTI Gene. The restriction sites in the mp18 cloning linkers in both the pRIT20 and pRIT30 series, shown in Fig. 3B, are all unique to each fusion plasmid, respectively. This facilitates the cloning of any gene fragment into the vector systems. In addition, *Not*I, a restriction enzyme recognizing 8 base pairs and therefore most likely unique in all constructions, has been introduced 5' of the promoter in both plasmid series. This site facilitates the recloning of the expression fragment into other vectors, or interconversions between pRIT20 and pRIT30 types of constructions.

As an example of the construction and expression of an SPA fusion protein, a fusion to a mutant form of bovine pancreatic trypsin inhibitor (BPTI) was selected. BPTI is a protein of 58 amino acid residues.[39] It has three cross-linking disulfide bonds between cysteines 5 and 55, 14 and 38, and 30 and 51, respectively. The BPTI gene fragment that was expressed as an SPA fusion is a mutant in which the codons for cysteines 14 and 38 have been replaced by codons for alanine residues.[40] The BPTI gene was cloned into pRIT23 (Fig. 3), by utilizing a *Stu*I site in the BPTI gene (data not shown) which overlaps the codons for amino acids 1 and 2 of BPTI, into the *Sma*I site of pRIT23. The gene fusion maintains the reading frame from the *spa* gene into the BPTI gene. The fusion plasmid was designated pRIT:BPTI and a linear representation of the expression fragment is shown in Fig. 1D.

Method 2: Expression and Purification of Fusion Proteins from pRIT20 Plasmid Series

Expression of SPA-BPTI. The expression and purification of SPA fusion proteins from the pRIT20 series in *E. coli* are exemplified by the expression of the fusion protein of SPA and the mutant form of BPTI. The protein expressed from the parent plasmid (pRIT23) is used as the control. The plasmid vectors pRIT23 and pRIT:BPTI were hosted in *E. coli* W3110*tonA*,[41] but most *E. coli* hosts tested are compatible with SPA expression. The expression levels of fusion proteins from this type of vector vary from 3 to 200 mg/liter in a shake flask (Table I). In a fermentor, expression levels exceeding 1 g/liter have been obtained from similar plasmids (B. N., unpublished).

[39] S. Anderson and I. B. Kingston, *Proc. Natl. Acad. Sci. U.S.A.* **80,** 6838 (1983).
[40] C. Berman Marks, H. Naderi, P. A. Kosen, I. D. Kuntz, and S. Anderson, *Science* **235,** 1370 (1987).
[41] C. Nan Chang, M. Rey, B. Bouchner, H. Heyneker, and G. Gray, *Gene* **55,** 189 (1987).

In *E. coli,* the expression of SPA fusions can be performed either by using overnight cultures or by growing the cultures for a shorter period of time. Overnight cultures typically express higher levels of fusion proteins as the production continues into stationary growth phase. In the SPA–BPTI experiment, the expression was performed for a shorter period of time. The growth conditions used will give an expression level of approximately 5–10 mg/liter and the quality of the fusion protein is generally good, even for the expression of proteolytically sensitive products. The SPA-BPTI fusion protein was recovered from the growth medium, but for most products, leakage to the medium is not as efficient in a shake flask and an osmotic shock procedure is needed to isolate the fusion protein from the periplasmic space.[2]

The procedure used to isolate and purify SPA–BPTI from the growth medium is as follows.

1. Seed cultures of 5 ml were grown for 16 hr at 30° in LB (10 g/liter tryptone, 5 g/liter yeast extract, and 10 g/liter NaCl) supplemented with 0.1% (w/v) glucose and 250 mg/liter ampicillin.

2. The cells were washed by centrifugation and resuspended in 1 ml of fresh LB medium. This washing procedure is performed as the medium of the seed culture contains β-lactamase which could otherwise rapidly degrade the ampicillin used to select for the plasmid. In a case where the volume of the inoculum would be less than 1% of the final volume, the wash of the inoculum would not be necessary.

3. Each washed cell suspension was used as inoculum to 25 ml of fresh medium (LB, 0.1% glucose, and 250 mg/liter ampicillin) preheated to 37°. The cultures were grown for 4 hr at 37°. The harvest will, under these conditions, be performed early in stationary growth phase.

4. The cultures were harvested by first cooling to 0° using an ice/water bath followed by centrifugation at 5° at 10,000 g for 20 min. The media were subsequently filtered through a 0.45-μm filter and 25 μl of each medium was precipitated with 100 μl of acetone and saved for the SDS–PAGE analysis.

5. The filtered media were loaded by gravity feed to two separate Pasteur pipet columns packed with 0.4 ml of IgG Sepharose 6 Fast Flow (Pharmacia, Sweden). Each gel matrix was equilibrated with 5 ml of TST [50 mM Tris-HCl, pH 7.4, 150 mM NaCl, and 0.05% (v/v) Tween 20] prior to the loading of the medium. The flow rate through the column is not a major concern as the binding of SPA to IgG is fast[4] (linear flow rates of up to 150 cm/hr have been used on a large scale using the IgG Sepharose 6 Fast Flow resin).

6. The columns were washed with 2 × 15 ml of TST followed by 5 ml of 1 mM CH$_3$COONH$_4$ (ammonium acetate). The latter wash is performed

to lower the buffer capacity of the gel matrix and to get the eluted materials exclusively in volatile buffers. The lowering of the buffer capacity might not be necessary in this experiment as a step elution was performed, but is crucial in a large-scale process to minimize the elution volume (the elution volume of SPA fusions from a saturated IgG Fast Flow Sepharose column could be as little as 10% of the total gel matrix volume under optimal conditions[4]). In the experiment with the SPA–BPTI fusion protein, the two columns were step eluted with 1 ml of 0.5 M CH$_3$COOH titrated to pH 3.3 using CH$_3$COONH$_4$. The eluted materials were lyophilized and one-eighth of each eluate was analyzed by SDS–PAGE (Fig. 4) using standard conditions.[42]

The SDS gel in Fig. 4 shows that the purified protein from cells harboring pRIT23 has a band corresponding in size to the expected value of 35.9 kDa deduced from the amino acid sequence of the mature protein. The protein purified from the culture harboring pRIT:BPTI is larger in size, in the range of the expected value for a full-length protein (41.0 kDa), and the fusion protein shows trypsin inhibitory activity (data not shown). Each medium contains several other proteins in which the SPA protein is only a minor protein component.

If the product is not excreted to the medium, which is often the case, an osmotic shock procedure should be performed to release the SPA fusion product among the periplasmic proteins from the *E. coli* cells. The osmotic shock procedure presented is in essence as was described by Randall and Hardy:[43] (1) Resuspend the cell pellet (from step 4 above) in 5 ml of ice-cold sucrose buffer (0.5 M sucrose, 0.1 M Tris-HCl, pH 8.2, and 1 mM EDTA) and incubate in an ice/water bath for 10 min. (2) Add 80 μl of lysozyme solution (10 mg/ml in H$_2$O) and immediately thereafter add 5 ml of ice-cold H$_2$O. Mix by vortexing. (3) Incubate in ice/water bath for 5 min. (4) Add 180 μl of 1 M MgSO$_4$ to stabilize the spheroplasts. (5) Centrifugate at 10,000 g for 20 min. (6) Load the clear supernatant to the IgG column and continue the protein purification as described above.

Method 3: Expression and Purification of SPA Fusion Proteins from pRIT30 Plasmid Series

The plasmids pRIT31, 32, and 33 are designed for intracellular expression of SPA fusion proteins in *E. coli*. The DNA region 5′ of the *spa* coding sequence contains two separate promoters: the P$_R$ promoter from phage λ situated directly 5′ of the *spa* structural gene and the *lac*UV5 promoter[36] further 5′ of the P$_R$ promoter. It has been shown[28] that this dual-promoter

[42] U. K. Laemmli, *Nature (London)* **227**, 680 (1970).
[43] L. L. Randall and S. J. S. Hardy, *Cell* **46**, 921 (1986).

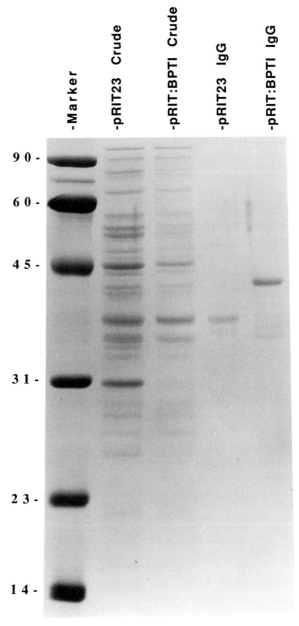

FIG. 4. Electrophoresis in 15% polyacrylamide gels in the presence of SDS of an expression experiment of the SPA–BPTI fusion. Crude refers to 25 μl of growth medium loaded after an acetone precipitation. IgG refers to proteins purified by IgG affinity chromatography as described in Method 2. Molecular weight markers are shown as $\times 10^{-3}$.

alignment can be used to regulate the expression so that three different levels of expression are obtained: (1) the repressed state, (2) induction with IPTG (lsopropyl-β-D-thiogalactopyranoside), which induces only the lacUV5 promoter, and (3) induction of the P_R promoter, e.g., by transferring the culture to a nonpermissive temperature for a temperature-sensitive cl repressor.[34] The relative expression values for the three different levels of expression are about 1:10:50, and the highest expression level often results in refractor body formations.[28] The induction of the P_R and the lacUV5 promoters simultaneously does not result in any further protein expression compared to induction of the P_R promoter alone.[28]

All gene fusion constructions into the pRIT30 plasmids are preferentially made in an *E. coli* host having a cl gene from phage λ [e.g., *E. coli* 514(λ)[34]] in which the P_R promoter is repressed. The expression initiated from the P_R promoter is performed in a host harboring a temperature-sensitive cl repressor expressed from the cl857 gene, e.g., *E. coli* K12$\Delta H_l trp$, cl857.[34]

To expand the number of easily available hosts harboring the cl or cl857 genes, two new plasmids were recently constructed.[28] The plasmids are based on the origin of replication from the low-copy-number plasmid R6K, which makes them compatible with the pRIT30 series of plasmids, and can thus be hosted simultaneously in *E. coli*. One R6K-derived plasmid has the cl gene cloned and is designated pRITcl, and the other has the cl857 gene cloned and is designated pRITcl857. The plasmids are selected for by resistance to kanamycin (40 mg/liter). These new plasmids are very useful and are used by cotransforming them with the pRIT30 expression plasmid to any *E. coli* host of choice.[28]

Expression from a pRIT30 type of plasmid, initiated from the P_R promoter, is performed by first growing a culture of the production plasmid in an *E. coli* cl857 host at a permissive temperature for the mutant repressor (30°) to form a cell mass without growth inhibitory expression, and then transferring the culture to a nonpermissive temperature (42°) in which the transcription from the promoter is induced. The procedure recommended is the following:

1. Grow an overnight culture at 30° in LB supplemented with 250 mg/liter ampicillin and 0.1% glucose. Kanamycin (40 mg/liter) is added if pRITcl857 is used to introduce the cl857 gene.

2. Inoculate fresh medium using the overnight culture, e.g., using 1% of the final growth volume.

3. Incubate at 30° until the absorbance of the culture at 580 nm (OD_{580}) has reached 0.8–1.2; this will take 3–4 hr using a 1% inoculum.

4. Dilute the culture with equal volume of fresh growth medium preheated to 54° and incubate for an additional 1–3 hr at 42°.

5. Harvest and recover the soluble intracellular proteins, e.g., by sonication followed by centrifugation at 5° at 20,000 g for 20 min. If the product is precipitated in a refractor body, procedures to solubilize the precipitated proteins from the pellet would be necessary.[10,11,37] We recommend, however, that the weaker induction described below is performed instead to try to avoid the precipitation. It should, however, be mentioned that the formation of the SPA fusion refractile has been shown to protect some proteolytically sensitive products from degradation.[11] In these cases, the refractor body formation is an advantage for the recovery of the product, and the SPA fusion should therefore be purified from the pellet. (Methods for solubilizing SPA fusion refractiles are not described here; see Refs. 10 and 11 for recommended methods.)

6. Pass the clear crude extract over an IgG Sepharose 6 Fast Flow column. To avoid compression of gel resin during loading of the typically viscous extract, the linear flow rate should not exceed 15 cm/hr. Wash and elute the column as was described for the purification of secreted products in Method 2.

The slightly weaker induction of the SPA fusion is accomplished by inducing transcription only from the lacUV5 promoter. This may be applicable when the expression from the strong P_R promoter results in refractor body formations. When inducing transcription only from the lacUV5 promoter, the plasmid is hosted in a cl E. coli host overproducing the lac repressor from the lacI^Q gene, e.g., JM103.[28] The cl gene could be introduced using pRITcl (as described above). The cell growth is performed at 37°, and the lacUV5 promoter is induced by the addition of IPTG to 1 mM: (1) Grow an overnight culture at 37° of the E. coli host harboring the expression plasmid based on the pRIT30 series in LB supplemented with 0.1% glucose and 250 mg/liter ampicillin. Kanamycin (40 mg/liter) is added if pRITcl is used to introduce the cl gene. (2) Inoculate fresh medium using 1% of the total culture volume from the overnight culture. (3) Incubate at 37° until the OD_{580} has reached 0.8–1.2. (4) Add IPTG to 1 mM and incubate for an additional 1–3 hr at 37°. (5) Harvest and recover the intracellular proteins by sonication and purify the SPA fusions using the IgG Sepharose as described above and in Method 2.

Acknowledgments

The SPA fusion system was developed at the Department of Biochemistry at the Royal Institute of Technology, Stockholm and we wish to thank all members of the department participating in the project. Specifically, we wish to acknowledge Mathias Uhlén and Tomas Moks for their major contributions. The SPA project has additionally received valuable suggestions and advice from outside collaborators, who are gratefully acknowledged: Lennart Philipson (EMBL, Heidelberg), Martin Lindberg (BMC, Uppsala), and KabiGen AB. Furthermore, the authors would like to thank Stephen Anderson, Cara Berman Marks, Paul Carter, and Mark Hurle for critical comments on the manuscript. Paul Carter is also acknowledged for permitting reference to results prior to their publication. Carol Morito is greatly acknowledged for the construction of Figs. 1 and 3. Finally, we thank KabiGen AB, the Swedish National Board of Technical Development, and the Swedish Natural Science Research Council for financial support in several projects which have made the development of the SPA fusion concept possible.

[14] Expression as *trpE* Fusion

By Daniel G. Yansura

Although expression of a fusion protein is not as elegant as a more direct approach, often it is the easiest and quickest means of achieving expression of a particular protein. For many small proteins or peptides, fusions appear to be the only practical way to accumulate useful levels of protein within the cell against losses due to proteolysis. This idea was first established for expression of somatostatin, a small 14-amino acid peptide;[1] however, much larger proteins may also benefit from this principle.[2] A second advantage of a fusion over direct expression is that problems due to poor translation initiation can be essentially avoided.[3] Overcoming such problems can be tedious and, for many situations, a fusion protein may satisfy all the particular needs of the experimenter.

Over the last several years, *trpE* or *trpLE* fusions have been used quite successfully, with more than 30 proteins or peptides reportedly expressed. The high level of expression of most of these fusions can be attributed to two factors: the relatively strong[4] but controllable *trp* promoter, and the insoluble nature of the *trpE* or *trpLE* fusions, which usually leads to proteolytic protection within refractile bodies. In terms of expression, *trpE* and *trpLE* fusions appear to be quite comparable, the main difference being in the level of transcriptional control. While the *trpE* fusions are under both repressor and attenuation control, the *trpLE* fusions retain only repressor control. The *trpLE* fusions have deletions between the end of the leader and distal parts of the *trpE* polypeptide and have been derived

[1] K. Itakura, T. Hirose, R. Crea, A. D. Riggs, H. L. Heyneker, F. Bolivar, and H. W. Boyer, *Science* **198**, 1056 (1977).
[2] A. R. Davis, D. P. Nayak, M. Ueda, A. L. Hiti, D. Dowbenko, and D. G. Kleid, *Proc. Natl. Acad. Sci. U.S.A.* **78**, 5376 (1981).
[3] B. E. Schoner, H. M. Hsiung, R. M. Belagaje, N. G. Mayne, and R. G. Schoner, *Proc. Natl. Acad. Sci. U.S.A.* **81**, 5403 (1984).
[4] H. A. de Boer, L. J. Comstock, D. G. Yansura, and H. L. Heyneker, *in* "Promoter Structure and Function" (R. L. Rodriguez and M. J. Chamberlain, eds.), p. 462. Praeger, New York, 1982.

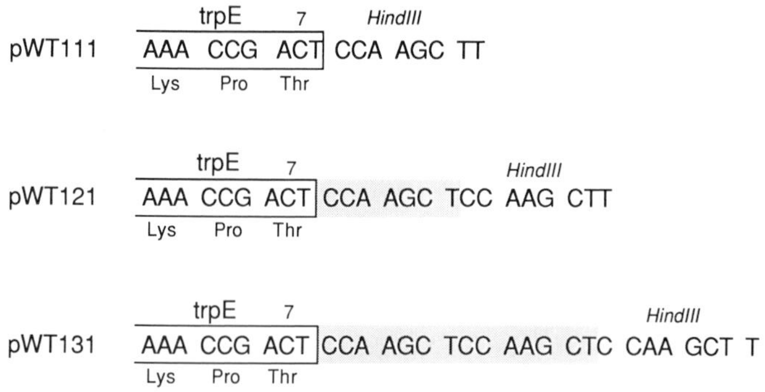

```
                    trpE      7      HindIII
    pWT111    AAA  CCG  ACT  CCA  AGC  TT
              Lys  Pro  Thr


                    trpE     7                HindIII
    pWT121    AAA  CCG  ACT  CCA  AGC  TCC  AAG  CTT
              Lys  Pro  Thr


                    trpE     7                          HindIII
    pWT131    AAA  CCG  ACT  CCA  AGC  TCC  AAG  CTC  CAA  GCT  T
              Lys  Pro  Thr
```

FIG. 1. Partial coding sequence of the *trpE* protein followed by *Hind*III restriction sites in three different reading frames for expression vectors pWT111, pWT121, and pWT131.[16] Numbers represent the amino acid positions of the *trpE* protein.

from the deletions *trpΔLE 1413* and *trpΔLE* 1417.[5] A more detailed review on the *trp* promoter and its use in expressing cloned genes was published by Nichols and Yanofsky in 1983.[6]

Designing Fusion

There are two major considerations when designing a *trpE* or *trpLE* fusion: the length of *trpE* polypeptide and the junction between the *trpE* polypeptide and the protein or peptide of interest. The length of *trpE* can be a significant factor in determining the levels of expression. In general, the longer *trpE* fusions give higher levels of expression.[7,8] This may not be just a matter of increasing the mass of the total fusion, but probably involves better precipitation of the longer fusions within the cell, thus avoiding proteolysis. Comparison of relatively short versus long *trpE* lengths demonstrates this tendency whereby the shorter fusions are more soluble than the larger ones.[9,10] The length of *trpE* or *trpLE* polypeptide necessary to achieve maximum expression is probably in the range of 100–300 amino acids. Besides expression levels, however, producing a more soluble fusion may be a more important factor, depending on the ultimate use of the fusion. If one is interested in some enzymatic activity associated with the protein of interest, a shorter, more soluble fusion will almost certainly be more useful. For maximal enzymatic activity in gen-

[5] G. F. Miozzari and C. Yanofsky, *J. Bacteriol.* **133,** 1457 (1978).

[6] B. P. Nichols and C. Yanofsky, this series, Vol. **101,** p. 155.

[7] R. Derynck, A. B. Roberts, M. E. Winkler, E. Y. Chen, and D. V. Goeddel, *Cell* **38,** 287 (1984).

[8] P. R. Szoka, A. B. Schreiber, H. Chan, and J. Murthy, *DNA* **5,** 11 (1986).

[9] N. Tanese, M. Roth, and S. P. Goff, *Proc. Natl. Acad. Sci. U.S.A.* **82,** 4944 (1985).

[10] C. A. Chlan, C. Coulter, and L. T. Feldman, *J. Virol.* **61,** 1855 (1987).

```
     trpE     517   EcoRI     PstI        EcoRI
  CAT GCA CAG GAA TTC TGC AGA ATT C
   His   Ala   Gln
```

FIG. 2. Partial coding sequence for the *trpLE* protein of plasmid pNCV with the *Eco*RI and *Pst*I restriction sites available for cloning. The number represents the amino acid position in the *trpE* protein and not that of the LE. Note that this LE consists of the first nine amino acids of the *trp* leader fused to amino acids 339–517 of the *trpE* polypeptide.

eral, the lengths of *trpE* polypeptide in the fusions have been in the range of 17–42 amino acids,[9–11] although much larger fusions also retained some activity.

The junction between the *trpE* polypeptide and the protein of interest can be designed to allow cleavage of the fusion at this point. There are a number of cleavable fusions reported using both chemical[7,8,12] and enzymatic[13,14] means. Because many of the *trpE* fusion proteins are soluble only in the presence of chaotropic agents, cleavage by chemical means, if feasible, is usually preferred. Enzymatic cleavage, however, can be used in some cases if solubility problems can be overcome. Usually, the shorter *trpE* fusions are more amenable to this type of cleavage.

Plasmids and DNA Construction

The *trp* promoter and the gene encoding the fusion protein are usually carried on a multicopy plasmid for maximum expression. $TrpR^+$ strains of *Escherichia coli* should be used in plasmid construction and expression experiments to help ensure plasmid stability;[6] strains HB101 and MM294 have been commonly used. There are a number of plasmids that can be used as a source of the *trp* promoter and *trpE* or *trpLE* coding sequences. One simply ligates the gene for the protein of interest, in frame, to some point in the *trpE* gene, using a convenient restriction site. Synthetic DNA is now readily available and can be used to link nonoverlapping restriction sites in the correct reading frame and also provide the desired junction. The complete nucleotide sequence of the *trp* operon has been published,[15] and can be used to generate restriction sites within the *trpE* gene.

[11] I. Sadowski and T. Pawson, *Oncogene* **1**, 181 (1987).
[12] T. C. Furman, J. Epp, H. M. Hsiung, J. Hoskins, G. L. Long, L. G. Mendelsohn, B. Schoner, D. P. Smith, and M. C. Smith, *Bio/Technology* **5**, 1047 (1987).
[13] G. Allen, M. D. Winther, C. A. Henwood, J. Beesley, L. F. Sharry, J. O'Keefe, J. W. Bennett, R. E. Chapman, D. E. Hollis, B. A. Panaretto, P. Van Dooren, R. W. Edols, A. S. Inglis, P. C. Wynn, and G. P. M. Moore, *J. Biotechnol.* **5**, 93 (1987).
[14] T. Imai, T. Cho, H. Takamatsu, H. Hori, M. Saito, T. Masuda, S. Hirose, and K. Murakami, *J. Biochem.* **100**, 425 (1986).
[15] C. Yanofsky, T. Platt, I. P. Crawford, B. P. Nichols, G. E. Christie, H. Horowitz, M. Van Cleemput, and A. M. Wu, *Nucleic Acids Res.* **9**, 6647 (1981).

<u>trpE</u>   323 *SmaI*   *BamHI*   *SalI*   *PstI*   *HindIII*
ATT GAG ATC CCC GGG GAT CCT CTA GAG TCG ACC TGC AGC CCA AGC TTA

FIG. 3. Sequence of the multiple cloning region of plasmid pATH2.[18] The number represents the amino acid position of the *trpE* protein.

Some plasmids, the so-called *trpE* expression vectors, have been rather useful in constructing specific fusion expression plasmids. These vectors have convenient restriction sites engineered at various distances into the *trpE* gene, often with more than one coding phase available for cloning. For short *trpE* fusions, vectors pWT111, pWT121, and pWT131 have been constructed.[16] These contain the *trp* promoter, leader, and coding sequence for the first seven amino acids of the *trpE*, followed by a *Hin*dIII restriction site in all three reading frames (Fig. 1). For intermediate length fusions, pNCV[17] which carries the deletion *trpΔLE1413*,[5] can be used. This plasmid has *Eco*RI and *Pst*I restriction sites, engineered at the end of the coding sequence for 188 amino acids of *trpLE*. This *trpLE* consists of the first nine amino acids of the leader fused to amino acids 339–517 of the *trpE* polypeptide (Fig. 2). For longer *trpE* fusions, the vector pATH2 has been extensively used.[18] This vector contains the *trp* promoter, leader, and coding sequence for the first 323 amino acids of *trpE*, followed by the multiple cloning region of pUC12 (Fig. 3).

trpE and *trpLE* fusion expression plasmids can be maintained in *trpR+* strains provided sufficient tryptophan is present in the media. In minimal media, cultures containing these plasmids should normally be supplemented with tryptophan to approximately 20 μg/ml.

trp Induction

Induction of the *trpE* or *trpLE* fusion generally involves tryptophan starvation of bacterial cultures containing the expression plasmids. Because some plasmids are unstable in their host, it is advisable to start with bacterial cells recently transformed with the appropriate expression plasmids. Such cells are usually grown overnight to saturation in a rich medium, such as LB,[19] containing the appropriate antibiotic. The overnight

[16] W. Tacon, N. Carey, and S. Emtage, *Mol. Gen. Genet.* **177**, 427 (1980).
[17] T. Maniatis, E. F. Fritsch, and J. Sambrook, *in* "Molecular Cloning: A Laboratory Manual," p. 426. Cold Spring Harbor Laboratory, Cold Spring Harbor, New York, 1982.
[18] C. L. Dieckmann and A. Tzagoloff, *J. Biol. Chem.* **260**, 1513 (1985).

1 2

FIG. 4. SDS–polyacrylamide gel analysis of the expression of a *trpLE* fusion. *Escherichia coli,* strain W3110, containing plasmid pBR322 and a plasmid encoding a *trpLE*–relaxin A chain fusion, was grown overnight in LB. The cultures were diluted 100-fold into minimal medium containing 3-indoleacrylic acid and grown for 7 hr. Samples of these cultures equivalent to 1 ml with an absorbance at 600 nm of 1 were then removed for analysis. The pelleted cells were resuspended in 250 μl of 0.1 M 2-mercaptoethanol, 2% (w/v) SDS, 60 mM Tris (pH 6.8), 10% (v/v) glycerol, heated to 95° for 2 min, and loaded onto an SDS–polyacrylamide gel. After electrophoresis, the gel was stained with Coomassie brilliant blue. Lane 1 shows the control culture containing pBR322. Lane 2 shows the culture containing the *trpLE*–relaxin A chain fusion with an arrow pointing to the fusion protein band.

culture is then diluted 25 to 50-fold into M9[19] with 0.5% (w/v) casamino acids and the appropriate antibiotic. After 1–2 hr growth, 3-indoleacrylic acid[20] is added to a final concentration of 25 μg/ml. Growth is then continued for 2–5 hr, at which time samples may be removed to assay for the particular fusion protein.

Samples removed from induced cultures are usually analyzed by SDS–polyacrylamide gels (Fig. 4). Expression levels of *trpE* or *trpLE* fusions are usually quite high, and the fusion can then be visualized after staining with Coomassie brilliant blue. For many fusions, the appearance of a new protein band on the gel after staining is the only way to verify and estimate the level of expression. One should therefore include on the gel samples of

[19] J. H. Miller, *in* "Experiments in Molecular Genetics," p. 431. Cold Spring Harbor Laboratory, Cold Spring Harbor, New York, 1972.
[20] W. F. Doolittle and C. Yanofsky, *J. Bacteriol.* **95,** 1283 (1968).

induced cultures containing the parent *trpE* or *trpLE* expression plasmid as a control. If the fusion protein is to be used to raise antibodies against the protein of interest, one can purify the fusion by preparative SDS gel electrophoresis once the prospective protein band has been identified.[21-23]

Comments

Occasionally, one cannot successfully express a protein of interest as a *trpE* or *trpLE* fusion. Usually, the bacterial cultures for these fusions stop growing completely when induced, and no new protein band can be seen on an SDS–polyacrylamide gel analysis of a total cell lysate. The hepatitis B surface antigen, for example, when fused to 190 amino acids of *trpLE,* displayed this type of expression problem. Removal of the last 24 amino acids of the hepatitis surface antigen, however, restored expression to a high level.[24] It is not clear what the nature of this problem is, but for many fusions such trimming of the protein of interest may be an acceptable alternative.

[21] L. K. Pape, T. J. Koerner, and A. Tzagoloff, *J. Biol. Chem.* **260,** 15362 (1985).
[22] D. G. Kleid, D. Yansura, B. Small, D. Dowbenko, D. M. Moore, M. J. Grubman, P. D. McKercher, D. O. Morgan, B. H. Robertson, and H. L. Bachrach, *Science* **214,** 1125 (1981).
[23] I. Sadowski, J. C. Stone, and T. Pawson, *Mol. Cell. Biol.* **6,** 4396 (1986).
[24] D. Yansura and D. G. Kleid, unpublished observation (1980).

[15] Engineering *Escherchia coli* to Secrete Heterologous Gene Products

By Joan A. Stader and Thomas J. Silhavy

The development of recombinant DNA technology has revolutionized the way we approach research problems in almost all biological systems, but its use in applied technology still favors only a few organisms and cell types. Certainly in terms of prokaryotes, *Escherichia coli* is one of the most favored organisms for many applications because of the ease with which the genome can be manipulated. Controlling factors such as promoter efficiency and regulation, transformation efficiency, protease levels, and the secretion of protein into the growth medium, is within our current capability. Still, despite all the progress, in many instances success is achieved empirically and the practice can best be described as an art. In this article, we attempt to bring to light some of the problems encountered in secreting heterologous proteins from *E. coli* and discuss some of the successes that have been achieved to date.

In some contexts, the term "protein secretion" has been used loosely in studies with *E. coli* to describe localization of proteins into the periplasm, the outer membrane, or the medium. In this article, we restrict our usage of the term "secretion" to mean localization of a protein into the growth medium. Localization to other extracytoplasmic compartments of the bacteria will be referred to as "export."

Why Secrete Proteins in *Escherichia coli?*

The ability to secrete heterologous proteins from *E. coli* is desirable for a number of reasons. First, industrial fermentations aimed at the production of a particular protein generally give a higher yield of product when the purification scheme is simplified. Because *E. coli* normally does not secrete protein, engineering a heterologous protein for secretion can allow for the use of continuous cultures producing little else but that protein in culture supernatants. In this regard, export of the protein in question from the cytoplasm can also simplify purification, albeit to a lesser degree, relative to purification from whole-cell extracts.

Second, export or secretion can lead to recovery of products with the correct primary amino acid structure. Many proteins translated in *E. coli* retain the initiating methionine while those from eukaryotic cells do not. The presence of fMet on a protein that normally does not contain it may result in a decrease in biological activity or may cause the protein to act as an immunogen if introduced into the original organism. By engineering a functional signal sequence onto the gene, the fMet will be removed during the proteolytic processing event that accompanies translocation of the protein across the membrane.

Third, biological activity is dependent on whether or not the protein has been correctly folded, and certain proteins may assemble poorly, if at all, in the cytoplasm of *E. coli*. For example, many proteins of commercial interest contain disulfide bonds and the *E. coli* cytoplasm does not seem to be favorable for disulfide bond formation.[1] However, it is known that secretion of the protein, or export into the periplasm, can provide a natural environment for this process.[1-3]

Fourth, many proteins produced in *E. coli* are unstable owing to proteolytic degradation. Secretion of the protein can relieve this problem by physically separating the protein from the harmful proteases. Here again, export to the periplasm can help, since the class of proteases present in this

[1] S. Pollitt and H. Zalkin, *J. Bacteriol.* **153**, 27 (1983).
[2] H. M. Hsiung, N. G. Mayne, and G. W. Becker, *Bio/Technology* **4**, 991 (1986).
[3] M. Better, C. P. Chang, R. R. Robinson, and A. H. Horwitz, *Science* **240**, 1041 (1988).

compartment differs from those present in the cytoplasm[4] and mutants defective in a major periplasmic protease (DegP) have been isolated.[4a]

Problems Encountered in Secreting Proteins from *Escherichia coli*

The major problems encountered in attempting to export or secrete foreign proteins from *E. coli* are inefficient or incomplete translocation of the molecule across the inner membrane, insufficient capacity of the localization machinery, and proteolytic degradation. These problems, separately or in combination can decrease yields substantially. In addition, other factors, including anomalous folding and absent or incorrect posttranslational modification, can decrease the biological activity of the molecule in question. Most of these problems are minimal when the protein expressed is a naturally secreted bacterial exoprotein. Thus, many bacterial toxins, extracellular proteases, kinases, lyases, and nucleases have been completely or partially secreted in biologically active form with reasonable yields. When the protein is of eukaryotic origin, secretion is generally more problematic, even if the protein has evolved to be translocated through a membrane.

Efforts to export or secrete small eukaryotic molecules such as growth factors and hormones have met with greater success than those of large, complex, and multimeric polypeptides. The explanation is not clear, but it seems that the endoplasmic reticulum (ER), which is lacking in *E. coli,* plays an active role in carrying out essential posttranslational modifications, folding, and assembly. For example, *E. coli* is incapable of carrying out N- and O-linked glycosylations, a posttranslational modification that normally occurs in the rough ER of eukaryotic cells. In addition, certain complex secreted proteins may also require the assistance of accessory factors present in the ER, such as BiP (an HSP70 homolog),[5-7] to achieve a functional tertiary or quarternary structure. Indeed, certain temporal and spatial requirements normally satisfied in the ER may not be duplicated in *E. coli.*

The precise nature of intragenic export or secretion signals such as the signal sequence is not yet clearly defined. Accordingly, it is not possible to graft such signals onto a particular gene and guarantee efficient export or secretion of the desired product. In certain cases, success is spectacular (see below), thus demonstrating the overall feasibility of the method. Far too often, however, results are disappointing and yields are low.

[4] K. H. S. Swamy and A. L. Goldberg, *J. Bacteriol.* **149**, 1027 (1982).
[4a] K. L. Strauch, K. Johnson, and J. Beckwith, *J. Bacteriol.* **171**, 2689 (1989).
[5] D. G. Bole, L. M. Hendershot, and J. F. Kearney, *J. Cell Biol.* **102**, 1558 (1986).
[6] S. Munro and H. R. B. Pelham, *Cell* **46**, 291 (1986).
[7] M. J. Gething, K. McCammon, and J. Sambrook, *Cell* **46**, 939 (1986).

If the protein is not normally secreted, it may contain sequences that are incompatible with the cellular export apparatus, regardless of whether the protein is eukaryotic or prokaryotic. Attaching a signal sequence to some of these proteins has resulted in killing the host cell. The classic example of the jamming phenomenon is β-galactosidase fusion proteins that carry functional export signals.[8-10] Again, because this incompatibility is not understood in molecular terms, accurate predictions are not possible.

Proteolysis is a major problem in the expression of heterologous proteins in *E. coli* in general. As noted above, export or secretion can provide a solution. In certain cases, however, attempts to secrete the protein can exacerbate the problem.[11,12] Apparently, the occurrence of proteolysis is very much dependent on the conformation characteristics of the protein itself, and targeting it to the export machinery may or may not provide a solution.

Overloading the export machinery may result from inefficient secretion of a foreign protein or because the protein is expressed at levels that simply exceed cellular capacity. Overloading prevents efficient export of normal *E. coli* proteins and the foreign protein causing the precursor forms of these molecules to accumulate in the cytoplasm.[2] Often, the accumulation of foreign proteins in the cytoplasm leads to the induction of a stress response, which in turn may increase the levels of active proteases in the cell.[13] Cell death can result if the export block is sufficiently severe.

To develop a better appreciation for the problems encountered in the export or secretion of foreign proteins in *E. coli* and their potential solutions, an understanding of the salient features of the organism's protein targeting pathways will be useful. The following is a summary of what is currently known about these various pathways in *E. coli*. We also describe some attempts at secreting foreign proteins that we think have contributed to our body of knowledge regarding this area of genetic engineering.

Protein Targeting Pathways in *Escherichia coli*

As noted above, *E. coli* does not normally secrete proteins into the growth media efficiently and it is not clear that the organism can do so without added genetic information or damage to the cell envelope. How-

[8] S. D. Emr, M. Schwartz, and T. J. Silhavy, *Proc. Natl. Acad. Sci. U.S.A.* **75**, 5802 (1978).
[9] S. D. Emr and T. J. Silhavy, *J. Mol. Biol.* **141**, 63 (1980).
[10] P. J. Bassford, Jr., T. J. Silhavy, and J. R. Beckwith, *J. Bacteriol.* **139**, 19 (1979).
[11] R. Gentz, Y. Kuys, C. Zwieb, D. Taatjes, H. Taatjes, W. Bannwarth, D. Stueber, and I. Ibrahimi, *J. Bacteriol.* **170**, 2212 (1988).
[12] K. L. Strauch and J. Beckwith, *Proc. Natl. Acad. Sci. U.S.A.* **85**, 1576 (1988).
[13] K. Ito, Y. Akiyama, T. Yura, and K. Shiba, *J. Bacteriol.* **167**, 201 (1986).

ever, certain pathogenic variants and strains carrying genes from other gram-negative bacteria can secrete proteins specifically and with high efficiency, and these specialized secretion pathways offer promising opportunities for the genetic engineer. Because some of the specialized systems utilize components of the general protein export pathway, it is necessary to summarize here the relevant features of normal protein export.

General Pathway

The *E. coli* cell envelope consists of two concentric membranes (the inner and outer membranes) separated by an aqueous gel (the periplasm) containing the peptidoglycan cell wall. Proteins destined for the periplasm and the outer membrane are thought to utilize all or part of the general pathway (for reviews, see Refs. 14 and 15). A protein must contain a signal sequence in order to be translocated across the inner membrane lipid bilayer. The signal sequence lies at the N terminus of the polypeptide and is proteolytically removed during translocation. This processing reaction is carried out by one of two signal peptidases that reside on the periplasmic side of the inner membrane. Signal peptidase I, the product of the *lep* gene, cleaves signal sequences preferentially on the C-terminal side of an Ala-X-Ala recognition site.[16-18] Signal peptidase II, the product of the *lsp* gene, removes signal sequences from lipoproteins (polypeptides that have been modified by a glycerol fatty acid at the cysteine residue occupying position 1 of the mature sequence.[18a]

In addition to the signal sequence and modified cysteine in lipoproteins, parts of the mature sequence of the protein are necessary, at least in certain cases, for efficient translocation from the cytoplasm.[19-22] It is not clear what role(s) these mature sequences play in the process. In certain cases, components of the export machinery recognize and interact with these sequences and this interaction prevents stable folding prior to translocation across the lipid bilayer and perhaps serves to pilot the precursor to the membrane-bound components of the export machinery.[23-23b]

[14] K. Baker, N. Mackman, and I. B. Holland, *Prog. Biophys. Mole. Biol.* **49**, 89 (1987).
[15] L. L. Randall, S. J. S. Hardy, and J. R. Thom, *Annu. Rev. Microbiol.* **41**, 507 (1987).
[16] G. von Heijne, *Eur. J. Biochem.* **133**, 17 (1983).
[17] D. Perlman and H. O. Halvorson, *J. Mol. Biol.* **167**, 391 (1983).
[18] P. B. Wolfe, W. Wickner, and J. M. Goodman, *J. Biol. Chem.* **258**, 12073 (1983).
[18a] F. Yu, H. Yamada, K. Daishima, and S. Mizushima, *FEBS Lett.* **173**, 264 (1984).
[19] M. N. Hall, M. Schwartz, and T. J. Silhavy, *J. Mol. Biol.* **156**, 93 (1982).
[20] J. Tommassen, H. van Tol, and B. Lugtenberg, EMBO J. **2**, 1275 (1983).
[21] C. S. Hoffman and A. Wright, *Proc. Natl. Acad. Sci. U.S.A.* **82**, 5107 (1985).
[22] B. A. Rasmussen and T. J. Silhavy, *Genes Dev.* **1**, 185 (1987).
[23] D. N. Collier, V. A. Bankaitis, J. B. Weiss, and P. J. Bassford, *Cell* **53**, 273 (1988).
[23a] C. A. Kumamoto and P. M. Gannon, *J. Biol. Chem.* **263**, 11554 (1988).
[23b] M. Watanabe and G. Blobel, *Cell* **58**, 695 (1989).

Besides the signal peptidases, additional extragenic loci have been shown to specify components of the general targeting pathway (for review, see Ref. 24). The *sec* genes (A,[25] B,[26] D,[27] E,[28] and Y[29]) were identified through the characterization of mutations that cause the accumulation of precursor forms of localized proteins. The *prl* genes (A,[30] C,[30] D,[31] and G[31a]) were identified as suppressors of signal sequence mutations. Following the initial selections, it was found that the *secY* and *prlA* mutations map to the same gene,[29,32-34] which we will refer to as *prlA;* also, recent data suggest that *secA* and at least certain alleles of *prlD* are allelic,[34a] and that *secE* and *prlG* are allelic as well.[31a,34b] To date, more is known about PrlA, SecA, and SecB than the other gene products primarily because of available biochemical data verifying their role in the export process.[23-23b,34c,d]

The *prlA* gene is essential and is located in the *spc* operon,[32] which encodes several ribosomal proteins. PrlA, a hydrophobic protein of 49 kDa, is not part of the ribosome but is located in the inner membrane.[35] The *prlA4* allele was originally isolated as a suppressor of a signal sequence mutant in *lamB*[30] and later shown to suppress other signal sequence mutations both in *lamB* and in other localized proteins,[36,37] suggesting interaction with signal sequences. The *prlA4* allele shows no detectable phenotype with wild-type proteins carrying wild-type signal sequences and may function by having broadened specificity. The location of PrlA in the membrane suggests that it may have a role in the translocation of proteins across

[24] S. A. Benson, M. N. Hall, and T. J. Silhavy, *Annu. Rev. Biochem.* **54**, 101 (1985).

[25] D. B. Oliver and J. Beckwith, *Cell* **25**, 765 (1981).

[26] C. A. Kumamoto and J. Beckwith, *J. Bacteriol.* **154**, 253 (1983).

[27] C. Gardel, S. Benson, J. Hunt, S. Michaelis, and J. Beckwith, *J. Bacteriol.* **169**, 1286 (1987).

[28] P. D. Riggs, A. I. Darman, and J. Beckwith, *Genetics* **118**, 371 (1988).

[29] K. Ito, M. Wittekind, M. Nomura, K. Shiba, T. Yura, A. Miura, and H. Nashimoto, *Cell* **32**, 789 (1983).

[30] S. D. Emr, S. Hanley-Way, and T. J. Silhavy, *Cell* **23**, 79 (1981).

[31] V. A. Bankaitis and P. J. Bassford, Jr., *J. Bacteriol.* **161**, 169 (1985).

[31a] J. Stader, L. J. Gansheroff, and T. J. Silhavy, *Genes Dev.* **3**, 1045.

[32] J. Shultz, T. J. Silhavy, M. L. Berman, N. Fiil, and S. D. Emr, *Cell* **31**, 227 (1982).

[33] D. P. Cerretti, D. Dean, G. R. Davis, D. M. Bedwell, and M. Nomura, *Nucleic Acids Res.* **11**, 2599 (1983).

[34] K. Shiba, K. Ito, T. Yura, and D. P. Cerretti, *EMBO J.* **3**, 631 (1984).

[34a] J. D. Fikes and P. J. Bassford, Jr., *J. Bacteriol.* **171**, 402 (1989).

[34b] P. J. Schatz, P. D. Riggs, A. JacQ, M. J. Fath, and J. Beckwith, *Genes Dev.* **3**, 1035 (1989).

[34c] J. P. Fandl and P. C. Tai, *Proc Natl. Acad. Sci. U.S.A.* **84**, 7448 (1987).

[34d] R. J. Cabelli, L. Chon, P. C. Tai, and D. B. Oliver, *Cell* **55**, 683 (1988).

[35] K. Ito, *Mol. Gen. Genet.* **197**, 204 (1984).

[36] S. D. Emr and P. J. Bassford, *J. Biol. Chem.* **257**, 5852 (1982).

[37] S. Michaelis, H. Inouye, D. Oliver, and J. Beckwith, *J. Bacteriol.* **154**, 366 (1983).

this barrier, and recent genetic studies support this view.[37a] In addition, these studies suggest that PrlA function may be rate-limiting for protein export in wild-type cells.

The *secA* gene is essential and encodes a 102-kD protein[38,39] that is found both in the cytoplasm and associated with the inner membrane.[38] Regulation of *secA* expression is controlled by signals coming from the general targeting pathway.[38] Thus, when export is compromised by growing conditional mutants under nonpermissive conditions for example, synthesis of SecA increases. There are data suggesting that SecA may interact both with PrlA[40,41] and with the precursor form of the localized protein.[34a,42] Biochemical studies reveal ATPase activity and thus SecA may function in energy coupling.[42a,b]

The nonessential *secB* gene was originally identified in a selection for decreased protein localization.[26] The gene encodes a 12-kD cytoplasmic protein which is required for efficient localization of a subset of outer membrane and periplasmic proteins.[43] SecB does not seem to interact with signal sequences[44] but has been shown to be dependent on sequences within the mature protein, and this interaction can prevent folding of the precursor into an export-incompetent conformation.[23–23b]

Biochemical analysis of protein export using the technique of *in vitro* protein translocation has revealed several additional cytoplasmic factors that facilitate the translocation reaction.[45–47] For the most part, these factors have not been purified nor has precise function been assigned. One notable exception is Trigger Factor, a 68-kD protein that also has antifolding and perhaps piloting activities.[46] The gene specifying Trigger Factor has not yet been identified.

[37a] K. L. Bieker and T. J. Silhavy, *Proc. Natl. Acad. Sci. U.S.A.* **86**, 968 (1989).
[38] D. B. Oliver and J. Beckwith, *Cell* **30**, 311 (1982).
[39] M. G. Schmidt, E. E. Rollo, J. Grodberg, and D. B. Oliver, *J. Bacteriol.* **170**, 3404 (1988).
[40] E. R. Brickman, D. B. Oliver, J. L. Garwin, C. Kumamoto, and J. Beckwith, *Mol. Gen. Genet.* **196**, 24 (1984).
[41] J. P. Fandl, R. Cabelli, D. Oliver, and P. C. Tai, *Proc. Natl. Acad. Sci. U.S.A.* **85**, 8953 (1988).
[42] K. Baker, N. Mackman, M. Jackson, and I. B. Holland, *J. Mol. Biol.* **198**, 693 (1987).
[42a] K. Cunningham, R. Lill, E. Crooke, M. Rice, K. Moore, W. Wickner, and D. Oliver, *EMBO J.* **8**, 955 (1989).
[42b] R. Lill, K. Cunningham, L. A. Brondage, K. Ito, D. Oliver, and W. Wickner, *EMBO J.* **8**, 961 (1989).
[43] C. A. Kumamoto and J. Beckwith, *J. Bacteriol.* **163**, 267 (1985).
[44] N. J. Trun, J. Stader, A. Lupas, C. Kumamoto, and T. J. Silhavy, *J. Bacteriol.* **170**, 5928 (1988).
[45] M. Müller and G. Blobel, *Proc. Natl. Acad. Sci. U.S.A.* **81**, 7737 (1984).
[46] E. Crooke and W. Wickner, *Proc. Natl. Acad. Sci. U.S.A.* **84**, 5216 (1987).
[47] Q. Weng, L. Chen, and P. C. Tai, *J. Bacteriol.* **170**, 126 (1988).

The general targeting pathway described above for *E. coli* is analogous in many ways to the eukaryotic protein secretion pathway originally identified by Palade (see review, Ref. 48) and described in greater detail by the "signal hypothesis."[49] This mechanistic similarity is most graphically demonstrated by the observed recognition of eukaryotic secreted proteins by the export machinery of *E. coli* and vice versa.[50-52] Indeed, it is this evolutionary conservation of mechanism that lead initially to the concept that *E. coli* could serve as a vehicle for the synthesis and export of eukaryotic secretory proteins.

Hemolysin Secretion Pathway

Hemolysin is secreted by certain pathogenic strains of *E. coli* that infect the urinary tract of humans or intestinal tract of animals.[53] The hemolysin determinant, which is plasmid borne, consists of four genes, *hlyC, hlyA, hlyB,* and *hlyD*. The *hlyA* gene encodes hemolysin, a 107-kDa protein which is efficiently secreted into the medium. Secretion is dependent on *hlyB* and *hlyD* and independent of *hlyC*, which is required for a posttranslational modification to activate hemolysin. Unlike all of the other exoproteins that will be discussed here, hemolysin does not contain an N-terminal signal sequence and does not seem to rely in any way on the general targeting pathway. In contrast, hemolysin contains a sequence at its C terminus[54] which has been shown by mutational inactivation to be required for its secretion. The sequence is not proteolytically removed in the same way that N-terminal sequences are, nor does it seem to function in a manner similar to the C terminus of proteases such as IgA₁ that are described in a following section. This sequence apparently targets the hemolysin to an export machinery dedicated exclusively to its secretion. The machinery is composed of the three other *hly* gene products. Two polypeptide products (66 and 46 kD) of the *hlyB* gene result from alternative translational starts within an overlapping reading frame[55,56] and contain a conserved sequence corresponding to an ATP-binding site. The gene products have been found to reside in the inner membrane.[56] The *hlyD*

[48] G. Palade, *Science* **189,** 347 (1975).
[49] G. Blobel and B. Dobberstein, *J Cell Biol.* **67,** 835 (1975).
[50] T. H. Fraser and B. J. Bruce, *Proc. Natl. Acad. Sci. U.S.A.* **75,** 5936 (1978).
[51] K. Talmadge, S. Stahl, and W. Gilbert, *Proc. Natl. Acad. Sci. U.S.A.* **77,** 3369 (1980).
[52] M. Müller, I. Ibrahimi, C. N. Chang, P. Walter, and G. Blobel, *J. Biol. Chem.* **257,** 11860 (1982).
[53] R. A. Welch, R. Hull, and S. Falkow, *Infect. Immun.* **42,** 178 (1983).
[54] L. Gray, N. Mackman, J. M. Nicaud, and I. B. Holland, *Mol. Gen. Genet.* **205,** 127 (1986).
[55] T. Felmlee, S. Pollitt, and R. Welch, *J. Bacteriol.* **163,** 94 (1985).
[56] N. Mackman, J. M. Nicaud, L. Gray, and I. B. Holland, *Mol. Gen. Genet.* **201,** 529 (1985).

gene product is a 53-kDa protein that also is a component of the machine, and has tentatively been localized primarily to the inner membrane with residual amounts in the outer membrane fraction.[54,56]

The mechanism for secretion of hemolysin is not yet clear. Earlier reports stated that there was a periplasmic intermediate in the pathway,[57] but more recent studies have revealed that this is probably not the case.[54] A current model (B. Holland, personal communication) suggests that a structural domain within the C-terminal 50 amino acids of HlyA is recognized by the secretion-specific HlyB and D proteins followed by the ATP-dependent extrusion of HlyA through a transenvelope channel.

Escherichia coli Enterotoxins

Escherichia coli produces a heat-labile (LT) and a heat-stable (ST) enterotoxin. Since these molecules act outside the organism, some mechanism for exit must exist. However, as the data presented below indicate, this mechanism does not involve a specific secretion pathway.

LT is a multimeric complex composed of one A subunit and five B subunits.[58-60] Each subunit contains an N-terminal signal sequence[61-63] and Hirst *et al.*[64] showed that *E. coli* localizes both subunits to the periplasm most likely via the general export pathway. Yamamoto and Yokota[65] demonstrated that only about 10% of the LT is released from *E. coli* into the culture medium and that it is necessary to have the B subunit present for LT release (i.e., B subunit alone can b released, but A subunit requires B subunit for release). Hirst *et al.*[64] speculate that conditions present in the intestine (low pH, low oxygen, bile salts) may bring about the leakage of LT out of the periplasm.

The cholera toxin of *Vibrio cholerae* is highly homologous to LT at the level of the DNA,[61,66] protein complex composition,[58,67,68] and enzyme activity in the mammalian host intestine.[68] In fact, antibody to cholera

[57] N. Mackman and I. B. Holland, *Mol. Gen. Genet.* **196**, 129 (1984).
[58] J. D. Clements and R. A. Finklestein, *Infect. Immun.* **24**, 760 (1979).
[59] D. M. Gill, J. D. Clements, D. C. Robertson, and R. A. Finkelstein, *Infect. Immun.* **33**, 677 (1981).
[60] S. L. Kunkle and D. C. Robertson, *Infect. Immun.* **25**, 586 (1979).
[61] W. S. Dallas and S. Falkow, *Nature (London)* **288**, 499 (1980).
[62] E. K. Spicer and J. A. Noble, *J. Biol. Chem.* **257**, 5716 (1982).
[63] T. Yamamoto, T. A. Tamura, M. Ryoji, A. Kaji, T. Yokota, and T. Takano, *J. Bacteriol.* **152**, 506 (1982).
[64] T. R. Hirst, L. L. Randall, and S. J. S. Hardy, *J. Bacteriol.* **157**, 637 (1984).
[65] T. Yamamoto and T. Yokota, *J. Bacteriol.* **150**, 1482 (1982).
[66] E. K. Spicer, W. M. Kavanaugh, W. S. Dallas, S. Falkow, W. H. Konigsberg, and D. E. Schafer, *Proc. Natl. Acad. Sci. U.S.A.* **78**, 50 (1981).
[67] D. M. Gill, *Biochemistry* **15**, 1242 (1976).
[68] W. S. Dallas and S. Falkow, *Nature (London)* **277**, 406 (1979).

toxin also precipitates LT. Interestingly, *V. cholerae* is much more efficient at secreting LT than is *E. coli*. Hirst and Holmgren[69] reported 65 to 85% of the enterotoxin was found extracellularly in *V. cholerae* during exponential growth. Conversely, Pearson and Mekalanos[70] cloned and expressed the genes for the cholera toxin A and B subunits in *E. coli* and found that the toxin secretion was very inefficient but parallels that observed for LT. This comparison seems to suggest that *V. cholerae* possesses a secretion function not present in *E. coli*. Consistent with this possibility, a mutant of *V. cholerae* is found to accumulate cholera toxin in the periplasm,[71] and this mutation maps outside the toxin structural genes. When LT is expressed in the mutant strain, it also is not secreted.[72]

The ST enterotoxin contains a single subunit of small size (about 5 kDa) that is also synthesized with a classical N-terminal signal sequence[73] that accumulates in the periplasm in a latent form and is released in the media with an efficiency of 70%.[74] Although the mechanism of ST secretion is not known, it may be dependent on the small size of the molecule.

Colicin Secretion

Colicins are a class of bacteriocins encoded on naturally occurring plasmids. In addition to a colicin gene, the plasmid generally carries genes encoding an immunity protein and a release (or "lysis") protein which is required for colicin release. Some colicins become complexed with the immunity protein[75,76] in the cytoplasm[77] prior to release. Both the colicin and the immunity proteins lack N-terminal signal sequences and are synthesized on free ribosomes. The lysis protein is a lipoprotein located in the membrane and synthesized with a cleavable signal sequence.[78,79]

We do not consider the release of colicins into the medium to be true secretion because it is nonspecific and associated with envelope damage. The presence of the lysis protein in the cell envelope activates a phospholipase which results in the release of free fatty acids and subsequent changes in outer membrane permeability.[80]

[69] T. R. Hirst and J. Holmgren, *J. Bacteriol.* **169,** 1037 (1987).

[70] G. D. N. Pearson and J. J. Mekalanos, *Proc. Natl. Acad. Sci. U.S.A.* **79,** 2976 (1982).

[71] R. K. Holmes, M. L. Vasil, and R. A. Finkelstein, *J. Clin. Invest.* **55,** 551 (1975).

[72] R. J. Neill, B. E. Ivins, and R. K. Holmes, *Science* **221,** 289 (1983).

[73] M. So and B. J. McCarthy, *Proc. Natl. Acad. Sci. U.S.A.* **77,** 4011 (1980).

[74] L. M. Guzman-Verduzco, R. Fonseca, and Y. M. Kupersztoch-Portnoy, *J. Bacteriol.* **154,** 146 (1983).

[75] K. Schaller and M. Nomura, *Proc. Natl. Acad. Sci. U.S.A.* **73,** 3989 (1976).

[76] K. S. Jakes and N. D. Zinder, *Proc. Natl. Acad. Sci. U.S.A.* **71,** 3380 (1974).

[77] D. Cavard, A. Bernadic, J.-M. Dagis, and C. Lazdunski, *Biol. Cell* **51,** 79 (1984).

[78] A. P. Pugsley, *Mol. Gen. Genet.* **211,** 335 (1988).

[79] S. T. Cole, B. Saint-Joanis, and A. P. Pugsley, *Mol. Gen. Genet.* **198,** 465 (1985).

[80] A. P. Pugsley and M. Schwartz, *EMBO J.* **3,** 2393 (1984).

The manner in which the colicins cross the cytoplasmic membrane remains a mystery, but it has been suggested[78] that the lysis protein may be directly involved. Mutant colicins that accumulate in the cytoplasm in the presence of the lysis gene exist,[81,82] but it has not been determined whether specific "signals" or tertiary structure requirements are responsible for the specificity of translocation across the inner membrane.

How Do We Define Secretion in *Escherichia coli?*

As evidenced by the lengthy and sometimes confusing literature regarding the secretion of enterotoxins and colicins in *E. coli,* it seems useful to list several criteria which we believe are necessary to document true secretion of an authentic protein product.

1. The recombinant vector used to direct expression of the foreign protein must be shown to contain the gene only. As we have seen, *E. coli* can secrete some proteins only if additional genetic information is provided. In many cases, such genes are closely linked to and are cloned together with the gene of interest. To understand how and why the protein is secreted, it is important to determine if special accessory proteins are involved.

2. Secretion of a recombinant protein must be shown to be specific (i.e., does not result from leakage or cell lysis). This can be accomplished by measuring the distribution of marker proteins in each of the cellular fractions and comparing it to the distribution of the secreted protein. In doing this, one should avoid using enzymatic activity as the sole assay for secreted proteins. Enzymatic activity may be an accurate assay for the presence of extracellular enzymes in the culture supernatant, but the precursor form of the molecule present in other cellular fractions may have little or no activity.

3. It must be demonstrated that appearance of the protein in the culture supernatant does not result from the release of vesicles containing the molecule. This can be determined by treating the supernatant with protease. If the protein is in a vesicle, it will be protected, and if it is free, it will be degraded. If the protein is associated with the outside surface of a vesicle, degradation may occur. Therefore, the result should also be confirmed by high-speed centrifugation, which will pellet the vesicles while the free protein remains in the supernatant.

4. If the protein contains a signal sequence, or if it is processed by another proteolytic event, it is necessary to determine by amino acid

---

[81] M. Mock and M. Schwartz, *J. Bacteriol.* **136,** 700 (1978).
[82] M. Yamada and A. Nakazawa, *Eur. J. Biochem.* **140,** 249 (1984).

sequence analysis and/or peptide mapping that the precursor has been correctly processed.

Reports describing the secretion from *E. coli* of exoproteins from many different sources have appeared, but we are not aware of any studies that have fulfilled rigorously all the criteria listed above.

Prokaryotic Exoprotein Secretion in *Escherichia coli*

Heterologous bacterial proteins that are secreted when expressed in *E. coli* fall into two general classes. Both types contain a typical signal sequence and thus utilize components of the general targeting pathway. The first requires the presence of additional proteins together with the export machinery of *E. coli*. The second class is composed of so-called "self-secreting" proteins. When expressed in *E. coli* they are secreted without the participation of additional gene products. This is remarkable, because *E. coli* normally does not secrete proteins. The explanation for this apparent exception may be that these proteins participate actively in their own secretion. Therefore, in a sense, they provide their own additional functions.

Exoproteins That Require the General Targeting Pathway and Accessory Proteins

The best characterized protein of this class is the pullulanase of *Klebsiella pneumoniae,* the product of the *pulA* gene. This starch-debranching enzyme is synthesized with an N-terminal signal sequence which is processed, and the mature polypeptide is fatty acylated at the amino-terminal cysteine of the mature protein in both *K. pneumoniae* and *E. coli*.[83] Pullulanase is expressed and localized on the outer face of the outer membrane and progressively released into the growth medium. d'Enfert *et al.*[84] have shown that the release of pullulanase into the medium is not accompanied by proteolytic cleavage as in the case of exoproteases, which is described below. To secrete pullulanase in *E. coli* correctly, the minimum amount of chromosomal DNA containing *pulA* that needs to be present is 18.8 kilobases (kb).[85] In addition to containing *pulA*, the plasmid insert contains regions on both sides of *pulA* that direct the production of eight other polypeptides. Some or all of these polypeptides are necessary for secretion of pullulanase. When the genes are mutated or absent, pullu-

---

[83] A. P. Pugsley, C. Chapon, and M. Schwartz, *J. Bacteriol.* **166,** 1083 (1986).

[84] C. d'Enfert, A. Ryter, and A. P. Pugsley, *EMBO J.* **6,** 3531 (1987).

[85] C. d'Enfert and A. P. Pugsley, *Mol. Microbiol.* **1,** 159 (1987).

lanase is correctly processed and modified but is found both in the inner and outer membranes and is inaccessible to external pullulan substrate.[83] The *pulCDEFG* operon lies upstream of *pulA* and is transcribed in the opposite direction.[85a] Like *pulA,* the *pulCDEFG* operon is under maltose regulation in *K. pneumoniae* and *E. coli,* and thus appears to be specialized. The *pulS* gene, which encodes a 12-kDa outer membrane lipoprotein, lies downstream of *pulA* and has its own promoter which is not under maltose regulation.[85a] It is not clear whether the *pulS* gene product functions in a general way in protein secretion in *K. pneumoniae* or if this differential regulation is required to control the temporal expression of PulS with respect to pullulanase and the *pulCDEFG* gene products.

Pugsley and co-workers have used *pulA–phoA* fusions to determine if the pullulanase system can be used to direct secretion of other proteins, in this case alkaline phosphatase. Although secretion was not observed, these investigators did succeed in placing PhoA sequences on the *E. coli* cell surface.[84] The ability to direct protein sequences to the surface may have interesting applications.

Several of the exoproteases of *Erwinia chrysanthemi* have been characterized and shown to be secreted in *E. coli.* These are the PRT protease[86] and proteases A, B, and C.[87] Here again, the regions of the cloned DNA are large and appear to contain additional genes that participate in the secretion process. Although the mechanism of secretion in these cases is not understood, it is clearly different from pullulanase since these proteins are not lipoproteins.

Many proteins that are synthesized with signal peptides and secreted in other organisms have been expressed in *E. coli* and found to be processed but incompletely secreted. In most cases, these proteins are found in the periplasm. An extensive list of these proteins can be found in a recent review by Pugsley.[88] At this point, we can only speculate on the possibility that, like the pathways mentioned here, special accessory proteins are required for secretion.

"Self-Secreting" Proteins

The strongest evidence for self-secreting proteins involves exoproteases that seem to contain a C-terminal domain that directs the preprotein to an outer membrane location. Once localized to the outer membrane, the

[85a] C. d'Enfert and A. P. Pugsley, *J. Bacteriol.* **171,** 3673 (1989).

[86] F. Barras, K. K. Thurn, and A. K. Chatterjee, *FEMS Microbiol. Lett.* **34,** 343 (1986).

[87] C. Wandersman, P. Delepelaire, S. Letoffe, and M. Schwartz, *J. Bacteriol.* **169,** 5046 (1987).

[88] A. P. Pugsley, *in* "Protein Transfer and Organelle Biogenesis," p. 607. Academic Press, New York, 1988.

enzymatically active protease will then cleave itself, freeing the enzyme from the membrane-bound C-terminal region. Two genetic studies on organisms that secrete a variety of exoproteins support the evidence for proteases being self-secreting. One study[89] reported the isolation of *out* mutants *Er. chrysanthemi* that were defective in the secretion of pectate lyases and endocellulases but not for at least one exoprotease. Another study[90] described *xcp* mutants of *Pseudomonas aeruginosa* that were defective in secreting alkaline phosphatase, phospholipase, staphylolytic enzyme, and elastase but were still able to secrete alkaline protease.

Neisseria gonorrhoeae is a pathogenic gram-negative bacterium that secretes extracellular proteases specific for human immunoglobulin A (IgA). Halter *et al.*[91] have shown that the *N. gonorrhoeae* MS11 Iga$_1$ protease can be efficiently produced and secreted in *E. coli*. It was demonstrated that the plasmid insert contained only the gene encoding the protease. Characterization of the secretion of IgA$_1$ protease in both organisms revealed that secretion and full activation of the protein involve at least four processing events and that the first two are required for secretion of a propeptide.[92] One of the processing events is the removal of a 27-amino acid N-terminal signal sequence which is presumably carried out by signal peptidase I. N-Terminal amino acid sequencing of the mature enzyme revealed that this processing was correct. The other secretion-specific processing reaction is believed to be the result of an autoproteolytic event which removes a 411-amino acid peptide from the C-terminal end. This C-terminal peptide has been localized to the outer membrane. The reported distribution of IgA$_1$ protease in the cell (96% in the growth supernatant, 4% in the periplasm, and 0% in the cytoplasm) was based solely on an enzyme assay. Inactive precursor in the cytoplasm, periplasm, or outer membrane would not have been detected. The fact that this information is lacking does not contradict the authors' conclusion that secretion was specific, because extracellular fractions, which do contain active enzyme, were high in protease activity and low in activity for the cytoplasmic and periplasmic markers. Therefore, one can conclude from this work that *E. coli* is capable of selectively secreting IgA$_1$ protease but the level of efficiency is unknown.

Yanagida *et al.*[93] reported on the secretion of *Serratia marcescens* extracellular serine protease in *E. coli*. The gene is carried on a 3.7-kb

[89] T. Andro, J. Chambost, A. Kotoujansky, J. Cattaneo, Y. Bertheau, F. Barras, F. Van Gijsegem, and A. Coleno, *J. Bacteriol.* **160,** 1199 (1984).
[90] B. Wretlind and O. R. Pavlovskis, *J. Bacteriol.* **158,** 801 (1984).
[91] R. Halter, J. Pohlner, and T. F. Meyer, *EMBO J.* **3,** 1595 (1984).
[92] J. Pohlner, R. Halter, K. Beyreuther, and T. F. Meyer, *Nature (London)* **325,** 458 (1987).
[93] N. Yanagida, T. Uozumi, and T. Beppu, *J. Bacteriol.* **166,** 937 (1986).

plasmid insert which was shown to contain only one open reading frame. This exoprotein appears to be the product of a double cleavage resulting in removal of a 28-amino acid N-terminal signal sequence and a 637-amino acid C-terminal peptide. Both of these processing events were confirmed by amino acid sequencing. This group showed by mutation that the large C-terminal portion of the proenzyme is essential for extracellular production of the enzyme. Secretion of the protease was not accompanied by leakage of host periplasmic enzymes and therefore seems to be specific. Efficiency of secretion was not measured in this study.

It is conceivable that one of these protease systems could be modified by gene fusion techniques to allow secretion of targeted sequences. No doubt experiments addressing this possibility are in progress.

Reports of other self-secreting proteins that are not proteases have appeared.[94-97] However, the data currently available do not fulfill the criteria established to document true secretion and, accordingly, an assessment of the potential of these systems must await further studies.

Export and/or Secretion of Eukaryotic Proteins

In the following paragraphs, we present some examples of eukaryotic proteins that have been exported or secreted in *E. coli*. Unfortunately, no standard approach emerges as a result of these successes, but they serve to highlight factors that can potentially hamper secretion.

OmpF - β-Endorphin

Nagahari *et al.*[98] constructed a fusion containing the *E. coli* OmpF signal sequence, the first eight amino acids of mature OmpF, four amino acids of β-lipotropin, and the entire human β-endorphin gene. On expression of this chimera, it was found that *E. coli* efficiently and specifically secreted the correctly processed peptide product. The amount of OmpF - β-endorphin in the medium was 290 times as much as that in the periplasm, and there was a negligible amount in the cytoplasm. In contrast, β-lactamase and alkaline phosphatase were found exclusively in the periplasm. The rationale behind this construction was to use the signal sequence and part of the mature sequence of a bacterial outer membrane

[94] A. Collmer and D. B. Wilson, *Bio/Technology* **2**, 594 (1983).
[95] H. Malke and J. J. Ferretti, *Proc. Natl. Acad. Sci. U.S.A.* **81**, 3557 (1984).
[96] S. Clegg and B. L. Allen, *FEMS Microbiol. Lett.* **27**, 257 (1985).
[97] T. K. Ball, P. N. Saurugger, and M. J. Benedik, *Gene* **57**, 183 (1987).
[98] K. Nagahari, S. Kanaya, K. Munakata, Y. Aoyagi, and S. Mizushima, *EMBO J.* **4**, 3589 (1985).

protein in an effort to make the chimera enter the outer membrane protein localization pathway. The hydrophilic nature of the protein would then cause it to be released into the medium. Apparently the peptide was secreted for different reasons, because deletion of the OmpF mature sequence did not interfere with the secretion of the peptide. It seems that, in this case, one must consider the possibility that the small size of the peptide (MW 4800) may have been a factor. Perhaps the peptide is secreted in a manner similar to the heat-stable (ST) enterotoxin (MW 5000) as described above. Regardless of mechanisms, results with this construct suggest that efficient secretion of small peptides in *E. coli* may be simply attained.

Human Growth Hormone

Three groups have been able to export the human growth hormone (hGH) to the periplasm of *E. coli*. This hormone is a single-chain polypeptide of 191 amino acids that contains two intramolecular disulfide bonds. Gray *et al.*[99] compared the efficiency of export of hGH directed by either its own signal sequence or the *E. coli* PhoA signal sequence. Results are comparable: the hGH signal sequence directed 76% of the protein into the periplasm, and the PhoA signal sequence directed 82% into the periplasm. This shows that, at least in certain cases, *E. coli* can recognize efficiently a eukaryotic signal sequence. Moreover, amino acid sequence analysis demonstrates that correct processing occurs in both cases and the disulfide bonds are correctly formed; thus the structure of the product obtained is authentic.

Hsiung *et al.*[2] obtained the secretion of hGH by utilizing the secretion cloning vector, pIN-III-OmpA.[100] This vector contains the OmpA signal sequence under the regulation of a hybrid *lpp–lac* promoter and was employed to achieve 10- to 20-fold higher yields than Gray *et al.*[99] with essentially the same distribution among the cellular fractions. In the absence of the inducer isopropyl-β-D-thiogalactoside (IPTG), hGH was estimated to be approximately 6% of the total cell protein and 30% of the total periplasmic protein. The approximate distribution of hGH among the various cellular fractions as determined by immunoblot analysis was as follows: 72% was present in the periplasm as mature protein; 22% was in the cytoplasm and membrane fractions both as precursor and mature protein; and 6% was a lower molecular weight periplasmic species. Induc-

---

[99] G. L. Gray, J. S. Baldridge, K. S. McKeown, H. L. Heyneker, and C. N. Chang, *Gene* **39**, 247 (1985).
[100] J. Ghrayeb, H. Kimura, M. Takahara, H. Hsiung, Y. Masui, and M. Inouye, *EMBO J.* **3**, 2437 (1084).

tion with IPTG increased hGH expression, but the additional protein was found to be in the precursor form in the cell, suggesting that export was rate-limiting. Accordingly, this may provide an example in which the export capacity of the cell has been exceeded. The utility of protein export is evidenced by a simple two-step purification of the periplasmic hGH that resulted in protein of 90% purity.

Kato et al.[100a] took a novel approach to obtaining the controlled release of hGH from the periplasm by cloning the gene coding for hGH into a pEAP excretion vector.[100b,c] These vectors contain a *kil* gene under the control of a promoter derived from an alkalophilic *Bacillus* strain. The weak expression of the *kil* gene from this plasmid in *E. coli* leads to the permeabilization of the outer membrane and subsequent release of the periplasmic proteins. The scheme succeeded in producing a yield of 11.2 mg hGH per liter of culture.

Hsuing et al.,[100d] using a similar approach to that of Kato et al., obtained 4.5 μg/ml authentic hGH in the culture medium. Their system involved the use of two compatible plasmids, one carrying the *ompA – hGH* construction used in their earlier work[2] and the other carrying the gene for the bacteriocin release protein (BRP). Both genes are under the control of the *lpp – lac* promoter/operator system and are expressed on induction with IPTG. The chimeric pOmpA – hGH precursor is synthesized and exported through the general export pathway into the periplasm, and the concurrent presence of BRP leads to leakage of authentic hGH into the medium. This approach was apparently successful because the copy number of the plasmid expressing BRP was lower than that of the plasmid expressing hGH, and the coordinated expression of both genes prevented premature lysis of the host. The hGH was not the only protein released, but it was easily purified to 98% homogeneity in one step by reversed-phase column chromatography.

In all of the above constructions, the recombinant hGH produced and purified was correctly processed and the disulfide bonds were correctly formed. Hsiung et al.[2] also showed by circular dichroic spectroscopy that structurally their recombinant product was virtually identical to that of authentic hGH. This result is important because it demonstrates that correct folding and disulfide bond formation can occur in *E. coli* when the molecule is translocated to the periplasm.

[100a] C. Kato, T. Kobayashi, T. Kudo, T. Furusato, Y. Murakami, T. Tanaka, H. Baba, T. Oishi, E. Ohtsuka, M. Ikehara, T. Yanagida, H. Kato, S. Moriyama, and K.Horikoshi, *Gene* **54**, 197 (1987).
[100b] T. Kudo, C. Kato, and K. Horikoshi, *J. Bacteriol.* **156**, 949 (1983).
[100c] C. Kato, T. Kobayashi, T. Kudo, and K. Horikoshi, *FEMS Microbiol. Lett.* **36**, 31 (1986).
[100d] H. M. Hsiung, A. Cantrell, J. Luirink, B. Oudega, A. J. Veros, and G. W. Becker, *Bio/Technology* **7**, 267 (1989).

Human Epidermal Growth Factor

Oka *et al.*[101] cloned the cDNA encoding human epidermal growth factor (hEGF) into a secretion vector containing the *E. coli* alkaline phosphatase signal sequence and found that correctly processed product was localized in the periplasm. However, only about 3% of the protein recovered from the periplasm was hEGF. Because the product was shown to have biological activity, it was assumed that three disulfide bridges per molecule were correctly formed. These results strengthen the hypothesis that export may be critical for the recovery of an active protein. Interestingly, they found that there was close to a 5-fold difference in levels of hEGF produced from the same construction in two different *E. coli* strains. They also showed that lowering the temperature from 37 to 30° for the duration of a 9-h induction doubled their yield. Our inability to explain such effects underscores our lack of knowledge.

Human Superoxide Dismutase

Recently, Takahara *et al.*[102] have shown that human superoxide dismutase (hSOD) can be exported to the *E. coli* periplasm with high efficiency when the structural gene is inserted into the pIN-III-OmpA secretion cloning vector mentioned above. Following a 6-hr induction, the hSOD was approximately 7% of the total cell protein and 64% of the periplasmic protein as determined by densitometric scanning of an SDS–polyacrylamide gel. Much of the hSOD was also found in the cytoplasm, but the levels were not quantitated. Authentic hSOD is acylated at the amino terminus and requires Cu^{2+} and Zn^{2+} for full activity. The periplasmic species was demonstrated to have the correct amino-terminal sequence but did not contain the acylated amino-terminal group. Moreover, in enzymatic assay, the periplasmic form was found to have a lower specific activity than the cytoplasmic form. The activity of the periplasmic species was reportedly increased by the addition of Cu^{2+}, resulting in a specific activity of approximately half of that of the cytoplasmic form.

The work with hSOD provides the first example of successful export of a eukaryotic cytoplasmic enzyme. This approach may prove quite useful as many proteins of commercial interest may be detrimental if produced and contained inside the bacteria in active form. Whether or not this method can be generalized is not yet apparent, given only one example. However, it should be noted that hSOD has a molecular mass of only 15 hDa. Recall that *E. coli* cannot export sequences of β-galactosidase, a cytoplasmic enzyme of 116 kDa.

[101] T. Oka, S. Sakamoto, K.-I. Miyoshi, T. Fuwa, K. Yoda, M. Yamasaki, G. Tamura, and T. Miyake, *Proc. Natl. Acad. Sci. U.S.A.* **82**, 7212 (1985).
[102] M. Takahara, H. Sagai, S. Inouye, and M. Inouye, *Bio/Technology* **6**, 195 (1988).

Mouse-Human Fab Proteins

Fab proteins result from proteolytic cleavage of immunoglobulin G (IgG) molecules are roughly one-third the size of an IgG. The Fab protein consists of two polypeptides, each containing two intrachain disulfide bonds, that assemble into a heterodimer containing one interchain disulfide bond. Previous attempts at obtaining active Fab protein ended in failure because the protein was not engineered for secretion and failed to fold properly in the cytoplasm.[103,104] Better *et al.*[3] have recently described a novel method that allows production of active Fab in *E. coli*. First, site-directed mutagenesis was employed to place a termination codon in the IgG heavy-chain cDNA, thus shortening the length of the coding region to that of the Fab fragment. Second, the homologous signal sequences were removed and replaced with the *pelB* (pectate lyase exoprotein) signal sequence from *Erwinia carotovora*. Finally, both genes were positioned downstream from the inducible *Salmonella typhimurium araB* promoter. In this orientation, both genes were encoded on one dicistronic message, thus ensuring that both proteins were translated in close proximity.

The ability of *E. coli* to fold and assemble a heterodimeric protein properly with intra- and interchain disulfide bonds is very encouraging. Clearly, secretion of the Fab contributed to this success, and it is possible that cotranscription of the two genes was also a factor.

When cultures of *E. coli* cells containing the recombinant operon were induced for 4 hr to express the Fab proteins, most of the protein appeared in the periplasm (M. Better, personal communication). However, after a 16-hr induction, 90% of the secreted protein was in the growth medium.[3] Cytoplasmic levels of Fab protein were not measured, nor were levels of cytoplasmic and periplasmic markers in the culture supernatant. One-step purification yielded 2 mg per liter of culture that was 90% pure product. As measured by their assays, this Fab protein seemed to be correctly folded and biologically active.

The reasons for such successful secretion are not entirely clear; but, judging from the fact that the protein was localized primarily to the periplasm after 4 hr of induction, it seems likely that the Fab protein was leaking into the growth medium after the longer induction periods. Even though the origin of the signal sequence was a secreted bacterial protein, it probably was only responsible for translocating the protein across the cytoplasmic membrane and not the outer membrane. Rather, we suspect

[103] M. A. Boss, J. H. Kenten, C. R. Wood, and J. S. Emtage, *Nucleic Acids Res.* **12**, 3791 (1984).
[104] S. Cabilly, A. D. Riggs, H. Pande, J. E. Shively, W. E. Holmes, M. Rey, L. J. Perry, R. Wetzel, and H. L. Heyneker, *Proc. Natl. Acad. Sci. U.S.A.* **81**, 3273 (1984).

that high-level expression over prolonged periods of time results in damage to the cell envelope, allowing the protein to escape to the media. Such effects have been observed previously, even with prokaryotic proteins.[105]

Novel Approach for Secretion of Foreign Proteins

In the examples cited above, secretion of a foreign eukaryotic protein was obtained not so much by design, but rather by default, i.e., secretion occurred by some passive mechanism. *Escherichia coli* does possess a system which specifically directs highly efficient secretion, i.e., hemolysin, and properties of this system are such that it can be adapted by techniques of gene fusion for the directed secretion of foreign proteins.

The hemolysin secretion system, described in an earlier section, has been shown to be capable of secreting a tribrid protein with high efficiency and selectivity.[106] It was shown that a 56-kDa chimera, containing 10 amino acids from the amino terminus of β-galactosidase fused to 300 amino acids of OmpF (not including its signal sequence) followed by the 218 carboxyl terminus amino acids of HlyA, was secreted in a HlyB- and HlyD-dependent manner. Secretion was shown to occur without the loss of cytoplasmic or periplasmic proteins. As mentioned previously, the hemolysin is a large protein (107 kDa) that is extremely hydrophilic. The significance of this result for genetic engineering purposes is 3-fold: (1) It does not require the presence of an N-terminal signal sequence, and therefore recombinant proteins will not complete with other cellular proteins for essential components of the general secretion pathway. (2) The system has no apparent size limitation, at least within the range of 23–107 kDa. (3) Internal hydrophobic sequences and sequences that are normally membrane localized do not cause the retention of the protein in the cell envelope, suggesting the potential of this system to secrete proteins that have not evolved to be translocated across biological membranes. Although efficient selective secretion might be possible using the Hly system, the required C-terminal HlyA sequence remains attached once the protein is in the growth medium, resulting in a nonauthentic protein.

Concluding Remarks

The technology for secretion of heterologous proteins from *E. coli* is in its infancy. Within the last 8–10 years, we have been able to establish that,

---

[105] G. Georgiou, J. N. Telford, M. L. Shuler, and D. B. Wilson, *Appl. Environ. Microbiol.* **52,** 1157 (1986).
[106] N. Mackman, K. Baker, L. Gray, R. Haigh, J.-M. Nicaud, and I. B. Holland, *EMBO J.* **6,** 2835 (1987).

in most cases, a protein that has evolved to be secreted through a biological membrane can be exported through the *E. coli* inner membrane into the periplasm. Poor efficiency and yields remain as major obstacles to the economic feasibility of exploiting this technology. The next major challenge is to learn how to make proteins cross the outer membrane. We have described cases where extracellular secretion has apparently been achieved in *E. coli* by leakage from the periplasm. From these studies, some observations regarding this type of secretion have been found and can be briefly summarized:

1. Strain. Marked differences in expression, secretion, and protein yields have been reported among various *E. coli* strains. No systematic study has yet been done to establish which is the best strain, but it is likely that no ideal strain will be found that is optimal for all purposes.

2. Growth temperature. Differences in both secretion efficiency[101] and protein solubility[107,108] have been shown to exist as a function of growth or induction temperature. These differences may be a result of alteration of the protein structure leading to insolubility or greater protease sensitivity. Some of the problems may arise following shifts up to 42° if a vector with a temperature-sensitive repressor is used. Such conditions will induce the heat-shock response and may lead to the insoluble association of the recombinant protein with host proteins.[108]

3. Length of induction. If the recombinant protein is controlled by an inducible promoter, the length of induction may be an important factor in secretion. This was found to be the case with the mouse–human Fab fragment[3] secretion described in a previous section. Generally, in older cultures, more lysis occurs. Therefore, when comparing secretion levels as a function of induction time, one should determine whether there is an accompanying change in the specificity of secretion.

4. Cotranscription. While only one example exists,[3] it is worth noting that, when secreting heterodimeric molecules, it may be helpful to construct a dicistronic message. It has been suggested[3] that if synthesis and secretion of the two polypeptides are temporally and spatially close, they are more likely to assemble correctly. A direct test of this hypothesis would be informative.

It seems that the greatest potential for solving the problem of crossing the outer membrane lies with specific secretion pathways. More needs to be learned about the manner in which exoproteases are secreted and what

---

[107] C. H. Schein and M. H. M. Noteborn, *Bio/Technology* **6**, 291 (1988).

[108] M. Piatak, J. A. Lane, W. Laird, M. J. Bjorn, A. Wang, and M. Williams, *J. Biol. Chem.* **263**, 4837 (1988).

the limits of tolerance are for substituting protease moieties with recombinant protein sequences. We also need to develop a better understanding of the hemolysin system since this pathway may become a means of secreting proteins that cannot be secreted by other pathways. The motivation of the genetic engineer to achieve secretion of heterologous proteins in *E. coli* ensures that progress will be made.

Acknowledgments

The authors would like to thank P. Bassford, P. C. Tai, D. Oliver, and B. Holland for communicating data prior to publication and T. Hirst, T. Pugsley, and M. Better for helpful discussions. The authors would also like to thank S. Benson and M. Igo for critically reading the manuscript and S. DiRenzo for typing.

[16] Refolding of Recombinant Proteins

By Tadahiko Kohno, David F. Carmichael, Andreas Sommer, and Robert C. Thompson

Proteins that are normally made in the cells of higher organisms are now routinely made in microorganisms by cloning the gene together with appropriate signals for transcription and translation. However, the protein made is often inactive because the polypeptide chain does not readily adopt its native conformation in these organisms. This result is not to be taken as evidence against the view that the primary structure of a protein determines its conformation. More reasonably, it indicates that the protein will only fold to its correct conformation under the appropriate conditions, and that these conditions are not met in the microorganism. The failure to fold properly may even induce the protein to form a precipitate, commonly known as an inclusion body, within the microorganism.

To induce the recombinant protein to assume its active conformation, two broad strategies have evolved. The most elegant, and intellectually pleasing, strategy is to recreate in the microorganism the conditions prevailing in the higher cell. Most frequently, this corresponds to cotranslationally secreting the protein to an oxidizing environment, either the periplasmic space or the extracellular medium. This strategy is remarkably successful for some proteins.[1,2] However, when some eukaryotic proteins are designed to secrete cotranslationally in microorganisms, the proteins appear to poison the microorganism.[2] For this reason, an additional

[1] J. A. Stader and T. J. Silhavy, this volume [15].
[2] T. Kohno *et al.*, manuscript in preparation.

strategy has been employed in which the protein is first made in a disordered state in the microorganism and is subsequently folded to its native state *in vitro.* This strategy, and the methods employed, are derived almost exclusively from the large volume of work on the *in vitro* refolding of denatured proteins that was initiated by Anfinsen and collaborators in the early 1960s.[3] However, few reports describe the refolding of recombinantly produced proteins. In this article, we illustrate this strategy by describing examples of folding two highly disulfide-bonded human proteins from the disordered state present in *Escherichia coli.*

The process of refolding of the recombinant protein, to be feasible for commercial application, should be fast and cheap and give a high yield of an active molecule. The strategy used employs three essential steps. In the first of these, the insoluble protein aggregate from the microorganism is effectively solubilized and partially purified. In the second, the solubilized protein is brought into an environment which favors the native, active structure. Finally, since it is too much to expect the folding to proceed in 100% yield, it is necessary to find some method to remove the residual, wrongly folded protein.

Solubilization of Inactive Proteins

The solubilization of insoluble aggregates can be achieved by various denaturing reagents. Low or high pH can solubilize some inclusion bodies, presumably by increasing electrostatic repulsions between protein molecules, thereby destabilizing aggregates.[4] High temperature, detergents, and high concentration of inorganic salts or organic solvents will also solubilize some aggregates.[5] Some proteins, such as DNase I, can be refolded after treatment with a detergent SDS.[6] However, in some cases, detergents may bind strongly to the protein,[7] inhibiting subsequent refolding. The most commonly used solubilizing agents are high concentrations of water-soluble organic solutes such as urea or guanidine-HCl. These are often used in the presence of reducing agents like mercaptoethanol or dithiothreitol (DTT) to break aggregates that are stabilized by disulfide bonds. The solubilized proteins are often partially purified by ion-exchange chroma-

---

[3] C. J. Epstein, R. F. Goldberger, and C. B. Anfinsen, *Cold Spring Harbor Symp. Quant. Biol.* **28,** 439 (1963).

[4] J. S. Emtage, S. Angal, M. T. Doel, T. J. R. Harris, B. Jenkins, G. Lilley, and P. A. Lowe, *Proc. Natl. Acad. Sci. U.S.A.* **80,** 3671 (1983).

[5] I. Bikel, T. M. Roberts, M. T. Bladon, R. Green, E. Amann, and D. M. Livingston, *Proc. Natl. Acad. Sci. U.S.A.* **80,** 906 (1982).

[6] S. A. Lacks and S. S. Springhorn, *J. Biol. Chem.* **255,** 7467 (1980).

[7] C. Tanford, *Adv. Protein Chem.* **23,** 121 (1968).

tography or other conventional methods at this stage, prior to the refolding process. For some proteins it is not absolutely required that the protein be purified prior to the refolding; however, recovery of the active molecule seems to be higher when this is done.

Refolding of Proteins

In general, the refolding of proteins is initiated by complete unfolding of the protein of interest. Very often this will have been accomplished by the conditions needed to dissolve the aggregates. However, if this is not the case, strong denaturants, such as urea or guanidine-HCl, and reducing agents are added at this stage. If no disulfide bonds are present in the final conformation, it would probably be sufficient to remove the denaturing agent slowly to allow the protein to fold correctly. However, when disulfide bonds are to be formed, it is frequently beneficial to form them *during* the course of refolding since they stabilize the native structure and thereby influence the process.

The formation of disulfide bonds during the refolding of a protein can be done by an air oxidation,[8] including oxidation by air catalyzed by the presence of trace metal ions,[9] or by the presence of a mixture of reduced and oxidized thiol compounds,[9] or by disulfide isomerase.[10] Air oxidation is a slow process. However, the addition of the thiol–disulfide exchanger increases the rate of reactivation. Refolding buffers containing conjugate thiol : disulfide pairs such as oxidized DTT and reduced DTT, oxidized glutathione (GSSG), and glutathione (GSH), cystine and cysteine, or cystamine and cysteamine yield similar results in some cases.[9] The appropriate ratio of oxidized and reduced thiols will vary and must be determined experimentally.

The formation of protein aggregates during the refolding is sometimes observed and will reduce the recovery of the active molecule. This aggregation may be limited by the presence of a low concentration of urea or guanidine-HCl during the refolding. But even when this is true, it is sometimes necessary to use very low concentrations of protein ($\sim 10 \, \mu g/ml$) to avoid the formation of insoluble aggregates. The pH of the solution may affect protein aggregation as well as the rate of disulfide interchange. Optimal pH of buffer must therefore be determined empirically to maximize the recovery of an active molecule. The removal of denaturants used

[8] C. B. Anfinsen, E. Haber, M. Sela, and F. H. White, *Proc. Natl. Acad. Sci. U.S.A.* **47,** 1309 (1961).
[9] P. Saxena and D. B. Wetlaufer, *Biochemistry* **9,** 5015 (1970).
[10] D. F. Carmichael, J. E. Morin, and J. E. Dixon, *J. Biol. Chem.* **252,** 7163 (1977).

can be done either by dialysis or by dilution in an appropriate buffer. The strategy we have adopted is first to oxidize all the free —SH groups to intermolecular disulfide bonds with a suitable low-molecular-weight thiol and then to remove the denaturing agent, relying on the fact that the favorable energy of folding will drive the formation of intramolecular disulfide bonds. A small amount of free thiol is incorporated into the reaction to catalyze disulfide bond interchange.

Isolation and Characterization of Correctly Folded (Active) Proteins

Once the proteins are folded in biologically active conformation, the active molecules have to be separated from incorrectly folded molecules. Affinity chromatography is one of the best methods for removing inactive molecules.

The biological activity of the refolded proteins should be identical to that of native proteins. Purity of the refolded protein can often be assessed by the combination of the following techniques: (1) sodium dodecyl sulfate–polyacrylamide gel electrophoresis (SDS–PAGE), (2) reversed-phase chromatography (RP-HPLC), and (3) ion-exchange chromatography (FPLC). However, it is generally not possible to show that all possible refolding isomers will separate from the authentic protein. Besides examining the N-terminal amino acid sequence of the refolded proteins, conformational identity between the recombinant proteins and human-derived proteins could be determined by peptide mapping of those proteins without prior reduction.

Example 1. Refolding of Secretory Leukocyte Protease Inhibitor

Secretory leukocyte protease inhibitor (SLPI) is a protease inhibitor originally isolated from human tissues. SLPI inhibits various serine proteases, which include leukocyte elastase, cathepsin G, and trypsin. This inhibitor is secreted protein and has been found in various secretory fluids such as those from the parotid glands.[11] It consists of 107 amino acids and contains 16 cysteine residues,[11] all of which are present in disulfide bonds in the native protein. The cDNA and genomic sequence of the gene for SLPI are available,[12] and the gene can be expressed in high yield in prokaryotic systems such as *E. coli*.[13] However, the protein produced in *E. coli* is inactive, probably due to the incorrect conformation. In order to obtain functional SLPI from *E. coli,* we learned to solubilize and refold the inactive molecules.

---

[11] R. C. Thompson and K. Ohlsson, *Proc. Natl. Acad. Sci. U.S.A.* **83**, 6692 (1986).
[12] G. L. Stetler, M. T. Brewer, and R. C. Thompson, *Nucleic Acids Res.* **14**, 7883 (1986).
[13] S. Eisenberg *et al.,* manuscript in preparation.

Procedure

Initial Purification. An *E. coli* strain which produces recombinant SLPI (rSLPI) at about 2% of cell protein was grown in a 10-liter fermenter at 37°. Growth was monitored by optical density. At 10 OD, SLPI synthesis was induced by adding 0.5 mM IPTG. After 5 hr of induction, the cells were harvested at 4° and were resuspended in 2× volume of 50 mM Tris-HCl, pH 7.5. The cell suspension was processed through a Gaulin mill two times and centrifuged at 900 g for 60 min. The resultant pellet was solubilized at 25° with 50 mM Tris-HCl, pH 7.5, containing 10 M urea and 50 mM 2-mercaptoethanol. Insoluble cell debris was removed by centrifugation at 14,000 g for 60 min. The supernatant was loaded onto an SPC25 column (18 × 25 cm) which was previously equilibrated with 50 mM Tris-HCl, pH 7.5. The column was washed extensively with the same buffer to remove unbound proteins. The rSLPI was eluted with 50 mM Tris-HCl, pH 7.5, buffer containing 0.3 M guanidine-HCl. This column step yielded greater than 80% homogeneous SLPI preparation. This eluate was concentrated to between 0.5 and 50 mg per milliliter by ultrafiltration prior to refolding.

Refolding. One liter of this concentrated SLPI was mixed with 1 liter of 6 M guanidine-HCl and incubated at room temperature for 30 min. After the end of 30 min, 1.23 g of solid DTT was added and gently stirred. The incubation was continued for 1 hr at 25°, then 114 ml of 250 mM cystine solution (in 0.5 N NaOH) was added to the above mixture. After a 10-min incubation, this mixture was added over a period of 5 min to 10 liters of 50 mM Tris solution containing 5.3 mM cysteine. This diluted material was incubated for up to 16 hr at 25°. The antichymotrypsin activity and antitrypsin activity of SLPI appeared rapidly when assayed by the method described by Thompson and Ohlsson,[11] and the refolding was almost complete after 2 hr of incubation (Fig. 1).

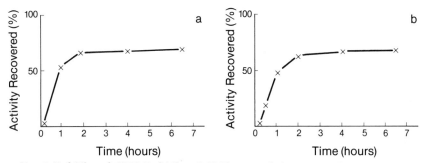

FIG. 1. Refolding of rSLPI. Refolding of rSLPI was carried out as described in the text. At the indicated time after refolding started, samples were drawn and assayed for (a) antichymotrypsin and (b) antitrypsin activities.

Final Purification. After refolding, the material was purified on an affinity column. The affinity resin is composed of anhydrochymotrypsin bound to Affi-Gel 10 (Bio-Rad Laboratories, Richmond, CA). Anhydrochymotrypsin was prepared by the method described by Ako *et al.*[14] and was coupled to the Affi-Gel 10 using the manufacturer's protocol. The above refolding mixture was loaded onto the affinity column (5 × 18 cm) and washed extensively with 50 m*M* sodium phosphate buffer, pH 8.0. Active SLPI was eluted by 50 m*M* sodium phosphate-HCl, pH 2.0. Judging by HPLC (C_8) chromatography (Fig. 2) and SDS–PAGE, the low-pH eluate contains over 90% pure SLPI. The major contaminant appears to be a dimer. The recovery of the pure, active SLPI by this refolding method was about 50%, regardless of the initial protein concentration used. Evidence supporting the identity of refolded rSLPI and parotid-derived SLPI was obtained by measuring their K_d values for human leukocyte elastase. The recombinant SLPI has a K_d of 1.9×10^{10} *M* and that for parotid SLPI is 2×10^{10} *M*.

Escherichia coli-produced rSLPI can be refolded to form an active molecule without an initial purification. However, the final yield of the active molecule was about 5 to 10%, based on a Western blot technique.

Example 2. Refolding of Recombinant Tissue Inhibitor of Metalloproteases (rTIMP)

TIMP is an inhibitor of many mammalian metalloproteases. It is a single polypeptide chain of 184 amino acids in length. In addition, the primary translation product has a 23-amino acid N-terminal extension signaling its secretory destination.[15] The mature inhibitor has 12 cysteine residues, all of which are apparently engaged in intrachain disulfide linkages.[16] Although TIMP is normally glycosylated, this is not required for its activity.[17]

Procedure

Initial Purification. The 184-amino acid polypeptide chain can be produced in *E. coli* as an inactive, insoluble protein found predominantly in cytoplasmic inclusion bodies containing approximately 50% pure TIMP. In order to purify the protein further, the inclusion bodies were solubilized in 8 *M* urea, 50 m*M* Tris-HCl, pH 7.5, containing 0.1 *M* 2-mercaptoeth-

---

[14] A. Ako, R. J. Fester, and C. A. Ryan, *Biochemistry* **13**, 132 (1974).
[15] D. F. Carmichael, A. Sommer, R. C. Thompson, D. C. Anderson, C. G. Smith, H. G. Welgus, and G. P. Stricklin, *Proc. Natl. Acad. Sci. U.S.A.* **83**, 2407 (1986).
[16] G. P. Stricklin and H. G. Welgus, *J. Biol. Chem.* **258**, 12252 (1983).
[17] G. Stricklin, unpublished.

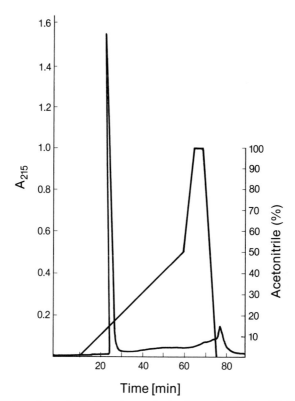

FIG. 2. HPLC pattern of affinity column eluate. A sample of low-pH eluate from the anhydrochymotrypsin–Affi-Gel 10 column was loaded on an HPLC column (Synchrom). The rSLPI was eluted with a 0.1% trifluoroacetic acid–acetonitrile gradient (0–50% acetonitrile over 50 min) at a flow rate of 1 ml/min.

anol. The insoluble material was removed by centrifugation at 30,000 g for 5 hr. The supernatant fraction was applied to a carboxymethyl-cellulose column (Whatman CM-23) in this same buffer. Four grams of soluble protein was loaded onto a 7.0 × 18 cm column. The column was washed in 50 mM Tris-HCl, pH 7.5, containing 50 mM 2-mercaptoethanol and finally eluted by including 0.25 M NaCl in the wash buffer. Fractions were analyzed for the presence of TIMP by Western blot techniques. TIMP-containing fractions were pooled, then used in refolding reactions.

Refolding. Refolding of the partially purified protein was effected by the stepwise addition of reagents and dilution of the denaturant from the protein solution. All additions were made at 0–4°. First, 2-mercaptoethanol was added to a final concentration of 140 mM in 6.0 M urea, 50 mM Tris-HCl, pH 7.5. The protein was further diluted in this solution to a final concentration of 1 mg/ml as determined by Bio-Rad protein assay kit.

Four volumes of 250 mM cystamine in 6.0 M urea, 50 mM Tris-HCl, pH 7.5, were then added to the protein solution. Finally, the urea was diluted to 0.3 M with the gradual addition of 50 mM Tris-HCl, pH 9.0. The pH of the final solution was approximately 8.3. This solution was allowed to stand at 4° overnight.

Final Purification. Following an overnight incubation, the solution was then concentrated 20-fold by pressurized ultrafiltration on an Amicon YM10 filter. The buffer solution was then exchanged either by dialysis or diafiltration for 50 mM Tris-HCl, pH 7.5. The protein at this pH was then purified by cation-exchange chromatography from contaminating proteins left in the solution and also from certain inactive conformers of the inhibitor protein. The chromatography was performed using Whatman CM-52 carboxymethyl-cellulose. One gram of protein in solution was applied to a 2.5 × 15 cm column and the column was washed with 50 mM Tris-HCl, pH 7.5. Adsorbed proteins were eluted from the column with a linear gradient from 0 to 200 mM NaCl in 50 mM Tris-HCl, pH 7.5. Two peaks of inhibitor protein are partially resolved under these conditions (Fig. 3). Fractions from the smaller peak contain inactive TIMP as well as some degradation products. The main peak contains the active inhibitor. Pooled fractions from this peak provide active inhibitor with a 40% recovery through the refolding and subsequent steps.

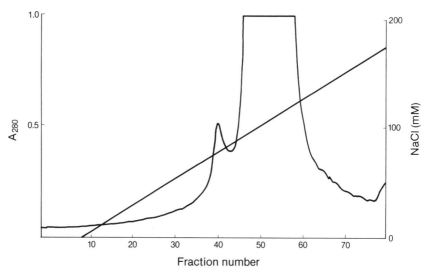

FIG. 3. CM-52 chromatogram of the refolding mixture. Both peaks contain full-length TIMP, indistinguishable by SDS–PAGE, but only the major peak contains active inhibitor.

Purity of the TIMP as determined by SDS – PAGE analysis is greater than 90%, and the activity has been shown to be equivalent on a molar basis to the inhibitor produced by human fibroblasts. In addition, the protein produced in prokaryotic cells behaves identically to the natural inhibitor in terms of its stability to extremes of pH, proteases, and thermal denaturation.[17]

Section III

Expression in *Bacillus subtilis*
Article 17

A. Intracellular Production
Article 18

B. Secretion
Article 19

C. Inducible Expression of Regulatory Proteins
Article 20

[17] Expression of Heterologous Genes in *Bacillus subtilis*

By DENNIS J. HENNER

There are several reasons why *Bacillus subtilis* might be considered useful as a host for expression of a protein of interest. It is a well-studied prokaryote, with extensive genetic systems, and well-understood physiology; there is a wealth of studies on transcription and translation in it. Additionally, a great deal of experience has been gained by growing various *Bacillus* species for industrial production. Over the years, a number of expression systems have been constructed which are reasonably simple to use. Perhaps the greatest advantage of *B. subtilis* is its ability to secrete proteins directly into the culture medium; however, as discussed below, this may have both positive and negative aspects.

Secretion

A major rationale for using *B. subtilis* as a host for expression of proteins is its ability to secrete proteins directly into the culture medium, accumulating them to high levels, and in a relatively pure state. A further advantage of secretion is that the proteins made are often correctly folded, disulfide-bonded, and soluble. However, the ability of an expression system to secrete proteins efficiently is not well understood. Even attempts to secrete heterologous proteins in the better characterized host *Escherichia coli* are not uniformly successful. Certain proteins seem to be efficiently secreted, while other, even closely related, proteins are secreted poorly. In our experience, most prokaryotic proteins are secreted efficiently from both organisms, while the behavior of eukaryotic proteins in a secretion system cannot be predicted. The underlying mechanism(s) for this failure of some proteins to be efficiently secreted has not been well documented in either *B. subtilis* or *E. coli*. One report demonstrated the accumulation of large amounts of the precursor of human interferon α_2 in a detergent-insoluble fraction in *B. subtilis* when an attempt was made to express it with the α-amylase signal sequence.[1] Our attempts to find similar material accumulating using human growth hormone failed to detect any significant accumulation of precursor material. The "competence" of different prokaryotic organisms to secrete heterologous proteins remains open to investigation.

[1] C. H. Schein, K. Kashiwagi, A. Fujisawa, and C. Weissmann, *Bio/Technology* **4**, 719 (1986).

Proteases and Protein Stability

Bacillus subtilis secretes several proteases into the culture medium which have the potential to degrade other proteins that are also secreted to the medium. This problem has been well documented.[2,3] The genes for both of the major proteases have been isolated and used to create deletions in the chromosome of *B. subtilis*.[4,5] Strains carrying such deletions are viable; however, such strains exhibited only modest improvements in the stability of heterologous proteins. There are several other minor proteases, at least two of which have been cloned and chromosomal deletions created.[6,7] The elimination of all the proteases will probably lead to improvements in the stability of heterologous proteins in the culture medium.

Another stability problem which should be distinguished from proteolysis is the stability of the protein to the physical conditions in the culture medium. We found that several proteins of interest to us were very unstable in sterile culture media during agitation. This instability was exhibited by a rapid aggregation and/or loss of reactivity in an immunological assay and may be caused by surface denaturation resulting from the agitation and aeration found during fermentation. Denaturation of the protein could also lead to increased susceptibility to proteolysis.

Plasmid Instability

There have also been a large number of anecdotal reports of plasmid instability in *B. subtilis*. However, in our hands, plasmid instability has never led to practical problems. The only plasmids that exhibited unacceptable levels of deletions or rearrangements were those which had a region repeated on the plasmid, or those which constitutively expressed a protein at high levels. Both problems can be avoided by taking care in construction of plasmids so as to avoid duplicating a region of DNA and by using promoters that are regulated.[8]

Another simple way to avoid plasmid instability in *B. subtilis* is to integrate the expression system into the chromosome.[9] It is very simple to

[2] I. Ulmanen, K. Lundstrom, P. Lehtovaara, M. Sarvas, M. Ruohonen, and I. Palva, *J. Bacteriol.* **162**, 176 (1985).

[3] K. Nakamura, T. Furusato, T. Shiroza, and K. Yamane, *Biochem. Biophys. Res. Commun.* **128**, 601 (1985).

[4] M. L. Stahl and E. Ferrari, *J. Bacteriol* **158**, 411 (1984).

[5] M.Y. Yang, E. Ferrari, and D. J. Henner, *J. Bacteriol* **160**, 15 (1984).

[6] A. Sloma, A. Ally, D. Ally, and J. Pero, J. Bacteriol. **170**, 5557 (1988).

[7] M. Ruppen and E. Ferrari, personal communication (1988).

[8] S. F. J. Le Grice, this volume [18].

[9] C. W. Saunders, B. J. Schmidt, M. S. Mirot, L. D. Thompson, and M. S. Guyer, *J. Bacteriol.* **157**, 718 (1984).

integrate nonreplicative plasmids into the chromosome by homologous recombination, and a number of vectors have been designed for this purpose. These integrated plasmids are replicated by the chromosomal replication machinery, and are very stable.

Expression vectors for secretion of proteins that is based on the *B. amyloliquefaciens* alkaline protease (subtilisin) promoter and signal sequence are discussed in [18]. These vectors have been successfully used to express both prokaryotic and eukaryotic proteins. The vectors are reasonably simple to use, and constructions can be made in either *E. coli* or *B. subtilis*. This expression system would be appropriate for the expression of most prokaryotic secreted proteins, and is also worth considering for eukaryotic proteins which can be demonstrated to be reasonably stable in the culture media of the *B. subtilis* host strain.

A series of vectors which allow the regulated expression of proteins in *B. subtilis* is described in [19]. Although primarily designed for intracellular expression, this system might be also utilized for secreted proteins. I am not aware of any proteins which have been successfully expressed in *B. subtilis* but not in *E. coli,* and I suspect that *E. coli* will be the first choice of most investigators. However, there has not been a detailed comparison of the two organisms as hosts, and the vectors described here can be used in either organism, making such comparisons easy to do. This system should also be useful to researchers working with *B. subtilis* who wish to overproduce proteins in this organism.

A brief description of two plasmids which can be used in one step to insertionally inactivate a gene in *B. subtilis* and place that same gene under the control of a tightly regulated promoter is presented in [20]. These plasmids would be of use for studies of gene regulation and expression in *B. subtilis.*

[18] Regulated Promoter for High-Level Expression of Heterologous Genes in *Bacillus subtilis*

By STUART F. J. LE GRICE

Introduction

A desirable feature of prokaryotic vector design is the ability to control tightly expression of a gene whose product might be detrimental to the host. In *Escherichia coli,* the most commonly employed systems have incorporated transcriptional control mediated through either the *lac,*[1] *trp,*[2]

---

[1] H. A. de Boer, L. J. Comstock, and M. Vasser, *Proc. Natl. Acad. Sci. U.S.A.* **80,** 21 (1983).
[2] B. P. Nichols and C. Yanofsky, this series, Vol. 101, p. 155.

or bacteriophage λ *C1* repressors.[3] When *Bacillus subtilis* is considered as a host for high-level expression of heterologous genes, utilization of similar regulatory elements would likewise be advantageous. Until recently, progress in this area was hampered by the lack of well-characterized gram-positive repressible systems. In light of this, an alternative would be to dissect the control elements of a gram-negative system and optimize these for utilization in a gram-positive host. This approach was first accomplished by Yansura and Henner,[4] who demonstrated inducible gene expression in *B. subtilis* mediated via modified *E. coli lac* regulatory elements.

During our preliminary studies with gram-positive expression vectors, we discovered that promoters of the *E. coli* bacteriophage T5 were efficiently recognized by the gram-positive transcriptional machinery.[5,6] One such promoter, P_{N25},[7] had previously been fused to a synthetic *lac* operator fragment, generating the regulatable promoter $P_{N25/O}$, and successfully utilized in an *E. coli* expression system.[8] We therefore assumed that insertion of this regulatable promoter and a functional *lacI* gene into a gram-positive vector would generate an inducible expression system for *B. subtilis*. This article describes two expression systems displaying these features. The first of these is a single plasmid pREP9 which contains both a regulatable promoter and a *lacI* gene modified for expression in *B. subtilis*. The second system comprises a *B. subtilis* strain which constitutively expresses the *E. coli lac* repressor from a derivative of the *Staphylococcus aureus* plasmid pE194.[9] This latter strain (designated BR : BL1) can be transformed with a compatible expression vector containing the regulatable promoter. In either the single-or dual-plasmid system, gene expression is tightly controlled and can be rapidly derepressed by addition of the inducer isopropyl-β-D-thiogalactopyranoside (IPTG). The usefulness of these systems in high-level production of heterologous proteins is illustrated in this article. Furthermore, the utilization of the dual-plasmid system to produce an enzymatically active heterologous protein is exemplified by the production of human immunodeficiency virus, type 1 (HIV-1) reverse transcriptase in *B. subtilis*.

[3] M. Rosenberg, Y.-S. Ho, and A. Shatzman, this series, Vol. 101, p. 123.

[4] D. G. Yansura and D. J. Henner, *Proc. Natl. Acad. Sci. U.S.A.* **81**, 439 (1984).

[5] U. Peschke, V. Beuck, H. Bujard, R. Gentz, and S. F. J. Le Grice, *J. Mol. Biol.* **186**, 547 (1985).

[6] S. F. J. Le Grice, R. Gentz, U. Peschke, D. Stüber, V. Beuck, and H. Bujard, *in* "Bacillus Molecular Genetics and Biotechnology Applications" (A. T. Ganesan and J. A. Hoch, eds.), p. 433. Academic Press, New York, 1986.

[7] R. Gentz and H. Bujard, *J. Bacteriol.* **10**, 70 (1985).

[8] D. Stüber, I. Ibrahimi, D. Cutler, B. Dobberstein, and H. Bujard, *EMBO J.* **3**, 3143 (1984).

[9] S. Horinouchi and B. Weisblum, *J. Bacteriol.* **150**, 804 (1982).

Elements of a *Lac*-Based Inducible System for *Bacillus subtilis*

Promoter/lac Operator Element

Stüber *et al.*[8] illustrated that the *E. coli* bacteriophage T5 promoter P_{N25}, when fused to a synthetic *lac* operator sequence, allowed inducible gene expression in *E. coli*. Our own work with gram-positive promoters suggested that an AT-rich region in the immediate upstream vicinity of the canonical −35 hexamer, as found in the *B. subtilis* chromosomal *veg* and bacteriophage SPO1 promoters,[10] might influence promoter strength. A similar AT-rich region is present at the same position of *E. coli* bacteriophage T5 promoters,[5] and we subsequently showed that T5 promoters could be efficiently recognized by the vegetative RVA polymerase of *B. subtilis*.[5,6] Thus, the T5 promoter/*lac* operator element $P_{N25/O}$, previously designed for high-level gene expression in *E. coli*, conveniently served as one component of a *lac*-based inducible system for *B. subtilis*. The DNA sequence and relevant features of this promoter/operator element used in our vectors are illustrated in Fig. 1.

lac Repressor Functional in Bacillus subtilis

The second component of a *lac*-based inducible system is a *lac* repressor gene capable of functioning in *B. subtilis*. Since neither the *lacI* promoter nor the ribosome binding site was expected to be utilized in *B. subtilis*, it was necessary to replace these with equivalent gram-positive regulatory elements. This was carried out in a stepwise manner as presented in Fig. 2. We made use of a *lacI* gene containing a *Hind*III site in the immediate vicinity of the initiator methionine codon to remove the original promoter and ribosome binding site. The *lacI* coding region was initially placed under the transcriptional control of the relatively weak *B. subtilis* promoter P_{vegII}.[12,13] Subsequently, we introduced a synthetic ribosome binding site, adapted for gram-positive utilization, between the *vegII* promoter and the initiator methionine of the *lacI* gene. Finally, this modified *lacI* cassette was tailored with a transcriptional terminator (T1)[14] to

[10] C. P. Moran III, N. Lang, S. F. J. Le Grice, G. Lee, M. Stephens, A. L. Sonenshein, J. Pero, and R. L. Losick, *Mol. Gen. Genet.* **186,** 339 (1982).
[11] M. D. Barkley and S. Bourgeois, *in* "The Operon" (J. H. Miller and W. S. Reznikoff, eds.), p. 177. Cold Spring Harbor Monograph Series, Cold Spring Harbor, New York, 1978.
[12] S. F. J. Le Grice and A. L. Sonenshein, *J. Mol. Biol.* **162,** 551 (1982).
[13] S. F. J. Le Grice, C.-C. Shih, F. Whipple, and A. L. Sonenshein, *Mol. Gen. Genet.* **204,** 229 (1986).
[14] J. Brosius, T. J. Doll, D. D. Sleeter, and H. F. Noller, *J. Mol. Biol.* **148,** 107 (1981).

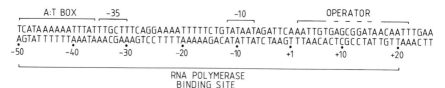

FIG. 1. DNA sequence and pertinent features of the regulatable *E. coli* bacteriophage T5 promoter–*lac* operator element $P_{N25/O}$ employed in our vectors for gene expression in *B. subtilis*. The sequence from −50 to +26 is indicated. In addition to the canonical hexamers around −10 (Pribnow box) and −35, the A:T box spanning −35 to −50, which possibly contributes to the utilization of this *E. coli* bacteriophage promoter in *B. subtilis*, is overlined. A "core" *lac* operator sequence[11] (i.e., *lac* repressor binding site) is fused in the vicinity of the transcription initiation site.

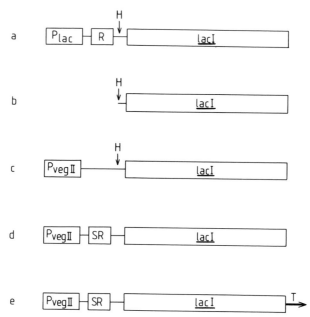

FIG. 2. Stepwise manipulation of the *E. coli lacI* gene for utilization in *B. subtilis*. The *lacI* gene (a), with a *Hind*III site (H) preceding the coding region, was separated from its own promoter (P_{lac}) and ribosome binding site (R) by *Hind*III digestion (b). In Step c, a fusion was made between the *lacI* coding region and the *B. subtilis* promoter P_{vegII}. A synthetic ribosome binding site (SR), functional in *B. subtilis*, was subsequently added between P_{vegII} and the *lacI* coding region (d). The *lacI* cassette after d is now functional in *B. subtilis* (d). Finally, to restrict transcription from P_{vegII} to the *lacI* cassette, a transcriptional terminator (T) was inserted 3′ to the *lacI* coding sequence (e).

ensure the transcription from P_{vegII} would be confined to the *lacI* cassette. Following these manipulations, we generated a vector, pREP4, from which the *E. coli lac* repressor was constitutively expressed in *B. subtilis* (data not shown).

Inducible Expression in *Bacillus subtilis* from Plasmid pREP9

The general features of pREP9, a bifunctional shuttle vector mediating inducible gene expression in either *E. coli* or *B. subtilis,* are outlined in Fig. 3A. In the major expression cassette, contained between the *Xho*I (X) and *Xba*I (Xb) sites, the *E. coli* chloramphenicol acetyltransferase *(cat)* gene[15] has been connected, via a multicloning site, to a synthetic ribosome binding site (P^x). Transcription of this gene is mediated by the promoter/ operator element $P_{N25/0}$ (P/O). Since the same plasmid constitutively synthesizes the *lac* repressor, transcription from P_{N25} is repressed in the absence of IPTG. A transcriptional terminator follows the *cat* gene to ensure that high-level transcription from P_{N25} does not proceed into the plasmid replication region, since this might have a deleterious effect on stability. Controlled CAT expression in *B. subtilis* from pREP9 can be achieved as follows.

1. Transformation: Plasmid pREP9 is introduced into competent cultures of *B. subtilis* BR151 *(trp, met, lys)* according to the method of Contente and Dubnau.[16] Although we predominantly use this method, protoplast transformation[17] or electroporation[18] can also be employed. Recombinant clones are selected on LB agar plates containing 10 μg/ml kanamycin at 37°.

2. Induction: Recombinant clones containing pREP9 are grown, with vigorous shaking, in LB medium containing 10 μg/ml kanamycin to mid-logarithmic phase ($A600 = 0.8$). Induction of protein synthesis is accomplished by adding IPTG (from a stock solution of 100 mg/ml) to a final concentration of 400μg/ml. Samples are removed over the next 4 hr for analysis.

3. Sample preparation: 200 μl of culture is removed and centrifuged at room temperature 12,000 rpm for 1 min. The cell pellets are resuspended in 20 μl of 15% (w/v) sucrose in 50 mM Tris-HCl, pH 7.2. Four microliters of a 5 mg/ml lysozyme solution (in 0.25 M Tris-HCl, pH 7.2) is added, and

[15] R. Marcoli, S. Iida, and T. A. Bickle, *FEBS Lett.* **110,** 11 (1980).
[16] S. Contente and D. Dubnau, *Mol. Gen. Genet.* **167,** 251 (1979).
[17] S. Chang and S. N. Cohen, *Mol. Gen. Genet.* **168,** 111 (1979).
[18] N. M. Calvin and P. C. Hanawalt, *J. Bacteriol.* **170,** 2796 (1988).
[19] H. Towbin, T. Staehelin, and J. Gordon, *Proc. Natl. Acad. Sci. U.S.A.* **76,** 4350 (1979).

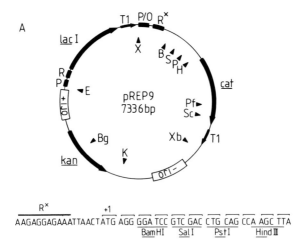

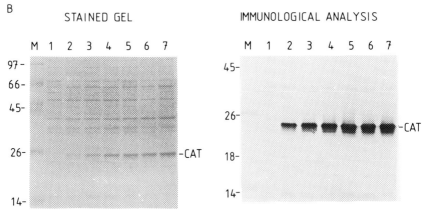

FIG. 3. (A) Structure of the bifunctional shuttle vector pREP9, determining regulatable gene expression in *B. subtilis*. In the *lacI* expression cassette [contained within the *Eco*RI (E) and *Xho*I (X) sites], P is the vegII promoter previously described; in the expression cassette [contained within the *Xho*I (X) and *Xba*I (Xb) sites], P/O is the regulatable promoter–operator element $P_{N25/O}$. Both promoters are linked to synthetic ribosome binding sites (R,R^x). The synthetic ribosome binding site R^x is followed by a multicloning site for the restriction endonucleases *Bam*HI (B), *Sal*I (S), *Pst*I (P), and *Hind*III (H); the DNA sequence below the plasmid diagram illustrates the reading frame from the ribosome binding site R^x through the multicloning site and is delineated into triplet codons. Gram-positive and -negative origins of replication are depicted ori+ and ori−, respectively. Genes conferring resistance to kanamycin (pUB110 derived) and chloramphenicol (*E. coli* derived[15]) are abbreviated *kan* and *cat*, respectively. Both the *lacI* and *cat* genes are flanked by the transcriptional terminator T1.[14] Additional unique restriction sites in the vector are abbreviated as follows: Pf, *Pfl*MI; Sc, *Sca*I; K, *Kpn*I (*Asp*718); Bg, *Bgl*II. (B) Regulatable expression of the *E. coli cat* gene in *B. subtilis* strain BR151 containing plasmid pREP9. In the stained gel analysis, lane 1 represents a cell lysate prior to addition of IPTG (i.e., preinduction). Following addition of IPTG, samples were withdrawn for analysis after 15, 30, 45, 60, 120, and 180 min (lanes 2–7,

the suspension incubated at 37° for 5 min. Thirty six microliters of SDS–polyacrylamide gel sample buffer is added, and the solution is heated for 10 min at 100°.

4. Electrophoresis: 10 μl of the cell lysate is fractionated through discontinuous 12.5 (w/v) SDS–polyacrylamide gels containing a 3.3% (w/v) stacking gel. Following electrophoresis, the gels are stained and destained according to standard procedures.

The results of such an experiment with pREP9 are presented in Fig. 3B. Prior to addition of IPTG, no CAT protein was observed on either a stained gel or by an immunological assay with polyclonal antibodies against CAT, which is at least an order of magnitude more sensitive than detection by Coomassie blue staining. The latter result illustrates clearly that, in the presence of *lac* repressor, transcription from P_{N25} is abolished. Addition of IPTG to the culture results in immediate derepression and high-level CAT production. After 2–3 hr of induction, CAT represents the major cellular protein, and levels approaching 15% of the total cell protein can be achieved.

Introduction of Foreign Genes into pREP9

In pREP9, the *E. coli* CAT protein is translated from an efficient synthetic ribosome binding site. A multicloning site in the immediate downstream vicinity of the initiator methionine residue (Fig. 3A) allows cloning of entire or fragmented genes. By addition of correctly sized molecular linkers, such genes or gene fragments can be expressed as amino-terminal fusions to CAT, allowing the possibility of detecting the fusion protein with CAT antibodies. Alternatively, if antibodies to the desired gene product are available, inserted DNA fragments may be so tailored to contain a translational stop codon. Since pREP9 is a bifunctional vector, it can be transformed into an *E. coli* host, from which large amounts of starting material for genetic manipulation can be rapidly isolated. Finally, it is worthwhile making a comment on utilization of the polylinker site of pREP9. Many researchers attempt to increase the versatility of their expression systems by adding new unique restriction sites to a polylinker. However, it is also possible to enhance the versatility of a multiple cloning

respectively). M, Molecular weight markers ($\times 10^{-3}$). In the immunological analysis, the same samples were similarly fractionated by SDS–polyacrylamide gel electrophoresis, then transferred to nitrocellulose by the method of H. Towbin *et al.*[19] Immunological detection was accomplished using a polyclonal antibody to chloramphenicol acetyltransferase (CAT) and a colorimetric assay. Lanes are denoted as in the stained gel. In both the stained gel and immunological analysis, the migration position of CAT is indicated.

site simply by making use of compatible cohesive ends shared by two or more restriction endonucleases. As an example, the cohesive end produced by *Bam*HI digestion of pREP9 is compatible with those produced by *Mbo*I, *Bgl*II, or *Bcl*I digestion. Similarly, DNA fragments produced by *Xho*I digestion can be introduced at the *Sal*I site, and those by *Nsi*I digestion at the *Pst*I site. Thus, without increasing the length of the polylinker region, sites for eight different restriction endonucleases can be utilized.

Dual-Plasmid Repressible System for *Bacillus subtilis*

Although the single plasmid repressible system afforded by pREP9 functions in *B. subtilis,* we were concerned that, at the time of entry of the recombinant plasmid into competent cells, RNA polymerase molecules might recognize the strong promoter P_{N25} in preference to P_{vegII}. Under such circumstances, a situation of transient derepression is possible and could conceivably have a destabilizing effect. One means of overcoming this problem would be to develop a strain of *B. subtilis* which constitutively expressed the *lac* repressor, and use this as recipient for expression plasmids containing the regulatable promoter.

The *S. aureus* plasmid pE194,[9] conferring erythromycin resistance, is capable of replicating in *B. subtilis*. We elected to transfer the *lacI* cassette of plasmid pREP9 into pE194. Previous reports showed that DNA might be inserted at the unique *Pst*I site of pE194 without disturbing either the replication or erythromycin resistance functions. We therefore excised the *lacI* cassette from pREP9 by *Eco*RI and *Xho*I digestion, converted the cohesive ends to blunt ends with DNA polymerase Klenow fragment, added *Pst*I linkers, and inserted this fragment into the *Pst*I site of pE194. The resulting plasmid, designated pBL1 (Fig. 4A), was transformed into *B. subtilis,* selecting at 32° (due to the temperature-sensitive replicon of pE194[20]) on 10 μg/ml erythromycin. By this approach, we generated the recombinant *B. subtilis* strain BR:BL1, which constitutively expressed *lac* repressor. Since the pE194 replicon contained pBL1 and the pUB110 replicon[21] on our shuttle vectors were compatible, it was possible to transform competent BR:BL1 with any pUB110-based plasmid containing the regulatable $p_{N25/0}$ element, selecting for recombinant clones on 10 μg/ml erythromycin (encoded on pBL1) and kanamycin (encoded on the pUB110-based vector) at 32°. As an example, we transformed BR:BL1 with the shuttle vector p602/20, generating the strain BR:BL1–p602/20 (Fig. 4A). Within p602/20, the *E. coli cat* gene is replaced with a gene

[20] J. Scheer-Abramowitz, T. Gryczan, and D. Dubnau, *Plasmid* **6,** 67 (1981).

[21] S. D. Ehrlich, *Proc. Natl. Acad. Sci. U.S.A.* **74,** 1680 (1977).

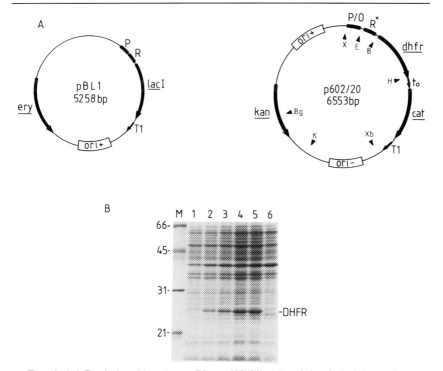

FIG. 4. (A) Dual-plasmid system, pBL1–p602/20, determining inducible synthesis of mouse DHFR in *B. subtilis.* Construction of the *lacI*-containing plasmid pBL1 has been outlined in the text. In p602/20, the reading frame at the *Bam*HI site connecting the synthetic ribosome binding site (R^x) to the *dhfr* coding sequence is the same as for plasmid pREP9, i.e., -G-G-A-T-C-C-. Notations and restriction site abbreviations are as in the legend to Fig. 3A. An additional transcription terminator, t_o [M. Rosenberg, B. De Crombrugghe, and R. Musso, *Proc. Natl. Acad. Sci. U.S.A.* **73,** 717 (1976)], is present on p602/20 between the *dhfr* and *cat* genes. On pBL1, *ery* is the erythromycin resistance gene. (B) Inducible DHFR synthesis in the recombinant *B. subtilis* strain BR:BL1–p602/20. The induction experiment was performed as described for pREP9, with the exception that the culture was grown at 32° in L broth containing 10 μg/ml of both erythromycin and kanamycin. Time points 1–5, at which samples were withdrawn for analysis, represent 0 (i.e., preinduction), 30, 60, 120, and 180 min, respectively. Lane 6 is a lysate of cells following prolonged growth in the *absence* of IPTG. The migration position of DHFR is indicated.

coding for mouse dihydrofolate reductase (*dhfr*[22]). The procedures for transformation and analysis of recombinant clones are similar to those previously described, with the following exceptions: (1) Recipient cells were competent BR:BL1, which were prepared at 32° in the presence of 10 μg/ml erythromycin. (2) Following introduction of p602/20, selection

[22] J. H. Nunberg, R. J. Kaufman, A. C. Y. Chang, S. N. Cohen, and R. T. Schimke, *Cell* **19,** 355 (1980).

of recombinant clones was at 32° on 10 μg/ml of both kanamycin and erythromycin.

Figure 4B illustrates inducible dihydrofolate reductase (DHFR) expression in the *B. subtilis* strain BR:BL1–p602/20. Prior to addition of IPTG, we observed no DHFR on a stained gel. Addition of IPTG led to rapid accumulation of DHFR, estimated to be in excess of 10% of the total cellular protein. Figure 4B also shows DHFR levels in BR:BL1–p602/20 following prolonged growth in the absence of IPTG (Fig. 4B, lane 6), demonstrating that, also in the dual-plasmid system, the $P_{N25/0}$ promoter element is very tightly regulated in the absence of IPTG.

As indicated in Fig. 4A, a *Bam*HI site lies between the initiator methionine residue and the coding region of *dhfr*. The reading frame within this *Bam*HI site is as in pREP9 (i.e., <u>GGA</u>-<u>TCC</u>); thus, by addition of appropriately sized *Bam*HI, *Bgl*II or *Bcl*I linkers to the DNA clone in question, it is possible to fuse a foreign coding sequence to DHFR. Alternatively, if the foreign coding sequence provides its own translational stop codon, it can be expressed individually.

Regulated Expression of Enzymatically Active Reverse Transcriptase in *Bacillus subtilis*

To determine whether enzymatically active heterologous protein could be produced in *B. subtilis* from our expression systems, we used the dual-plasmid system to express the HIV-1 reverse transcriptase. As illustrated in Fig. 5A, the major portion of the HIV-1 polymerase *(pol)* open reading frame[23] was introduced as a translational fusion to the *E. coli cat* gene. This *pol* fragment encodes a protease, reverse transcriptase, and a truncated endonuclease (here as a CAT fusion). We used plasmid p602/18 as the vector, which does not contain the *lacI* gene now supplied by plasmid pBL1. The *pol*-encoding shuttle vector, pRTL11 (Fig. 5A), was introduced into strain BR:BL1, selecting for transformants at 32° on LB agar containing 10 μg/ml erythromycin and kanamycin. When a culture of the recombinant clone BR:BL1–pRTL11 was induced with IPTG and immunoreactive polypeptides in the cell lysate analyzed with a pool of HIV-1 positive sera,[24] we observed accumulation of a high-molecular-weight polypeptide (Fig. 5B), indicative of the *pol*-encoded precursor poly-

---

[23] L. Ratner, W. Haseltine, R. Patarca, K. J. Livak, B. Starich, S. F. Josephs, E. R. Doran, J. A. Rafalski, E. A. Whitehorn, K. Baumeister, L. Ivanoff, S. R. Petteway, Jr., M. L. Pearson, J. A. Leutenberger, T. S. Papas, J. Ghrayeb, N. T. Chang, R. C. Gallo, and F. Wong-Staal, *Nature (London)* **313**, 277 (1985).

[24] U. Certa, W. Bannwarth, D. Stüber, R. Gentz, M. Lanzer, S. F. J. Le Grice, F. Guillot, I. Wendler, G. Hunsmann, H. Bujard, and J. Mous, *EMBO J.* **5**, 3051 (1986).

protein. After prolonged induction, the precursor polyprotein disappeared, and was replaced by polypeptides of molecular mass 66 and 51 kDA. Since these latter polypeptides correspond in size to the expected size of the HIV-1 reverse transcriptase polypeptides, the results presented in Fig. 5B suggest that the HIV-1 *pol* polyprotein is properly processed in *B. subtilis.* Once again, no immunoreactive polypeptides were observed prior to IPTG induction, showing tight regulation of the N25 promoter. Using a rapid purification method outlined in more detail elsewhere,[25] we isolated HIV-1 reverse transcriptase from strain BR : BL1 – pRTL11. The following steps were used in the purification: (1) lysozyme – Triton X-100 lysis at 10°, (2) high-speed centrifugation to remove cell debris, (3) overnight dialysis of the soluble fraction, (4) elimination of nucleic acids by streptomycin sulfate precipitation, and (5) ion-exchange chromatography on DEAE-Sephacel.

Figure 5C illustrates that, under these conditions, the p66 and p51 reverse transcriptase polypeptides did not bind to the DEAE-Sephacel, providing a very convenient means of partial purification (approximately 90% of the soluble protein is retained on the column). Finally, Fig. 5D shows that the partially purified enzyme preparation displayed high levels of reverse transcriptase activity.

Inducible *Bacillus subtilis* System Employing Integrated Copy of *lacI* Gene

Although both the single- and dual-plasmid-inducible systems work well in *B. subtilis,* each system has disadvantages. For example, pREP9 is somewhat large (7.3 kb), and increasing its size via insertion of foreign DNA could affect plasmid stability. In the dual-plasmid system, the expression vector is smaller, but recombinant clones must be grown at 32°. Integration of the *lacI* gene into the *B. subtilis* chromosome would overcome both problems. One consideration in following this approach is to ensure sufficient levels of *lac* repressor from an integrated gene. In preliminary experiments, we have constructed a *lacI* cassette with a stronger transcriptional signal. This cassette on a multicopy plasmid produces *lac* repressor levels in *B. subtilis* approaching 5 – 10% of the total cellular protein. This cassette was introduced into the *B. subtilis* chromosome via a bacteriophage φ105 cloning vehicle recently constructed by Errington *et al.*[26] *Bacillus subtilis* containing the integrated *lacI* gene can be transformed with plasmid p602/20 described earlier and again displays inducible DHFR synthesis. (S. F. J. Le Grice, unpublished data).

[25] S. F. J. Le Grice, V. Beuck, and J. Mous, *Gene* 55, 95 (1987).
[26] J. Errington, *in* "Bacillus Molecular Biology and Biotechnology Applications" (A. T. Ganesan and J. A. Hoch, eds.), p. 217. Academic Press, New York, 1986.

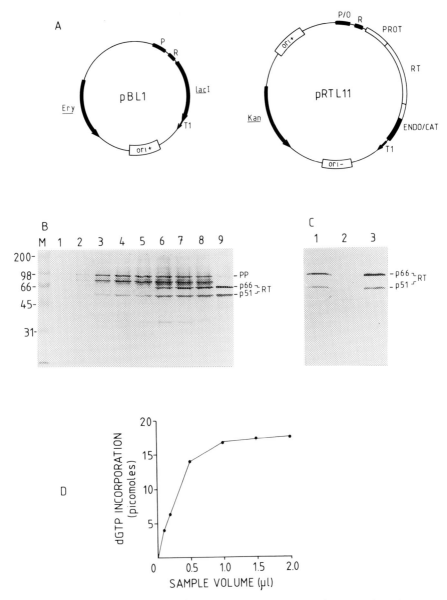

FIG. 5. (A) Dual-plasmid repressible system, pBL1–pRTL11, for expression of a truncated HIV-1 *pol* open reading frame in *B. subtilis.* Plasmid pBL1 is as before. On plasmid pRTL11, a *Bgl*II–*Xmn*I *pol* fragment was cloned into a related expression vector, p602/19. The cloning translationally fused the *pol* endonuclease portion to the *E. coli* CAT protein (ENDO/CAT). The other *pol*-encoded proteins, protease and reverse transcriptase, are abbreviated PROT and RT, respectively. (B) Induction of reverse transcriptase from the HIV-1 *pol* precursor polyprotein in a culture of the *B. subtilis* strain BR:BL1–pRTL11. The culture was grown and induced at 32° in a similar manner to the BR:BL1–p602/20 system. Lane 1 is the preinduction sample. Lanes 2–8 represent samples removed after 15, 30, 45, 60, 120, 180, and 240 min, respectively. An immunological analysis is presented, wherein the source

Construction of Inducible Secretion Vectors

Although another section of this volume will deal with secretion vectors for *B. subtilis* in considerably more detail, it is nonetheless worthwhile to mention that the regulatable system outlined here can also be modified to secrete foreign proteins. In both the single- and dual-plasmid systems described above we have inserted a cassette containing the *Bacillus amyloliquifaciens* α-amylase signal peptide[27] linked to the mature portion of the *E. coli* β-lactamase gene.[28] When cultures of *B. subtilis* containing the recombinant plasmids are grown in a minimal medium and induced during logarithmic growth, we observe accumulation of mature β-lactamase in the culture supernatant.

Concluding Remarks

Rather than give explicit details of vector construction, I have attempted to give a brief description of the systems available, together with their application. The single- and dual-plasmid systems presented here have been constructed so that entire expression cassettes can be mobilized. For example, the modified *lacI* cassette can be removed from pREP9 by *Eco*RI and *Xho*I digestion, or from the plasmid pBL1 by *Pst*I digestion. Similarly, the cassette conferring inducible CAT expression can be excised from pREP9 by *Xho*I and *Xba*I digestion. This flexibility in the expression vectors should permit those interested to modify further or produce their own gram-positive expression systems. This could be especially worthwhile with the *lacI* cassette, which could be stably integrated into the *B. subtilis* chromosome in an alternative manner to that described here.

[27] I. Palva, R. F. Petterson, N. Kalkkinen, P. Lehtovaara, M. Servas, H. Söderlund, K. Takkinen, and L. Kääriäinen, *Gene* 15, 43 (1981).
[28] J. T. Kadonaga, P. E. Gautier, D. R. Strauss, A. D. Charles, M. D. Edge, and J. R. Knowles, *J. Biol. Chem* 259, 2149 (1984).
[29] P. Dhaese, C. Hussey, and M. Van Montagu, *Gene* 32, 181 (1984).
[30] M. S. Osbourne, R. J. Craig, and D. Rothstein, *J. Bacteriol.* 163, 1101 (1985).

of antibody was a pool of HIV-1-positive sera. PP refers to the *pol* precursor polyprotein. Lane 9 contains a sample of purified p66/p51 HIV-1 reverse transcriptase (RT). (C) Partial purification of HIV-1 reverse transcriptase, produced from the *B. subtilis* recombinant strain BR:BL1–pRTL11 by DEAE-Sephacel ion-exchange chromatography. An immunological analysis is presented, using again pooled HIV-1-positive sera as antibody source. Lane 1, DEAE-Sephacel flow-through material (i.e., non-DEAE-Sephacel-binding proteins); lane 2, proteins eluted with a buffer containing 0.2 *M* NaCl; lane 3, purified p66/p51 HIV-1 reverse transcriptase. (D) Activity profile for HIV-1 reverse transcriptase isolated from *B. subtilis* strain BR:BL1-pRTL11. The assay here, performed on the DEAE-Sephacel-purified enzyme, is based on the ability of reverse transcriptase to incorporate [^{32}P]dGTP into polynucleotide, using a poly(rC)oligo(dG) template–primer system.[25]

In the systems presented, foreign genes are expressed during logarithmic growth, taking advantage of the observation that the *E. coli* bacteriophage T5 promoter P_{N25} is recognized by the vegetative ($\sigma55$) RNA polymerase of *B. subtilis*. Since there is rapid interchange of σ factors in *B. subtilis* at later growth stages,[31] it is unlikely that P_{N25} and P_{vegII} are recognized by the alternative forms of RNA polymerase. If it were necessary to produce foreign proteins at later growth stages, the system described here could be modified by replacing the vegII and N25 promoters with counterparts utilized by other forms of *B. subtilis* RNA polymerase. Alternatively, these growth-stage-specific promoters could be placed in the immediate 5' vicinity of P_{vegII} and $P_{N25/0}$. Such a "twin" promoter would ensure constant synthesis of *lac* repressor. Furthermore, Deutchle *et al.* have recently demonstrated that the *lac* operator sequence can be separated from the promoter and still prevent RNA polymerase from entering into productive transcription.[32] Thus, transcription from a growth-stage-specific promoter placed 5' to the $P_{N25/0}$ element would still be controlled by the *lac* operator.

The plasmids and strains described in this article, together with complete nucleotide sequence and restriction maps, can be made available upon request.

Acknowledgments

I wish to thank Ursula Peschke and Verena Beuck for their assistance in early expression vector studies. The gifts of various expression cassettes from D. Stuber, R. G. Gentz, and J. Knowles are also acknowledged, as well as helpful comments on the manuscript from Jan Mous and Oktavian Shatz. Finally, I wish to thank Professor H. Bujard, whose excellent work on T5 promoters and belief in *B. subtilis* were instrumental in generation of the early expression systems.

[31] R. Losick and J. Pero, *Cell* **25**, 585 (1981).
[32] U. Deutchle, R. Gentz, and H. Bujard, *Proc. Natl. Acad. Sci. U.S.A.* **83**, 4134 (1986).

[19] System for Secretion of Heterologous Proteins in *Bacillus subtilis*

By VASANTHA NAGARAJAN

Introduction

Secretion of heterologous proteins from *Bacillus subtilis* has several attractive properties. The secreted protein is usually soluble and active. Most importantly, secreted proteins from *B. subtilis* are located in the growth medium, in contrast to *Escherichia coli,* where the majority of secreted proteins are localized in the periplasmic space. The protein can usually be recovered from the medium with relatively few contaminating proteins, simplifying further steps in the purification. Several vectors designed for the secretion of heterologous proteins from *B. subtilis,* based on the genes for several secreted proteins (α-amylase, alkaline protease, neutral protease, and levansucrase), have been described.[1-5]

This article describes a secretion vector derived from the alkaline protease (subtilisin) gene (*apr[BamP]*) of *Bacillus amyloliquefaciens.* The promoter, translation initiation, and signal sequences are all provided by the *apr[BamP]* gene. The vector also provides origins of replication and selectable markers for both *E. coli* and *B. subtilis,* allowing constructions to be made in either organism. The *E. coli* origin and ampicillin resistance gene are derived from pBR322, while the pC194 chloramphenicol resistance gene and origin of replication provide replication functions and selection for *B. subtilis.*

Description of *apr[BamP]*-Based Secretion Vector

Although alkaline protease is synthesized as a preproprotease,[6,7] the first 30 amino acids are sufficient to function as a signal sequence. In order to simplify construction of heterologous gene fusions, a *Bam*HI recognition sequence was created by site-directed mutagenesis two codons from the end of the signal (pre) peptide coding region (Fig. 1). Introduction of this *Bam*HI site resulted in the addition of two amino acids (Asp and Pro) in the pro peptide sequence. The addition of these two amino acids did not interfere with the expression or activity of the alkaline protease.[3] The insertion of a heterologous gene at this *Bam*HI site results in the inactivation of the *apr[BamP]* gene. The ability of the alkaline protease signal peptide to translocate heterologous proteins was demonstrated with staphylococcal protein A and bovine pancreatic ribonuclease A (RNase) as described below. A diagram of pGX2134, an *E. coli–B. subtilis* shuttle vector containing the *apr* gene with the introduced *Bam*HI site is shown in Fig. 2.

[1] Palva *et al., Proc. Natl. Acad. Sci. U.S.A.* **79,** 5582 (1982).
[2] Palva *et al., Gene* **22,** 229 (1983).
[3] N. Vasantha and L. D. Thompson, *J. Bacteriol.* **165,** 837 (1986).
[4] Honjo *et al., J. Biotechnol.* **4,** 63 (1986).
[5] Joyet *et al.,* in "Bacillus Molecular Genetics and Biotechnology Applications" (A. T. Ganesan and J. A. Hoch, eds.), p. 479. Academic Press, New York, 1986.
[6] N. Vasantha *et al., J. Bacteriol.* **159,** 811 (1984).
[7] S. D. Power, R. M. Adams, and J. A. Wells, *Proc. Natl. Acad. Sci. U.S.A.* **83,** 3096 (1986).

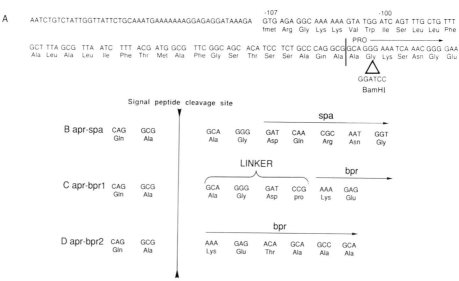

FIG. 1. The entire DNA sequence of *apr[BamP]* has been published.[6] (A) DNA sequence of the signal peptide coding region and location of the introduced *Bam*HI site. DNA sequences across the fusion junction of (B) *apr-spa*(pGX2136), (C) *apr-bpr1*(pGX2211), and (D) *apr-bpr2*(pGX2214) are also shown.

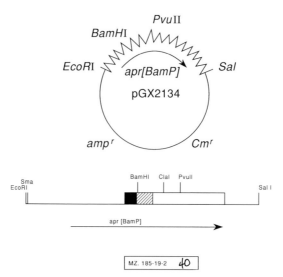

FIG. 2. pGX2134 is an *E. coli-B. subtilis* shuttle vector containing the *apr[BamP]* sequence on an *Eco*RI–*Sal*I fragment. A linear map of the *Eco*RI–*Sal*I fragment with the relevant restriction sites is shown below. ■, pre; ▨, pro; □, mature coding sequence of *apr[BamP]*.

Construction of Hybrid Gene Fusions

Plasmid pGX2912 is a previously described plasmid with the protein A gene that has a *Bcl* site at the twenty-second and twenty-third codons of mature protein A gene *(spa)* and a *Pvu*II sites following the gene.[8] pGX2134 was digested with *Bam*HI and *Pvu*II and ligated to the *Bcl*I–*Pvu*II fragment containing the *spa* gene from pGX2912. The ligation mixture was transformed into *E. coli* and the desired plasmid (pGX2136) was isolated. *Bacillus subtilis* strain GX4935 *(trpC2 metB10 lys-3 ΔaprE Δnpr)* was transformed with pGX2136, and colonies containing the proper plasmid expressing the protein A gene were screened by colony immunoassay (see below).

A *Bam*HI site was created preceding the mature sequence of bovine pancreatic ribonuclease *(bpr)* by site-directed mutagenesis of the *bpr* gene which had been subcloned into M13mp19.[9] The replicative form of this M13 phage was isolated, and a *Bam*HI–*Pvu*II fragment containing the *bpr* gene was isolated and ligated into pGX2134 which had also been digested with *Bam*HI and *Pvu*II. The ligation was done at a high DNA concentration (100 μg/ml) to generate long concatemers. Both *E. coli* and *B. subtilis* were transformed with the ligated DNA. Two out of the 12 *E. coli* transformants screened had the correct plasmid based on restriction analysis and were designated as pGX2211 and pGX2212. Site-directed mutagenesis was performed on pGX2211 (an *E. coli–B. subtilis*–M13 shuttle vector containing *apr–bpr1*) to delete the 12 nucleotides (Fig. 1) between the end of the signal peptide coding region and mature *bpr*. The resulting plasmid carrying *apr–bpr2* was designated pGX2214. *Bacillus subtilis* GX4935 was transformed with pGX2211 and pGX2214 by the competent cell transformation protocol of Gryczan *et al.*[10] Selection was for chloramphenicol resistance on TBAB plates supplemented with 5 μg/ml of chloramphenicol. Transformants were screened for RNase activity by colony screening as described below.

Once positive colonies expressing protein A or RNase were identified, plasmid DNA was isolated and its structure verified by restriction analysis. The bacteria carrying the correct plasmid were stored by resuspension in Spizzen minimal medium [0.2% (w/v) ammonium sulfate, 0.14% (w/v) dibasic potassium phosphate, 0.6% (w/v) monobasic potassium phosphate, 0.1% (w/v) sodium citrate] plus 20% (v/v) glycerol, frozen in a dry ice–ethanol bath, and maintained at −70°.

[8] S. R. Fahnestock and K. E. Fisher, *J. Bacteriol.* **165,** 796 (1986).
[9] N. Vasantha and D. Filpula, *Gene* **76,** 53 (1989).
[10] T. S. Gryczan, S. Contente, and D. Dubnau, *Mol. Gen. Genet.* **177,** 459

Colony Screening

Transformants were patched onto two filters overlaid on TBAB [tryptose blood agar base (Difco, Detroit, MI), 33 g/liter] + Cm (5 μg/ml) plates.[11] The top filter was a cellulose acetate membrane filter [Schleicher & Schuell (Keene, NH) OE67] and the bottom filter was a nitrocellulose filter (Schleicher & Schuell BA85). The plates were incubated for 17 hr at 37°. The bacteria were retained on the cellulose acetate filter and secreted proteins passed through the cellulose acetate filter and bound to the nitrocellulose filter below. The nitrocellulose filter was removed, and processed as follows:

1. Incubate the filter for 1 hr in 10 ml of TS buffer [10 mM Tris-hydrochloride, pH 8.0, 0.9% (w/v) sodium chloride] containing 3% (w/v) bovine serum albumin (BSA) at room temperature.

2. Add 10 μl of rabbit anti-RNase serum and incubate the filter for 1 hr at 25°.

3. Wash three times with TS buffer to remove the antibody.

4. Add 5 μl of horseradish peroxidase conjugate of goat anti-rabbit IgG (Bio-Rad, Richmond, CA) to 10 ml of TS buffer and incubate the filter for 1 hr at room temperature.

5. Wash with TS buffer three times and develop using 4-chloro-1-naphthol and hydrogen peroxide. (Dissolve 36 mg of 4-chloro-1-naphthol in 12 ml of methanol and add 48 ml of TS buffer. Mix thoroughly and add 120 μl of 30% hydrogen per oxide. Use immediately and do not store.)

RNase Activity Screen

The nitrocellulose filter can also be screened for RNase activity by overlaying the filter on an RNA-agarose plate [1% (w/v) yeast RNA in 50 mM Tris-HCl, pH 7.0, 10 mM EDTA, and 1% (w/v) agarose] for 4 to 8 hr and incubated at 37°. The filter is removed and the plate is treated with 0.1 M sulfuric acid; zones of clearing indicate RNase activity.

Analysis of Expression of Heterologous Protein

There are several methods to determine how efficiently a heterologous protein is secreted from *B. subtilis*. Immunoblot analysis of a culture supernatant cannot determine whether a protein is present in the growth medium due to secretion or cell lysis. Proteolysis of the accumulated protein in the culture supernatant can complicate the interpretation of the results of an immunoblot. Therefore, pulse–chase experiments can be used to determine the efficiency of secretion of a protein and its stability.

---

[11] C. W. Saunders, *et al., J. Bacteriol.* **169,** 2917 (1987).

Protocol for Pulse – Chase Analysis

1. Start overnight cultures of the desired strain and a negative control strain on a TBAB + Cm plate from the frozen stock and incubate at 30°.

2. Inoculate the strains into 20 ml of medium S7 [10 mM ammonium sulfate, 50 mM potassium phosphate, pH 7.0, 2 mM MgCl$_2$, 0.7 mM CaCl$_2$, 50 μM MnCl$_2$, 5 μM FeCl$_3$, 1 μM ZnCl$_2$, 2 μM thiamin, 1% glucose, 20 mM L-glutamate (adjusted to pH 7.0 with KOH), tryptophan (50 μg/ml), methionine (50 μg/ml), and lysine (50 μg/ml)] in a 125-ml flask such that the absorbance of the initial inoculum is OD$_{600}$ = 0.05. No antibiotic is added because the shift down and addition of antibiotic results in a long growth lag.

3. Grow at 37° with aeration at 120 rpm; doubling time of the cultures is normally 60 to 85 min; if longer, add 0.001% yeast extract.

4. At an absorbance of OD$_{600}$ = 0.5, spin 2.0 ml of the culture for 5 min at 6000 rpm at room temperature.

5. Wash the cells with 2 ml of prewarmed synthetic medium lacking methionine.

6. Resuspend the cells in 2 ml of prewarmed synthetic medium lacking methionine.

7. After a 10-min incubation at 37° with aeration, add 50 – 100 μCi of L-[^{35}S]methionine. After 3 min, add 30 μl of chase solution (1 mg of puromycin per milliliter, 5 mg of methionine per milliliter). Withdraw samples (500 μl) after 2, 5, and 10 min, and add them to a microfuge tube containing 25 μl of BSA (5 μg/ml). Spin for 20 sec at room temperature in a microfuge to separate the cell and supernatant fractions. Precipitate proteins in both the cell pellet and supernatant with trichloroacetic acid [5% (v/v) final concentration].

9. After 1 hr at 4°, spin down the precipitate at 4° and wash with 80% (v/v) acetone.

10. Air dry the acetone pellet and add 50 μl of lysozyme (1 mg/ml) in buffer A [50 mM Tris – 10 mM EDTA + 1 mM phenylmethylsulfonyl fluoride (PMSF)] to the cell pellet and buffer A alone to the supernatant fraction.

11. Vortex thoroughly and incubate at 37° for 10 min; add 15 μl of 4× sample preparation buffer (Laemmli)[12] to all samples, vortex, and boil for 5 min.

12. Spin at room temperature for 15 min in a microfuge and transfer the supernatant to a microfuge tube. Process the supernatant for immunoprecipitation using any standard protocol.[3] Resuspend the immunoprecipitate in 30 μl of Laemmli sample buffer[12] and run 10 μl per lane on a polyacrylamide gel.

[12] U. K. Laemmli, *Nature (London)* **227**, 680 (1970).

Demonstration of Synthesis of Precursor Proteins

Pulse–chase experiments do not always reveal the presence of a precursor protein due to the rapid processing of the precursor. Precursors for secreted proteins can be observed by inhibiting protein translocation. In *E. coli,* precursors of secreted proteins have been observed by labeling cells in the presence of 2-phenylethanol (PEA), which disrupts the proton-motive force.[13] Precursor species for protein A, human serum albumin, ribonuclease, and β-lactamase have been observed by labeling *B. subtilis* in the presence of PEA.[3,9,11,14] Figure 3 shows results from a culture of *B. subtilis* cells containing pGX2214 labeled with L-[35S]methionine in the presence of varying amounts of PEA. Cultures treated with 0 to 0.6% PEA showed a protein with an apparent mobility of 14 kDa, identical to that of mature RNase, in both the cell and supernatant fractions. The cultures treated with 0.6% PEA showed an additional protein with an apparent mobility of 17 kDa in the cellular fraction. Prolonged exposure of the fluorogram revealed a 17-kDa protein in the cellular fraction of cells treated with 0.8% PEA (not shown). However, the 17-kDa protein was never observed in the supernatant fraction. The expected size of the precursor protein would be 3 kDa larger than the mature protein due to the signal sequence, and thus the 17-kDa cell-associated protein is likely to be the unprocessed precursor protein.

Determination of Signal Peptide-Processing Site

It is important to confirm that a secreted protein has the correct N-terminal amino acid sequence. One method is first to purify the protein and then determine its N-terminal amino acid sequence. Another simple method to determine the N-terminal sequence is to radiolabel the protein with one amino acid and then determine the position of labeled residue in the sequence. For example, lysine is the third and fourth residue in the alkaline protease signal peptide and the first and eighth residue in bovine pancreatic ribonuclease (Fig. 1A, D). A culture of *B. subtilis* carrying pGX2214 was labeled with [3H]lysine and processed as described below. The majority of the counts released by the Edman degradation were in cycles one and eight. These results are consistent with correct processing of the signal peptide, indicating that *B. subtilis* can secrete ribonuclease with the same N terminus as the native enzyme.

[13] C. J. Daniels, *et al. Proc. Natl. Acad. Sci. U.S.A.* **78,** 5396 (1981).
[14] N. Vasantha and L. D. Thompson, *Gene* **49,** 23 (1986).

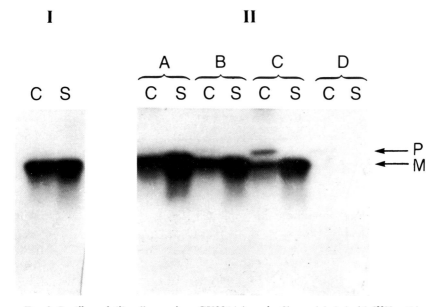

FIG. 3. *Bacillus subtilis* cells carrying pGX2214 *(apr–bpr2)* were labeled with [³⁵S]methionine for 3 min. (I) Control culture with no PEA; (II) PEA-treated culture. PEA at 0.2% (A), 0.4% (B), 0.6% (C), and 0.8% (D) was present during methionine starvation. The cellular (C) and supernatant (S) fractions were separated and immunoprecipitated. P, precursor; M, mature.

N-Terminal Analysis of Protein by Radiolabeling

Steps 1 through 4 are as described previously for pulse-labeling experiments.

5. Wash the cells with 2 ml of prewarmed S7 medium lacking lysine.

6. Resuspend the cells in 1 ml of S7 medium lacking lysine and containing 2 mM PMSF (a protease inhibitor is necessary to prevent proteolysis of the secreted protein). Incubate for 10 min with aeration (at 37°).

7. Add [³H]lysine (200 μCi) and label for 5 min.

8. Add 30 μl of chase solution (1 mg/ml puromycin, 5 mg/ml lysine).

9. After 10 min, separate the cell and supernatant fractions and immunoprecipitate as described above.

10. The immunoprecipitate is dissolved in 40 μl of 2% SDS, boiled for 5 min, and centrifuged in a microfuge for 15 min. The supernatant is removed and 5 μl is analyzed by polyacrylamide gel electrophoresis to ensure that only a single species is present.

11. The rest of the solubilized protein (35 μl) is diluted to 100 μl with water, 800 μl of 80% cold acetone is added, and the protein is precipitated for 17 hr at −20°.

12. The precipitated protein is centrifuged for 15 min in a microfuge at

TABLE I
MEDIUM OPTIMIZATION[a]

Medium	Cell density	Protein A (mg/liter)
Pen assay + $MnCl_2$	4.8	3
2% yeast extract	4.3	12
GY medium	9.7	32
Medium A	10.2	300

[a]Cell density (absorbance at 600 nm) and amount of
protein A [S. Lotdahl, B. Guss, M. Uhlen, L. Philipson,
and M. Lindberg, *Proc. Natl. Acad. Sci. U.S.A.* **80**, 697
(1983)] were determined in a 12-hr culture.

4° and the supernatant is gently aspirated. The pellet is washed again with
80% acetone to remove all traces of SDS.

13. Dissolve the precipitate in 30% acetic acid and subject it to automated Edman degradation.

Medium Optimization

The accumulation of the proper product will be a balance between the
final cell density, the specific productivity, and the degradation of the
product. *Bacillus subtilis* strain GX4935 containing pGX2143 (*apr–spa* in
pPL703) was grown in the following media:

Medium A (tryptone, 33 g; yeast extract, 20 g; NaCl, 7.4 g; Na_2HPO_4,
8 g; KH_2PO_4, 4 g; casamino acids, 20 g; glucose, 10 g; $MnCl_2$,
0.06 mM; and NaOH to pH 7.5)
Pen assay broth + 50 μM $MnCl_2$
2% yeast extract
Glycerol–yeast (GY) extract medium (50 mM potassium phosphate,
2 mM $MgCl_2$, 0.7 mM $CaCl_2$, 50 μM $MnCl_2$, 5 μM $FeCl_3$, 1 μM
$ZnCl_2$, 2 μM thiamin, 1% glycerol, 1% yeast extract, and 20 mM
sodium glutamate)

The highest accumulation of protein A was in medium A (Table I).
Some proteolysis of the protein A was still observed, even though the host
strain has deletions of the two major extracellular proteases.[15,16] Although
the yield of product was lower, proteolysis of the protein A was less in a
glycerol–yeast extract medium. Optimization of the medium for each
heterologous protein would probably be worthwhile.

[15] M. Y. Yang, E. Ferrari, and D. J. Henner, *J. Bacteriol.* **160**, 15 (1984).
[16] S. R. Fahnestock and K. E. Fisher, *Appl. Environ. Microbiol.* 379 (1987).

Acknowledgments

I thank Mark Guyer, Stephen Fahnestock, Charles Saunders, and Ethel N. Jackson for critical discussions during the course of this work and Leo Thompson for the construction of *apr-spa*. This work was performed at Genex Corporation, Gaithersburg, MD 20877.

[20] Inducible Expression of Regulatory Genes in *Bacillus subtilis*

By DENNIS J. HENNER

Introduction

The ability to manipulate the expression of a gene of interest easily can be a valuable means to regulate gene expression. In *Bacillus subtilis,* the homologous recombination of nonreplicative plasmids into the chromosome makes it particularly simple to inactivate genes, to produce gene duplications, and to put any gene under the control of a regulated promoter. By this means, one can study the consequences of the loss of expression of an essential gene, or the inappropriate expression of a gene product. There are many examples of such studies in other organisms. For example, a system devised to put the *Escherichia coli* signal peptidase under control of the *araC* promoter was used to show that the signal peptidase was essential for cell viability.[1] In yeast, the plasma membrane ATPase was put under the control of a galactose-inducible promoter, and it was shown that the cells were not viable unless inducer was present.[2] This article describes a very simple system for placing genes under control of an inducible promoter in *B. subtilis.*

Design of Regulatory System

As previously described in this volume by Le Grice,[3] the simplest approach to an inducible promoter system in *B. subtilis* was to import one from *E. coli.* The hybrid promoter that we constructed, designated *spac-1,* contains the RNA polymerase recognition site from an early promoter of the *B. subtilis* phage SPO-1 and the *lac* operator.[4] The *lacI* gene, encoding the *lac* repressor, was placed under the control of the *Bacillus licheniformis* penicillinase transcriptional and translational control signals to ensure expression in *B. subtilis.*[4]

[1] R. E. Dalbey and W. Wickner, *J. Biol. Chem.* **260,** 15925 (1985).
[2] A. Cid, R. Perona, and R. Serrano, *Curr. Genet.* **12,** 105 (1987).
[3] S. F. J. Le Grice, this volume, [18].
[4] D. Yansura and D. J. Henner, *Proc. Natl. Acad. Sci. U.S.A.* **81,** 439 (1984).

Figure 1 shows the structure of the integration plasmids that have been constructed. These plasmids have the pBR322 origin of replication and ampicillin resistance gene. A *cat* gene derived from plasmid pC194 allows selection for integration in *B. subtilis* on plates containing chloramphenicol. The polylinkers following the two plasmids provide a variety of sites to allow insertion of DNA fragments.

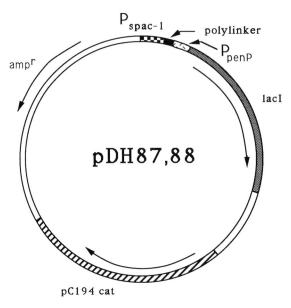

pDH87

HindIII			Xbal	Sall	Pstl	Sphl
AAGCTTAAGGAGGTGATCTAGAGTCGACCTGCAGGCATGC						

pDH88

HindIII	Smal	Xbal	Hpal	BglII
AAGCTTTTCCCGGGTTTCTAGATTTGTTAACTTAGATCT				

Clal	Sphl
TTATCGATTTGCATGC	

FIG. 1. Structure of plasmids pDH87 and pDH88. The locations of the *spac-1* promoter, polylinker, *lacI* gene, and pC914 *cat* gene are as shown and are approximately to scale. The rest of the plasmid is derived from pBR322 and contains the origin of replication and ampicillin resistance gene (*amp*r). The sequences of the polylinkers of the two plasmids are shown below, along with the relevant restriction sites. All of the restriction sites listed, with the exceptions of the *Pst*I site in pDH87 and the *Hpa*I site in pDH88, are unique in each plasmid.

Placing *Bacillus subtilis spo0A* Gene under Inducible Control

The *B. subtilis spo0A* gene is essential for sporulation, and mutations at this locus block sporulation at a very early stage.[5] An approximately 210-bp fragment of the *spo0A* gene was isolated and ligated behind the *spac-1* promoter in the polylinker. This fragment contains about 30 nucleotides preceding the *spo0A* initiation codon, which includes the natural ribosome binding site, and extends through the first 62 of the 267 codons of the gene.[6] The fragment does not contain the *spo0A* promoter. The plasmid carrying the *spo0A* fragment was designated pDH86. As shown in Fig. 2, integration by a crossover event within the *spo0A* gene fragment on the plasmid results in a truncated copy of the *spo0A* gene behind the natural *spo0A* promoter, and a complete copy of the gene behind the *spac-1* promoter. A point of caution must be raised, the copy number of the integrated plasmid can be higher than one, especially if higher levels of chloramphenicol are used. In such cases, the *spac-1* promoter would also drive the expression of the truncated protein. There is always a possibility that the truncated protein could have an unanticipated effect.

Transformation of *Bacillus subtilis* with Integrative Plasmids

1. Grow the recipient strain overnight at 30° on a TBAB (tryptose, blood, agar base; Difco, Detroit, MI) plate.

2. Resuspend cells from the overnight plate into stage 1 medium at an OD_{600} of approximately 0.1. Incubate with good aeration at 37° for 4–5 hr.

3. Add 0.1 ml of the stage 1 culture to 0.9 ml of stage 2 medium containing 5 μl of the appropriate plasmid from any standard miniprep protocol or approximately 0.5 μg of DNA.

4. Continue incubation at 37° for 60–90 min.

5. Plate 0.1-ml aliquots on TBAB plates supplemented with chloramphenicol at 5 μg/ml.

6. Streak individual transformants for single colonies on TBAB plates supplemented with chloramphenicol at 5 μg/ml.

Control of Sporulation by IPTG

As shown in Table I, a strain carrying pDH86 now shows isopropyl-β-D-thiogalactopyranoside (IPTG)-inducible sporulation. At the highest IPTG concentration, the sporulation frequency of the 168/pDH86 is indistinguishable from its parent. Intermediate levels of IPTG give intermediate

[5] J. A. Hoch, *Adv. Genet.* **18,** 69 (1976).

[6] F. A. Ferrari, K. Trach, D. LeCoq, J. Spence, E. Ferrari, and J. A. Hoch, *Proc. Natl. Acad. Sci. U.S.A.* **82,** 2647 (1985).

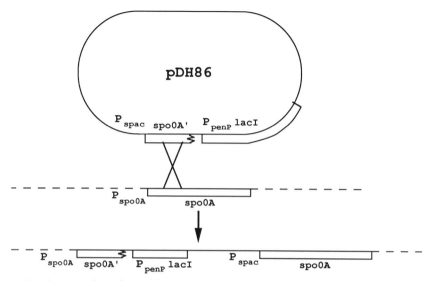

FIG 2. Integration of pDH86 by homologous recombination. The plasmid sequence is indicated by solid lines, the chromosomal sequences by dotted lines. The boxes indicate the *spo0A* and *lacI* coding regions, and the jagged ends of the *spo0A'* indicate a truncated copy of the gene. The positions of the promoters are shown by P_{spac} and P_{spo0A}. The figure is not to scale.

levels of sporulation, and the uninduced cultures have sporulation frequencies six orders of magnitude below that of the fully induced cultures. Even this low level of sporulation indicates some leakiness of the control system, as a strain with a deletion of the *spo0A* gene can produce no spores at all. A further point to be noted is that the integration event shown in Fig. 1 is a reversible process, and that cells that lose the integrated plasmid would score as Spo+ and chloramphenicol sensitive. Although no such recombinants were seen in this experiment, they can be easily avoided by maintaining selective pressure for chloramphenicol.

Induction Ratio of *spac-1* Promoter

The above experiment with the *spo0A* gene demonstrates that this system can be used to regulate a biological response. The induction ratio of the system cannot be easily determined, since the role of the *spo0A* gene product in sporulation has not been determined, nor how much of the gene product is necessary for function. To determine the induction ratio, a similar plasmid was constructed using the *E. coli lacZ* gene. Plasmid pDH90 contains a *spoVG'-'lacZ* fragment that has the *spoVG* ribosome binding site and initiation codon, and the first 275 codons of the *lacZ* gene.

TABLE I
INDUCTION OF SPORULATON WITH IPTG OF
STRAIN 168/pDH68

IPTG [M]	Spores/ml[a]	
	168	168/pDH86
0	1×10^8	1×10^2
5×10^{-5}	7.1×10^7	9.5×10^2
10^{-4}	8×10^7	2.1×10^4
5×10^{-4}	5.7×10^7	3.5×10^7
10^{-3}	7.5×10^7	8×10^7

[a]Cultures were grown in liquid SM medium at 37° to late logarithmic phase and then diluted into 5 ml of prewarmed SM medium containing the indicated concentrations of IPTG. After 24 hr at 37° with aeration, 0.6 ml of chloroform was added and the tubes were vortexed vigorously for 10 sec. Serial dilutions of the cultures were made in fresh SM medium and aliquots plated on SM plates. After incubation at 37° for 36 hr, the plates were scored. None of the 168/pDH86 colonies scored as Spo[+] was genotypically spo0A[+].

A recipient strain that had a single copy of the *lacZ* gene under the control of the *aprE* promoter was used.[7] Transformation was carried out as above, and the induction of β-galactosidase at different concentrations of IPTG was determined as shown in Table II. The induction ratio for the system was at least 200-fold. Since the parent strain BG125 has a small background level of β-galactosidase activity, the induction ratio of 200 is a minimal estimate of the true induction ratio of the *lacZ* gene.

Media

　　10× MG (minimal glucose) salts: Add per 100 ml: 14 g of K_2HPO_4, 6 g of KH_2PO_4, 2 g of $(NH_4)_2SO_4$, and 1 g of $Na_3C_6H_5O_7 \cdot 2H_2O$ (trisodium citrate)

　　Stage 1 transformation medium: 1× MG salts, 2.5 mM $MgSO_4$, 0.5% (w/v) glucose, 5×10^{-3} mM $FeCl_3$, 2×10^{-2} mM $MnCl_2$, 0.02% (w/v) casamino acids, 0.02% (w/v) yeast extract, 50 μg of required amino acids per milliliter, and 10 μg of required bases per milliliter

[7] E. Ferrari, D. J. Henner, M. Perego, and J. A. Hoch, *J. Bacteriol.* **170**, 289 (1988).

TABLE II
IPTG INDUCTION OF β-GALACTOSIDASE

IPTG [M]	β-Galactosidase (units/mg)[a]		
	BG125	pDH90A	pDH90B
10^{-3}	0.4	118	102
10^{-4}	ND	27	20
10^{-5}	ND	0.4	0.5
0	0.4	0.3	0.5

[a] Cultures of BG125 (trpC2, hisA1, thr-5) or BG4089/pDH90 (trpC2, hisA1, thr-5, amyE::[ermC, aprE'-'lacZ]) were grown at 37° in 2YT medium to the late logarithmic phase of growth and diluted to an OD_{600} of approximately 0.1 in the same medium containing the indicated concentrations of IPTG. The cultures were grown at 37° with aeration until an OD_{600} of 1.5. Samples were harvested in triplicate for the determination of β-galactosidase activity and the values shown are the averages of these determinations. The pDH90A and pDH90B are two independent transformants of plasmid pDH90 into strain BG4089.

Stage 2 transformation medium: $1\times$ MG salts, 5.0 mM MgSO$_4$, 0.5% glucose, 5×10^{-3} mM FeCl$_3$, 2×10^{-2} mM MnCl$_2$, 0.01% casamino acids, and 10 μg of required amino acids or bases per milliliter

SM (Schaeffer's medium): Add per liter: 8 g of nutrient broth, 1 g, of KCl, 0.125 g of MgSO$_4$, pH to 7 and autoclave. Add 1 ml per liter: 1 mM Fe$_2$(SO$_4$)$_3$, 1 M Na$_2$SO$_4$, and 10 mM MnCl$_2$

2YT medium: Add per liter: 16 g of tryptone, 10 g of yeast extract, and 5 g of NaCl

Acknowledgments

I would like to acknowledge Alan Grossman, who first used an integrative plasmid in an identical fashion as described here for controlled expression of the B. subtilis spo0H gene and shared his data prior to its publication. Plasmid pDH87 described here is identical to plasmid pAG58 constructed by him. I would also like to thank Jim Hoch for analyzing the sporulation of 168/pDH86.

Section IV

Expression in Yeast
Article 21

A. Plasmid Vectors
Articles 22 and 23

B. Inducible Expression Cassettes
Articles 24 through 26

C. Constitutive Expression Cassettes
Articles 27 through 29

D. Translation
Article 30

E. Proteases
Articles 31 through 33

F. Protein Secretion and Modification
Articles 34 through 37

[21] Heterologous Gene Expression in Yeast

By Scott D. Emr

In making the choice of an appropriate experimental organism for heterologous gene expression, one must carefully consider the unique advantages that each of the available bacterial and eukaryotic expression systems offer. The goal of this volume is to help the individual researcher to make this choice in as an informed a manner as presently is possible. The following set of articles describes in practical terms the manipulations and design considerations that are necessary to achieve high-level gene expression in yeast. In addition, they offer helpful information as to the benefits one expression system has over another depending on the nature of the heterologous protein one has chosen to express.

Yeast have proved to be useful experimental organisms for the expression and study of a wide variety of eukaryotic gene products both for basic research and for industrial and pharmaceutical applications. As they are unicellular microorganisms, many of the manipulations commonly used in bacteria can also be readily applied to yeast. Yeast have a rapid growth rate and can be grown to very high cell densities, they can be propagated on simple defined media, they can be transformed with a variety of either self-replicating or integrating plasmid vectors, and a wide range of simple genetic, molecular and biochemical techniques have been developed for use in these organisms. The yeast *Saccharomyces cerevisiae* can be propagated vegetatively as one of three different cell types: **a** haploid cells or α haploid cells, or these can mate and give rise to the third cell type, the a/α diploid. Diploid cells, when starved for nitrogen, undergo meiosis and form tetrads containing four haploid spores. This well-developed genetics has permitted the simple construction of yeast strains that are ideally suited for high-level gene expression and gene product isolation.

Unlike bacteria, yeast, being eukaryotic, possess much of the complex cell biology typical of multicellular organisms, including a highly compartmentalized intracellular organization and an elaborate secretory pathway which mediates the secretion and modification of many host proteins. Use of the yeast expression system thus affords one a broader range of potential applications than presently is possible with bacterial expression systems.

Vectors and Expression Cassettes

During the past 10 years, there has been a dramatic increase in our understanding of the organization and roles specific sequence elements play in controlling yeast gene expression and plasmid maintenance. Many of these sequence elements now exist in the form of simple movable cassettes that can be mixed and matched in a manner that leads to predictable patterns of gene expression. The replication and stable maintenance of yeast plasmid vectors are ensured by the presence of other sequence elements. These include sequences either from the endogenous multicopy 2-μm plasmid of yeast that confer stable high-copy maintenance, or from a yeast *ARS* element (autonomous replicating sequence—presumed to correspond to a chromosomal origin of replication) that can be added together with a yeast chromosomal centromere sequence to ensure stable low-copy maintenance of the plasmid vector. It also is possible to direct site-specific integration of a plasmid vector into the yeast genome, where it can be stably replicated and maintained as part of a host chromosome. Because plasmids can be amplified and isolated with greater ease from *Escherichia coli,* each of the yeast plasmid vectors is designed as a composite shuttle vector that includes sequences which permit the selection and maintenance of the plasmids in both host systems.

Sequence elements that promote efficient transcription and translation of a foreign gene together with sequences that lead to precise transcription termination are described in detail in articles 24–30. A large variety of low- and high-copy number plasmid vectors and a choice of several constitutive and regulated promoters that can be used in these vectors are presented. Importantly, these articles also offer helpful information as to the benefits of one expression system over another, depending on the kind of recombinant protein being produced. It should be noted that expression levels similar to the highest levels seen in the *E. coli* expression system have been achieved in yeast. *S. cerevisiae,* has been used successfully to express high levels of several different types of proteins, including soluble cytoplasmic proteins, integral membrane proteins, and secreted proteins.

Secretion and Modification

Like other eukaryotes, yeast have a complex secretory pathway. Because of the tremendous purification advantage associated with secreting a foreign protein into the growth medium, considerable effort has been devoted to setting up expression systems that permit high-level secretion of heterologous proteins in yeast (see [34] and [35] in this volume). These proteins are often modified with Asn-linked oligosaccharides. Yeast possess the necessary machinery for the addition and processing of these N-linked carbohydrates. In the article by Ballou, yeast mutants are described with defects in the glycosylation pathway which permit one to regulate the extent of glycosyl processing to which a secreted protein will be subjected. These mutants often grow as well as wild-type yeast strains, and all of the various vector and promoter systems that are presented here can easily be introduced and propagated in these mutant yeast backgrounds.

Protein Stability

Achieving high-level transcription and translation of a heterologous gene in either yeast or *E. coli* is nearly guaranteed at present; however, recovering large amounts of the desired gene product is anything but a guarantee. Some protein products are rapidly degraded either during or shortly after synthesis. Others are lost during cell breakage and subsequent purification. Proteolysis therefore represents one of the most significant barriers to heterologous gene expression in any organism. In yeast, the vacuole contains several endo- and exoproteinases. These enzymes gain access to heterologous gene products expressed in the cytoplasm when the yeast cells are harvested and lysed. Many of the genes that encode these proteases have been cloned and null mutants have been constructed. In the article by Jones [31], the characteristics and uses of these mutants are described. Use of these mutants has dramatically improved the yields of a number of gene products that initially were lost during their purification. No simple genetic or biochemical tricks yet exist that can assure the stabilization of proteins that are degraded rapidly after their synthesis in the cytoplasm. Secretion of the protein into the media where only low levels of proteases have been detected is one answer. However, many nonsecreted proteins do not transit through the secretory pathway efficiently enough (even when the appropriate secretion signals are added to the protein) to make this a reasonable solution to the problem. Recent work on the ubiquitin-mediated pathway for cytoplasmic protein turnover in yeast may help to overcome this problem. The article by Wilkinson [32] describes the ubiquitin-dependent proteolysis system in yeast and offers some useful suggestions that may enable the researcher to minimize protein degradation mediated by this pathway.

A final related problem results from the expression of certain recombinant proteins that turn out to be highly toxic to the cell. The only possible solutions here are to (1) switch to a different expression organism, (2) target the toxic protein into an organelle where its toxicity is no longer manifested, or (3) select a yeast mutant that is insensitive to the toxic effects of the protein.

The articles in this section make it clear that yeast represent excellent host systems for heterologous gene expression. They by no means replace *E. coli* as the first choice for recombinant gene expression. However, in cases where the gene product of interest is incompatible with expression in *E. coli* or must be secreted and posttranslationally modified, yeast represent the next best expression system.

[22] Propagation and Expression of Cloned Genes in Yeast: 2-μm Circle-Based Vectors

By ALAN B. ROSE and JAMES R. BROACH

A number of strategies exist for introducing cloned genes into yeast. These strategies fall into three general categories, which yield quite different cellular copy levels and different stability values for the cloned gene. The first strategy is to link the cloned gene to a centromere, either by integrating it into a chromosome or by inserting it into a YCp (centromere-based) plasmid. Such an approach yields very stable persistence of the cloned gene in yeast, but at a per-cell copy level of only one to two.[43,149] * A second strategy, using ARS-based vectors, yields high per-cell copy levels of the cloned gene but at the cost of stability: such plasmids are generally rapidly lost from the culture in the absence of selection.[65,126,150] The third strategy, propagating the cloned gene on a 2-μm circle-based vector, combines the advantages of the other two approaches, namely, high cellular copy levels and stable propagation.[65,126] Thus, for situations in which maximum expression of the cloned gene is desired—either for production of the gene product or identification of high-copy suppressors—use of 2-μm circle-based vectors represents the optimum strategy.

In this article, we provide a practical guide for using 2-μm circle-based vectors. This includes a short description of the 2-μm plasmid itself, highlighting the salient features of its copy control and segregation systems. We then discuss separately various classes of 2-μm vectors, providing in each section a reasonably comprehensive list of currently available vectors, a discussion of the situations in which the class of vector is appropriate, and a summary of the rigorous, as well as anecdotal, data on the expected behavior of these classes of vectors. Finally, we describe several novel vector systems for specialized uses, such as ultra-high copy propagation, inducible high copy levels, or promoter dissection. In general, this article concentrates on the pragmatic aspects of 2-μm vector use. More detailed considerations of why the various vectors behave as they do can be obtained from several other recent, and not so recent, reviews.[4,12,26,31] Perusal of these other reviews is recommended before embarking on a program of 2-μm vectorology.

---

* References are listed at the end of this article, rather than as footnotes.

Yeast Plasmid 2-μm Circle

Plasmid Properties and Structure

The yeast plasmid 2-μm circle is a small double-stranded DNA plasmid present in the nuclei of most *Saccharomyces* strains at approximately 60–100 copies per haploid genome (cf. Refs. 25, 167). The 2-μm circle confers no apparent selective advantage to cells in which it is resident.[64] Nonetheless, the plasmid achieves stable maintenance even in the absence of positive selection by elaborating a set of interconnected systems dedicated to its persistence in the population. As described below, the persistence of the plasmid is due at least in part to its ability to distribute its copies between both progeny cells following cell division and to amplify its relative copy level in those cells that receive a diminished number of copies.[63,96,103,126]

The structural organization of the plasmid (Fig. 1) is integral to its mode of persistence. The plasmid consists of two unique regions separated by two regions of 599 base pairs (bp) each that are precise inverted repeats of each other.[80] The product of the largest plasmid coding region, *FLP*, catalyzes recombination at specific sites lying near the center of the inverted repeats.[28,30] The result of this recombination event is inversion of the two unique regions with respect to each other. A single origin of replication lies at the junction between one of the repeats and the large unique region.[24,32,92] In addition, the plasmid contains four extended open coding regions, each of which is transcribed into a discrete polyadenylated RNA.[27,153]

Mechanics of Plasmid Persistence

Central to the stability of the plasmid is its ability to ensure its transmission to both progeny cells during cell division. This equipartitioning process is promoted by the trans-acting products of two plasmid coding regions, designated *REP1* and *REP2*.[37,96,103] In addition, equipartitioning requires the integrity of a series of direct tandem repeats of a 62-bp element lying near the origin of replication. This collection of repeats, designated *REP3* or *STB*, is active only in cis and appears to function as a centromere-like element in the partitioning process.[97,103] Partitioning activity resides solely in the repeat elements. However, flanking sequences, primarily the region between the *Pst*I and *Hpa*I sites, are required for partitioning activity in some vector constructions. The transcription termination site within this flanking domain serves to isolate the partitioning site and protect it from transcriptional inactivation.[127] The mechanism of plasmid

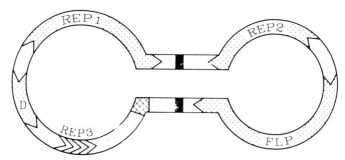

FIG. 1. The yeast plasmid 2-μm circle. A diagram of the genomic organization of the yeast plasmid is shown, drawn to distinguish the inverted repeats (horizontal lines) from unique sequences (circular regions). On the diagram are indicated the locations of the open coding regions (dotted regions, arrows signifying the 5' to 3' orientation), the origin of replication (cross-hatched region), the *FLP* recombination target site (solid region), and the cis-acting stability locus (chevrons).

partitioning is currently not known, although several models for this process have been proffered.[173]

A second facet of the plasmid's stability is its capacity for copy-number amplification, a process that may serve to correct any downward fluctuation in copy levels that could arise from imprecise segregation of plasmid molecules during mitosis. Plasmid amplification occurs by a novel mechanism, first proposed by Futcher,[63] in which *FLP* recombination between the repeats promotes a shift in replication from θ to rolling circle. Under steady-state conditions, replication initiates at the single origin in each plasmid, proceeds in a bidirectional manner through a θ intermediate, and terminates to yield two copies from each parent plasmid. Under conditions of low plasmid copy, *FLP*-mediated intramolecular recombination within the θ intermediate can shift a plasmid into rolling circle replication to yield a number of progeny plasmids from a single parent. This shift to rolling circle replication allows the plasmid to increase its relative copy level in those cells harboring few plasmid molecules.[166]

Control of Plasmid Copy Levels

The 2-μm circle maintains a copy level of approximately 60–100 copies per haploid genome. Since *FLP* protein is required for amplification, one would expect *a priori* that regulation of plasmid copy control would have to be intimately associated with regulation of *FLP* activity. In confirmation of this expectation, transcription of *FLP* is repressed more than 100-fold by the concerted action of the products of *REP1* and *REP2*.[128,136,147] This regulatory scheme yields a credible copy control system. In cells with high copy levels, intercellular *REP1* and *REP2* protein concentrations are sufficient to maintain repression of *FLP* activity and, accordingly, to inhibit amplification. However, in cells acquiring a reduced level of plasmids, the diminished gene dosage of *REP1* and *REP2* yields reduced concentrations of these proteins, which in turn provokes derepression of *FLP* activity and induction of amplification. Thus, regulation of *FLP* expression by *REP1* and *REP2* provides for induction of amplification when copy level is low and repression of amplification when copy number is high. Additional features of the regulation of plasmid gene expression, including antirepressor activity of the *D* gene product and autoregulation of *REP1* expression by the *REP1/REP2* complex,[128,147] provide for fine tuning of the copy control system.

Despite the elaborate systems in place to ensure stable copy control and efficient propagation, copy level is not rigidly maintained in every cell. Plasmid copy levels vary from cell to cell, assuming a fairly broad distribution around the mean value. Although rectification mechanisms tend to drive copy number toward this mean value, cell-to-cell variation persists for sufficiently long times to be measurable within clonal populations.[65] The consequence of this "sloppy" copy control process is that selection pressures can enrich for cells with either extreme of copy level. Thus, if high copy levels of a 2-μm derived plasmid are deleterious to the cell, then the population will be enriched for cells with lower copy levels, and the average per-cell copy level will be lower than that for nondeleterious plasmids.

Standard 2-μm Circle-Based Vectors

Composition

A variety of vectors designed for propagation of cloned fragments in yeast and based on the 2-μm circle are described in Table I. These vectors include plasmids, such as pCV20 and pCV21, that carry the entire 2-μm circle genome, as well as those, such as YEp13 and YEp24, that incorporate only a portion of the genome. Almost all of the 2-μm vectors encompass the *REP3* locus, which lies between the *Hpa*I site and *Ava*I site in the large unique region.[97,103,127] Since the 2-μm circle-promoted partitioning requires the products of the *REP1* and *REP2* genes, those plasmids that do not carry *REP1* and *REP2* themselves have to be propagated in a [cir+] strain or in some other strain in which these proteins are present.[96,103] Most of the constructs carry the *REP3* locus as the 2.2-kilobase (kb) *Eco*RI fragment or the 2.1-kb *Hin*dIII fragment from the B form of 2-μm circle. These fragments also span the plasmid origin of replication and one of the inverted repeats, which encompasses a *FLP* recombination target site.

TABLE I
STANDARD 2-μm CIRCLE-BASED VECTORS[a]

Single Yeast Marker

Name[b]	Yeast marker	E. coli genes	Parental plasmid	2-μm genes[c]	Other DNA	Size (kb)	Cloning sites[d]	Comments	Ref.[e]
YEpADE8.1	ADE8	amp	pUC18	REP3		8.1	Q, B, F	BamHI–BglII ADE8 and 2.2-kb EcoRI 2-μm fragments in pUC18	L. Levin, T. Toda, S. Cameron
fdY7	HIS3	amp	fd106	REP3		10.9		For production of single-stranded DNA	146
pHV1	HIS3	amp	pUC	REP3		5.4	E, Q, M, S, F		P. Sass, J. Nikawa
pJM35	HIS3	amp	pBR322	REP2, REP3	λ cos	13.2	B, S	Cosmid made with the large EcoRI fragment from 2-μm A, the 1.7-kb BamHI HIS3 fragment, and a BglII–EcoRI fragment of λ containing cos in pBR322	116
pSRT1	HIS3	amp	pBR322		pSR1	13.5	Y	Made with pSR1, a 2-μm-like plasmid from Zygosaccharomyces rouxii. Transforms Z. rouxii and S. cerevisiae, but behaves like an ARS plasmid in S. cerevisiae. pSRT2 contains the isomeric form of pSR1, which is called pSR2	98, 99, 158
pYc1	HIS3	amp	pBR322	REP3	λ cos	10.0	B, E, S, Y	λ cos site in YEp6	87
pYF177	HIS3	amp	pBR322	REP3		8.2	A, S	Small EcoRI 2-μm B in pBR322 with HIS3 in BamHI site	117
pYF88	HIS3	amp	pBR322	REP2, REP3		10.0	S	1.7 kb HIS3 in pYF85	117, 151
pYF92	HIS3	amp	pBR322	REP1, REP2, REP3, D		12.7	S	1.7 kb HIS3 in pYF86	151
YEp421	HIS3	amp	pBR322	REP3		7.7	C, E, V, R, S, F	In vivo recombinant between YEp420 and YEp6	112
YEp430	HIS3	amp	pBR322	REP3		7.4	C, E, R, S, W, F, Y	In vivo recombinant between YEp420 and pRB328	112
YEp4-Sc2703	HIS3	tet	pBR322	REP1, REP2, REP3, FLP		17.8		YEp4 with a 7.1-kb SalI fragment containing HIS3	152
YEp6	HIS3	amp	pBR322	REP3		8.3	E, Y, S		20, 152
pC500	HIS4	amp	pJC75-58	REP1, REP2, REP3, FLP	cos site	27.5		Cosmid containing whole 2-μm and HIS4. pC501 differs only in orientation of HIS4 fragment	90
pBT1-1	LEU2	amp, tet	pBR322	REP3	λ cos	11.1	B	Cosmid with λ cos site in YEp13	124
pBT1-5	LEU2	amp	pBR322	REP3	λ cos	11.0	B, H	Cosmid derived from pBT1-1 by deleting the small HindIII fragment, leaving a unique HindIII site and inactivating tet	124
pBT1-8	LEU2	tet	pBR322	REP3	λ cos	12.7		Cosmid for cloning large DNA fragments	124

Plasmid	Marker	E. coli marker	Backbone	2-μm region	Special	Size (kb)	Sites	Description	Reference
pBT1-9	LEU2	tet	pBR322	REP3	λ cos	10.2	X	Cosmid derived from pBT1-8 by deleting a 2.5-kb XhoI fragment	124
pC4B	LEU2	amp	pBR322	REP3		7.2	B, S, H, A	pC4 has a unique SalI site while pC4B has this site converted to a unique BamHI site	29
pCV17	LEU2	amp	pBR322	REP2, REP3		10.5	S, B, U	Large EcoRI A form 2-μm fragment and LEU2 in pBR322	29
pCV20	LEU2	amp	pBR322	REP1, REP2, REP3, D		13.0	S, A, B	pBR322 and LEU2 in the EcoRI site in FLP of whole 2-μm. Available with a temperature-sensitive allele of REP1. A similar plasmid, pCV19, contains a different isolate of 2-μm that differs slightly in size and restriction sites	26, 29 B. Veit
pCV21	LEU2	amp	pBR322	REP1, REP2, REP3, FLP		13.0	S, A, B	pBR322 plus LEU2 in EcoRI site in large unique region of 2-μm B.	26
pCV7	LEU2	amp	pBR322	REP3		8.8	B, S, H	2.2-kb EcoRI 2-μm B and 2.2-kb LEU2 SalI–XhoI in pBR322. pCV7A has 2-μm sequences in the opposite orientation	26, 29
pDB248	LEU2	amp tet	pBR322	REP3		10.1	P, B, A, H	Also transforms Schizosaccharomyces pombe. pDB248' contains a tandem duplication of the 1.5-kb EcoRI fragment containing part of LEU2	8, 9
pEMBLYe30	LEU2	amp lacZα	pMB1	REP3	fl ori	9.5	B, H, M	The fl origin allows isolation of single-stranded DNA. Color screen for plasmids containing inserts	5
pEMBLYe31	LEU2	amp lacZα	pMB1	REP3	fl ori	8.1	M, B, S, P, H	Derived from pEMBLYe30 by deleting nonessential DNA between LEU2 and 2-μm sequences	39, 40
YEpL3	LEU2	amp lacZα	pMB1	REP3	fl ori	6.8	M, B, S, P, H	Derivative of pEMBLYe31. Has the fl origin for isolating single-stranded DNA. Color screen for plasmids containing inserts	40
pGT40	LEU2	amp	pBR322	REP3		8.4	B, S	Differs from pCV7 only in orientation of LEU2 and 2-μm fragments	160
pGT41	LEU2	amp	pBR322	REP1, REP2, REP3, FLP		12.5	S, B, A	Differs from CV21 only in orientation of LEU2 and 2-μm fragments	160
pHKB52	LEU2		pBR322	REP1, REP2, REP3, FLP		12.9	S	Entire 2-μm B as PstI fragment and LEU2 as SalI–XhoI fragment in pBR322, apparently inactivating both amp and tet	57
pJDB248	LEU2	tet	pMB9	REP1, REP2, REP3, FLP		13.6	B, S, A	Whole 2-μm A in pMB9	10, 11
pJH4	LEU2	amp lacZα	pUC8	REP3		6.2	P, S, B, M	Color screen for plasmids containing inserts	F. Volkert
pJZ1	LEU2	amp tet					B		144

239

(continued)

TABLE I *(continued)*

Name[a]	Yeast marker	E. coli genes	Parental plasmid	2-μm genes[c]	Other DNA	Size (kb)	Cloning sites[d]	Comments	Ref.[e]
pDB262	LEU2	Pr/tet	pBR322	REP3	λ c1 repressor	9.8	L, H	Cloning vector with a positive selection for inserts. The tet gene is under control of the λ Pr promoter. Inserts in the cloning sites destroy the c1 repressor, allowing expression of the tet gene. Also transforms S. pombe	172
pWH4	LEU2	Pr/tet amp	pBR322	REP3	λ c1 repressor	10.6	L, H	Cloning vector with a positive selection for inserts, derived from pDB262. Inserts in the cloning sites destroy the c1 repressor, allowing expression of the tet gene, which is under the control of the λ Pr promoter. Also transforms S. pombe	172
pWH5	LEU2	Pr/tet amp	pBR322	REP3	λ c1 repressor	10.6	L, M, H	Derived from pWH4. Inserts in the cloning sites destroy the tet gene, allowing expression of the tet gene, which is under the control of the λ Pr promoter. Also transforms S. pombe	172
pYF90	LEU2	amp	pBR322	REP2, REP3		14.0	Y		151
pYF91	LEU2	amp	pBR322	REP2, REP3		13.5	B, H, Y		151
pYK71	LEU2	amp	pBR322	REP3	T7 φ10 promoter terminator	11.0	B	YEp13 containing the T7 promoter φ10 and a T7 transcriptional terminator, allowing isolation of RNA homologous to inserts in the BamHI site	R. Sternglanz, W.-K. Eng
pYK72	LEU2	amp	pBR322	REP3	T7 promoter	11.0	B	YEp13 containing the T7 promoter φ10 inserted as a BamHI–BglII fragment into the BamHI site, allowing isolation of RNA homologous to inserts	R. Sternglanz, W.-K. Eng
pYT7810	LEU2		pBR325	REP3		8.0	H, B, S	Unclear which 2-μm and pBR325 genes are present	148
YEp13	LEU2	amp tet	pBR322	REP3		10.7	B, U, Y	Small EcoRI 2-μm B form and 4.0-kb PstI LEU2 in pBR322. pPL200 is a derivative in which the LEU2 SalI site has been filled in, leaving a unique SalI site	33, 107
YEp13S	LEU2	amp tet	pBR322	REP3		10.7	B, S, H	Made from YEp13 by inverting the SalI–XhoI fragment containing LEU2, destroying both sites and leaving a unique SalI site in tet. Also known as YEp213	26
pPL116	LEU2	amp	pBR322	REP3		10.6	B, H, U, Y	YEp13 with the 135-bp HindIII fragment deleted	P. Lagosky
pPL160	LEU2	amp	pBR322	REP3		10.7	B, S, H	YEp13 with the SalI site in tet filled in. pPL199 is YEp13 with both SalI sites destroyed	P. Lagosky
YEp13M4	LEU2	amp	pUC19	REP3		5.8	Q, M, B, S, P, F, H, A, L	Has sequencing primer upstream of cloning sites	P. Sass, J. Nikawa

YEp20	LEU2	amp	pBR322	REP1, REP2, no ori		B, S	10.6	Large 2-μm B EcoRI fragment, SalI–XhoI LEU2, in pBR322	20
YEp21	LEU2	amp	pBR322	REP3		B, S	8.8	2.2-kb EcoRI 2-μm B, 2.2-kb SalI–XhoI LEU2, in pBR322	20
YEp351	LEU2	amp lacZα	pUC18	REP3		Q, M, B, X, S, P, F, H	5.6	Color screen for plasmids containing inserts	83
pMCB974	LEU2	amp lacZα	pUC18	REP3		E, Q, K, M, B, X, S, P, F	5.6	Derivative of YEp351 lacking the HindIII site. Color screen for plasmids containing inserts	J. Schultz, M. Carlson
YEp423	LEU2	amp	pBR322	REP3		B, H, J, R, S, F	8.1	In vivo recombinant between YEp420 and YEp21	112
YEplac181	LEU2	amp	pUC19	REP3		H, F, P, S, X, B, M, K, Q, E	5.7	The KpnI and EcoRI sites were removed from the LEU2 gene by site-directed mutagenesis. The XbaI site in 2-μm was removed by filling in. Color assay for plasmids containing inserts	69
YEpLEU2.1	LEU2	amp	pUC119	REP3	M13 ori	Q, M, B, S, P, H	7.3	2.2-kb SalI–XhoI Leu2 and 1.9-kb PstI–HindIII 2-μm fragments in pUC119. The M13 ori allows easy isolation of single-stranded DNA	L. Levin
YEpM7	LEU2	amp	pUC19	REP3		Q, M, B, S, P, F, H, L, A	6.0		P. Sass, J. Nikawa
YEp-DE	LEU2	amp lacZα	pUC18	REP3		Q, M, B, S, F	7.3	Color screen for plasmids containing inserts	102
YEp426	LYS2	amp	pBR322	REP3		C, H, S, F	11.0	In vivo recombinant between YEp425 and pBR322	112
YEp620	LYS2	tet	pBR322	REP3		S, U	11.7		7
pYTH4	SUP4.0	amp	pBR322	REP3		M, G	11.8	Can select for by the suppression of chromosomal ochre mutations. One of a series of plasmids consisting of the SUP4.0 gene, pBR322, and various 2-μm fragments	156
pFL45	TRP1	amp lacZα	pUC19	ori only		E, Q, K, M, B, S, P, F, H	4.0	Probably unstable, since it lacks REP3. Color screen for plasmids containing inserts. The TRP1 fragment lacks EcoRI, HindIII, and AccI sites	F. Lacroute
pRC4	TRP1	amp kan	pKC7	REP3	ARSI	B, M, S	10.3	Derived from pRC3 by insertion of a BamHI–SmaI adaptor into the tet BamHI site, creating a BamHI–SmaI–BamHI site and destroying tet	L. Hartwell
pRE1052	TRP1	amp	pBR322	REP3	ARSI	E	5.2		159
pTRP584	TRP1	amp	pUC	REP3		B, Y, U	4.4		120
pTV3	TRP1	amp		REP3		E, Q, K, S, F, B			P. Sass, J. Nikawa
pYE	TRP1	amp	pBR322	REP1, REP2, REP3, FLP		B	11.5	Derived from pGT41 by replacing LEU2 with TRP1	12, 14, 175
YEp427	TRP1	amp tet	pBR322	REP3	ARSI	B, C, R, S, F	7.4	In vivo recombinant between YEp420 and YRp7	112
YEp432	TRP1	amp	pBR322	REP3		B, E, R, S, F	6.4	In vivo recombinant between YEp420 and pRB315	112

241

(continued)

TABLE I (continued)

Name[b]	Yeast marker	E. coli genes	Parental plasmid	2-μm genes[c]	Other DNA	Size (kb)	Cloning sites[d]	Comments	Ref.[e]
YEp9T	TRP1	amp tet	pBR322	REP3		7.2	E, B		41
YEplac112	TRP1	amp lacZα	pUC19	REP3		4.9	H, F, P, S, X, B, M, K, Q, E	The XbaI, HindIII, and PstI sites were removed from the TRP1 gene by site-directed mutagenesis. The XbaI site in 2-μm was removed by filling in. Color screen for plasmids containing inserts	
G18	URA3		pCR1	REP1, REP2, REP3, D		20.9	A	Derivative of PTY39 with URA3 in the HindIII site of the kan gene, inactivating kan. G23 has URA3 in the opposite orientation	15, 68
pAB18	URA3	amp	pBR322	REP1, REP2, REP3, D		11.9	B, J, S, M	2-μm A in YIp5 via ClaI sites	125
pCGS40	URA3	amp tet	pBR322	REP3		7.1	E, C, H, B, S, M, F	1.5-kb HindIII–HpaI 2-μm B in the PvuII site of YIp5	70
pEMBLYe23	URA3	amp lacZα	pMB1	REP3	f1 ori	7.4	B, H, S	Color screen for plasmids containing inserts. The f1 origin of replication enables easy isolation of single-stranded DNA. HindIII fragment from YEp24 containing URA3 and 2-μm sequences in pEMBL9	5, 39
pEMBLYe24	URA3	amp lacZα	pMB1	REP3	f1 ori	7.4	B, H, S	Similar to pEMBLYe23, but with the 2-μm–URA3 fragment in the opposite orientation	5
pFL2	URA3	amp	pBR322	REP3		7.6	A, B, S, U, C	2.2-kb EcoRI 2-μm B and 1.1-kb HindIII URA3 in pBR322. Plasmids pFL1, pFL3, and pFL4 differ from pFL2 only in orientation of the inserts. pFL1 also available with the ClaI site destroyed	42, F. Messenguy
pFL44	URA3	amp lacZα	pUC19	ori only		4.4	E, Q, K, M, B, X, S, P, F, H	Probably unstable, since it lacks REP3. Color assay for plasmids containing inserts. URA3 has PstI and SmaI sites destroyed. Also available with EcoRI site destroyed	F. Lacroute, F. Messenguy
pG63	URA3	amp	pBR322	REP3		7.5	A, B, S, E	1.1-kb HindIII URA3 in pG6 (EcoRI–HindIII 2-μm B in pBR322). A derivative, pG63-11, also has a HindIII cloning site	67
pGN621	URA3	amp tet	pBR322	REP3		7.8	E, C, J, B, F, S, U	2.2-kb EcoRI 2-μm B in the SmaI site of YIp5	J. Boeke
pHCG3	URA3	amp	pHC79	REP3	cos site	10.0	H, B, S, E, A, G	Cosmid derived from pHC79 and pG63-11	48, 67
pJDB110	URA3	amp	pAT153	REP3		6.7	E, B, S		11

Plasmid	Marker	Selection	Backbone	REP3/ori	ARS	Size (kb)	Sites	Description	Reference
pSEY8	URA3	amp lacZα	pBR322			6.8	E, M, B, S, H	Color screen for plasmids containing inserts	55
pTB199	URA3	amp tet	pBR322	REP3		7.7	E, C, J, B, F, S, M, D, X, A, U	2.2-kb EcoRI fragment of 2-μm B in the BalI site of YIp5. Derivatives have a unique HindIII site in tet (pTB220) or in 2-μm sequences (pTB221). pTB224 is pTB220 with the unique EcoRI site destroyed	L. Prakash
pTG807	URA3	amp tet	pBR322	ori only		6.3	B, E, S,U	Probably unstable, since it lacks REP3	110
pUV2	URA3	amp	pUC	REP3		4.8	E, Q, K, B, S, F, H		P. Sass, J. Nikawa
pYK2060	URA3	amp tet	pBR322	REP3	ARS1	7.9	B, S, E	Derivative of YRp16 containing REP3 from pJDB219	103
YEp103	URA3	amp	pBR322	REP3		7.7	E, H, B, S, U	YIp5 containing the EcoRI–HindIII fragment spanning the 2-μm origin and REP3	46
YEp16	URA3	amp tet	pBR322	REP3		7.7	B, S, H		26
YEp24	URA3	amp	pBR322	REP3		7.8	B, L, C, A, J, U, S, M, F, X	2.2-kb EcoRI 2-μm B and 1.2-kb URA3 in pBR322. YEp24-B has the BamHI site destroyed	20, 73
YEp352	URA3	amp lacZα	pUC18	REP3		5.2	E, Q, K, M, B, X, S, P, F, H	Color screen for plasmids containing inserts	83
YEp352B	URA3	amp	pUC18	REP3		4.9	B	YEp352 with the multiple cloning site replaced by a BamHI linker	A. Tzagoloff
YEp352C	URA3	amp	pUC18	REP3		4.9	Q	YEp352 with the multiple cloning sites replaced by a SacI linker	A. Tzagoloff
YEp352E	URA3	amp	pUC18	REP3		4.9	E	YEp352 with the multiple cloning sites replaced by an EcoRI linker	129
YEp352H	URA3	amp	pUC18	REP3		4.9	H	YEp352 with the multiple cloning sites replaced by a HindIII linker	A. Tzagoloff
YEp352P	URA3	amp	pUC18	REP3		4.9	P	Yep352 with the multiple cloning sites replaced by a PstI linker	A. Tzagoloff
YEp352S	URA3	amp	pUC18	REP3		4.9	F	YEp352 with the multiple cloning site replaced by a SphI linker	A. Tzagoloff
YEp420	URA3	amp tet	pBR322	REP3		7.1	E, C, H, B, F, S, R	HindIII–HpaI from 2-μm B in the PvuII site of YIp5. Previously called β72. pRB308 contains the 2-μm sequences in the opposite orientation. pRB381 lacks the SalI site	53, 112 D. Botstein, A. Hoyt
YEp429	URA3	amp	pBR322	REP3		6.8	B, G, C, E, H, R, S, M, F	In vivo recombinant between YEp420 and pPL7	112
YEplac195	URA3	amp lacZα	pUC19	REP3		5.2	H, F, P, S, X, B, M,	The PstI site was removed from the URA3 gene by site-directed mutagenesis. The XbaI site in	69

(continued)

TABLE I (continued)

Name[b]	Yeast marker	E. coli genes	Parental plasmid	2-μm genes[c]	Other DNA	Size (kb)	Cloning sites[d]	Comments	Ref.[a]
pGT38	LEU2	amp	pBR322	REP3	CEN3	10.4	K, Q, E; B, S	2-μm was removed by filling in. Color screen for plasmids containing inserts. SalI–XhoI LEU2, BamHI–HindIII CEN3, and small EcoRI fragment of 2-μm B in pBR322	160
pGT39	LEU2	amp	pBR322	REP1, REP2, REP3, FLP	CEN3	14.5	B, S	SalI–XhoI LEU2, BamHI–HindIII CEN3, and entire 2-μm B in pBR322	160
pScT6	LEU2	amp	pBR322	REP1, REP2, REP3, D	Tetrahymena telomere	19.5		Linear plasmid able to transform both S. cerevisiae and S. pombe	75
YCp2micron	LEU2	amp	pBR322	REP3	CEN5	12.2	H, Y	1.5-kb BamHI CEN5 fragment in YEp13	113
Two Yeast Markers									
pA3	ADE3 URA3	amp	pBR322	REP1, REP2, REP3, D		18.7	B, S	Can be used in a color assay for plasmid stability in an ade2 ade3 strain. Available with a temperature-sensitive allele of REP1	B. Veit
pABADE8	ADE8 LEU2	amp	pBR322	REP3		13.4	B	Could be used as a constitutive (ADC1 promoter) expression vector if the BamHI fragment containing the ADE8 gene were deleted	3
TLC-1	CAN1 LEU2	amp tet	pBR322	REP3	λ cos	15.2	B, U, Y	CAN1 gene in YEp13	33
pJM52	HIS3 LEU2	amp	pBR322	REP2, REP3		15.4	B	pJM35 with LEU2	116
YEp422	HIS3 URA3	amp	pBR322	REP3		8.9	C, E, R, S, M, Z, F	In vivo recombinant between YEp420 and YEp6	112
YEp431	HIS3	amp	pBR322	REP3		8.6	C, E, R, S, M, F, Y	In vivo recombinant between YEp420 and pRB328	112
YEp13a	LEU2 SUP53	amp tet	pBR322	REP3		10.7	B, U, Y	YEp13 with an amber supressor, made by mutation of the tRNA 3^su normally found in the LEU2-containing PstI fragment of YEp13	61
pRB28	LEU2 URA3	amp	pBR322	REP3		14.3	S	5.5-kb BamHI URA3 fragment in YEp21. The large URA3 fragment could be useful for integrations	D. Botstein
pRB37	LEU2 URA3	amp	PBR322	REP3		12.7	B, S	3.9-kb BglII–BamHI URA3 fragment in YEp21. Large URA3 fragment may be useful for integrations. pRB38 has the URA3 fragment in the opposite orientation	D. Botstein
YEp424	LEU2 URA3	amp	pBR322	REP3		9.3	B, H, J, R, S, M, F	In vivo recombinant between YEp420 and YEp21	112

244

Plasmid	Yeast markers	Bacterial markers	Vector	2μm	Other	Size (kb)	Selectable markers	Description	Ref.
pDA6200	LYS2 URA3	amp	pBR322	REP3		18.2	S, C	LYS2 gene in YEp24 BamHI site	7
YEp425	LYS2 URA3	amp	pBR322	REP3		12.1	C, H, S, M, F	In vivo recombinant between YEp420 and pRB506	112
pYK2074	TRP1 URA3	amp	pBR322	REP3	ARS1	8.5	B, S	Derivative of pYK2060 containing the TRP1 gene. Differs from pYK2075 in orientation of TRP1 fragment	103
YEp428	TRP1 URA3	amp tet	pBR322	REP3	ARS1	8.6	B, C, R, S, M, F	In vivo recombinant between YEp420 and YRp7	112
YEp433	TRP1 URA3	amp	pBR322	REP3		7.6	B, E, R, S, M, F	In vivo recombinant between YEp420 and pRB315	112
Dominant Selectable Markers									
YEp24cry1	cry1 URA3	amp	pBR322	REP3		12.8	B, S	YEp24 containing the cry1 gene, conferring resistance to cryptopleurine. In high copy number, the normally recessive cryptopleurine resistance allele (cry1) can supress the wild-type CRY1 allele	108
pPL241	DFR1 LEU2	amp	pBR322	REP3		11.8		pPL200 (see YEp13) containing a 1.1-kb DraI–PstI fragment encoding the yeast dihydrofolate reductase gene. Confers resistance to methotrexate	107
pJM81	HIS3 LEU2	amp	pBR322	REP3	HSV-TK, λcos	16.4	B	Cosmid containing the HSV-TK gene, which can be selected for in yeast, in addition to the LEU2 and HIS3 genes. Derivative of pJM52	116
pJM94	HIS3 LEU2	amp	pBR322	REP3	HSV-TK	14.2	B, S	Similar to pJM81, but derived from pJM35	116
pADA1	LEU2	amp	pBR322	REP3	Rdhfr	12.9	B, S	Rdhfr (dihydrofolate reductase) gene conferring methotrexate resistance under control of the ADC1 promoter in pAAH5	121
R5H	PBS2 URA3	amp	pBR322	REP3		15.6	U, S, B, K	7.8-kb PBS2 gene, conferring polymyxin B resistance, in the BamHI site of YEp24	19
YEpPBS2	PBS2 URA3	amp	pBR322	REP3		9.5	B, S	3.2-kb DraI–SacI PBS2 gene, conferring polymyxin B resistance, in the PvuII–SalI sites of YEp24	G. Boguslawski
pJR41	TUN URA3	amp	pBR322	REP3		13.1	S	Tunicamycin resistance gene from yeast in YEp24	137
YEp36	CUP1 LEU2 URA3	amp tet	pBR322	REP3		9.7	U, Y	Copper metallothionine gene in YEp13	34
pUT302	URA3	ble amp	pBR322		ARS1			Contains the ble gene (conferring resistance to phleomycin) from Tn5 in pEX-2 under control of the CYC1 promoter	66
pCH100	TRP1	cat amp	pBR322	REP3		8.3	S	Contains cat gene from pBR328 under control of the ADC1 promoter in a derivative of pYcDE-2. Confers resistance to chloramphenicol	76
pYT11		cat	pBR325	REP1, REP2,		11.7	B, S	Whole 2-μm in pBR325 by PstI site. Confers	59

(continued)

TABLE I *(continued)*

Name[b]	Yeast marker	E. coli genes	Parental plasmid	2-μm genes[c]	Other DNA	Size (kb)	Cloning sites[d]	Comments	Ref.[e]
		tet		*REP3, FLP*				resistance to chloramphenicol. Differs from pYT14 only in orientation	
pYT11-LEU2	*LEU2*	*cat*	pBR325	*REP1, REP2, REP3, FLP*		13.9	B, S, A	*SalI–XhoI LEU2* fragment in pYT11. pYT14-LEU2 differs only in orientation of 2-μm	45
pLG89	*URA3*	*hyg amp*		*REP3*	*ARS1*	10.3	X	The *hyg* gene (confers resistance to hygromycin B) under control of the *CYC1* promoter in pEX-2	71
pGH-1	*URA3*	*hyg kan amp*	pBR322	*REP3*		11.4		Contains two dominant selectable genes, encoding resistance to G418 and hygromycin B. Made by inserting *kan* into pLG89	58
fdY5		*kan*	fd106	*REP3*		11.3		For production of single-stranded DNA. The *kan* gene is selectable in *E. coli* and confers G418 resistance in yeast	146
PTY39		*kan*	pCR1	*REP1, REP2, REP3, D*		19.8	A	2-μm B in pCR1 via *FLP EcoRI* site. The *kan* gene is selectable in *E. coli* and confers G418 resistance in yeast	68
CVneo	*LEU2*	*kan amp*	pBR322	*REP1, REP2, REP3, D*		13.9	S, B, A	The *kan* gene from Tn5 in CV20. The *kan* gene is selectable in *E. coli* and confers G418 resistance in yeast	75
pNF1	*URA3*	*kan amp*	pBR322	*REP3*			S	YEp24 with *kan* in the *BamHI* site. The *kan* gene is selectable in *E. coli* and confers G418 resistance in yeast	131

246

Plasmid	Yeast marker	E. coli markers	E. coli replicon	2-μm	ARS	Size (kb)	Cloning sites	Description	Ref.
pNF2	URA3	kan amp	pBR322	REP3			B, S	Derived from pNF1. SalI sites flank the BamHI site, so that fragments cloned into the BamHI site can be removed with SalI. The kan gene is selectable in E. coli and confers G418 resistance in yeast	131
pMC3G	URA3	kan amp lacZ	pBR322	REP3		27.0		Has duplicated kan genes and lacZ, lacY, and lacA. The kan gene is selectable in E. coli and confers G418 resistance in yeast	169
pYE13G	LEU2	kan amp tet	pBR322	REP3		18.0		The kan gene is selectable in E. coli and confers G418 resistance in yeast	169
pRC2	TRP1	kan amp tet	pKC7	REP3	ARS1	10.5	B, M	2-μm ori from YEp13 in pRC1 (1.45-kb EcoRI TRP1–ARS1 fragment in pKC7). The kan gene is selectable in E. coli and confers G418 resistance in yeast	60
pRC3	TRP1	kan amp tet	pKC7	REP3	ARS1	10.3	B, S	Derived from pRC2 by deleting a 200-bp SmaI–SalI fragment, leaving a unique SalI cloning site. The kan gene is selectable in E. coli and confers G418 resistance in yeast	60

[a] Three groups of standard 2-μm circle-based vectors are listed. The first group consists of those plasmids carrying a single yeast selectable marker that will complement recessive alleles in the host strain. The second group consists of those plasmids carrying two such selectable markers. The third group consists of those plasmids that carry a dominant marker for selection in yeast. In the first two groups, the plasmids are listed alphabetically by the yeast selectable marker. In the third group the plasmids are listed alphabetically first by E. coli selectable marker and second by yeast selectable marker.

[b] Plasmids with multiple designations are listed only by the one most commonly used.

[c] Unless otherwise noted, all plasmids carry the 2-μm circle origin of replication in addition to the listed genes. If the space is blank, the particular 2-μm circle genes present could not be determined from the available plasmid maps.

[d] Cloning sites: A, HpaI; B, BamHI; C, ClaI; D, NdeI; E, EcoRI; F, SphI; G, BglI; H, HindIII; I, NarI; J, NheI; K, KpnI; L, BclI; M, SmaI; N, NotI; O, NcoI; P, PstI; Q, SacI; R, NruI; S, SalI; T, StuI; U, PvuII; V, EcoRV; W, ScaI; X, XbaI; Y, XhoI; Z, SpeI.

[e] References to individuals are personal communications.

In addition to *REP3*, 2-μm circle-based vectors comprise a yeast ARS element, a yeast selectable marker, and bacterial vector sequences. An origin of replication or ARS element is essential for propagation of the vectors in yeast. Although most of the original vectors used the 2-μm circle ARS, this ARS confers no distinctive properties to the vectors. Addition of the *REP3* domain to a plasmid containing any ARS element yields a vector that is as stable in yeast as those based on the 2-μm circle origin of replication.[97,103]

Every vector carries a yeast selectable marker, which is required both to obtain initial transformants with the vector and to provide continual selection to maintain the vector in the culture. As is evident from Table I, 2-μm circle-based vectors are available with a wide variety of selectable markers, encompassing all of the markers routinely present in strains used in standard yeast molecular genetics. In addition, a number of vectors are available with dominant selectable markers, such as G418 or chloramphenicol resistance. Obviously, such vectors can be used to introduce cloned genes into strains lacking any of the coterie of standard recessive alleles.

Finally, to facilitate construction and preparation of plasmid DNA, all the vectors contain sequences to permit selection and propagation in *Escherichia coli*. Most of the initial vectors were based on the bacterial plasmid pBR322. However, more recently developed vectors have used pUC plasmids, which permit higher plasmid yields from bacteria, the cloning versatility of polylinkers, and a monitoring system to detect insertion of cloned sequences.

Vector Stability

The singular feature of the 2-μm circle-based vectors is their ability to engage the 2-μm partitioning system. This renders them more stable during mitotic growth than vectors based solely on ARS elements. That is, ARS plasmids exhibit a mitotic loss rate of approximately 1/10 cells per generation. The standard 2-μm circle-based vectors listed in Table I are, on average, an order of magnitude more stable than ARS plasmids, with a loss rate of approximately $1/10^2$ cells per generation.[65,126]

Although 2-μm circle-based vectors are more stable than ARS-based vectors, none of the standard vectors achieves the stability of the 2-μm plasmid itself. That is, most of the 2-μm circle-based vectors are at least two orders of magnitude less stable during mitotic growth than authentic 2-μm circle.[65] As a consequence, long-term propagation of the plasmid-bearing strain in the absence of selection for retention of the plasmid yields a steady decline in the proportion of plasmid-bearing cells and a reduction in the average per-cell copy number. Under growth conditions selective for the plasmid, plasmid-bearing cells usually comprise between 60% and 95% of the population, depending on the particular plasmid and selectable marker used.[96,126] The reason for the reduced stability of 2-μm vectors *vis-à-vis* 2-μm itself is not readily apparent. The explanation most consistent with available data suggests that the primary defect is reduced replication or amplification potential caused merely by the presence of the exogenous sequences carried in the vector.[4]

As discussed above, stability and copy number are not properties rigidly fixed by the vector but, rather, exhibit considerable plasticity. As a consequence, a number of factors influence the stability of 2-μm vectors. First, inclusion of additional sequences *per se*—even genomic yeast sequences —tends to reduce stability. For instance, plasmid TLC1, which consists of plasmid YEp13 plus 4 kb of DNA spanning the yeast *CAN1* gene, is twice as unstable as plasmid YEp13 itself.[65] Similarly, vectors constructed from pUC plasmid tend to be somewhat more stable than the equivalent plasmid based on pBR322.[69] Thus, as a general rule of thumb, the smaller the plasmid, the better the stability. A second obvious factor influencing stability is the nature of the sequences cloned. Addition of a gene whose presence is deleterious to the host strain yields diminished stability and reduced average copy levels in the population, even under growth conditions selective for the plasmid.[12] As a consequence, inducible promoters are essential for optimum expression of genes whose products exhibit even mild inhibition of cell growth. A third factor influencing stability is copy level. In general, plasmids with increased copy levels exhibit increased stability. Vectors designed for high copy levels, and thus enhanced stability, are described later in this article.

Vector Copy Levels

Most of the standard 2-μm circle-based vectors whose copy levels have been determined have been found to be maintained in [cir⁺] strains at between 10 and 40 copies per cell. This value has been obtained by direct measurement of plasmid copy levels[96,103] as well as by determination of the specific activity of various enzyme products of plasmid-borne genes. For instance, the specific activity of β-isopropylmalate dehydrogenase, the product of *LEU2*, in a strain harboring plasmid pYT14-*LEU2* is 30 times that of a strain containing a single copy of the gene.[91] Similarly, the specific activity of orotidine-5′-phosphate decarboxylase, the product of the *URA3* gene, is approximately 30 times higher in a strain containing YEp24 than in a strain containing a single chromosomal copy of the gene.[67,138]

It should be noted that most standard vectors achieve this level of amplification, even though they contain only a single *FLP* recombination site and are, accordingly, incapable of amplifying via the inversional recombination system described in the preceding section. Exactly how this occurs is not known, although a number of different processes undoubtedly contribute to achieving the moderately high copy levels of these standard vectors. These likely include hitchhiking to high copy number by recombination with endogenous 2-μm circles as well as concentration of plasmid molecules by asymmetric segregation in conjunction with selection for plasmid-bearing cells. This issue is discussed in depth in a separate review.[4]

As noted in the previous section, vector copy levels are not uniform from cell to cell, but rather assume a distribution in the population whose mean is the average copy level per cell. Thus, at any one time, some cells in the population will contain a large number of plasmid molecules, others will have very few or none, and most will have levels near the mean. For many of the uses to which these vectors are applied, this is not a significant issue. In some cases, though, such as genetic selections or screens involving plasmid-borne markers, one has to keep in mind that selection can enrich for either end of the spectrum.

Most of the factors that were noted above to reduce stability also reduce average copy levels. This is true not only because plasmids with reduced stability are present in fewer cells, but also because the general distribution of cellular copy levels is shifted downward. In addition, for those plasmids carrying the entire 2-μm circle genome, cellular copy levels are also influenced by plasmid incompatability. That is, the copy level of the vectors in a [cir⁺] strain is less than that in a [cir⁰] strain, presumably because such 2-μm circle vectors share the same copy control system as the endogenous plasmids.

Applications

Cloning. Standard 2-micron circle vectors were originally developed as cloning vehicles for isolation of specific genes by complementation following transformation of appropriately marked yeast strains.[33] This is still a viable application of these vectors. Several partial *Sau*3A genomic yeast banks on YEp13 and YEp24 are available, from which a number of specific genes have been recovered. In addition, technology for generating such a bank using any of the standard vectors is quite straightforward and readily achievable in any molecular genetic laboratory.[114]

Despite historical precedent and the availability of representative clone banks, two considerations have diminished the appeal of using 2-μm-based vectors for cloning by complementation. First, since the vector, and the fragment cloned into it, are present at multiple copies in a preponderance of the cells in any transformant, any gene whose overexpression is deleterious to the cell will not be readily recovered from such a bank. As a consequence, certain genes, such as *KAR1,* could not be isolated from a 2-μm vector-based gene bank even though they are readily recovered from genomic banks carried on centromeric vectors.[140,141] Second, several investigators have found that their mutation of interest could be complemented by overexpression of a heterologous gene as well as by the wild-type copy of the desired gene.[157] Although this facet can be turned to advantage as described below, it adds a complication to the cloning process that is avoided using centromeric vectors.

Although the preceding discussion refers exclusively to isolation of yeast genes, several specialized 2-μm-based yeast vectors extend the applicability of complementing yeast mutations to isolation of genes from a wide variety of species. In these systems, random cDNA molecules from any source are cloned directly into a vector between signals for expression in yeast.[115] In this manner, cDNA libraries from essentially any organism can be screened directly in yeast for activities that will complement specific yeast mutations.

Overexpression for Protein Production. To a first approximation, propagation of a cloned yeast gene on a standard 2-μm circle vector yields an increase in production of the encoded protein commensurate with increased gene dosage. That is, 10- to 30-fold higher specific activity of a protein is routinely obtained when its cognate gene is propagated on a 2-μm circle vector than when it is present at single copy in the genome.[67,138] While this enhancement is not usually sufficient for commercial applications, it does provide a very facile method for increased production from a gene of interest that requires very little knowledge of the specific structure of the cloned gene nor substantial engineering of the construct. The 10- to 30-fold enrichment often provides the critical boost required for visualizing the product of the gene, either through enzymatic or immunological means. Thus, as a first pass at overexpression of a cloned gene, standard 2-μm vectors are ideal.

Certain limitations preclude this method from being generally applicable to every cloned gene. Obviously, if expression of the gene is limited not by gene dosage but by availability of a specific transcription or translation factor, then increased gene dosage will not yield increased expression. In addition, as discussed above, if the gene product is deleterious to the cell, then propagation on a 2-μm circle vector will not yield the expected increase in copy number and, accordingly, will not yield an increase in production. In this latter case, though, use of inducible copy level plasmids as described below might circumvent this restriction.

Overexpression for Genetic Suppression. Recently, 2-μm circle-promoted overexpression has been exploited as a means of genetic selection. That is, a number of investigators have used 2-μm circle vectors to identify genes whose overexpression alters the phenotype of a particular yeast strain. For example, Meeks-Wagner *et al.*[118] were able to isolate specific genes whose overexpression on 2-μm circle vectors diminished efficiency of chromosome segregation in yeast. Similarly, F. Winston and colleagues (personal communication) used 2-μm circle vectors to show that disproportionate expression of histone pairs yielded suppression of certain Ty insertion mutations. Finally, Toda *et al.*[157] isolated one of the genes encoding the catalytic subunit of cAMP-dependent protein kinase from a 2-μm vector-based random genomic library as a high-copy suppressor of a *cdc25* mutation. Given these examples, 2-μm vector-promoted enhanced expression should become as standard a procedure for analysis of specific problems in yeast molecular genetics as second-site repressor analysis is now. One obvious but noteworthy benefit of this approach is that identification of the suppressor locus and cloning it are one and the same step.

Vectors for High Copy Propagation

Vectors Based on pJDB219

Most ultra-high-copy yeast vectors currently in use are based on plasmid pJDB219.[10] A number of such vectors are listed in Table II. The salient structural features of pJDB219-derived plasmids is the hybrid 2-μm-*LEU2* allele adventitiously formed in the initial construction of pJDB219. Plasmid pJDB219 contains a fragment of yeast DNA spanning the *LEU2* gene — isolated from randomly sheared, total genomic DNA — inserted at the *Pst*I site in the *D* gene of the 2-μm circle. The fragment spans the entire *LEU2* coding region but lacks the *LEU2* promoter.[57] The *LEU2* gene is inserted in pJDB219 in the opposite orientation to *D* and, accordingly, transcription of the gene is most likely driven by the *REP1/REP2*-repressible promoter for a 1950n transcript, located near *REP3*.[147,153] Expression of this truncated *LEU2* gene, as measured by specific activity of the encoded enzyme, is several orders of magnitude less than that of the chromosomal *LEU2* gene.[57] Accordingly, the gene carried on the plasmid has been designated *leu2-d* to emphasize its diminished activity.

Plasmid pJDB219, and derivatives of it containing the *leu2-d* allele, persist in yeast cells at approximately 200–300 copies per haploid genome, an order of magnitude higher than that observed with most standard 2-μm circle-based vectors.[65] Several lines of evidence suggest that this high copy propagation is a consequence of the requirement for high copy numbers of the *leu2-d* allele to achieve leucine prototrophy in a *leu2⁻* strain. First, the copy level of plasmid pJDB219 drops rapidly from 200 copies per cell to approximately 50 copies per cell following shift of a plasmid-bearing strain from selective to nonselective medium (J. Broach, unpublished observations). Second, the presence of a normal *LEU2* gene on a pJDB219-based plasmid abolishes the high-copy properties of the plasmid. For example, the copy level of plasmid pSI5, which carries the *leu2-d* allele derived from plasmid pJDB219, persists at copy levels of approximately 200 per cell in [cir⁺] strains. However, plasmid pSI3, which is identical to plasmid pSI5 except that it carries a normal *LEU2* gene in addition to the pJDB219 *leu2-d* allele, is maintained at only 20 copies per cell in the same [cir⁺] strain.[26] In a similar vein, V. Zakian and colleagues (personal communication) have examined the behavior of a plasmid carrying both the *leu2-d* allele and the *URA3* locus. They find that only a small proportion of the Ura⁺ transformants of a *leu2 ura3* strain obtained with this plasmid are also Leu⁺, whereas all of the Leu⁺ transformants are also Ura⁺. From all these observations, we conclude that the high cellular copy levels of pJDB219 result from selection for high copy numbers of the *leu2-d* gene. In contrast to these observations, though, some investigators have noted that certain plasmids carrying the *leu2-d* allele attain high copy number even when selection for leucine prototrophy is not applied.[6] The reason for the discrepancy between this observation and those noted above is unknown. From a practical point of view, though, applying selection for leucine prototrophy will certainly ensure high copy propagation.

The use of plasmid pJDB219 or its derivatives for propagation of exogenous genes in yeast is reasonably straightforward. Genes of interest can be cloned into the bacterial moiety of the plasmid in any of a number of unique restriction sites. The resulting plasmid can be transformed by protoplast or lithium acetate procedures into any [cir⁺] *leu2* strain, selecting for leucine prototrophy. Transformants arise in 2 to 3 days and grow rapidly under selective or nonselective conditions.

For vectors such as pSI4 or pJDB219 itself, which carry the entire 2-μm circle genome, [cir⁰] strains are preferred hosts. Recombination between the pJDB219 vector and an endogenous 2-μm circle in [cir⁺] strains can yield a plasmid that carries the *leu2-d* marker but lacks the bacterial vector moiety and any gene cloned into it.[49] Since this recombinant plasmid is more stable than pJDB219—even more so if the additional genes carried confer a selective disadvantage—the gene of interest could be rapidly lost from the population. Thus, for the most part, use of [cir⁺] strains is best avoided for propagation of exogenous genes. A facile procedure for generating [cir⁰] strains is described in the next section. Transformants of [cir⁰]

TABLE II
High-Copy Vectors[a]

Name	Yeast markers	E. coli genes	Parental plasmid	2-μm genes	Other DNA	Size (kb)	Cloning sites[b]	Comments	Ref.[c]
pET13.1	CUP1, leu2-d	amp	pBR322	REP3		11.6	B	The CUP1 gene confers resistance to copper sulfate	81
pMP81	leu2-d	kan	pCR1	REP3		11.9	B, S, P, A	EcoRI fragments from pJDB219 in pCR1	56
pCI/1	leu2-d	amp, tet	pBR322	REP1, REP2, REP3			H, B, S, P	Differs from pSI4 only in orientation of pBR322 with respect to 2-μm sequences	22, 94
pJDB207	leu2-d	amp, tet	pAT153	REP3		6.9	S, B, P	3.3-kb fragment from pJDB219 containing leu2-d and REP3 cloned into pAT153. Same as pMA40	11, 21
pJDB209	leu2-d	amp	pAT153	REP3		6.8	B, S, A		11
pJDB219	leu2-d	tet	pMB9	REP1, REP2, REP3		12.4		Whole 2-μm B in pMB9. Very high copy number	10, 11
pL623	leu2-d	amp	pBR325	REP1, REP2, REP3					57
pMH158	leu2-d	amp, tet	pJRD158b	REP3		7.2	H, B, F, S, Q, P, L, G, X, U, Y, W, T, A, D		47, 82
pMCB803	leu2-d	amp	pJRD158b	REP3		6.0	D, H, X, U, Y, T, A	Deletion of BamHI–BglII of pMH158, removing BamHI, SphI, SalI, XmaIII, SacI, PstI, MluI, BclI, and BglII sites, and leaving the KpnI site in LEU2 unique	J. Schultz, M. Carlson
pMCB967	leu2-d	amp	pJRD158b	REP3		7.2	D, B, F, S, Q, P, L, G, X, U, Y, W, T, A	Derivative of pMH158 lacking the HindIII site	J. Schultz, M. Carlson
pMP78-1	leu2-d	amp, cat	pBR325	REP3		8.7	B, S, P	HindIII fragment of pJDB219 in pBR325	56, 57, 88, 89
pMP80-3	leu2-d	amp, tet	pBR325	REP3		8.7	B, S, P	EcoRI fragments from pJDB219 containing leu2-d in the EcoRI site of pBR325	88
pMP80-4	leu2-d	amp, tet	pBR325	REP3			H	EcoRI fragments from pJDB219 in pBR325	89

Plasmid	Yeast marker	Bacterial marker	Backbone	REP genes	Other	Size (kb)	Sites	Comments	Ref.
pPL262	leu2-d	amp	pUC19	REP1, REP2, REP3		9.9	Y, P, S, B	Derivative of pCl/1 with pBR322 sequences replaced with pUC19. Several closely related plasmids that differ in cloning sites are available	P. Lagosky
pPL269	leu2-d	amp	pAT153	REP1, REP2, REP3		11.2	I, Y, Q, K, S	Made from pCl/1 by replacing pBR322 sequences with pAT153 and a polylinker. Several closely related plasmids that differ in cloning sites are available	P. Lagosky
pPM40	leu2-d	amp	pBR322	REP1, REP2, REP3, FLP		9.7	AatII, A	Slightly more stable than pJDB219	P. McCabe
pROG5	leu2-d	amp tet	pBR322	REP3		7.6	B, P	3.2-kb HindIII fragment from pJDB219 in pBR322	44
pSI4	leu2-d	amp tet	pBR322	REP1, REP2, REP3		11.9	B, S, P, A	Derivative of pJDB219 with pBR322 substituted for pMB9 sequences	26
pSI5	leu2-d	amp	pBR322	REP3		8.3	B, S	3.9-kb HindIII from pJDB219 containing leu2-d, REP3, and origin in pBR322	31
pY43	leu2-d	amp	pMB1	REP3	fl ori	7.6	E, M, B, S, P	Very high copy number. The fl ori allows isolation of single-stranded DNA	6
YEp437	leu2-d HIS3	amp	pBR322	REP1, REP2, REP3		13.9	S	In vivo recombinant between pSI4 and YEp6	112
pJDB210	leu2-d URA3	amp	pAT153	REP3		7.9	B, S	pJDB211 differs only in orientation of URA3 fragment	11
pKS2-20	leu2-d URA3	amp tet	pBR322	REP3		8.6	B	Formed by in vivo ligation of HindIII fragments of pJDB219 and YIp5. Several other similar plasmids were recovered from this same ligation	154
pYK2024	leu2-d URA3	amp tet	pBR322	REP1, REP3	ARS1	11.4	B, S	4.4-kb XbaI fragment containing leu2-d, REP1, and REP3 from pJDB219 in the XbaI site of YRp16	103
YEp436	leu2-d URA3	amp tet	pBR322	REP1, REP2, REP3		13.3	B, J, S, M	In vivo recombinant between pSI4 and YIp5	112
pIA2	TRP1	kan	pHSS6	REP1, REP2, REP3, D		9.3	N, B, C	Inducible very-high-copy-number plasmid. In circle 0 strains containing GAL10–FLP, this plasmid can be driven to very high copy number by inducing FLP with galactose	P. Hieter
pIA1	URA3	kan	pHSS6	REP1, REP2, REP3, D		8.8	N, C, B	Inducible very-high-copy-number plasmid. In circle 0 strains containing GAL10–FLP, this plasmid can be driven to very high copy number by inducing FLP with galactose	P. Hieter

[a] Plasmids designed for very-high-copy-level propagation in yeast are listed alphabetically by yeast selectable marker.

[b] Restriction site abbreviations are as in Table I, footnote d.

[c] References to individuals are personal communications.

strains obtained with *leu2-d* vectors take a week or so to appear. However, once obtained, the strains grow quite well on selective medium and plasmid loss is very rare. In either case, *leu2-d* vectors persist in transformants indefinitely at high copy levels as long as selection is maintained. In addition, the vectors persist in most cells in the population for many generations following a shift to nonselective medium. This property makes the plasmids ideal for use in large-scale fermentation situations, in which continual selection for the plasmid is not always practical.

FLP-Induced High-Copy Propagation

As described in the first section of this article, the site-specific recombinase encoded by *FLP* promotes amplification of 2-μm circle. This amplification process is suppressed at normal plasmid copy levels by limiting *FLP* activity in the cell, through *REP1/REP2*-mediated repression of *FLP* gene expression. An obvious prediction of this model is that increased *FLP* expression during steady-state propagation of the plasmid should yield further amplification and result in an increase in the steady-state level of the plasmid. This prediction has been tested and confirmed.[128,136,147] Using a galactose-inducible *FLP* gene fusion, we provided additional *FLP* activity to a cell containing a normal complement of 2-μm circle. We measured the copy level of the plasmid in these cells after growth either in the absence of additional *FLP* activity or under conditions yielding high-level expression of the *FLP* construct. Plasmid copy levels were measured by ethidium bromide staining of restriction-digested genomic DNA fractionated by agarose gel electrophoresis. Results with several such strains are shown in Fig. 2. As is evident, the copy level of the plasmid in the absence of additional *FLP* activity is substantially less than that of the 150–200 copies of the rDNA repeats—consistent with previously reported values of 50 copies per haploid genome. In contrast, the plasmid copy number maintained in the presence of additional *FLP* activity is approximately two times higher than that of the rDNA repeats, or approximately 200–400 copies per cell. Thus, we can conclude that *FLP* activity is limiting for amplification during steady-state propagation of 2-μm circle, and that infusion of additional activity yields a dramatic increase in steady-state copy levels.

Similar *FLP*-induced increase in copy number can be achieved with 2-μm circle-based vectors. P. Hieter and colleagues (personal communication) have constructed a set of general purpose, amplifiable vectors (Fig. 3). These plasmids consist of the entire 2-μm circle genome, interrupted within the *FLP* coding region by vector sequences from plasmid pHSS6 plus a yeast selectable marker. Plasmid pHSS6 was constructed by Seifert *et al.*[145] and carries a kanamycin resistance gene and an origin of replica-

GLU   GAL   GLU   GAL   GLU   GAL

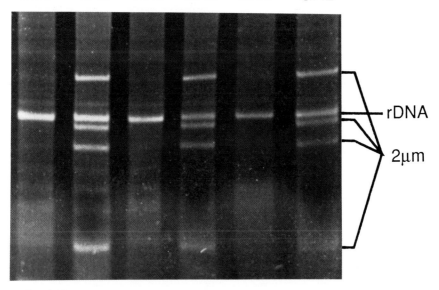

FIG. 2. Increased *FLP* expression induces high plasmid copy number. Ethidium bromide staining pattern of *Ava*I-digested genomic DNA isolated from *S. cerevisiae* strain GF4 grown either on YEP plus 2% glucose (Glu) or YEP plus 2% galactose (Gal). Each pair of tracks represents DNA obtained from a different isolate of the same strain. Bands corresponding to rDNA and to 2-μm circle DNA are indicated.

tion from pMB8, for selection and propagation in *E. coli*. The vector also carries a polylinker, to facilitate insertion of cloned sequences, bracketed by *Not*I sites. Once a sequence of interest has been cloned into any of the polylinkers sites, it can be readily moved among any of the related plasmids as a *Not*I fragment.

The advantage of using this system, rather than one based on plasmid pJDB219 are 2-fold. First, such a system could be used with a wide variety of preexisting plasmids and vectors. This would minimize the amount of plasmid reconstruction needed to generate a high-copy vector and would permit use of the full panoply of current selectable markers. Second, such a system could be tailored to permit high copy propagation even in nonselective conditions. All that would be required would be expression of the chromosomal *FLP* gene, by growth either on galactose-containing medium, if currently available *GAL10 – FLP* strains were used, or on any rich medium if a strain containing a constitutively expressed *FLP* gene were used. By coupling this vector to a plasmid-borne marker that was selective even on rich medium — using a currently available essential gene — then high copy, stable propagation could be achieved under optimal growth and fermentation conditions.

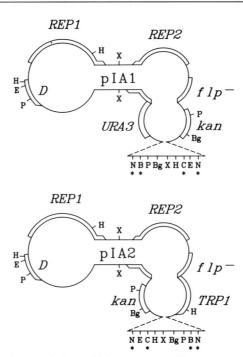

FIG. 3. Vectors for *FLP*-induced high copy levels. Shown are plasmids designed for *FLP*-induced high copy propagation. When propagated in a strain carrying an integrated copy of *GAL10–FLP*, the plasmids maintain a copy level of 10–20 when grown on noninducing carbon sources, such as glucose or raffinose, and a copy level of 200–400 when grown on galactose-containing medium. Plasmids pIA1 and pIA2 were constructed by inserting either the *Hin*dIII fragment spanning the *URA3* gene or the *Eco*RI to *Bgl*III fragment spanning the *TRP1* gene into the *Sma*I site of plasmid pHSS6 (145) to yield plasmids pHSS6-URA and pHSS6-Trp, respectively. *Bal*I plus *Eco*RV-digested 2-μm circle genomic DNA was then inserted into the *Sma*I site of pHSS6-URA and into the *Stu*I site of pHSS6-TRP to yield plasmids pIA1 and pIA2, respectively. The positions of various 2-μm, yeast, and plasmid genes are indicated, as are the locations of a number of restriction sites. Those sites in the polylinker designated with asterisks are suitable for insertion of exogenous DNA. Restriction sites: B, *Bam*HI; Bg, *Bgl*II; C, *Cla*I; E, *Eco*RI; H, *Hin*dIII; N, *Not*I; P, *Pst*I; X, *Xba*I.

Method for Isolating Strains Free of 2-μm Circle

Both of the procedures described above require strains devoid of 2-μm circle as hosts for propagating the various high-copy vectors. Prior methods for isolating such strains are based on competition with plasmid pJDB219 and are generally hit-or-miss propositions.[26] We have developed a simple procedure for isolating strains lacking 2-μm circles, based on our observation that 2-μm circle strains in which *FLP* is overexpressed are substantially inhibited for growth. The procedure involves introducing a *GAL10–*

FLP construct, either on a plasmid or integrated into a chromosome, into the strain of interest, plating the strain on selective media containing galactose, and isolating fast-growing papillae from the lawn of cells. Such galactose-resistant cells are present in approximately $1/10^4$ in a culture and are thus readily recovered. Most of the galactose-resistant cells arise as a result of losing 2-μm circles from the strain. However, a small proportion arise by inactivation of the *GAL10–FLP* construct, either by loss or by mutation. These can be distinguished readily, either by Southern analysis of DNA isolated from the galactose-resistant clones or by scoring galactose sensitivity of diploids obtained by crossing the isolates to a strain carrying the *GAL10–FLP* construct.

Procedure 1. Generating [cir⁰] Strains with Integrated GAL10–FLP

1. Transform a *leu2* strain to leucine prototrophy on synthetic complete glucose plate lacking leucine (SD-Leu) with 5 μg of *Bst*EII-cut pFV17 plasmid DNA.

2. Purify transformants by streaking for single colonies on SD-Leu plates.

3. Streak several purified colonies onto YEPG [1% (w/v) yeast extract, 2% (w/v) peptone, 2% (w/v) galactose] plate and incubate for 3 days. If the original strain is *gal2, gal7* or *gal10,* then add 2% (w/v) raffinose or 2% (v/v) glycerol plus 2% (w/v) acetate to the YEPG plate.

4. After 4 days of incubation, streaks should consist predominantly of small colonies with a few larger colonies (see Fig. 4). Purify several of the larger colonies by streaking on YEPG.

5A. Isolate genomic DNA from 5-ml cultures of individual isolates. Analyze the DNA by the method of Southern after digestion with *Eco*RI and probe with labeled YEp13 or YEp24 DNA. The autoradiogram should show hybridization to the single-copy *LEU2* or *URA3* gene, and no hybridization to 2-μm circle-specific fragments. Use DNA from known [cir⁺] and [cir⁰] strains as controls.

5B. As an alternative or in addition to 5A, cross individual isolates to the following tester strains: Y22 (**a** *leu2 ura3 his3 trp1* [cir⁺]), Y23 (**a** *leu2 ura3 his3 trp1* [cir⁰]), and Y894 (**a** *leu2::LEU2-GAL10-FLP ura3 his3 trp1* [cir⁰]), selecting and purifying diploids on appropriate media. Streak diploids on YEPG and incubate for 3 days. If the isolate is [cir⁰], then the diploids with Y23 and Y894 should yield large colonies and that with Y22 should yield small.

Procedure 2. Generating [cir⁰] Strains Lacking GAL10-FLP

1. Transform a *leu2* strain to leucine prototrophy on synthetic complete glucose plate lacking leucine (SD-Leu) with 5 μg of YEp51–*FLP* plasmid DNA.

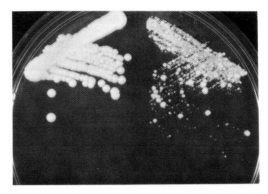

FIG. 4. Selection of plasmid-free strains by *FLP* overexpression. *Saccharomyces cerevisiae* strains Y23 (**a** *leu2 ura3 his3 trp1* [cir⁰], left-hand side) and Y22 (**a** *leu2 ura3 his3 trp1* [cir⁺], right-hand side) were transformed to Leu⁺ with *Bst*EII–cut pFV17 plasmid DNA, which carries a *GAL10–FLP* fusion. Transformants were purified on synthetic glucose plates lacking leucine, and then a single colony from each was inoculated into a 5-ml YEP plus galactose culture and grown overnight at 30°. A loopful of cells from each culture was streaked on one-half of a YEP plus galactose plate, and the plate was incubated at 30° for 4 days. As evident, the [cir⁰] strain (Y23) yielded all large colonies, whereas the [cir⁺] strain (Y22) yielded predominantly very small colonies and a few large colonies. Most of the large colonies from the [cir⁺] strain now lack endogenous 2-μm circles.

2. Purify transformants by streaking for single colonies on SD-Leu plates.

3. Streak several purified colonies onto synthetic complete galactose medium minus leucine plate and incubate for 3 days. If the original strain is *gal2, gal7,* or *gal10,* then add 2% raffinose or 2% glycerol plus 2% acetate to the plate.

4. After 3 days of incubation, streaks should consist predominantly of small colonies with a few larger colonies. Purify several of the larger colonies by streaking on YEPG.

5. Replicate the YEPG plate onto SD-Leu. Identify a Leu⁻ clone and recover the corresponding clone from the YEPG master plate. Proceed as in step 5 in Procedure 1.

Vectors for Expression of Cloned Genes in Yeast

A large number of 2-μm circle vectors suitable for expression of cloned genes in yeast are listed in Table III. Such vectors are available with a variety of constitutive or inducible promoters and a wide selection of yeast markers. Some of the vectors are designed to yield translational fusions with or without secretory signals, whereas others yield transcriptional fusions. Issues pertaining to promoter selection, optimization of yield of the

TABLE III
EXPRESSION VECTORS[a]

Name	Pro-moter	Secretion signal	Termi-nator	Yeast marker	2-μm genes	E. Coli genes	Parental plasmid	Other DNA	Size (kb)	Cloning sites[b]	Comments	Ref.[c]
							Constitutive Promoters					
pAAH5	ADC1		ADC1	LEU2	REP3	amp	pBR322		12.5	H	YEp13 with the ADC1 promoter and terminator inserted in the BamHI site	3, 121
pACF301	ADC1		CYC1	LEU2		amp				Y	Contains the first 6 codons of the ADC1 structural gene before the XhoI linker	85
pAD1	ADC1			LEU2	REP3	amp	pUC19		7.8	H, P, S, M, Q	Multiple cloning site between the ADC1 promoter and a sequencing primer binding site	P. Sass, J. Nikawa
pAD4	ADC1		ADC1	LEU2	REP3	amp	pUC19		8.2	H, P, S, M, Q		P. Sass, J. Nikawa
pAH5	ADC1		FLP	LEU2	REP3	amp	pBR322		11.8	H	HindIII cloning site is 12 bp upstream of the ADC1 ATG	3
pAH9	ADC1		FLP	LEU2	REP3	amp	pBR322		11.8	H	Very similar to pAH5, but the HindIII site is downstream of the ADC1 ATG. Plasmids pAH10 and pAH21 differ in the precise location of the HindIII site, so that the HindIII site is in a different reading frame in each plasmid	3
pAXα11	ADC1		MATα1	LEU2	REP3	amp	pBR322		13.6	Y	XhoI site is 10 bp upstream of the ADC1 ATG	3
pADH040-2	ADC1			leu2-d	REP3	amp	pBR322		7.0		The amp gene lacks its own promoter, but is expressed in yeast from the ADC1 promoter. Originally used to study expression of β-lactamase in yeast	57, 135
pADH322-2	ADC1			leu2-d	REP3	amp	pBR322		7.2		There is a nonunique EcoRI site adjacent to the ADC1 promoter	135
pMA56	ADC1		FLP	TRP1	REP3	amp	pBR322		8.3	E		3, 57, 168
pMAC561	ADC1		CYC1	TRP1	REP3	amp	pBR322		8.8	E, K		115

(continued)

TABLE III *(continued)*

Name	Promoter	Secretion signal	Terminator	Yeast marker	2-μm genes	E. Coli genes	Parental plasmid	Other DNA	Size (kb)	Cloning sites[b]	Comments	Ref.[c]
pWY4	ADC1		ADC1	TRP1		amp	pBR322			E		95
pYcDE8	ADC1		CYC1	TRP1	REP3	amp	pBR322		6.4	E, K, M, B		104
pYcDE-2	ADC1		CYC1	TRP1	REP3	amp	pBR322		7.5	E	A derivative, pCH98, also has a BglII cloning site flanked by EcoRI sites	76
pRIT10774	ARG3		ARG3	LEU2	REP3	amp	pBR322		13.1	B		35
pRIT10779	ARG3		ARG3	LEU2	REP3	amp	pBR322		13.1	O, B	Very similar to pRIT10774, but with the BamHI site just downstream of the ARG3 ATG, which is contained within an NcoI site	35
pEX-2	CYC1		CYC1	URA3		amp	pBR322	ARS1	8.9	B, M		66, 71
pEX-4	CYC1		CYC1	URA3	REP3	amp, kan	pBR322	ARS1	11.0	B	Derived from pEX-2 by adding the kan gene, which confers G418 resistance in yeast	58
pAC1	ENO1		ENO1	LEU2	REP3	amp	pBR322		10.7	S, H		93
pMAC101	ENO1		ENO1	LEU2	REP3	amp	pBR322		10.7	H	Derived from pAC1	162
pYE8	GAPDH		FLP	TRP1	REP3	amp	pBR322	ARS1 IFN		E	This plasmid contains an interferon gene that can be removed as an EcoRI fragment, producing a more general expression vector that lacks an ATG	143
pGPD-1	GPD		TRP1	TRP1	REP1, REP2, REP3, FLP	amp	pBR322			B	TRP1 gene acts as yeast marker and transcription terminator	13, 14
pGPD-2	GPD		PGK	TRP1	REP1, REP2, REP3, FLP	amp	pBR322			B	Similar to pGPD-1, but has the PGK terminator instead of the TRP1 terminator	13, 14
pMA230	PGK			LEU2	REP3	amp	pBR322		8.0	B	One of a series of plasmids named pMA22aδ	51, 161
pMA301	PGK			LEU2	REP3	amp	pBR322		8.4	G	Similar to pMA230, but with a BglII cloning site	119
pMA91	PGK		PGK	LEU2	REP3	amp	pBR322		9.5	G	Derived from pMA301 by adding the PGK terminator	16, 119

YEpIPT	PGK	FLP	TRP1	REP3	amp	pBR322		8.6	E		84, 86
pMA36	TRP1		LEU2	REP3	amp	pBR322		8.9	B	Lacks the TRP1 ATG	50
Inducible Promoters											
YEp434	GAL1		LEU2	REP3	amp	pBR322		8.7	B	In vivo recombinant between pGM65 and YEp21	112
YEp435	GAL1		LEU2, URA3	REP3	amp	pBR322		9.9	B	In vivo recombinant between pGM65 and YEp21	112
pLG12	GAL1		leu2-d	REP3	amp	pBR322	GAL4	9.9	B	GAL1 expression vector containing the GAL4 gene. Made by inserting a 390-bp promoter fragment into the BAMHI site of pLG1/4Δ60, recreating the BAMHI site and a functional GAL4 gene	A. Buchman, R. Freedman
pLG1/4Δ60	GAL1		leu2-d	REP3	amp	pBR322	GAL4	9.6	B	GAL1 promoter with a BamHI site 10 bp upstream of the mRNA start site fused to the GAL4 gene, which lacks its own promoter. Unless the fragment inserted in the BamHI site carries a promoter, GAL4 will not be expressed	A. Buchman, R. Freedman
pYMS-3	GAL1	ADC1	leu2-d	REP1, REP2, REP3	amp	pBR322			B		170
YEpG3	GAL1	CDC28	leu2-d	REP3	amp	pBR322		10.0	B	Derived from pMA91	S. Reed
pYRF101	GAL1		leu2-d, URA3	REP3	amp	pBR322	GAL4	10.7	B	1.2-kb HindIII URA3 fragment in pLG1/4Δ60	R. Freedman
pYRF102	GAL1		leu2-d, URA3	REP3	amp	pBR322	GAL4	11.1	B	1.1-kb HindIII URA3 fragment in pLG12	R. Freedman
pCGE329	GAL1		URA3	REP3	amp	pBR322		7.6	Y	Does not contain the GAL1 mRNA start site	17
pGAL-X/P	GAL1		URA3	REP3	amp	pBR322		7.6	Y, Q, U, E, B		D. Garfinkel
pMAC105	GAL1-ENO1	ENO1	LEU2	REP3	amp	pBR322		10.2	H	Hybrid promoter composed of the GAL1-10 UAS fused to the ENO1 transcription initiation region	P. McCabe
YEp213-G	GAL1, GAL10	FLP	LEU2	REP3	amp	pBR322		11.5	B, S	0.8-kb BamHI-SalI GAL1-GAL10 promoter fragment in YEp213 (see YEp13S) allowing the choice of using either the GAL1 or the GAL10 promoter	R. Sternglanz, J. Mullen

(continued)

TABLE III (continued)

Name	Promoter	Secretion signal	Terminator	Yeast marker	2-μm genes	E. Coli genes	Parental plasmid	Other DNA	Size (kb)	Cloning sites[b]	Comments	Ref.[c]
pGM65	GAL1 GAL10			URA3	REP3	amp	pBR322		7.7	B, E	Divergent GAL1–GAL10 promoters in YEp420. Either promoter can be used with the BamHI site (GAL1) or the EcoRI site (GAL10)	112
YEp57	GAL10		HIS3	HIS3	REP3	amp	pBR322	leu2 (680-bp deletion)	8.6	S	1.8-kb BamHI HIS3 fragment in the BclI site of pAR38	4
YEp51	GAL10		FLP	LEU2	REP3	amp	pBR322		7.6	S	Galactose-inducible expression vector. Can use BamHI, HindIII, BclI, and SpeI sites downstream of the SalI site for directional cloning	31
YEp52	GAL10		FLP	LEU2	REP3	amp	pBR322		6.9	H	Derived from YEp51 by deleting the 757-bp HindIII–SalI fragment	31
YEp55	GAL10		FLP	LEU2	REP3	amp	pBR322		7.4	B, S	Derivative of YEp51 containing the GAL10 ATG. Available with the BamHI site in all three reading frames	4
YEp62	GAL10		lacZ	LEU2	REP3	amp lacZ	pBR322		10.3	S, B, M	For galactose-inducible production of trihybrid GAL10–ORF–lacZ fusion proteins. These proteins can be studied or purified using the lacZ moiety	31
YEpBam	GAL10		FLP	LEU2	REP3	amp	pBR322		6.9	S, B	Derivative of YEp51	F. Tamanoi
YEpUC	GAL10		FLP	LEU2	REP3	amp	pBR322		6.9	E, M, B, P	Contains an ATG upstream of the cloning sites. Made by placing the polylinker of pUC8 into the SalI–HindIII sites of YEp51	F. Tamanoi
YEp61	GAL10		lacZ	leu2-d	REP3	amp lacZ	pBR322			M	For making trihybrid proteins. Contains the GAL10 ATG upstream of the SmaI cloning site, followed by the lacZ gene. Open reading frames inserted	

into the SmaI site will produce a galactose-inducible GAL10–ORF–lacZ fusion protein

Plasmid	Promoter									Description	Ref.
YEp56	GAL10	LYS2	LYS2	REP3	amp	pBR322	leu2 (680-bp deletion)	11.6	S	4.8-kb HindIII fragment containing LYS2 in the HindIII site of pAR38	4
YEp54	GAL10	FLP	TRP1	REP3	amp	pBR322	ARS1 leu2 (inactive)	8.3	S	1.45-kb EcoRI TRP1–ARS1 fragment in the EcoRI site of pAR38	4
YEp53	GAL10	URA3	URA3	REP3	amp	pBR322	leu2 (680-bp deletion)	8.0	S	1.2 kb HindIII URA3 fragment in the HindIII site of pAR38	4
pEMBLyex2	GAL10–CYC1	FLP	leu2-d URA3	REP3	amp	pMB1	f1 ori	8.9	M, B, S, P, H	Very similar to pEMBLyex4	6, 40
pEMBLyex4	GAL10–CYC1	FLP	leu2-d URA3	REP3	amp	pMB1	f1 ori	8.8	M, B, X, S, P, H	Contains a hybrid GAL10–CYC1 promoter and the f1 origin of replication, allowing easy isolation of single-stranded DNA	40, 128
pEMBLyexT3	GAL10–CYC1	FLP	leu2-d URA3	REP3	amp	pMB1	f1 ori T3 promoter	9.0	M, B, X, S, P, H	Derivative of pEMBLyex4 containing the T3 promoter 3' to the polylinker. Can be used to make RNA probes	J. Murray
pEMBLyexT7	GAL10–CYC1	FLP	leu2-d URA3	REP3	amp	pMB1	f1 ori T7 promoter	8.9	M, B, X, S, P, H	Derivative of pEMBLyex4 with the T7 promoter at the 3' end of the polylinker. Can be used to make RNA probes	J. Murray
pAM81	PHO5		LEU2		amp	pBR322	ARS1	12.0	S	Differs from pAM82 only in the cloning site	142
pAM82	PHO5		LEU2		amp	pBR322	ARS1	12.0	Y	Derived from pAT77 by Bal31 deletion from the SalI site	122, 130
pAT77	PHO5		LEU2		amp	pBR322	ARS1	12.3	S	Contains the first 83 bp of the PHO5 coding sequences	122
pJDB207/PHO5	PHO5	TRP1	leu2-d	REP3	amp	pAT153		8.6	H	Contains the entire PHO5 gene	78
pYE4	PHO5	FLP	TRP1	REP3	amp	pBR322		9.1	E	Lacks the PHO5 ATG	106
pYE7	PHO5		TRP1	REP3	amp	pBR322	ARS1 IFN		E	This plasmid contains an interferon gene that can be removed as an EcoRI fragment, producing a more general expression vector that lacks an ATG	143
p1A1RI-15	PHO5	TRP1	TRP1 URA3	REP3	amp	pBR322			E	Derivative of p1A1 containing the TRP1 gene	133

(continued)

TABLE III (continued)

Name	Promoter	Secretion signal	Terminator	Yeast marker	2-μm genes	E. Coli genes	Parental plasmid	Other DNA	Size (kb)	Cloning sites[b]	Comments	Ref.[c]
pPV1	PHO5		URA3	URA3	REP3	amp	pHC79	λ cos	9.8	H	Cosmid containing the PHO5 promoter. Derived from pHCG3	48
pPV2	PHO5		URA3	URA3	REP3	amp	pHC79		6.5	H	Derived from pPV1 by deleting the cos site	48
SB10-2	PHO5			URA3	REP3	amp	pBR322		8.1	K, S	636-bp BamHI–SalI PHO5 promoter in the BamHI–SalI sites of YEp24. Similar to p1A1	133
Secretion												
pαADH2	ADH2	MFα1	FLP	TRP1	REP3	amp	pBR322				K	Derived from pEMBLyex2 by the addition of a synthetic oligonucleotide encoding the leader peptide of the Kluyveromyces lactis killer toxin gene	134
YEpsec1	GAL10-CYC1	k1	FLP	leu2-d URA3	REP3	amp	pMB1	f1 ori	9.0	M, B, S, P, H			6
pJC1-5	MFα1	MFα1	MFα1	LEU2	REP3	amp	pBR322		10.1	H	Contains the MFα1 promoter and leader sequence and one copy of the α-factor structural gene	165
pTG881	MFα1	MFα1	PGK	leu2-d URA3	REP3	amp	pBR322			S, H	Contains the α-factor coding sequences which can be removed with HindIII	109
YEp70αT	MFα1	MFα1	FLP	TRP1	REP3	amp	pBR322		8.2	E	Contains the α-factor prepro sequence and three more codons before the EcoRI cloning site	84

pMFα8	MFα1	MFα1	TRP5	TRP1	REP3	amp	pBR322	ARS1	10.0	T, Y		120
pSEY210	MFα1	MFα1		URA3	REP3	amp	pBR322	SUC2		H	Contains the SUC2 gene fused to the promoter and prepro-leader sequences from the MFα1 gene. The SUC2 sequences can be removed as a HindIII fragment, creating a general expression–secretion vector	54
pTG1819	PGK	MFα	PGK	LEU2 URA3	REP3	amp	pBR322			H, S, F, G	Contains the α-factor coding sequences which can be removed with HindIII	109
pNW033	PGK	pGKL1 97K Killer	TRP1	TRP1		amp			6.7	B, M		159
pIA1	PHO5	PHO5		URA3	REP3	amp	pBR322		7.3	K, S	Fragment containing 555 bp of the promoter and the first 29 codons of PHO5 and a 1.4-kb fragment of 2-μm in YIp5	79
pCGS681	TPI	SUC2		URA3		amp				O		123

[a] Plasmids designed for expression of cloned genes in yeast are listed in three groups. The first group comprises vectors in which expression is driven by a constitutive yeast promoter. The second group comprises vectors containing an inducible yeast promoter. The third group are those vectors designed to promote secretion of the product of the expressed gene. In each group, vectors are listed alphabetically first by promoter designation and then by selectable marker designation. No separation is made in either of the first two groups between those vectors that yield transcription fusions and those that yield translation fusions, although the comments appended to each vector usually indicate which type of fusion is obtained.

Unless otherwise noted, all plasmids carry the 2-μm circle origin of replication in addition to the listed genes. If the space is blank, the particular 2-μm circle genes present could not be determined from the available plasmid maps.

[b] Restriction site abbreviations are as listed in Table I, footnote d.

[c] References to individuals are personal communications.

desired protein, targeting to the secretory pathway, and patterns of glycosylation and other posttranslational modification are discussed in detail in other articles in this volume. Additional considerations regarding the use of inducible promoters and expression in yeast in general can be found elsewhere.[12,31] Accordingly, the following comments focus only on factors pertinent to propagating these vectors in yeast.

First, as discussed in the previous sections, stability and copy number can be affected by the particular sequences cloned into a 2-μm-based vector. Specifically, expression of a gene whose product is inhibitory to cell growth yields reduced copy levels and stability of the vector construct. The extent to which copy number and stability are reduced and the rapidity with which such diminished persistence is manifested are a function of the degree of toxicity of the product. However, even minimal toxicity will, with time, yield diminished vector integrity. Thus, use of inducible promoters to drive synthesis of the even minimally toxic products is imperative in order to achieve maximal expression of the product.

Second, as the copy number of a particular expression vector increases, transcription and translation factors required for synthesis of the desired product can become limiting. In those promoter systems in which genes encoding specific activators have been identified, these genes can be amplified to preclude their being limiting. For instance, to ensure adequate levels of the transcriptional activator encoded by GAL4 for expression from vectors using the GAL1 or GAL10 promoter, some GAL vectors incorporate a copy of the GAL4 gene. As a more general solution, plasmid pGAL4, a 2-μm circle vector carrying a copy of GAL4, can be copropagated in a strain containing any of the GAL expression vectors to enhance the levels of the activator in that strain. We have routinely observed a 2- to 3-fold increase in the level of expression of a product of a gene carried on a GAL expression vector when pGAL4 is also present in the strain over that obtained when pGAL4 is absent. Analogous measures, using a cloned copy of the PHO4 gene, should be effective in enhancing expression from PHO5-based vectors.

Vectors for Specialized Applications

A number of vectors based on 2-μm circle have been designed for specialized uses. Several of these are listed in Table IV and their applications are discussed in this section.

Analysis of cis-Acting Expression Elements

Several vectors have been developed that can be used to determine whether a particular segment of DNA encompasses an element, such as an enhancer or a terminator, active in affecting gene expression. These vectors carry a reporter gene encoding a protein that is readily assayed in yeast, such as β-galactosidase or chloramphenicol acetyltransferase. The vector is constructed so that the reporter gene lacks one or more elements required for expression in yeast, as well as a suitably positioned restriction site to permit insertion of test sequences. Fragments can then be cloned into the vector and assessed for their ability to compensate for the missing element, following transformation into an appropriate yeast strain.

The prototypic vector for such analysis is plasmid pLG670-Z, which consists of a *CYC1–lacZ* fusion including the *CYC1* TATA domain but lacking the *CYC1* upstream activation site (UAS) elements.[72] The vector alone yields no *lacZ* expression. An exogenous fragment inserted into the vector can be tested for enhancerlike activity by measuring the level of β-galactosidase activity in a strain harboring the vector-plus-test-fragment construct. Similar vectors have been designed that lack other components required for expression. These include UAS deletions from other yeast promoters, UAS + TATA deletions from a variety of promoters, and *SUC2* signal peptide deletions. These vectors provide a means of identifying sequences that provide any of these missing functions. An additional set of vectors use reporter genes whose activities can be selected in yeast. Accordingly, specific sequences fulfilling the requisite function can be identified and retrieved from banks of randomly cloned fragments.

The clear advantage of using 2-μm circle-based vectors for analyzing transcriptional elements is that the enhanced copy levels promoted by the plasmid amplify the experimental signal, as compared with analysis of the element at single copy in the chromosome. However, one caveat to this approach is that an element on a multicopy plasmid may yield a different response than it would at single copy in the chromosome. This could be due to the abnormal context of the element within the plasmid or simply to the presence of the multiple copies of the promoter of interest. A second caveat is that an element that affected copy number or stability of the plasmid, either directly or indirectly, would yield an inaccurate measure of its functional activity. To date, neither of these potential problems has been documented. However, with any new system, one should establish experimentally that the activity of the element as measured on the high-copy plasmid provides an accurate reflection of the element in its normal, single-copy extent.

Strain Manipulation

Several vectors are available that promote specific modifications or alterations of yeast strains. Plasmids carrying the yeast *HO* gene can be used to convert haploid **a** strains to α strains and vice versa. Another set of

TABLE IV
SPECIALIZED VECTORS[a]

Name	Yeast markers	2-μm genes	E. coli genes	Parental plasmid	Other DNA	Size (kb)	Cloning sites[b]	Comments	Ref.[c]
						Cis-Acting Signals			
pIT213	URA3	REP3	amp 'hyg kan	pBR322		10.7	B, S	Allows screening for promoter fragments, since the hygromycin resistance gene lacks a promoter	101
pEP100	LEU2		amp 'cat	pBR327			S	For studying promoters using the cat gene from Tn9	2
pJM102	LEU2	REP3	amp	pBR322	HSV-TK λcos		G	For studying promoters using the HSV-TK gene	116
pMC1587	LEU2	REP3	amp 'lacZ	pBR322		16.0	M, B	For studying promoters using the lacZ gene	36
pYT760	LEU2	REP2, REP3	amp 'lacZ	pBR322		15.6	H, B	For studying promoters using the lacZ gene	111
pRB45	LEU2 URA3	REP3	amp 'lacZ	pBR322		20.0	B	For studying promoters using the lacZ gene	139
pJRB10	leu2-d	REP3	amp 'lacZ	pBR322			B	For studying promoters using the lacZ gene	31
pSL24	leu2-d	REP3	amp 'lacZ	pBR322		12.9	B	For studying promoters using the lacZ gene	77
YEpZ36	leu2-d	REP3	amp 'lacZ	pBR322		9.5	H, B	For studying promoters using the lacZ gene	132
pMC2010	TRP1	REP3	amp 'lacZ	pBR322		8.0	E, M, B	For studying promoters using the lacZ gene	36
pCGS139	URA3	REP3	amp 'lacZ			9.7	E	For studying promoters using the lacZ gene	54
pMC1585	URA3	REP3	amp 'lacZ	pBR322		13.4	B	For studying promoters using the lacZ gene	36
pRY140	URA3	REP3	amp 'lacZ	pBR322		9.5	M, B	For studying promoters using the lacZ gene	174
pSEY101	URA3	REP3	amp 'lacZ			9.7	E, M, B	For studying promoters using the lacZ gene. Derivative of pCGS139 lacking the SmaI site in URA3 and the SalI and XhoI sites at the 3' side of lacZ	52,55
pLG670-Z	URA3		amp 'lacZ	pBR322			B	For studying promoters using the lacZ gene. Derived from pLG669-Z by deleting the XhoI fragment containing the CYC1 UAS	2,72
YEp363	LEU2	REP3	amp 'lacZ	pUC8		8.4	E, M, B, S, P, H	For studying promoters using the lacZ gene. Plasmids YEp364 and YEp365 differ only in reading frame of cloning sites with respect to lacZ	129

270

Plasmid	Marker	Replication	Resistance	Vector	Insert	Size (kb)	Sites	Description	Ref.
YEp366	LEU2	REP3	amp 'lacZ	pUC18		8.4	E, Q, K, M, B, X, S, P, F, H	For studying promoters using the lacZ gene. Plasmids YEp367 and YEp368 differ only in reading frame of cloning sites with respect to lacZ. Plasmids YEp366R, YEp367R, and YEp368R have the multiple cloning sites in the other orientation with respect to lacZ	129
YEp353	URA3	REP3	amp 'lacZ	pUC8		8.0	E, M, B, S, P, H	For studying promoters using the lacZ gene. Plasmids YEp354 and YEp355 differ only in reading frame of cloning sites with respect to lacZ	129
YEp356	URA3	REP3	amp 'lacZ	pUC18		8.0	E, Q, K, M, B, X, S, P, F, H	For studying promoters using the lacZ gene. Plasmids YEp357 and YEp358 differ only in reading frame of cloning sites with respect to lacZ. Plasmids YEp356R, YEp357R, and YEp358R have the multiple cloning sites in the other orientation with respect to lacZ	129
pHZ18	URA3	REP3	amp	pBR322	CYC1–rp51–lacZ	11.1	H	For studying splicing. The plasmid has the GAL10 UAS and CYC1 transcription start site fused to a fragment of the rp51 gene containing an intron, followed by the lacZ gene	155
pJT24	URA3	REP3	amp 'lacZ	pBR322	GAL10 UAS	10.3	Y, M, B	For studying transcription start signals. Galactose regulated β-galactosidase production will occur if a promoter and adjacent transcribed sequences are inserted into the cloning sites	62
pSEY303	URA3	REP3	amp	pBR322	'SUC2	9.0	E, M, B, S, H	For making fusions to invertase in order to study intracellular location signals	55
Strain Manipulation									
pGTyH3	URA3	REP3	amp	pBR322	GAL1-Ty		G	For chromosomal amplification of genes. Genes are inserted into the Ty element and transcription is induced with galactose. The Ty elements transpose, causing an increasing in the chromosomal copy number of the gene. Has been used with inserts as large as 2.7 kb	18
pLB318	TRP1	FLP, no ori	amp	pBR322	ARS1	9.2		Contains FLP under control of the ADC1 promoter on a YRp-type plasmid	164
YEp51-FLP	LEU2	REP3, FLP	amp	pBR322		7.7	S	For galactose-inducible FLP overexpression from an episomal plasmid	166
pFV17	LEU2	FLP, no ori	amp	pBR322		6.8		For galactose-inducible FLP overexpression. This is an integrating plasmid containing GAL10:FLP	166
pGAL4	URA3	REP3	amp	pBR322	GAL4	10.9	E	For increasing expression from the GAL1 or GAL10 promoters	M. Marshall
p82-6B + LEU	LEU2	REP1, REP2, REP3, FLP, D	tet	pMB9				For introducing 2-μm into cir° strains	163

(continued)

TABLE IV (continued)

Name	Yeast markers	2-μm genes	E. coli genes	Parental plasmid	Other DNA	Size (kb)	Cloning sites[b]	Comments	Ref.[c]
pPS13	URA3	2 repeats only	amp	pIC19R		5.7	E, B, S, P, H, G	For making gene disruptions. Polylinkers flank two direct repeats of the 2-μm 599-bp repeats, separated by the URA3 gene. The cassette is used to make gene disruptions. Inducing FLP expression causes loss of URA3 and one repeat, allowing URA3 to be used for further manipulations	R. Seale
YEpHO	LEU2	REP3	amp	pBR322	HO	16.7		HO endonuclease gene in YEp13. Can be used to change the mating type of a strain	100
pGALHO	LEU2	REP3	amp	pBR322	GAL10–HO	8.4		HO endonuclease gene in YEp51. For galactose-inducible switching of mating type	J. Broach
pMR848	URA3	REP3	amp tet	pSC101		7.5	E, H, B, S, M	Low copy number in E. coli, allowing isolation of yeast genes that may be toxic to E. coli at high copy numbers	M. Rose
p2-1A	TRP1	REP3	amp	pBR322	ODC	8.9	D, F, M	pYcDE8 containing a fragment of DNA from Schwanniomyces occidentalis which encodes orotidine-5'-phosphate decarboxylase (ODC) under control of the ADC1 promoter. Complements mutations in the Saccharomyces cerevisiae URA3 gene	104
pADE	TRP1	REP3	amp	pBR322	S. occidentalis ADE2	9.4		pYcDE8 containing the ADE2 complementing gene from S. occidentalis between the ADC1 promoter and the CYC1 terminator. Complements ade2 mutations in both S. occidentalis and Saccharomyces cerevisiae	105, R. Klein

Vector Construction

Name	Yeast markers	2-μm genes	E. coli genes	Parental plasmid	Other DNA	Size (kb)	Cloning sites[b]	Comments	Ref.[c]
pAR38		REP3	amp	pBR322	leu2 (680-bp deletion)	6.8	S, F, B, J, H, L, X, A, R, E	For making GAL10 expression vectors. The SalI site is adjacent to the GAL10 promoter, and the other sites can be used to insert the yeast marker of choice. Derived from YEp51 by deleting a 680-bp BstEII–BstXI fragment from the LEU2 coding region and a small HindIII fragment	4

272

Plasmid	Marker	REP	amp/lacZ	Backbone	Special	Physiological Response		Comments	References
UTX25	LEU2	REP3	tet	pBR322		9.7		Digestion with SalI releases a 6.5-kb fragment containing LEU2 and 2-μm sequences, allowing easy conversion of intergrating or ARS plasmids into 2-μm–LEU2 vectors. Derived from YEp13 by digestion with PstI and religating, deleting a small fragment and inverting the remaining piece	**R. Needleman, D. Finkelstein**
UTX45	TRP1							For easy isolation of a SalI fragment containing 2-μm and TRP1 for conversion of integrating plasmids to 2-μm–TRP1 plasmids. UTX46 is similar, but the 2-μm–TRP1 DNA is isolated as a BglII fragment	**R. Needleman, D. Finkelstein**
pYT760.ADH1	LEU2	REP2, REP3	amp 'lacZ	pBR322		17.1	H	Control plasmid to determine the activity of the ADC1 promoter	111
pLG669-Z	URA3	REP3	amp 'lacZ	pBR322		12.0	B	CYC1 promoter fused to the lacZ gene, with a BamHI site at the junction. Derived from YEp24	73
pLGSD5	URA3	REP3	amp	pBR322	CYC1–lacZ fusion	9.9	B	GAL10 promoter in place of the CYC1 UAS in pLG669-Z. Makes galactose-inducible β-galactosidase. Has been used to estimate plasmid copy number	38,74
pRY121	URA3	REP3	amp 'lacZ	pBR322		11.5	B, S	GAL1–lacZ fusion plasmid derived from pLG669. Also has a SalI site by the GAL10 promoter. A similar plasmid, pRY131, has 1.0 kb of nonfunctional tet sequences (and the SalI site) deleted	171,174
pIT210	LEU2	REP3	amp 'lacZ	pBR322				HSP1–lacZ fusion for studying the heat-shock response	23
pJH5	LEU2 URA3	REP3	amp lacZα	pUC8	CEN4	9.8	B, S	Inducible high-copy-number vector for studying genes lethal in high copy. When a strain containing GAL10:FLP and this plasmid is induced, the FLP promotes recombination that removes URA3 and CEN4, allowing the copy number to rise from one copy per cell to more than 20	F. Volkert

a 2-μm circle-based plasmids designed for specialized purposes are listed in four categories. These are (1) plasmids for analysis of cis-acting expression signals, (2) plasmids for strain manipulation, (3) plasmids to facilitate vector construction, and (4) plasmids for monitoring various in vivo physiological responses. A short description of the specific purpose for which each plasmid has been designed is provided in the Comments column.

b Restriction site abbreviations are as in Table I, footnote d.

c References to individuals are personal communications.

plasmids can be used to cure [cir$^+$] strains of 2-μm plasmids as described in the previous section. With due concern for symmetry, other vectors have been constructed that allow introduction of wild-type 2-μm circles into [cir^0] strains. Plasmid pGTyH3 can be used, in conjunction with a Ty cassette, to introduce multiple copies of a sequence stably into the genome by random integration. Finally, a cassette consisting of direct repeats of the *FLP* target site bracketing the yeast *URA3* gene can be excised from plasmid pPS13 and inserted into cloned yeast sequences to allow repetitive, omega-style inactivations in a single strain in a manner first suggested by Alani *et al.*[1] Specific directions for optimum use of these vectors can be obtained from the reference appended to each listing in Table IV.

Vector Construction

Three plasmids listed in Table IV facilitate vector construction. Two of these contain a readily transferable cassette, comprising a yeast selectable marker and 2-μm circle replication and stability loci, that can be used to convert an integrating plasmid into a replicating plasmid. The third plasmid is a marker-free expression vector into which novel selectable markers can be inserted as required.

Monitoring Physiological Responses

Plasmid pIT210 consists of a translational fusion of the heat-shock responsive gene *HSP1* to the *E. coli lacZ* gene. This construct allows a quantitative as well as a visual metric of a strain's response to heat shock. Similarly, plasmid pLG669-Z can be used to monitor glucose repression and/or heme sufficiency in various strains.

Acknowledgments

We are very grateful to all of our colleagues who responded promptly and generously to our request for published and unpublished information on 2-μm-based plasmids. The usefulness of this article is a direct consequence of their timely contributions.

Research from our laboratory described in this article was supported by NIH grant GM34596.

References

[1] E. Alani, L. Cao, and N. Kleckner, *Genetics* **116**, 541 (1987).
[2] M. R. Altherr and R. L. Rodriguez, *in* "Vectors: A Survey of Molecular Cloning Vectors and Their Uses" (R. L. Rodriguez and D. T. Denhardt, eds.), p. 405. Butterworth, Stoneham, Massachusetts, 1988.
[3] G. Ammerer, this series, Vol. 101, p. 192.
[4] K. A. Armstrong, T. Som, F. C. Volkert, A. Rose, and J. R. Broach, *in* "Yeast Genetic

Engineering" (P. J. Barr, A. J. Brake, and P. Valenzuela, eds.), p. 165. Butterworth, Stoneham, Massachusetts, 1989.

[5] C. Baldari and G. Cesareni, *Gene* **35**, 27 (1985).

[6] C. Baldari, J. A. H. Murray, P. Ghiara, G. Cesareni, and C. L. Galeotti, *EMBO J.* **6**, 229 (1987).

[7] D. A. Barnes and J. Thorner, *Mol. Cell. Biol.* **6**, 2828 (1986).

[8] D. Beach and P. Nurse, *Nature (London)* **290**, 140 (1981).

[9] D. Beach, M. Piper, and P. Nurse, *Mol. Gen. Genet.* **187**, 326 (1982).

[10] J. D. Beggs, *Nature (London)* **275**, 104 (1978).

[11] J. D. Beggs, *in* "Molecular Genetics in Yeast" (D. Von Wettstein, A. Stenderup, M. Kielland-Brandt, and J. Friis, eds.), p. 383 (Alfred Benzon Symposium, Vol. 16). Munksgaard, Copenhagen, 1981.

[12] G. A. Bitter, K. K. Chen, A. R. Banks, and P.-H. Lai, *Proc. Natl. Acad. Sci. U.S.A.* **81**, 5330 (1984).

[13] G. A. Bitter and K. M. Egan, *Gene* **32**, 263 (1984).

[14] G. A. Bitter, K. M. Egan, R. A. Koski, M. O. Jones, S. G. Elliott, and J. C. Giffin, this series, Vol. 153, p. 516.

[15] H. Blanc, C. Gerbaud, P. P. Slonimski, and M. Guerineau, *Mol. Gen. Genet.* **176**, 335 (1979).

[16] M. W. Bodmer, S. Angal, G. T. Yarranton, T. J. R. Harris, A. Lyons, D. J. King, G. Pieroni, C. Riviere, R. Verger, and P. A. Lowe, *Biochem. Biophys. Acta* **909**, 237 (1987).

[17] J. D. Boeke, D. J. Garfinkel, C. A. Styles, and G. R. Fink, *Cell* **40**, 491 (1985).

[18] J. D. Boeke, H. Xu, and G. R. Fink, *Science* **239**, 280 (1988).

[19] G. Boguslawski and J. O. Polazzi, *Proc. Natl. Acad. Sci. U.S.A.* **84**, 5848 (1987).

[20] D. Botstein, S. C. Falco, S. E. Stewart, M. Brennan, S. Scherer, D. T. Stinchcomb, K. Struhl, and R. W. Davis, *Gene* **8**, 17 (1979).

[21] B. A. Bowen, A. M. Fulton, M. F. Tuite, S. M. Kingsman, and A. J. Kingsman, *Nucleic Acids Res.* **12**, 1627 (1984).

[22] A. J. Brake, J. P. Merryweather, D. G. Coit, U. A. Heberlein, F. R. Masiarz, G. T. Mullenbach, M. S. Uerda, P. Valenzuela, and P. J. Barr, *Proc. Natl. Acad. Sci. U.S.A.* **81**, 4642 (1984).

[23] C. Brazzell and T. D. Ingolia, *Mol. Cell. Biol.* **4**, 2573 (1984).

[24] B. J. Brewer and W. L. Fangman, *Cell* **51**, 463 (1987).

[25] J. R. Broach, *in* "The Molecular Biology of the Yeast *Saccharomyces:* Life Cycle and Inheritance" (J. N. Strathern, E. W. Jones, and J. R. Broach, eds.), p. 455. Cold Spring Harbor Laboratory, Cold Spring Harbor, New York, 1981.

[26] J. R. Broach, this series, Vol. 101, p. 307.

[27] J. R. Broach, J. F. Atkins, C. McGill, and L. Chow, *Cell* **16**, 827 (1979).

[28] J. R. Broach, V. R. Guarascio, and M. Jayaram, *Cell* **29**, 227 (1982).

[29] J. R. Broach, V. R. Guarascio, M. H. Misiewicz, and J. L. Campbell, *in* "Molecular Genetics in Yeast" (D. Von Wettstein, A. Stenderup. M. Kielland-Brandt, and J. Friis, eds.), p. 227 (Alfred Benzon Symposium, Vol. 16). Munksgaard, Copenhagen, 1981.

[30] J. R. Broach and J. B. Hicks, *Cell* **21**, 501 (1980).

[31] J. R. Broach, Y.-Y. Li, L.-C. C. Wu, and M. Jayaram, *in* "Experimental Manipulation of Gene Expression" (M. Inouye, ed.), p. 83. Academic Press, New York, 1983.

[32] J. R. Broach, Y. Y. Li, J. Feldman, M. Jayaram, J. Abraham, K. A. Nasmyth, and J. B. Hicks, *Cold Spring Harbor Symp. Quant. Biol.* **47**, 1165 (1982).

[33] J. R. Broach, J. N. Strathern, and J. B. Hicks, *Gene* **8**, 121 (1979).

[34] T. R. Butt, E. Sternberg, J. Herd, and S. T. Crooke, *Gene* **27**, 23 (1984).

[35] T. Cabezon, M. De Wilde, P. Herion, R. Loriau, and A. Bollen, *Proc. Natl. Acad. Sci. U.S.A.* **81**, 6594 (1984).

[36] M. J. Casadaban, A. Martinez-Arias, S. K. Shapira, and J. Chou, this series, Vol. 100, p. 293.

[37] A. M. Cashmore, M. S. Albury, G. Hadfield, and P. A. Meacock, *Mol. Cell. Biol.* **203**, 154 (1986).

[38] S. E. Celniker, K. Sweder, F. Srienc, J. E. Bailey, and J. L. Campbell, *Mol. Cell. Biol.* **4**, 2455 (1984).

[39] G. Cesareni, *in* "Vectors: A Survey of Molecular Cloning Vectors and Their Uses" (R. L. Rodriguez and D. T. Denhardt, eds.), p. 103. Butterworth, Stoneham, Massachusetts, 1988.

[40] G. Cesareni and J. A. H. Murray, *in* "Genetic Engineering" (J. K. Setlow, ed.), Vol. 9, p. 135. Plenum, New York, 1987.

[41] C. Y. Chen, H. Oppermann, and R. A. Hitzeman, *Nucleic Acids Res.* **12**, 8951 (1984).

[42] M.-R. Chevallier, J.-C. Bloch, and F. LaCroute, *Gene* **11**, 11 (1980).

[43] L. Clarke and J. Carbon, *Nature (London)* **287**, 504 (1980).

[44] L. E. Clarke, R. F. Gibson, R. F. Sherwood, and N. P. Minton, *J. Gen. Microbiol.* **131**, 897 (1985).

[45] J. D. Cohen, T. R. Eccleshall, R. B. Needleman, H. Federoff, B. A. Buchferer, and J. Marmur, *Proc. Natl. Acad. Sci. U.S.A.* **77**, 1078 (1980).

[46] K. G. Coleman, H. Y. Steensma, D. B. Kabak, and J. R. Pringle, *Mol. Cell. Biol.* **6**, 4516 (1986).

[47] J. Davison, M. Heusterspreute, and F. Brunel, this series, Vol. 153, p. 34.

[48] P. Dehoux, V. Ribes, E. Sobczak, and R. E. Streeck, *Gene* **48**, 155 (1986).

[49] M. J. Dobson, A. B. Futcher, and B. S. Cox, *Curr. Genet.* **2**, 193 (1980).

[50] M. J. Dobson, M. F. Tuite, J. Mellor, N. A. Roberts, R. M. King, D. C. Burke, A. J. Kingsmar, and S. M. Kingsman, *Nucleic Acids Res.* **11**, 2287 (1983).

[51] M. J. Dobson, M. F. Tuite, N. A. Roberts, A. J. Kingsman, and S. M. Kingsman, *Nucleic Acids Res.* **10**, 2625 (1982).

[52] M. G. Douglas, B. L. Geller, and S. D. Emr, *Proc. Natl. Acad. Sci. U.S.A.* **81**, 3983 (1984).

[53] H. J. Drabkin and U. L. RajBhandary, *J. Biol. Chem.* **260**, 5596 (1985).

[54] S. D. Emr, R. Schekman, M. C. Flessel, and J. Thorner, *Proc. Natl. Acad. Sci. U.S.A.* 80, 7080 (1983).

[55] S. D. Emr, A. Vassarotti, J. Garrett, B. L. Geller, M. Takeda, and M. G. Douglas, *J. Cell Biol.* **102**, 523 (1986).

[56] E. Erhart and C. P. Hollenberg, *Curr. Genet.* **3**, 83 (1981).

[57] E. Erhart and C. P. Hollenberg, *J. Bacteriol.* **156**, 625 (1983).

[58] J. F. Ernst and R. K. Chan, *J. Bacteriol.* **163**, 8 (1985).

[59] H. J. Federoff, J. D. Cohen, T. R. Eccleshall, R. B. Needleman, B. A. Buchferer, J. Giacalone, and J. Marmur, *J. Bacteriol.* **149**, 1064 (1982).

[60] J. Ferguson, J. C. Groppe, and S. I. Reed, *Gene* **16**, 191 (1981).

[61] D. A. Fischhoff, R. H. Waterston, and M. V. Olson, *Gene* **27**, 239 (1984).

[62] H. M. Fried, H. G. Nam, S. Loechel, and J. Teem, *Mol. Cell. Biol.* **5**, 99 (1985).

[63] A. B. Futcher, *J. Theor. Biol.* **119**, 197 (1986).

[64] A. B. Futcher and B. S. Cox, *J. Bacteriol.* **154**, 612 (1983).

[65] A. B. Futcher and B. S. Cox, *J. Bacteriol.* **157**, 283 (1984).

[66] A. Gatignol, M. Baron, and G. Tiraby, *Mol. Gen. Genet.* **207**, 342 (1987).

[67] C. Gerbaud, C. Elmerich, N. T. de Marsac, N. Chocat, M. Charpin, M. Guerineau, and J.-P. Aubert, *Curr. Genet.* **3**, 173 (1981).

[68] C. Gerbaud, P. Fournier, H. Blanc, M. Aigle, H. Heslot, and M. Guerineau, *Gene* **5**, 233 (1979).

[69] R. D. Gietz and A. Sugino, *Gene* **74**, 527 (1988).

[70] C. G. Goff, D. T. Moir, T. Kohno, T. C. Gravius, R. A. Smith, E. Yamasaki, and A. Taunton-Rigby, *Gene* **27**, 35 (1984).

[71] L. Gritz and J. Davies, *Gene* **25**, 179 (1983).

[72] L. Guarente, this series, Vol. 101, p. 181.

[73] L. Guarente and M. Ptashne, *Proc. Natl. Acad. Sci. U.S.A.* **78**, 2199 (1983).

[74] L. Guarente, R. R. Yocum, and P. Gifford, *Proc. Natl. Acad. Sci. U.S.A.* **79**, 7410 (1982).

[75] A. M. Guerrini, F. Ascenzioni, C. Tribioli, and P. Donini, *EMBO J.* **4**, 1569 (1985).

[76] C. Hadfield, A. M. Cashmore, and P. A. Meacock, *Gene* **45**, 149 (1986).

[77] D. C. Hagen and G. F. Sprague, Jr., *J. Mol. Biol.* **178**, 835 (1984).

[78] R. Haguenauer-Tsapis and A. Hinnen, *Mol. Cell. Biol.* **4**, 2668 (1984).

[79] S. D. Hanes, V. E. Burn, S. L. Sturley, D. J. Tipper, and K. A. Bostian, *Proc. Natl. Acad. Sci. U.S.A.* **83**, 1675 (1986).

[80] J. L. Hartley and J. E. Donelson, *Nature (London)* **286**, 860 (1980).

[81] R. C. A. Henderson, B. S. Cox, and R. Tubb, *Curr. Genet.* **9**, 133 (1985).

[82] M. Heusterspreute, J. Oberto, V. Ha-Thi, and J. Davison, *Gene* **34**, 363 (1985).

[83] J. E. Hill, A. M. Myers, T. J. Koerner, and A. Tzagoloff, *Yeast* **2**, 163 (1986).

[84] R. A. Hitzeman, C. N. Chang, M. Matteucci, L. J. Perry, W. J. Kohr, J. J. Wulf, J. R. Swartz, C. Y. Chen and A. Singh, this series, Vol. 119, p. 424.

[85] R. A. Hitzeman, F. E. Hagie, H. L. Levine, D. V. Goeddel, G. Ammerer, and B. D. Hall, *Nature (London)* **293**, 717 (1981).

[86] R. A. Hitzeman, D. W. Leung, L. J. Perry, W. J. Kohr, H. L. Levine, and D. V. Goeddel, *Science* **219**, 620 (1983).

[87] B. Hohn and A. Hinnen, in "Genetic Engineering II" (J. K. Setlaw and A. Hollander, eds.), p. 169. Academic Press, New York, 1980.

[88] C. P. Hollenberg, in "Extrachromosomal DNA" (D. J. Cummings, P. Borst, I. B. Dawid, S. M. Weissman, and C. F. Fox, eds.), p. 325 (ICN–UCLA Symposium 15). Academic Press, New York, 1979.

[89] C. P. Hollenberg, B. Kustermann-Kuhn, V. Mackedonski, and E. Erhart, in "Molecular Genetics in Yeast" (D. Von Wettstein, A. Stenderup. M. Kielland-Brandt, and J. Friis, eds.), p. 341 (Alfred Benzon Symposium, Vol. 16). Munksgaard, Copenhagen, 1981.

[90] S. Holmberg, J. G. L. Petersen, T. Nilsson-Tillgren, and M. C. Kielland-Brandt, *Carlsberg Res. Commun.* **44**, 269 (1979).

[91] Y.-P. Hsu and G. B. Kohlhaw, *J. Biol. Chem.* **257**, 39 (1982).

[92] J. A. Huberman, L. D. Spotila, K. A. Nawotka, S. El-Assouli, and L. R. Davis, *Cell* **51**, 473 (1987).

[93] M. A. Innis, M. J. Holland, P. C. McCabe, G. E. Cole, V. P. Wittman, R. Tal, K. W. K. Watt, D. H. Gelfand, J. P. Holland, and J. H. Meade, *Science* **228**, 21 (1985).

[94] M. Irani, W. E. Taylor, and E. T. Young, *Mol. Cell. Biol.* **7**, 1233 (1987).

[95] M. A. Jabbar, N. Sivasubramanian, and D. P. Nayak, *Proc. Natl. Acad. Sci. U.S.A.* **82**, 2019 (1985).

[96] M. Jayaram, Y.-Y. Li, and J. R. Broach, *Cell* **34**, 95 (1983).

[97] M. Jayaram, A. Sutton, and J. R. Broach, *Mol. Cell. Biol.* **5**, 2466 (1985).

[98] A. Jearnpipatkul, H. Araki, and Y. Oshima, *Mol. Gen. Genet.* **206**, 88 (1987).

[99] A. Jearnpipatkul, R. Hutacharoen, H. Araki, and Y. Oshima, *Mol. Gen. Genet.* **207**, 355 (1987).

[100] R. Jensen, G. F. Sprague, Jr., and I. Herskowitz, *Proc. Natl. Acad. Sci. U.S.A.* **80**, 3035 (1983).

[101] K. R. Kaster, S. G. Burgett, and T. D. Ingolia, *Curr. Genet.* **8**, 353 (1984).

[102] M. I. Khan, D. J. Ecker, T. Butt, J. A. Gorman, and S. T. Crooke, *Plasmid* **17**, 171 (1987).

[103] Y. Kikuchi, *Cell* **35**, 487 (1983).

[104] R. D. Klein and L. L. Roof, *Curr. Genet.* **13,** 29 (1988).
[105] R. D. Klein and P. G. Zaworski, *in* "Yeast Strain Selection" (C. P. Panchal, ed.), in press. Dekker, New York, 1990.
[106] R. A. Kramer, T. M. DeChiara, M. D. Schaber, and S. Hilliker, *Proc. Natl. Acad. Sci. U.S.A.* **81,** 367 (1984).
[107] P. A. Lagosky, G. R. Taylor, and R. H. Haynes, *Nucleic Acids Res.* **15,** 10355 (1987).
[108] J. C. Larkin and J. L. Woolford, Jr., *Nucleic Acids Res.* **11,** 403 (1983).
[109] G. Loison, A. Findeli, S. Bernard, M. Nguyen-Juilleret, M. Marquet, N. Riehl-Bellon, D. Carvallo, L. Guerra-Santos, S. W. Brown, M. Courtney, C. Roitsch, and Y. Lemoine, *Bio/Technology* **6,** 72 (1988).
[110] G. Loison, M. Nguyen-Juilleret, S. Alouani, and M. Marquet, *Bio/Technology* **4,** 433 (1986).
[111] S. Lolle, N. Skipper, H. Bussey, and D. Y. Thomas, *EMBO J.* **3,** 1383 (1984).
[112] H. Ma, S. Kunes, P. J. Schatz, and D. Botstein, *Gene* **58,** 201 (1987).
[113] G. T. Maine, P. Sinha, and B.-K, Tye, *Genetics* **106,** 365 (1984).
[114] T. Maniatis, E. F. Fritsch, and J. Sambrook, "Molecular Cloning: A Laboratory Manual." Cold Spring Harbor Laboratory, Cold Spring Harbor, New York, 1982.
[115] G. L. McKnight and B. L. McConaughy, *Proc. Natl. Acad. Sci. U.S.A.* **80,** 4412 (1983).
[116] J. B. McNeil and J. D. Friesen, *Mol. Gen. Genet.* **184,** 386 (1981).
[117] J. B. McNeil, R. K. Storms, and J. D. Friesen, *Curr. Genet.* **2,** 17 (1980).
[118] D. Meeks-Wagner, J. S. Wood, B. Garvik, and L. H. Hartwell, *Cell* **44,** 53 (1986).
[119] J. Mellor, M. J. Dobson, N. A. Roberts, M. F. Tuite, J. S. Emtage, S. White, P. A. Lowe, T. Patel, A. J. Kingsman, and S. M. Kingsman, *Gene* **24,** 1 (1983).
[120] A. Jiyajima, M. W. Bond, K. Otsu, K. Arai, and N. Arai, *Gene* **37,** 155 (1985).
[121] A. Miyajima, I. Miyajima, K.-I. Arai, and N. Arai, *Mol. Cell. Biol.* **4,** 407 (1984).
[122] A. Miyanohara, A. Toh-e, C. Nozaki, F. Hamada, N. Ohtomo, and K. Matsubara, *Proc. Natl. Acad. Sci. U.S.A.* **80,** 1 (1983).
[123] D. T. Moir and D. R. Dumais, *Gene* **56,** 209 (1987).
[124] D. W. Morris, J. D. Noti, F. A. Osborne, and A. A. Szalay, *DNA* **1,** 27 (1981).
[125] G. T. Mullenbach, A. Tabrizi, R. W. Blacher, and K. S. Steimer, *J. Biol. Chem.* **261,** 719 (1986).
[126] A. W. Murray and J. W. Szostak, *Cell* **34,** 961 (1983).
[127] J. A. H. Murray and G. Cesareni, *EMBO J.* **5,** 3391 (1986).
[128] J. A. H. Murray, M. Scarpa, N. Rossi, and G. Cesareni, *EMBO J.* **6,** 4205 (1987).
[129] A. M. Myers, A. Tzagoloff, D. M. Kinney, and C. J. Lusty, *Gene* **45,** 299 (1986).
[130] Y. Nakamura, T. Sato, M. Emi, A. Miyanohara, T. Nishide, and K. Matsubara, *Gene* **50,** 239 (1986).
[131] L. Naumovski and E. C. Friedberg, *Gene* **22,** 203 (1983).
[132] J. Oberto and J. Davison, *Gene* **40,** 57 (1985).
[133] S. A. Parent, C. M. Fenimore, and K. A. Bostian, *Yeast* **1,** 83 (1985).
[134] V. Price, D. Mochizuki, C. J. March, D. Cosman, M. C. Deeley, R. Klinke, W. Clevenger, S. Gillis, P. Baker, and D. Urdal, *Gene* **55,** 287 (1987).
[135] G. Reipen, E. Erhart, K. D. Breunig, and C. P. Hollenberg, *Curr. Genet.* **6,** 189 (1982).
[136] A. Reynolds, A. Murray, and J. Szostak, *Mol. Cell. Biol.* **7,** 3566 (1987).
[137] J. Rine, W. Hansen, E. Hardeman, and R. W. Davis, *Proc. Natl. Acad. Sci. U.S.A.* **80,** 6750 (1983).
[138] M. Rose and D. Botstein, *J. Mol. Biol.* **170,** 883 (1983).
[139] M. Rose, M. J. Casadaban and D. Botstein, *Proc. Natl. Acad. Sci. U.S.A.* **78,** 2460 (1981).
[140] M. Rose and G. R. Fink, *Cell* **48,** 1047 (1987).
[141] M. D. Rose, P. Novick, J. H. Thomas, D. Botstein, and G. R. Fink, *Gene* **60,** 237 (1987).

[142] T. Sato, S. Tsunasawa, Y. Nakamura, M. Emi, F. Sakiyama, and K. Matsubara, *Gene* **50**, 247 (1986).

[143] M. D. Schaber, T. M. DeChiara, and R. A. Kramer, this series, Vol. 119, p. 416.

[144] L. D. Schultz and J. D. Friesen, *J. Bacteriol.* **155**, 8 (1983).

[145] H. S. Seifert, E. Y. Chen, M. So, and F. Heffron, *Proc. Natl. Acad. Sci. U.S.A.* **83**, 735 (1986).

[146] H. Singh, J. J. Bieker, and L. B. Dumas, *Gene* **20**, 441 (1982).

[147] T. Som, K. A. Armstrong, F. C. Volkert, and J. R. Broach, *Cell* **52**, 27 (1988).

[148] P. P. Stepien, R. Brousseau, R. Wu, S. Narang, and D. Y. Thomas, *Gene* **24**, 289 (1983).

[149] D. T. Stinchcomb, C. Mann, and R. W. Davis, *J. Mol. Biol.* **158**, 157 (1982).

[150] D. T. Stinchcomb, K. Struhl, and R. W. Davis, *Nature (London)* **282**, 39 (1979).

[151] R. K. Storms, J. B. McNeil, P. S. Kilandekar, G. An, J. Parker, and J. D. Friesen, *J. Bacteriol.* **140**, 73 (1979).

[152] K. Struhl, D. T. Stinchcomb, S. Scherer, and R. W. Davis, *Proc. Natl. Acad. Sci. U.S.A.* **76**, 1035 (1979).

[153] A. Sutton and J. R. Broach, *Mol. Cell. Biol.* **5**, 2770 (1985).

[154] K. Suzuki, Y. Imai, I. Yamashita, and S. Fukui, *J. Bacteriol.* **155**, 747 (1983).

[155] J. L. Teem and M. Rosbash, *Proc. Natl. Acad. Sci. U.S.A.* **80**, 4403 (1983).

[156] D. Y. Thomas and A. P. James, *Curr. Genet.* **2**, 9 (1980).

[157] T. Toda, S. Cameron, P. Sass, M. Zoller, and M. Wigler, *Cell* **50**, 277 (1987).

[158] A. Toh-e, S. Tada, and Y. Oshima, *J. Bacteriol.* **151**, 1380 (1982).

[159] M. Tokunaga, N. Wada, and F. Hishinuma, *Biochem. Biophys. Res. Commun.* **144**, 613 (1987).

[160] G. Tschumper and J. Carbon, *Gene* **23**, 221 (1983).

[161] M. F. Tuite, M. J. Dobson, N. A. Roberts, R. M. King, D. C. Burke, S. M. Kingsman, and A. J. Kingsman, *EMBO J.* **1**, 603 (1982).

[162] J. N. Van Arsdell, S. Kwok, V. L. Schweickart, M. B. Ladner, D. H. Gelfand, and M. A. Innis, *Bio/Technology* **5**, 60 (1987).

[163] B. E. Veit and W. L. Fangman, *Mol. Cell. Biol.* **5**, 2190 (1985).

[164] D. Vetter, B. J. Andrews, L. Roberts-Beatty, and P. D. Sadowski, *Proc. Natl. Acad. Sci. U.S.A.* **80**, 7284 (1983).

[165] G. P. Vlasuk, G. H. Bencen, R. M. Scarborough, P.-K. Tsai, J. L. Whang, T. Maack, M. J. F. Camargo, S. W. Kirsher, and J. A. Abraham, *J. Biol. Chem.* **261**, 4789 (1986).

[166] F. C. Volkert and J. R. Broach, *Cell* **46**, 541 (1986).

[167] F. C. Volkert and J. R. Broach, *in* "The Biochemistry and Molecular Biology of Industrial Yeasts" (G. G. Stewart, I. Russell, R. D. Klein, and R. R. Hiebsch, eds.), p. 145. CRC Press, Boca Raton, Florida, 1987.

[168] F. Watts, C. Castle, and J. Beggs, *EMBO J.* **2**, 2085 (1983).

[169] T. D. Webster and R. C. Dickson, *Gene* **26**, 243 (1983).

[170] D. Wen and M. J. Schlesinger, *Proc. Natl. Acad. Sci. U.S.A.* **83**, 3639 (1986).

[171] R. W. West, Jr., R. R. Yocum, and M. Ptashne, *Mol. Cell. Biol.* **4**, 2467 (1984).

[172] A. Wright, K. Maundrell, W.-D. Heyer, D. Beach, and P. Nurse, *Plasmid* **15**, 156 (1986).

[173] L. C. C. Wu, P. A. Fisher, and J. R. Broach, *J. Biol. Chem.* **262**, 883 (1987).

[174] R. R. Yocum, S. Hanley, R. West, Jr., and M. Ptashne, *Mol. Cell. Biol.* **4**, 1985 (1984).

[175] K. M. Zsebo, H.-S. Lu, J. C. Fieschko, L. Goldstein, J. Davis, K. Duker, S. V. Suggs, P.-H. Lai, and G. A. Bitter, *J. Biol. Chem.* **261**, 5858 (1986).

[23] Manipulating Yeast Genome Using Plasmid Vectors

By TIM STEARNS, HONG MA, and DAVID BOTSTEIN

The yeast *Saccharomyces cerevisiae* has proved to be a popular and successful experimental system for biological research. It is in many ways a typical eukaryote, sharing the most fundamental aspects of cell biology with the cells of multicellular organisms, yet it stands alone among eukaryotic organisms in the ease with which it can be experimentally manipulated. The current high status of yeast as an experimental system is in large part due to the work of the many geneticists who recognized early on the virtues of this organism. Features such as a life cycle with essentially isomorphous haploid and diploid phases, the formation of tetrads in meiosis, a short generation time, and the ability to grow in defined media formed the fundamentals of yeast methodology, and are still important today.

For those interested in expression of heterologous genes, yeast holds particular advantages, evident in the articles in this volume devoted to the topic. The sequence elements required for expression of a number of genes have been intensely studied and are well characterized. From these studies have come a variety of useful promoters, both regulated and constitutive, that make it relatively easy, to a first approximation, to achieve expression of a foreign gene in yeast. In addition, study of the yeast secretory pathway has led to a knowledge of both the *cis*-acting sequences necessary for secretion and the components that make up the secretion apparatus.

Two particular advances in yeast technology have directly led to the current high level of experimental tractability: first, the advent of DNA transformation[1,2] and the ensuing proliferation of vector systems; second, the demonstration that recombination of transforming DNA with the chromosome usually occurs via homology[1] and that the recombination can be "directed."[3] We will describe the range of yeast vectors available and discuss methods for manipulating both the yeast genome and yeast vectors using homologous recombination. Some of these methods have been reviewed in a previous volume of this series.[4]

---

[1] A. Hinnen, J. B. Hicks, and G. R. Fink, *Proc. Natl. Acad. Sci. U.S.A.* **75,** 1929 (1978).

[2] J. D. Beggs, *Nature (London)* **275,** 104 (1978).

[3] T. L. Orr-Weaver, J. W. Szostak, and R. J. Rothstein, *Proc. Natl. Acad. Sci. U.S.A.* **78,** 6354 (1981).

[4] R. J. Rothstein, this series, Vol. 101, p. 202.

METHODS IN ENZYMOLOGY, VOL. 185

Yeast Vector Systems

Yeast vectors are of four general types. These differ in the manner in which they are maintained in yeast cells. YIp (*y*east *i*ntegrating *p*lasmid) vectors lack a yeast replication origin, so must be propagated as integrated elements in a yeast chromosome, usually in a single copy per genome. YRp (*y*east *r*eplicating *p*lasmid) vectors have a chromosomally derived autonomously replicating sequence (ARS) and are propagated as medium-copy-number, autonomously replicating, unstably segregated plasmids. YCp (*y*east *c*entromere *p*lasmid) vectors have both a replication origin and a centromere sequence and are propagated as low-copy-number, autonomously replicating, stably segregated plasmids. YEp (*y*east *e*pisomal *p*lasmid) vectors have a fragment of the yeast 2-μm plasmid and, are propagated as high-copy-number, autonomously replicating, irregularly segregated plasmids.

All of these plasmids are "shuttle vectors" that allow propagation in *Escherichia coli* for convenient manipulation and large-scale preparation of their DNA. We will concentrate on YIp and YCp plasmids, as YEp plasmids are reviewed elsewhere in this volume,[5] and YRp plasmids are now rarely used because of their extreme instability.

Yeast Vector Components

There are two components shared by all yeast vectors: a backbone derived from a bacterial plasmid, allowing selection of *E. coli* transformants and subsequent propagation of the DNA; and a yeast selectable marker, allowing selection of yeast transformants. YCp, YRp, and YEp plasmids have additional components required for autonomous replication in yeast.

The backbone of most yeast vectors is derived from the *E. coli* plasmid pBR322.[6] This plasmid has a bacterial replication origin that allows maintenance at high copy number, and two selectable antibiotic resistance genes. The *bla* gene encodes resistance to the β-lactam ampicillin, and the *tet* gene encodes resistance to tetracycline. In addition, pBR322 has a number of unique restriction sites for cloning DNA of interest. Because almost all yeast vectors share this common backbone of pBR322 DNA, they are all homologous to each other over some portion of their length. This homology is both a blessing, in that it can be exploited in the creation of new plasmids *in vivo,* as discussed later, and a curse, as homologous

---

[5] A. B. Rose and J. R. Broach, this volume [22].

[6] F. Bolivar, R. L. Rodriguez, P. J. Greene, M. C. Betlach, H. L. Heyneker, H. W. Boyer, J. H. Crosa, and S. Falkow, *Gene* **2**, 95 (1977).

TABLE I

YEAST SELECTABLE MARKERS AND THEIR
COGNATE MUTATIONS

Selectable marker gene	Stable yeast mutation	Stable bacterial mutation
HIS3	his3-Δ200	hisB463
LEU2	leu2-3,112	leuB6
LYS2	lys2-201	—
TRP1	trp1-Δ901	trpC9830
URA3	ura3-52	pyrF::Tn5

recombination can occur among virtually all members of this family of plasmids, even when unwanted.

The yeast selectable marker allows selection of yeast cells that have received the plasmid by transformation. There are a number of markers available, the most common being yeast genes that complement a specific auxotrophy. For example, the *LEU2* gene encodes β-isopropylmalate dehydrogenase[7] and complements the leucine auxotrophy of a *leu2⁻* mutant. Of equal importance to the selectable marker is the cognate chromosomal mutation that causes the auxotrophy. Ideally, this mutation must be completely recessive, and nonreverting; the commonly used *leu2-3,112* mutation is a double frameshift mutation that reverts extremely rarely ($< 10^{-10}$) and is completely complemented by the wild-type *LEU2* gene.[1] Additionally, four of the most widely used selectable markers (*HIS3, LEU2, TRP1,* and *URA3*) can be selected in *E. coli*. Table I lists commonly used pairs of yeast selectable markers and chromosomal mutations and the bacterial mutations that are complemented by the yeast genes.

Two yeast selectable markers have the additional advantage of both positive and negative selection. For both *URA3* and *LYS2* the positive selection is for complementation of an auxotrophy, and the negative selection is for ability to grow on medium containing a compound that inhibits the growth of cells expressing the wild-type function. For the *URA3* gene the negative selection is growth in medium containing 5-fluoroorotic acid;[8] for the *LYS2* gene the negative selection is growth in medium containing α-aminoadipate.[9] Thus, cells that have lost the selectable marker can be selected from a population; this is very useful, as many experiments that

[7] B. Ratzkin and J. Carbon, *Proc. Natl. Acad. Sci. U.S.A.* **74,** 487 (1977).

[8] J. D. Boeke, F. Lacroute, and G. R. Fink. *Mol. Gen. Genet.* **197,** 345 (1984).

[9] B. B. Chatoo, F. Sherman, T. A. Fjellstedt, D. Mehnert, and M. Ogur, *Genetics* **93,** 51 (1979).

we discuss below require that one be able to find the rare cells that have undergone a recombination or loss event to remove the plasmid DNA sequences. Unfortunately, the *LYS2* gene is quite large, and contains sites within the coding sequence for many of the most common enzymes used in plasmid construction.[10] There are ways to get around this limitation, however, by using *in vivo* homologous recombination to construct plasmids with this marker (see below).

A Starter Set of Yeast Vectors

For most experiments we use the *URA3* gene as the selectable marker, resorting to other genes only when the experiment requires more than one marker. In addition to the advantage of negative selection noted above, the *URA3* gene is small and lacks sites for the most commonly used restriction enzymes. A trio of plasmids that contain the *URA3* gene are particularly useful (Fig. 1). These are the integrating vector YIp5 and two vectors derived from it: YEp420, derived by addition of a fragment of the 2-μm plasmid that contains the replication origin, and YCp50, derived by addition of *ARS1* and *CEN4*. With these three plasmids all of the experiments described here can be performed.

Yeast Transformation

Before describing methods for manipulating the genome of yeast, a protocol for transforming yeast in a reproducible manner must be presented. The following is a distillation of various modifications to the lithium acetate method of Ito *et al.*[11] and works well with most laboratory strains. Transformation by this protocol consistently yields 1,000–10,000 transformants/μg of DNA. This protocol can be scaled up or down to provide competent cells for one or many transformation experiments. Approximately 2×10^8 cells are needed for each transformation, so about 10 ml of 2×10^7 cells/ml culture is sufficient per transformation. For example, for five transformations, grow 50 ml of culture to a density of 2×10^7 cells/ml. The transforming DNA can be from *E. coli* minipreps as well as from CsCl-purified preparations. DNA that is digested with a restriction enzyme(s) prior to transformation need not be purified again; we routinely heat-kill the enzyme(s) and use the reaction mixture to transform yeast. In many of the examples that follow, the desired transforming DNA is a restriction fragment of a plasmid. The fragment can be purified

[10] D. A. Barnes and J. Thorner, *Mol. Cell. Biol.* **6**, 2828 (1986).
[11] H. Ito, Y. Fukuda, K. Murada, and A. Kimura, *J. Bacteriol.* **153**, 163 (1983).

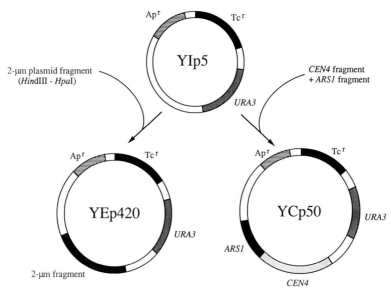

FIG. 1. Three commonly used yeast vectors that employ the *URA3* gene as a selectable marker. YIp5 was constructed by inserting the *URA3* gene into the *Ava*I site of pBR322. YEp420 was derived from YIp5 by insertion of a *Hind*III–*Hpa*I fragment of the yeast 2-μm circle into the YIp5 *Pvu*II site. YCp50 was derived from YIp5 by insertion of both *ARS1* and *CEN4* into the YIp5 *Pvu*II site.

from away from the plasmid backbone if desired, but this is often not necessary, such as when the backbone fragment cannot give transformants under the selection condition used. All steps are carried out at room temperature, except where noted.

Yeast Lithium Acetate Transformation Protocol

Solutions required:

LiAc buffer: 100 mM Lithium acetate, 10 mM Tris, 1 mM EDTA, pH 8.0

Carrier DNA: commercial chick blood, salmon sperm, or calf thymus DNA made up as a 10 mg/ml stock solution and sonicated to reduce viscosity

Transforming DNA: usually 0.1–1.0 μg of circular or linear plasmid DNA per transformation, in a volume of 20 μl or less

PEG buffer: 40% Polyethylene glycol (PEG) 3350 (Sigma Chemical Co., St. Louis, MO, #P-3640) in 100 mM lithium acetate, 10 mM Tris, 1 mM EDTA, pH 8.0

1. Grow cells to approximately 2×10^7/ml.

2. Harvest by centrifugation, and wash once with water.

3. Resuspend cells in LiAc buffer so that total volume is 0.01 of original culture volume (i.e., 0.5 ml for 50 ml of culture).

4. Incubate for 30 min at 30° with agitation (any type of agitation that keeps cells from settling to the bottom is sufficient).

5. Split cells into 100-μl aliquots in sterile 1.5-ml Eppendorf tubes; one aliquot for each transformation. Add 5 μl of carrier DNA to each tube. Add transforming DNA to each tube. Vortex briefly.

6. Incubate for 30 min at 30° with agitation.

7. Add 700 μl of PEG buffer, vortex.

8. Incubate for 60 min at 30° with no agitation required.

9. Heat shock for 5 min at 42°.

10. Spin in Eppendorf centrifuge for approximately 5 sec, pour off supernatant.

11. Resuspend cell pellet in 200 μl of sterile water and plate by spreading on selective medium.

Notes: Transformants should be purified by picking single colonies from the transformation plate and streaking for single colonies on selective plates.

Frozen competent yeast cells can be prepared after step 4 by adding sterile glycerol to 15% (v/v) and freezing at $-70°$. When needed, these cells are thawed on ice and used in the protocol as for fresh cells, starting at step 5; the transformation efficiency of such cells is usually reduced 2- to 5-fold from fresh cells.

On occasion, a particular strain does not transform well by the lithium acetate method, or an experiment requires a very high frequency of transformation; in these cases, the more efficient, but more tedious, spheroplasting method can be used.[1,12]

Integrating Plasmids: Uses of Homologous Recombination

Integrating plasmids lack a yeast origin of replication; the only way they can be propagated in yeast is by integration into the host genome. Recombination of introduced DNA with the chromosome in yeast occurs extremely efficiently, for all practical purposes exclusively by homologous recombination;[1] (the "illegitimate recombinants" found by Hinnen *et al.* turned out to be homologous crossovers within a repeated element). A very useful technique allows directed integration by homology[3]: this is particularly important when the plasmid contains more than one yeast gene,

---

[12] F. Sherman, G. R. Fink, and C. W. Lawrence, "Methods in Yeast Genetics." Cold Spring Harbor Laboratory, Cold Spring Harbor, New York, 1979.

namely, any instance in which a yeast gene has been cloned into a vector with a yeast selectable marker. Cleavage in one of these regions of homology with the genome greatly stimulates the frequency of integration in the immediate vicinity of the cleavage site. Thus, the investigator, using YIp plasmids, can manipulate the integration of exogenous DNA with considerable exactitude. A list of commonly used YIp plasmid vectors is given in Table II.

The simplest form of experiment using YIp vectors is integration into the yeast genome using the homology in the cloned fragment, directed, if desired, to the chromosomal locus of the cloned fragment by cleaving within the cloned sequences. This results in a direct duplication of the cloned DNA in the chromosome with vector sequences between the repeats (Fig. 2). The duplication event has increased the copy number of the cloned fragment from one to exactly two in the genome. If the cloned fragment contains a functional gene, then the copy number of that gene has been increased, and the effect of the increased gene dosage can be

TABLE II
Yeast Integrating Plasmid (YIp) Vectors

Plasmid	E. coli markers	Yeast markers	Comments[a]	Ref.[b]
YIp1	amp[r]	HIS3		1
YIp5	amp[r], tet[r]	URA3		1
YIp25	tet[r]	HIS4		1
YIp26	amp[r]	URA3, LEU2	All orientations of markers available	1
YIp30	amp[r]	URA3	Available in both orientations	1
YIp32	amp[r]	LEU2	Available in both orientations	1
pRB61	amp[r]	HIS3		2
pRB290	amp[r], tet[r]	URA3	YIp5 without HindIII site	3
pRB315	amp[r]	TRP1	No ARS1	4
pRB328	amp[r]	HIS3	Polylinker (1) flanks HIS3	4
pRB331	amp[r]	URA3	Polylinker (1) flanks URA3	2
pRB388	amp[r], tet[r]	URA3	13-kb URA3 fragment	2
pRB502	amp[r], tet[r]	URA3	Polylinker (2) flanks URA3	2
pRB506	amp[r]	LYS2	Polylinker (1) flanks LYS2	2
pRB758	amp[r]	LEU2	Polylinker (3) flanks LEU2	2
pRB759	amp[r]	URA3	Polylinker (3) flanks URA3	2

[a]The polylinkers are (1) pPL7 (J. Mullins, personal communication; sites are EcoRI–ClaI–HindIII–XbaI–BglII–MboII–PstI–BamHI), (2) pT7-2 (United States Biochemical Corporation), (3) pUC19 (Yanisch-Perron et al., 1985).

[b]1, D. Botstein, S. C. Falco, S. E. Stewart, M. Brennan, S. Scherer, D. T. Stinchcomb, K, Struhl, and R. W. Davis, Gene 8, 17 (1979); 2, this laboratory (Dept. of Biology, Massachusetts Institute of Technology); 3, Schatz et al. (1986); 4, H. Ma, S. Kunes, P. J. Schatz, and D. Botstein, Gene 58, 201 (1987).

assayed. It is important to realize also that this strain now carries a selectable genetic marker, *URA3* in this case, at the chromosomal locus of the cloned DNA; this construction can be used to map the locus in relation to other genetic markers, even if the cloned fragment has no function. This duplication is quite stable in comparison to autonomously replicating plasmids; the integrated sequences excise at a frequency of approximately 1×10^{-3} – 1×10^{-4}.

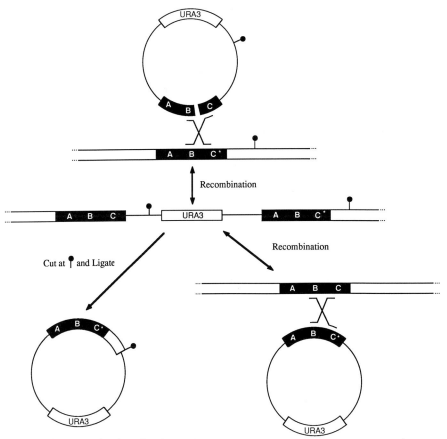

FIG. 2. Recombination of an integrating plasmid with the yeast genome. The plasmid and chromosome differ at position C (C versus C*). The recombination event is directed by cleaving the transforming DNA between B and C, leading to a direct duplication in the chromosome. If genomic DNA is isolated from this transformant, cleaved at an appropriate restriction site, and ligated, then a region of the chromosome, including the chromosomal C*, can be recovered in *E. coli*. Alternatively, recombination between the duplicated sequences will lead to recovery of cells that have looped out the plasmid sequences, occasionally leaving the plasmid C behind in the chromosome.

Of course, any such construction must be checked by physical means, such as restriction digestion and Southern hybridization, to determine that the desired event is actually what took place, and that only a single copy of the plasmid has integrated. In particular, one must be aware of the possibility that integration has taken place at the chromosomal locus of the selectable marker, as it also provides homology for recombination. Indeed, there are situations where this event is desired, and can be accomplished by cleaving the plasmid in the selectable marker sequences rather than the insert sequences. An example of this would be the introduction of a foreign gene into the yeast genome; as the foreign DNA provides no homology for recombination with the yeast chromosome, integration could be directed to the locus of the selectable marker.

Excision of the inserted DNA from the chromosome also takes place by homologous recombination. Therefore, recombination between the duplicated sequences in Fig. 2 will result in complete excision of the plasmid DNA sequences, including the cloned fragment, leaving the chromosome as it was before integration. This *in vivo* recombination event can take place at any point within the homology created by the duplicated segments. Thus, if the cloned fragment contains a mutation made *in vitro*, integration and subsequent excision will, at some frequency, lead to the replacement of the wild-type sequence with the mutant sequence, leaving the chromosome otherwise wild-type. This kind of perfect replacement of mutant for wild type (often called "transplacement") is the best way of testing the *in vivo* consequences of mutations made *in vitro*. As discussed above, the *URA3* and *LYS2* genes are particularly useful for this type of experiment as both positive and negative selections are possible. Thus, to isolate the rare cells that have undergone the excision event, one can grow the strain under nonselective conditions for a few generations to allow the recombination event to occur, then plate them under conditions that select against the function of the selectable marker gene. The colonies that appear can then be checked for the presence of the mutant phenotype.

Integration of YIp plasmids can also be exploited to clone mutations made *in vivo* so that their structure can be analyzed *in vitro*. In this case, a wild-type gene is integrated into a mutant allele, creating a duplication. If genomic DNA from this strain is then prepared, and cleaved with an appropriate restriction enzyme, the mutant DNA can be isolated by circularizing the cut DNA by ligation and selecting in *E. coli* for the vector marker (Fig. 2).[13,14]

The examples thus far have involved integration of an intact copy of a

[13] F. Winston, F. Chumley, and G. R. Fink, this series, Vol. 101, p. 211.
[14] G. S. Roeder and G. R. Fink, *Cell* 21, 239 (1980).

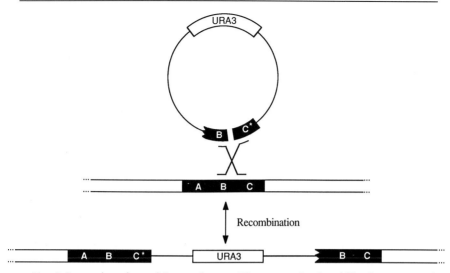

FIG. 3. Integration of a partial copy of a gene. The gene on the plasmid has been truncated such that it lacks one end, and it differs from the chromosome at C. Integration of this plasmid, directed by cleaving between B and C*, will lead to a direct duplication in the chromosome in which only one of the two copies of the gene is active. Because the integration was directed to occur between B and C*, the functional copy contains C*.

gene at its chromosomal locus; other useful approaches have been developed that make use of integration of an incomplete copy of a gene. By constructing the integrating plasmid so that the gene of interest lacks its 5' end, and integrating the plasmid at the locus, a partial gene duplication is created (Fig. 3). Only one of the copies is functional in this construct, thus the locus has been genetically marked without an accompanying change in gene dosage. Of course, this could also be done with a construct lacking the 3' end of the gene. If the plasmid contains a mutation, then, depending on the location of the mutation with respect to the restriction enzyme site used for directing recombination, the functional copy of the gene may contain the mutation. Thus, this is a useful method for isolating recessive mutations in one step; integration of the plasmid in the correct way will lead to expression of the mutant phenotype. This has been exploited in isolating conditional mutations in a number of genes: *ACT1*,[15] *TOP2*,[16] and *TUB2*.[17]

Homologous integration of a partial copy of a gene can also be used to create null mutations in a single step. If the integrating plasmid construct

[15] D. Shortle, P. Novick, and D. Botstein, *Proc. Natl. Acad. Sci. U.S.A.* **81,** 4889 (1984).
[16] C. Holm, T. Goto, J. C. Wang, and D. Botstein, *Cell* **41,** 553 (1985).
[17] T. C. Huffaker, J. H. Thomas, and D. Botstein, *J. Cell Biol.* **106,** 1997 (1988).

contains an internal fragment of the gene, then, after recombination, the gene is split into two inactive fragments (Fig. 4A), assuming, of course, that neither truncated gene fragment has activity. Other methods for producing null mutations exist, the most common being single-step transplacement[4] in which yeast is transformed with a linear DNA fragment that contains a gene or gene fragment into which a selectable marker has been inserted. The free ends of the fragment direct integration to the homologous chromosomal locus, resulting in direct replacement of the chromosomal sequences (Fig. 4B). For both of these methods of gene disruption it is important to know that the mutation is actually a null mutation and does not retain partial function or gain novel function, and that the construction does not interfere with surrounding genes.

One of the major advantages of yeast as a genetic organism is that there are stable haploid and diploid phases of the life cycle. Thus, null alleles of uncharacterized genes can be constructed as described above in a diploid strain; this diploid can then be induced to undergo meiosis to yield four haploid progeny. Two of these will be wild-type, and two will contain the mutation. If the two containing the mutation fail to grow, then the mutation is lethal. If they are viable, the phenotype of the null mutation can be studied. Again, the presence of the construction in the transformant must be confirmed by physical means.

Centromere Plasmids

Yeast centromere plasmids (YCp) stably replicate as autonomous plasmids. YCp vectors share the selectable markers of YIp vectors, but differ from YIp vectors in two important ways: they have an element that functions in yeast as replication origin, and they have a yeast centromere sequence that allows fairly regular mitotic and meiotic segregation. The replication origin sequence is termed ARS (autonomously replicating sequence); the most widely used YCp vectors contain *ARS1,* which is found immediately adjacent to the *TRP1* gene of yeast.[18] The centromere sequence is termed CEN;[19] *CEN4,* the centromere of chromosome IV, is commonly used. Interestingly, as few as 125 base pairs (bp) of authentic centromere DNA are required for proper mitotic and meiotic segregation of chromosomes, although the fragment commonly used in yeast vectors contains considerably more adjacent sequence. Table III is a listing of some available YCp vectors. We have included only those that are derived from

[18] D. T. Stinchcomb, K. Struhl, and R. W. Davis, *Nature (London)* **282,** 39 (1979).
[19] L. Clarke and J. Carbon, *Nature (London)* **287,** 504 (1980).

A

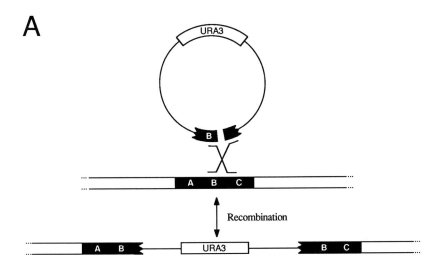

B

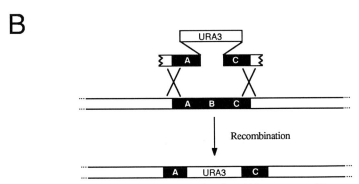

FIG. 4. Methods of gene disruption. In A, the plasmid-borne copy of the gene is truncated such that it lacks both ends. Integration of this construct into the chromosome leads to two nonfunctional copies of the gene, one lacking its 5' end, the other its 3' end. In B, a portion of the gene has been replaced with the *URA3* coding sequence. Integration of a linear fragment of DNA containing the insertion, homologous to the chromosome at both ends, leads to replacement of the chromosomal locus with the disrupted version. No plasmid sequences are left in the chromosome.

YCp50, the most frequently used centromere vector, as these are the ones of which we have had experience.

The presence of a yeast centromere on a plasmid does not ensure that the plasmid will remain single copy in the yeast cell. This is because plasmids bearing a centromere sequence are not as stable as natural linear

TABLE III

YEAST CENTROMERE PLASMID (YCp) VECTORS

Plasmid	E. coli markers	Yeast markers	Comments	Ref.[a]
YCp50	amp^r, tet^r	URA3	CEN4 and ARS1	1
YCp401	amp^r	LYS2, URA3	Derived from YCp50	2
YCp402	amp^r	LEU2	Derived from YCp50	2
YCp403	amp^r	HIS3	Derived from YCp50	2
YCp404	amp^r	HIS3, URA3	Derived from YCp50	2
YCp405	amp^r	LYS2	Derived from YCp50	2
YCp406	amp^r	URA3	Derived from YCp50	2
YCp407	amp^r	HIS3	Derived from YCp50	2
YCp408	amp^r	HIS3, URA3	Derived from YCp50	2
YCp409	amp^r	LEU2	Derived from YCp50	2
YCp410	amp^r	TRP1	Derived from YCp50	2
YCp411	amp^r	TRP1, URA3	Derived from YCp50	2
YCp412	amp^r	URA3	YCp50 with pUC19 polylinker	3

[a] 1, C. Mann (see Ma et al., 1987, for map); 2, H. Ma, S. Kunes, P. J. Schatz, and D. Botstein, Gene **58**, 201 (1987); 3, this laboratory (Dept. of Biology, Massachusetts Institute of Technology).

chromosomes. For example, chromosome V is lost approximately once in 10^5 divisions,[20] while YCp vectors typically are lost approximately once in 10^2 divisions.[21] This lowered fidelity of mitotic transmission results in some cells receiving two copies of the plasmid at mitosis, rather than one (actually, only about 20% of the loss events result in a daughter getting two copies; most often, one daughter receives one copy, and the other receives none). If cells carrying more than one copy of the plasmid have a growth advantage, then they may dominate the culture. It has been shown however, that a large excess of centromere-bearing plasmids causes sickness and inviability, presumably because the extra centromeres interfere with the normal process of chromosome segregation,[22] so it can be assumed that the copy number of a sequence on a YCp vector in yeast will be on the order of one to a few copies.

YCp vectors are true shuttle vectors in that they can replicate freely in both S. cerevisiae and E. coli. Indeed, the shuttle can be quite rapid: plasmid DNA from standard E. coli "minipreps" can be used for transformation of yeast, and DNA from several yeast DNA miniprep procedures can be used to transform bacteria. The most rapid of these yeast minipreps

[20] L. H. Hartwell, S. K. Dutcher, J. S. Wood, and B. Garvik, Rec. Adv. Yeast Mol. Biol. **1**, 28 (1982).

[21] D. T. Stinchcomb, C. Mann, and R. W. Davis, J. Mol. Biol. **158**, 157 (1982).

[22] B. Futcher and J. Carbon, Mol. Cell. Biol. **6**, 2213 (1986).

involves breaking cells open by agitation with glass beads, and requires only a few minutes.[23] We present here a version of this protocol that we have successfully used.

Yeast DNA Miniprep for Shuttling Plasmids

Materials

> Lysis buffer: 2% (v/v) Triton X-100, 1% (w/v) SDS, 100 mM NaCl, 10 mM Tris, 1 mM EDTA, pH 8.0
>
> Glass beads: Obtained from Sigma Chemical Co., (St. Louis, MO; Cat. No. G-9268, 425–600 μm glass beads); prepared by soaking in concentrated nitric acid for 1–2 hrs, washing extensively in distilled water, and baking until dry in a 200° oven
>
> PCI: A 25:24:1 mixture of phenol:chloroform:isoamyl alcohol

Method

1. Grow 2-ml overnight cultures in a medium that selects for presence of the plasmid.

2. Fill a 1.5-ml microfuge tube with culture and spin for 5 sec to collect the cells. Pour off the supernatant.

3. Add 50 μl of lysis buffer and 50 μl of PCI. Then add 0.2 g of glass beads.

4. Vortex at top speed for 2 min.

5. Spin for 5 min in a microfuge.

6. Use ~ 1 μl of the aqueous phase to transform competent *E. coli* cells.

Using *E. coli* cells that give approximately 10^6 transformants/μg DNA, one can expect on the order of 10–30 transformants when transferring centromere plasmids from yeast to *E. coli,* and 100–300 transformants when transferring 2-μm-based plasmids. The number of transformants does not increase in proportion to the amount of aqueous phase added; no more than 3 μl should be added to the transformation.

A typical experiment with YCp vectors is to insert a DNA fragment into the plasmid and to determine the phenotype of yeast carrying this fragment at low copy number. This is very useful when subcloning a DNA fragment to determine the region required for complementation of a recessive mutant phenotype. An extension of this is to clone a gene by complementation of a mutant phenotype. In this case, a library of random yeast DNA inserts is made in the centromere vector, then the library is transformed into the mutant yeast. Cells that receive a plasmid carrying the wild-type gene will be able to grow at the restrictive condition. The use of a

[23] C. S. Hoffman and F. Winston, *Gene* **57**, 267 (1987).

centromere vector for library construction is advantageous because the cloned wild-type gene is present in approximately the same copy number as when at its chromosomal locus. Indeed, some genes cannot be cloned from libraries made in YEp vectors because they are lethal in high copy number: *KAR1*,[24] *TUB2*,[25] and *ACT1* (unpublished observation). An excellent centromere–vector yeast DNA library in *E. coli*[26] is available from the American Type Culture Collection. This library consists of *Sau*3A partial fragments of ~ 15 kb inserted into the *Bam*HI site of YCp50. Once strains carrying complementing plasmids have been isolated, the plasmids can be examined by shuttling them from yeast to *E. coli,* and using physical means to determine their relationship to each other. To confirm that a given plasmid clone contains the authentic gene, and not a suppressor, the insert can be cloned into a YIp vector, integrated into the genome at the locus of the cloned fragment using the homology of the fragment to direct integration, and crossed to a strain bearing a mutation in the gene to be cloned. If the cloned sequence is the authentic gene, then tetrad analysis should show complete genetic linkage between the locus of the cloned fragment and the locus of the mutant gene.

Centromere plasmids can also be used to clone mutations from the chromosome onto a plasmid. The previously discussed method of doing this, using YIp vectors, requires that DNA be isolated from a transformant, cleaved, and recircularized by ligation. Using centromere vectors, this cloning can be done directly.[27] This is accomplished by introducing a linear centromere plasmid containing a gap that covers the region of the mutation into a mutant strain. *In vivo,* the homology on the ends of the gapped plasmid direct repair DNA synthesis such that the gap is repaired from the chromosomal sequence containing the mutation. The plasmid resulting from this repair can be transferred to *E. coli* as described above and analyzed.

Construction of New Vectors by *in Vivo* Homologous Recombination

One of the limitations of *in vitro* recombinant DNA techniques is that the researcher is often at the mercy of available restriction sites when making new constructions. This is especially apparent in yeast research because genetic analysis of any given gene now routinely involves moving the gene onto plasmids of different copy number and different selectable

[24] M. D. Rose and G. R. Fink, *Cell* **48,** 1047 (1987).

[25] J. H. Thomas, Ph.D., thesis, Massachusetts Institute of Technology (1984).

[26] M. D. Rose, P. Novick, J. H. Thomas, D. Botstein, and G. R. Fink, *Gene* **60,** 237 (1987).

[27] T. Orr-Weaver and J. Szostak, *Proc. Natl. Acad. Sci. U.S.A.* **80,** 4417 (1983).

markers. A recently described alternative to the *in vitro* approach takes advantage of *in vivo* homologous recombination to create new plasmids.[28]

This *in vivo* method can, in principle, be used to construct any plasmid with a yeast selectable marker and a yeast replication origin. Briefly, the method is based on the observation that two linear DNA molecules with overlapping homologies at or near the ends of the molecules can recombine efficiently after cotransformation into yeast to regenerate a circular molecule. In practice, one of the linear molecules is a plasmid linearized at a restriction site, the other is a DNA fragment that bears homology to the plasmid, but cannot itself circularize to form a replicating plasmid.

In the simplest case, a new selectable marker is introduced onto a plasmid either as a simple insertion or with concomitant loss of the original marker (Fig. 5). The homology for recombination is provided by the pBR322 backbone sequences common to most yeast plasmids. Because the linearized plasmid itself can give rise to transformants at a very low frequency by *in vivo* ligation, a small amount of this DNA is used in the transformation. Typically, a DNA mixture containing approximately 0.1 μg of a linearized plasmid DNA and 1.0 μg of a DNA restriction fragment is used to transform competent yeast cells, and the new marker is selected. We have observed that the frequency of transformation is quite high in these circumstances in comparison to transformation with closed-circular DNA controls. Plasmid DNA from the transformants is then isolated and shuttled to *E. coli* so that the structure of the plasmids can be determined.

The location of the cleavage site in the linearized plasmid and the extent of homology that the fragment has with the linearized plasmid determine the types of recombinants that will be recovered. If the cleavage site is outside of the selectable marker present on the linearized plasmid, then it is possible for that selectable marker to be maintained in the recombinational repair event. If the cut is made in the selectable marker gene of the linearized plasmid, then that selectable marker will be lost and replaced by the new marker. Similarly, if the ends of the restriction fragment carrying the new selectable marker do not cover the region of the selectable marker of the linearized plasmid, then the recombinant plasmid will contain both markers. If the ends of the restriction fragment do cover the site of the linearized plasmid's selectable marker, then the recombinant plasmid may or may not have the original marker, depending on the location of the recombination event (Fig. 5).

This method is also applicable to cases where the selectable marker itself is not exchanged. It may be necessary, for example, to exchange

[28] H. Ma, S. Kunes, P. J. Schatz, and D. Botstein, *Gene* **58**, 201 (1987).

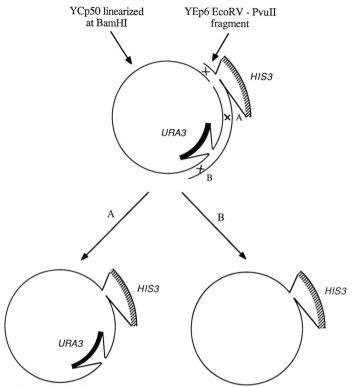

FIG. 5. Construction of new plasmids by *in vivo* recombination. In this example, YCp50 linearized at the *Bam*HI site is cotransformed into yeast with a fragment of YEp6. The *HIS3* marker of YEp6 is selected in the transformation. Because the YEp6 fragment cannot itself replicate, most transformants contain plasmids derived by recombination between YCp50 and the fragment. If one end of the recombination event takes place at A, then the recombinant plasmid will contain both the *URA3* and *HIS3* markers. If the recombination event takes place at B, then the plasmid will contain only *HIS3*.

alleles of the insert gene on the plasmid, from wild type to mutant or vice versa. In such a case, the plasmid is cleaved in the insert sequences to create a gap that covers the change in the new allele. This is then cotransformed with a fragment containing the new allele, and the marker of the original plasmid is selected. Because ligation of the plasmid to form a circle *in vivo* is infrequent compared to the frequency of recombination between the two cotransformed DNA fragments to form a functional plasmid, the majority of transformants will have the allele replacement. In this case, it is important that there be a deletion of the locus in the chromosome so that the plasmid cannot be repaired from the chromosomal copy.

In some cases, a restriction site needed to linearize the plasmid at the desired position is not available, or is not unique in the plasmid. In such a case, the circular plasmid can be cotransformed, uncut, with the linear fragment, and the selectable marker of the linear fragment selected. This method will only work if there is a selectable marker on the linear fragment, as otherwise there is no selection for the recombination event.

Summary

The vectors and techniques described here enable one to manipulate the yeast genome to meet specific needs. Genes can be cloned, and the clone used to delete the wild-type gene from the chromosome, or replace it with mutant versions. Mutants derived by classical methods, such as mutagenesis of whole cells, or by reversion of a phenotype, can be cloned and analyzed *in vitro.* Yeast genes and foreign genes can either be inserted into autonomously replicating plasmid vectors that are reasonably stable or integrated into a yeast chromosome where they are maintained at one copy per genome. The combination of these techniques with the characterized promoter systems available in yeast make it possible to express almost any gene in yeast. Once this is achieved, the entire repertoire of yeast genetics is available to probe the function of the gene, or to engineer the expression in useful ways.

[24] Regulated *GAL4* Expression Cassette Providing Controllable and High-Level Output from High-Copy Galactose Promoters in Yeast

By Lawrence M. Mylin, Kathryn J. Hofmann, Loren D. Schultz, and James E. Hopper

The *Saccharomyces* galactose-inducible promoters *GAL1, GAL7,* and *GAL10* have been extensively exploited for regulated and high-level production of foreign and nonforeign proteins in yeast. The inducibility of these promoters is based on a physical interplay between the GAL4 protein, a transcriptional activator, and the GAL80 protein, an antagonist of GAL4 protein activity.[1,2] The fact that the *GAL1, GAL7,* and *GAL10* promoters can be rapidly switched from a near-zero *off* state to a very high *on* state by addition of galactose to a nonglucose-containing growth medium makes these promoters especially attractive. In particular, when

[1] N. F. Lue, D. I. Chasman, A. R. Buchman, and R. D. Kornberg, *Mol. Cell. Biol.* **7,** 3446 (1987).
[2] M. Johnston, *Microbiol Rev.* **51,** 458 (1987).

METHODS IN ENZYMOLOGY, VOL. 185

high-level production of the desired protein is toxic to the yeast cell, the use of a regulated promoter is imperative. In such cases of toxicity, and where a *GAL* promoter is used, the cell culture is grown to a high cell density prior to the addition of galactose.

A limitation of the utility of the *GAL* promoters for providing the highest possible levels of a desired protein is realized when the desired GAL promoter–gene fusion construct is carried on plasmids which propagate to very high copy number in yeast. The structure and utility of such plasmids have been described elsewhere by others.[3,4] In theory, maximizing the recombinant GAL promoter–gene construct copy number ought to increase production of the desired protein. This is not realized in practice since the extremely low abundance of the GAL4 protein is rate-limiting for maximal induction of even the normal complement of *GAL* promoters in wild-type yeast.[2,5] This limitation is accentuated in yeast cells containing multiple copies of a plasmid bearing a *GAL* promoter construct.[6] Increasing the constitutive production of the GAL4 protein is not an acceptable solution to this problem since an excess of GAL4 protein overrides GAL80 protein function and leads to constitutive *GAL* promoter expression.[5,7] Consequently, the desired inducibility of the system is lost.

To service a high number of *GAL* promoters without loss of desired induction control, we have constructed a strain of *Saccharomyces cerevisiae* which carries both the chromosomal wile-type *GAL4* gene and an expression cassette consisting of the *GAL4* structural gene fused to the *GAL10* promoter *(GAL10p–GAL4)* integrated into the chromosome at the *HIS3* locus. In the absence of galactose, the normally low level of GAL4 protein expressed from the wild-type *GAL4* locus is maintained in an inactive state by the GAL80 protein. Expression of the *GAL1, GAL7,* and *GAL10* promoters is nondetectable. In the presence of galactose and absence of glucose a 60-fold increase in the cellular abundance of GAL4 protein occurs. This increase in GAL4 protein drives higher expression from multiple, plasmid-linked copies of *GAL* promoters than is achieved in the same strain lacking the *GAL10p–GAL4* expression cassette.[8]

[3] J. R. Broach, this series, Vol. 101, p. 307.

[4] S. Rosenberg, P. J. Barr, R. C. Najarian, and R. A. Hallewell, *Nature (London)* **312,** 77 (1984).

[5] S. A. Johnston and J. E. Hopper, *Proc. Natl. Acad. Sci. U.S.A.* **79,** 6971 (1982).

[6] S. M. Baker, S. A. Johnston, J. E. Hopper, and J. A. Jaehning, *Mol. Gen. Genet.* **208,** 127 (1987).

[7] S. A. Johnston, M. J. Zavortink, C. Debouck, and J. E. Hopper, *Proc. Natl. Acad. Sci. U.S.A.* **83,** 6553 (1986).

[8] L. D. Schultz, K. J. Hofmann, L. M. Mylin, D. L. Montgomery, R. W. Ellis, and J. E. Hopper, *Gene* **61,** 123 (1987).

We report here the detailed construction method, the function verification, and usage protocol for the *GAL10p–GAL4* expression cassette. As packaged here, the cassette can be readily integrated into the *HIS3* chromosomal locus of *S. cerevisiae* and related yeasts or conveniently transferred to a variety of centromere (CEN)-based yeast vectors. Alternatively, a stripped-down version can be easily tailored for integration into a genomic locus other than *HIS3*.

Overall Construction Strategy

The first stage in the construction entailed the fusion of the *GAL10* promoter to a site 5' to the initiator ATG of the *GAL4* coding sequence and the subsequent insertion of this recombinant molecule into *S. cerevisiae HIS3* gene sequences carried on a plasmid (Fig. 1). In the second stage, the entire ensemble of the *GAL10p–GAL4* fusion embedded in *his3* was moved as a *BAM*HI fragment to a plasmid suitable for subsequent insertion of the *Saccharomyces cerevisiae URA3* gene into a site 3' to the *GAL4* coding region. The movement of the URA3 gene into that site yielded a new plasmid, pKHint-C (Fig. 2), which provides a conveniently movable and selectable 5'-*his3–GAL10p–GAL4–URA3–his3*-3' cassette for targeting the *GAL10p–GAL4* recombinant sequence into the yeast genome. The third stage entailed enzymatic cleavage of pHKint-C at the *Bam*HI sites within the *his3* sequences, and integrative transformation of yeast with this digested DNA to yield a yeast bearing the *GAL10p–GAL4–URA3* cassette inserted into the chromosomal *HIS3* locus (Fig. 3).

Materials and Methods

Yeast Strains and Media

Sc252 (21R) is a GAL4 revertant of strain 21 *(gal4-2 ura3-52, leu2-3 leu2-112 ade1 MEL1).*[5] Sc340 is genetically identical to strain Sc252 except that it contains the regulated GAL4 overproducing cassette integrated at the *his3* locus, constructed as described here. Sc413 is genetically identical to strain Sc252 except that the *GAL4* gene initiator ATG and the majority of *GAL4* coding sequences have been replaced with the *S. cerevisiae LEU2* gene. This strain produces no detectable GAL4 protein or transcript.[9]

Yeast strains lacking plasmids were grown in succinate/NaOH-buffered synthetic complete media containing either 2% (w/v) dextrose (for inocula-

[9] L. M. Mylin, J. P. Bhat, and J. E. Hopper, *Genes Devel.* **3**, 1157 (1989); W. Bajwa and J. E. Hopper, unpublished results (1987).

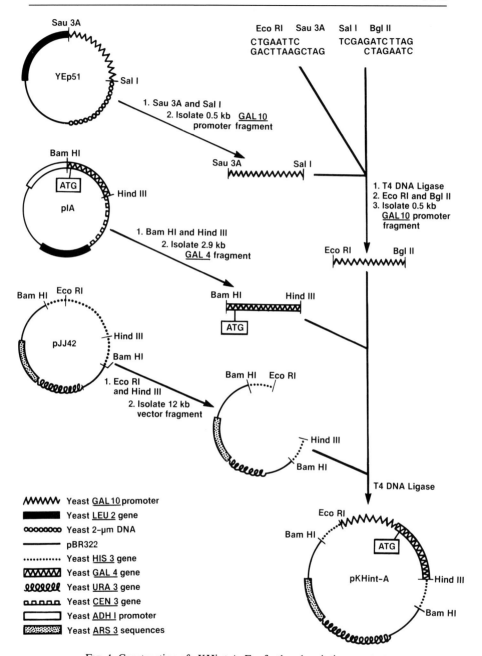

FIG. 1. Construction of pKHint-A. For further description, see text.

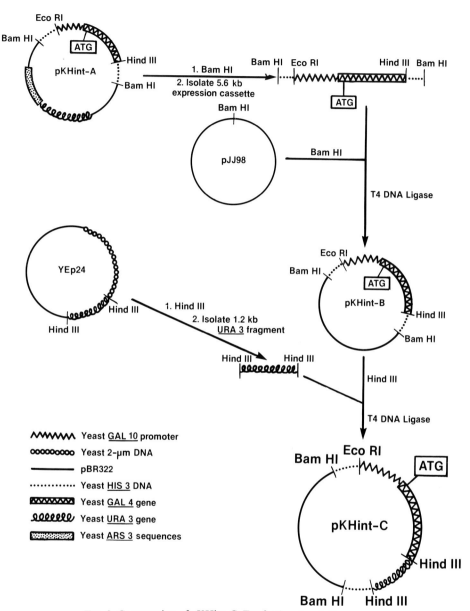

FIG. 2. Construction of pKHint-C. For further description, see text.

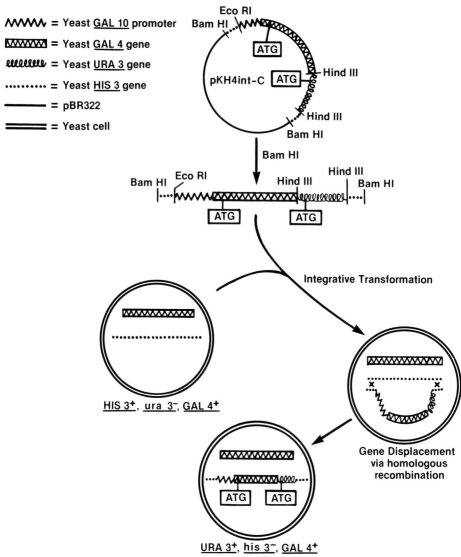

FIG. 3. Targeted integration of the *GAL4* expression cassette into the yeast chromosomal *HIS3* locus. For further description, see text.

tion) or 3% (v/v) glycerin, 2% (v/v) potassium lactate, pH 5.7,[10] and 0.05% (w/v) dextrose (for growth prior to induction of *GAL4* overproduction). Buffered synthetic complete media contained per liter of final solution:

[10] T. E. Torchia, R. W. Hamilton, C. L. Cano, and J. E. Hopper, *Mol. Cell. Biol.* **4**, 1521 (1984).

7.2 g of yeast nitrogen base lacking amino acids and ammonium sulfate [Difco (Detroit, MI) 0335-15], 6.5 g of ammonium sulfate [Sigma (St. Louis, MO) A-55132], 10 g of succinic acid, 6 g of sodium hydroxide, 100 mg each of adenine, uracil, tryptophan, histidine-HCl, arginine, and methionine, 150 mg each of tyrosine, lysine, and isoleucine, 250 mg of phenylalanine, 1.0 g of threonine, 750 mg of valine, 500 mg of aspartic acid, 300 mg of leucine, and the appropriate carbon sources.

Transformation and Isolation

Yeast Transformation. Yeast transformation was carried out as described by Schultz *et al.*[11]

Escherichia coli Transformation. Competent *E. coli* cells were purchased from Bethesda Research Laboratories (Gaithersburg, MD), and were transformed according to the supplier's procedure.

Plasmid Isolation. Plasmids were isolated from *E. coli* transformants by an alkaline lysis method detailed elsewhere.[12]

Preparation of Oligodeoxynucleotides

Oligonucleotides were synthesized by phosphoramidite chemistry using the Applied Biosystems Model 380A automated synthesizer according to the Applied Biosystems directions. The oligonucleotides were desalted by chromatography on Sephadex G-50 (superfine) in 0.1 M triethylammonium bicarbonate using a modification of the procedure recommended by the manufacturer. Partial purification was achieved by selecting only that fraction of the oligodeoxynucleotide population eluting at the leading edge of the chromatogram. The resulting preparation was used without further purification.

Growth and Induction of GAL4 Overexpression

Stationary cultures of yeast strains Sc413, Sc252, and Sc340 grown in complete, succinate-buffered liquid media containing 2% glucose (w/v) were diluted $\frac{1}{200}$ into noninducing media (as above but with 3% glycerin, 2% potassium lactate, pH 5.7, and 0.05% glucose as carbon sources) and agitated vigorously at 30° until culture densities reached $A_{600} = 0.4$. At this point, zero time samples were removed, and galactose added to a final concentration of 2%. Incubation continued at 30°, with additional culture samples removed at 2 and 4 hr after addition of galactose.

[11] L. D. Schultz, J. Tanner, K. J. Hofmann, E. A. Emini, J. H. Condra, R. E. Jones, E. Kieff, and R. W. Ellis, *Gene* **54**, 113 (1987).
[12] T. Maniatis, E. F. Fritsch, and J. Sambrook "Molecular Cloning: A Laboratory Manual." Cold Spring Harbor Laboratory, Cold Spring Harbor, New York, 1982.

Cell Extract Preparation

Culture aliquots were chilled on ice for 10 min, pelleted at 4°, washed one time each with cold water and breaking buffer (lacking protease inhibitors, reducing agent, and glycerol), and stored as cell pellets in 1.5-ml microfuge tubes at −70° for less than 2 weeks. Breaking buffer is essentially buffer A(50) used by Bram and Kornberg.[13]

Cell samples (5×10^8 cells) were allowed to thaw for 30 min on ice. Pellets were resuspended in 0.3 ml AT(50) each. [AT(50) in which Tris-HCl replaces HEPES–NaOH as the buffering agent.] Following addition of an equal volume of 0.45-mm glass beads, the cells were lysed with six 20-sec pulses of a vortex mixer interrupted by equal periods of cooling on ice. Cell lysates were removed from the beads, the beads rinsed with a second 0.2-ml aliquot of AT(50), and recovered volumes combined. The unfractionated lysates were then incubated at 100° for 5 min after addition of 0.25 volumes of electrophoresis sample buffer concentrate (5% SDS, 50% sucrose, 50 mM Tris-Cl, pH 8.0, 5 mM EDTA, 160 mM DTT). The extracts were centrifuged for 5 min in an Eppendorf centrifuge at room temperature, and the supernatants saved for protein estimation and Western blotting analysis.

Protein determination was performed according to Peterson[14] using bovine serum albumin as a standard and excluding deoxycholate from the trichloroacetic acid precipitation step.

Western Blotting Analysis

Samples containing 20 μg of protein were fractionated by SDS–PAGE[15] on 5%/7.5% discontinuous gels using a Hoefer SE 250 apparatus. Proteins were electrophoretically transferred to nitrocellulose membranes [Schleicher and Schuell (Keene, NH) 0.45 μm pore size] for 3 hr at 200 mA constant current in a Hoefer TE-22 chamber. The transfer buffer contained 25 mM Tris-Cl, 192 mM glycine.

Immunodetection was performed according to the method of Towbin[16] with the following modifications. Incubations were performed in 50 mM sodium phosphate (made by 20-fold dilution of 1 M stock prepared by addition of mono- to dibasic for pH of 7.2), 145 mM NaCl (PBS). Blocking solution was PBS containing 10% (w/v) Carnation nonfat dry milk.[17]

[13] R. Bram and R. Kornberg, *Proc. Natl. Acad. Sci. U.S.A.* **82,** 43 (1985).

[14] G. L. Peterson, *Anal. Biochem.* **83,** 346 (1977).

[15] U. K. Laemmli, *Nature (London)* **227,** 680 (1970).

[16] H. Towbin, T. Staehelin, and J. Gordon, *Proc. Natl. Acad. Sci. U.S.A.* **76,** 4350 (1979).

[17] D. A. Johnson, J. W. Gautsch, J. R. Sportsman, and J. H. Elder, *Gene Anal. Tech.* **1,** 3 (1984).

Blocking was performed overnight at 4°. Antibody incubations were performed in blocking solution containing 0.3% (v/v) Tween 20 (Sigma)[18] in a total volume of 10 ml per blot (enclosed in seal-a-meal bags). Washing solution was PBS containing 0.3% (v/v) Tween 20. Rabbit antiserum raised against the full-length GAL4 protein produced in *E. coli*[7] was used as primary antibody at 1 : 1000 dilution.[9] The GAL4 protein used as antigen was isolated by preparative SDS–PAGE and electroelution followed by acetone precipitation prior to injection mixed with RIBI MPL + TDM + CWS Emulsion (RIBI Immunochem Research, Inc., Hamilton, MT).[19] Alkaline phosphatase conjugated affinity-purified goat anti-rabbit IgG (Cooper Biomedical, West Chester, PA) was used as the second antibody at a 1 : 1000 dilution. The alkaline phosphatase color reaction was performed according to Billingsley *et al.*[20] and an additional wash in Tris buffer lacking the chromogenic substrates was included following the final set of phosphate-buffered washes. The dried blot was scanned using a Beckman DU-8B spectrophotometer equipped with a slab gel scanning system. Comparison of peak areas was performed using the integrating software package supplied with the Compuset module.

Construction Details

YEp51[21] was digested with *Sau*3A + *Sal*I, and the 0.5 kilobase (kb) *GAL10* promoter fragment was isolated. Two oligodeoxynucleotide adapters were synthesized and ligated to the 0.5-kb fragment, thus converting the 5' *Sau*3A and the 3' *Sal*I ends to *Eco*RI and *Bgl*II ends, respectively, following digestion with the appropriate enzymes. pIA carries the *GAL4* structural gene and contains a *Bam*HI site centered at 30 base pairs (bp) upstream of the *GAL4* translational initiation codon plus a *Hin*dIII site downstream of the *GAL4* transcriptional terminator.[7] pIA was digested with a *Bam*HI + *Hin*dIII, and the 2.9-kb *Bam*HI–*Hin*dIII fragment bearing *GAL4* was isolated. YRp14-Sc3121 (Δ26)[22,23] (shown as pJJ42 in Fig. 1), which contains the yeast *HIS3* gene, was digested with *Eco*RI + *Hin*dIII, which cleave within the *HIS3* promoter and coding region, respectively. The resulting 12-kb DNA fragment was isolated and ligated simultaneously to the 0.5- and 2.9-kb fragments described above to yield

[18] B. Batteiger, W. S. Newhall, V, and R. B. Jones, *J. Immunol. Methods* **55,** 297 (1982).

[19] L. M. Mylin and J. E. Hopper, unpublished results (1987).

[20] M. L. Billingsley, K. R. Pennypacker, C. G. Hoover, and R. L. Kincaid, *BioTechniques* **5,** 22 (1987).

[21] J. R. Broach, Y. Li, L. C. Wu, and M. Jayaram *in* "Experimental Manipulation of Gene Expression" (M. Inouye, ed.), p. 83. Academic Press, New York, 1983.

[22] K. Struhl, *Proc. Natl. Acad. Sci. U.S.A.* **79,** 7385 (1982).

[23] K. Struhl, *Cold Spring Harbor Symp. Quant. Biol.* **47,** 901 (1982).

pKHint-A (Fig. 1). The 4.7-kb *Bam*HI fragment of pKHint-A-bearing the *GAL10p–GAL4* expression cassette inserted into the *HIS3* gene was then cloned into the *Bam*HI site of pJJ98,[6] a modified pBR322 lacking the *Hin*dIII site, to yield pKHint-B. The 1.2-kb *Hin*dIII fragment bearing the yeast *URA3* gene was isolated from YEp24[24] and cloned into the unique *Hin*dIII site of pKHint-B which is located beyond the 3' end of *GAL4*. The resulting plasmid, pKHint-C, contains the complete 5.8-kb 5'-*his3*-*GAL10p-GAL4-URA3-his3*-3' cassette (Fig. 2). pKHint-C was digested with *Bam*HI, and the digest products were used to transform *S. cerevisiae* strain Sc252. Ura⁺ transformants were selected and then screened for His⁻. The replacement of chromosomal internal *HIS3* gene sequences with the 5'-*his3*-*GAL10p-GAL4-URA3-his3*-3' cassette (Fig. 3) occurs by homologous recombination.[25] Integration was confirmed by Southern blot analysis of restriction digests of genomic DNA.[26]

Results

Increased production of GAL4 protein occurs in a regulated manner in yeast strain Sc340 which contains the integrated *GAL10p–GAL4* cassette (Fig. 4). Following 4 hr of induction, the GAL4 protein level in Sc340 was increased 60-fold over that found in the isogenic wild-type strain Sc252. Prior to induction, both GAL4 mRNA[8] and protein levels (Fig. 4) were equivalent in the two strains.

The presence of the *GAL10p–GAL4* cassette has been shown previously to increase production from a heterologous gene controlled by the *GAL10* promoter carried on a high-copy yeast vector.[8] Expression of Epstein-Barr viral glycoprotein was enhanced 10-fold in the integrant strain Sc340 over wild-type strain Sc252. In that case, the expression kinetics of the viral proteins paralleled that of the GAL4 protein, reaching maximum levels between 4 and 8 hr after addition of galactose to the culture.

Comments

Yeast strains bearing the *GAL10p–GAL4* expression cassette either integrated[8,19] or on a CEN vector[27] cannot grow either on plates containing galactose as the sole carbon source, or on plates containing noninducing

[24] D. Botstein, S. C. Falco, S. E. Stewart, M. Brennan, S. Scherer, D. T. Stinchcomb, K. Struhl, and R. W. Davis, *Gene* **8**, 17 (1979).
[25] R. J. Rothstein, this series, Vol. 101, p. 202.
[26] E. M. Southern, *J. Mol. Biol.* **98**, 503 (1975).
[27] X. Yu and J. E. Hopper, unpublished results (1988).

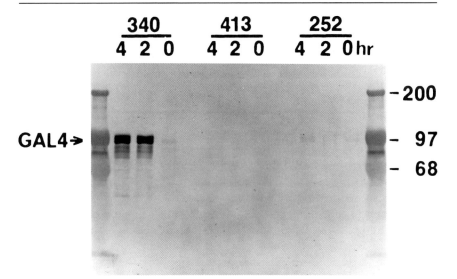

FIG. 4. Western blot analysis of GAL4 protein levels. GAL4 protein was detected by Western blotting of extracts (20 μg protein per lane) made from uninduced (0 hr) and induced (2, 4 hr) yeast strains Sc252 (wild type), Sc413 (*gal4Δ*), and Sc340 (*GAL4* overproducer) as described in the text. The position of the *GAL4* doublet is indicated. Densitometric comparison of the *GAL4* doublet in the Sc340 and Sc252 4-hr lanes indicates a 60-fold overproduction. Outside lanes contain prestained molecular weight standards ($\times 10^{-3}$) (Bethesda Research Labs, Lot #73101). The slower migrating *GAL4* band includes two distinct forms, one of which has been shown to be a phosphorylated form of *GAL4*.[9]

carbon sources (glycerol and lactic acid) supplemented with lower levels of galactose. This toxicity is not dependent on the presence of the *GAL1*, *GAL7*, or *GAL10* gene products.[19] In fact, this toxicity can be used as a preliminary indication that a functional cassette has been successfully introduced, provided the recipient yeast strain contains functional alleles of all *GAL* regulatory genes. A similar galactose toxicity was observed previously by Laughon *et al.*[28] in yeast strains bearing a *GAL1p–GAL4* fusion on the multicopy vector YEp13, although the kinetics of *GAL4* mRNA accumulation and disappearance observed with their *GAL1p–GAL4* fusion differed markedly from what we observe in the single-copy *GAL10p–GAL4* integrant.[8] Although the basis for the kinetic differences is unknown, there are several differences between their experiment and ours that could serve as possible explanations: (1) GAL4 was expressed under structurally distinct promoters; (2) the copy number of expression cassettes

[28] A. Laughon, R. Driscoll, N. Wills, and R. F. Gesteland, *Mol. Cell. Biol.* **4**, 268 (1984).

in the cells differed; (3) the strains in which the cassettes were expressed differed; and (4) our strains contained a large excess of *GAL10* promoters on pKH4-related vectors that could have potentially provided a sink for excess GAL4 protein at early times of induction.

Since the *GAL10p–GAL4* cassette works when either integrated into a chromosome or carried on a CEN vector, it has potential utility in yeasts other than *S. cerevisiae* providing they carry *GAL* regulatory functions analogous to those in *S. cerevisiae*. For transfer of the expression cassette to such yeasts, one need simply transfer either the 3.4-kb *GAL10p–GAL4* *Eco*RI–*Hind*III fragment from pKHint-A or the 5.8-kb 5'-*his3-GAL10p-GAL4-URA3-his3*-3' *Bam*HI fragment from pKHint-C to a suitable CEN vector for subsequent transformation into the desired yeast. Alternatively, these fragments can be integrated into the genome using the *his3* sequences as outlined here, providing the recipient carries homologous *HIS3* sequences.

In summary, *S. cerevisiae* strain Sc340 and/or the readily movable *GAL10p–GAL4* cassette should prove generally useful for optimized and regulated expression of *GALp*–gene fusions carried on high-copy vectors in yeast.

[25] Expression of Heterologous Proteins in *Saccharomyces cerevisiae* Using the *ADH2* Promoter

By Virginia L. Price, Wayne E. Taylor, William Clevenger, Marlis Worthington, and Elton T. Young

Introduction

The yeast alcohol dehydrogenase 2 gene *(ADH2)* is one of many yeast genes whose expression is regulated by glucose repression. The transcription of *ADH2* is undetectable when yeast are grown on glucose and is derepressed to a level representing about 1% of soluble cellular protein when yeast are grown on a nonfermentable carbon source.

The *ADH2* promoter has been extensively analyzed by both site-specific mutagenesis and deletion analysis to reveal two cis-acting regulatory components (upstream activation sequences, or UAS) mediating derepression.[1,2] Either UAS element alone is sufficient to confer glucose-regulated expression on a heterologous promoter; both UAS elements together act

[1] D. R. Beier, A. T. Sledziewski, and E. T. Young, *Mol. Cell. Biol.* **5,** 1743 (1986).

[2] J. Shuster, J. Yu, D. Cox, R. V. L. Chan, M. Smith, and E. T. Young, *Mol. Cell. Biol.* **6,** 1894 (1986).

synergistically to confer maximum expression on the promoter while keeping it tightly repressed on glucose.[3]

A protein encoded by the gene *ADR1* has been identified as a positive activator of *ADH2* transcription.[4] The ADR1 protein contains two zinc fingers[5] enabling it to bind specifically to UAS1,[6] a perfect 22 base pair. (bp) inverted repeat. Glucose repression of *ADH2* is due to a lack of positive activation by ADR1 rather than due to a negative mechanism acting through DNA-binding repressor proteins.[7] Recent data suggest that ADR1 may be a phosphoprotein whose phosphorylation state could be regulated by glucose repression.[8]

The *ADH2* promoter is very amenable for use in the expression of heterologous genes, not only because it provides a strong transcriptional start signal, but also because transcription from the *ADH2* promoter is tightly repressed by glucose in the growth medium.[1,4] Cultures can be grown to a high density in the presence of glucose, which keeps the promoter repressed. When the glucose in the medium is depleted by metabolism, the promoter is derepressed to a high level. This procedure avoids changing the growth medium, adding inducing compounds, or changing the temperature in order to induce the promoter. Possible cyto-toxic effects of the heterologous protein are avoided since initial cell growth occurs in the absence of the protein. In addition, the cloned regulatory gene, *ADR1,* can be used to enhance expression from the *ADH2* promoter (see below). In this article, we describe the construction of various *ADH2* promoter plasmids for expression of heterologous proteins in yeast as well as for secretion into the culture medium.

Construction of *ADH2* Promoter Plasmids

A 4.7-kilobase-pair (kb) *Bam*HI–*Eco*RI DNA fragment containing the *ADH2* structural gene was subcloned into pBR322 at the *Bam*HI and *Eco*RI sites (p*ADH2*, Fig. 1A).[9] This fragment contains 1.2 kb of DNA sequences 5′ to the *ADH2* coding region and 2.7 kb of DNA 3′ to the *ADH2* coding region, and includes both the *ADH2* promoter and a tran-scription terminator. To isolate the promoter fragment from the *ADH2* structural gene for the insertion of heterologous coding sequences, a series

[3] J. Yu, M. S. Donoviel, and E. T. Young, *Mol. Cell. Biol.* **9,** 34 (1989).

[4] C. L. Denis, M. Ciriacy, and E. T. Young, *J. Mol. Biol.* **148,** 355 (1981).

[5] T. Hartshorne, H. Blumberg, and E. T. Young, *Nature (London)* **320,** 283 (1986).

[6] A. Eisen, W. E. Taylor, H. Blumberg, and E. T. Young, *Mol. Cell. Biol.* **8,** 4552 (1988).

[7] M. Irani, W. E. Taylor, and E. T. Young, *Mol. Cell. Biol.* **7,** 1233 (1987).

[8] C. L. Denis and C. Gallo, *Mol. Cell. Biol.* **6,** 4026 (1986).

[9] J. R. Broach, this series, Vol. 101. p. 307.

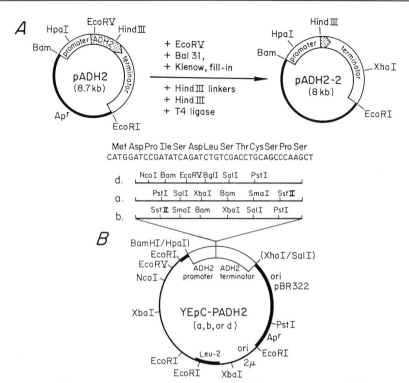

FIG. 1. *ADH2* promoter plasmids. (A) Plasmid pADH2 contains the *ADH2* structural gene, promoter, and terminator cloned into the *Bam*HI and *Eco*RI sites of pBR322, as described in the text. This plasmid was used to generate pADH2-2 by *Bal*31 deletions. pADH2-2 contains the *ADH2* promoter sequence, a *Hind*III site at position −1 to the ADH2 initiation codon, the C-terminal 279 bp of the *ADH2* structural gene, and the *ADH2* transcription terminator. (B) Plasmid YEpC-pADH2a, b, and d are yeast expression plasmids derived from pC1/1 (see text and Ref. 9) that contain a polylinker following the *ADH2* promoter. YEpC-pADH2a and b contain the pUC13 polylinker in both orientations. YEpC-PADH2d contains a polylink encoding the initiation codon ATG. The nucleotide sequence of polylinker d is shown above line d, which depicts the restriction sites.

of *Bal*31 deletions were made as shown in Fig. 1A. Plasmid pADH2 was digested with *Eco*RV, which cleaves uniquely in the plasmid at nucleotide 69 in the *ADH2* structural gene, and *Bal*31 deletions were generated. DNA polymerase (Klenow fragment) was used to fill in overhanging ends and *Hind*III linkers were ligated onto the *Bal*31-generated ends. The DNA was then digested with *Hind*III to remove excess linkers and to cut the *ADH2* gene at its unique *Hind*III site at amino acid 254. Ligation of the resulting mixture created deletions extending leftward (5′) variable distances from the *Eco*RV site and containing a common 3′ end point defined by the *Hind*III site in the *ADH2* structural gene.

The deletions of interest were of two categories: (1) A deletion to -2 relative to the *ADH2* translation initiation ATG, which therefore included the promoter and 5′ untranslated region of *ADH2* (plasmid pADH2-2). The presence of the natural mRNA start site and 5′ untranslated region may aid in the translation of heterologous proteins. (2) Deletions to $+19$, $+26$, and $+27$ bp (relative to the *ADH2* ATG) provide, in addition to the above elements, the ATG initiation codon and a restriction site (*Hind*III) in three reading frames to fuse heterologous coding sequences lacking translational starts. These plasmids are designated pADH2 + 19, pADH2 + 26, and pADH2 + 27, respectively (not shown).

Construction of Yeast Expression Plasmids Containing ADH2 Promoter

The *ADH2* promoter was inserted into the yeast – *Escherichia coli* shuttle vector pC1/1 for expression in yeast. pC1/1 is a multicopy 2-μm plasmid that was constructed by insertion of pBR322 into the entire 2-μm form B plasmid at the *Eco*RI site which maps at 0 bp.[9] In this vector (see Fig. 1B), the selectable marker is the defective (promoterless) *LEU2d* gene, which was inserted into the blunted *Pst*I site of the 2-μm plasmid.[10] When pC1/1 is transformed into a cir⁰ yeast strain (lacking endogenous 2-μm plasmid) and grown selectively in leucine-deficient medium, the plasmid copy number reaches very high levels of 400–500 copies per cell.[7] When the *ADH2* gene was cloned into this plasmid, the ADHII protein was overproduced about 20-fold, reaching approximately 20% of total cellular protein (unpublished observations).

To generate a cloning site downstream of the *ADH2* promoter, a set of plasmids was constructed by first inserting the polylinker from pUC13 into the *Hind*III site of plasmid pADH2-2. Then, the DNA fragment from the *Hpa*I site (900 bp 5′ to the polylinker) to the *Xho*I site (1.45 kb 3′ to the polylinker) containing the *ADH2* promoter, polylinker, and transcription terminator was inserted into the *Sal*I site of vector pC1/1.

The addition of the *ADH2* promoter and terminator to pC1/1 generated plasmids YEpC-PADH2a and YEpC-PADH2b containing the polylinker in the two orientations illustrated in Fig. 1B with unique restriction sites *Sma*I, *Bam*HI, and *Sal*I. This expression vector does not have an ATG codon within the 5′ untranslated region of the polylinker. Therefore, it can be used for the cloning and expression of an intact gene having its own initiation codon.

[10] J. D. Beggs, in "Molecular Genetics in Yeast" (D. Von Wettstein, A. Stenderup, M. Kielland-Brandt, and J. Friis, eds.), p. 383 (Alfred Benzon Symposium, Vol. 16). Munksgaard, Copenhagen, 1981.

A third yeast expression vector was made that is identical to YEpC-PADH2a and b except that the polylinker region is replaced with a different polylinker containing an initiator ATG codon at its 5′ end (YEpC-PADH2d, Fig. 1B). This polylinker contains an *Nco*I site (not unique) at the ATG followed by the unique restriction sites *Bam*HI, *Bgl*II, and *Sal*I. Such a vector can be used to express coding regions lacking an initiator methionine.

In all of the expression vectors described, the polylinker is located between the promoter and the transcription terminator of *ADH2*. This ensures that transcription initiated at the *ADH2* promoter will not continue around the plasmid, possibly interfering with plasmid replication.

Construction of Yeast Expression Plasmid Allowing Regulated Secretion of Heterologous Proteins

A yeast expression vector was also constructed that contained the *ADH2* promoter joined to the leader peptide from the secreted pheromone α-factor. This allows secretion of heterologous proteins whose transcription is regulated by the *ADH2* promoter. This yeast–*E. coli* shuttle vector, designated pαADH2[11] (Fig. 2A), contains most of pBR322 (from the *Sph*I site at nucleotide 562, not regenerated in the vector, to the *Eco*RI site at nucleotide 4361), which includes the origin of replication and the ampicillin resistance marker. Sequences from yeast include (1) the *TRP1* gene as a selectable yeast marker derived from plasmid YRp7;[12] (2) the 2-μm origin of replication contained on the *Pst*I to *Spe*I fragment obtained from plasmid YEp13.[9] This fragment also contains a transcription termination signal which is used to terminate transcripts from the *ADH2* promoter. Unlike plasmid pC1/1, plasmid pαADH2 must be transformed into a cir$^+$ yeast strain, as it does not carry the entire 2-μm plasmid genome; (3) the *ADH2* promoter fragment from the *Bam*HI site (not regenerated) to the *Hin*dIII site from plasmid pADH2-2; and (4) a DNA fragment coding for the 85-amino acid signal peptide derived from the α-factor gene to allow secretion of proteins. The α-factor signal peptide was fused to the *ADH2* promoter by means of a synthetic oligodeoxynucleotide that joined the *Hin*dIII site in the *ADH2* promoter to the *Pst*I site at nucleotide 24 in the α-factor leader sequence (regenerating the first eight amino acids of the α-factor leader). The *Hin*dIII site is not regenerated at the junction of the promoter and leader sequence.

[11] V. Price, D. Mochizuki, C. March, D. Cosman, M. Deeley, R. Klinke, W. Clevenger, S. Gillis, P. Baker, and D. Urdal, *Gene* **55**, 287 (1987).
[12] K. Struhl, D. Stinchcomb, S. Scherer, and R. Davis, *Proc. Natl. Acad. Sci. U.S.A.* **76**, 1035 (1979).

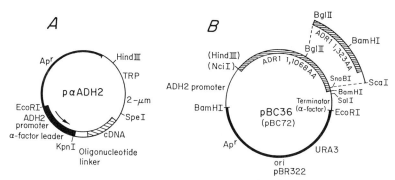

FIG. 2. Yeast expression plasmids. (A) The yeast–*Escherichia coli* shuttle vector pαADH2 is described in the text and in Ref. 11. This vector allows regulated secretion of heterologous proteins via the *ADH2* promoter fused to the α-factor leader sequence. (B) The YIp5-derived integrating vectors pBC36 and pBC72 allow overproduction of *ADR1* in yeast. pBC36 contains the amino-terminal 1068 amino acids of *ADR1* fused to the *ADH2* promoter while pBC72 contains the entire *ADR1* coding region fused to the *ADH2* promoter as described in the text. The gene encoding *ADR1* is followed, in both cases, by a transcription terminator derived from the α-factor gene. [A. Brake, J. Merryweather, D. Coit, U. Heberlein, F. Masiarz, G. Mullenbach, M. Urdea, P. Valenzuela, and P. Barr, *Proc. Natl. Acad. Sci. U.S.A.* **81**, 4642 (1984)].

An Asp-718 (*Kpn*I) restriction site was introduced at nucleotide-237 in the α-factor leader sequence to facilitate its fusion to heterologous genes. (The T residue at nucleotide-241 was changed to a C residue by site-directed mutagenesis.) A polylinker was also inserted into the vector between the Asp-718 site and the *Spe*I site in the 2-μm sequence, placing it between the α-factor leader and the transcription termination signal in 2-μm DNA. cDNAs are cloned into the vector by digesting with Asp-718, which cleaves five amino acids from the C-terminus of the α-factor leader, and a site in the polylinker. The leader peptide is then fused in-frame to the cDNA of interest using a synthetic oligodeoxynucleotide to regenerate the 3′ end of the α-factor leader (the amino acids Pro-Leu-Asp-Lys-Arg) and any N-terminal sequences needed in the cDNA. (For a detailed description of the proteolytic processing of the α-factor leader, see Refs. 13 and 14.)

Expression of Heterologous Proteins Using ADH2 Promoter

The expression of several cDNAs encoding mammalian cytokines was tested using vector pαADH2 (Fig. 2A). These proteins, listed in Table I, are normally secreted hormones and therefore have an endogenous signal

[13] D. Julius, R. Schekman, and J. Thorner, *Cell* **36**, 309 (1984).
[14] A. Brake, J. Merryweather, D. Coit, U. Heberlein, F. Masiarz, G. Mullenbach, M. Urdea, P. Valenzuela, and P. Barr, *Proc. Natl. Acad. Sci. U.S.A.* **81**, 4642 (1984).

TABLE I
SECRETION OF CYTOKINES FROM YEAST USING
THE *ADH2* Promoter

Cytokine[a]	Cytokines (μg/ml) secreted[b]
Murine GM-CSF	50 ± 10
Human GM-CSF	40 ± 10
Bovine IL-2	60 ± 10
Human IL-2	5 ± 2

[a] GM-CSF, Granulocyte–macrophage colony-stimulating factor; IL-2, interleukin 2.
[b] Cytokine (μg/ml) present in conditioned yeast medium after derepression. Values were determined either by radioimmunoassay or by silver-stained SDS–polyacrylamide gel electrophoresed material compared to a purified protein standard.

sequence. For their expression and secretion in yeast, the endogenous signal was removed and the mature coding region of the cDNA was fused in-frame with the α-factor leader, as described above. Proteolytic processing by the *KEX2* gene product[12,13] releases mature protein into the culture medium.

For these studies, yeast strain XV2181 (a/α, *trp1-1/trp1-1*) was used; however, other *trp1* yeast strains (i.e., X2181 or DBY 746) obtainable from the Yeast Genetic Stock Center (University of California, Berkeley, CA) work similarly. Yeast were transformed with plasmid pαADH2 containing the cDNA for several cytokines (Table I) by either the spheroplasting procedure[15] or the lithium acetate procedure.[16] Transformants were selected for on YNB-trp plates [0.67% (w/v) yeast nitrogen base (Difco, Detroit, MI) 0.5% (w/v) casamino acids, 2% (w/v) glucose, and 20 μg/ml each of adenine and uracil]. Maintenance of plasmid-containing strains and growth of cultures for inoculum were done in selective (YNB-trp) medium. Typically, cultures were grown for optimal expression of heterologous protein in rich medium (YPD, consisting of 1% yeast extract, 2% peptone, 1% glucose, and 80 μg/ml each of adenine and uracil). Derepression of the *ADH2* promoter was obtained by exhaustion of glucose in the medium. Cultures in shake flasks were grown for 18–24 hr (stationary phase) such that glucose was depleted from the medium for 10–12 hr before harvest. Despite some plasmid loss during growth in nonselective

[15] A. Hinnen, J. Hicks, and G. Fink, *Proc. Natl. Acad. Sci. U.S.A.* **75**, 1929 (1978).
[16] F. Sherman, G. Fink, and J. Hicks, *in* "Laboratory Course Manual for Methods in Yeast Genetics," p. 121. Cold Spring Harbor Laboratory, Cold Spring Harbor, New York, 1986.

medium, we have observed better protein expression in YPD than in synthetic selective medium.

Table I lists several different cytokines expressed in yeast using the *ADH2* promoter, and the amount of protein secreted into the culture medium. The amount of heterologous protein secreted ranged from 5 to 70 μg/ml. The inducibility of the *ADH2* promoter is illustrated in Fig. 3, using the expression of murine granulocyte macrophage colony-stimulating factor (MuGM-CSF) as an example. Depletion of glucose in the medium coincides with the appearance of MuGM-CSF in the culture medium. The expression of protein from the *ADH2* promoter continues for several hours after derepression.

Overproduction of ADR1: Effects on *ADH2* Promoter

ADR1 is required for positive activation of the *ADH2* promoter and its availability is a limiting factor in *ADH2* transcription.[7] This is also true when the *ADH2* promoter is present on a multicopy plasmid (ADR1 is

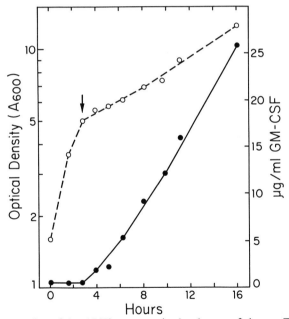

FIG. 3. Derepression of the *ADH2* promoter in the absence of glucose. The dashed line represents the growth (A_{600}) of *Saccharomyces cerevisiae* transformed with plasmid pαADH2 containing the cDNA encoding the cytokine murine GM-CSF. The culture was grown as described in the text. The arrow indicates the point at which glucose was no longer detectable in the culture medium (Diastix glucose test strips, Miles Laboratories, Elkhart, IN). The solid line shows the amount of secreted murine GM-CSF in the culture medium as determined by radioimmunoassay.

normally provided by a single-copy chromosomal gene). When ADR1 is overexpressed, the chromosomal *ADH2* locus is constitutively expressed and derepression to a much higher level occurs.[7,17] The overproduction of ADR1 and its effect on transcription from the *ADH2* promoter carried on a plasmid were therefore of interest.

The expression of ADR1 from its own promoter is relatively weak, based on analyses of ADR1 mRNA and β-galactosidase activity of an *ADR1*–β-galactosidase fusion protein.[17] One means of increasing ADR1 levels in yeast would be to express the gene from a very strong promoter. However, overexpression of ADR1 on glucose-containing medium allows the *ADH2* promoter to partially escape glucose repression. Hence, we sought a regulated promoter in order to maintain a low level of ADR1 during glucose repression. It was thought that the *ADH2* promoter itself could achieve this and, additionally, place the overexpression of ADR1 under glucose control. Placing the expression of ADR1 under control of the *ADH2* promoter also might create a form of runaway "autoregulation" of ADR1 synthesis: the more ADR1 protein made, the more positive activation of the promoter, thus more ADR1 made, etc. In order to initiate the cycle, the chromosomal copy of *ADR1* is retained under the control of its own promoter, which allows a level of *ADR1* to be present on glucose that can be posttranslationally modified to initiate the cycle.

An integrating plasmid was constructed using the yeast integrating vector YIp5[12] and inserting the 1.2-kb *ADH2* promoter fragment (from the *Bam*HI site to the *Hin*dIII site from plasmid pADH2-2) followed by the *ADR1* gene and the transcription termination sequences derived from the α-factor gene. Figure 2B illustrates two vectors that were made. pBC36 contains a 3.3-kb DNA fragment of the *ADR1* gene from the *Nci*I site, which includes 120 bp 5′ to the initiation ATG codon, and DNA encoding the N-terminal 1068 amino acids (out of 1323 total amino acids) to the *Bam*HI site. This portion of the ADR1 protein is functional in activating the *ADH2* promoter.[17] A second vector, pBC72, is identical to pBC36 except it includes the entire *ADR1* coding region on a 6-kb *Nci*I fragment.

Plasmid pBC36 was integrated into the yeast chromosome at the *ADR1* locus by first linearizing the plasmid with *Bgl*II, which cleaves at nucleotide-1922 in the *ADR1* gene (relative to the ATG). Transformation of yeast strain XV617 *(MATa, ste5, his6, leu2, trp1, ura3)* was done, selecting for Ura⁺ transformants. Total DNA from four transformants was isolated and subjected to Southern blot analysis, which confirmed the integration of pBC36 at the *ADR1* locus in all four cases (data not shown). This yeast strain was designated XV36. pBC72, on the other hand, was integrated

[17] H. Blumberg, T. Hartshorne, and E. T. Young, *Mol. Cell. Biol.* **8,** 1868 (1988).

876 bp upstream of the *ADH2* coding region by targeting to the *Mst*II site. Again, four Ura⁺ transformants were subjected to Southern blot analysis and the site of integration at the *ADH2* locus was verified. This strain was designated XV72.

The glucose regulation of ADR1 mRNA synthesis was demonstrated for strain XV36 by S1 and Northern blot analysis (data not shown). The *ADR1* mRNA derived from the gene fusion was undetectable during repressed growth, and was derepressed to a very high level when glucose was absent, consistent with the *ADR1* gene being under the control of the *ADH2* promoter. Comparison of the ADR1 mRNA levels in the parent strain, XV617, and in strain XV36 also showed that ADR1 mRNA is greatly overproduced in strain XV36.

To assess the effects of excess ADR1 protein on expression from the *ADH2* promoter, the endogenous ADHII levels in strains XV617, XV36, and XV72 were compared. Strains XV36 and XV72 showed a 4-fold and a 5.5-fold increase in ADHII levels, respectively, over the parent strain, XV617 (not shown).

The effect of excess ADR1 on heterologous gene expression from the *ADH2* promoter was determined in a strain containing an integrated copy of the *ADH2* promoter fused to the *lacZ* gene. Plasmid pADH2-*lacZ* (Fig. 4) is a YIp5-derived integrating plasmid containing the *ADH2* promoter and DNA sequences corresponding to the N-terminal 23 amino acids of *ADH2* (to the *Eco*RV site) fused to the *lacZ* gene from plasmid pMC1871.[18] This plasmid was integrated at the *ADH2* locus in strain XV36, which overexpresses ADR1 mRNA, and in strain XV617 to serve as a control. Comparison of the regulated expression of β-galactosidase in the two strains showed a 4- to 10-fold increase in β-galactosidase levels in strain XV36 on derepression (Fig. 4). Under repressed conditions (growth on glucose), β-galactosidase activity was very low, and similar to the activity of the parent strain, XV617, containing the *ADH2–lacZ* gene fusion. These results indicate that an excess of ADR1 protein can enhance expression of foreign proteins from the *ADH2* promoter and that the excess of positive activator can be regulated by glucose repression of the *ADH2* promoter. The level of activity obtained from a single copy of the ADH2 promoter and excess ADR1 provided by the same promoter is similar to the level of activation provided by a single-copy of the normal *ADR1* gene and the *ADH2* promoter on a high-copy plasmid. The advantages of the former system are 2-fold: All genes are stably integrated into a chromosome and the *ADH2* promoter remains tightly repressed.

[18] M. Casadaban, A. Martinez-Arias, S. Shapira, and J. Chou, this series, Vol. 100, p. 293.

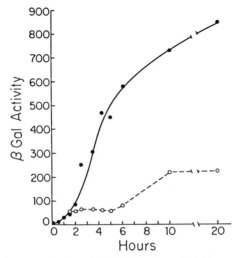

FIG. 4. Regulated overproduction of ADR1 stimulates *ADH2*-promoted expression of an ADH2–β-galactosidase fusion protein. Strains XV617 and XV36 contain an *ADH2–lacZ* gene fusion integrated at the *ADH2* locus. Strain XV36 contains an *ADHR1* plasmid integrated at the *ADR1* locus. Strains were grown in YPD and derepressed by shifting to complete medium containing ethanol (2%) and glycerol (3%) as carbon sources. Samples were taken for β-galactosidase (βGal) assays at the times indicated. β-Galactosidase activity is shown as the change in A_{420} per minute per unit absorbance at A_{600}, multiplied by 1000.

It was hoped that the ADR1 runaway system would lead to even higher levels of expression from a high-copy-number plasmid containing the *ADH2* promoter driving a heterologous gene. We tested this idea using the *ADH2* gene itself on a high-copy plasmid. The *ADH2* activities in strain XV36 or XV72 containing YEpADH2 were no higher than the activity in strain XV617 with the plasmid (30,000 U/mg after depression). This level is so high already that other factors may become limiting when both ADR1 and the *ADH2* promoter are present in large quantities. Moreover, high levels of ADR1 seem to lead to toxicity during derepression, which may limit the ultimate level of expression.

Acknowledgments

This work was supported by Immunex Corporation, Merck, Sharpe and Dohme Research Laboratories, and The National Institutes of Health.

[26] Induced Expression Using Yeast Copper Metallothionein Promoter

By TINA ETCHEVERRY

Different promoters have been utilized to express recombinantly engineered genes in yeast. The promoters from highly expressed genes such as phosphoglycerate kinase (PGK) and alcohol dehydrogenase (ADH) have been used to express the human interferon genes,[1,2] the α-factor expression system has been used to direct the synthesis of epidermal growth factor,[3] and the triose-phosphate isomerase promoter has been used to express the human insulin precursor.[4] The efforts on heterologous gene expression in yeast have led to the pharmaceutical production and licensing of hepatitis B surface antigen using the glycerol-3-phosphate dehydrogenase promoter.[5]

However, in the process of producing pharmaceuticals from microorganisms, it is often desirable to separate the growth phase of cells from product synthesis. A regulated promoter allows for the directed transcription of a foreign gene independent of cell metabolism during rapid cell growth. Promoters have been developed for regulated synthesis of gene products; e.g., the *GAL1* promoter,[6] which is controlled by carbon source, and the *PHO5* promoter,[7] which is induced by phosphate limitation. These methods, although effective on a small scale, might be inefficient for large-scale production processes. Therefore, development of different promoters for unique applications becomes necessary.

The inducible metallothionein (MT) promoter from yeast has been used for the expression of several different proteins with reasonably high

[1] R. A. Hitzeman, F. E. Hagie, H. L. Levine, D. V. Goeddel, G. Ammerer, and B. D. Hall, *Nature (London)* **293**, 717(1981).

[2] R. A. Hitzeman, D. W. Leung, L. J. Perry, W. J. Kohr, H. L. Levine, and D. V. Goeddel, *Science* **219**, 620(1983).

[3] A. J. Brake, J. P. Merryweather, D. G. Coit, U. A. Heberlein, F. R. Masiarz, G. T. Mullenbach, M. S. Urdea, P. Valenzuela, and P. Barr, *Proc Natl. Acad. Sci, U.S.A.* **81**, 4642 (1984).

[4] L. Thim, M. T. Hansen, K. Norris, I. Hoegh, E. Boel, J. Forstrom, G. Ammerer, and N. P. Fiil, *Proc. Natl. Acad. Sci. U.S.A.* **83**, 6766 (1986).

[5] W. J. McAleer, E. B. Buynak, R. Z. Maigetter, D. E. Wampler, W. J. Miller, and M. R. Hilleman, *Nature (London)* **307**, 178 (1984).

[6] C. G. Goff, D. T. Moir, T. Kohno, T. C. Gravius, R. A. Smith, E. Yamasaki, and A. Taunton-Rigby, *Gene* **27**, 35 (1984).

[7] R. A. Kramer, T. M. DeChiara, M. D. Schaber, and S. Hilliker, *Proc. Natl. Acad. Sci. U.S.A.* **81**, 367 (1984).

expression levels. This promoter is rapidly induced by the addition of copper ions to the media. The gene was initially cloned and characterized by Fogel and Welch[8] and the DNA sequence published by Karin *et al.*[9] and Butt *et al.*[10] The gene resides at the *CUP1* locus and its product has been termed both Cu-chelatin and Cu-metallothionein. Thiele and Hamer have analyzed the regulatory control elements using this promoter fused to the *Escherichia coli* galactokinase gene.[11] Described here are the methods for the construction of an inducible promoter for heterologous gene expression and the induction conditions used to maximize transcription from this promoter.

General Outline of Methods

The MT promoter constructions described here contain between 230 and 452 base pairs (bp) of DNA originating upstream of the *CUP1* coding sequence. The constructions have been designed with restriction sites available so that the promoter can be inserted proximal to any coding region in a cassette fashion. This promoter region contains the TATA box and metal regulatory elements sufficient for high-level gene expression with a translationally effective mRNA leader sequence.

The first step of the method involves selection of the appropriate shuttle vector and construction of the plasmid. Various yeast vectors are now available, containing Ura, Leu, His, and/or Trp selectable markers with either autonomously replicating sequences generating high-copy-number plasmids or centromere-containing plasmids to limit copy number. The *CUP1* promoter has been engineered using synthetic linkers altering the 10 bp proximal to the natural translation start signal resulting in *Xba*I and *Eco*RI recognition sites.

The second step is to test host strains for their copper resistance levels. The relationship between the induction protocol and strain tolerance to copper is critical. Inappropriately high levels of copper will reduce translation levels by poisoning the cells. Low levels of copper will not induce maximal expression as copper is chelated and thus removed by the endogenous metallothionein. The third step of the method requires transforming these strains with the constructed vectors using standard spheroplasting methods of transformation.[12]

[8] S. Fogel and J. W. Welch, *Proc. Natl. Acad. Sci. U.S.A.* **79,** 5342 (1982).

[9] M. Karin, R. Najarian, A. Haslinger, P. Valenzuela, J. Welch, and S. Fogel, *Proc. Natl. Acad. Sci. U.S.A.* **81,** 337 (1984).

[10] T. R. Butt, E.J. Sternberg, J. A. Gorman, P. Clark, D. Hamer, M. Rosenberg, and S. T. Crooke, *Proc. Natl. Acad. Sci. U.S.A.* **81,** 3332 (1984).

[11] D. J. Thiele and D. H. Hamer, *Mol. Cell. Biol* **6,** 1158 (1986).

[12] T. L. Orr-Weaver, J. W. Szostak, and R. J. Rothstein, this series, Vol. 14, p. 228.

Finally, one needs to optimize the induction protocol. This is related to the growth conditions. Essentially, the cells are grown under selective conditions with the addition of copper either as a single shot or multiple-pulsed additions to induce the promoter and express the product.

Materials

Strains

Yeast strain 20B-12 (α *trp1 pep4-3*) is available from the Berkeley Stock Center. Strain TE412 (α *trp1 ura3-52*) was obtained by a cross between 20B-12 and SF648-5D (*a leu2 ura3-52 ade2-1 his4-580 prc1-1*) generously donated by R. Schekman. The *CUP1* disruption strain TE412Δ6 (α *trp1 ura3-52 cup1* Δ/*URA3*) has been described previously,[13] with the *URA3* marker replacing metallothionein-coding sequences at the *CUP1* locus. Strain TE412Δ2 is identical to strain TE412Δ6 except that the *URA3* displacement removed only two of the three metallothionein repeated elements.

Media

Minimal medium contains 0.67% (w/v) yeast nitrogen base without amino acids (Difco-Detroit, MI), 2% (w/v) glucose, (w/v) and 0.5% (w/v) casamino acids. Agar plates contain minimal medium solidified with 3% (w/v) Phytagar (GIBCO Laboratories, Grand Island, NY). Copper sulfate ($CuSO_4$) was stored as a 100 mM sterile stock and added to cultures as needed. High-density cultures require supplementation with a buffered feed solution containing 20% glucose and 1 M sodium glycerol phosphate.

Construction of Promoter

Gene Isolation

The metallothionein gene can be isolated from a yeast DNA bank prepared from *Sau*3AI digested genomic DNA. Most laboratory yeast strains contain tandemly repeated metallothionein genes; each gene is contained within a 1320-bp *Sau*3AI fragment. The multiple genomic copies provide for an enriched source of the gene. The *CUP1* gene is isolated by transforming a copper-sensitive strain, such as TE412, with the genomic bank. The methods for yeast transformation have been well described.[12] On transformation, one selects for complementation of a selectable marker on the plasmid (such as Trp1) and protection against toxic levels of copper. The strain TE412 is resistant to 0.1 mM copper (Table I)

TABLE I
RELATIONSHIP OF LEVEL OF COPPER RESISTANCE IN HOST
STRAIN AND INDUCTION CONDITIONS

Host strain	Number of *CUP1* repeat units	Level of resistance (mM)	Induction range (mM)
20B-12	>6	0.6	0.2–0.5
TE412	3	0.1	0.03–0.1
TE412Δ2	1	0.03	0.01–0.03
TE412Δ6	0	0.01	0.01

but sensitive to 0.3 mM copper. This higher level of copper, when included in the Trp⁻ transformation plates, will select for cells carrying the *CUP1* gene on the plasmid. The coselection for copper resistance and Trp⁺ can result in false positives in some strain backgrounds. This difficulty can be avoided by first selecting for Trp⁺ transformants and screening these colonies for increased copper resistance.

In the process of cloning the gene described here, an unidentified *Sau*3AI fragment was cloned upstream of the copper regulatory region, adjacent to the 1320 bp of *CUP1* DNA. This additional fragment of DNA does not affect the regulation or expression of the gene and was retained in the constructions of pTE432 (Fig. 1) for the convenience of unique *Cla*I and *Sal*I sites. The actual construction has been previously described.[13] The promoter can be isolated from the genomic fragment by a *Sau*3AI and *Rsa*I digest, yielding a 428-bp fragment. This piece of DNA contains the metal regulatory sequences, the mRNA cap site, the TATA box, and associated transcription signals. It lacks 28 bp of DNA immediately adjacent to the initiating ATG. These are provided by the synthetic linkers shown in Fig. 1.

Linker Addition

Complementary strands containing *Eco*RI sticky ends and the *Rsa*I site 28 bp upstream of the ATG were chemically synthesized. These oligonucleotides provide 18 bp of natural sequence and alter the 10 bp adjacent to the transcription initiation site, adding the convenience of *Xba*I and *Eco*RI cloning sites. These restriction sites allow for the insertion of a coding region of interest. The synthetic linker can include other restriction sites; *Sac*I, *Sma*I, and *Bam*HI have been used at this position. The general idea is to connect the *Rsa*I site in the promoter to the gene of interest, keeping the sequences as conserved as possible. The process of linker addition involves

[13] T. Etcheverry, W. Forrester, and R. Hitzeman, *Bio/Technology* **4**, 726 (1986).

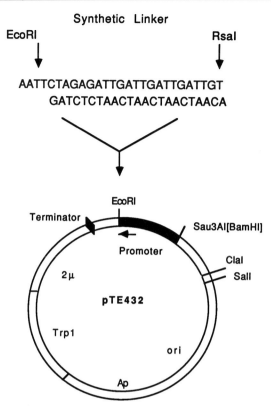

Synthetic Linker

EcoRI   RsaI

AATTCTAGAGATTGATTGATTGATTGT
GATCTCTAACTAACTAACTAACA

EcoRI

Terminator

Sau3AI[BamHI]

Promoter

ClaI

2 μ   SalI

pTE432

Trp1

ori

Ap

FIG. 1. Diagram of the expression vector pTE432. pTE432 contains the *CUP1* promoter (dark band) and the 2-μm terminator (diamond) in addition to the 2-μm replication sequences, the *Trp1* selective marker, and the *E. coli* ampicillin marker and origin of replication. The *Eco*RI site was introduced using synthetic linkers (top).

attaching these synthetic strands to the *Sau*3AI–*Rsa*I promoter fragment and inserting this reconstructed promoter into the shuttle vector by standard ligation procedures. The construction is shown in Fig. 1. The vector is now ready for insertion of the gene to be expressed.

Variations of This Promoter: 5′ and Internal Deletions

Shorter promoters were constructed to determine the minimal region required for strong and regulated gene expression. The promoter of 452 bp ending at the *Sau*3AI site functions normally and is independent of plasmid orientation or flanking DNA. Shorter promoters were generated by Bal-31 digestion of upstream sequences retaining 301 and 251 bp of *CUP1* DNA. These shorter promoters are affected by the context of the plasmid.

The association of pBR322 DNA proximal to the copper regulatory elements in these shorter plasmids results in constituitive expression. Both shorter promoters still contain the trans-acting regulatory protein binding sites, but no longer require copper for induction. Therefore, the integrity of the promoter is influenced by sequences beyond the defined protein-binding regions.

The cis-acting elements which bind trans-acting protein factors have been designated UAS_P and UAD_D[11] and lie at positions -173 to -204 and -213 to -246 with respect to the ATG. Deletion mapping of these regions have shown that both elements are required for maximal gene expression (unpublished data, this laboratory). The UAS sequences contain an inverted repeat of the element TCTTTTGCT. Attempts to use synthetic UAS elements have failed to produce high, regulated gene expression (unpublished data, this laboratory), suggesting that multiple elements are involved.

Selection of Copper Levels for Induction

Selection of copper levels for induction depends on the copper resistance level of the host. The endogenous *CUP1* gene plays a major role in MT promoter regulation. The MT protein is a scavenger of free copper and neutralizes the toxic effect of copper by chelating the metal. The stoichiometric ratio of copper to MT has been measured at 8 mole equivalents.[14] Yeast responds to selective pressure of copper in the environment by amplifying the *CUP1* locus. This amplification occurs by a homologous recombinational event with the frequency of 3×10^{-6} events per generation.[15] The laboratory strains related to S288C or X2180 contain an average of six repeated copies of this gene at the *CUP1* locus. In general, these strains can tolerate 0.3–0.6 mM copper when grown on plates. In liquid, this translates to a slightly lower level of 0.2–0.5 mM copper. The relationship of copper resistance of the host strain to the range of copper for induction is described in Table I. As the number of *CUP1* repeats in the genome decreases, the level of copper required for induction also decreases.

The best method for determining the range of copper concentration for the induction is to replica transfer fresh cells from minimal media agar plates to a series of plates containing 0.03, 0.1, 0.3, or 0.6 mM copper. Cells which exhibit poor growth at 0.3 mM copper but good growth at 0.1 mM (for example, strain TE412 in Table I) should be induced in liquid culture

[14] D. R. Winge, K. B. Nielson, W. R. Gray, and D. H. Hamer, *J. Biol. Chem.* **260**, 14464 (1985).
[15] M. T. Etcheverry, *in* "Genome Rearrangement" (M. Simon and I Herskowitz, eds.), p. 221. Alan R. Liss, New York, 1985.

within the copper range of $0.03-0.1$ mM. Higher copper levels will start to inhibit the metabolic machinery of the cell. It may not inhibit growth as much as reduce the expression levels from the MT promoter.

Induction Timing and mRNA Levels

Different copper feeding strategies have been used to try to maximize expression. Shake flask cultures which grow to an optical density of $10-15$ do not need complicated copper feeding routines. Copper sulfate is added to the growth medium when cells reach an OD_{600} of 0.3. No additional copper is required if the final harvest is within 24 hr. It is critical that the copper levels have been adjusted to the tolerance levels of the cells.

An alternative method is to induce the culture late in logarithmic phase of growth or after the cells have reached stationary phase. This procedure separates the growth of the culture from the production phase. Yield of product are lowered by this method, especially if the material is to be secreted and requires efficient intracellular transport and secretion. However, this is ideal for any protein which has been found to be detrimental to cell growth. It also conserves on metabolic substrates and machinery required for high-level protein production. Metabolic substrates are directed to cell growth initially and diverted to product synthesis late in growth phase by the addition of copper.

The mRNA profile for induction in the late growth phase is shown in Fig. 2. A modification of plasmid p432 was used in a time course to follow mRNA levels following copper induction. The human serum albumin (HSA)[13] gene contained on an *Eco*RI cassette was inserted into the plasmid as a marker to follow expression. Late-log phase cells ($OD_{600} = 10-15$) were induced with 0.3 mM copper sulfate and aliquots were removed at 0, 5, 10, 30, 60, and 180 min postinduction. A rapid harvest method was used at each time point to stop cell metabolism. A 10 OD sample was pelleted for 10 sec in a microfuge and placed directly on dry ice. After all the samples were collected, they were resuspended in room-temperature RNA extraction buffer. This buffer contains 0.1 M NaCl, 50 mM Tris-HCl, pH 7.5, 10 mM EDTA, and 5% (w/v) SDS. One-half volume of 0.45-mm glass beads was added. The glass beads had been acid washed, neutralized, and heat-baked to prevent ribonuclease contamination. This extraction mixture was vortexed for 5 min with an equal volume of phenol : chloroform : isoamyl alcohol (50 : 50 : 1). This treatment was repeated two times and then the RNA was precipitated by the addition of 0.1 volume of 5 M NaCl and 2.5 volumes of cold 95% (v/v) ethanol. After 2 hr at $-20°$, the samples were spun at 10,000 rpm for 15 min, resuspended in 0.2 M

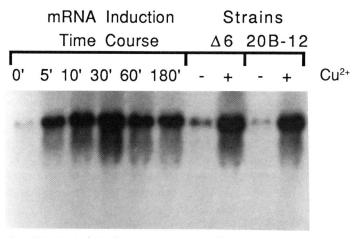

FIG. 2. mRNA levels following copper induction. The time course of mRNA induction was measured in yeast strain 20B-12 following the addition of 0.3 mM Cu^{2+}. Cell samples were removed at the times (minutes) indicated. RNA was extracted as described in the text. The mRNA bound to nitrocellulose was probed with *HSA*-specific sequences. The four lanes to the right exhibit the levels of *HSA* mRNA before and after 60 min of copper induction (as indicated) in the *CUP1* disruption strain TE412Δ6 and in the copper-resistant strain 20B-12.

sodium acetate, pH 5.2, and reprecipitated with ethanol. Ten micrograms of RNA was added to a loading solution of 50% formamide, 2.2 M formaldehyde, 20 mM MOPS, pH 7.0, and heated to 90° for 2 min. These RNA samples were subjected to electrophoresis on a denaturing gel (1% agarose, 6% formaldehyde, 20 mM MOPS) for 6 hr at 80 V. Induction of the metal-regulated promoter as measured by Northern analysis was found to be rapid. Maximal levels of mRNA were reached within 10 min of copper addition (Fig. 2), and remained constant over the 3-hr period.

mRNA levels were also measured for the *CUP1* disruption strain TE412Δ6. This strain is highly sensitive to copper since it lacks any endogenous Cu-metallothionein. Expression levels of HSA-specific mRNA in this strain were compared to levels in the copper-resistant strain 20B-12. Both strains were grown to late-log phase (OD$_{600}$ = 12) and samples removed prior to copper addition. Strain TE412Δ6 was then induced with 0.0.1 mM copper, whereas strain 20B-12 was induced with 0.3 mM. After 60 min, another sample was removed for RNA extractions as described above. The results are shown in Fig. 2. Final mRNA levels cannot be distinguished between the two strains. The *CUP1* disruption strain does express more HSA in the absence of added copper since small amounts of copper are present in the trace elements of the minimal medium. This is difficult to avoid because copper is an essential element for electron-transfer proteins and other enzymes. However, there are benefits to using the disruption strains for copper-regulated gene expression. As shown in Fig. 2., mRNA levels remain high in this strain even though the copper concentration was reduced 30-fold. This may be a consideration for product recovery when one is concerned about possible copper contamination.

Induction during High-Cell-Density Growth

High cell densities can be reached by modifying the standard shake flask protocol. By limiting the glucose available to the cell, ethanol production is minimized. This allows the cells to continue log-phase growth to optical densities (OG_{600}) of 50 or more. We have developed a method of growing cells in minimal medium plus 0.5% casamino acids by supplementing growth with pulsed additions of buffered feed containing 20% glucose and 1 M sodium-glycerol phosphate. Initially, cells are inoculated from a fresh overnight culture into minimal medium containing 2% glucose at a concentration of 2×10^7 cells/ml. Cells are grown with high aeration (300 rpm in a Lab-Line Environ Shaker) at 30° for up to 150 hr. The pH and glucose levels are measured twice a day by removing 1-ml samples and testing this aliquot with pH paper and glucose reagent strips (Diastix; Miles Laboratories, Naperville, IL). Small amounts of buffered feed (0.02 volume) are added to the culture after the initial glucose is consumed. This maintains the pH at 4.0 to 5.5 and limits the glucose to less than 1%. It is best to be conservative and add the buffered feed once or twice a day until the optical density exceeds $OD_{600} = 25$. At this time, the cells will rapidly consume the glucose and can tolerate four pulses of buffered feed daily without producing ethanol. Copper is present throughout the entire growth period. Copper (0.1 mM) is included in the initial growth medium and aliquots of copper in 0.1 mM increments are added daily to a maximum level of 0.5 mM.

mRNA Analysis of Different Genes

Different homologous and heterologous (nonyeast) proteins have been tested for expression in yeast using the *CUP1* promoter. Levels of gene transcripts were measured for the yeast PGK gene[13] and the human tissue plasminogen activator (tPA).[16] These expression vectors were constructed by R. Kastelein (PGK) and A. Singh (tPA) in similar fashion to the HSA-encoding plasmid. An engineered *Eco*RI site was placed preceding

[16] D. Pennica, W. E. Holmes, W. J. Kohr, R. N. Harkins, G. A. Vehar, C. A. Ward, W. F. Bennett, E. Yelverton, P. H. Seeburg, H. L. Heyneker, D. V. Goeddel, and D. Collen, *Nature (London)* **301**, 214 (1983).

HSA   PGK   tPA

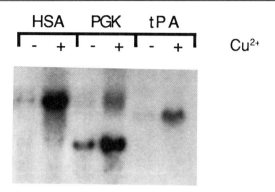

FIG. 3. Levels of transcripts of *HSA, PGK,* and *tPA* genes under the direction of the *CUP1* promoter. mRNA was isolated from cultures of yeast strain 20B-12 before and after 60 min of copper treatment as indicated above each lane. The Northern blot was probed with the 27-bp synthetic fragment (shown in Fig. 1) common to all sequences.

the initiator ATG for both genes. These gene cassettes were then inserted into the plasmid p432 at the *Eco*RI site so that the mRNA initiation site and 5′ untranslated region originated from the *CUP1* promoter and were identical in each case. A synthetic probe of 27 bp complementary to this region (upper strand in Fig. 1) was then used to measure RNA levels. The results of this Northern analysis are shown in Fig. 3. Both genes show a low level of expression in the absence of copper. Induction with 0.3 mM copper at a cell density of $OD_{600} = 12$ produces a rapid increase in mRNA level. Transcriptional levels varied between the different genes approximately 5-fold, whereas translational discrepancies between the genes were even greater (data not shown). Although strong transcription is the initial critical step for high-level gene expression, product levels are equally dependent on translational and other posttranscriptional events.

Conclusions

The *CUP1* promoter can be engineered to express a variety of genes under copper regulation. The product can be induced during log phase of cell growth or after the cells reach stationary phase. This decision will depend on the quantity of product desired (early induction) or the quality of product (late induction for toxic or labile proteins). The induction ratio (induced mRNA/uninduced mRNA) and the absolute level of mRNA depend on the vector chosen and the product expressed. High-copy vectors will produce more total mRNA and yield higher product, but will also have higher relative expression levels in the absence of copper. The level of copper used for induction will depend on the strain that is used. This is empirically determined by selecting a level of copper which is lower than the toxic threshold for the cell.

The methods described here can be used to facilitate the expression of cloned genes for laboratory applications or pharmaceutical production. The ease of induction and the strength of the promoter make this *CUP1* system an excellent method for expressing heterologous genes in yeast.

Acknowledgments

I wish to thank Wayne Forrester for his help in gene constructions during the course of this work, Anne Lucas and Robert Hamilton for development of the high-cell-density protocol, and Mike Covarrubias for assistance in the preparation of this manuscript.

[27] High-Efficiency Yeast Expression Vectors Based on the Promoter of the Phosphoglycerate Kinase Gene

By S. M. KINGSMAN, D. COUSENS, C. A. STANWAY, A. CHAMBERS, M. WILSON, and A. J. KINGSMAN

Introduction

The *PGK* (phosphoglycerate kinase) gene is one of the most efficiently expressed genes in yeast as the protein and mRNA comprise about 5% of total cell proteins and mRNA. The *PGK* promoter has been manipulated to produce a series of high-efficiency expression vectors which have been used to express a number of different eukaryotic proteins. These include human interferons,[1,2] calf chymosin,[3] immunoglobulins,[4] wheat α-amylase,[5] and HIV antigens.[6] The *PGK* gene is regulated to some degree by carbon source; levels of expression are low on acetate or pyruvate[1] (J. Tsang, unpublished observation) and are increased on glucose. The difference in levels is, however, only about 30-fold. The levels of expression on acetate or pyruvate are therefore still too high to allow the expression of

[1] M. F. Tuite, M. J. Dobson, N. A. Roberts, R. M. King, D. C. Burke, S. M. Kingsman, and A. J. Kingsman. *EMBO J.* **1,** 603

[2] R. Derynck, A. Singh, and D. V. Goeddel, *Nucleic Acids Res.* **11,** 1819 (1983).

[3] J. Mellor, M. J. Dobson, N. A. Roberts, M. F. Tuite, A. J. Kingsman, and S. M. Kingsman, *Gene* **24,** 1 (1983).

[4] C. R. Wood, M. A. Boss, J. H. Kenton, J. E. Calvert, N. A. Roberts, and J. S. Emtage, *Nature (London)* **314,** 446 (1985).

[5] S. J. Rothstein, C. M. Lazarus, W. E. Smith, D. C. Baulcombe, and A. A. Gatenby, *Nature (London)* **308,** 662 (1984).

[6] S. E. Adams, K. M. Dawson, K. Gull, S. M. Kingsman, and A. J. Kingsman, *Nature (London)* **329,** 68 (1987).

toxic proteins such as ricin. Derivatives of the basic *PGK* expression vectors have been prepared to allow regulated expression. All of the *PGK* expression vectors are based on a derivative of the endogenous multicopy plasmid, the 2-μm circle. The 2 μm-plasmid is normally present at 100–200 copies per cell. The levels of heterologous gene expression directed by the *PGK* promoter on these high-copy plasmids are usually about 1–5% of total cell protein. This is lower than the theoretical yields and lower than yields of the homologous protein, PGK, expressed from the same vector.[7] The major reason for this seems to be that transcription is less effective in the heterologous configuration because essential sequences in the *PGK* coding region are missing.[8] Many proteins are, however, turned over more rapidly than PGK and there is some influence of codon usage on mRNA levels.[9]

PGK Expression Vectors

An outline of the *PGK* promoter is shown in Fig. 1 (see Refs. 10 and 11 for sequence). It extends for 500 base pairs (bp) upstream from the initiating ATG and a number of key sequences have been identified. These include the upstream activator (UAS) which is located between −473 and −422.[12] The essential sequences are three repeats of the sequence CTTCC and the AC sequence which has homology to the RAP1/GRF binding site and to the RPG box in ribosomal protein genes.[13] An additional protein which we call M binds to nucleotides −495 to −520 in a carbon source-dependent manner, but it is not essential for activation.[14] There is also a functional heat-shock element in the promoter which mediates a transient 3-fold increase in expression after heat shock.[15] In addition, there is a TATA box which is not essential for efficient expression but is involved in

---

[7] J. Mellor, M. J. Dobson, N. A. Roberts, A. J. Kingsman, and S. M. Kingsman, *Gene* **33**, 215 (1985).

[8] J. Mellor, M. J. Dobson, A. J. Kingsman, and S. M. Kingsman, *Nucleic Acids Res.* **15**, 6243 (1987).

[9] A. Hoekema, R. A. Kastelein, M. Vasser, and H. A. deBoer, *Mol. Cell. Biol.* **7**, 2914 (1987).

[10] M. J. Dobson, M. F. Tuite, N. Roberts, A. J. Kingsman, and S. M. Kingsman, *Nucleic Acids Res.* **10**, 2625 (1982).

[11] R. A. Hitzeman, F. E. Hagie, J. S. Hayflick, C. Y. Chen, P. H. Seeburg, and R. Derynck, *Nucleic Acids Res.* **10**, 7791 (1982).

[12] J. Ogden, C. A. Stanway, S. Y. Kim, J. Mellor, S. M. Kingsman, and A. J. Kingsman, *Mol. Cell. Biol.* **6**, 4335, (1986).

[13] A. Chambers, Tsang, J., C. A. Stanway, A. J. Kingsman, and S. M. Kingsman *Mol. Cell. Biol.* **9**, 5516 (1989).

[14] J. Stanway, J. Mellor, J. E. Ogden, A. J. Kingsman, and S. M. Kingsman, *Nucl. Acids Res.* **15**, 6243 (1987).

[15] P. W. Piper, B. Curran, M. W. Davies, K. Hirst, A. Lockheart, J. Ogden, C. A. Stanway, A. J. Kingsman, and S. M. Kingsman, *Nucleic Acids Res.* **16**, 1333 (1988).

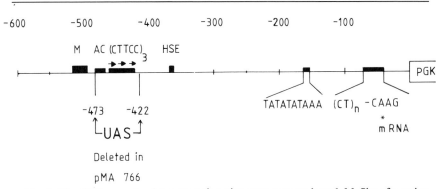

FIG. 1. The *PGK* promoter. Important cis-active sequences are boxed. M, Site of a major protein binding that may modulate expression but is not essential. AC, Activator core, a major component of the upstream activator sequence (UAS) that has homology to sequences in other promoters. HSE, Heat-shock response element. Three direct repeats of the sequences CTTCC are indicated.

determining the transcription start site[12] (unpublished observations) and a pyrimidine-rich tract which may be involved in determining the efficiency of initiation.[10]

The *PGK* coding region was removed by *Bal*31 deletion to nucleotide -2 and a synthetic *Bgl*II linker was added. The *PGK* gene was reconstructed by adding back the 3′ flanking region from the *Bgl*II site at position $+1157$ within the coding region to a *Hind*III site downstream of the transcription termination site. This produces a *PGK* expression cassette with the *PGK* promoter and terminator regions flanking a unique *Bgl*II expression site. This *Hind*III cassette was inserted into the *Hind*III site of the 2-μm shuttle vector pMA3a (originally called pMA3[10]). This shuttle vector has the 3.2 kilobase (kb) double *Eco*RI fragment containing *LEU2* from pJDB219[16] inserted into the *Eco*RI site of pBR322. This *LEU2* gene lacks a promoter and is referred to as *LEU2-d*. It is thought that the poor expression of this gene creates a selective advantage for cells that have a higher copy number of the plasmid. The final expression vector, called pMA91, is shown in Fig. 2.

To create a regulatable version of the *PGK* vector the UAS-*PGK* was deleted and replaced with the UAS from the *GAL1-10* gene.[17] The expression of the *GAL1-10* gene is repressed on glucose and induced by galactose by about 2000-fold. The UAS-*GAL* was isolated as a 144-bp *Rsa*I to *Alu*I fragment. This was converted to a *Bgl*II to *Bam*HI fragment by inserting it into the *Sma*I site of a derivative of pUC8 which had the *Eco*RI site converted to a *Bgl*II site (Fig. 3). This portable *GAL* UAS fragment was inserted into a deletion derivative of the *PGK* promoter called pMA766

[16] J. D. Beggs, *Nature (London)* **275**, 104 (1978).
[17] R. W. West, R. R. Yocum, and M. Ptashne, *Mol. Cell. Biol.* **4**, 2467 (1984).

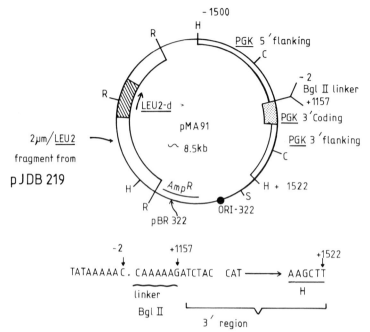

FIG. 2. The high-efficiency constitutive expression vector pMA91. Restriction sites H, HindIII; R, EcoRI; s, SalI; C, ClaI. The sequence flanking the expression site is shown. The figure is not to scale.

that lacks the UAS. pMA766 lacks nucleotides −422 to −473 and was created from joining a 5′ and 3′ Bal31 deletion series as described previously.[12] The PGK coding sequence was then removed by Bal31 deletion to nucleotide −7 and a synthetic BglII linker as described for pMA91 was added. The PGK gene was reconstructed by the addition of a PGK terminator fragment that extends from a synthetic BamHI site at codon 414 of PGK to the HindIII site downstream of the transcription stop. In this case the HindIII site was converted to a SalI site and digestion of the resulting plasmid and religation removed the sequences between this site and the SalI site in pBR322 to leave a unique SalI site. The final vector called pKV49 is shown in Fig. 3. The provision of a transcription terminator in both pMA91 and pKV49 is important to ensure the production of short discrete transcripts which are more stable than long transcripts.[7,21]

Both pMA91 and pKV49 are maintained at a copy number of 50 to 100 molecules per cell even in the absence of selection. In some cases, the addition of a heterologous coding region reduces both the copy number and stability of the plasmid. It is always essential to confirm the copy number of an expression plasmid to ensure that low protein yields are not

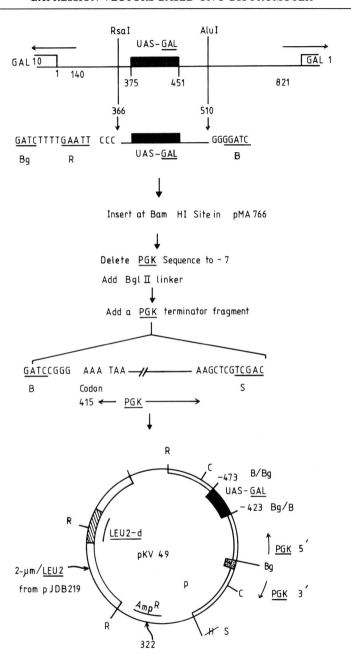

FIG. 3. Construction of a galactose-regulated derivative of the *PGK* promoter.

due to loss of plasmid. In general, strains containing expression plasmids are grown with selection for the *LEU2* gene. Very rarely, plasmid rearrangements or deletions are observed.[18] This can be checked by an appropriate restriction digest on DNA that has been prepared for copy number analysis (see below). Rearrangements usually only result when there is extreme selection against the expression configuration. The use of a regulated vector usually overcomes these problems because the expression of toxic proteins is delayed until the end of the culture period.

Methods

Media

Yeast selective medium (synthetic complete, SC): 6.7 g of Difco yeast nitrogen base without amino acids, 10 to 20 g of glucose per liter. For pyruvate as a carbon source, use 20 g/liter. For galactose as a carbon source, use 10 g liter. An amino acid mix can be prepared containing all the amino acids except the selected one. Alternatively, only individual amino acids required by the strains need be added.

Yeast rich medium: 20 g of peptone, 10 g of yeast extract, 20 g of glucose per liter (see Ref. 19).

Yeast Strains and Transformation

The yields of heterologous proteins can be markedly influenced by the host strain, but there are no good rules for predicting the most effective strain for a given product. The following strains have been useful in our laboratory, but we have not undertaken a comprehensive survey.

MD40-4c: α, *ura2 trp1 leu2-3 leu2-112 his3-11 his3-15*
MT302-1C: α, *ura-2 his3-15 leu2-3 leu2-112 arg5-6 ade1 trp1 pep4-3*
BJ2168: α, *leu2-3 leu2-112 ∧ trp1 ura3-52 prb1-1122 pep4-3 prc1-407* (obtained from Dr. E. Jones)
DBY745: α, *ade1-100 leu2-3 leu2-112 ura3-52*

MD40-4C will not grow on galactose and probably has a permease defect. This strain is not used for the pKV49 plasmids. The protease deficiencies of MT302-1C and BJ2168 may be important but they do not always give high yields. Plasmids are introduced into these strains by the standard polyethylene glycol (PEG)-mediated transformation procedure as

18 G. J. Belsham, D. G. Barker, and A. E. Smith, *Eur. J. Biochem.* **156**, 413 (1986).
19 D. C. Hawthorne and R. K. Mortimer, *Genetics* **45**, 1085 (1960).

described by Hinnen *et al.*[20] In the case of toxic proteins, no transformants are obtained with pMA91 on glucose or with pKV49 selected on galactose. It is standard procedure to select pKV49 transformants on glucose so that they are not expressing any foreign proteins during recovery from the transformation. Some variation in expression levels is seen between transformants, this is particularly marked when DBY745 is the host strain. It is standard practice to select two independent transformants for analysis.

Preparation of Heterologous Coding Sequences

It is usually advisable to trim the heterologous coding sequence to remove 5' and 3' flanking DNA. The *PGK* translation initiation environment is optimal when there is a low GC environment. It might be further optimized by creating the authentic environment, but these effects on translation efficiency appear to be marginal. This only becomes significant when maximum yields for minimum costs are required and 2-fold effects are then critical. The longer the transcript the more unstable it is likely to be,[21] and so trimming the 3' end is a sensible step to ensure maximum RNA levels.

Storage of Transformants

As soon as transformants have been checked glycerol stocks are established. The strain is inoculated into selective media at a high inoculum and grown overnight. An equal volume of sterile glycerol is added and, after thorough mixing, the culture is stored at $-70°$. Transformants on plates are rarely stored for more than 4 weeks. This is because selection for rearranged plasmids can occur. When repeated analyses are required, the glycerol stocks are used or a freshly transformed strain is produced.

Growth of Transformants

The doubling time of strains transformed with the vectors alone is about 2.5 hr on selective medium. When heterologous genes are expressed, these doubling times are usually increased. Doubling times range from about 3 hr as seen when interferon α is expressed to 12 hr when influenza virus hemagglutinin is expressed. Usually, there is an intermediate value of about 4 hr. Longer doubling times are difficult to work with because of the selection pressure for plasmid rearrangements. In these cases a regulated vector should always be used.

---

[20] A. Kinnen, J. B. Hicks, and G. R. Fink, *Proc. Natl. Acad. Sci. U.S.A.* **75,** 1929 (1978).
[21] K. Zaret and F. Sherman, *J. Mol. Biol.* **177,** 107 (1984).

To ensure reproducibility we use a standard inoculation protocol. An inoculum is taken from a single colony on a fresh (less than 1 week old) plate and resuspended in medium, counted, and used to inoculate a culture at 5×10^4 ml^{-1}. The culture is grown with vigorous orbital shaking to a density of $5-7 \times 10^6$ ml^{-1}.

Galactose Induction

The optimum conditions for induction seem to vary with the particular protein being expressed. We have used two protocols. Cells are grown in SC containing glucose to 4×10^6 cells ml^{-1}, washed in SC containing galactose (SC-GAL), and then resuspended in the same volume of SC-GAL for $12-16$ hr. Alternatively, cells are inoculated at 1×10^7 cells ml^{-1} into SC plus 0.3% (w/v) glucose and 1% (w/v) galactose and grown for 24 hr. It is best to analyze proteins during a time course of induction to select an optimum for each particular product.

Constitutive expression can be achieved with pKV49 by growing on 1% galactose throughout the culture period.

Labeling Cells

Cells are grown to 5×10^6 ml^{-1} in selective medium and then harvested and resuspended in fresh SC lacking either methionine or cysteine. After 1 hr at 30° cells are harvested and resuspended at 1×10^8 cells ml^{-1} in fresh medium containing 100 μCi ml^{-1} [^{35}S]Met (1450 Ci mmol^{-1}, Amersham, UK) or [^{35}S]Cys for 30 min. Cells are then chilled, washed once in phosphate buffer, and lysed as described below.

Preparation of Protein Extracts

Cells are grown to 5×10^6 ml^{-1}. All subsequent procedures are at 4° unless stated otherwise. Fifty milliliters of cells is harvested and washed in 6 ml of water and then in 1.5 ml of sodium phosphate buffer (25 mM, pH 7.0) containing 20 μl of 50 mM phenylmethylsulfonyl fluoride (PMSF) in an Eppendorf tube. The cell pellet is finally resuspended in 100 μl of phosphate buffer and 5 μl of PMSF. Sterile glass beads (40 mesh; BDH, UK) are added up to the meniscus; this requires about 8 g. The cells are lysed by vigorous vortexing for 1 min and then centrifuged to remove beads and insoluble material. A further 100 μl of resuspended cells is added to the beads and vortexing and centrifugation are repeated. The supernatants are pooled spun at 1200 g at 4° for 15 min, and the supernatant stored at -20°. This is referred to as the soluble fraction. Alternatively, the whole mixture is collected after allowing the glass beads to settle briefly. This is a total cell extract. The debris can be collected by washing the glass bead pellet with buffer, removing the supernatant, and then recentrifuging to collect the pellet. This is the "insoluble" fraction.

A typical protein profile is shown in Fig. 4. The coding region for human serum albumin (HSA) was inserted into pMA91 and pKV49 and the resulting vectors were introduced into DBY745. Cells were grown on glucose or on galactose and labeled with 50 μCi ml^{-1} [^{35}S]cysteine. Total protein extracts were prepared and analyzed by 10% SDS–PAGE. HSA was produced in high yields by the strain transformed with pMA91-HSA grown on glucose, but very little if any HSA was detected in the strain transformed with pKV49-HSA when grown on glucose. However, when the strains were grown on galactose, there were high yields with both the vectors, indicating that the *GAL* UAS derivative of the *PGK* promoter was tightly regulated by galactose. The yields of HSA are about 5% of total cell protein.

Preparation of RNA

Cells are grown as described above for protein extracts, washed twice in RNA extraction buffer (LET: 100 mM LiCl, 1 mM EDTA, (100 mM Tris, pH 7.4) and resuspended in LET at 5×10^9 cells ml^{-1}. The cells are lysed by the addition of glass beads and vortexing for 30 sec. One hundred microliters of phenol saturated with TNE (10 mM Tris, pH 7.4, 0.14 M NaCl, 1 mM EDTA) is added and the mixture is vortexed for 30 sec. Twenty microliters of 10× TNE/1% SDS, 100 μl of chloroform, and 100 μl of H$_2$O are added and the mixture is shaken vigorously for 1 min. The phases are separated by centrifugation at room temperature and the aqueous phase is reextracted four times with a 1:1 phenol:chloroform mix. The aqueous phase is precipitated with ethanol, and the pellet is washed in 80% ethanol and resuspended in water.[22] This RNA preparation can then be treated with RNase-free DNases (BRL, Gaithersburg, MD) and proteinase K by standard procedures (see Ref. 23 for further details), but this is not always necessary for simple Northern blot analysis.

Copy Number Determination

Total yeast DNA is prepared from each of the transformants according to the method of Cryer *et al.*[24] The DNA is digested with *Eco*RI, fraction-

[22] M. Dobson, J. Mellor, A. M. Fulton, N. A. Roberts, B. Bowen, S. M. Kingsman, and A. J. Kingsman, *EMBO J.* **3**, 1115 (1984).

[23] M. J. Dobson, M. F. Tuite, J. Mellor, N. A. Roberts, R. M. King, D. C. Burke, A. J. Kingsman, and S. M. Kingsman, *Nucleic Acids Res.* **11**, 2287 (1983).

[24] D. R. Cryer, R. Eccleshall, and J. Marmur, *in* "Methods in Cell Biology" (D. Prescott, ed.), Vol. 12, p. 39. Academic Press, New York, 1975.

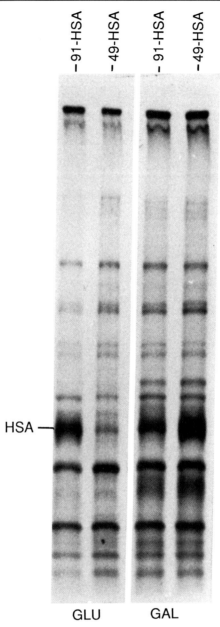

FIG. 4. Expression of human serum albumin in yeast. [^{35}S]Cysteine-labeled proteins from a total cell extract from yeast strain DBY745 containing either pMA91 or pKV49 carrying the HSA coding sequence were fractionated by gel electrophoresis. Strains are grown on glucose or galactose as indicated.

ated by agarose gel electrophoresis, and transferred to nitrocellulose by the procedure of Southern.[25] Filters are hybridized with a labeled probe that is common to either vector and with a probe that is specific for ribosomal DNA labeled to the same specific activity. A typical example is shown in Fig. 5. DNA from yeast transformed with pKV49-HSA and pMA91-HSA grown in glucose or galactose was hybridized with a *PGK* fragment isolated from pMA27[3] and a 1.8 kb *Eco*RI fragment of ribosomal DNA from plasmid pYIRG12.[26] This detects vector sequences and the chromosomal rDNA repeats. As these are known to be present in about 140 copies, the vector copy number can be established by comparison. In this case, the vectors are present in similar copy numbers, although consistently lower numbers of pKV49-HSA are obtained on galactose than on glucose. We have other observations that suggest that high levels of transcription often result in a reduction in plasmid copy number, although this has not been rigorously tested in our laboratory.

Conclusions

The *PGK*-based vectors have been used to express a wide range of different coding sequences from mammalian genomes, eukaryotic viruses, plants, and fungi. In most cases, proteins have been produced but the yields can be variable. Using the regulated vector pKV49 it is possible to obtain proteins at 1–5% of total cell protein in the majority of cases even when the product is toxic to the cell. It has not yet been possible to increase the efficiency of *PGK* promoter-directed heterologous transcription per template to approach that of homologous transcription.

In initial studies with a new coding sequence, it is essential to verify that the plasmid is present at a high copy number and that the correct mRNA is being produced. The hunt for the protein then begins with a Coomassie stain of total protein or of the soluble and insoluble fractions. Some proteins, e.g., HSA, are only detected in large amounts in the cell debris, although they are readily solubilized by additional procedures. Western blotting is more reliable for detecting specific proteins. It is possible that the protein may be turned over very rapidly, but it may be detectable after a very short pulse labeling with either cysteine or methionine, depending on which is most represented in the coding sequence. It is rare to find an extremely unstable protein, so if there is plenty of specific RNA and still no protein recheck the vector construction, particularly sequences of synthetic linkers. Once there is a hint of protein, it is possible to optimize expression

[25] E. Southern, *J. Mol. Biol.* **153,** (1975).
[26] T. D. Petes, L. M. Hereford, and K. G. Konstantin, *J. Bacteriol.* **134,** 295 (1978).

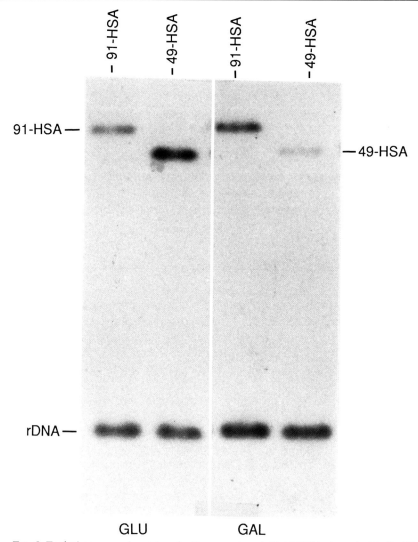

FIG. 5. Typical copy number determination experiment. Total DNA cleaved with *Eco*RI and hybridized with a mixture of rDNA- and pBR322-specific probes was subjected to Southern blotting. The DNA is from strains transformed with either pMA91 or pKV91 containing the *HSA* gene and grown on glucose or galactose.

by varying the growth stage of harvesting, varying the induction protocol, changing host strains, and mutagenizing production strains to select high producers. These mutant host strains are usually optimized for a particular product and so, unfortunately, this is not a generic approach.

Acknowledgments

Research into promoter structure and function and heterologous gene expression has been supported by grants from Celltech, Delta Biotechnology, Science and Engineering Research Council, and Agriculture and Forestry Research Council.

[28] Glyceraldehyde-3-phosphate Dehydrogenase-Derived Expression Cassettes for Constitutive Synthesis of Heterologous Proteins

By STEVEN ROSENBERG, DORIS COIT, and PATRICIA TEKAMP-OLSON

The 5′ flanking regions of several of the glycolytic genes of *Saccharomyces cerevisiae* have been used to drive the transcription of heterologous genes in yeast expression vectors. Glyceraldehyde-3-phosphate dehydrogenase (GAPDH) is one of the most abundant proteins in yeast, constituting as much as 5% of the soluble protein in glycolytically growing cells.[1] There are three *GAPDH* genes which are expressed to different extents.[2-5] The most prevalent GAPDH protein is derived from the *TDH3* or *GAP491* gene described by McAlister and Holland,[5] and the promoter from this gene is the one we and others have used to direct the high-level constitutive expression of heterologous genes. Examples of proteins expressed using this system are hepatitis B surface and core antigens,[6-8] wild-type and mutant

[1] E. G. Krebs, *J. Biol. Chem* **200**, 471 (1953).

[2] J. P. Holland, and J. Holland, *J. Biol. Chem.* **254**, 9839 (1979).

[3] J. P. Holland and M. J. Holland, *J. Biol. Chem.* **255**, 2596 (1980).

[4] J. P. Holland, L. Labienciec, C. Swimmer, and M. J. Holland, *J. Biol. Chem.* **258**, 5291 (1983).

[5] L. McAlister and M. J. Holland, *J. Biol. Chem.* **260**, 15019 (1985).

[6] G. A. Bitter and K. M. Egan, *Gene* **32**, 263 (1984).

[7] P. Valenzuela, D. Coit, R. Hallewell, U. Heberlein, O. Laub, A. Medina, S. Rosenberg, and P. Tekamp-Olson, *in* "Medical Virology III" (M. de la Maza and E. M. Peterson, eds.), p. 313. Elsevier, New York, 1984.

[8] P. J. Kniskern, A. Hagopian, D. Montgomery, P. Burke, N. R. Dunn, K. J. Hofmann, W. J. Miller, and R. W. Ellis, *Gene* **46**, 135 (1986).

α_1-antitrypsins (α_1-AT),[9] human superoxide dismutase (hSOD),[10] and epidermal growth factor.[11] In at least two of these cases, hSOD and core antigen, the levels of expression are more than 30% of the soluble yeast protein, comparable to levels seen in the best *Escherichia coli* expression systems.

We describe here the methods used to construct expression "cassettes" containing both 5' and 3' flanking sequences from the *GAP491* gene. Although the major advantage of this system is the promoter, which is likely among the strongest RNA polymerase II promoters in yeast, the 3' flanking sequences can also contribute substantially to expression levels. In addition, unusual effects of promoter length and plasmid flanking sequences on gene expression from the *GAPDH* promoter constructs are discussed.

Isolation of *GAP491* Gene

Restriction endonucleases and DNA modifying enzymes were obtained from New England Biolabs (Boston, MA) or Bethesda Research Labs (Gaithersburg, MD) and were used according to the manufacturer's instructions. Standard DNA manipulations were done according to Maniatis *et al.*[12] Holland and Holland reported the sequence of 150 base pairs (bp) of 5' flanking sequence as well as the entire coding sequence for the *GAP491* gene;[2,3] Bitter and Egan extended this to include 675 bp of 5' flanking sequence.[6] We prepared poly(A)$^+$ RNA from yeast strain A364A (*Mata, ade 1, ade2, ura1, his7, lys2, tyr1, gal1*) grown glycolytically and constructed a cDNA library inserted into the dG-tailed *Pst* site of pBR322. The cDNA was made using oligo(dC) tailing of the cDNA using terminal transferase for annealing into the vector. First-strand cDNA was labeled and used to screen 1000 transformants at low C_0t values. Positives were analyzed by restriction mapping and those consistent with the *GAPDH* gene map were subjected to sequence analysis. One clone was found to contain a full-length cDNA of ~1200 bp.

This cDNA was labeled and used as probe to screen a yeast genomic library in λ Charon 28.[13] The yeast *GAPDH* gene from one of these

[9] J. Travis, M. Owen, P. George, R. Carrell, S. Rosenberg, R. A. Hallewell, and P. J. Barr, *J. Biol. Chem.* **260**, 4384 (1985).

[10] R. A. Hallewell, R. Mills, P. Tekamp-Olson, R. Blacher, S. Rosenberg, F. Otting, F. R. Masiarz, and C. J. Scandella, *Bio Technology* **5**, 363 (1987).

[11] M. S. Urdea, J. P. Merryweather, G. T. Mullenbach, D. Coit, U. Heberlein, P. Valenzuela, and P. J. Barr, *Proc Natl. Acad. Sci. U.S.A.* **72**, 1317 (1983).

[12] T. Maniatis, E. F. Fritsch, and J. Sambrook, *in* "Molecular Cloning: A Laboratory Manual." Cold Spring Harbor Laboratory, Cold Spring Harbor, New York, 1982.

[13] F. R. Blattner, B. G. Williams, A. E. Blechl, K. Denniston-Thompson, H. E. Faber, L. Furlong, D. J. Gronwald, D. O. Kiefer, D. D. Moore, J. W. Schumm, E. L. Sheldon, and O. Smithies, *Science* **196**, 161 (1977).

genomic clones was subcloned into pBR322 either as a 2.1-kb *Hind*III fragment (pGAP-1) or a 3.5-kb *Bam*HI fragment (pGAP-2), and was subsequently shown to correspond to the *GAP491* gene of Holland and Holland.[2,3]

Preparation of *GAPDH* Promoter and Terminator Fragments

The sequence of the 385 bp of 5' flanking sequence from −1060 to −675 of the *GAP491* gene which has not been previously published, is shown in Fig. 1. We have found three differences in the 5' flanking sequence from −675 to +1 from that reported by Holland and Holland and Bitter and Egan[2,6]: T for C at −8, G for T at −185, and a G insertion at −358. These differences are likely due to strain polymorphisms or to sequencing artifacts.

In order to construct *GAPDH* promoter fragments, the 460-bp *Hin*fI fragment encompassing the initiation codon was isolated from pGAP-1 and treated with *Bal*31 exonuclease so as to remove 50–100 bp. After repairing the ends with Klenow polymerase and deoxynucleoside triphosphates (dNTPS), *Hind*III linkers (CAAGCTTG) were ligated on, the mixture digested with *Hind*III, gel isolated, and ligated into the *Hind*III site of pBR322. One clone, pGAP128, had retained the sequences from −390 to −27, while another, pGAP396, contained sequences from −375 to −3.

```
-1060
AAGCTTACCA GTTCTCACAC GGAACACCAC TAATGGACAC AAAATTCGAAA TACTTTGACC
HindIII
-1000
CTATTTTCGA GGACCTTGTC ACCTTGAGCC CAAGAGAGCC AAGATTTAAA TTTTCCTATG

-940
ACTTGATGCA AATTCCCAAA GCTAATAACA TGCAAGACAC GTACGGTCAA GAAGACATAT

-880
TTGACCTCTT AACTGGTTCA GACGCGACTG CCTCATCAGT AAGACCCGTT GAAAAGAACT

-820
TACCTGAAAA AAACGAATAT ATACTAGCGT TGAATGTTAG CGTCAACAAC AAGAAGTTTA

-760
ATGACGCGGA GGCCAAGGCA AAAAGATTCC TTGATTACGT AAGGGAGTTA GAATCATTTT

-700                        -675
GAATAAAAAA CACGCTTTTT CAGTTCGA
                         TaqI
```

FIG. 1. Sequence of the 5' flanking region of the *GAPDH491* gene from the *Hind*III site at −1060 to the *Taq*I site at −675 is shown. The sequence was determined using the chain terminator method after cloning the 1060-bp *Hind*III fragment of pGAP347 in M13mp8 and mp9. The sequence from −675 to +1 has been reported previously.[2,6]

The former was flush ended with Klenow and dNTPS and inserted into the *Sma*I site of ptac5,[14] adding 7 bp of the linker sequence GATCCCC to the 5' end of the fragment, yielding plasmid ptac5GAP128. This was digested with *Bam*HI and *Taq*I, and a 372-bp fragment comprising the promoter sequence from −390 to −26 and the linker sequence was isolated. This fragment was ligated to a mixture of complementary oligonucleotides with *Taq*I and *Sal*I overhangs, whose sequence is shown below, encoding the region from −26 to −3 of the promoter and a small polylinker ending with a *Sal*I site:

*Taq*I   *Nco*I   *Eco*RI   *Sal*I

CGAATAAACACACATAAACAAACAACCATGGGAATTCGTTAGG

After gel isolation, the resulting *Bam–Sal* fragment was subcloned between the *Bam* and *Sal* sites of pBR322 to yield plasmid pGAPNRS.

A longer *GAPDH* promoter fragment was isolated and engineered as follows. pGAP1 was digested with *Hha* and *Hind*III and the 700-bp fragment corresponding to nucleotides −1060 to −360 was isolated. pGAP396 was digested with the same two enzymes and the 360-bp fragment corresponding to −360 to −3 was isolated. These fragments were ligated together, the mixture digested with *Hind*III, and then inserted into the *Hind*III site of pBR322 to yield plasmid pGAP347 containing a 1060-bp *GAPDH* promoter. This promoter was subsequently further engineered to preserve the *GAPDH* sequence just 5' of the ATG, as in pGAPNRS (see above).

A 900-bp *Sal–Bam* fragment corresponding to the last 10 amino acids of the *GAP491* gene and 870 bp of 3' flanking sequence was isolated from plasmid pGAP2. This was subsequently used as a *GAPDH* "terminator" fragment in all constructions.

Construction of *GAPDH* Expression Cassette Plasmids

Given the large size and complex restriction maps of the yeast–*E. coli* shuttle vectors we utilize, such as pC1/1,[15] we chose to develop a series of subcloning vectors based on pBR322. These vectors would enable one to clone a gene of interest between the transcription signals derived from the *GAP491* gene. The expression "cassette" thus constructed, consisting of the promoter, gene, and terminator, could then be easily moved into the shuttle vectors and subsequently transformed into yeast.

[14] R. A. Hallewell, F. M. Masiarz, R. C. Najarian, J. P. Puma, M. R. Quiroga, A. Randolph, R. Sanchez-Pescador, C. J. Scandella, B. Smith, K. S. Steimer, and G. T. Mullenbach, *Nucleic Acids Res.* **13,** 2017 (1985).
[15] S. Rosenberg, P. J. Barr, R. C. Najarian, and R. A. Hallewell, *Nature (London)* **312,** 77 (1984).

The plasmid pBR322 was digested with *Eco*RI and *Sal*I, the sites filled-in with Klenow polymerase and dNTPS, and *Bam* linkers (GGATCCGGATCC) ligated on. The mixture was then digested with *Bam*HI and the 3.7-kb linear fragment isolated, religated, and transformed into *E. coli*. The resulting plasmid, pPRΔR1-*Sal*, has a single *Bam*HI site and is missing the 650 bp of the tetracycline resistance *(tet)* gene between the *Eco*RI and *Sal*I sites. Neither of these sites was regenerated by the *Bam*HI linker used. Plasmid pGAPNRS was digested with *Bam* and *Sal* and the ~400-bp promoter and linker fragment isolated. This was ligated to the *Sal–Bam* terminator fragment described above, and the mixture was digested with *Bam* and inserted into the *Bam* site of pBRΔR1-*Sal* to yield plasmid pPGAP.[9] This plasmid was subsequently modified to yield the plasmids pPGAP2 and pPGAP3; these differ from the original (Fig. 2). The linker sequence for these two plasmids is

        *Nco*I   *Eco*RI   *Bgl*II

        CCATGGGAATTCGTTAGGTCGAGATCTCGAC

They differ in that the cassettes can be exised with *Bam*HI (pPGAP2) or *Sal*I (pPGAP3). We thus have a set of plasmids into which a gene can be easily cloned between the sequences required for efficient transcriptional initiation and termination in yeast.

Effects of Promoter Length and Orientation on Expression

We first utilized the *GAPDH* promoter to drive transcription of the hepatitis B surface antigen gene, the first antigen used successfully in a recombinant DNA-based human vaccine.[16,17] We constructed two expression cassettes containing either the long (1060 bp) or short (400 bp) promoters fused to a 840-bp surface antigen gene encompassing the coding sequence and 120 bp of 3′ untranslated sequence. For these constructs, we used a fragment of the yeast *ADC1* gene 3′ untranslated region as a terminator,[18] instead of the *GAPDH* fragment described above. Subsequent studies showed these fragments to be functionally equivalent. The expression cassettes were inserted into the unique *Bam*HI site in the tetracycline resistance gene of pC1/1.[15] This shuttle vector is pJDB219 in which the bacterial sequences are pBR322 instead of pMB9.[19] Four result-

[16] P. Valenzuela, M. Quiroga, J. Zaldivar, P. Gray, and W. J. Rutter, *in* "Animal Virus Genetics" (B. N. Fields, R. Jaenisch, and C. F. Fox, eds.), p. 55. Academic Press, New York, 1980.

[17] W. J. McAleer, E. B. Buynak, R. Z. Maigetter, D. E. Wampler, W. J. Miller, and M. R. Hilleman, *Nature (London)* **298,** 347 (1984).

[18] J. Bennetzen and B. D. Hall, *J. Biol. Chem.* **257,** 3018 (1982).

[19] J. D. Beggs, *Nature (London)* **275,** 104 (1978).

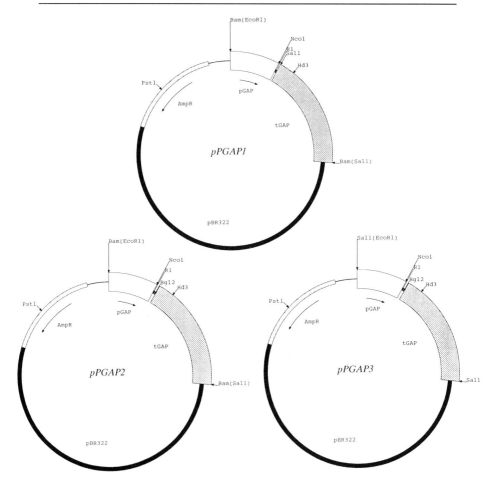

Fig. 2. Structure of plasmids pPGAP1, 2, and 3. Each consists of a 3.7-kb portion of pBR322 into which the 400-bp *GAPDH* promoter and 915-bp *GAPDH* terminator sequences have been inserted as described in the text. They differ by which restriction enzymes can mobilize the expression cassettes (*Bam*HI or *Sal*I) and by the sites in the linker between the promoter and terminator.

ing plasmids, containing expression cassettes in the two possible orientations with respect to the tetracycline resistance gene, were obtained (Fig. 3). We transformed yeast strain 2150-2-3 (*Mat **a**, leu2, ade1 nib6*[+] [cir°]) with the plasmids, selected leucine prototrophs, grew expression cultures in YEPD (yeast extract, peptone, dextrose) medium, and assayed for surface antigen particle production in cell lysates.[20,21] The amount of particle was determined with an enzyme-linked immunosorbent assay (ELISA) (Aus-

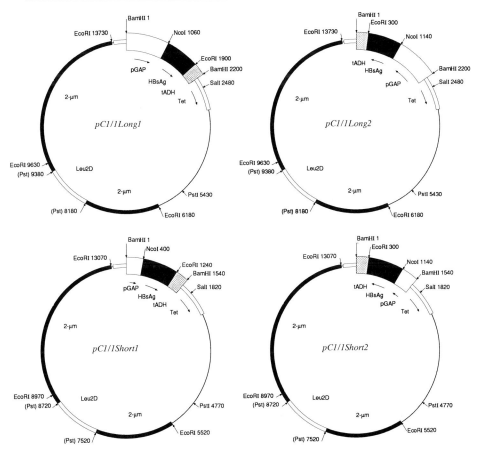

FIG. 3. Long and short *GAPDH* promoter plasmids for expression of hepatitis B surface antigen. The four plasmids are derivatives of pC1/1,[15] into which a *Bam*HI expression cassette has been inserted. The plasmids differ in that pC1/1Long1 and 2 have the 1060-bp promoter, while pC1/1Short1 and 2 use the 400-bp promoter. In addition, pC1/1Long1 and Short1 have transcription of the expression cassette in the same orientation as the *tet* gene, whereas their counterparts are in the anti-*tet* direction.

zyme II; Abbott Laboratories, North Chicago, IL) and the results are summarized in Table I.

For the long promoter plasmids, expression is essentially independent of orientation, and corresponds to ~ 0.1% of the soluble protein. This result confirms that of Bitter and Egan,[6] that no more than 675 bp of 5′ flanking sequence are required for efficient transcription. When the shorter

[20] A. Hinnen, J. B. Hicks, and G. R. Fink, *Proc. Natl. Acad. Sci. U.S.A.* **75**, 1929 (1978).
[21] N. N. Bradford, *Anal. Biochem.* **72**, 248 (1976).

TABLE I

EXPRESSION OF HEPATITIS B SURFACE ANTIGEN FROM
SHORT AND LONG GAPDH PROMOTER CONSTRUCTS[a]

Promoter (bp)	Orientation[b]	Transformant	HSaAg (μg/mg protein)
1060	+	1	0.90
1060	−	1	0.98
		2	0.80
400	+	1	0.01
		2	0.02
400	−	1	1.55
		2	1.3

[a]Yeast transformants were obtained using the method of Hinnen et al.[20] Independent transformants were grown in leucine selective medium to saturation and then diluted 1:25 into YEPD medium. Expression cultures were grown to mid-log phase ($A_{650\ nm} = 1-2$), harvested, and lysed with glass beads. Surface antigen activity was determined as described in the text; protein concentration was measured by the Coomassie blue dye binding method.[21]
[b]Orientation of the direction of transcription of the expression cassettes with respect to the tetracycline resistance gene. + indicates the same orientation as the *tet* gene.

promoter (400 bp) is used in the same orientation as the *tet* gene, expression is reduced 50- to 100-fold as compared to the long promoter constructs. Surprisingly, this defective expression is restored in the anti-*tet* orientation, where levels at least as high as for the long promoter constructs are obtained.

Table II shows that a similar orientation effect is seen in the expression of α_1-antitrypsin using the pPGAP1 expression cassette.[9] The anti-*tet* orientation yields ~ 10-fold more active inhibitor than the *tet* orientation plasmid, and Western analysis shows that this accurately reflects the total amount of α_1-AT related protein synthesized (S. Rosenberg, unpublished observations). Thus, these effects are not limited to hepatitis B surface antigen expression, and they suggest that eliminating the *GAPDH* promoter sequence upstream of -400 deletes or truncates a sequence necessary for maximum transcription, at least in one orientation.

In order to test this directly, we isolated RNA from 2150-2-3 cells transformed with the long and short promoter–HBsAg constructs in the *tet* orientation. Samples were blotted in duplicate and hybridized with either a 3.2-kb *Eco*RI fragment encompassing the entire HBsAg coding sequence,[16] or a 900-bp *Xba–Sal* fragment of the *GAPDH-491* gene.[2] The

TABLE II

EXPRESSION OF α_1-ANTITRYPSIN FROM SHORT
GAPDH PROMOTER CONSTRUCTS[a]

Promoter orientation[b]	Transformant	α_1-AT[c]
+	1	0.17
	2	0.11
−	1	0.70
	2	1.2

[a]Expression plasmids containing the α_1-AT cDNA fused to the NcoI site in plasmid pPGAP1 were constructed as described in Travis et al.[9] The expression cassettes were inserted into the BamHI site of pC1/1. Yeast strain AB103.1 (Mat a, leu2-3 leu2-112 ura3-52 his4-580 pep4-3 [cir⁰]) was transformed to leucine prototrophy, transformants were grown in selective medium, and then diluted 1:25 into YEPD for expression.

[b]The orientation of the promoter in pC1/1 is indicated with respect to the tetracycline resistance gene.

[c]The activity of the α_1-AT was measured by human leukocyte elastase inhibition as described in Rosenberg et al.[15]

results of these Northern blots are shown in Fig. 4. Clearly, the steady-state level of HBsAg mRNA is greatly reduced in the short promoter construct as compared to the full-length sequence. In addition, the amount of *GAPDH* mRNA from the chromosomal genes is markedly decreased in the long but not the short promoter or control plasmid transformants. These results suggest competition between the chromosomal *GAPDH* genes and the plasmid copies containing the long promoter for a limiting transcription factor. It seems likely, given the structure of other yeast promoters,[22] that the short promoter deletes or inactivates a *GAPDH-491* upstream activation site (UAS). We have shown that this is likely since functional hybrid promoters utilizing the *Gal1, Gal10, PHO5,* and *ADH2* UAS regions with the *GAPDH* short promoter can be constructed (S. Rosenberg and P. Tekamp-Olson, unpublished, and[23]). These hybrids yield regulated promoters of comparable strength to the intact long promoter.

[22] L. Guarente, *Cell* **36**, 799 (1984).
[23] L. S. Cousens, J. S. Shuster, C. Gallegos, L. Ku, M. M. Stempien, M. S. Urdea, R. Sanchez-Pescador, A. Taylor, and P. Tekamp-Olson, *Gene* **61**, 265 (1987).

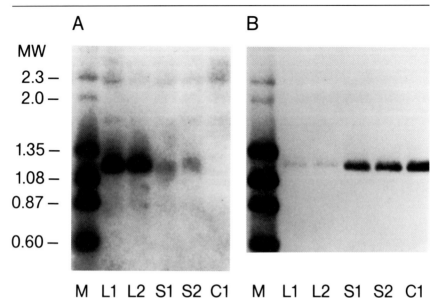

Fig. 4. Northern blot analyses of *GAPDH* promoter-driven plasmid transformants. Plasmids pC1/1Long1 and Short1 and the pC1/1 vector were transformed into yeast strain 2150-2-3 and leucine prototrophs selected. Total RNA was prepared from transformants grown in leucine selective media, and 8 μg of each RNA preparation was electrophoresed, transferred to filters, and hybridized with a probe specific for HBsAg (A) or GAPDH (B). Lanes M contain molecular weight markers ($\times 10^{-3}$), lanes C1 contain pC1/1 controls, lanes L1 and L2 are from pC1/1Long 1 transformants, and lanes S1 and S2 are from pC1/1Short1 transformants.

What then is the role of the vector sequences in stimulating short promoter expression? We have shown that the vector sequences required for this effect are not directly flanking the *GAPDH* promoter sequences (S. Rosenberg and P. Tekamp-Olson, unpublished), and that this phenomenon is observed with other truncated constitutive yeast promoters as well. A similar phenomenon has been mentioned by Guarente with respect to the *CYC1* promoter.[24] Although vector sequence "artifacts" of this type interfere with the study of the regulation of yeast genes, they can be very useful in the expression of heterologous genes.

Conclusion

We have described the construction of expression cassette plasmids utilizing the strong *GAPDH-491* promoter. These cassettes, when transferred to shuttle vectors and transformed into yeast, can yield high levels of

[24] L. Guarente, this series, Vol. 100, p. 181.

heterologous protein production, ranging from 0.1 to > 30% of the soluble cell protein. Dramatic effects of vector sequences on promoter strength have been observed in complete 2-μm vectors such as pC1/1, suggesting that it is always a good idea to examine flanking sequence effects on expression. Other glycolytic gene promoters are likely to show similar effects. Although efficient transcription of a cloned gene is a necessary prerequisite for high-level expression, it is by no means sufficient. Other factors, including protein toxicity necessitating regulated expression, strain effects, and protein folding and stability, need to be taken into account in order to address carefully the problems in heterologous protein expression.

Acknowledgments

We thank Pablo Valenzuela, Ulrike Heberlein, Rick Najarian, Qui Lim Choo, and Carol Gallegos for contributions to this work and the support of all our colleagues at Chiron.

[29] Superimposition of Temperature Regulation on Yeast Promoters

By ANDRZEJ Z. SLEDZIEWSKI, ANNE BELL, CARLI YIP, KIMBERLY KELSAY, FRANCIS J. GRANT, and VIVIAN L. MACKAY

The expression of heterologous proteins in the yeast *Saccharomyces cerevisiae* has become a routine procedure. Nevertheless, various aspects of this complex process are poorly understood and need further clarification. Transcription and transcriptional regulation have received a lot of attention[1] since it was postulated that an abundance of specific messenger RNA is necessary for efficient expression of foreign proteins in yeast.[2,3] However, the data indicate that yeast strains in which strong constitutive promoters were employed to express certain foreign proteins at high levels can be at a selective disadvantage relative to low- or nonexpressing strains.[4,5] The technology of regulated promoters, which was expected to circumvent this problem, is often difficult to transfer from the laboratory to large-scale

---

[1] C. Y. Chen, H. Oppermann, and R. A. Hitzeman, *Nucleic Acid Res.* **12**, 8951 (1984).

[2] M. F. Tuite, M. J. Dobson, N. A. Roberts, R. M. King, D. C. Burke, S. M. Kingsman, and A. J. Kingsman, *EMBO J.* **1**, 603 (1982).

[3] G. A. Bitter and K. M. Egan, *Gene* **32**, 263 (1984).

[4] V. Price, D. Mochizuki, C. J. March, D. Cosman, M. C. Deeley, R. Klinke, W. Clevenger, S. Gillis, P. Baker, and D. Urdal, *Gene* **55**, 287 (1987).

[5] V. L. MacKay, C. Yip, A. Bell, and A. Z. Sledziewski, unpublished observations.

fermentation.[6-8] Another problem encountered by several laboratories is that attempts to produce high levels of heterologous proteins through the yeast secretory pathway by using strong promoters and high-copy-number vectors frequently decreases rather than increases overall yields or quality of the final product.[9,10] For many products, it would be advantageous to obtain a range of messenger RNA levels from one expression cassette in order to assess product quality and quantity at various mRNA levels. There is also clearly a need for a yeast promoter system which can be easily regulated in large-scale fermentation, as well as in the laboratory. The temperature-regulated promoters described in this article offer both of these advantages, and thus temperature regulation has potential for use in the large-scale fermentation of heterologous proteins. Cultures with temperature-regulated expression units can be grown rapidly to high cell density at the restrictive temperature (with the promoter off), then synthesis of the heterologous protein can be induced by a simple temperature shift without additions to, or modifications of, the culture medium. Furthermore, as shown here, intermediate levels of expression can be achieved by adjusting the growth temperature between the extremes of the fully restrictive and permissive temperatures. This promoter "rheostat" may be particularly useful in creating production strains with an optimal balance between the expression level of heterologous protein and the secretory capacity of the host.

Principle of Method

Mating type α cells of the yeast *S. cerevisiae* contain but do not express a class of genes that are transcribed only in mating-type **a** cells. Promoters of these **a**-specific genes contain a 31-bp operator sequence[11,12] that binds the repressor protein encoded by the *MATα2* gene expressed only in α cells.[11] It has recently been demonstrated that another protein, GRM, which binds to the *MATα2* operator, is a necessary component of the *MATα2* repressor complex and is present in all three yeast cell types: **a**, α, and **a**/α diploids.[13] The *MATα2* operator sequence also serves as an up-

[6] A. Miyanohara, A. Toh-e, C. Nozaki, F. Hamada, N. Ohtomo, and K. Matsubara, *Proc. Natl. Acad. Sci. U.S.A.* **80**, 1 (1983).

[7] R. A. Kramer, T. M. DeChiara, M. D. Schaber, and S. Hilliker, *Proc. Natl. Acad. Sci. U.S.A.* **81**, 367 (1984).

[8] S. Rosenberg, P. J. Barr, R. C. Najarian, and R. A. Hallewell, *Nature (London)* **312**, 77 (1984).

[9] R. A. Smith, M. J. Duncan, and D. T. Moir, *Science* **229**, 1219 (1985).

[10] J. F. Ernst, *DNA* **5**, 483 (1986).

[11] A. D. Johnson and I. Herskowitz, *Cell* **42**, 237 (1985).

[12] A. M. Miller, V. L. MacKay, and K. Nasmyth, *Nature (London)* **314**, 598 (1985).

[13] C. A. Keleher, C. Goutte, and A. D. Johnson, *Cell* **53**, 927 (1988).

stream activation site (UAS) for **a**-specific genes.[14,15] *MATα2* repression can be imposed on other genes by the insertion of the operator sequence into their promoters, which offers the possibility of constructing a set of yeast promoters that would be subject to mating-typing regulation.

Promoters containing the *MATα2* operator sequence can be expressed at one temperature and repressed at another temperature if the synthesis or activity of the *MATα2* repressor protein can be made temperature dependent. For example, a temperature-sensitive mutation in the *MATα2* gene[16] allows low-level transcription of the **a**-specific gene *STE2* in α cells grown at the restrictive temperature.[17] Alternatively, temperature-dependent synthesis of the *MATα2* repressor can be achieved in **a** cells carrying a temperature-sensitive mutation in any of the four *SIR* (silent information regulator) genes whose products are required for the repression of *HMRa* and *HMLα* loci,[18,19] which are exact, but not expressed copies of the *MATa* and *MATα* information present at the mating-type locus. As illustrated in Fig. 1, the *sir3-8* mutation renders the Sir3 protein inactive at 35°; hence, the *HMRa* and *HMLα* silent loci are no longer repressed, leading to the synthesis of several mating-type regulatory proteins, including the *MATα2* repressor. Consequently, a promoter containing the *MATα2* operator(s) should be repressed at 35°. At the permissive temperature, 25°, the Sir3 protein is functional, which blocks transcription of *HMRa* and *HMLα* and prevents the synthesis of the *MATα2* repressor, thereby allowing expression from promoters controlled by the *MATα2* system.

Materials and Reagents

Yeast Strains

The *S. cerevisiae* strain XK1-C2 (*MATa sir3-8 leu2-3,112 trp1 ura3 Δpep4::URA3*) was derived by a standard genetic cross of a suitably marked strain. Strains bearing most of the mutations in this strain are available from the Yeast Genetics Stock Center (Berkeley, CA). The one-step gene replacement method of Rothstein[20] was used to obtain the deletion/gene disruption mutation for *pep4::URA3*. The wild-type

[14] J. W. Kronstad, J. A. Holly, and V. L. MacKay, *Cell* **50**, 369 (1987).
[15] A. Bender and G. F. Sprague, Jr., *Cell* **50**, 681 (1987).
[16] T. R. Manney, P. Jackson, and J. Meade, *J. Cell Biol.* **96**, 1592 (1983).
[17] A. Hartig, J. Holly, G. Saari, and V. L. MacKay, *Mol. Cell. Biol.* **6**, 2106 (1986).
[18] I. Herskowitz and Y. Oshima, *in* "The Molecular Biology of the Yeast *Saccharomyces:* Life Cycle and Inheritance" (J. N. Strathern, E. W. Jones, and J. R. Broach, eds.), p. 181. Cold Spring Harbor Laboratory, Cold Spring Harbor, New York, 1981.
[19] J. Rine and I. Herskowitz, *Genetics* **116**, 9 (1987).
[20] R. J. Rothstein, this series, Vol. 101, p. 202.

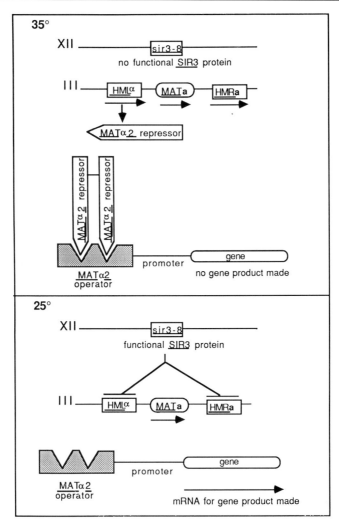

FIG. 1. Representation of temperature-dependent regulation of hybrid promoters. See text for explanation.

PEP4[21,22] gene was cloned and sequenced, and the availability of restriction sites determined internal regions of the promoter and/or the coding sequence that could be replaced with exogenous DNA sequences encoding

[21] G. Ammerer, C. P. Hunter, J. H. Rothman, G. S. Saari, L. A. Valls, and T. H. Stevens, *Mol. Cell. Biol.* **6,** 2490 (1986).

[22] C. A. Woolford, L. B. Daniels, F. J. Park, E. W. Jones, J. N. Van Arsdell, and M. A. Innis, *Mol. Cell. Biol.* **6,** 2500 (1986).

gene products for which there was a selection or screen. The *URA3* gene permits a direct selection of Ura$^+$ transformants in a host strain with a nonreverting *ura3* mutation.

Plasmids

pMVR1, a plasmid containing the *TPI1* promoter was constructed at ZymoGenetics from plasmids pTPI1C10[23] and pFATPOT (ATCC 699). Two plasmids containing *ADH2* promoters with small deletions, *ADH2-7* and *ADH2-10,* were obtained from E. T. Young (University of Washington).[24] Plasmid pDBC3T1 is a yeast–*Escherichia coli* shuttle vector containing the *S. cerevisiae TRP1* and *CEN3* sequences and pBR322 sequences. Because of its stability and low copy number in yeast (1–2 copies per genome), experimental variation due to copy number fluctuation is minimized.

Yeast Media

Standard yeast media were used throughout.[25] YEPD [1% (w/v) yeast extract, 2% (w/v) peptone, 2% (w/v) glucose, 60 mg/liter adenine, 40 mg/liter leucine ± 1.8% agar] is a rich complex medium used for routine growth and maintenance. Synthetic, chemically defined media contained 0.67% yeast nitrogen base (Difco, Detroit, MI) without amino acids, 2% glucose, appropriate nutritional supplements[25] ± 1.8% agar. Yeast strains transformed with plasmids derived from vector pDBC3T1 were grown either on synthetic medium lacking tryptophan or on YNB + CAA medium which contains 0.67% Difco-yeast nitrogen base without amino acids, 0.5% casamino acids (which is nearly devoid of tryptophan and therefore selective for presence of the *TRP1* gene on the vector), and either 5% glucose or 2% (v/v) ethanol as the carbon source. Unless noted, yeast strains were grown at 30°.

Enzymes

Glusulase (Du Pont) was used to prepare spheroplasts for yeast transformation. Restriction endonucleases and other DNA-modifying enzymes were purchased from standard suppliers and used according to their specification.

[23] T. Alber and G. Kawasaki, *J. Mol. Appl. Genet.* **1**, 419 (1982).
[24] J. Shuster, J. Yu, D. Cox, R. V. L. Chan, M. Smith, and E. Young, *Mol. Cell. Biol.* **6**, 1894 (1986).
[25] F. Sherman, G. R. Fink, and C. W. Lawrence, "Methods in Yeast Genetics." Cold Spring Harbor Laboratory, Cold Spring Harbor, New York, 1979.

General Plasmid Constructions

Plasmids were constructed by standard methods of DNA preparation, restriction digestion, DNA fragment isolation, and ligation. Oligonucleotides were synthesized with an Applied Biosystems 380A synthesizer, using β-cyanoethylphosphoramidites and CPG-LCAA solid support. See text below for details of specific constructions.

DNA Sequence Analysis

All DNA sequencing was performed by the dideoxy chain termination method.[26] The MTR series promoters were subcloned into M13 vectors[27] and sequenced with a universal primer. Double-stranded DNA fragments (*Bgl* II – *Bam*HI) of the SXR series promoters were sequenced directly with a synthetic primer near the linker site. Some orientations of linkers (head to head, tail to tail) are capable of forming stable hairpin structures which result in sequencing gel artifacts (i.e., compressions).

Transformations

Escherichia coli RR1 and *S. cerevisiae* strains were transformed by established procedures as described previously.[28,29]

β-Galactosidase Assays

Plate Assays[30]. Single yeast colonies or streaks were lifted onto a nitrocellulose filter by placing the filter for 10 min on the top of the master plate. The filter was peeled off and dipped, with the colonies up, into a tray with liquid nitrogen for about 10–15 sec. (The nitrocellulose becomes very fragile and requires great care while handling.) The filter was then transferred, colonies up, into the top of a petri dish containing a Whatman filter presoaked with 3–4 ml of Z buffer [0.06 M Na$_2$HPO$_4$, 0.04 M NaH$_2$PO$_4$, 0.01 M KCl, 1.8 mM MgSO$_4$, 0.27% (v/v) 2-mercaptoethanol, pH 7.0] and 0.1 ml of 2% (v/v) 5-bromo-4-chloro-3-indolyl-β-D-galactoside (XGal) solution. A petri plate bottom was placed over the nitrocellulose and the whole filter sandwich was incubated at 37° until the blue color develops (from 5 min to 2 hr, depending on promoter strength). The blue colonies were easily identified and the comparable colonies on the master plate further analyzed.

[26] F. Sanger, S. Nicklens, and A. R. Coulson, *Proc. Natl. Acad. Sci. U.S.A.* **74**, 5463 (1977).
[27] J. Messing, this series, Vol. 101, p. 20.
[28] T. Maniatis, E. F. Fritsch, and J. Sambrook, "Molecular Cloning: A Laboratory Manual." Cold Spring Harbor Laboratory, Cold Spring Harbor, New York, 1982.
[29] J. D. Beggs, *Nature (London)* **275**, 104 (1978).
[30] M. Rose and D. Botstein, this series, Vol. 101, p. 167.

Liquid Assays[31]. After growth to the appropriate culture density, cells were harvested and washed two times with ice-cold sterile H_2O. The cells were resuspended in Z buffer and broken by vortexing with glass beads. The cell debris was removed by centrifugation for 15 min in a microfuge and the supernatant fluid was assayed for β-galactosidase activity. Z buffer (1 ml) was supplemented with 0.2 ml of a 4 mg/ml solution of *o*-nitrophenyl-β-D-galactoside (ONPG) and preincubated at 30° for 10 min before addition of 0.01–0.1 ml extract. After an additional incubation for 1–100 min, the reaction was stopped by the addition of 0.5 ml of 1 M Na_2CO_3 and the OD_{420} determined for each sample. β-Galactosidase activity was expressed as milliunits per milligram protein, where 1 milliunit of enzyme activity produces 1 picomole of *o*-nitrophenyl-β-D-galactoside per minute at 30°, pH 7.0. Each number in Table I represents the mean of the number of assays for that sample with a standard deviation of 15%. If β-galactosidase activity of a sample was very low, the standard deviation could be somewhat higher.

Construction of Temperature-Regulated Promoters

Temperature regulated variants can be constructed either for constitutive promoters (e.g., the promoter from the *TPI1* gene that encodes triosephosphate isomerase)[23] or for regulated promoters (such as the glucose-repressible promoter from the *ADH2* gene for alcohol dehydrogenase II).[32] One or more copies of a synthetic oligonucleotide with the sequence of the *MATα2* operator[11,12] can be inserted at an appropriate site in the promoter and the number of copies and the orientation of the oligonucleotides determined for each insertion event by restriction analysis and DNA sequencing. Temperature regulation of the variant promoters can then be assessed by ligating them to a gene/cDNA encoding a product for which there exists a direct, quantitative assay (e.g., the *E. coli lacZ* gene[33]). Plasmids bearing these test constructions are then transformed into appropriate yeast host strains for analysis.

Identification of Insertion Site for MATα2 Operators within Promoter of Choice

In any yeast promoter, the *MATα2* operator should be inserted upstream of the TATAA box to achieve efficient temperature regulation. Since the *MATα2* operators are found approximately 135 to 200 bp upstream of the TATAA box sequences in the promoters of a-specific genes,

[31] J. Miller, "Experiments in Molecular Genetics." Cold Spring Harbor Laboratory, Cold Spring Harbor, New York, 1972.
[32] D. R. Beier and E. T. Young, *Nature (London)* **300,** 724 (1982).
[33] L. Guarente and M. Ptashne, *Proc. Natl. Acad. Sci. U.S.A.* **78,** 2199 (1981).

we suggest placing the *MATα2* oligonucleotides within this range in a hybrid promoter. For stringent temperature regulation of regulated promoters with a clearly identifiable UAS, the *MATα2* operator(s) should be placed between the TATAA box of the promoter and its UAS. However, such placement might affect the original regulatory mode of the promoter. Another option is to place the *MATα2* operator(s) upstream of the promoter's UAS. This site should not influence normal regulation of the promoter but may influence the efficiency of temperature regulation. These "rules" have been derived from experiments performed in our laboratory with the constitutive *TPI1* promoter and the glucose-repressible *ADH2* promoter, as well as from published results employing the *CYC1* promoter.[11,12]

Important features of both the constitutive *TPI1* promoter and the glucose-repressible *ADH2* promoter are depicted in Fig. 2. The *TPI1*[23] promoter has two putative TATAA boxes, one at position −146 and another at −81 (where +1 is the first nucleotide coding for *TPI1* mRNA). A unique *Sph*I site, located upstream of these *TPI1* TATAA boxes at position −198, was chosen as the insertion site of the *MATα2* operator sequences within the *TPI1* promoter. The structure of the *ADH2* promoter[24,32] is far more complex. There is only one TATAA box, which is located at position −105, but there are two important regulatory sequences: a stretch of 20 As starting at −167 bp and a strong UAS at −215 bp. We wanted to insert the *MATα2* elements between −105 (the TATAA box) and −167 (the first regulatory element) but the wild-type *ADH2* promoter does not contain any useful restriction sites within this region. However, during extensive deletion analysis of the *ADH2* promoter in E.T. Young's laboratory, several mutations in the 5′ flanking region of the *ADH2* gene were generated. Two of these mutants, *ADH2-10* and *ADH2-7* (Fig. 2), were used in our experiments. The *ADH2-10* promoter contains an *Xho*I linker which replaces a small deletion between −164 and −146 bp, thereby creating an insertion site between the TATAA box and the stretch of As. The *ADH2-7* promoter was deleted between −367 and −276 bp and an *Xho*I linker was inserted in its place, creating an insertion site upstream of the *ADH2* UAS.

Preparation of MATα2 Operator Sequences

The *MATα2* operator in the *STE2* gene was synthesized as a pair of oligonucleotides:

5′TCGAGTCATGTACTTACCCAATTAGGAAATTTACATGG 3′
3′ CAGTACATGAATGGGTTAATCCTTTAAATGTACCAGCT 5′

When annealed, these 38-bp oligonucleotides form ZC609, which has *Xho*I

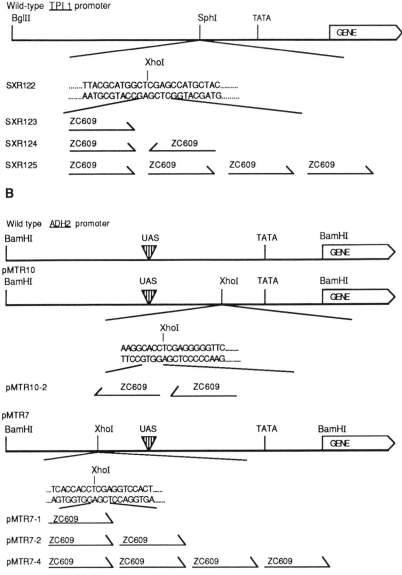

FIG. 2. (A) An *Xho*I site was introduced into a unique *Sph*I site in the *TPI1* promoter by linker insertion. Synthetic *MATα2* operator sequences were inserted into the *Xho*I site. Shown are the orientation, number of operator sequences inserted, and relative position of insertion with respect to the start of the gene and the TATA box (TATA). (B) An *Xho*I linker was inserted into the *ADH2* promoter following brief deletion analysis (see text). In pMTR10 the *Xho*I site is between the upstream activation sequences (UAS) and the TATA box (TATA). In pMTR7 the *Xho*I site is upstream of the UAS. The orientation, number, and relative position of insertion of the *MATα2* sequences are shown. See text for additional information.

compatible ends, but only one end will recreate a *Xho*I site when cloned into *Xho*I–cut DNA.

The *MATα2* oligonucleotides were phosphorylated in separate reactions containing 20 pmol of the oligonucleotide in 50 μl containing 50 mM Tris-HCl, pH 7.6, 10 mM MgCl$_2$, 5 mM dithiothreitol, 0.1 mM spermidine, 1 mM ATP, and 5 units of polynucleotide kinase for 30 min at 37°. The phosphorylated oligonucleotides were annealed by mixing them 1:1, incubating for 15 min at 65°, and allowing to cool slowly to room temperature. Annealed oligonucleotides were kept at 4° until needed.

Insertion of MATα2 Operators into Promoters

It is convenient if the region chosen for the insertion of the *MATα2* element(s) contains a restriction site unique within the plasmid carrying the promoter. In our experiments, the *Xho*I site was unique in both plasmids carrying the *ADH2* promoter variants. The unique *Sph*I site present in the *TPI1* promoter was converted to a unique *Xho*I site by ligating an *Sph*I–*Xho*I adaptor (GCTCGAGCCATG) into the *Sph*I site in order to use the same *MATα2* oligonucleotides for insertion into the *TPI1* promoter.

Dephosphorylation of the vector is crucial since this lowers the background of self-ligated plasmid DNA. Plasmids carrying the promoter of interest were digested with *Xho*I, and then dephosphorylated at 37° for 30 min in a 100-μl reaction containing 50 mM Tris-HCl, pH 9.0, 1 mM MgCl$_2$, 0.1 mM ZnCl$_2$, 1 mM spermidine, 2–5 pmol of DNA dissolved in 20 μl of 10 mM Tris-HCl, pH 8.0, and 0.1 unit of calf intestinal alkaline phosphatase. The dephosphorylation reaction was terminated by the addition of 10 μl of 10% (w/v) SDS and incubation of the sample at 70° for 15 min. The sample was then extracted with phenol:chloroform:isoamyl alcohol (25:24:1) and ethanol precipitated. In a test ligation, if the background is reduced to 1% we assume the dephosphorylation reaction was successful.

The molar ratio of oligonucleotide to vector is an important consideration when setting up a ligation reaction. A 1:1 ratio results in primarily one *MATα2* element inserted in the promoter, whereas a 5:1 (oligonucleotide:vector) ratio usually produces an insertion of one to four *MATα2* elements. Our typical ligation reaction contains 50 fmol of dephosphorylated vector DNA, 250 fmol of phosphorylated oligonucleotide, 50 mM Tris-HCl, pH 7.4, 10 mM MgCl$_2$, 10 mM dithiothreitol, 1 mM spermidine, 1 mM ATP, 1 mg/ml bovine serum albumin, and 1 unit of T4 polynucleotide ligase in a total reaction volume of 20 μl. The reaction is incubated overnight at 8° before transforming into competent RR1 cells.

Screening for Clones Containing MATα2 Oligonucleotides

Bacterial clones containing *MATα2* elements are identified by colony hybridization. Then, based on signal intensity from the hybridization, clones are selected for restriction analysis to determine the number and orientation of the *MATα2* sequences inserted. This information is then confirmed by DNA sequencing.

Quick Colony Hybridization Procedure

1. On each plate lay a filter (Whatman #541) for 10 min. As a negative control, a filter is placed on a plate with colonies carrying the plasmid with no *MATα2* elements inserted. Mark filters and plates asymmetrically.

2. Peel filters off and wash them at room temperature for 10 min in 0.5 *M* NaOH and three times for 10 min each in 2× SSC (20× SSC contains, per liter, 175.3 g of NaCl and 88.2 g of sodium citrate, pH 7.0).

3. Air dry filters on a sheet of Whatman 3MM filter paper before baking them for 30 min at 80°.

4. Prepare hybridization mix (about 15 ml per 12 100-mm filters). The hybridization mix contains 6x SSC, 1x Denhardt's solution (50x Denhardt's solution contains, per liter, 5 g of Ficoll, 5 g of polyvinylpyrrolidone, and 5 g of bovine serum albumin), 20 μg/ml yeast tRNA, 0.05% (w/v) sodium pyrophosphate, and the probe. The *MATα2* oligonucleotide is labeled in a phosphorylation reaction as described above with one modification: instead of 1 m*M* ATP, 75–150 μCi of [γ-^{32}P]ATP is used per reaction. The filters are hybridized for 2 hr at 42°.

5. After hybridization, wash the filters on a shaking platform for 5 min in 6x SSC at room temperature, twice for 15 min in 6x SSC at 37°, and once for 10 min in 6x SSC at 50° before placing against X-ray film. Usually, 1–2 hr is long enough to receive a clear signal, although an overnight exposure may bring out differences in intensities that relate to the number of *MATα2* elements inserted.

DNA Preparation. Plasmid DNA is prepared from the colonies identified in the colony hybridization according to the method of Birnboim[34] and is analyzed to determine the orientation and copy number of the inserted *MATα2* operators. For example, the *ADH2-7* promoter has two *Bam*HI sites located 774 bp upstream and 346 bp downstream of the unique *Xho*I site (the insertion site for the *MATα2* operators). If one cleaves the plasmid containing the wild-type *ADH2-7* promoter with both *Bam*HI and *Xho*I restriction endonucleases, the digest will yield two bands of interest, 774 and 346 bp, as well as a large band containing the rest of the plasmid DNA. If one copy of the synthetic *MATα2* operator is ligated into

---

[34] H. C. Birnboim, this series, Vol. 100, p. 243.

the *Xho*I site of the *ADH2-7* promoter, orientation of the insert can be determined by digestion with *Bam*HI and *Xho*I. If the oligonucleotide is inserted such that the *Xho*I site is proximal to the 5′ end of the promoter, the digest will produce bands of 774 and 384 bp. If the operator is inserted in the opposite orientation, the digest will yield bands of 812 and 346 bp. The plasmid DNA is digested as described above and subject to electrophoresis on 3% NuSieve agarose gels which allow separation of low-molecular-weight DNA. One must be very careful when handling these gels as they are fragile and break easily.

After restriction analysis, clones were selected for sequencing. DNA sequencing confirmed the sequence of the *Xho*I linker inserted into the *TPI1* promoter which converted the *Sph*I site to *Xho*I (pSXR122). Plasmids pSXR123, pSXR124, and pSXR125 were established to have one, two, and four copies, respectively, of the *MATα2* operator inserted into the *Xho*I site of the *TPI1* promoter. pMTR7 contains the *ADH2-7* promoter with no *MATα2* sequences, whereas pMTR7-1 has one, pMTR7-2 has two, and pMTR7-4 has four operators inserted. Similarly, pMTR10 contains the *ADH2-10* promoter with no *MATα2* sequences inserted and pMTR10-2 has two copies of the operator. The orientation of these inserts is shown in Fig. 2.

Testing Temperature-Regulated Promoters

The efficiency of temperature regulation of these hybrid promoters can be tested by fusing them with a gene encoding a protein that is easy to assay, e.g., the *lacZ* gene.[33] Such fusions can be cloned into yeast vectors and transformed into a yeast strain carrying the *MAT*a and *sir3-8* loci. The yeast transformants are then grown at various temperatures to assess the efficiency of temperature regulation on the expression of β-galactosidase. Results with the *SXR* and the *MTR* promoters are presented in Table I. β-Galactosidase activity is completely repressed in the XK1-C2[pSXR124] and XK1-C2[pSXR125] transformants grown at 35° (Table IA). When grown at 25°, these transformants express β-galactosidase at wild-type levels (i.e., like the XK1-C2[pSXR122] transformants grown at any temperature). In contrast, XK1-C2[pSXR123] clones exhibit partially repressed β-galactosidase activity when grown at 35° or 30°. It appears that full repression of a strong promoter like *TPI1* requires the insertion of at least two copies of the *MATα2* operator.

Temperature regulation of the *MTR* promoters is complicated by the fact that the *ADH2* promoter is already regulated by glucose. We were interested in determining if the insertion of *MATα2* operator elements would affect normal glucose repression of the *MTR* promoters. The

TABLE I
β-GALACTOSIDASE ACTIVITY PRODUCED AT DIFFERENT TEMPERATURES IN YEAST
TRANSFORMED WITH PLASMIDS BEARING VARIANT PROMOTERS[a]

Plasmid	Copies of MAT α2 operator	Temperature		
		35°	30°	25°
A. pSXR plasmids (*TPI1* promoter variants)[b]				
pSXR122	0	105	125	129
pSXR123	1	14	38	189
pSXR124	2	2	4	129
pSXR125	4	1	2	115

Plasmid	Copies of MAT α2 operator	Temperature					
		35°		30°		25°	
		Glu	EtOH	Glu	EtOH	Glu	EtOH
B. pMTR plasmids (ADH2 promoter variants)[c]							
pMTR7	0	16	676	14	694	9	830
pMTR7-1	1	12	1378	13	922	10	1414
pMTR7-2	2	2	216	2	294	11	956
pMTR7-4	4	1	76	1	78	7	435
pMTR10	0	16	548	14	572	8	634
pMTR10-2	2	6	88	13	136	125	702

[a]Reprinted with permission from A. Z. Sledziewski, A. Bell, K. Kelsay, and V. L. MacKay, *BioTechnology* 6, 411 (1988).
[b]Cells were grown in YNB + CAA containing 5% glucose at the temperature indicated.
[c]Cells were grown in YNB + CAA containing either 5% glucose (Glu) or 2% ethanol (EtOH) as carbon source.

XK1-C2[pMTR] transformants were grown in medium containing either glucose (to assess temperature regulation when the promoter is repressed) or ethanol (to assess temperature regulation of the fully derepressed promoter). The *MTR* promoters containing *MATα2* elements are still repressed by glucose except *MTR10-2* at 25° (Table 1B). It has been demonstrated that the *MATα2* operator contains a UAS element and it is possible that in the *MTR10-2* promoter the UAS of the *MATα2* operator is able to override glucose regulation of the promoter. However, the *MTR10-2* promoter is still repressed by glucose and/or temperature at 30° and 35° when the transformants are grown in glucose. *MATα2* repression was not sufficient for complete repression of β-galactosidase synthesis when the pMTR transformants were grown in media containing ethanol. Several of the *MTR* promoters exhibit intermediate derepression levels. Thus, temperature, site of insertion within the promoter, and number of *MATα2* opera-

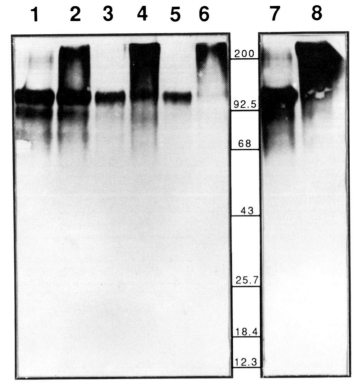

FIG. 3. Temperature-dependent hyperglycosylation of exported barrier protein. Yeast strain XCY42-28B (*MATa sir3-8 leu2-3,112 trp1 ura3 Δmnn9::URA3 lys2 ade2 tyr1? gal2? cry1?*) was transformed with plasmid pSW24 and either the vector pM111 (lanes 1, 3, and 5) or its derivative pZY66 (lanes 2, 4, and 6). Plasmid pSW24 is derived from the yeast–*E. coli* shuttle vector YEp13 [J.R. Broach, J. N. Strathern, and J. B. Hicks, *Gene* **8**, 121 (1979)], which has the *S. cerevisiae LEU2* gene for selection in yeast and contains a modified form of the *S. cerevisiae BAR1* gene [V. L. MacKay, S. K. Welch, M. W. Insley, T. R. Manney, J. Holly, G. S. Saari, and M. L. Parker, *Proc. Natl. Acad. Sci. U.S.A.* **85**, 55 (1988)]. *BAR1* encodes a 563-amino acid, heavily glycosylated protein that is exported to the culture medium. The modified *BAR1* gene in pSW24 has had it promoter replaced by the promoter from the *S. cerevisiae ADH1* gene (alcohol dehydrogenase I) and its C-terminal 61 amino acids replaced with an oligonucleotide encoding the substance P antigen [S. Munro and H. R. B. Pelham, *EMBO J.* **3**, 3087 (1984)]. Plasmid pM111 (constructed at ZymoGenetics) is also a yeast–*E. coli* shuttle vector that contains *TRP1* and *CEN3* sequences. Plasmid pZY66 is derived from pM111 and contains the *SXR–MNN9* expression cassette. For transformation of *mnn9* cells, synthetic media contained 2.5% agar and 1 *M* sorbitol in the top agar and 1 *M* sorbitol in the plates. Media for growth of *mnn9* mutants were supplemented with 0.5–1 *M* sorbitol. Yeast transformants were grown at the temperature indicated in selective medium lacking leucine and tryptophan for 30–36 hr. After removal of the cells, the exported barrier protein was precipitated from the supernatant fluid [V. L. MacKay, S. K. Welch, M. W. Insley, T. R. Manney, J. Holly, G. S. Saari, and M. L. Parker, *Proc. Natl. Acad. Sci. U.SA.* **85**, 55 (1988)], denatured, and subjected to electrophoresis on 10% polyacrylamide gels containing sodium lauryl sulfate [U. K. Laemmli, *Nature (London)* **227**, 680 (1970)]. The proteins separated by electrophoresis were transferred to nitrocellulose by electrophoretic transfer and immunoblotted, using a rat monoclonal antibody against the substance P antigen (NC1/34

tors inserted all seem to affect the levels of expression from these promoters (Table I).

Use of Temperature-Regulated Promoters

Our strategy was to use the temperature-regulated promoters to prevent overglycosylation of homologous and heterologous proteins secreted by yeast. Although the early steps of asparagine-linked glycosylation (through the assembly of core oligosaccharide in the endoplasmic reticulum) are quite similar in yeast and mammalian cells, yeast cells frequently elongate core oligosaccharide by the addition of a long mannose outer chain and side chains.[35] This elongation is blocked in cells bearing a *mnn9* mutation,[35] but the mutant cells exhibit slower growth and osmotic sensitivity relative to wild-type cells. The ability to regulate *MNN9* gene expression might permit both healthy culture growth and the secretion of proteins with homogeneous, limited glycosylation.

We have cloned and sequenced the *MNN9* gene and subsequently replaced its promoter with the *SXR* promoter from plasmid pSXR124. A centromeric plasmid bearing this expression unit was transformed into an appropriate *MATa sir3-8 mnn9::URA3* yeast strain (see legend to Fig. 3). The strain was simultaneously cotransformed with plasmid pSW24, which carries a modified form of the yeast *BAR1* gene that has eight potential asparagine-linked glycosylation sites. Cultures of a single transformant were grown at 25°, 30°, and 35° for 48 hr before harvest and then barrier protein was precipitated from the culture supernatant. Immunoblot analysis of the secreted barrier protein (Fig. 3) indicated that, at 25° and 30°, by far the predominant species was a hyperglycosylated form that migrated with an apparent molecular weight of > 200,000. At 35°, however, the majority of the barrier protein comigrated with that secreted by *mnn9* cells that were not transformed with the *SXR–MNN9* construction. We have not yet attempted optimization of this strategy under fermentation conditions.

The immunoblot results support the hypothesis that, at an intermediate temperature (30°), the temperature-regulated promoter from pSXR124 is expressed at a low level in each cell. As shown in Table I, β-galactosidase

---

[35] C. E. Ballou, *in* "The Molecular Biology of the Yeast *Saccharomyces:* Metabolism and Gene Expression" (J. N. Strathern, E. W. Jones, and J. R. Broach, eds.), p. 361. Cold Spring Harbor Laboratory, Cold Spring Harbor, New York, 1982.

HL, Accurate Chemical & Scientific, Westbury, NY), as described [S. Munro and H. R. B. Pelham, *EMBO J.* **3**, 3087 (1984)]. Lanes 1 and 2, 35°; lanes 3 and 4, 30°; lanes 5 and 6, 25°; lane 7, *mnn9* strain transformed with pSW24 alone and grown at 30°; lane 8, isogenic *MNN9* (wild-type) strain transformed with pSW24 and grown at 30°. Sizes of protein standards are indicated in kilodaltons.

expression from pSXR124 was nearly 97% repressed at 30° compared to 25°. These data, however, could not distinguish between two hypotheses: (1) all of the cells were 97% repressed for β-galactosidase expression or (2) 97% of the cells were completely repressed and 3% were fully expressed. If the latter were true, then one would predict that only 3% of the barrier protein secreted at 30° would be hyperglycosylated. Clearly, a much higher percentage of the protein migrates in a heterogeneous, highly glycosylated form (Fig. 3), indicating that, at intermediate temperatures, the promoter has intermediate strength in each cell.

We have demonstrated the construction of temperature-regulated variants of two strong yeast promoters (*TPI1* and *ADH2*) using the *MATα2* repression system. We have shown that both promoter series express β-galactosidase in a temperature-dependent fashion. We have also shown that these promoters can be used to regulate the expression of the *MNN9* gene and thus regulate the extent of glycosylation of proteins secreted by yeast.

[30] Sequence and Structural Requirements for Efficient Translation in Yeast

By THOMAS F. DONAHUE and A. MARK CIGAN

Introduction[1]

It is widely accepted that the ease with which one can manipulate the simple eukaryotic yeast *Saccharomyces cerevisiae* makes this organism particularly attractive as a host for the overexpression of a variety of higher eukaryotic proteins that have potential commercial or therapeutic value. However, recent studies,[2,3] which show that the synthesis of mammalian transcriptional factors in yeast will direct the expression of yeast genes that contain the appropriate 5' transcriptional signals indicate that the genetics of this organism may have unlimited potential to resolve detailed molecular interactions of mammalian gene regulation and cell biology. Typically, heterologous gene expression involves placing a cDNA under the control of a strong yeast promoter that will direct the synthesis and overproduce the encoded mRNA. However, in some cases, overexpression of a mRNA does not result in a coordinate increase of the translated gene product.[4,5]

[1] This work was supported by P.H.S. grant GM32263 and in part by the Searle Scholars Program of The Chicago Community Trust awarded to T.F.D.
[2] M. Schena and K. R. Yamamoto, *Science* **241**, 965 (1988).
[3] K. Lech, K. Anderson, and R. Brent, *Cell* **52**, 179 (1988).
[4] R. A. Smith, M. J. Duncan, and D. T. Moir, *Science* **229**, 1219 (1985).
[5] R. C. Scarpulla and S. H. Nye, *Proc. Natl. Acad. Sci. U.S.A.* **83**, 6352 (1986).

Based on recent studies in yeast, one possible explanation for these reduced levels of protein synthesis is that intrinsic properties of these "foreign mRNAs" or features introduced in the mRNA as a result of the heterologous construction may not be compatible with efficient translational requirements in yeast. It is therefore the intent of this article to point out those features of yeast mRNA that afford efficient translation, and relate these features to the general properties of higher eukaryotic messages in order to maximize their translation in yeast.

Translation Initiation

Comparative, genetic, and direct mutational analyses of translational initiator regions of yeast genes are all consistent with a basic mechanism of translation initiation as proposed by the ribosomal scanning model[6] for higher eukaryotes. Namely, the ribosome binds the 5′ end of mRNA and migrates the leader until it encounters the first AUG codon, whereby protein synthesis commences. However, the ability to initiate translation of a yeast mRNA efficiently does not precisely parallel the sequence and structural requirements for efficient initiation of mammalian genes. Below we discuss the similarities and differences between the yeast and mammalian studies and the variability in features of the initiator regions of yeast genes that afford efficient initiation. Most of the information discussed about yeast translation initiation stems from mutational studies that have been conducted primarily at the *CYC1* and *HIS4* initiation regions. In the former, the effects of mutations were quantitated by measuring the *in vivo* levels of the encoded iso-1-cytochrome *c* and in the latter by quantitation of β-galactosidase activity in mutant *his4 – lacZ* fusion strains. In addition, comparative analysis is based on a recent compilation of initiator region from 131 yeast genes.[7]

Differences in Yeast Leader Lengths Have No Effect on Translation Initiation Efficiency

Comparative analysis shows that the distance from the 5′ end of yeast mRNA to the start codon ranges from 11 to 591 nucleotides (nts), with most (70%) having leaders that range from 20 to 60 nts.[7] The average yeast leader is 52 nts in length.[7] This is in close agreement with the 3 to 572 nts range reported for higher eukaryotes,[8] with the majority falling between 20 and 80 nts and having an average length of 60 nts.

[6] M. Kozak, *Cell* **15**, 1109 (1978).
[7] A. M. Cigan and T. F. Donahue, *Gene* **59**, 1 (1987).
[8] M. Kozak, *Nucleic Acids Res.* **12**, 857 (1984).

Mutational studies in yeast suggest that variability in leader length does not play any significant role in controlling the efficiency of translation initiation, at least within the limits of the range apparent to the majority of genes. Deletion and duplication of the normal leader length of the *HIS4* gene in yeast from 39 to 115 nts result in 100% levels of wild-type initiation.[9] A study at the *CYC1* locus suggests that a leader distance as small as 11 nts is still capable of supporting normal initiation events.[10] Thus, natural variation in yeast leader lengths is tolerated by the translational process, consistent with the interpretation that the scanning ribosome migrates any leader length until it encounters an AUG start codon.

Secondary Structure in Yeast Leader Region Virtually Abolishes Translation Initiation

Insertion mutagenesis studies of mammalian genes[11,12] suggest that localized stem–loop structures in the 5' noncoding region of mRNA block translation initiation either by preventing the ribosome from binding the 5' end of the message or its migration toward the AUG start codon. Insertion studies at *HIS4* also demonstrate that secondary structure is inhibitory to the initial steps of protein synthesis in yeast when placed either near the 5' end of the message, at an intermediate position between the 5' end and the first AUG, or closer toward the AUG start codon.[9] A similar inhibitory effect was observed at the *CYC1* gene when secondary structure was inserted slightly 5' to the AUG start codon.[10]

One difference between the yeast and mammalian studies is that translation initiation in yeast would appear to be particularly sensitive to secondary structure. Based on studies at *HIS4*, regions of dyad symmetry that contribute a ΔG value of -20 kcal reduce *HIS4* expression to approximately 1–5% of wild-type levels, depending on its position in the leader.[9] To put these studies into perspective, the region of dyad symmetry inserted in the *HIS4* leader constitutes a *Bam*HI linker flanked by *Eco*RI restriction sites (5'-GGAATTCCCGGATCCGGGAATTCC-3'). Similarly, iso-1-cytochrome *c* expression was reduced to 5% of wild-type levels upon introduction of a shorter region of dyad symmetry (5'-GAATTCGTTAAC-GAATTC-3') immediately 5' to the *CYC1* AUG start codon.[10] Interestingly, the calculated ΔG value for the insert at *CYC1* was only -7.6 kcal. In contrast, inserts of similar lengths or ΔG values (-20 to -30 kcal) have virtually no inhibitory effect on translation initiation when introduced into the leader of the herpes simplex thymidine kinase[11] or rat

[9] A. M. Cigan, E. K. Pabich, and T. F. Donahue, *Mol. Cell. Biol.* **8**, 2964 (1988).
[10] S. B. Baim and F. Sherman, *Mol. Cell. Biol.* **8**, 1591 (1988).
[11] J. Pelletier and N. Sonenberg, *Cell* **40**, 515 (1985).

preproinsulin[12] genes as measured from extracts of transfected cells. Only when inserts of -50 to -60 kcal are used is significant inhibition of expression of these mammalian genes observed. These differences in the effects of inserts on translation initiation between the yeast and mammalian studies do not appear to be attributed to the overall secondary structure of the leader, as the ΔG values of the wild-type *HIS4* and *CYC1* leaders region are -4.7 and -6.3 kcal,[7] respectively, in comparison to -64 kcal calculated for the herpes thymidine kinase leader.[11] Instead, the yeast translation initiation process appears to be more sensitive to intramolecular and perhaps intermolecular base pair interactions in the leader. This is consistent with the general properties of yeast leader regions, being extremely rich in A nucleotides (between the -25 and -1 position, 46% A) while G nucleotides are underrepresented (12%), and lacking significant secondary structure in the 5' noncoding region (based on 85 full-length leaders from yeast genes: ΔG range, $+1.6$ to -41.5; average, -7 kcal).[7] Based on our inspections, these features are more pronounced than vertebrate mRNAs suggesting that intrinsic properties of the latter, namely, G-, C-, or G+C-rich leaders or restriction sites containing dyad symmetry introduced during cDNA constructions, could have an inhibitory effect on the translational expression of foreign genes in yeast. The reduced ability to initiate translation of a foreign gene can be easily accessed by primer extension analysis.[9] These studies at *HIS4* show that dyad symmetry does not result in a reduction in mRNA levels but rather results in premature termination of the priming reaction at the site of the dyad symmetry inserted in the mRNA.

No Simple Sequence Can Override Importance of First AUG Codon near 5' End of Yeast Message in Establishing Start Site of Translation

The site of translation initiation in 95% of yeast mRNA corresponds to the first AUG codon nearest the 5' end of the message.[7] Simple inspection or comparison[7] of the initiator regions from yeast mRNAs does not identify any extensive homology at a fixed distance from the AUG that would resemble a ribosomal binding site complementary to the 3' end of 18S rRNA as a consensus eukaryotic counterpart to the prokaryotic Shine–Dalgarno sequence.[13] One sequence, 5'-CACACA-3', has been observed[7] at 10% of yeast leader regions at a relatively fixed position upstream from the AUG start codon (-25 to -10 position), as well as implicated by point mutation to be involved in translation initiation at *CYC1*.[14] However, 5'-CACACA-3' does not appear to be significant to the general mechanism

[12] M. Kozak, *Proc. Natl. Acad. Sci. U.S.A.* **83**, 2850 (1986).

[13] J. Shine and L. Dalgarno, *Proc. Natl. Acad. Sci. U.S.A.* **71**, 1342 (1974).

[14] J. I. Stiles, J. W. Szostak, A. T. Young, R. Wu, S. Cosaul, and F. Sherman, *Cell* **25**, 277 (1981).

of translation initiation in yeast, as insertion of this sequence in the leader region of the *HIS4* message[9] and its deletion from the *CYC1*[10] leader region neither stimulate nor reduce translational expression. Thus, yeast mRNAs like mammalian mRNAs do not appear to require a ribosomal binding site or sequence element of any type to mediate the general mechanism of translation.

Comparative[8] and mutational[15] studies of mammalian initiator regions suggest that the sequence context 5'-CACC<u>A</u>UGG-3' surrounding an AUG start codon is the optimal initiator region for protein synthesis. Alterations of this optimal sequence context at the preproinsulin gene have been shown to result in a 20-fold lowering of the translational efficiency at the AUG start codon which leads to ribosomal bypass of the initiator region.[15] In contrast, the yeast concensus start region 5'-A$\frac{A}{Y}$A$\frac{A}{U}$AAUGUCU-3' differs from the higher eukaryotic consensus with the exception of a similar high bias for an A nucleotide at the -3 position (75%, yeast; 79%, higher eukaryotes).

Aside from demonstrating that the AUG triplet is the only codon that can function as the site of initiation, elegant genetic studies at the *CYC1*[16] and *HIS4*[17] genes in yeast have suggested that sequence context does not play a significant role in the translation initiation process. This has been confirmed by more direct mutational studies at *HIS4*.[9] Of simple base changes 5' to the *HIS4* AUG start codon, only a change from the favorable A nucleotide to the unfavorable U nucleotide, as indicated by comparative studies,[7] results in a decrease in *HIS4* expression.[9] However, this mutation results in no more than a 2-fold reduction in initiation at the AUG start codon.[9] Furthermore, changing the *HIS4* sequence context 5'-AAUAAUGG-3' to 5'-CACCAUGG-3', the optimal mammalian concensus start region, has no stimulatory effect on translation initiation at *HIS4*,[9] as suggested by the absence of this sequence at 131 yeast initiator regions.[7]

Other mutational analyses at *HIS4*[9] suggest that the concensus sequence 3' to the AUG start codon (5'-AUG<u>UCU</u>-3') is insignificant to translation initiation efficiency. Instead, this bias, as suggested by comparative studies, seems to reflect a nonrandom distribution of the $+2$ amino acid position and the preferred codon usage of these amino acids in yeast genes.[7] Therefore, with the exception of an A nucleotide at the -3 position, genetic and mutational studies in yeast indicate that the first AUG codon in the message is efficiently recognized as the start signal for transla-

[15] M. Kozak, *Cell* **44**, 283 (1986).
[16] F. Sherman, J. W. Stewart, and A. M. Schweingruber, *Cell* **20**, 215 (1980).
[17] T. F. Donahue and A. M. Cigan, *Mol. Cell. Biol.* **8**, 2955 (1988).

tion. This would agree with the effects of introducing AUG codons in the leader region upstream and out-of-frame with the *CYC1*[16] and *HIS4*[17] coding region. In each case, these upstream AUG codons preclude initiation at the downstream AUG. Although in some constructions residual levels of translation initiation are observed at the downstream AUG, these residual levels have not been observed to constitute more than 10–20% of wild-type levels of expression.[9]

In summary, the general mechanism of translation initiation in yeast is consistent with the basic features of the ribosomal scanning model, suggesting that the yeast translation initiation process will afford efficient expression of mRNAs derived from higher eukaryotes. However, in contrast to mammalian studies, regions of dyad symmetry or the presence of upstream AUG codons despite the immediate sequence context effectively inhibit gene expression at the level of initiation of translation. Therefore, these features should be avoided in heterologous cDNA constructions to maximize the expression of "foreign mRNAs" in yeast.

Translation Elongation and Termination

Although the genetics of yeast has been used extensively to study the mechanism of elongation of translation in eukaryotes,[18] due to the complexity of this process a clear picture has not emerged as to the important sequence and structural features that aid the elongation process, and virtually no data exist of the same type in mammalian organisms for comparative analysis. However, some observations have been made that would suggest that the preferred sequence context observed in yeast may be optimal for expression of yeast proteins,[19,20] and this bias differs from the preferred codon usage bias in higher eukaryotes.[21] One study, which employs systematic substitution of least-preferred for preferred codons in the yeast phosphoglycerate kinase *(PGK)* gene, suggests that the preferred yeast context is important for maximal expression.[22] However, this study does employ substitution of regions of the *PGK* by synthetic oligonucleotides containing restriction sites with dyad symmetry. Based on observations at the *CYC1* gene, regions of dyad symmetry may also inhibit the elongation process during protein synthesis.[23] Thus, the codon usage analysis of *PGK* may be subject to other complex interpretations.

[18] K. Chakraburtty and A. Kamath, *Int. J. Biochem.* **20**, 581 (1988).
[19] J. L. Bennetzen and B. D. Hall, *J. Biol. Chem.* **257**, 3026 (1982).
[20] P. M. Sharp, T. M. F. Tuohy, and K. R. Mosurski, *Nucleic Acids Res.* **14**, 5125 (1986).
[21] T. Maruyama, T. Gojobori, A. Aota, and T. Ikemura, *Nucleic Acids Res.* **14**, r151 (1986).
[22] A. Hoekema, R. A. Kastelein, M. Vasser, and H. A. deBoer, *Mol. Cell. Biol.* **7**, 2914 (1987).
[23] S. B. Baim, D. F. Pietras, D. C. Eustice, and F. Sherman, *Mol. Cell. Biol.* **5**, 1839 (1985).

Finally, we have analyzed the termination region of 131 yeast genes. In essence, there is no clear consensus that emerges that would suggest that a specific sequence element is required for termination of translation in yeast genes other than a termination codon of which there is no particular preference for UAA, UAG, or UGA at this position.

[31] Vacuolar Proteases in Yeast *Saccharomyces cerevisiae*

By ELIZABETH W. JONES

Interest in vacuolar proteases of the yeast *Saccharomyces cerevisiae,* for those involved in gene expression technology, takes the form of trying to eliminate the proteases and being sure that one has done so. The former task is accomplished genetically; the latter through a combination of genetic and biochemical tests.

The complement of proteases found in the lumen of the vacuole includes two endoproteinases, protease A (PrA) and protease B (PrB); two carboxypeptidases, carboxypeptidase Y (CpY) and carboxypeptidase S (CpS); and two aminopeptidases, the 600K aminopeptidase I (API) and the Co^{2+}-dependent aminopeptidase.[1,2]

Most of the genetic analysis has concentrated on the two endoproteinases, PrA and PrB, and the two carboxypeptidases, CpY and CpS. Plate tests or microtiter well tests have been developed that allow one to test directly for activity of PrB, CpY, and CpS and indirectly for activity of PrA in colonies. In this article, I first present the plate and well tests for assaying protease activities of colonies, followed by assays that allow quantitation of activity levels in cell-free extracts, and end with guidelines for designing useful protease-deficient strains and a list of strains that we have sent to the Yeast Genetics Stock Center (MCB/Biophysics and Cell Physiology, University of California, Berkeley, CA 94720).

Genetic Analyses

Requirements of Tests

Several requirements must be met if one is to succeed in assaying activities of particular intracellular enzymes in colonies. The first condition is that the enzyme gain access to the externally supplied substrate.

---

[1] E. W. Jones, *Annu. Rev. Genet.* **18**, 233 (1984).
[2] T. Achstetter and D. H. Wolf, *Yeast* **1**, 139 (1985).

This is accomplished either by permeabilizing the cells in a colony with a solvent or by establishing conditions that result in lysis of cells. The second requirement is that, where lysis is employed, conditions be established that free the enzyme from its naturally occurring, intracellular, polypeptide inhibitor.[1,2] This is accomplished, for PrB, by including sodium lauryl sulfate (SDS) in the overlay. The third requirement is to find a substrate or condition that tests for one enzyme activity only. How this is accomplished for each enzyme will be given in the procedure for that enzyme.

Protease B

Protease B activity in colonies can be assayed using an overlay test. Initially, we developed a procedure for cells grown on YEPD plates [20 g Difco (Detroit, MI) Bacto-peptone, 10 g Difco yeast extract, 20 g dextrose, 13–20 g agar, according to brand, per liter] that necessitated use of a lysis mutation to cause release of intracellular PrB from cells. We have since realized that growth of cells on YEPG plates (20 g Difco Bacto-peptone, 10 g Difco yeast extract, 50 g glycerol, 13–20 g agar, according to brand, per liter) obviates use of a lysis mutation, since some lysis occurs when cells are grown on this medium. The substrate is particulate Hide Powder Azure (HPA), for PrB is the only protease in *S. cerevisiae* that catalyzes cleavage and solubilization of this substrate.[3]

HPA Overlay Test for PrB Activity.[4,5] *Principle.* Protease B, which is freed from cells by lysis and from its inhibitor by the SDS present in the overlay, solubilizes the particles of Hide Powder Azure in the overlay, uncovering the colony and surrounding it with a clear halo. Mutant colonies remain covered.

Reagents

Sodium lauryl sulfate (SDS): 20% (w/v) in 0.1 M Tris-HCl pH 7.6
Cycloheximide: Sterile solution at 5 mg/ml
Penicillin G–streptomycin: Use a sterile solution containing 5000 Units/ml penicillin base and 5000 μg/ml streptomycin base
0.6% agar, molten, held at 50°
Hide Powder Azure (Calbiochem, San Diego, CA): Hide Powder Azure is pulverized by homogenization (VirTis homogenizer) of 200 ml of a slurry (100 mg/ml 95% ethanol) in a 500-ml flask for

[3] R. E. Ulane and E. Cabib, *J. Biol. Chem.* **249,** 3418 (1974).
[4] G. S. Zubenko, A. P. Mitchell, and E. W. Jones, *Proc. Natl. Acad. Sci. U.S.A.* **76,** 2395 (1979).
[5] C. M. Moehle, M. W. Aynardi, M. R. Kolodny, F. J. Park, and E. W. Jones, *Genetics* **115,** 255 (1987).

5 min at 40,000 rpm or by sonication of a slurry of the same proportions. Aliquots containing 50–100 mg are transferred to sterile 13 × 100 mm tubes, centrifuged (5 min at 1650 g), and the supernatants are discarded. The pellets are washed with 2.5–3 ml of sterile water, repelleted, and the supernatants discarded.

Procedure. Add 0.2 ml of cycloheximide solution, 0.2 ml of penicillin–streptomycin solution, and 0.1 ml of 20% SDS to a HPA pellet, vortex, then add 4 ml of molten agar. Vortex to mix and resuspend particles. Streaks or replica plates of cells grown for 2 days at 30° on YEPG agar are overlaid with the molten agar cocktail, with pouring along the length of the stripes rather than across. After the agar solidifies, the plates are incubated at 34–36° for 8 hr to 2 days, depending on the properties of the strains. (Some strains lyse well and are difficult to score at later times.) Plates can (and should) be incubated upside down, so long as the medium will absorb the moisture in the overlay. It is important that a seal does *not* form between the lid and base of the petri dish.

Utility. This test works very well for following mutations in the PrB structural gene *PRB1,* and less well for pleiotropic mutations like the *pep* mutations. In using *prb1* mutations or in constructing strains, be aware that, although *prb1* homozygotes sporulate, the asci may be very small (the size of a normal spore) and that superimposition of heterozygosity for *pep4* in such *prb1* homozygotes may prevent sporulation.[6]

Carboxypeptidase Y

Carboxypeptidase Y activity in colonies can be assessed by using an overlay test that relies on the esterolytic activity of the enzyme. The substrate is *N*-acetyl-DL-phenylalanine β-naphthyl ester (APE), whose cleavage, in colonies anyway, is catalyzed only by CpY.

APE Overlay Test for CpY Activity.[7] *Principle.* Dimethylformamide present in the initial overlay permeabilizes cells on the surface of colonies. CpY within cells catalyzes cleavage of the ester. The product β-naphthol reacts nonenzymatically with the diazonium salt Fast Garnet GBC to give an insoluble red dye; Cpy$^+$ colonies are red; Cpy$^-$ colonies are yellow or pink.

Reagents

N-Acetyl-DL-phenylalanine β-naphthyl ester (Sigma, St. Louis, MO): Make a solution of 1 mg/ml dimethylformamide
0.6% agar, molten, held at 50°

[6] G. S. Zubenko and E. W. Jones, *Genetics* **97,** 45 (1981).
[7] E. W. Jones, *Genetics* **85,** 23 (1977).

Fast Garnet GBC (Sigma)
0.1 M Tris-HCl, pH 7.3 – 7.5

Procedure. Replica plate strains to thick YEPD plates (40 – 45 ml/100-mm plate). Grow for 3 days at 30°. To form the overlay mix, add 2.5 ml of the ester solution to 4 ml of molten agar in a 13 × 100 mm tube. Vortex or cover with Parafilm and invert 3 or 4 times until the schlieren pattern disappears. After the bubbles exit, pour the contents over the surface of colonies or stripes (along, not across, stripes; colonies must be covered). After 10 min (or after the agar is hard), carefully flood the surface of the agar with 4.5 – 5 ml of a solution of Fast Garnet GBC (5 mg/ml 0.1 M Tris-HCl, pH 7.3 – 7.5). Do not tear the agar overlay during the flooding. The Fast Garnet GBC solution must be made *immediately* before use, for diazonium salts are very unstable in solution. We use Fast Garnet GBC from Sigma and store it in the freezer. Watch the color develop and pour off the fluid when Cpy$^+$ colonies turn red (several to a few minutes). If color development takes longer than 5 – 10 min, replace your bottle of Fast Garnet GBC. Do the test at room temperature. The color is not stable, but is more stable if the plates are placed in the cold in the dark.

Utility. The APE test has wide utility for following many mutations that reduce protease activity so long as CpY activity is among the activities reduced as a consequence of the mutation. The APE test can be used for following mutations in the CpY structural gene *PRC1,* and for following the pleiotropic *pep* mutations, including *pep4-3*.[7] If an *ade1* or *ade2* mutation is segregating in the cross, the red pigmentation problem can be circumvented by growing cells on YEPG or on YEPD supplemented with 100 μg/ml adenine sulfate. Petites give aberrant phenotypes in the test (but see well test below).

PEP4 is the structural gene for the PrA precursor.[8,9] PrA activity is essential for proper maturation of several vacuolar hydrolases, including PrB, CpY, one or more RNase species, the 600K vacuolar aminopeptidase API, and the repressible species of alkaline phosphatase.[10-12] We have cloned and sequenced several of the pleiotropic *pep4* mutations (including

[8] C. A. Woolford, L. B. Daniels, F. J. Park, E. W. Jones, J. N. Van Arsdell, and M. A. Innis, *Mol. Cell. Biol.* **6,** 2500 (1986).

[9] G. Ammerer, C. Hunter, J. Rothman, G. Saari, L. Valls, and T. Stevens, *Mol. Cell. Biol.* **6,** 2490 (1986).

[10] B. A. Hemmings, G. S. Zubenko, A. Hasilik, and E. W. Jones, *Proc. Natl. Acad. Sci. U.S.A.* **78,** 435 (1981).

[11] E. W. Jones, G. S. Zubenko, and R. R. Parker, *Genetics* **102,** 665 (1982).

[12] B. Mechler, M. Müller, H. Müller, and D. H. Wolf, *Biochem. Biophys. Res. Commun.* **107,** 770 (1982).

pep4-3) as well as the allelic *pra* mutations[13] that are not fully pleiotropic.[14] Pleiotropic *pep4* mutations like *pep4-3* usually prove to be nonsense mutations that totally eliminate PrA activity and greatly reduce levels of all hydrolase activities, including CpY, in whose maturation PrA participates.[13] For this reason, one can follow many mutations that eliminate PrA activity (the pleiotropic *pep4* mutations) by means of this APE test. Indeed, the more devastating the effect of the *pep4* mutation on PrA activity, the easier it is to follow by the APE test. And, of course, utilization of *pep4* mutations, because of their effects on hydrolase maturation, results in reduction or elimination of several protease activities simultaneously, including PrA, PrB, CpY, and API (but not CpS; the Co^{2+}-dependent aminopeptidase was not tested) as well as RNase activity.[11]

In working with pleiotropic *pep4* mutations (in crosses or in constructing alleles), be aware that *pep4* homozygotes do not sporulate[6] and that the mutations show phenotypic lag.[15] To circumvent phenotypic lag, streak spore clones for single colonies on YEPD plates (quadrants suffice). Do the APE test. Stab negative colonies through the agar overlay and make a new master plate. Then carry out the usual analyses for other markers. If this procedure is followed, the overlays should be made with sterile solutions and the colonies should be stabbed soon after the test, for the cocktail will kill cells.

Well Test for CpY Activity.[8] *Principle.* Dimethylformamide present in the solution permeabilizes cells. Cleavage of the amide bond in *N*-benzoyl-L-tyrosine *p*-nitroanilide (BTPNA) is catalyzed only by CpY to give the yellow product *p*-nitroaniline. This test works for petites as well as for grandes, and is unaffected by *ade1* and *ade2* mutations.

Reagents

> *N*-Benzoyl-L-tyrosine *p*-nitroanilide (Sigma): Make a solution of 2.5 mg/ml dimethylformamide
> 0.1 *M* Tris-HCl, pH 7.5: Use a sterile solution

Procedure. Mix 4 volumes of buffer to 1 volume BTPNA solution. Distribute 0.2 ml into wells in a 96-well microtiter test plate. Cells are transferred into the solution by rotating an applicator stick in the fluid after dipping the sterile stick into a colony grown on a YEPD plate. (Alterna-

[13] E. W. Jones, C. A. Woolford, C. M. Moehle, J. A. Noble, and M. A. Innis, *in* "Cellular Proteases and Control Mechanisms" (Proceedings UCLA Symposium), p. 141. Alan R. Liss, New York, 1989.

[14] E. W. Jones, G. S. Zubenko, R. R. Parker, B. A. Hemmings, and A. Hasilik, in "Molecular Genetics in Yeast" (D. von Wettstein, J. Friis, M. Kielland-Brandt, and A. Stenderup, eds.), p. 182 (Alfred Benzon Symposium, Vol. 16). Munksgaard, Copenhagen, 1981.

[15] G. S. Zubenko, F. J. Park, and E. W. Jones, *Genetics* **102**, 679 (1982).

tively, a 48-prong replicator can be used, taking care to adjust individual wells for colonies that may not transfer effectively with the technique.) Cover the wells and incubate overnight at $34-37°$. Cpy$^+$ cells give yellow fluid in the wells; Cpy$^-$ cells produce no color.

Utility. This test can be used to follow *prc1* mutations as well as pleiotropic *pep* mutations, like *pep4*, that result in CpY deficiency.

Carboxypeptidase S

Carboxypeptidase S activity in colonies can be assessed by using a well test that incorporates a coupled assay that detects release of free leucine from the blocked dipeptide carbobenzoxyglycyl-L-leucine (Cbz-Gly-Leu). Carboxypeptidase Y, which also catalyzes this cleavage, is inactivated by preincubation with phenylmethylsulfonyl fluoride (PMSF), which reacts covalently to inactivate serine proteases like CpY.

Principle. The principle of the coupled assay, devised by Lewis and Harris[16] and adapted by Wolf and Weiser,[17] is shown below.

$$N\text{-Cbz-Gly-Leu} + H_2O \xrightarrow{\text{carboxypeptidase}} \text{Leucine}$$

$$\text{Leucine} + O_2 \xrightarrow{\text{L-amino-acid oxidase}} \text{Keto acid} + NH_3 + H_2O_2$$

$$H_2O_2 + o\text{-dianisidine} \xrightarrow{\text{peroxidase}} \text{Oxidized dianisidine}$$

Oxidized dianisidine is dark brown in color.

The amino acid generated must be a substrate for the L-amino-acid oxidase employed. Typically, snake venoms serve as sources. For *Crotalus adamanteus* (Eastern Diamondback rattlesnake) venom, leucine, isoleucine, phenylalanine, tyrosine, methionine, and tryptophan are good substrates; arginine, valine, and histidine are poor substrates and the other amino acids are not oxidized. We have successfully used venoms from *Crotalus atrox* (Western Diamondback rattlesnake) and *Bothrops atrox* (a viper). Cps$^+$ colonies give brown fluid in the wells; Cps$^-$ cells do not. Cells are permeabilized and CpY is inactivated by preincubation of cells in a solution containing Triton X-100 and PMSF.

Reagents

0.1% (v/v) Triton X-100 made 1 mg/ml in phenylmethylsulfonyl fluoride
0.2 M KPO$_4$, pH 7.0
50 mM MnCl$_2$
N-Cbz-Gly-Leu
Horseradish peroxidase, type I (Sigma)

[16] W. H. P. Lewis and H. Harris, *Nature (London)* **215**, 351 (1967).
[17] D. H. Wolf and U. Weiser, *Eur. J. Biochem.* **73**, 553 (1977).

L-Amino-acid oxidase, type VI (Sigma; actually crude dried venom from *Crotalus atrox*) or type II (Sigma; dried venom from *Bothrops atrox*)
o-Dianisidine dihydrochloride

Procedure. Cells are transferred into 50 μl/well of the Triton X-100/ PMSF solution in a 96-well microtiter test plate by applicator stick or multiprong replicator (see CpY procedure) after growth on a YEPD plate. Cover and let sit for 2 hr at room temperature. Add to each well 150 μl of the substrate mix made in the proportions: 1 ml of buffer, 0.01 ml of $MnCl_2$, 3.22 mg of Cbz-Gly-Leu (final concentration, 10 mM), 0.2 mg of peroxidase, 0.4 mg of amino-acid oxidase type VI or 1.2 mg of type II, 0.4 mg of dianisidine dihydrochloride. Cover and incubate at 37° for 17–18 hr. (We have not investigated other concentrations of venoms but know that these work.)

An example of this test for 20 tetrads from a cross is shown in Fig. 1. One parent of this cross carried mutations in *PRC1,* the structural gene for CpY, and *DUT1,* a gene required for production of CpS activity. The *prc1-407 dut1-1* strain was isolated based on its inability to use Cbz-Gly-

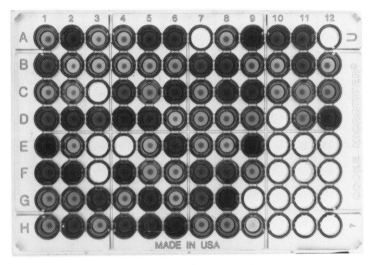

Fig. 1. Scoring for the presence or absence of carboxypeptidase S (CpS) activity in meiotic segregants using the well test for CpS activity. Parents of the cross were of genotype α *leu2 trp1 prc1-407* (well 9E) and **a** *ura3-52 leu2 prc1-407 dut1-1* (well 9F). The *dut1-1* mutation results in failure to produce CpS activity. Spores of a given tetrad are in columns: 1A–D, 2A–D, 1E–H, 2E–H, etc. The scoring for CpS activity for tetrads 1 and 2 is, respectively, −+−+ and +−−+. Blank wells such as 3C correspond to segregants that did not grow. Well 9H contains the reaction mixture but no cells. The Cps⁺ phenotype and ability to use Cbz-Gly-Leu as a nitrogen source cosegregated in all 20 tetrads.

Leu as a nitrogen source.[18] (Either CpY or CpS activity alone is sufficient to allow utilization of the dipeptide as a nitrogen source.) The diploid was homozygous for *prc1-407* and heterozygous for *dut1-1*. In this cross, segregants possessing CpS activity should retain the ability to use Cbz-Gly-Leu as a nitrogen source and give oxidation of *o*-dianisidine; those lacking CpS activity should do neither. The two characteristics cosegregated in all tetrads, as expected.

Utility. This test appears to work well for following mutations that result in CpS deficiency, whether or not the strain lacks CpY activity. We have used it particularly to follow *dut1-1*,[19] a mutation we suspect to be allelic to *cps1*.[18]

Protease A

We have found no general plate test to directly assess PrA activity that is satisfactory, although preliminary tests indicate that it may be possible to adapt the fluorescence-based assay given in the Biochemistry section of this article for a well test. (A plate test that necessitates that strains carry a lysis mutation has been described.[20] However, since total loss of function for PrA results in failure to activate a set of vacuolar hydrolase precursors, including that of CpY, many mutations in the PrA structural gene *PEP4* can be followed using the indirect test (APE test, see section Carboxypeptidase Y above) that detects the esterolytic activity of CpY, the processing and activation of which is dependent on PrA. This latter test is satisfactory for following the pleiotropic *pep4* mutations (Pra⁻ Prb⁻ Cpy⁻ . . .) that are of most utility in biotechnological applications.

Biochemical Analyses

Growth of Cells and Preparation of Extracts

For most of our assays of cell-free extracts for protease activities we employ extracts prepared according to the following protocol. Cells are grown to stationary phase in YEPD at 30° with vigorous shaking (usually 48–52 hr for our conditions of inoculation), harvested by centrifugation, washed once with distilled water and resuspended in 2 ml of 0.1 M Tris-HCl, pH 7.6, per gram of cells. The cells are broken (3 min) with 0.45-mm glass beads [40:60 to 50:50 (v/v) glass beads: cell suspension] in a Braun homogenizer (Braun-Melsungen, West Germany) without CO_2 cooling.

[18] D. H. Wolf and C. Ehmann, *J. Bacteriol.* **147,** 418 (1981).
[19] G. S. Zubenko, Ph.D. Thesis, Carnegie Mellon University, Pittsburgh, Pennsylvania, 1981.
[20] B. Mechler and D. H. Wolf, *Eur. J. Biochem.* **121,** 47 (1981).

After centrifugation for 30 min at 35,000 g in the cold, the supernatant is removed to a fresh tube on ice.

Enzymatic Assays for Protease A

Two assays for protease A activity will be described, the first based on release of peptides from hemoglobin, the second a fluorescence assay based on cleavage of a peptide.

Principle. The most commonly used assay for PrA activity measures the release of tyrosine-containing acid-soluble peptides from acid-denatured hemoglobin. Protease A is apparently the only protease to catalyze the reaction at acid pH. The procedure is based on that of Lenney et al.[21]

Reagents

2% (w/v) acid-denatured hemoglobin: Dissolve 2.5 g of hemoglobin (Sigma) in 100 ml of distilled water, and dialyze against three changes of 3 liters of water in the cold. Bring the pH to 1.8 with 1 N HCl. After 1 hr of incubation with stirring at 35°, bring the pH to 3.2 with 1 M NaOH and adjust the volume to 125 ml. Aliquots can be stored frozen for years

0.2 M glycine-HCl, pH 3.2

1 N perchloric acid

0.5 M NaOH

2% (w/v) Na_2CO_3 in 0.1 M NaOH

1% (w/v) $CuSO_4 \cdot 5H_2O$

2% (w/v) Sodium or potassium tartrate

Folin and Ciocalteu's phenol reagent (diluted 1:1 with water)

Procedure. The reaction mixture consists of 2 ml of a hemoglobin solution (prepared by mixing equal volumes of the 2% hemoglobin, pH 3.2, described above and 0.2 M glycine, pH 3.2) and 0.1 ml of cell-free extract (1–2 mg protein/incubation), with incubation at 37°. At 0, 15, and 30 min, 0.4-ml samples are removed to 0.2 ml of 1 N perchloric acid on ice and the tubes are shaken briefly. After centrifugation at 1650 g for 5 min, 0.1 ml of each sample is removed to 0.1 ml 0.5 M NaOH. Tyrosine-containing peptides in the neutralized 0.2-ml sample are determined with the Folin reagent according to Lowry et al.[22] To each 0.2-ml sample add 1 ml of a reagent consisting of 2% Na_2CO_3 in 0.1 M NaOH, 1% $CuSO_4 \cdot 5H_2O$, 2% sodium or potassium tartrate in the proportion of 100:1:1 (mix just before use). Incubate at room temperature for at least 10 min. Add 0.1 ml

[21] J. Lenney, P. Matile, A. Wiemken, M. Schellenberg, and J. Meyer, *Biochem. Biophys. Res. Commun.* **60**, 1378 (1974).

[22] O. Lowry, N. Rosebrough, A. Farr, and R. Randall, *J. Biol. Chem.* **193**, 265 (1951).

of the diluted phenol reagent and vortex immediately. After 30 min, determine absorbance at 750 nm.

Definition of Unit and Specific Activity. One unit corresponds to 1 μg of tyrosine per minute. The A_{750} of 1 μg of tyrosine is 0.058 in the Lowry assay performed as described. Using the change in absorbance for a 30-min incubation, the conversion to μg Tyr/min/mg protein is made by the following calculation:

$$\frac{\Delta A_{750}}{30} \times \frac{1}{0.058} \times \frac{2.1}{0.1} \times \frac{6}{4} \times \frac{1}{0.1 \text{ (mg protein/ml extract)}}$$
$$= (181)(\Delta A_{750})/(\text{mg protein/ml extract})$$

An abbreviated PrA assay can be used for segregants of crosses known to be segregating a *pra* mutation, where distinction between + and − is all that is sought. It can also be used for mutant screens if a 0 point is added. Extract (0.05 ml) is added to 1 ml of the hemoglobin solution (equal volumes of 2% hemoglobin and 0.2 *M* glycine, pH 3.2). Incubate at 37° for 30 min. Remove a sample to PCA and work up as described above.

An alternative, much more sensitive assay for PrA is available. We find it more satisfactory than the hemoglobin assay with respect to linearity, reproducibility, etc., but have much less experience with it. It was developed for assaying renin[23] but has been used for assaying PrA.[24] We have adapted it for use in crude extracts.

Principle. Protease A will catalyze cleavage at the Leu-Val bond of the octapeptide *N*-succinyl-L-arginyl-L-prolyl-L-phenylalanyl-L-histidyl-L-leucyl-L-leucyl-L-valyl-L-tyrosine 7-amido-4-methylcoumarin. After removal of the valine and tyrosine residues by aminopeptidase M, the fluorescence of 7-amino-4-methylcoumarin can be determined at 460 nm after excitation at 380 nm. PMSF is included to react covalently with and inactivate PrB, a serine protease that could also catalyze cleavage of the peptide.

Reagents

McIlvaine's buffer: 0.2 *M* Na_2HPO_4, 0.1 *M* citric acid. Adjust pH to 6.0 with NaOH or HCl

N-Succinyl-L-arginyl-L-prolyl-L-phenylalanyl-L-histidyl-L-leucyl-L-leucyl-L-valyl-L-tyrosine 7-amido-4-methylcoumarin (Sigma; MW 1300): Make a solution of 0.325 mg/ml dimethylformamide (0.25 m*M*)

7-Amino-4-methylcoumarin (Sigma; MW 175): Make a 0.4 mg/ml solution.

[23] K. Murakami, T. Ohsawa, S. Hirose, K. Takada, and S. Sakakibara, *Anal. Biochem.* **110**, 232 (1981).
[24] H. Yokosawa, H. Ito, S. Murata, and S.-I. Ishii, *Anal. Biochem.* **134**, 210 (1983).

Serially dilute this $1/100 \times 1/40$ (1/4000) to make a stock solution
for the standard curve

PMSF (Sigma): 0.2 M in 95% ethanol (34.8 mg/ml)

Aminopeptidase M (Sigma, L-0632): 0.5 mg/ml in McIlvaine's buffer.
(We have not explored whether cheaper and apparently more active
preparations like L-9876 or L-1503 will work, as they should, since
the original protocol for renin calls for only 50 mU.[23])

Procedure. Mix 90 or 85 μl of McIlvaine's buffer, pH 6.0, 10 μl of
peptide solution, and 10 or 15 μl of crude extract (pretreat 1 ml of extract
with 5 μl of 0.2 M PMSF for 2 hr at room temperature). Incubate for
15 min at room temperature. Immerse the tube in boiling water for 5 min
to stop the reaction. Cool. Add 5 μl of aminopeptidase M. After a 90-min
incubation at room temperature, add 1.61 ml of buffer (to dilute the
reaction 15-fold) and centrifuge the tubes at 1650 g for 5 min. Remove the
supernatant and determine the fluorescence in a fluorimeter, with excitation at 380 nm and emission at 460 nm. A standard curve is constructed
for 7-amino-4-methylcoumarin. The diluted stock (0.1 μg/ml) is mixed
with buffer as given below and the fluorescence is determined.

Stock (μl)	Buffer (ml)	Total nanomoles
350	1.375	0.2
175	1.550	0.1
88	1.637	0.05
44	1.681	0.025
22	1.703	0.0125
11	1.714	0.00625

For experimental samples, read nanomoles in the sample from the
standard curve (the volumes for the standard and the experimental sample
are the same). The samples and standards must be read in exactly the same
way (same slit width, etc.). Prepare the standard curve the same day.

Definition of Unit and Specific Activity. We have been using a unit to
equal 1 nmol/min and the specific activity to be units/milligram protein.

Crude extracts for this assay procedure are made as follows. The volume of cells sampled is 2500 ml/Klett. Pellet the cells, wash with water,
and freeze the pellet. Resuspend the pellet in 1.5 ml of 0.1 M Tris-HCl, pH
7.6, and transfer to a small Braun homogenizer tube (40–50% full of glass
beads). Add buffer to the top (fill tube completely) to prevent foaming.
Homogenize for 3 min at room temperature. Transfer to a long centrifuge
tube and centrifuge for 20 min at 25,000 g in the cold. Transfer the supernatant to a 1.5-ml microfuge tube. (The small homogenizer tubes are
about 35 mm high and are cut down from 12×75 mm tubes.)

We have preliminary evidence that suggests that this assay can be adapted for use in a well test for genetic analyses.

Protease B

Protease B is an endoproteinase that will solubilize particulate substrates like HPA and Azocoll. It is apparently the only enzyme in the yeast cell that can do so.[25]

Principle. Protease B catalyzes hydrolysis of the peptide bonds in Azocoll, resulting in release of the trapped red dye. Absorbance of the dye is read at 520 nm.[4]

Reagents

Azocoll (Calbiochem)
1% Triton X-100
0.1 M Tris-HCl, pH 7.6
20% SDS in 0.1 M Tris-HCl, pH 7.6

Procedure. Use 20 mg of Azocoll for each 0.54 ml of solution made by mixing the three listed solutions in the proportion 0.125 ml of Triton X-100, 0.375 ml of buffer, 0.040 ml of SDS. Once the Azocoll is thoroughly wetted, 0.54 ml of the suspension is transferred with a wide-bore pipetter (cut $\frac{1}{4}$ inch off of a 1-ml pipette tip) to a tube. Extract (0.2 ml) of a suitable dilution is added and the tube is placed in a 37° constant temperature block. At 1-min intervals, each tube is removed and shaken gently (do not vortex) to resuspend the Azocoll, and replaced in the block. Avoid leaving Azocoll on the tube walls. At the end of the 15-min incubation, the tubes are plunged into ice and 3.5 ml or 2 ml of ice-cold distilled water is added. Tubes are immediately centrifuged for 3–5 min at 1650 g and the supernatants are removed to fresh tubes. The absorbance of these supernatants is relatively stable. Absorbance is read at 520 nm. Dilutions are chosen such that the kinetics are linear with time and protein concentration. Best results are obtained if the ΔA_{520} is less than 0.3. When comparing different strains, the extracts should be diluted such that the protein concentrations are similar for all strains. For extracts made from stationary phase wild-type cells, about 0.1 mg of extract protein/assay is appropriate. Correction is made for a blank lacking extract. A 0-min time point is needed for *ade2* mutant strains.

Definition of Unit and Specific Activity. One unit of PrB activity is defined as a change in absorbance at 520 nm of 1.0 per minute for the 0.74 ml of reaction mixture as assayed at 37°. For a reaction stopped with

[25] E. Juni and G. Heym, *Arch. Biochem. Biophys.* **127,** 89 (1968).

3.5 ml of water and run for 15 min, the conversion to units/milligram protein is

$$\frac{\Delta A_{520}}{15} \times \frac{4.24}{0.74} \times \frac{1}{0.2 \text{ (mg protein/ml extract)}}$$
$$= (1.91)(\Delta A_{520})/(\text{mg protein/ml extract})$$

Carboxypeptidase Y

Carboxypeptidase Y will catalyze cleavage of esters, amides, and peptides. An assay based on its amidase activity can be employed for kinetic analyses[26] or modified to a fixed time point assay.[7]

Principle. Carboxypeptidase Y will catalyze the hydrolysis of N-benzoyl-L-tyrosine p-nitroanilide to give the yellow p-nitroaniline. Production can be followed by absorbance at 410 nm.

Reagents

> N-Benzoyl-L-tyrosine p-nitroanilide (BTPNA) (Sigma) (6 mM): Dissolve 2.43 mg in 1 ml of dimethylformamide (DMF)
> 0.1 M Tris-HCl, pH 7.6
> 1 mM HgCl₂
> 20% SDS in 0.1 M Tris-HCl, pH 7.6

Procedure. To a tube containing 0.40 ml of 0.1 M Tris-HCl, pH 7.6, and 0.1 ml of extract at 37° is added 0.1 ml of 6 mM BTPNA in dimethylformamide. After 30 min, 1.5 ml of 1 mM HgCl₂ is added to stop the reaction. If the extract being assayed has low activity (as is typical for our wild-type strains), 0.2 ml of 20% SDS, pH 7.6, is added and, after vortexing, the tubes are incubated at 70° until solubilization of the protein, as evidenced by clearing, ensues. Absorbance at 410 nm is determined. Permeabilized cells can be used as an enzyme source in this assay. We have used cells permeabilized with 0.1–0.2% Triton X-100 or with 10–20% DMF.

Definition of Unit and Specific Activity. One unit of activity corresponds to 1 μmol of p-nitroaniline produced per minute, assuming a molar absorbance of 8800. Corrections for absorbance due to substrate and protein are made. The conversion to units/milligram protein is

$$\frac{\Delta A_{410}}{30} \times \frac{2.3}{8800} \times \frac{10^3}{0.1 \text{ (mg protein/ml extract)}}$$
$$= (0.087)(\Delta A_{410})/(\text{mg protein/ml extract})$$

Carboxypeptidase Y levels can also be determined using the kinetic assay described below for CpS, but using Cbz-Phe-Leu as the substrate,

---

[26] S. Aibara, R. Hayashi, and T. Hata, *Agric. Biol. Chem.* **35**, 658 (1971).

since 95% of the hydrolytic activity toward this peptide is apparently due to CpY.[17]

Carboxypeptidase S

Carboxypeptidase S will catalyze hydrolysis of dipeptides that are blocked at the amino terminus.

Principle. Free leucine released from a peptide by CpS catalysis is oxidized by amino-acid oxidase. Reduction of the product hydrogen peroxide by horseradish peroxidase is coupled to oxidation of *o*-dianisidine, yielding brown oxidized dianisidine. The absorbance at 405 nm is followed. The kinetic assay was developed by Wolf and Weiser.[17]

Reagents

0.2 M KPO$_4$ buffer, pH 7.0, containing 0.5 mM MnCl$_2$
L-Amino-acid oxidase type I (Sigma)
Horseradish peroxidase type I (Sigma)
Cbz-Gly-Leu (20 mM): 6.44 mg/ml 0.2 M KPO$_4$ buffer, pH 7.0 [or carbobenzoxy-L-phenylalanyl-L-leucine (15 mM) in 0.2 M KPO$_4$ buffer, pH 7.0, for CpY]
o-dianisidine dihydrochloride: 2 mg/ml water
0.2 M PMSF in 95% ethanol

Procedure. A solution is made in the proportion 1 ml of phosphate buffer–MnCl$_2$, 0.25 mg of amino-acid oxidase, 0.4 mg of peroxidase. To 0.5 ml of this is added 0.5 ml of the peptide solution followed by 0.05 ml of the *o*-dianisidine solution and 0.05 ml of dialyzed extract. The mixture is incubated at 25° and the absorbance followed at 405 nm. To render the assay specific for CpS when Cbz-Gly-Leu is the substrate, extracts are preincubated for 2 hr at 25° with 0.1 mM PMSF to inactivate CpY. Use Cbz-Phe-Leu as the substrate if CpY activity is to be measured with this assay.

Definition of Unit and Specific Activity. One unit corresponds to production of 1 nmol of L-leucine per min and 0.1 μmol of leucine corresponds to a change in absorbance of 0.725 for this procedure. Specific activity is expressed as nmol L-leucine/min/mg extract protein.

Designing the Most Useful Protease-Deficient Strain

A great deal of evidence, both published (see Ref. 27 for compilation) and anecdotal, suggests that a major source of *in vitro* proteolytic artifacts is PrB, and that elimination of this protease by mutation results in stabilization of numerous proteins and, indeed, even changes the gross protein signature revealed by Coomassie blue staining of SDS–polyacrylamide gels after electrophoresis of extracts. It is worth remembering that PrB activity in extracts is activated both by heating[27,28] and by treatment with SDS,[4,27] two treatments generally employed in preparing samples for SDS–polyacrylamide gel electrophoresis.

The usual approach taken to reduce proteolytic degradation of a useful protein is to start with a strain that carries a *pep4* mutation. Because *pep4* mutations are pleiotropic and greatly reduce or eliminate activity for PrA, PrB, CpY, and API, they have proved to be useful starting points. However, while attempting to purify the 40K precursor to PrB from a *pep4*-bearing strain, C. M. Moehle found that the precursor became activated to PrB of mature size during one column purification step.[29] We have no information on whether the activation was autocatalytic or due to the activity of another protease. Nonetheless, it seems clear that strains that carry the precursor to PrB may not be ideal. A better starting point may well be a *pep4 prb1* double mutant that carries no PrB precursor.

We have constructed and lodged the following strains in the Yeast Genetics Stock Center, MCB/Biophysics and Cell Physiology University of California, Berkeley, CA 94720 for the use of interested parties.

BJ926 α *trp1* + *prc1-126 pep4-3 prb1-1122 can1 gal2*
      a + *his1 prc1-126 pep4-3 prb1-1122 can1 gal2*
BJ1984 ≡ 20B-12 α *trp1 pep4-3 gal2*
BJ1991 α *leu2 trp1 ura3-52 prb1-1122 pep4-3 gal2*
BJ2168 **a** *leu2 trp1 ura3-52 prb1-1122 pep4-3 prc1-407 gal2*
BJ2407α *leu2 trp1 ura3-52 prb1-1122 prc1-407 pep4-3 gal2*
      a *leu2 trp1 ura3-52 prb1-1122 prc1-407 pep4-3 gal2*
BJ3501 α *pep4*::*HIS3 prb1-Δ1.6R his3-Δ200 ura3-52 can1*
BJ3505 α *pep4*::*HIS3 prb1-Δ1.6R his3-Δ200 lys2-801 trp1-Δ101 ura3-52 gal2 can1*

The *prb1-1122* and *pep4-3* alleles are nonsense mutations (UAA[30] and UGA[13], respectively). The *pep4*::*HIS3* mutation is an insertion of a *Bam*HI fragment bearing *HIS3* into the *Hin*dIII site in *PEP4*.[8] The *prb1-Δ1.6R* mutation is a deletion of a 1.6-kb *Eco*RI fragment internal to the *PRB1* gene.[31]

[27] J. R. Pringle, in "Methods in Cell Biology" (D. B. Prescott, ed.), Vol. 12, p. 149. Academic Press, New York, 1975.
[28] R. E. Ulane and E. Cabib, *J. Biol. Chem.* **251**, 3367 (1976).
[29] C. M. Moehle, Ph.D. Thesis, Carnegie Mellon University, Pittsburgh, Pennsylvania, 1988.
[30] G. S. Zubenko, A. P. Mitchell, and E. W. Jones, *Genetics* **96**, 137 (1980).
[31] C. M. Moehle, M. W. Aynardi, M. R. Kolodny, F. J. Park, and E. W. Jones, *Genetics* **115**, 255 (1987).

[32] Detection and Inhibition of Ubiquitin-Dependent Proteolysis

By KEITH D. WILKINSON

As the number of recombinant proteins being expressed in the yeast system has grown, it has become apparent that it is important to understand the systems by which this organism deals with "foreign" proteins. This article describes one such system, the covalent attachment of the protein ubiquitin and the subsequent proteolysis of the conjugated protein. This system could adversely effect the yields of recombinant protein if the protein were a substrate for the ubiquitin-dependent proteolysis system. Because our knowledge of and experience with this system are so limited, the focus of this article is to familiarize the reader with the system and suggest approaches which might be employed to detect the ubiquitination of, and minimize the degradation of, recombinant proteins being expressed in yeast. Because of the similarity of the system in all eukaryotic cells, this information is also applicable to other expression systems described in this volume.

Ubiquitin-Dependent System

Ubiquitin is a highly conserved and universally distributed eukaryotic protein of 76 amino acids. It is unique in that it becomes covalently attached to a variety of intracellular proteins.[1] The linkage is an amide bond between the carboxyl terminus of ubiquitin and ϵ- and/or α-amino groups on other proteins. This conjugation functions as a posttranslational modification which appears to target abnormal or damaged proteins for intracellular proteolysis.[2] Ubiquitin conjugation has also been implicated as being involved in maintenance of chromatin structure, in the cellular response to stress and heat shock, and in the structure and function of cellular receptors such as the platelet-derived growth factor receptor and the T-cell homing receptor.[2]

The levels and types of conjugates formed intracellularly can be examined by sodium dodecyl sulfate–polyacrylamide gel electrophoresis (SDS–PAGE) using either ^{125}I-labeled ubiquitin[1] or antibodies raised against cross-linked ubiquitin.[3] By either technique, a large number of conjugates

---

[1] K. D. Wilkinson, M. K. Urban, and A. L. Haas, *J. Biol. Chem.* **255**, 7529 (1980).

[2] K. D. Wilkinson, *Anti-Cancer Drug Des.* **2**, 211 (1987).

[3] A. L. Haas and P. M. Bright, *J. Biol. Chem.* **260**, 12464 (1985).

of a wide molecular weight range is observed. If high levels of a substrate protein are added, one observes the formation of a characteristic ladder of conjugates[4,5] resulting from the conjugation of one, two, and three or more molecules of ubiquitin to the target protein. It appears from these studies that a high ratio of ubiquitin-to-protein is required in order to generate a good substrate for the proteases.

Figure 1 illustrates this cytoplasmic pathway of ubiquitination and proteolysis as it is currently known. This can be conveniently broken down into three steps: activation of ubiquitin, conjugation of ubiquitin, and degradation of ubiquitin conjugates. Each is briefly discussed below.

Activation of Ubiquitin

The formation of the amide bond upon conjugation requires the prior activation of the carboxyl terminus of ubiquitin. A ubiquitin activating enzyme (E1) has been isolated and characterized, and found to catalyze the adenylation of the carboxyl terminus of ubiquitin. The enzyme also contains a thiol ester site with an active-site cysteine which attacks the bound adenylate, releasing AMP and forming an intramolecular thiol ester between the activating enzyme and ubiquitin. A second molecule of ATP and ubiquitin can then be bound and converted to the adenylate. The fully charged enzyme thus contains a bound ubiquitin as the thiol ester, as well as a tightly bound ubiquitin adenylate.[6]

Conjugation of Ubiquitin

The transfer of the activated forms of ubiquitin to the target protein requires one or two more enzymes. The thiol ester form of ubiquitin is first transferred to a free cysteine on one of a family of ubiquitin carrier proteins (E2).[7] These proteins are analogous to the acyl-carrier protein of fatty acid synthesis. At least two of these carrier proteins can directly transfer the bound ubiquitin to amino groups on basic proteins such as cytochrome *c* or histone H2a.[8] In other instances, the transfer is catalyzed by a distinct ligase (E3) with specificity for denatured proteins. Multiple ubiquitin molecules can become attached to a single protein molecule, resulting in a ladder of ubiquitin–protein conjugates.

The specificity of this conjugation, and thus the degradation of cellular

[4] A. Hershko, E. Leshinsky, D. Ganoth, and H. Heller, *Proc. Natl. Acad. Sci. U.S.A.* **81**, 1619 (1984).

[5] R. Hough, G. Pratt, and M. Rechsteiner, *J. Biol. Chem.* **261**, 2400 (1986).

[6] A. L. Haas and I. A. Rose, *J. Biol. Chem.* **257**, 10329 (1982).

[7] A. Hershko, H. Heller, S. Elias, and A. Ciechanover, *J. Biol. Chem.* **258**, 8206 (1983).

[8] C. M. Pickart and I. A. Rose, *J. Biol. Chem.* **260**, 1573 (1985).

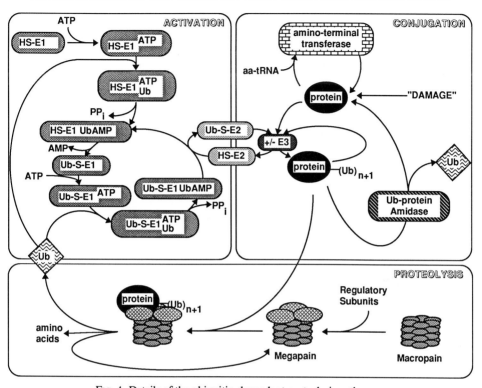

FIG. 1. Details of the ubiquitin-dependent proteolysis pathway.

proteins, is a complicated question. There is good evidence with at least some proteins that the nature of the amino-terminal amino acid is important in determining its degradation rate.[9] Recent work suggests that the conjugation of ubiquitin may involve three specific ubiquitin ligases with specificity for the amino-terminal residue of peptide chains.[10] Class I substrates have basic amino-terminal residues, class II substrates have bulky hydrophobic amino-terminal residues, and class III substrates contain amino-terminal residues which are neither bulky nor basic. It has been postulated that the rate of degradation of a protein with a stabilizing amino-terminal residue (class III) can be enhanced by arginylation, i.e., addition of arginine to the amino terminus to generate a more labile class I substrate. This reaction utilizes arginyl-tRNA and, presumably, a specific amino-terminal transferase.[11] This machinery offers a way in which the cell

[9] A. Bachmair, D. Finley, and A. Varshavsky, *Science* **234,** 179 (1986).
[10] Y. Reiss, D. Kaim, and A. Hershko, *J. Biol. Chem.* **263,** 2693 (1988).
[11] S. Ferber and A. Ciechanover, *Nature (London)* **326,** 808 (1987).

can modulate the stability of a normally stable protein in response to some metabolic signal.

Another modulation of a proteins' susceptibility to degradation is illustrated by the arrow labeled "damage" in Fig. 1. Thus, it is known that denatured proteins, oxidized proteins, and prematurely terminated polypeptide chains are preferred substrates for the ubiquitin-dependent system. The increase in degradation rate observed on damage may relate to the ability of the ligases (E3) to bind to denatured protein.[10]

Degradation of Ubiquitin Conjugates

Once ubiquitin conjugates are formed, they are subject to one of two fates: proteolysis of the protein and liberation of the attached ubiquitin, or deconjugation of ubiquitin to yield free ubiquitin and intact protein.[4] It is not clear what structural features of the conjugates are responsible for targeting conjugates to one or the other of these pathways.[12] It appears, however, that conjugates of damaged, denatured, or abnormal proteins are primarily degraded by proteolysis.[13]

Proteolysis of Conjugates. The specific proteolysis of ubiquitin–protein conjugates is accomplished by an ATP-dependent protease, presumably containing a specific binding site for ubiquitin. Preparations of this protease from reticulocytes resemble a previously identified ribonucleoprotein structure called the prosome which is found in all nucleated cells.[14] Two versions of this structure exist: a 19S version which is probably identical with the multicatalytic proteinase from rat, but does not show a dependence on ATP or ubiquitin, and a 26S version which has additional polypeptides associated with it, as well as a dependence on ATP and preference for ubiquitin conjugates as the substrate.[15] It appears that the larger version arises from the binding of regulatory subunits to the basic 19S structure, a cylinder consisting of four stacked rings or disks of subunits of molecular weight 25,000 to 32,000. It is possible that the additional polypeptides present in the larger complex confer the ubiquitin selectivity and that the role of ATP may be to translocate the peptide chain of the substrate through this structure. These two types of multimeric proteases have also been termed macropain (19S, 700 kDa) and megapain (26S, 1000 kDa) because of the apparent involvement of an essential thiol group. The products of the ubiquitin-dependent proteolysis are free amino acids and ubiquitin.

---

[12] M. J. Cox, A. L. Haas, and K. D. Wilkinson, *Arch. Biochem. Biophys.* **250**, 400 (1986).
[13] A. Hershko, E. Eytan, A. Ciechanover, and A. L. Haas, *J. Biol. Chem.* **257**, 13964 (1982).
[14] P. E. Falkenburg, C. Haass, P. M.-Kloetzel, B. Niedel, F. Kopp, L. Kuehn, and B. Dahlmann, *Nature (London)* **331**, 190 (1988).
[15] R. Hough, G. Pratt, and M. Rechsteiner, *J. Biol. Chem.* **262**, 8303 (1987).

Deconjugation of Ubiquitin. The known specificity of the ubiquitin-conjugating system makes it clear that most proteins can act as substrates for ubiquitination. This is also demonstrated by the large number and variety of ubiquitin–protein conjugates observed *in vitro* and *in vivo*.[1,3] In contrast, however, few proteins have been shown to be substrates for this system. This suggests that some conjugates are not metabolized by the proteases of the system. The alternative fate of ubiquitin–protein conjugates is removal of the nascent ubiquitin molecule. This is accomplished by one or more of a family of ubiquitin carboxyl-terminal hydrolases. These enzymes, ubiquitinyl-protein amidases, possess a binding site for ubiquitin and an active-site thiol required for activity.[16] An enzyme activity of approximately 38 kDa is responsible for deconjugating ubiquitin from ubiquitin–histone conjugates,[17] and an activity of approximately 200 kDa[4,5] is less specific and appears able to deconjugate ubiquitin from a wide variety of conjugates. We have suggested that these activities can act to proofread the structure of a ubiquitin–protein conjugate and release ubiquitin from conjugates with nativelike structures.[12] If, however, the protein is damaged or denatured, a conformational change in ubiquitin is triggered and these conjugates become preferred substrates for the protease complex. This mechanism allows the cell to "randomly sample" the conformation of cellular proteins by attaching ubiquitin, and efficiently degrade only those that are damaged.

Detecting Ubiquitin-Dependent Degradation

If proteolysis of an expressed protein is a problem, it would be useful to be able to evaluate if the degradation of that specific gene product was occurring through the ubiquitin-dependent pathway. In general, this can be examined in two different ways: by detection of the cleavage intermediate, the ubiquitin–protein conjugate, and by a demonstration that known inhibitors of the pathway prevent the observed degradation. It is possible that most proteins are ubiquitinated to some extent; therefore, the detection of protein–ubiquitin conjugates does not in itself indicate that the ubiquitin-dependent pathway is quantitatively the most important degradative event. The combination of detection of ubiquitin conjugates and demonstration that inhibition patterns are consistent with involvement of this system would be strong presumptive evidence that it is the major pathway of degradation.

[16] A. N. Mayer and K. D. Wilkinson, *Biochemistry* **28,** 166 (1989).
[17] F. Kanda, D. E. Sykes, H. Yasuda, A. A. Sandberg, and S. Matsui, *Biochem. Biophys. Acta* **870,** 64 (1986).

Immunological Detection of Conjugates Formed in Vivo

The detection of ubiquitin–protein conjugates is complicated by the presence of conjugates between ubiquitin and endogenous yeast proteins. Therefore, it is necessary to have an antibody against the protein of interest for the detection and immunoisolation of the newly expressed protein and any ubiquitin conjugates of it. The appropriate controls to establish the specificity of the antibody and immunostaining procedures are also necessary; i.e., all experiments would include samples from the parental yeast strain transfected with the control vector or an unrelated gene.

Antiprotein Antibodies. The most convenient way to detect specific ubiquitin conjugates is to lyse the cells directly into SDS sample buffer, followed by SDS–PAGE, then to transfer to nitrocellulose for immunodetection using the specific antibody against the recombinant protein. The hallmark of ubiquitin conjugation is the appearance of the characteristic ladder of cross-reacting species. These will appear on a Western blot as a series of bands spaced at intervals of approximately 8 kDa above the native molecular mass of the recombinant protein. The steady-state level of these conjugates may be quite low,[9] especially if the degradation of such conjugates is rapid.

Alternatively, if the levels of expression are too low to detect the protein directly, it is possible to pulse label the cells with [^{35}S]methionine, immunoisolate the protein and any conjugates, and detect the ladder of conjugates by radioautography of SDS gels of the immunoprecipitate.[9]

Antiubiquitin Antibodies. To confirm unambiguously that the ladder is the result of ubiquitin conjugation, the specific protein conjugates can be immunoisolated, and subjected to SDS–PAGE and Western blotting with antibody against cross-linked, SDS-denatured ubiquitin.[3] We have found that the published preparation of antigen and the antigenic response in rabbits are extremely reproducible. Antibodies made in our laboratory are identical in specificity and sensitivity to those originally reported.[3] The affinity purification step yields approximately 0.3 mg of affinity-purified IgG per milliliter of serum.

Immunological Staining Procedures. We have used a variation of the original immunostaining procedure[3] with good results in our laboratory. It is necessary to heat-treat the nitrocellulose filters after transfer by immersing them in boiling water for 30 min. This exposes immunological determinants in the conjugates and results in a much improved signal. The filters are blocked by incubation for 1 hr at room temperature in buffer A [20 mM Tris-buffered saline, 0.05% (v/v) Tween 20, 5% (w/v) milk solids] containing 0.1% (w/v) NaN$_3$. Affinity-purified primary antibody is added to a final concentration of 10 μg/ml in buffer A containing 0.1% NaN$_3$ and incubated for 3 hr at room temperature or overnight at 4°. The filter is next washed three times for 15 min each with buffer A and incubated with the detection reagent (see below) in buffer A for 3 hr at room temperature. After four washes of 15 min each in buffer B (20 mM Tris-buffered saline, 0.05% Tween 20) the filter is ready for signal development. We have used two different detection reagents. With ^{125}I-labeled staphylococcal-A protein, (5 × 10⁴ dpm/ml, 1.5 ng/ml) the linear range of detection is from 5 to 500 fmol/mm² of immunogen in 24 to 48 hr of exposure using Kodak X-OMAT AR film and Cronex intensifier screens. With horseradish peroxidase conjugated to goat anti-rabbit IgG antibodies (Sigma A-6154, 1/1000 dilution), color development is achieved by incubation with 0.33 mg/ml diaminobenzidine and 0.6% (v/v) H$_2$O$_2$ in buffer B. After a maximum of 5 min at room temperature, color development is stopped by washing the filters with water. Longer exposures to the substrate result in markedly nonlinear staining intensities. With this procedure the linear range of response is from 0.25 to 1 pmol of ubiquitin conjugate.

Use of Inhibitors to Detect Degradation in Vitro

Studies on the degradation of the expressed protein after lysis of the cells can also be useful in determining if the ubiquitin-dependent system is involved in the degradation of the recombinant protein. This is largely because the ubiquitin-dependent system is still active upon disruption of the cell, and a wide variety of inhibitors and biochemical interventions can be applied in extracts to define the nature of the degradation observed.

In this approach, cells are centrifuged, suspended in four volumes of cold homogenization buffer (50 mM Tris-HCl, pH 7.6, 5 mM MgCl$_2$, 5 mM ATP, 5 mM dithiothreitol, and 10 mM phosphocreatine), and broken by vortexing with glass beads at 4°. Cellular debris is removed by centrifugation at 13,000 g for 10 min at 4°, and the stability of the soluble protein of interest is examined in the presence and absence of various inhibitors of the ubiquitin-dependent system. Because of the high energy of activation for this proteolysis system, degradation should be minimal at low temperatures. Incubations are initiated by warming to 37° and adding 1 unit/ml of creatine phosphokinase. It is important to include the ATP-generating system in these incubations, since the endogenous ATP is rapidly depleted under these conditions. Techniques utilized to detect the protein of interest and measure its degradation in the extracts will vary with the nature of the protein. The only requirements are that the method be sufficiently sensitive, specific, and at least semiquantitative. These could include immunochemical, enzymatic, or affinity chromatographic approaches.

Table I lists some of the conditions which have been shown to inhibit ubiquitin-dependent proteolysis. All of this information is derived from

TABLE I

INHIBITORS OF UBIQUITIN-DEPENDENT PROTEOLYSIS AND SITE(S) OF ACTION[a]

Treatment	Substrate	Inhibition (%)	Target	Ref.
Analogs and cofactors				
ATP depletion	Lysozyme	100	E1, protease	b
Antiubiquitin (3 mg/ml)	Lysozyme	78	Ub depletion	b
Ubiquitin aldehyde	Lysozyme	60	Ub depletion	c
Iodoubiquitin (10 mM)	rcm-BSA	50	Protease	d
Chaotropic agents				
Oleate (125 mM)	LLVY-MCA	84	Protease	e
SDS (0.01%)	LLVY-MCA	98	Protease	e
NaCl (0.5 M)	Lysozyme	90	Protease, other?	f
Competitive amines				
Methylamine (125 mM)	Lysozyme	80	Competition for Ub?	f
Dansylcadavarine (10 mM)	Lysozyme	100	Competition for Ub?	f
Lysine (100 mM)	Lysozyme	70	Competition for Ub?	f
Metals or chelators				
o-Phenanthroline (1 mM)	Lysozyme	20	Zinc chelation	f
(5 mM)	Lysozyme	95		b
EGTA (0.1 mM)	Lysozyme	35	Calcium chelation	f
EDTA (5 mM)	Lysozyme	100	Magnesium chelation	f
Vanadate (2 mM)	Lysozyme	69	Unknown	f
Manganese (10 mM)	Lysozyme	45	Unknown	f
Miscellaneous				
Hemin (25 mM)	BSA	50	Proteases	g
(0.1 mM)	Lysozyme	90		f
(0.2 mM)	Lysozyme	100		b
Ribonuclease (6 mg/ml)	rcm-BSA	80	tRNA?	h
(60 mg/ml)	Lysozyme	0		
Acylated proteins (0.4 mg/ml)	rcm-BSA	50	E3	i
Phenylalanylalanine (2 mM)	ox-RNase	0	E3	j
	β-Lacto	90	E3	j
Lysylalanine (2 mM)	ox-RNase	65	E3	j
	β-Lacto	0	E3	j
Aurin tricarboxylate (0.2 mM)	Lysozyme	90	Unknown	f
Protease inhibitors				
PMSF (2 mM)	Lysozyme	25	Unknown	f
Chymostatin (0.25 mM)	LLVY-MCA	95	Proteases	e
TLCK (0.5 mM)	Lysozyme	50	Unknown	f
TPCK (0.15 mM)	Lysozyme	50	Unknown	f
(0.5 mM)	Lysozyme	90		
Thiol reagents				
N-Ethylmaleimide (10 mM)	Lysozyme	100	E1,E2	b
p-Hydroxymecuriphenyl-sulfonyl acid (2.5 mM)	LLVY-MCA	95	Protease	e

[a] Unless otherwise indicated, the inhibition was measured in protein degradation assays using reticulocyte fraction II. Ub, Ubiquitin; ATP, adenosine triphosphate; SDS, sodium dodecyl

the reticulocyte system, but indications are that a very high degree of homology exists between the systems in yeast and other eukaryotic cells. It is obvious that few of these inhibitors are absolutely specific for the ubiquitin system. Therefore, identification of a protein as a substrate for the ubiquitin system by this approach is limited to demonstrating patterns of inhibition similar to those compiled here.

Minimizing Ubiquitin-Dependent Degradation

If the studies suggested above demonstrate that the recombinant protein is being degraded via the ubiquitin-dependent system, a number of approaches can be taken to minimize this unwanted complication.

Expression as Secreted Protein

The evidence in most systems suggests that the conjugation and net proteolysis of the conjugates are restricted to abnormal or damaged proteins in the cytoplasm. Thus, expression of the desired recombinant protein as a fusion protein with a secretion signal attached is a viable alternative. This approach assumes that the fusion protein will be translocated into the secretory apparatus before it becomes conjugated to ubiquitin and that its expression is not lethal.

Acute Expression

Another approach would be to express the recombinant protein under control of an inducible promoter which could be turned on for a brief period of expression just before harvesting the cells. If enough protein is synthesized rapidly, it would be expected to saturate the capacity of the

sulfate; EGTA, ethylene glycol bis(β-aminoethyl ether)-N,N,N',N'-tetraacetic acid; EDTA, ethylenediaminetetraacetic acid; PMSF, phenylmethylsulfonyl fluoride; TLCK, tosyllysyl chloromethyl ketone; TPCK, tosylphenylalanyl chloromethyl ketone; BSA, bovine serum albumin; rcm-BSA, reduced, carboxymethylated bovine serum albumin; ox-RNase, oxidized RNase; β-Lacto, β-lactoglobulin; LLVY-MCA, succinylleucylleucylvalyltyrosylmethylcoumarin amide.

[b] A. Hershko, E. Leshinsky, D. Ganoth, and H. Heller, *Proc. Natl. Acad. Sci. U.S.A.* **81**, 1619 (1984).
[c] A. Hershko and I. A. Rose, *Proc. Natl. Acad. Sci. U.S.A.* **84**, 1829 (1987).
[d] M.J. Cox, A.L. Haas, and K.D. Wilkinson, *Arch. Biochem. Biophys.* **250**, 400 (1986).
[e] R. Hough, G. Pratt, and M. Rechsteiner, *J. Biol. Chem.* **262**, 8303 (1987).
[f] R. Hough, G. Pratt, and M. Rechsteiner, *J. Biol. Chem.* **261**, 2400 (1986).
[g] A. L. Haas and I. A. Rose, *Proc. Natl. Acad. Sci. U.S.A.* **78**, 6845 (1981).
[h] S. Ferber and A. Ciechanover, *J. Biol. Chem.* **261**, 3128 (1986).
[i] E. Breslow, R. Daniel, R. Ohba, and S. Tate, *J. Biol. Chem.* **261**, 6530 (1986).
[j] Y. Reiss, D. Kaim, and A. Hershko, *J. Biol. Chem.* **263**, 2693 (1988).

ubiquitin-dependent system, with the result that most of the expressed protein would be prevented from being conjugated to ubiquitin.

Genetic Constructs

Under some circumstances, it may be possible to minimize the complications due to ubiquitination by carefully choosing the genetic constructs utilized.

Cloning Backgrounds. Recent evidence suggests that the primary source of ubiquitin during periods of stress is the polyubiquitin gene *UBI4*.[18] This heat-shock gene codes for several head-to-tail repeats of the monomeric protein and, when deleted from yeast, allows normal vegetative growth. During periods of stress, however, these mutants do poorly and show a lower level of free ubiquitin compared to wild type.[18] Thus, expression of the recombinant protein in this background would be expected to diminish the capacity of the cells to conjugate ubiquitin to it, resulting in a greater yield of intact protein.

A second genetic locus that should be amenable to manipulation is the gene coding for the ubiquitin-activating enzyme (E1). A temperature-sensitive cell cycle mutant from a mouse mammary carcinoma cell line *(ts-85)*[19] has been identified as a mutation in the structural gene for E1. These mutants fail to conjugate ubiquitin or progress past early G_2 at the nonpermissive temperature. Identification or construction of such a mutant in yeast should allow expression of recombinant proteins under conditions where E1 is unable to activate ubiquitin (high temperature) and thus prevent conjugation.

Some information is available concerning the genes for the family of ubiquitin carrier proteins (E2). The yeast DNA repair gene, *RAD6*, encodes a 20-kDa E2 which is responsible for conjugation of ubiquitin to histones H2a and H2b.[20] Further work on these and other components of the E2 subfamily should result in mutants defective in the conjugation of ubiquitin to other classes of proteins, and may allow one to select an expression background where conjugation of ubiquitin to the desired recombinant protein is defective.

Finally, there is no information about the genes encoding the polypeptides of the proteases. However, the recent identification of this complex[15] should allow cloning and genetic manipulation of these enzymes as well.

---

[18] D. Finley, E. Ozkaynak, and A. Varshavsky, *Cell* **48,** 1035 (1987).
[19] D. Finley, A. Ciechanover, and A. Varshavsky, *Cell* **37,** 43 (1984).
[20] S. Jentsch, J. P. McGrath, and A. Varshavsky, *Nature (London)* **329,** 131 (1987).
[21] T. R. Butt, M. I. Khan, J. Marsh, D. J. Ecker, and S. T. Crooke, *J. Biol. Chem.* **263,** 16364 (1988).

Amino-Terminal Residues. One very important recent finding in yeast is that the identity of the amino-terminal residue can have a profound influence on the half-life of a protein, presumably reflecting the ability of the ubiquitin-dependent system to degrade it.[9] Thus, β-galactosidase with an unrelated 23-amino acid leader sequence exhibited a half-life of approximately 2 min with an amino-terminal Arg and over 20 hr with amino-terminal Met, Ser, Ala, Thr, Val, or Gly. The identity of the amino-terminal residue of a recombinant protein can be controlled by expressing the protein with the desired amino-terminal residue as a carboxyl-terminal extension of ubiquitin. The potent processing enzyme activity in yeast efficiently removes the ubiquitin and generates the recombinant protein with any amino acid except proline in the first position.[9,21] Subsequently, this protein is processed by the normal cellular machinery, and in at least some cases, substitution at the amino terminus can have a profound influence on the rate at which it is conjugated and degraded by the cells.

Conclusions

It is clear that the posttranslational conjugation of ubiquitin can have a profound influence on the structure and stability of proteins. Further, if an expressed protein is recognized as abnormal, this system will participate in the degradation of the protein with adverse effects on the yield of the desired product. In such cases, it is vital to identify this complication and minimize it. Techniques are well in hand to identify the problem, and an ever-increasing genetic knowledge of the system is opening the way to intervention. One note of caution should, however, be expressed to temper this optimistic view. The ubiquitin-dependent degradation system is universal and highly conserved in eukaryotic cells and we still do not understand the entirety of its functions. Thus, there may be a practical limit to how much we can cripple the system in the interest of limiting the degradation of recombinant proteins.

[33] Cotranslational Amino-Terminal Processing

By RICHARD L. KENDALL, RYO YAMADA, and RALPH A. BRADSHAW

Proteins are subject to a variety of co- and posttranslational processing steps that direct or determine function, location, and turnover in both prokaryotic and eukaryotic cells.[1] This article focuses on cotranslational processing events that occur at the N termini of nascent polypeptide chains and their effect on heterologous gene expression and protein stability. An understanding of these processes and the degree to which they can be presently manipulated is an important aspect of the production of natural and recombinant proteins.

Cotranslational Modifications

Prokaryotic translation is initiated with *N*-formylmethionine-tRNA binding to the 30S ribosomal subunit,[2] directed by the germane codon of the mRNA. However, during translation, both the *N*-formyl group and, in most but not all proteins, the methionine residue are removed quickly by a deformylase and a methionine aminopeptidase (MAP), respectively. This processing occurs in a stepwise fashion, with the formyl group removed first followed by the hydrolysis of the methionine.[3] The deformylase is highly specific for *N*-formylmethionine and will not remove the formyl group from other *N*-formylamino acids or from *N*-formylmethionine where the methionine has been oxidized.[4] Further, this enzyme will not deacetylate acetylated dipeptides.[3] MAP is less specific; this specificity is apparently mainly dependent on the identity of the adjacent or penultimate amino acid.[3,5]

Stepwise cotranslational processing also occurs in eukaryotic cells. In these cases, translation is initiated with an unmodified methionine, and processing of the nascent N terminus consists of methionine hydrolysis and/or acylation. The removal of the methionine, also by a methionine aminopeptidase, depends on the penultimate residue[6,7] and occurs when

[1] R. A. Bradshaw, *Trends Biochem. Sci.* **14**, 276 (1989).
[2] D. B. Wilson and H. M. Dintzis, *Proc. Natl. Acad. Sci. U.S.A.* **66**, 1282 (1970).
[3] J. Adams, *J. Mol. Biol.* **33**, 571 (1968).
[4] M. Takeda and R. Webster, *Proc. Natl. Acad. Sci. U.S.A.* **60**, 1487 (1968).
[5] C. Flinta, B. Persson, H. Jornvall, and G. Heijne, *Eur. J. Biochem.* **154**, 193 (1986).
[6] F. Sherman, J. W. Stewart, and S. Tsunasawa, *BioEssays* **3**, 27 (1986).
[7] S. Huang, R. C. Elliott, P. S. Liu, R. K. Koduri, J. Weickmann, J. H. Lee, L. C. Blair, P. Ghosh-Dastidar, R. A. Bradshaw, K. M. Bryan, B. Einarson, R. L. Kendall, K. Kolacz, and K. Saito, *Biochemistry* **26**, 8242 (1987).

the growing nascent chain is between 15 and 20 residues in length.[8,9] This leads to the production of ribosome-bound intermediates with and without the initiator methionine which can be subsequently modified by an N^α-acetyltransferase (NAT), or in a few special cases, by an N-myristoyltransferase (NMT), on the α-amino group.[1] NAT is reported to modify the N termini of growing polypeptide chains of about 25 to 70 residues.[10] Thus, the cotranslational processing of newly synthesized proteins by these enzymes can produce four classes of N terminals in eukaryotes, those with and without the initiator methionine residue and those with and without acylation. The purposes of these initial modifications are not fully understood but appear to be related, at least in part, to protein turnover.

It is important to note that these modifications, although affecting all proteins, are only manifested in proteins that are *not* further processed by signal peptidases. In those cases, the N terminus of the mature protein reflects an internal sequence and will rarely be modified on α-amino groups. However, if such proteins are expressed from nonsecretory constructs, i.e., with the signal peptide sequences deleted from the corresponding gene (or cDNA), the resulting protein may be isolated with the initiator methionine, an N^α-acetyl group or both. The converse, although less frequently encountered, would apply to an intracellular protein expressed with a secretory construct. In that case, other groups such as unwanted disulfides and glyco moieties may also be a problem.[1]

Effect of State of N Terminus on Function and Stability

The state of the N terminus of proteins, particularly those that occur in the cytoplasm and its contiguous compartments, may be an important consideration in the expression of eukaryotic proteins. This is reflected in the fact that about 80% of cytosolic proteins in mammalian cells are acetylated,[11] but only about 50% in *Saccharomyces fragilis* and *Neurospora crassa*,[12] and virtually none at all in prokaryotes.[13] N-Terminal modifications may affect biological function, stability and folding, translocation, and degradation, and can therefore influence gene (cDNA) design and choice of expression vector. A brief description of these problems follows.

The biological function of proteins may be altered if the N terminus is important for that function (or it affects the folding of the protein). The

[8] R. Jackson and T. Hunter, *Nature (London)* **227**, 672 (1970).
[9] A. Yoshida, S. Watanabe, and J. Morris, *Proc. Natl. Acad. Sci. U.S.A.* **67**, 1600 (1970).
[10] A. Pestana and H. C. Pitot, *Biochemistry* **14**, 1404 (1975).
[11] J. L. Brown and W. K. Roberts, *J. Biol. Chem.* **251**, 1009 (1976).
[12] J. L. Brown, *Int. Congr. Biochem. Abstr.* **11**, 90 (1979).
[13] J. P. Waller, *J. Mol. Biol.* **7**, 483 (1963).

N$^{\alpha}$-acetylation of melanocyte-stimulating hormone (MSH) increases the melanotropic effect of that molecule,[14] whereas this modification of β-endorphin suppresses its opioid activity.[15] The acetylation in these cases probably results from an acetyltransferase other than the cotranslational (ribosomal) enzyme, since these modifications occur posttranslationally following proteolytic processing. Another example of N-acetylation affecting the properties of a protein is the feline β-hemoglobin chain, where the oxygen affinity of the molecule is rendered insensitive to the modifying effect of organic phosphates.[16] Similar effects on the properties of hemoglobin molecules from other organisms have been noted. Thus, the unnatural removal of the initiator methionine or the introduction of a new blocking of the amino group could have an effect on the function of a recombinant protein.

The structure of the N terminus may also affect the stability of proteins either directly or by influencing degradation pathways. There is not much direct evidence for effect on stability, although observations such as that the unacetylated forms of the *Escherichia coli* ribosomal protein S5 and NADP-specific glutamate dehydrogenase from *Neurospora crassa*[17,18] are more thermosensitive than the acetylated forms of these proteins suggest that, in some cases, this may be important. Of greater significance is the effect on protein degradation. Jornvall made the observation that a majority of acetylated proteins were structural proteins.[19] He suggested that a possible role for acetylation is to protect the N terminus from degradation by aminopeptidases, thereby stabilizing these proteins (by decreasing turnover). Although, in general, acetylated and nonacetylated proteins in mammalian cells have the same turnover rates, there are clear exceptions to this rule. Johnson and colleagues have shown that the products of two alleles of hypoxanthine phosphoribosyltransferase which differ at the N terminus by an *N*-acetylalanine for proline substitution (and only at one other site) exhibit differential turnover.[20,21] In some cases, particularly with proteins exported from cells, posttranslation modification can protect against degradation and affect the lifetime of the agent. This can be of

[14] J. Ramachandran and C. H. Li, *Adv. Enzymol.* **29**, 391 (1967).

[15] D. G. Symth, D. E. Massey, S. Zakarian, and M. D. A. Finnie, *Nature (London)* **279**, 252 (1979).

[16] F. Taketa, A. G. Mauk, and J. L. Lessard, *J. Biol. Chem.* **246**, 4471 (1971).

[17] A. G. Cumberlidge and K. Isono, *J. Mol. Biol.* **131**, 169 (1979).

[18] M. A. M. Siddig, J. A. Kinsey, J. R. S. Fincham, and M. Keighren, *J. Mol. Biol.* **137**, 125 (1980).

[19] H. Jornvall, *J. Theor. Biol.* **55**, 1 (1975).

[20] G. G. Johnson and V. M. Chapman, *Genetics* **116**, 313 (1987).

[21] G. G. Johnson, W. A. Kronert, S. I. Bernstein, V. M. Chapman, and K. D. Smith, *J. Biol. Chem.* **263**, 9079 (1988).

marked practical significance for polypeptide therapeutics. The *in vitro* acetylation of the C-terminal octapeptide of cholecystokinin as protection against degradation by soluble proteases from smooth muscle tissues is a case in point.[22]

Recent studies by Varshavsky and colleagues have provided a basis at the molecular level for understanding the role of the N terminus of intracellular proteins in determining their stability.[23-25] Using an engineered ubiquitin – β-galactosidase fusion construct, which when expressed in yeast was quickly processed by "ubiquitinase" precisely at the junction of this fusion, producing a β-galactosidase molecule with a free N-terminal, they showed that the penultimate residue had a decisive effect on protein turnover. Site-directed mutagenesis was used to change the residue normally following the initiator methionine of the β-galactosidase portion, thus generating molecules of this protein with different free N terminals *in vivo;* the half-life of the β-galactosidase generated in this system depended on the N-terminal residue exposed. They found that stabilizing residues (producing proteins with half-lives of greater then 20 hr) are *generally* small (Ala, Ser, Thr, Gly, Val, and Met) and that destabilizing residues (producing proteins with half-lives of from 2 to 30 min) are large. They formulated the N-end rule based on these observations. Curiously, this specificity is quite similar to the proposed specificity of MAP.[6,7] The cotranslational MAP, by its specificity, removes the initiator methionine from proteins when the penultimate position is occupied by a stabilizing residue (except methionine), but not when it is a destabilizing residue.

While these observations provide a clear rationale for the function and specificity of MAP, such is not the case for the acylation reactions, particularly acetylation.[26] Both the specificity and the magnitude (or spectrum) of activity appear to vary somewhat from species to species, and there is generally less evidence to support a stabilizing role. In yeast, N-terminal alanine, serine, threonine, and glycine generally are acetylated[7] and the isolated enzyme showed a pattern consistent with this.[27] In addition, the initiator methionine was acetylated when the adjacent residue was aspartic acid, glutamic acid, or asparagine, also in keeping with the distribution

[22] M. Praissman, J. W. Fara, L. A. Praissman, and J. M. Berkowitz, *Biochim. Biophys. Acta.* **716**, 240 (1982).

[23] A. Bachmair, D. Finley, and A. Varshavsky, *Science* **234**, 179 (1986).

[24] V. Chan, J. W. Tobias, A. Bachmair, D. Marriott, D. J. Ecker, D. K. Gonda, A. Varshavsky, *Science* **243**, 1576 (1989).

[25] A. Bachmair and A. Varshavsky, *Cell* **56**, 1019 (1989).

[26] S. M. Arfin and R. A. Bradshaw, *Biochemistry* **27**, 7979 (1988).

[27] F. -J. S. Lee, L. -W. Lin, and J. A. Smith, *J. Biol. Chem.* **263**, 14948 (1988).

observed from an analysis of the protein sequence data base.[28] Penultimate proline, valine, and cysteine, from which the methionine is readily, if not completely removed, are rarely acetylated nor are the initiator methionine residues that protect the isoleucine, leucine, methionine, glutamine, lysine, arginine, histidine, tyrosine, phenylalanine, and tryptophan when they occur in the second position. Since these latter residues are not exposed cotranslationally, they are not presented as possible substrates.

The acetylation of the methionine residues preceding aspartic acid, glutamic acid, and asparagine renders these proteins potential substrates for an acetylmethionine-specific aminoacyl hydrolase.[29,30] Removal of this grouping intact exposes the aspartic or glutamic acid and allows for tRNA-dependent arginylation by arginine tRNA protein transferase.[31] (Asparagine exposed in this fashion must first be deamidated.) Since N-terminal arginine is among the most destabilizing residues of the N-end rule,[23] this "pathway," which is possible because of the combined specificities of MAP and NAT, allows for the degradation of a select, predetermined groups of proteins.[26]

A major technical problem associated with N-acylation is the refractiveness of such proteins to any N-terminal degradation process (chemical or enzymatic). N^{α}-Acetylation, which overwhelmingly occurs in the soluble protein fraction of eukaryotic cells,[11] cannot be reversed systematically or efficiently. Some deacylases have been described but their specificities are always highly restricted and applicable only to short peptides.[32,33] While this problem can eventually be overcome in whole sequence analyses by overlapping peptides and techniques that do not depend on a free α-amino group, such as mass spectroscopy, it is a severe impediment to characterization of synthetic proteins or matching real protein sequence data to predicted structures.

Overexpression of heterologous proteins in different hosts has, in some instances, resulted in the production of a heterogeneous population of molecules. In *E. coli,* mixtures of molecules retaining the N-formyl group

[28] H. P. C. Driessen, W. W. de Jong, G. I. Tesser, and H. Bloemendal, *CRC Crit. Rev. Biochem.* **18,** 281 (1985).

[29] F. Wold, *Trends. Biochem. Sci.* **9,** 256 (1984).

[30] G. Radhakrishna and F. Wold, *J. Biol. Chem.* **261,** 9572 (1986).

[31] R. L. Soffer, *in* "Transfer RNA: Biological Aspects" (D. Soll, J. N. Abelson, and P. Schimmel, eds.), p. 493. Cold Spring Harbor Laboratory, Cold Spring Harbor, New York, 1980.

[32] K. Kobayashi and J. A. Smith, *J. Biol. Chem.* **262,** 11435 (1987).

[33] W. M. Jones and J. M. Manning, *Biochem. Biophys. Res. Commun.* **126,** 933 (1985).

and the initiator methionine have been reported.[34,35] Likewise, overexpression in yeast has led to proteins being incompletely processed by both MAP and NAT.[6] Small- and large-scale preparations of proteins by this method may require processing to produce a homogeneous population of molecules. Chemical modification of methionine or N-acylmethionine is possible by treatment with cyanogen bromide (CNBr); however, this method is not gentle and if there are any internal methionines the molecule will be cleaved at those positions. One approach to this problem is enzymatic: two groups[35,36] have cloned the MAP gene from *E. coli* and *Salmonella,* and, in one study, purified MAP was used to remove the N-terminal methionine from human interleukin-2 (IL-2) produced in *E. coli* without any other alterations to the molecule. A related strategy is to overexpress MAP and the protein of interest in the same cell, thus ensuring complete removal of initiator methionine *in situ.* Potentially, the gene encoding the deformylase enzyme from *E. coli* could be used in a similar fashion.

Strategies for Producing Desired N-Terminal Structure

With the developing appreciation of how different translocation processes affect protein structure,[1] it is possible to manipulate expression systems so as to alter certain features of the desired product. For example, by constructing secretory vectors (with signal peptides), one can introduce disulfide bonds through natural processes, whereas a nonsecretory construct will produce a reduced (and unusually denatured product).

The specificities of MAP and NAT can also be utilized to produce proteins with varying N termini, and thus potentially affect their properties and stability. The same N termini can also be produced by selective processing of fusion proteins using either chemical or enzymatic reagents. Finally, modifications can be manipulated in some cases by blocking the acylation reaction and by removing N termini enzymatically. The latter will probably be of principal value in overproduction situations. A similar approach to that used to remove N^α-acetyl groups is not practical at this time.

[34] S. Tsunasawa, K. Yutani, K. Ogasahara, M. Taketani, N. Yasuoka, M. Kadudo, and Y. Sugino, *Agric. Biol. Chem.* **47,** 1393 (1983).
[35] A. Ben-Bassat, K. Bauer, S.-Y. Chang, K. Myambo, A. Boosman, and S. Chang, *J. Bacteriol.* **169,** 751 (1987).
[36] C. G. Miller, K. L. Straunch, A. M. Kukral, J. L. Miller, P. J. Wingfield, G. J. Mazzei, R. C. Werlen, P. Graber, and N. R. Movva, *Proc. Natl. Acad. U.S.A.* **84,** 2718 (1987).

Amino-Terminal Structure Selection

Substrate Specificities. For intracellular proteins (nonsecretory gene constructions), the combined activity of NAT and MAP can produce four classes of proteins: those with and without methionine and those with and without N-acetyl groups. The assignments observed with the expression of a simple synthetic gene in yeast are listed in Fig. 1.[7] The methionine distribution is in excellent agreement with that predicted by Sherman *et al.*,[6] and seems highly constant over a wide range of organisms. Accordingly, the retention of the initiator methionine (without subsequent modification) can be safely predicted to occur with any of 10 amino acids in the penultimate position. By substituting aspartic acid or glutamic acid, the methionine will most likely be acetylated as well. It is noteworthy that the stability of this latter class is likely governed (at least to some degree) by aminoacyl hydrolyase(s) capable of removing N^α-acetylmethionine. Little is presently known about their function or activity. They appear to be localized in the cytoplasm.

In a similar fashion, initiator methionine will be removed efficiently from proteins with penultimate alanine, glycine, serine, threonine, cysteine, proline, and valine. The first four of these residues are also likely to be acetylated. If this is desirable, alanine and serine appear to be the best (most reliable) substrates for NAT. Although found to occur naturally, synthetic substrates with glycine and threonine in the N-terminal position have proved to be poor substrates for rat liver NAT *in vitro* (R. Yamada and R. A. Bradshaw, unpublished observations). It has also been observed that the second and third positions (counting the penultimate residue as the first position) can be dramatic in their effect on NAT activity (R.

	Met (−)	Met (+)
N^α–acetyl (−)	C, P, V	F, H, I, K, I M, Q, R, W, Y
N^α–acetyl (+)	A, G, S, T	D, E, N

FIG. 1. Distribution of protein N termini resulting from the actions of MAP and NAT in eukaryotic cells. Abbreviations: A, Ala; C, Cys; D, Asp; E, Glu; F, Phe; G, Gly; H, His; I, Ile; K, Lys; L, Leu; M, Met; N, Asn; P, Pro; Q, Gln; R, Arg; S, Ser; T, Thr; V, Val; W, Trp; Y, Tyr. Plus and minus signs indicate presence or absence of a methionine and/or N^α-acetyl group.

Yamada and R. A. Bradshaw, unpublished observations), and probably account for the much larger number of exceptions from the acetylation pattern shown in Fig. 1.[5,27,28,37] Serine, alanine, threonine, aspartic acid, tyrosine, and phenylalanine have quite a positive effect in the second position, while glycine, lysine, arginine, asparagine, proline, valine, isoleucine, and tryptophan are rather inhibitory. All were measured with serine as the N-terminal residue. Most of the amino acids exerted much less effect in the third position. Glycine, aspartic acid, glutamic acid, and tryptophan were notably negative in their effect. Clearly, depending on whether acetylation is desired or not, these residues may be incorporated or avoided.

To achieve expression of a protein with unblocked N-terminal residues (with initiator methionine removed), valine in the penultimate position appears to be the most reliable choice. Both cysteine and proline can lead to only partial removal of the methionine.[6]

Cleavage Methods. Many proteins are expressed as "fusions" and, when the foreign protein is positioned to the N-terminal side (5'-end) of the desired protein, the method of cleavage will dictate the N terminus that will be formed. Clearly, products of this approach will not be acetylated.

The most common chemical method employed for this purpose is CNBr, which cleaves polypeptides at methionine in relatively strong acid solutions. However, the method is not selective, meaning the desired protein must be devoid of methionine naturally or be engineered to convert all the existing methionine residues to acceptable substitutes (leucine is commonly used). Of course, not all proteins will tolerate such substitutions. This approach does have the advantage that essentially any residue can be adjacent to the methionine (and thus end up as the N-terminal residue) although Met-Ser and Met-Thr bonds can be rather resistant and should probably be avoided for this use.

Two other chemical methods utilize the facile cleavage of Asn-Gly and Asp-Pro bonds, respectively. The former can be specifically cleaved by NH_2OH, while the latter is unstable in acid (70% formic acid overnight). As with CNBr, these sequences must be unique (with respect to the desired protein) and will render N termini of glycine and proline, respectively.

Enzymatic hydrolysis can also be used to process fusion proteins providing the susceptible bond is substantially more labile then the other bonds normally cleaved by that protease. This is usually accomplished by inserting a sequence assumed to be poorly structured (not highly folded) between the two major protein domains of the fusion complex. If proper folding is not achieved first, the yield of the desired protein is likely to be very low.

---

[37] J.-P. Boissel, T. J. Kasper, and H. F. Bunn, *J. Biol. Chem.* **263**, 8443 (1988).

Greater selectivity can be introduced by inserting sequences that are recognized by proteases normally functioning with high selectivity. The ubiquitin construct employed by Bachmair et al.[23] is an excellent example. The naturally occurring ubiquitinase readily removed the N-terminal ubiquitin extension immediately following synthesis. This method could easily be employed in expression systems by incorporating the C-terminal sequence of ubiquitin as a bridge between the fusion protein and the product sought and expressing the entire construct as a secretory protein. The ubiquitinase (as an extract) can be added to the purified fusion protein subsequently to produce the desired protein. Clearly, any N-terminal residue (unblocked) could be generated by the approach. Proteases, such as epidermal growth factor-binding protein (M. Blaber and R. A. Bradshaw, unpublished experiments) and factor Xa,[24] with well-defined specificities, are additional examples of enzymes suitable for use in this approach.

Inhibition of Acetylation

Two methods of synthesizing a protein with an unblocked N-terminus in a cell-free system have been described.[38,39] One such approach, reported by Palmiter,[38] takes advantage of the requirement for acetyl-CoA by N^α-acetyltransferase.[40] Acetyl-CoA is depleted from the translation system by citrate synthase with the addition of oxaloacetate.

$$\text{Acetyl-CoA} + \text{oxaloacetate} \xrightarrow{\text{citrate synthase}} \text{CoA} + \text{citrate}$$

The translation lysate is preincubated for 5 min at 26° with 1 mM oxaloacetate (final concentration) and 30 units of citrate synthase. Ammonium sulfate concentrations greater than 20 mM should be avoided as this will inhibit the translation reaction [citrate synthase supplied by Sigma as a 2.2 M (NH$_4$)$_2$SO$_4$ suspension].

Rubenstein et al.[41] used the above method to synthesize *Dictyostelium discoideum* actin in a reticulocyte translation system. They found routinely only 25–30% of the actin produced was nonacetylated. The addition of S-acetonyl-CoA (a nonreactive acetyl-CoA analog[40]) to lysates at concentrations as high as 100 μM produced similar levels of nonacetylated actin. However, when the acetyl-CoA trap and S-acetonyl-CoA (75 μM) were used together, 85% of the actin synthesized was found to be in the unacetylated form. Clearly, the most effective inhibition of NAT in cell-free systems is achieved by use of both the "trap" and the competitive inhibitor.

[38] R. D. Palmiter, *J. Biol. Chem.* **252**, 8781 (1977).
[39] P. Rubenstein and R. Dryer, *J. Biol. Chem.* **255**, 7858 (1980).
[40] J. A. Traugh and S. B. Sharp, *J. Biol. Chem.* **252**, 3738 (1977).
[41] P. Rubenstein, P. Smith, J. Deuchler, and K. Redman, *J. Biol. Chem.* **256**, 8149 (1981).

Isolation of MAP (from Prokaryotic Sources)

Recently, the purification of prokaryotic MAP from *Salmonella typhimurium*[36,42] and *E. coli*[35] has been reported. Briefly, *S. typhimurium* cells are washed in 20 mM sodium phosphate, pH 7.5, 5 mM MgCl$_2$, and 250 mM sucrose, then lysed by two passes through a French press at 124 mPa in 100 mM sodium phosphate, pH 7.0, and centrifuged at 10,000 g for 30 min. The resulting supernatant is centrifuged at 60,000 g for 60 min and filtered through a membrane with a 0.45 μm pore size. The extract is applied to a DEAE-Sepharose column equilibrated with 20 mM sodium phosphate buffer, pH 7.0. The MAP activity of interest elutes in the flow-through fractions. The pooled active fractions are adjusted to pH 8.0 with NaOH and applied to a second DEAE-Sepharose column equilibrated with 50 mM sodium phosphate, pH 8.0, and the activity eluted with a linear gradient of 0–0.2 M NaCl. The active fractions are again pooled, concentrated by ultrafiltration, and dialyzed against 5 mM sodium phosphate, pH 6.8, 0.5 mM dithiothreitol and then against 5 mM sodium phosphate. This material is applied to a hydroxyapatite column and eluted with a linear gradient from 5 to 200 mM sodium phosphate, pH 6.8. The pooled fractions are adjusted to 0.5 mM dithiothreitol, concentrated by ultrafiltration, and applied to an Ultragel AcA 54 column (95 cm $\times$ 5 cm) equilibrated with 50 mM sodium phosphate, pH 7.5, 1 mM sodium azide.

The active fractions are pooled, made 5% (w/v) in sucrose, and in this fashion can be stored at −70°.[42] The enzyme has been used to remove N-terminal methionines from recombinant IL-1β[36] and human IL-2 *in vitro*.[35]

Acknowledgments

Aspects of this work emanating from the authors' laboratory were supported by USPHS grant DK 32465. The expert assistance of Ms. Kisma Stepanich in preparing this manuscript is gratefully acknowledged.

[42] P. Wingfield, P. Graber, G. Turcatti, N. P. Movva, M. Pelletier, S. Craig, Rose, K. and C. G. Miller, *Eur. J. Biochem.* **180**, 23 (1989).

[34] α-Factor Leader-Directed Secretion of Heterologous Proteins from Yeast

By ANTHONY J. BRAKE

The development of a wide variety of useful plasmid vectors, promoters, and host strains, combined with the foundation of classical genetic and biochemical methods, has resulted in the emergence of the bakers' yeast *Saccharomyces cerevisiae* as the most widely used eukaryotic microorganism for the expression of heterologous proteins. Almost any gene product can now be expressed at some level in yeast, and there are many examples of foreign proteins being expressed at levels higher than any endogenous yeast protein.[1] With the well-developed methods available for the propagation of yeast to very high cell densities and in large volumes of inexpensive media, such recombinant yeast strains are ideal for the production of proteins on an industrial scale. On the other hand, readily available vectors and strains,[2] as well as simple equipment and media requirements, make the production of proteins for research use possible in virtually any biology or biochemistry laboratory. The secretion of heterologous proteins from yeast is particularly desirable, since the level of endogenous secreted proteins is quite low, thus greatly simplifying the purification of the desired protein. In addition, disulfide bonds are more likely to be properly formed in proteins passing through the secretory pathway than in proteins expressed in the reducing environment of the cytoplasm and then oxidized after cell breakage.

A number of different leader sequences have been employed to direct the secretion of proteins from *S. cerevisiae*. In a few cases, expression of the naturally occurring precursor forms of foreign proteins has resulted in the secretion of the correctly processed, mature proteins.[3-5] More commonly, fusions of leader sequences from yeast proteins such as invertase[6,7] have

[1] P. J. Barr, A. J. Brake, and P. Valenzuela, "Yeast Genetic Engineering." Butterworth, Stoneham, Massachusetts, 1989.

[2] S. A. Parent, C. M. Fenimore, and K. A. Bostian, *Yeast* **1**, 83 (1985).

[3] R. A. Hitzeman, D. W. Leung, L. J. Perry, W. J. Kohr, H. L. Levine, and D. V. Goeddel, *Science* **219**, 620 (1983).

[4] M. A. Innis, M. J. Holland, P. C. McCabe, G. E. Cole, V. P. Wittman, R. Tal, K. W. K. Watt, D. H. Gelfand, J. P. Holland, and J. H. Meade, *Science* **228**, 21 (1985).

[5] T. Sato, S. Tsunasawa, Y. Nakamura, M. Emi, F. Sakiyawa, and K. Matsubara, *Gene* **50**, 247 (1986).

[6] R. A. Smith, M. J. Duncan, and D. T. Moir, *Science* **229**, 1219 (1985).

[7] C. N. Chang, M. Matteucci, L. J. Perry, J. J. Wulf, C. Y. Chen, and R. A. Hitzeman, *Mol. Cell. Biol.* **6**, 1812 (1986).

been used to direct the secretion of heterologous proteins. The focus of this article is on the use of the leader sequence of the precursor of the yeast mating hormone α-factor, which has been the leader most widely used for the secretion of foreign proteins from yeast.[8]

Secretion in *Saccharomyces cerevisiae*

Saccharomyces cerevisiae has become an important model system for the study of protein transport secretion. Of particular importance has been the isolation of a large collection of *sec* mutants blocked at specific points in the secretory pathway at a nonpermissive temperature.[9] Considerable biochemical and genetic evidence demonstrates that the pathways of protein transport and secretion in yeast appear to have much in common with those in other eukaryotic organisms.[10,11] The large amount of research being carried out on protein transport and secretion in yeast continues to provide an increasingly detailed understanding of the underlying cell biology and biochemistry.

As in other eukaryotes, N-linked and O-linked carbohydrate chain addition occurs onto proteins transported to the cell surface or to the lysosome-like vacuole. This is an important consideration when attempting to secrete heterologous proteins containing sites (Asn-X-Thr/Ser) for asparagine-linked or O-linked carbohydrate addition. Although the structure of the core oligosaccharide transferred to Asn residues of yeast glycoproteins is identical to that found in mammalian glycoproteins, subsequent modifications of these chains are quite different in yeast.[12] Instead of the extensive "trimming" of the oligosaccharide core and subsequent addition of sialic acid, galactose, and N-acetylglucosamine as in mammalian cells, secreted yeast glycoproteins undergo less extensive trimming of the core followed by elongation with long "outer chains" of mannose residues (50 or more). As a result of these differences, heterologous glycoproteins secreted from yeast may be inactive or antigenically different from the natural proteins. This may be especially important in the case of proteins

---

[8] A. J. Brake, *in* "Yeast Genetic Engineering" (P. J. Barr, A. J. Brake, and P. Valenzuela, eds.), p. 269. Butterworth, Boston, Massachusetts, 1989.

[9] P. Novick, C. Field, and R. Schekman, *Cell* **21**, 205 (1980).

[10] R. Schekman and P. Novick, *in* "The Molecular Biology of the Yeast *Saccharomyces cerevisiae:* Metabolism and Gene Expression" (J. N. Strathern, E. W. Jones, and J. R. Broach, eds.), p. 361. Cold Spring Harbor Laboratory, Cold Spring Harbor, New York, 1982.

[11] R. Schekman, *Annu. Rev. Cell Biol.* **1**, 115 (1985).

[12] C. E. Ballou, *in* "The Molecular Biology of the Yeast *Saccharomyces cerevisiae:* Metabolism and Gene Expression" (J. N. Strathern, E. W. Jones, and J. R. Broach, eds.), p. 335. Cold Spring Harbor Laboratory, Cold Spring Harbor, New York, 1982.

destined for human therapeutic use.[13] Efforts have been made to eliminate or minimize this problem through the use of yeast mutants defective in the ability to add outer-chain oligosaccharides.[14,15]

Secreted yeast proteins also contain short mannooligosaccharide chains of one to four residues attached to serine and threonine residues. As in the case of mammalian glycoproteins, no defined amino acid sequence signal for this modification has been found. Heterologous proteins secreted from yeast have been found to be O-glycosylated at the same sites found in the natural products.[4,16,17]

Biosynthesis of α-Factor

Haploid *S. cerevisiae* cells of the α mating type secrete a 13-residue peptide required for efficient mating with cells of the opposite **a** mating type.[18] Like other peptide hormones, α-factor is derived from larger precursor polypeptides encoded by two structural genes, *MFα1* and *MFα2*.[19,20] The protein sequence deduced from the major structural gene *MFα1*, shown in Fig. 1, is a 165-residue polypeptide containing four repeats of the mature α-factor peptide, each preceded by a spacer peptide of 6–8 residues with the structure Lys-Arg-(Glu/Asp-Ala)$_{2-3}$. These repeats are preceded by an 83-residue leader sequence containing a hydrophobic signal sequence and three potential sites for Asn-linked oligosaccharide addition. The minor *MFα2* gene encodes a similar precursor polypeptide containing only two repeats of the mature α-factor peptide.

Immunochemical analysis of α-factor-related species, resulting from *in vitro* translation during passage through the secretory pathway in various *sec* mutants confirmed that α-factor is derived from larger, glycosylated

[13] M. A. Innis, *in* "Yeast Genetic Engineering" (P. J. Barr, A. J. Brake, and P. Valenzuela, eds.), p. 233. Butterworth, Stoneham, Massachusetts, 1989.

[14] D. T. Moir and D. R. Dumais, *Gene* **56**, 209 (1987).

[15] C. L. Yip, S. K. Welch, T. Gilbert, and V. L. MacKay, *Yeast* **4**, S457 (1988).

[16] J. F. Ernst, J.-J. Mermod, J. F. DeMarter, R. J. Mattaliano, and P. Moonen, *Bio/Technology* **5**, 831 (1987).

[17] J. N. Van Arsdell, S. Kwok, V. L. Schweichart, M. B. Ladner, D. H. Gelfand, and M. A. Innis, *Bio/Technology* **5**, 691 (1987).

[18] J. Thorner, *in* "The Molecular Biology of the Yeast *Saccharomyces cerevisiae:* Life Cycle and Inheritance" (J. N. Strathern, E. W. Jones, and J. R. Broach, eds.), p. 143. Cold Spring Harbor Laboratory, Cold Spring Harbor, New York, 1981.

[19] J. Kurjan and I. Herskowitz, *Cell* **30**, 933 (1982).

[20] A. Singh, E. Y. Chen, J. M. Lugovoy, C. N. Chang, R. A. Hitzeman, and P. W. Seeburg, *Nucleic Acids Res.* **11**, 4049 (1983).

[21] A. Brake, D. Julius, and J. Thorner, *Mol. Cell. Biol.* **3**, 1440 (1983).

[22] O. Emter, B. Mechler, T. Achstetter, H. Müller, and D. H. Wolf, *Biochem. Biophys. Res. Commun.* **116**, 822 (1983).

[23] D. Julius, R. Schekman, and J. Thorner, *Cell* **36**, 309 (1984).

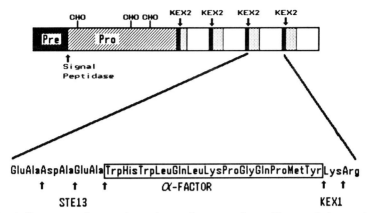

Fig. 1. Structure and processing pathway of prepro-α-factor. The translation product of the *MFα1* gene has three sites for Asn-linked oligosaccharide addition (CHO) and sites for endoproteolytic cleavage by signal peptidase and the Kex2 protease. An expanded view (below) of the peptide released by Kex2 cleavage shows sites for exoproteolytic processing by products of the *STE13* and *KEX1* genes.

precursor polypeptides.[21-23] The translocation and processing of prepro-α-factor have also been studied *in vitro*.[24-26] Processing of prepro-α-factor requires four different proteolytic activities. The availability of mutations in the appropriate genes as well as the corresponding cloned genes has resulted in α-factor being the first peptide hormone whose processing pathway has been genetically defined.[27]

Signal peptidase cleavage occurs between residues 19 and 20 of the prepro-α-factor.[28] The apparent lack of an intermediate resulting from signal peptide cleavage of the primary translation product appears to have been due to anomalous gel mobility of the two species. Signal peptide cleavage is blocked in *sec11* mutants, suggesting that *SEC11* may be the structural gene for signal peptidase.[29]

The glycosylated pro-α-factor is subsequently cleaved by an endoproteinase cleaving on the carboxyl side of the Lys-Arg sequence in the "spacer" peptide of each repeat. This cleavage is blocked in *kex2* mutants, resulting in a mating defect in *MATα* strains, failure to secrete active killer toxin peptide, and in the secretion of a hyperglycosylated form of pro-α-factor.[30,31] The *KEX2* gene was cloned on the basis of complementation of

[24] J. A. Rothblatt and D. Meyer, *Cell* **44,** 619 (1986).
[25] W. Hansen, P. B. Garcia, and P. Walter, *Cell* **45,** 397 (1986).
[26] M. G. Waters and G. Blobel, *J. Cell Biol.* **102,** 1543 (1986).
[27] R. S. Fuller, R. E. Sterne, and J. Thorner, *Annu. Rev. Physiol.* **50,** 345 (1988).
[28] M. G. Waters, E. A. Evans, and G. Blobel, *J. Biol. Chem.* **263,** 6209 (1988).
[29] P. C. Bohni, R. J. Deshaies, and R. W. Schekman, *J. Cell Biol.* **106,** 1035 (1988).
[30] M. J. Leibowitz and R. B. Wickner, *Proc. Natl. Acad. Sci. U.S.A.* **73,** 2062 (1976).
[31] D. Julius, A. Brake, L. Blair, R. Kunisawa, and J. Thorner, *Cell* **37,** 1075 (1984).

kex2 mutants and has been shown to be a membrane-bound, calcium-dependent serine protease homologous to subtilisin and related proteases.[32-34]

Following excision of each of the repeats from pro-α-factor, maturation to mature α-factor requires exoproteolytic processing at both the C and N termini. The *STE13* gene encodes a membrane-bound, heat-stable dipeptidylaminopeptidase, which is responsible for removal of Glu-Ala and Asp-Ala dipeptides from the N terminus of each repeat.[35] Mutations in this gene result in sterility of α haploids due to the defect in α-factor processing, with no other known phenotype.

Also required for maturation of α-factor is removal of the arginyl and lysyl residues at the C terminus of each of the first three repeats. The *KEX1* gene has been shown to encode a serine protease (of the trypsin family) with the required carboxypeptidase B-like specificity for C-terminal Lys and Arg residues on α-factor and killer toxin precursors.[36] The *KEX1* gene is essential for production of active killer toxin peptide. The failure of *kex1* mutants to show an α-specific sterile phenotype is apparently due to the fact that the α-factor species derived from the fourth repeat of prepro-α-factor does not possess C-terminal Lys and Arg residues, and that α-factor production at a reduced level (presumably about 25% that of an isogenic *KEX1* strain) is apparently sufficient for normal mating.

Expression in Yeast of Hybrid Proteins Containing the α-Factor Leader

In order to determine whether the leader sequence of prepro-α-factor contained sufficient sequence information for targeting α-factor for secretion and processing, and whether this could be utilized for a general secretion system, gene fusions were constructed which joined the prepro region to various other proteins. The first such hybrid protein reported contained a portion of the *MFα1* gene encoding the leader region of prepro-α-factor fused to a portion of the *SUC2* gene encoding the secreted yeast enzyme invertase.[37] Yeast transformants expressing this fusion exported active invertase to the cell surface. This report was soon followed by

---

[32] K. Mizuno, T. Nakamura, T. Ohshima, S. Tanaka, and H. Matsuo, *Biochem. Biophys. Res. Commun.* **156**, 246 (1988).

[33] R. S. Fuller, A. Brake, and J. Thorner, *Proc. Natl. Acad. Sci. U.S.A.* **86**, 1434 (1989).

[34] R. S. Fuller, A. J. Brake, and J. Thorner, *Science* **245**, 482 (1989).

[35] D. Julius, L. Blair, A. Brake, G. Sprague, and J. Thorner, *Cell* **32**, 839 (1983).

[36] A. Dmochowska, D. Dignard, D. Henning, D. Y. Thomas, and H. Bussey, *Cell* **50**, 573 (1987).

[37] S. D. Emr, R. Schekman, M. C. Flessel, and J. Thorner, *Proc. Natl. Acad. Sci. U.S.A.* **80**, 7080 (1983).

reports of the expression of fusions of the α-factor leader to proteins foreign to yeast. In these studies, the fusion partners were the human proteins epidermal growth factor (hEGF),[38] β-endorphin, a consensus interferon α (IFN-con$_1$),[39] and interferon α1 (IFN-α1).[40] These genes encoded none of the leader sequences present in the human precursor proteins and therefore must have been targeted for secretion and processing in yeast by the α-factor leader sequence, since direct expression of mature hEGF and IFN-α1 in yeast had resulted in no secretion of these proteins.[41,42] In fact, expression of hEGF as a fusion resulted in an increase in production level from ~30 μg/liter for internal expression to greater than 1 mg/liter for secretion. In the case of IFN-α1, secretion directed by the α-factor leader resulted in more efficient export and more precise processing than that seen for secretion directed by the naturally occurring IFN signal sequence. Since these reports, many other proteins have been successfully secreted using the α-factor leader system,[8] thus demonstrating its general utility, as well as providing insight into the biosynthesis and secretion of α-factor and other yeast proteins.

"Improvements" in α-Factor Secretion Systems

Although there is an increasing list of proteins which have been efficiently expressed and secreted from yeast using the α-factor leader, other proteins have proved refractory to efficient secretion and/or processing using the same expression. A number of laboratories have introduced modifications to overcome limitations or problems which have arisen in the basic α-factor expression system for secretion of various heterologous proteins.

Initial studies on expression of α-factor fusions took advantage of a convenient *Hin*dIII site in the MFα1 gene at the junction of the leader and the first α-factor repeat (Fig. 2). The resulting fusions thus contained spacer peptide sequences, requiring the action of the *STE13* gene product for complete maturation of the secreted product. It was found that a large

[38] A. J. Brake, J. P. Merryweather, D. G. Coit, U. A. Heberlein, F. R. Masiarz, G. T. Mullenbach, M. S. Urdea, P. Valenzuela, and P. J. Barr, *Proc. Natl. Acad. Sci. U.S.A.* **81,** 4642 (1984).

[39] G. A. Bitter, K. K. Chen, A. R. Banks, and P.-H. Lai, *Proc Natl. Acad. Sci. U.S.A.* **81,** 5330 (1984).

[40] A. Singh, J. M. Lugovoy, W. J. Kohr, and L. J. Perry, *Nucleic Acids Res.* **12,** 8927 (1984).

[41] M. S. Urdea, J. P. Merryweather, G. T. Mullenbach, D. Coit, U. Heberlein, P. Valenzuela, and P. J. Barr, *Proc. Natl. Acad. Sci. U.S.A.* **80,** 7461 (1983).

[42] R. A. Hitzeman, F. E. Hagie, H. L. Levine, D. V. Goeddel, G. Ammerer, and B. D. Hall, *Nature (London)* **293,** 717 (1981).

A

```
MFα1              GluGluGlyValSerLeuAspLysArgGluAlaGluAla
                  GAAGAAGGGGTATCTTTGGATAAAAGAGAGGCTGAAGCTT
                  CTTCTTCCCCATAGAAACCTATTTTCTCTCCGACTTCGAA
                                                    HindIII


pAB126            GluGluGlyValSerLeuAspLysArg
                  GAAGAAGGGGTATCTCTAGATAAAAGA
                  CTTCTTCCCCATAGAGATCTATTTTCT
                             XbaI
```

B

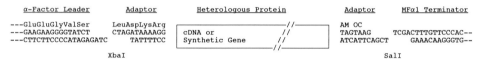

FIG. 2. Sequences around the junction of α-factor leader fusions. (A) DNA sequences and protein translation of the *MFα1* gene and the modification found in pAB126. (B) Strategy to clone a hypothetical heterologous gene, with blunt ends, into the α-factor leader vector pAB126, using synthetic oligonucleotide adaptors.

fraction of the secreted hEGF, IFN, and β-endorphin contained an N-terminal extension corresponding to the (Glu-Ala)$_n$ spacer sequence. This indicated that the *STE13*-encoded dipeptidylaminopeptidase is present in an amount insufficient to process the high levels of protein expressed by these synthetic genes. This had been encountered previously in strains overexpressing α-factor due to the presence of the MFα1 gene on a high-copy plasmid.[35] These strains secreted forms of α-factor similar to that secreted by *ste13* mutants. Two approaches have used to overcoming this problem for proteins secreted from yeast.

The first method was initially described for α-factor leader fusions of hEGF[38] and IFN-α1,[40] and involved site-directed mutagenesis of the fusion genes to delete the portion of these genes encoding the spacer (Glu-Ala)$_n$ dipeptides. Such modified genes were found to result in efficient secretion of the heterologous proteins at levels similar to that seen for the original, spacer-containing fusions. These results indicated that the spacer regions of prepro-α-factor were not essential for transport or processing by the *KEX2* protease. Subsequent α-factor leader fusions have thus usually utilized this direct joining of the heterologous protein to the Lys-Arg processing site of the α-factor leader.

The second approach to circumventing the limiting amount of the *STE13* dipeptidylaminopeptidase present in yeast host strains is to increase the expression of this enzyme by including the cloned *STE13* gene on the same plasmid as the α-factor leader fusion gene. Although there have been no reports of the use of this approach for heterologous proteins, a plasmid containing both the *MFα1* and *STE13* genes was reported to increase the amount of active α-factor over that produced from a plasmid containing only the *MFα1* gene.[43]

The expression of some α-factor leader fusions lacking spacer peptide sequences has resulted in the intracellular accumulation or secretion of unprocessed and partially processed forms, while the same fusions containing (Glu-Ala) spacers are efficiently processed by the *KEX2* protease. Zsebo *et al.*[44] carried out Western blot analysis of proteins secreted from yeast strains expressing α-factor leader–IFN-α1 fusions. Fusions lacking a spacer peptide resulted in secretion of considerable levels (> 50%) of unprocessed, heavily glycosylated fusion protein. This product is analogous to the hyperglycosylated pro-α-factor species secreted from *kex2* mutants.[31] Yeast transformants expressing similar fusions including a (Glu-Ala)$_2$ spacer efficiently secreted fully processed IFN. Thus, some fusions linked directly to the Lys-Arg processing site must be poor substrates for the *KEX2* protease. However, the presence of the charged spacer peptide may convert such fusions to good substrates. The expression of the *STE13* gene may have to be increased as described above to provide complete processing of such spacer-containing fusions.

Increased expression of the *KEX2* gene provides an alternative solution to poor processing at the Lys-Arg site. This approach has been employed to improve the secretion of correctly processed transforming growth factor α (TGF-α) from an α-factor leader–TGF-α fusion.[45] Expression of an α-factor leader-TGF-α fusion resulted in the secretion of both properly processed TGF-α and higher molecular weight species, which were presumably variously glycosylated forms of the uncleaved fusion protein. Insertion of the *KEX2* gene into the multicopy plasmid carrying the α-factor leader–TGF-α fusion gene resulted in the elimination of these unprocessed forms of the fusion protein with a corresponding increase in the secretion of properly processed TGF-α.

[43] D. Barnes, L. Blair, A. Brake, M. Church, D. Julius, R. Kunisawa, J. Lotko, G. Stetler, and J. Thorner, *Recent Adv. Yeast Mol. Biol.* **1**, 295 (1982).

[44] K. M. Zsebo, H.-S. Lu, J. C. Fieschko, L. Goldstein, J. Davis, K. Duker, S. V. Suggs, P.-H. Lai, and G. A. Bitter, *J. Biol. Chem.* **261**, 5858 (1986).

[45] P. J. Barr, H. L. Gibson, C. T. Lee-Ng, E. A. Sabin, M. D. Power, A. J. Brake, and J. R. Shuster, *in* "Industrial Yeast Genetics" (M. Korhola and H. Nevalainen, eds.), p. 139. Foundation for Biochemical and Industrial Fermentation Research, Helsinki, 1987.

As is the case when any yeast secretion system is used for the production of heterologous proteins, N-linked glycosylation of Asn residues in these proteins may result in undesirable differences between the yeast-produced and natural proteins, as discussed above. Site-directed mutagenesis of the sites for addition of N-linked oligosaccharide can be used to eliminate the production of glycosylated proteins entirely.[16,46,47] However, such modified genes will still result in production of proteins which may be antigenically distinct from the naturally occurring proteins. As discussed above, one may also use host yeast strains carrying mutations which reduce the amount of outer-chain mannose addition.[14,15,48,49] It is not currently possible, however, to produce glycoproteins in yeast with oligosaccharide structures identical to those found in mammalian cells.

In addition to defined modifications of the α-factor secretion system, one may screen, using either an immunological or enzymatic assay, for random mutations resulting in improved secretion efficiency. Such an approach resulted in the isolation of yeast mutants showing improved secretion efficiency of bovine prochymosin produced from invertase leader fusions.[6] Two of these mutants also showed increase secretion efficiency for α-factor leader–prochymosin fusions.

Finally, the transcription of α-factor leader gene fusions may be increased by substituting a stronger promoter, such as one of the promoters of the structural genes for the glycolytic enzymes, for that of the *MFα1*, or placed under control of an easily regulated promoter such as that of the *ADH2* gene encoding the repressible alcohol dehydrogenase.[50] Interestingly, it has been reported that the use of a weaker promoter results in increased secretion of insulin-like growth factor I (somatomedin C).[51]

Construction of α-Factor Fusions

A number of different cloning strategies have been used for construction of genes encoding fusions of the α-factor leader with heterologous proteins. Initial studies utilized a fortuitously positioned *Hin*dIII site at the

[46] V. L. MacKay, *in* "Biological Research on Industrial Yeasts" (G. G. Stewart, I. Russell, R. D. Klein, and R. R. Hiebsch, eds.), Vol. 2, p. 27. CRC Press, Boca Raton, Florida, 1986.
[47] A. Miyajima, K. Otsu, J. Schreurs, M. W. Bond, J. S. Abrams, and K. Arai, *Embo J.* **5**, 1193 (1986).
[48] P. K. Tsai, J. Frevert, and C. E. Ballou, *J. Biol. Chem.* **259**, 3805 (1984).
[49] P. W. Robbins, *in* "Biological Research on Industrial Yeasts" (G. G. Stewart, I. Russell, R. D. Klein, and R. R. Hiebsch, eds.), Vol. 2, p. 193. CRC Press, Boca Raton, Florida, 1986.
[50] J. R. Shuster, *in* "Biological Research on Industrial Yeasts" (G. G. Stewert, I. Russell, R. D. Klein, and R. R. Hiebsch, eds.), Vol. 2, p. 20. CRC Press, Boca Raton, Florida, 1986.
[51] J. F. Ernst, *DNA* **5**, 483 (1986).

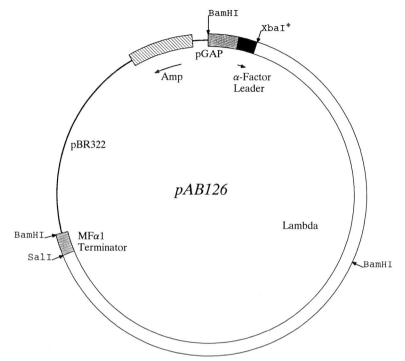

Fig. 3. Map of the plasmid vector pAB126. The asterisk indicates that the *Xba*I site is methylated, and thus resistant to cleavage, in *dam*+ strains of *Escherichia coli.*

junction of the α-factor leader and the mature α-factor sequence and a *Sal*I site 20 base pairs beyond the translational termination codon of MFα1, as shown in Fig. 2. Subsequently, various workers modified the prepro-α-factor coding region to include convenient restriction sites which allow precise fusions to be made directly to the Lys-Arg processing site using specific oligonucleotide adaptors.[38,44,52]

Described here is one such modified vector, pAB126, that is commonly used in our laboratory. A map of this plasmid is shown in Fig. 3. It contains a segment of the *MFα1* gene encoding the leader region, which has been modified to include an *Xba*I restriction site as shown in Fig. 2. This gene has been fused to the GAP promoter[53] which was derived from the *TDH3* gene, one of three structural genes for glyceraldehyde-3-phosphate dehydrogenase. A large (8.2 kb) *Xba*I–*Sal*I fragment of bacteriophage λ has been inserted into the *MFα1* gene and is replaced by a gene encoding a protein of interest.

[52] A. Miyajima, M. W. Bond, K. Otsu, K. Arai, and N. Arai, *Gene* 37, 155 (1985).
[53] S. Rosenberg, D. Coit, and P. Tekamp-Olsen, this volume [28].

In order to construct an α-factor leader fusion, pAB126 is digested with *Xba*I and *Sal*I and the 4.7-kb vector fragment isolated by preparative agarose gel electrophoresis. A cDNA or synthetic gene encoding the protein one wishes to have secreted is ligated to this vector after addition of synthetic oligonucleotide adaptors as shown in Fig. 2. The 5' *Xba*I adaptor is designed to include the *KEX2* processing site and be compatible with a blunt end or an overhang of a convenient restriction site in the heterologous gene, thus producing an in-frame fusion. The 3' *Sal*I adaptor should likewise ligate to an end produced by a restriction site near the 3' end of the heterologous gene, and should include a translational termination codon if one is not contained within the heterologous gene fragment.

Once a plasmid containing the correct gene fusion is isolated, the expression "cassette" is excised by digestion with *Bam*HI and the resulting fragment inserted into a *Bam*HI site of any appropriate yeast plasmid vector, most of which have unique *Bam*HI sites.[2] We routinely use the plasmid pAB24 shown in Fig. 4. The expression plasmid may be introduced by transformation of spheroplasts[54] or lithium-treated whole cells[55] of an appropriate yeast strain carrying either a *ura3* or *leu2* mutation. Because the *leu2-d* allele present in this vector is partially defective,[56,57] Leu$^+$ transformants cannot be obtained by the lithium transformation method, but must be obtained by spheroplast transformation or by leucine selection of transformants obtained initially by uracil selection. Because of this defective *leu2-d* allele, however, leucine selection results in extremely high plasmid copy number. Host strains defective for the major proteases, encoded by the *PRA1 (PEP4), PRB1,* and *PRC1,* are particularly useful to minimize proteolytic degradation of secreted products. Such strains can be obtained from the Berkeley Yeast Genetic Stock Center.[58]

Analysis of Secretion Products

Yeast transformants expressing the α-factor leader fusion are grown in the appropriate liquid selective medium.[59] Yeast cells are removed by centrifugation and the resulting culture supernatant analyzed by sodium dodecyl sulfate (SDS) gel electrophoresis.[60] Usually, the solution must be

[54] A. Hinnen, J. B. Hicks, and G. R. Fink, *Proc. Natl. Acad. Sci. U.S.A.* **75,** 1929 (1978).
[55] H. Ito, Y. Fukuda, and A. Kimara, *J. Bacteriol.* **153,** 163 (1983).
[56] J. D. Beggs, *Nature (London)* **275,** 104 (1978)
[57] E. Erhart and C. Hollenberg, *J. Bacteriol.* **156,** 625 (1983).
[58] Yeast Genetic Stock Center, Department of Biophysics and Medical Physics, University of California, Berkeley, California 94720.
[59] F. Sherman, G. R. Fink, and J. B. Hicks, "Methods in Yeast Genetics." Cold Spring Harbor Laboratory, Cold Spring Harbor, New York, 1986.
[60] U. K. Laemmli, *Nature (London)* **227,** 680 (1970).

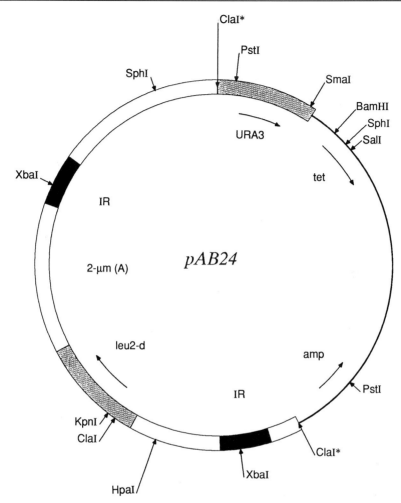

Fig. 4. Map of the yeast plasmid vector pAB24. This plasmid contains the A form of the yeast 2-μm plasmid (this volume [22]) cloned at the *Cla*I site of the *E. coli* vector pBR322, as well as the *S. cerevisiae LEU2* and *URA3* genes. The asterisks indicate that two of the *Cla*I sites are methylated in *dam*+ strains of *E. coli*.

concentrated to obtain sufficient concentrations to allow visualization of the secreted products. A useful general method involves precipitation by trichloroacetic acid (TCA). To the culture supernatant is added one-fourth volume of a solution of 50% TCA, 0.2% sodium deoxycholate (w/v). After 30 min at 0°, the precipitate is collected by centrifugation. The resulting pellet is washed first with 10% TCA, then with acetone, and dried under vacuum. This pellet can then be dissolved in SDS gel sample buffer and a sample representing up to 10 ml of culture electrophoresed. The resolved proteins may be visualized by staining with Coomassie blue, silver chloride,[61] or by Western blot[62] using antibodies directed to the heterologous protein. The cultures analyzed should include a control of the host strain transformed with the vector lacking an expression cassette. Alternatively, the secreted proteins may be concentrated by ultrafiltration or by chromatography using an ion-exchange or hydrophobic interaction resin.

In order to determine whether secreted products have been modified by Asn-linked carbohydrate, the concentrated secreted proteins are dissolved by boiling in a small volume of a solution of 1% (w/v) SDS, then diluted and digested with endoglycosidase H.[63] The resulting deglycosylated proteins are then analyzed as described previously. This will sometimes require concentration by one of the previously described methods to avoid undesirable salt effects during electrophoresis.

To examine products which may remain cell-associated, cells from the same culture are lysed by boiling in SDS sample buffer, then analyzed by SDS electrophoresis and Western blotting.

Alternatively, both secreted and cell-associated products can be analyzed by metabolically labeling yeast cells with radioactive amino acids or ^{35}S-labeled sulfate, then specifically immunoprecipitating proteins using antibodies directed against the heterologous protein.[23]

Conclusion

Expression systems based on the yeast α-factor leader have proved to be generally useful for directing the secretion of a wide variety of proteins of both commercial and research interest. Proteins originating from organisms ranging from bacteria to humans have been successfully secreted using the α-factor leader. These proteins vary in size from 14 amino acids (somatostatin[64]) to over 800 residues (Epstein-Barr virus envelope glycoprotein[65]), and include a number of proteins which have been refractory to efficient expression using other recombinant DNA systems. A number of these proteins have progressed to the stage of clinical trials, and, in fact, the first recombinant protein approved for use in food products, bovine chy-

[61] C. R. Merril, R. C. Switzer, and M. L. Van Keuren, *Anal. Biochem.* **105**, 361 (1980).
[62] W. N. Burnette, *Anal. Biochem.* **112**, 195 (1981).
[63] R. B. Trimble and F. Maley, *Anal. Biochem.* **141**, 515 (1984).
[64] Y. Bourbonnais, D. Bolin, and D. Shields, *J. Biol. Chem.* **263**, 15342.
[65] L. D. Schultz, J. Tanner, K. J. Hofmann, E. A. Emini, J. H. Condra, R. E. Jones, E. Kieff, and R. W. Ellis, *Gene* **54**, 113 (1987).

mosin, is being produced using the *S. cerevisiae* α-factor leader in another budding yeast, *Kluyveromyces lactis*.[66] Although efficient secretion using α-factor leader-based systems is commonly achieved, much more remains to be learned about the basic biology of α-factor secretion, and secretion in general, before optimization of these systems can be achieved.

[66] K. Rietveld, J. G. Bakhuis, N. J. Jansen in de Wal, R. W. van Leen, A. C. M. Noordermeer, A. J. J. van Ooyen, A. Schaap, and J. A. van den Berg, *Yeast* **4**, S163 (1988).

[35] Use of Heterologous and Homologous Signal Sequences for Secretion of Heterologous Proteins from Yeast

By RONALD A. HITZEMAN, CHRISTINA Y. CHEN, DONALD J. DOWBENKO, MARK E. RENZ, CHUNG LIU, ROGER PAI, NANCY J. SIMPSON, WILLIAM J. KOHR, ARJUN SINGH, VANESSA CHISHOLM, ROBERT HAMILTON, and CHUNG NAN CHANG

Introduction

The development of yeast genetic engineering has made possible the expression of heterologous genes and the secretion of their protein products from yeast. This article deals exclusively with the yeast *Saccharomyces cerevisiae* (bakers' yeast).

The advantages of secretion (export) of heterologous gene products are clearly exemplified by human serum albumin (HSA), and are discussed in this article. This 65-kDa blood protein has no N-linked glycosylation sites but does have 35 cysteines;[1] in the blood, this protein monomer has 17 disulfide linkages.[2] HSA is misfolded when produced intracellularly in yeast without its amino-terminal secretion peptide sequence. This conclusion is based on its insolubility, loss of greater than 90% of its antigenicity (as compared to human-derived HSA), and formation of large protein aggregates. Using its natural secretion signal to promote secretion into the media of yeast, HSA is soluble and appears to have the same disulfide linkages as the human blood-derived material. As a pharmaceutical product, which will be potentially used in gram amounts in humans, recombinant HSA will require such identity with the natural product. Other advantages of secreting a product into the growth media of yeast are proper

[1] R. M. Lawn, J. Adelman, S. C. Bock, A. E. Franke, C. M. Houck, R. C. Najarian, P. H. Seeburg, and K. L. Wion, *Nucleic Acids Res.* **9**, 6103 (1981).
[2] J. R. Brown, "Albumin Structure, Function, and Uses," p. 27. Pergamon, New York, 1977.

amino-terminal processing (no initiator methionine residue), the difficulty of preparing yeast extracts, the resistance of yeast cells to lysis, the increase in purity obtained by placing the product in an environment which contains only 0.5–1.0% of the total yeast protein, and the lack of toxic proteins in this environment.

Secretion of the type of HSA described above can be accomplished with homologous (yeast) or heterologous (nonyeast) secretion signals. Processing of the signal sequence (amino-terminal peptide) during the secretion process can be accomplished at the level of signal peptidase[3] or at the level of a peptide prosequence clipped off within the secretion pathway. Examples of these various options are discussed. Many heterologous proteins have been secreted so as to give the proper amino terminus on the mature protein; however, most proteins secreted by the procedures that are described here are secreted proteins in their normal (homologous) environment. Whether cytoplasmic (normally nonsecreted) proteins can be secreted by these methods for various obvious applications remains to be extensively tried, and many may be misfolded or trapped in compartments or membranes.

Materials, Strains, and Methodology

Most of the materials, strains, and methodology have been previously described or referenced in Volume 119 of *Methods in Enzymology* in an article by Hitzeman *et al.*[4] HSA amounts were assayed in media and in glass-beaded extracts of yeast cells using a standard ELISA[5] immunological technique with goat anti-HSA antiserum from Cappel. Somatostatin and gp120 assays were done in a similar fashion using antibody solutions prepared at Genentech, Inc. Yeast strain 30-4 is described in an accompanying article in this volume by Chisholm *et al.*[6]

Plasmid Components Used for Secretion of Heterologous Gene Products

Heterologous Gene

In order for heterologous (nonyeast) gene products to be secreted from yeast, one must first produce the mRNA of the isolated heterologous gene

[3] D. Perlman and H. O. Halvorson, *Cell* 25, 525 (1981).
[4] R. A. Hitzeman, C. N. Chang, M. Matteucci, L. J. Perry, W. J. Kohr, J. J. Wulf, J. R. Swartz, C. Y. Chen, and A. Singh, this series, Vol. 119, p. 424.
[5] A. Voller, D. E. Bidwell, and A. Barlett, "Enzymatic Immunoassays in Diagnostic Medicine; Theory and Practice." World Health Organization, 1976.
[6] V. Chisholm, C. Y. Chen, N. J. Simpson, and R. A. Hitzeman, this volume [37].

in yeast. This is done using yeast 5'- and 3'-flanking DNA sequences from isolated yeast genes to act as promoter and transcription termination signals. The protein-encoding portion from the heterologous gene is inserted between such control regions.[7] Cloned cDNAs are normally used since yeast intron excision involves different signals than, for example, human cell intron excision[8] and even the intron excision of other fungi.[9]

Yeast Origin of Replication for High Copy Number of Plasmid

The expression system described above is generally used on a plasmid as shown in Fig. 1. To obtain high expression, the 2-μm origin is used as well as *REP3* and the *FLIP* gene recombination site in adjacent DNA. The functional unit is 2 kilobase pairs (kb) of DNA between *Eco*RI and *Pst*I in Fig. 1.[10] This system, explained in a separate chapter by Rose and Broach,[11] leads to high plasmid copy number (30–60 copies/cell) which is fairly evenly distributed in a population of *CIR*$^+$ yeast cells. *CIR*$^+$ yeast contain natural 2-μm circle plasmid which is necessary for high copy number of the recombinant plasmid and contains at least two genes for necessary replication functions.

Genes for Selective Maintenance of Plasmid

The plasmid also contains the *TRP1*[12] and *URA3*[13] genes for transformation and selective maintenance of the plasmid in *trp1*$^-$, *ura3*$^-$, or *trp1*$^-$*ura3*$^-$ yeast strains. The *URA3* gene has the added advantage that a plasmid containing it can be cured from the yeast easily using 5-fluoroorotic acid. Many other selective markers are available (e.g., *LYS2*, which can be cured in a similar way using α-aminoadipic acid[14]). This general yeast expression–secretion plasmid also contains pBR322 functions, including an *Escherichia coli* origin of replication and a functional ampicillin resistance gene *(Ap^R)*, which allow for transformation and maintenance of the plasmid in *E. coli.*[15]

[7] R. A. Hitzeman, F. E. Hagie, H. L. Levine, D. V. Goeddel, G. Ammerer, and B. D. Hall, *Nature (London)* **293,** 717 (1981).

[8] S. M. Mount, *Nucleic Acids Res.* **10,** 459 (1982).

[9] M. A. Innis, M. J. Holland, P. E. McCabe, G. E. Cole, V. P. Wittman, R. Tai, K. W. Watt, D. H. Gelfard, J. P. Holland, and J. H. Mead, *Science* **228,** 21 (1985).

[10] J. L. Harley and J. E. Donelson, *Nature (London)* **286,** 860 (1980).

[11] A. B. Rose and J. R. Broach, this volume [22].

[12] G. Tsumper and J. Carbon, *Gene* **10,** 157 (1980).

[13] F. Fasiolo, J. Bonnet, and F. Lacroute, *J. Biol. Chem.* **56,** 2324 (1981).

[14] D. A. Barnes and J. Thorner, *Mol. Cell. Biol.* **6,** 2828 (1986).

[15] F. Bolivar, R. L. Rodriguez, P. Y. Green, M. C. Betlach, H. L. Heyneker, H. W. Boyer, Y. H. Crosa, and S. Falkow, *Gene* **2,** 95 (1977).

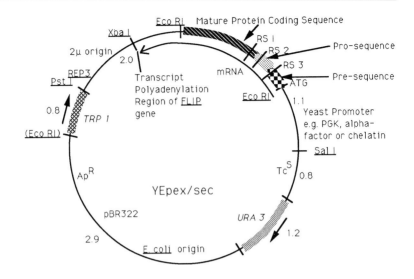

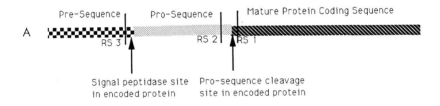

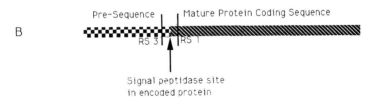

Fig. 1. General yeast expression–secretion plasmid for insertion of heterologous genes. Numbers denote kilobase pairs between junctions. Arrows parallel to genes denote transcription direction. A and B show an expanded version of the heterologous gene expressed with two options for secretion and proper processing of the protein produced.

Carrying Expression Unit on Low-Copy Plasmid or Integrated into Genomic DNA

If high expression is not required or is deleterious to the cell, plasmids containing a centromere with a chromosomal origin of replication[16] may be used to carry the expression system. These plasmids maintain low copy number, from 1–4 copies/cell. Plasmids which disrupt centromere function by an inducible promoter may also be used to raise and lower plasmid copy number at will.[17] Integrative plasmids may also be used if one or more copies of the expression system is desired in the chromosome.[18] Linear DNA integrations can be used to substitute an expression–secretion system for a particular region of DNA in the yeast genome using a method described by Rothstein.[19] A similar strategy has been used to incorporate several copies of an expression system at a targeted region of the yeast genome.[20]

Promoters Used for Heterologous Gene Expression

Many 5'-flanking sequences from various yeast genes have been used for expression and are discussed in various articles in this volume. However, the experiments discussed here have mainly used the yeast 3-phosphoglycerate kinase (PGK) gene promoter,[21,22] the yeast chelatin promoter,[23,24] and the yeast α-factor promoter.[25,26] In each case, sufficient 5'-flanking sequence is present to provide for optimal promoter function. Generally, one kilobase or more is necessary for optimal function since

[16] L. Clarke and J. Carbon, *Nature (London)* **287**, 504 (1980).

[17] L. Panzeri, I. Groth-Clausen, J. Shepard, A. Stotz, and P. Philippsen, *Chromosomes Today* **8**, 46 (1984).

[18] R. A. Hitzeman, C. Y. Chen, F. E. Hagie, J. M. Lugovoy, and A. Singh, "Recombinant DNA Products: Insulin, Interferon, and Growth Hormone," p. 47. CRC Press, Boca Raton, Florida, 1984.

[19] R. J. Rothstein, this series, Vol. 101, p. 202.

[20] A. Singh, C. N. Chang, J. M. Lugovoy, M. D. Matteucci, and R. A. Hitzeman, "Genetics: New Frontiers," p. 169. Oxford and IBH Publishing Co., New Delhi, Bombay, Calcutta, 1983.

[21] R. A. Hitzeman, F. E. Hagie, J. S. Hayflick, C. Y. Chen, P. H. Seeburg, and R. Derynck, *Nucleic Acids Res.* **10**, 7791 (1982).

[22] S. M. Kingsman, D. Cousens, C. A. Stanway, A. Chambers, M. Wilson, and A. J. Kingsman, this volume [27].

[23] M. Karin, R. Najarian, A. Haslinger, P. Valenzuela, J. Welch, and S. Fogel, *Proc. Natl. Acad. Sci. U.S.A.* **81**, 337 (1984).

[24] T. Etcheverry, this volume [26].

[25] J. Kurjan and I. Herskowitz, *Cell* **30**, 933 (1982).

[26] A. Singh, E. Y. Chen, J. M. Lugovoy, C. N. Chang, R. A. Hitzeman, and P. H. Seeburg, *Nucleic Acids Res.* **11**, 4049 (1983).

some promoter activation sequences[27] are quite far upstream of the ATG. The chelatin promoter results in 10 times more mRNA (when induced) than the PGK or α-factor promoters using the same test genes (data not shown). The chelatin promoter (chelp) is about 5X inducible on a 2-μm plasmid using 0.15–0.3 mM Cu^{2+} ion.[28] If more control is desired, one should use the *GAL1-10* promoter, which is essentially turned off during growth on glucose and turned on during growth on galactose.[29]

Transcription Termination or Processing Signals for Heterologous Genes

Many different 3'-flanking sequences have also been used. In the general plasmid shown in Fig. 1, the *FLIP* gene from 2-μm circle plasmid is used for termination and the transcript polyadenylation occurs near the *Xba*I site close to the 2-μm origin.[10,11] The PGK transcription termination region can also be used and sometimes results in higher steady states of mRNA than the *FLIP* terminator (data not shown). It is recommended that enough 3'-flanking region is used (usually ≥ 300 bp) and at least part of the normal yeast structural gene (to make sure junctional information has not been lost); however, a few exact constructions containing homologous 3'-flanking sequences immediately after the stop codon of the heterologous structural gene do function well.[30,31] These data suggest that the end of the normal yeast structural gene is not necessary for proper transcription termination and polyadenylation. Two mammalian cDNAs have 3'-flanking regions which are known to function in yeast [e.g., γ-IFN[32] and IFN-α2[18]]. Whether this is a universal phenomenon has not been explored, but, if an extensive 3'-region is not removed, a Northern analysis of the mRNA[33] made from a Fig. 1-type plasmid can easily determine whether the yeast or heterologous 3'-flanking sequence is being used. We generally remove most of the 3'-flanking sequence on the heterologous gene (≤ 30 bp remain). It has been previously published that the absence of a yeast 3'-flanking sequence on homologous[34] or heterologous[18] genes results in a 10-fold or more reduction in mRNA and protein. Transcripts become very long, ending at various regions downstream from the normal stop regions.

[27] K. Struhl, *Cell* **49**, 295 (1987).
[28] T. Etcheverry, W. Forrester, and R. Hitzeman, *Bio Technology* **4**, 726 (1986).
[29] M. Johnston and R. W. Davis, *Mol. Cell. Biol.* **4**, 1440 (1984).
[30] C. Y. Chen, H. Opperman, and R. A. Hitzeman, *Nucleic Acids Res.* **12**, 8951 (1984).
[31] C. Y. Chen and R. A. Hitzeman, *Nucleic Acids Res.* **15**, 643 (1987).
[32] R. Derynck, A. Singh, and D. V. Goeddel, *Nucleic Acids Res.* **11**, 1819 (1983).
[33] P. R. Dobner, E. S. Kawasaki, L. Y. Yu, and F. C. Bancroft, *Proc. Natl. Acad. Sci. U.S.A.* **78**, 2230 (1981).
[34] K. S. Zaret and F. Sherman, *Cell* **28**, 563 (1982).

Promoter–Heterologous Gene Junctions

Between the homologous promoter and the ATG of the secretion signal, we normally use the DNA sequence TCTAGAATTC, which contains *Xba*I and *Eco*RI restriction sites and is present in the transcript. This sequence is substituted for the first 10 bases at the junction between the homologous promoter and the ATG initiation coding sequence of its natural gene. We have determined that this sequence does not affect mRNA levels or protein production for the homologous promoter and its natural gene.[31] We have also used other linker sequences such as *Bam*HI and *Xho*I without effects on expression. Of course, ATGs in the linker prior to the correct ATG must be avoided since the linker sequence is included in the mRNA. Such linkers are placed at the 5' end of the heterologous structural gene using M13 mutagenesis[35] or by reconstruction with synthetic DNA from a restriction site within the structural gene.

Secretion Signal Sequences

Figure 1 also shows a general schematic representation of a presequence (secretion signal) used to secrete the desired mature protein. Two types of secretion signals have been used in such expression systems: those from heterologous genes and those from homologous genes. Heterologous signals have generally been used to secrete their natural gene products from yeast.[36] This is discussed in more detail in the following sections. Homologous signals have been more often chosen for hybrid expression units containing a natural yeast secretion signal and a heterologous gene (as shown in Fig. 1) to allow the production of the mature protein product in the fermentation media. In most cases, the signals have been obtained from the yeast invertase gene,[37] the α-factor genes,[25,26] an acid phosphatase gene,[38] and the killer toxin gene.[39] The nature of the processing of these proteins during the secretion process has been determined. The desired heterologous protein product is usually the one with the same amino terminus as that of the protein from its natural environment. This is accomplished by attaching the DNA junction of the signal peptidase[3] cleavage site directly to the DNA of the first encoded amino acid of the

[35] M. J. Zoller and M. Smith, *Nucleic Acids Res.* **10**, 6487 (1982).
[36] R. A. Hitzeman, D. W. Leung, L. J. Perry, W. J. Kohr, H. L. Levine, and D. V. Goeddel, *Science* **219**, 620 (1983).
[37] M. Carlson, R. Taussig, S. Kustu, and D. Botstein, *Mol. Cell. Biol.* **3**, 439 (1983).
[38] B. Meyhack, W. Bajwa, H. Rudolph, and A. Hinnen, *EMBO J.* **1**, 675 (1982).
[39] K. A. Bostian, V. E. Burn, S. Jayachandran, and D. J. Tipper, *Nucleic Acids Res.* **11**, 1077 (1983).

mature peptide-encoding sequence (as shown in Fig. 1B). The right side of the signal cleavage site seems to be very tolerant of different amino acids, and such a construct does give desired processing in many cases.[40] Processing can also be obtained at the prosequence level (as shown in Fig. 1A). The best example of this is the use of one of the α-factor genes of yeast. Processing occurs for the natural system after a Lys-Arg site at the end of the prosequence due to the action of the *KEX2* gene endopeptidase.[41] Various groups have determined that the attachment of the mature heterologous gene DNA after this Lys-Arg-encoding sequence results in correct processing of the mature protein even with many changes on the right side of the junction.[42-44]

Secretion Signal and Prosequence Junctions with Heterologous Genes

The exact connection of the signal sequence or the prosequence to the mature heterologous gene is done by means of convenient restriction sites shown in Fig. 1 as RS1, RS2, and RS3. Reconstruction is accomplished in this vector or in simpler vectors using synthetic DNA. Junctions can also be engineered using oligonucleotide mutagenesis[35] for loop-out of undesired sequences or insertion of useful restriction sites. A stretch of synthetic DNA linkers between the presequence or prosequence and the transcription termination region may be useful for the insertion of the heterologous genes. However, since exact in-frame (translational) constructions are desired without extraneously encoded amino acids, unwanted DNA sequences must then be removed by oligonucleotide mutagenesis.

Examples of Use of Heterologous and Homologous Secretion
  Signals in Yeast

Heterologous Signals

The first heterologous gene products were secreted from yeast using their natural secretion signal sequences.[36] Thus, DNA modifications within the structural gene were not necessary. These were cDNAs of human genes,

[40] C. N. Chang, M. Matteucci, L. J. Perry, J. J. Wulf, C. Y. Chen, and R. A. Hitzeman, *Mol. Cell. Biol.* **6**, 1812 (1986).
[41] D. Julius, A. Brake, L. Blair, R. Kunisawa, and J. Thorner, *Cell* **37**, 1075 (1984).
[42] A. J. Brake, J. P. Merryweather, D. G. Coit, U. A. Heberkin, F. R. Masiarz, G. T. Mullenbach, M. S. Urdea, P. Valenzuela, and P. J. Barr, *Proc. Natl. Acad. Sci. U.S.A.* **81**, 4642 (1984).
[43] A. Singh, J. M. Lugovoy, W. J. Kohr, and L. J. Perry, *Nucleic Acids Res.* **12**, 8927 (1984).
[44] G. A. Bitter, K. K. Chen, A. R. Banks, and P. H. Lai, *Proc. Natl. Acad. Sci. U.S.A.* **81**, 5330 (1984).

including interferon α (IFN-α), as well as leukocyte interferons IFN-α1 and IFN-α2. A hybrid signal encoded by DNA sequences from pre-IFN-α1 and pre-IFN-α2 was also used productively for the secretion of IFN-α2.

Figure 2 shows two of these amino-terminal sequences (structures 1 and 2) that result in the secretion of IFN-α1 and IFN-α2. Although 45–64% of the interferon in the media was properly processed between amino acids -1 and $+1$ (Gly-Cys) by the signal peptidase,[3] significant proportions were misprocessed between amino acids -4 and -3. A small amount (8%) of a different misprocessing also occurred in structure 2 (Fig. 2). These misprocessed forms suggest a difference in signal peptidase recognition for these signals in yeast versus human cells. The 50–70% of the interferon that remained cell-associated show similar levels of correct and incorrect processing and misprocessing.

Expression of the human growth hormone (hGH) cDNA[45] results in properly processed hGH in yeast media[18] (see Fig. 2, 3). This suggests that signal peptidase recognition is not flawed for this signal. However, only 10% of the expressed material is secreted while 90% of the hGH remains cell-associated and retains the entire signal sequence, suggesting that differences exist in the efficiency of signal recognition between yeast and human secretion pathways.

Another class of heterologous signals is represented by structure 4 (Fig. 2). The expression of the HSA cDNA[1] has been previously reported using the yeast chelatin promoter.[28] This human secretion signal works well in yeast, producing about 50% of the HSA free of the cell in yeast fermentation media. All of the HSA in the media is correctly processed at the prosequence cleavage site. These data are consistent with the data of Bathurst et al.,[46] who showed that human proalbumin obtained from human cells is cleaved in KEX2 yeast extract but not in a kex2⁻ yeast mutant extract. Thus, this processing is most likely done by the KEX2-encoded endoprotease and is exactly like the processing seen after the amino acid pair Lys-Arg in the prosequence of α-factor.[41] However, in proHSA the amide bond is cleaved after an Arg-Arg amino acid pair.

In the general yeast plasmid shown in Fig. 1, this pre- and prosequence of HSA results not only in HSA secretion with proper processing but also in the secretion and desired processing of other heterologous genes. Structure 5 in Fig. 2 is an example of this for human immunodeficiency virus

[45] D. V. Goeddel, H. L. Heyneker, T. Hozumi, R. Arentzen, K. Itakura, D. G. Yansura, M. J. Ross, G. Miozzari, R. Crea, and P. H. Seeburg, Nature (London) 281, 544 (1979).
[46] I. C. Bathurst, S. O. Brennan, R. W. Carrell, L. S. Cousens, A. J. Brake, and P. J. Barr, Science 235, 348 (1987).

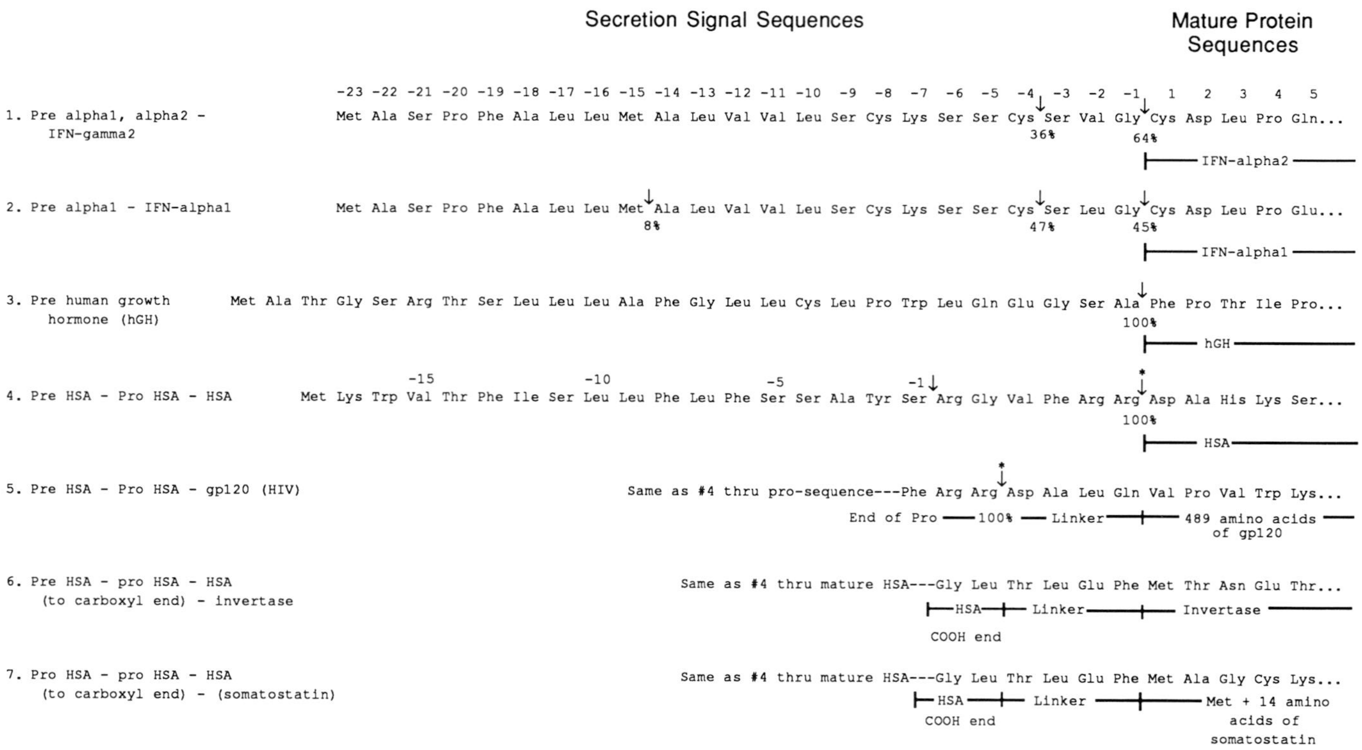

Secretion Signal Sequences

Mature Protein Sequences

   -23 -22 -21 -20 -19 -18 -17 -16 -15 -14 -13 -12 -11 -10 -9 -8 -7 -6 -5 -4 -3 -2 -1   1 2 3 4 5

1. Pre alpha1, alpha2 –
IFN-gamma2
Met Ala Ser Pro Phe Ala Leu Met Ala Leu Val Val Leu Ser Cys Lys Ser Ser Val Gly↓Cys Asp Leu Pro Gln...
                                    36%   64%   ——— IFN-alpha2

2. Pre alpha1 – IFN-alpha1
Met Ala Ser Pro Phe Ala Leu Leu Met Ala Leu Val Val Leu Ser Cys Lys Ser Ser Cys↓Ser Leu Gly Cys Asp Leu Pro Glu...
                     8%   47%   45% ——— IFN-alpha1

3. Pre human growth
hormone (hGH)
Met Ala Thr Gly Ser Arg Thr Ser Leu Leu Leu Ala Phe Gly Leu Leu Cys Leu Pro Trp Leu Gln Glu Gly Ser Ala↓Phe Pro Thr Ile Pro...
                                     100% ——— hGH

4. Pre HSA – Pro HSA – HSA
Met Lys Trp Val Thr Phe Ile Ser Leu Leu Phe Leu Phe Ser Ser Ala Tyr Ser Arg Gly Val Phe Arg Arg↓Asp Ala His Lys Ser...
                                       *   100% ——— HSA

5. Pre HSA – Pro HSA – gp120 (HIV)
Same as #4 thru pro-sequence---Phe Arg Arg Asp Ala Leu Gln Val Pro Val Trp Lys...
                                  *
                         End of Pro — 100% — Linker — 489 amino acids of gp120

6. Pre HSA – pro HSA – HSA
(to carboxyl end) – Invertase
Same as #4 thru mature HSA---Gly Leu Thr Leu Glu Phe Met Thr Asn Glu Thr...
                                    HSA—|—Linker—|—Invertase
                                        COOH end

7. Pro HSA – pro HSA – HSA
(to carboxyl end) – (somatostatin)
Same as #4 thru mature HSA---Gly Leu Thr Leu Glu Phe Met Ala Gly Cys Lys...
                                    HSA—|—Linker—|—Met + 14 amino acids of somatostatin
                                      COOH end

FIG. 2. Heterologous secretion signal sequences (structures 1–7) encoded by plasmids of the general construction shown in Fig. 1. Arrows without an asterisk designate signal peptidase cleavage sites observed by amino-terminus peptide sequencing. Arrows with an asterisk designate prosequence peptide cleavage sites.

(HIV) gp120.[47] HIV gp120. is secreted from wild-type yeast as a hyperglycosylated form containing the processing after the dibasic Arg-Arg residues in the prosequence. Therefore, HSA DNA sequences may be as useful as the homologous α-factor pre- and prosequences which have been used for secretion and processing of heterologous gene products (see next section). Furthermore, these HSA sequences are shorter (24 encoded amino acids versus 85 for α-factor) than the homologous signal from α-factor (prosequence contains three N-linked glycosylation sites). In fact, in a side-by-side comparison of α-factor and HSA pre/prosequences used for the secretion and processing of gp120, secretion levels produced were essentially identical.

Figure 3 shows the great potential of the expression–secretion unit encoding preHSA–proHSA–HSA (structure **4** Fig. 2) under the control of the chelatin promoter. A 10-liter fermentation run produces levels of HSA in the crude media as seen in lanes 3–5. Greater than 90% of the Coomassie blue-staining material is correctly processed HSA of 65,000 molecular weight. Interestingly, these high media levels seen in high-density fermentations are not seen in shake flasks. In shake flasks, most of the HSA remains cell-associated in the periplasm. Thus, if such yeast systems are used to obtain secreted proteins in media, vigorous fermentation conditions may be required for some proteins. It should also be noted that the percentage of material secreted varies from one heterologous gene production to another.

It is also possible to use the signal sequences of HSA to produce fusion proteins in yeast media. Structures **6** and **7** in Fig. 2 are examples of such fusions. These constructions are just like structure **4**; however, in this case, the carboxyl end encoding DNA of mature HSA has been modified to eliminate the translation stop and replace this DNA with a linker sequence for the amino acids Thr, Leu, Glu, and Phe. This linker DNA contains the sequence TCTAGAATTC, which contains the overlapping restriction sites *Xba*I and *Eco*RI. The *Eco*RI site has been used to connect the mature invertase-encoding DNA to HSA-encoding DNA in the proper reading frame. We modified the 5′ end of yeast invertase gene[48] to contain an *Eco*RI site immediately before the ATG of mature invertase. This particular construction uses the 3′-flanking sequence of the yeast invertase gene for transcription termination. Extremely large quantities of the fusion protein are secreted (comparable to structure **4** in Fig. 2) and much of this is found in the media and is glycosylated. Structure **7** was made in a similar

[47] L. A. Lasky, J. E. Groopman, C. W. Fennie, P. M. Benz, D. J. Capon, D. J. Dowbenko, G. R. Nakamura, W. M. Nunes, M. E. Renz, and P. W. Berman, *Science* **233**, 209 (1986).
[48] R. Taussig and M. Carlson, *Nucleic Acids Res.* **11**, 1943 (1983).

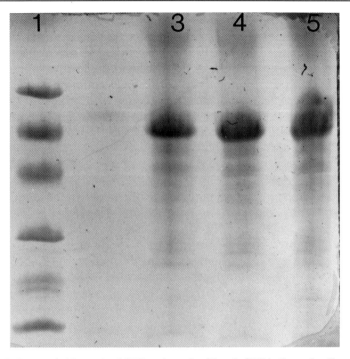

FIG. 3. Coomassie blue-stained SDS–polyacrylamide gel of HSA. One-microliter samples of crude media from yeast fermentation runs were applied to the SDS–polyacrylamide (10%) gel in lanes 3–5. Lanes 3–5 contain HSA (strain 30-4) produced by construct **4** (Fig. 2). Lane 1 has protein standards with the second down being bovine serum albumin, which migrates essentially identically to HSA produced and secreted by yeast.

fashion and contains the small coding region of somatostatin[49] at the 3′ end of mature HSA. Again, large amounts of the fusion protein are found in the media (comparable to structure **4** in Fig. 2 produced levels).

The use of such fusions as carriers of small peptides or as antigens shows great promise. Such constructions may be of general research use for obtaining portions of proteins or entire proteins in extremely pure form for antibody production and for functional studies. These fusions may also be advantageous when direct secretion signal constructions do not secrete the protein of interest.

Many other heterologous signal sequences have been used for secretion of their normally mature protein products (for review, see Ref. 50); however, there are few examples of processing determination or of the use of

[49] K. Itakura, T. Hirose, R. Crea, A. D. Riggs, H. L. Heyneker, F. Bolivar, and H. W. Boyer, *Science* **198**, 1056 (1977).

heterologous signals for the secretion of other heterologous gene products (hybrid secretion proteins). Almost every hybrid secretion gene has been made using a homologous signal and a heterologous mature protein encoding sequence, as is discussed in the next section. However, the authors wish to point out the usefulness of a highly characterized heterologous signal attached to another heterologous gene.

Homologous Signals

Figure 4 shows examples of how homologous (yeast) secretion signals can be used for the secretion and proper processing of heterologous gene products. Structure **8** demonstrates how the yeast invertase protein is processed during its secretion, driven by its amino-terminal peptide secretion signal sequence.[37] Knowing the natural junction sequence, we used this signal in structure **9** to secrete human interferon, IFN-α2, from yeast into the growth media.[40] Several other human gene products have also been secreted from yeast using this signal.[40] All these constructions resulted in properly processed protein in the media due to protein cleavage at the level of the signal peptidase. These results strongly suggest that the amino acid to the right of the signal peptidase cleavage site can be varied without affecting cleavage.

A variant of the invertase presequence (Fig. 4, structure **10**) results in secreted and properly processed IFN-α2.[40] This construction had an alanine missing in the signal as well as a methionine at the right-hand side of the signal peptidase junction. The significance of the missing alanine will not be discussed. However, the junctional change of methionine to the right of the cleavage site further shows a flexibility in signal peptidase recognition of hybrid junctions. Another demonstration of a junctional change is demonstrated by structure **11**. When this structure is expressed in yeast, protein secretion results and 100% properly processed HSA is found in the growth media. For this construction and the next, the yeast PGK promoter was used in the general yeast expression–secretion vector (Fig. 1).

Structure **12** (Fig. 4) is similar to structure **11** but has an insert of 6 amino acids (Arg Gly Val Phe Arg Arg) at the signal peptidase cleavage site of structure **11**. These additional amino acids are the prosequence encoded by the cDNA of HSA. Production and secretion for structures **11** and **12**, as well as **4**, are essentially identical when all other parts of the expression plasmid are kept the same. Structure **12** also gives correctly processed HSA in the yeast media. However, unlike previously discussed structures, this one contains a homologous secretion signal and a heterologous prose-

[50] R. C. Das and J. L. Shultz, *Biotechnol. Prog.* **3**, 43 (1987).

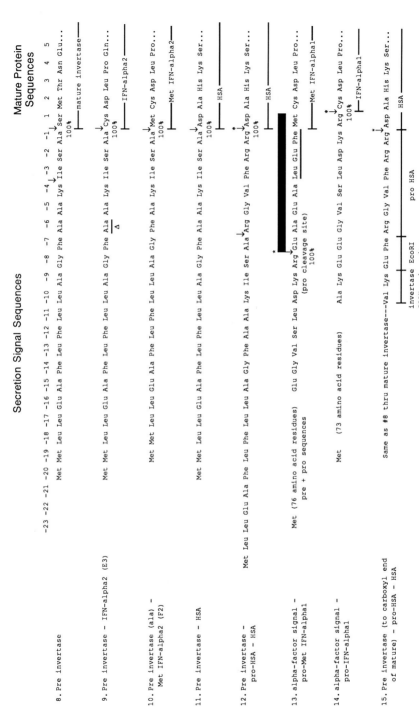

Fig. 4. Homologous secretion signal sequences (structures **8–15**) encoded by plasmids of the general construction shown in Fig. 1. Arrows identify peptide cleavage sites as described in the legend to Fig. 2. The dark bar represents the amino acid sequence removed in structure **14**. The Δ designation shows an amino acid deleted in structure **10**.

quence. Correct processing is at the level of the yeast *KEX2* gene endoprotease as described in the previous section. Thus, another option is available for the secretion and processing of a heterologous gene product.

Structure **15** reveals still another option for secretion and processing. This structure is similar to **8** (Fig. 4); however, at the carboxyl end encoding DNA of mature invertase, the translation stop codon has been replaced by an *Eco*RI restriction site using M13 oligonucleotide mutagenesis. An *Eco*RI site was also placed at the beginning of the prosequence of structure **4** (Fig. 2). Then modified structures **8** and **4** were placed together using these *Eco*RI sites to give structure **15**. Structure **15** encodes a procleavage site between mature invertase and mature HSA. It is interesting that, on expression of this gene in yeast, one obtains active invertase secreted as well as normal-size secreted HSA as determined by Western analysis. These results suggest that the HSA prosequence can be used at the carboxyl end of a normally secreted protein in yeast to get processing of a heterologous gene protein, which is being carried along through the secretion pathway. It will be interesting to see if such an expression–secretion system may be more effective than the attachment of pre- and prosequences alone in the secretion of some proteins (especially those that are normally not secreted in their natural environment).

Finally, we have saved the most frequently used secretion system for the last part of this discussion. This system is based on the yeast α-factor pre- and prosequences. Two α-factor genes have been cloned from yeast.[25,26] These genes produce the α mating pheromone, a small peptide which prepares the opposite mating type (**a**) for cell fusion and zygote formation.[51] The processing of the pheromone precursor from one of these gene products has been extensively characterized.[25,41,52] The signal sequence is attached to a large prosequence to give a combined pre/prosequence of 85 amino acids. This prosequence contains three N-linked glycosylation sites (unlike that of HSA). The prosequence, as well as intervening amino acid sequences between four tandem copies of the mature pheromone, are clipped by the *KEX2* gene product after Lys-Arg residues. The cleaved α-factor pheromones retain Glu-Ala-Glu/Asp-Ala at their amino-terminal ends and Lys-Arg at the carboxyl end. The amino-terminal sequences are then removed by a dipeptidylaminopeptidase which removes two amino acids at a time.[52,53] The Lys-Arg residues are removed by another protease encoded by *KEX1*;[52] however, this protease is irrelevant to the processing of heterologous gene products.

[51] I. Herskowitz and Y. Oshima, "The Molecular Biology of the Yeast *Saccharomyces:* Life Cycle and Inheritance," p. 181. Cold Spring Harbor Laboratory, Cold Spring Harbor, New York, 1981.
[52] D. Julius, L. Blair, A. Brake, G. Sprague, and J. Thorner, *Cell* **32**, 839 (1983).
[53] R. B. Wickner, *Genetics* **76**, 423 (1974).

Singh *et al.*[43] have described the use of structure **13** in Fig. 4 to obtain Met IFN-α1. The α-factor promoter was used in a Fig. 1-type plasmid. The pre- and prosequences of the α-factor gene were attached by means of an *Xba*I and *Eco*RI restriction site linker [encodes Leu-Glu-Phe, which allows for convenient in-frame (translational) connection of other genes as well]. It was not expected that this amino acid sequence would be removed during secretion of the protein. However, the *KEX2* protease cleavage at the procleavage site (after Lys-Arg) was expected, as well as the removal of Glu-Ala-Glu-Ala. All the yeast media interferon retained the Glu-Ala-Glu-Ala. Further studies suggest that, at high expression levels, the protease that performs this function may be limiting. In order to overcome this problem, structure **14** (Fig. 4) was made which places the Lys-Arg next to the Cys of mature IFN-α1. Correctly processed IFN-α1 was obtained at levels of about 50% of the media protein. Since this time, many constructions like this have been made with different amino acid sequences to the right of Lys-Arg and all have been fairly well processed. An accompanying article about α-factor discusses this in greater detail.[54]

Glycosylation of Heterologous Gene Products

Problem

When proteins travel through the secretion pathway of yeast, core glycosylation occurs at the endoplasmic reticulum and outer-chain glycosylation occurs in the Golgi bodies. One type of glycosylation happens at Asn residues within the amino acid sequence Asn-X-Ser/Thr (X is any amino acid) (see Fig. 5).[55,56] An accompanying article by Ballou[57] describes this glycosylation process in great detail.

However, when heterologous gene products are glycosylated during secretion from yeast, they are glycosylated somewhat differently than what normally occurs in the cell from which the gene was obtained. For example, most of the genes described in this article are from human cells where the glycosylation process is similar. However, the sugar content at Asn residues is somewhat different. Human cells have a very similar inner core but lack the extensive "flowering" of the large outer chain which can be repeated several times. This characteristic results in a gross difference

---

[54] A. J. Brake, this volume [34].
[55] C. E. Ballou, "This Molecular Biology of the Yeast *Saccharomyces:* Metabolism and Gene Expression," p. 335. Cold Spring Harbor Laboratory, Cold Spring Harbor, New York, 1982.
[56] P. K. Tsai, J. Frevert, and C. E. Ballou, *J. Biol. Chem.* **259**, 3805 (1984).
[57] C. E. Ballou, this volume [36].

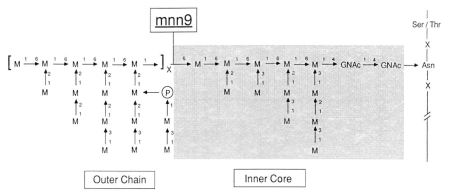

FIG. 5. Asparagine-linked glycosylation in yeast. The primary protein sequence is at the right. M refers to mannose residues, P is phosphate, and GNAc is N-acetylglucosamine. Only one type of mutant that affects outer-chain glycosylation is shown, *mnn9*.[56] Note that this structure has been modified since the submission of this chapter.[57]

between the sugar content of a heterologous human protein produced in yeast and mammalian cells. Other differences also exist between yeast and human glycosylation. The mannose content relative to other sugars in human glycosylated proteins is not as great as in yeast. Several other sugars are added within the inner core by a series of reactions that vary from one human cell type to another.

The extent of the difference between human and yeast glycosylation can be easily seen with expression of the *gp120* gene of HIV. This retrovirus envelope glycoprotein gene has been previously truncated by placing a premature translation stop within the hydrophobic region between *gp120* and *gp41*. This truncated protein has been shown to be secreted by CHO cells.[47] We have further modified the *gp120* gene as described earlier in this article (Fig. 2, structure 5). The pre- and prosequences of human HSA were attached to the first 489 amino acids of *gp120*. The general expression plasmid in Fig. 1 was used with the yeast chelatin promoter and the yeast PGK transcription terminator. This portion of the *gp120* gene should produce a secreted protein monomer of about 54,000 Da if it is not glycosylated. However, there are 24 possible N-linked glycosylation sites.

In mammalian cells, this results in a protein monomer of 120,000 Da due to extensive sugar addition. In yeast, this gene expression results in a hyperglycosylated form of gp120 of greater than 600,000 Da. This is shown by the Western blot[58] in Fig. 6. A reducing SDS–polyacrylamide gel was run with media from yeast expressing the *gp120* gene (lane 1) and from the same yeast strain containing a plasmid without this gene (lane 2). When the yeast medium material was treated with endo H (which removes the

[58] W. N. Burnette, *Anal. Biochem.* **112**, 195 (1981).

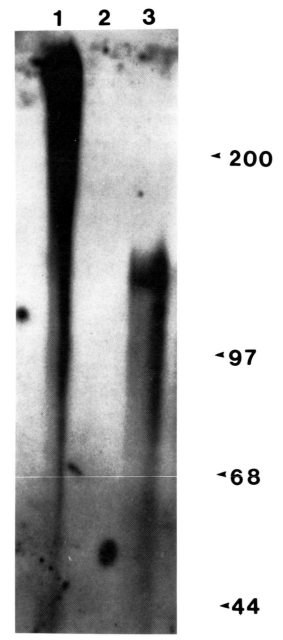

FIG. 6. HIV gp120 secreted by wild-type yeast and a *mnn9* mutant yeast. This is a Western blot[58] of a SDS–polyacrylamide (7%) gel. Lane 1 contains culture medium from yeast 20B-12 containing construction **5** (Fig. 2). Lane 2 is medium from yeast 20B-12 containing construction **4** (Fig. 2) (no gp120). Lane 3 is medium from yeast α *mnn9 trp1⁻* containing construction **5**. Antiserum against HIV gp120 was incubated with blot, after gel proteins were blotted onto nitrocellulose. Antibody binding to gp120 was visualized using ¹²⁵I-labeled protein A and autoradiography. Standard protein migration is as indicated (MW × 10⁻³).

sugars), the gp120 migrates at about 60,000 Da (data not shown). This hyperglycosylated form of gp120 does not bind to the CD4 receptor[59] and binds poorly to polyvalent and monovalent antibody columns (antibodies to the mammalian cell-produced material[47]).

Partial Solution

A *mnn9* mutant[57] yeast characterized by Tsai *et al.*[56] (kindly provided by Clinton Ballou) behaves as shown in Fig. 5. The mutant lacks an active α-1$\rightarrow$6-mannosyltransferase that initiates outer-chain glycosylation. Therefore, the *mnn9* mutant produces an N-glycosylated protein which totally lacks the outer chain.

We crossed this *mnn9* mutant with strain S191[60] to obtain an α *mnn9 trp1* spore. This strain was transformed with the *gp120* gene expression plasmid described above and the production of gp120 protein in the yeast medium was examined. In Fig. 6, lane 3, the medium from this strain is shown by Western analysis to produce a smear at around 120,000 Da. This is identical to the mammalian-produced material. When compared by SDS–gel electrophoresis, it is difficult to notice any difference between the two. In fact, with storage, some of the same-size degradation products appear in both (data not shown).

Therefore, relative sugar addition to this protein containing an extremely high number of glycosylated sites is essentially identical with respect to size in this yeast mutant and in mammalian cells. Unfortunately, this drop in size does not significantly improve binding to CD4 (the receptor protein), although binding to polyvalent antibody columns is greatly improved. The lack of binding to CD4 suggests that there may still be folding differences between yeast and mammalian cell-produced gp120. Whether this is related to inner-core glycosylation differences is unknown.

Recent experiments by Moir and Dumais[61] which describe the expression and secretion of human α-1-antitrypsin suggest similar conclusions. They found that α-1-antitrypsin secreted by the wild-type yeast was hyperglycosylated (3-N-glycosylation sites) in comparison to human cell-derived material. However, when it was produced in the *mnn9* yeast, it was slightly smaller than the natural protein.

Recently, we have cloned the wild-type *MNN9* gene using the extremely sensitive nature of *mnn9* yeast to the antibiotic hygromycin B. We have verified the identity of the gene by several criteria and have sequenced the structural gene and its flanking regions. This gene can now be conve-

[59] D. H. Smith, R. A. Byrn, S. A. Marsters, T. Gregory, J. E. Groopman, and D. J. Capon, *Science* **238**, 1704 (1987).
[60] Yeast strain S191, a *trp1 met15* from Arjun Singh, unpublished results.
[61] D. T. Moir and D. R. Dumais, *Gene* **56**, 209 (1987).

niently deleted in the yeast genome (e.g., in a *ura3⁻* yeast strain using a linear DNA containing *URA3* flanked by *MNN9* sequences). Such deletions are useful since we have previously had some trouble with reversion of the original *mnn9* mutant.

Concluding Comments

If one intends to produce and secrete a heterologous gene product using yeast, several options are available to accomplish this. If it is a secreted protein in its normal cell environment, one can first try its own amino-terminal secretion signal. Many have been functional; however, the amount of protein produced, the amount secreted, and the processing which occurs are all variable from one gene product to another. If the heterologous signal does not function in any of the desired aspects mentioned above, homologous or heterologous signals that are known to function in a somewhat predictable fashion in yeast may be tried. This article has described how such hybrid constructions can be made and the many options available.

[36] Isolation, Characterization, and Properties of *Saccharomyces cerevisiae mnn* Mutants with Nonconditional Protein Glycosylation Defects

By CLINTON E. BALLOU

Introduction

Three classes of mutant have been obtained that are defective in protein glycosylation in yeast, one based on selection by [³H]mannose-killing and called *alg* (asparagine-linked glycosylation defective) mutants,[1] one based on changes in cell density owing to accumulation of membrane material and called *sec* (secretion defective) mutants,[2] and one based on changes in cell surface carbohydrate structure and called *mnn* (mannan defective) mutants.[3] The first two selections are designed to detect conditional mutants, whereas the last selects nonconditional mutants because the cells must grow in order to express the change in phenotype. The *alg* mutants are affected mostly in early steps involving synthesis of the dolichol-linked precursor oligosaccharide and often lead to underglycosylation of proteins. The *sec* mutants are defective in the processing of glycoproteins through

[1] T. C. Huffaker and P. W. Robbins, *Proc. Natl. Acad. Sci. U.S.A.* **80**, 7466 (1983).
[2] P. Novick, C. Field, and R. Schekman, *Cell* **21**, 205 (1980).
[3] W. C. Raschke, K. A. Kern, C. Antalis, and C. E. Ballou, *J. Biol. Chem.* **248**, 4660 (1973).

the secretory pathway, although some are similar or identical to *alg* mutants. The *mnn* mutants are mostly concerned with synthesis of the outer-chain portion of N-linked oligosaccharides, although some affect glycosylation of the core and of O-linked oligosaccharides.[4] This article is concerned only with *mnn* mutants and will describe their isolation, characterization, and properties.

Detection of *mnn* Mutants

The cell surface of *Saccharomyces cerevisiae* is covered by mannan, which is actually the carbohydrate part of mannoproteins, a complex of N- and O-linked glycopeptides that, in strain X2180, is characterized by the presence of terminal $\alpha 1 \rightarrow 3$-linked mannose units that are very immunogenic and by mannobiosylphosphate units that bind alcian blue dye.[4] Antisera raised in rabbits will agglutinate X2180 cells, and *mnn* mutants are selected for the loss of agglutinability by such sera following mutagenesis with ethylmethane sulfonate. Some mutants of this class also lose the ability to bind alcian blue dye, the consequence of a defect that eliminates the acceptor sites for phosphorylation.[5,6]

Some wild-type strains, such as *S. cerevisiae* A364A, lack terminal $\alpha 1 \rightarrow 3$-linked mannose and have the phenotype of a *mnn1* mutant.[7] Such strains bind alcian blue more strongly than strain X2180, and they also bind well to QAE-Sephadex beads at pH 3.[8] The mannosylphosphate structure is also very immunogenic in rabbits, although the mannobiosylphosphate group loses this specificity and reacts only with sera against $\alpha 1 \rightarrow 3$-linked mannose.[9] There are at least four types of mannosylphosphorylation sites in *S. cerevisiae* mannoproteins, the side chains in the outer-chain part of N-linked oligosaccharides,[10] two sites in the core[11,12]

[4] C. E. Ballou, *in* "The Molecular Biology of the Yeast *Saccharomyces:* Metabolism and Gene Expression" (J. N. Strathern, E. W. Jones, and J. R. Broach, eds.), p. 335. Cold Spring Harbor Laboratory, Cold Spring Harbor, New York, 1982.

[5] C. E. Ballou, K. A. Kern, and W. C. Raschke, *J. Biol. Chem.* **248**, 4667 (1973).

[6] E. M. Karson and C. E. Ballou, *J. Biol. Chem.* **253**, 6484 (1978).

[7] C. E. Ballou and W. C. Raschke, *Science* **184**, 127 (1974).

[8] W. D. Sikkema, L. Ballou, J. Letters, and C. E. Ballou, *Fed. Proc., Fed. Am. Soc. Exp. Biol.* **41**, 755 (1982).

[9] L. Rosenfeld and C. E. Ballou, *J. Biol. Chem.* **249**, 2319 (1974).

[10] T. R. Thieme and C. E. Ballou, *Biochemistry* **10**, 4121 (1971).

[11] C. Hashimoto, R. E. Cohen, W.-J. Zhang, and C. E. Ballou, *Proc. Natl. Acad. Sci. U.S.A.* **78**, 2244 (1981).

[12] L. M. Hernandez, L. Ballou, E. Alvarado, B. L. Gillece-Castro, A. L. Burlingame and C. E. Ballou, *J. Biol. Chem.* **264**, 11849 (1989).

and on the nonreducing terminus in the *mnn10* mutant.[13] Phosphorylation of the O-linked oligosaccharides has not been demonstrated. Because mutants of the *mnn7* class are clumpy, tend to lyse, and allow periplasmic enzymes to leak into the medium, any one of these phenotypes could be used to detect a new *mnn* mutant type.

An important use of *mnn* mutants is to elucidate the structure of the N-linked carbohydrate chains in *S. cerevisiae* (Fig. 1) and to gain some ideas regarding the biosynthetic pathway for their assembly and regulation.[14] The outer chain is highly branched in the wild type, and loss of the side chains in the *mnn2* mutant provides the important evidence that the yeast polymannose chain is differentiated into a core oligosaccharide, identical in structure to the high-mannose chain of other eukaryotes, and an attached $\alpha1 \rightarrow 6$-linked outer chain.[15] The exposed $\alpha1 \rightarrow 6$-linked outer chain of the *mnn2* mutant is immunogenic, and rabbit antiserum against this strain is used to isolate mutants that have lost part or all of the outer chain (the *mnn7* class), such as the *mnn9* strain that lacks an outer chain and the *mnn10* strain that has an outer chain of 10–12 mannose units.[16]

In studies with the *mnn* mutants, it is often helpful to use strains with multiple *mnn* lesions. The *mnn1* defect eliminates all terminal $\alpha1 \rightarrow 3$-linked mannose attached to $\alpha1 \rightarrow 2$-linked mannoses in the core, outer chain, and on O-linked oligosaccharides. Terminal $\alpha1 \rightarrow 3$-linked mannose is immunogenic in rabbits, so elimination of this structural feature may be an important consideration when expressing mammalian glycoprotein genes in yeast for production of materials for clinical use. When *mnn1* is present in the *mnn9* background, the cells make uniform N-linked oligosaccharides with 10 mannoses, although some are still phosphorylated.[17] If the *mnn4* or *mnn6* mutation is added, the level of phosphorylation is reduced about 90%. Interestingly, the *mnn2* defect is not expressed in presence of *mnn9* even though one $\alpha1 \rightarrow 2$-linked mannose is added during the processing of the core oligosaccharide.[18] On the other hand, in the *mnn10* background, *mnn2* eliminates branching of the shortened outer chain, with the exception of the nonreducing end stop signal.[19]

[13] L. M. Hernandez, L. Ballou, E. Alvarado, P.-K. Tsai and C. E. Ballou, *J. Biol. Chem.* **264,** 13648 (1989).
[14] T. Nakajima and C. E. Ballou, *Proc. Natl. Acad. Sci. U.S.A.* **72,** 3912 (1975).
[15] T. Nakajima and C. E. Ballou, *J. Biol. Chem.* **249,** 7684 (1974).
[16] L. Ballou, R. E. Cohen, and C. E. Ballou, *J. Biol. Chem.* **255,** 5986 (1980).
[17] P.-K. Tsai, J. Frevert, and C. E. Ballou, *J. Biol. Chem.* **259,** 3805 (1984).
[18] P. K. Gopal and C. E. Ballou, *Proc. Natl. Acad. Sci. U.S.A.* **84,** 8824 (1987).
[19] L. Ballou, E. Alvarado, P.-K. Tsai, A. Dell, and C. E. Ballou, *J. Biol. Chem.* **264,** 11857 (1989).

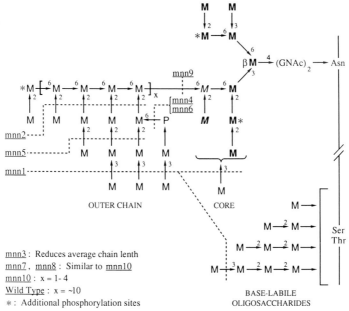

FIG. 1. Chemotypes of mannoprotein mutants. The revised structure of *S. cerevisiae* X2180 mannoprotein carbohydrate chains[12] with an indication of the effects of *mnn* mutations is shown. In the *mnn2* or *mnn2 mnn10* mutant mannoprotein, the italicized mannose in the core is absent and is located at the nonreducing end of the outer chain. In the *mnn9* or *mnn2 mnn9* mannoproteins, the outer chain is absent and the italicized mannose is present. Thus, addition of this $\alpha 1 \rightarrow 2$-linked mannose is not regulated by the *MNN2* locus. Some wild-type N-linked oligosaccharides on external invertase and all of those on vacuolar carboxypeptidase Y have structures identical to those formed in the *mnn9* mutant, which suggests that the mannoproteins made in this mutant bypass the compartment in which the outer chain is added or else the $\alpha 1 \rightarrow 2$-linked mannose acts as a stop signal to prevent elongation of the backbone. Outer-chain phosphorylation is reduced by the *mnn4* and *mnn6* mutations, and additional sites of phosphorylation (indicated by asterisks) are observed in *mnn1 mnn9* and *mnn1 mnn2 mnn10* cells and may be present in wild-type mannoproteins.

Properties of Individual *mnn* Mutants

mnn1. This mutation leads to synthesis of N- and O-linked chains that lack terminal $\alpha 1 \rightarrow 3$-linked mannose attached to mannose in $\alpha 1 \rightarrow 2$ linkage, thus eliminating the major immunogenic determinant of the mannoprotein.[3] The mutation does not affect synthesis of the $\alpha 1 \rightarrow 3$ linkages in the lipid-linked precursor, in which the sugars are attached to $\alpha 1 \rightarrow 6$-linked mannoses. Since extracts of the *mnn1* mutant lack $\alpha 1 \rightarrow 3$-mannosyltransferase activity with artificial acceptors composed of $\alpha 1 \rightarrow 2$-linked mannose, the mutation may affect the structural gene for this enzyme.[14] The *mnn1* locus is centromere-linked and has been mapped to chromosome V near the *ura3* locus.[20] Other strains, such as *Saccharomyces italicus,* have a second $\alpha 1 \rightarrow 3$-linked mannose attached to the mannose whose addition is controlled by the *MNN1* locus.[7] The addition of this mannose is controlled by a different gene.

mnn2. This mutation eliminates branching of the outer-chain portion of mannoproteins that occurs by addition of mannose in $\alpha 1 \rightarrow 2$ linkage.[15] Extracts of *mnn2* cells show a normal level of $\alpha 1 \rightarrow 2$-mannosyltransferase activity with artificial acceptors, which suggests that the defect is in a regulatory gene;[21] but so many different $\alpha 1 \rightarrow 2$ linkages occur in the mannoprotein that it is possible there are multiple transferase activities and that the assay does not reveal the absence of one of them. The *mnn2* mutation exposes a new immunogenic determinant, the unsubstituted $\alpha 1 \rightarrow 6$-linked backbone of the outer chain, and simultaneously reduces reaction with the $\alpha 1 \rightarrow 3$-mannosyl-specific antiserum, even though the core and O-linked carbohydrate chains still carry mannose in such linkage.[16]

mnn3. This strain leads to a general shortening of the carbohydrate chains of the mannoproteins.[22] Like *mnn1,* it affects both N- and O-linked oligosaccharides, with the latter consisting exclusively of mannobiose units in *mnn3* mannoproteins.

mnn4 and mnn6. The two mutations reduce the extent of substitution of the N-linked carbohydrate chains with mannosylphosphate units, and they both appear to be regulatory defects. In the *mnn1* strain, the mannosylphosphate unit is the most important immunogenic determinant, whereas in absence of *mnn1* these units are substituted with an $\alpha 1 \rightarrow 3$-linked mannose and are transformed in specificity.[9] The *mnn6* defect is recessive,[6] but *mnn4* is dominant under certain conditions as determined by the dye-binding assay. Because phosphate provides the major charge to the cell wall, binding of the basic phthalocyanin dye alcian blue provides a simple way to assess the extent of phosphorylation of the mannoprotein.[23] Haploid *mnn4* or diploid *mnn4/mnn4* cells fail to bind dye when grown on either YEPD (1% yeast extract, 2% peptone, 2% glucose) or YEPD + 0.5 M KCl, whereas *mnn4/X2180* or *mnn4/mnn6* diploid cells do not bind dye when grown on YEPD but do bind dye when grown on YEPD + KCl. Thus, *mnn4* is dominant in the heterozygous diploids when grown without KCl but is recessive when grown in the presence of KCl or sorbitol.[24]

[20] D. L. Ballou, *J. Bacteriol.* **123,** 616 (1975).

[21] A. J. Parodi, *J. Biol. Chem.* **254,** 8343 (1979).

[22] R. E. Cohen, L. Ballou, and C. E. Ballou, *J. Biol. Chem.* **255,** 7700 (1980).

[23] J. Friis and P. Ottolenghi, *C. R. Trav. Lab. Carlsberg* **37,** 327 (1970).

[24] L. M. Hernandez, L. Ballou, and C. E. Ballou, unpublished results (1989).

mnn5. This defect prevents addition of the second $\alpha 1 \rightarrow 2$-linked mannose to the side chains of the outer chain, a phenotype that could be explained by a defect in an $\alpha 1 \rightarrow 2$-mannosyltransferase II.[22] Unlike *mnn1* and *mnn3*, this mutation has no effect on the O-linked oligosaccharides.

mnn7 Class. This mutant class originally included *mnn7, mnn8, mnn9*, and *mnn10*, and the members are defective in elongation of the outer chain, a phenotype that is characterized by a reduced agglutinability of the cells with $\alpha 1 \rightarrow 6$-mannosyl-specific antiserum when in the *mnn1 mnn2* background.[16] Recently, *mnn7* and *mnn8* have been shown to be allelic.[19] The *mnn9* strain is the most dramatically altered and makes all of its mannoproteins with oligosaccharides that have structures identical to those found on carboxypeptidase Y in wild-type cells.[17] In the *mnn1 mnn9* strain, the oligosaccharides are mainly $Man_{10}GlcNAc_2$ units along with a small amount of $Man_{11}GlcNAc_2$ units, which have the following structures.

$\alpha Man \rightarrow^6 \alpha Man \rightarrow^6 \beta Man \rightarrow^4 \beta GlcNAc \rightarrow^4 \beta GlcNAc \rightarrow Asn$
$\uparrow^2 \qquad \uparrow^3 \qquad \uparrow^3$
$\alpha Man \quad \alpha Man \quad \alpha Man \,^6\!\leftarrow \alpha Man$
$\qquad\qquad\qquad \uparrow^2 \qquad \uparrow^2$
$\qquad\qquad\qquad \alpha Man \quad \alpha Man$
$\qquad\qquad\qquad \uparrow^2$
$\qquad\qquad\qquad \alpha Man$

$\alpha Man \rightarrow^6 \alpha Man \rightarrow^6 \beta Man \rightarrow^4 \beta GlcNAc \rightarrow^4 \beta GlcNAc \rightarrow Asn$
$\uparrow^2 \qquad \uparrow^3 \qquad \uparrow^3$
$\alpha Man \quad \alpha Man \quad \alpha Man \,^6\!\leftarrow \alpha Man$
$\qquad\qquad \uparrow^2 \qquad \uparrow^2 \qquad \uparrow^2$
$\qquad\qquad \alpha Man \quad \alpha Man \quad \alpha Man$
$\qquad\qquad\qquad \uparrow^2$
$\qquad\qquad\qquad \alpha Man$

The *mnn1 mnn2 mnn10* strain makes oligosaccharides with the structure

$\alpha Man \rightarrow^6 \alpha Man \rightarrow^6 \beta Man \rightarrow^4 \beta GlcNAc \rightarrow^4 \beta GlcNAc \rightarrow Asn$
$\uparrow^2 \qquad \uparrow^3 \qquad \uparrow^3$
$\alpha Man \quad \alpha Man \quad \alpha Man(^6\!\leftarrow \alpha Man)_x \,^6\!\leftarrow \alpha Man$
$\qquad\qquad\qquad \uparrow^2 \qquad\qquad \uparrow^2$
$\qquad\qquad\qquad \alpha Man \qquad\qquad \alpha Man$
$\qquad\qquad\qquad \uparrow^2$
$\qquad\qquad\qquad \alpha Man$

in which x is $4-14$.[19] A notable feature of this oligosaccharide is that it is branched at the end and has one less $\alpha 1 \rightarrow 2$-linked mannose in the core than does the *mnn1 mnn2 mnn9* oligosaccharide. It is postulated that this side-chain mannose may serve as a structural block to further elongation of the outer chain. Strain *mnn8* differs from *mnn10* in that it makes N-linked oligosaccharides with a wider range of sizes, where $x = 4-19$. The *mnn9* defect is dominant to all other members of this class in the haplophase, whereas *mnn10* is partially dominant to *mnn8*.

Isolation of *mnn* Mutants

Three procedures have been used, namely, agglutination with specific antisera, fluorescence-activated cell sorting of cells labeled with fluorescent antibody or lectin, and selective binding to anion-exchange resin.

Preparation of Specific Antiserum

Antiserum can be prepared by two procedures

Procedure 1.[25]   Rabbits are injected in the marginal ear vein three times weekly for 4 weeks with a suspension of whole yeast cells that are killed by heating at 70° for 1 hr. Each 1-ml injection contains about 3×10^7 cells, or about 1 mg (dry weight) of cells in 0.9% (w/v) NaCl. One week after the last injection, the rabbits are bled, the blood is allowed to clot overnight at room temperature, the clot is removed, and the serum is collected by centrifugation. The serum is filter-sterilized and stored at 4°.

Procedure 2.[3]   Rabbits are injected intradermally over three sites on the back with a total of 1 ml of an emulsion of 2 mg of lyophilized heat-killed yeast cells in Freund's incomplete adjuvant. After 1 month, the animals receive three intravenous injections at 2-day intervals, with each injection containing 1 mg of dry cells suspended in 1 ml of sterile 0.9% NaCl. One week after the last injection, the rabbits are bled. The two immunization procedures give comparable results and yield antisera with titers of about 1 : 1280. To increase specificity for terminal $\alpha 1 \rightarrow$ 3-linked mannose, the serum raised against *S. cerevisiae* S288C or X2180 cells is adsorbed with *Kloeckera brevis* or *S. cerevisiae mnn1* cells to remove antibodies against the mannosylphosphate group and terminal $\alpha 1 \rightarrow$ 2-linked mannose. Ten milliliters of anti-*S. cerevisiae* S288C serum is added to 1 g (wet weight) of heat-killed *K. brevis* cells in 25 ml of saline, and the mixture is left overnight at room temperature with 0.2% (w/v) sodium azide as a preservative. The cells are removed by centrifugation and the adsorption is repeated in the same manner. Finally, the serum is collected, dialyzed against saline, concentrated to 10 ml by pressure dialysis, and sterilized by filtration. Alternatively, sera are diluted in saline and passed through an affinity column prepared by coupling purified mannoprotein of appropriate structure to cyanogen bromide-activated Sepharose beads. The antibodies that pass through the column are recovered and those that bind to the column are eluted with 0.1 M acetic acid.[16] Antiserum with specific-

[25] C. E. Ballou, *J. Biol. Chem.* **245**, 1197 (1970).

ity for the α-mannosylphosphate group is raised against *K. brevis* or *S. cerevisiae mnn1* cells, while serum specific for $\alpha 1 \rightarrow$ 6-mannosyl units is prepared by immunization with *S. cerevisiae mnn1 mnn2* cells.

Isolation of mnn Mutants by Agglutination[3]

Mutagenesis of *S. cerevisiae* haploid strains X1280-1A *ade1* and X1280-1B *ade2* is accomplished with ethylmethane sulfonate. For each strain, 50 μl of mutagen is added to approximately 2×10^8 cells suspended in 1 ml of sterile 0.1 M sodium phosphate buffer, pH 7.0. After the suspension has shaken at 30° for 1 hr, 0.2 ml of the mutagenized culture is transferred to 8 ml of sterile 5% (w/v) sodium thiosulfate, which terminates the reaction.

The mutagenized cells are centrifuged, resuspended in 5 ml of complete medium, and grown at 30° for 2 days. Cells with mannoprotein containing the $\alpha 1 \rightarrow$ 3 linkage are agglutinated by the addition of 0.5 ml of sterile adsorbed *S. cerevisiae* antiserum. After 1 hr at room temperature, the cells are resuspended by gentle shaking. The clusters of agglutinated cells settle rapidly, and after 5 min only very small clusters and nonagglutinated cells remain in suspension. One-half of the supernatant is removed and centrifuged, and the cells are resuspended in fresh medium and grown for 24 hr. The agglutination and transfer processes are repeated twice, each time enriching for mutants lacking the $\alpha 1 \rightarrow$ 3 linkage. The resulting supernatant is plated onto complete medium and, after growth for 4 days, colonies exhibiting the pink pigmentation of the parent strains are picked and screened for agglutination with the adsorbed antiserum. Isolates from each strain that show decreased agglutinability are used for further study.

Isolation of mnn Mutants by Cell Sorting[26]

Kluyveromyces lactic Y-58 has a mannoprotein carbohydrate structure that is very similar to *S. cerevisiae,* but it differs in having α-linked *N*-acetylglucosamine units attached to some of the side chains. This hexosamine binds wheat germ agglutinin and stimulates the formation of a specific antiserum in rabbits. Mutagenesis of *K. lactis* leads to the formation of strains with defects in synthesis of the glucosamine-containing side chains.

Kluyveromyces lactis Y-58a cells, grown to a density of 4×10^8 cells/ ml, are mutagenized with ethylmethane sulfonate. The reaction is terminated by transferring 0.1 ml of this solution into 5 ml of 5% sodium thiosulfate. The mutagenized cells are centrifuged, suspended in 0.6 ml of

[26] R. H. Douglas and C. E. Ballou, *J. Biol. Chem.* **255,** 5975 (1980).

YEPD medium, and 0.1-ml samples are used to inoculate 1-ml YEPD medium cultures. The cultures are incubated at 30° until the cells have undergone two divisions (approximately 8 hr). Portions of 5×10^6 cells are centrifuged, and the cells are washed with sterile 0.9% NaCl and labeled by incubating them for 1 hr with 25 μl of fl-WGA (0.25 mg/ml) in sterile saline. The labeled cells are washed twice with 1 ml of 0.9% NaCl and then separated on a modified Becton-Dickinson FACS II instrument equipped with a Spectra-Physics 171 laser operated at 488 nm with 1 W energy. Cells that show less than 10% of the maximum fluorescence intensity are sorted directly onto YEPD medium – agar plates. The plates are incubated at 30° for 48 hr, and all of the resulting clones are picked and screened for fl-WGA binding and agglutination by specific antisera.

For testing of lectin binding, cells grown overnight at 30° in 5 ml of YEPD medium are harvested, washed twice with saline, and then suspended in 0.1 ml of saline containing 0.25 mg/ml of fl-WGA. After a 1-hr incubation at 25°, the cells are pelleted, washed twice, suspended in saline, and the fluorescence emission spectrum is determined using a Spex Fluorolog fluorimeter. Binding of the fl-WGA to cells is also monitored visually with a Zeiss photomicroscope II equipped with Epifluorescence condenser IIIRS.

For screening many clones, a portion of cells from a YEPD medium – agar culture is suspended in 0.1 ml of saline in the well of a microtiter plate. The cells are pelleted by centrifugation at 3000 rpm in an International PR-2 centrifuge equipped with swinging microtiter plate holders. Removal of the supernatant liquid is accomplished by inverting the plate rapidly once. The pelleted cells are then suspended in 20 μl of fl-WGA (0.25 mg/ml) in saline. After a 30-min incubation at 25°, the cells are washed three times with saline, transferred into 3.0 ml of saline, and the fluorescence emission at 520 nm is measured upon excitation at 460 nm. Cell number is estimated from light scattering at 660 nm.

Isolation of mnn2 Transformants by Cell Sorting[27]

A haploid *S. cerevisiae mnn2 ura3* strain is transformed by the spheroplast method with the CsCl-purified yeast genomic DNA of a library constructed in the vector YEp24, and *URA* transformants are patched onto YM plates supplemented with 0.003% (w/v) adenine and 0.002% (w/v) histidine. Approximately 5000 patched transformants are replicaplated onto fresh plates and grown for 2 days at 30°. Cells are harvested by adding 0.5 ml of sterile water to the plates, scraping the plates with a sterile plate-spreader, and collecting the cell suspension. Cells are washed with 0.9% NaCl, resuspended by vortexing and sonication, and quantitated by

---

[27] C. P. Devlin, Doctoral Dissertation, University of California, Berkeley, Berkeley, CA, 1987.

absorbance. One A_{600} unit of cells is mixed with fluorescein-labeled *mnn4* ($\alpha 1 \rightarrow$ 3-specific) or *mnn2* ($\alpha 1 \rightarrow$ 6-specific) Fab′ fragments. After the cell pellet is washed, it is resuspended in 1.0 ml of potassium phosphate buffer by vortexing and sonication in an ultrasonic bath.

A Becton-Dickinson fluorescence-activated cell sorter (FACS IV) is flushed with 70% ethanol for 30 min and sterile saline for 10 min before the fluorescein-labeled yeast cells are added. Fluorescence excitation at 488 nm is obtained with an argon laser (Spectra-Physics, 2W) operated at 0.4 W. Light scattering and fluorescence emission of individual cells are measured by separate photomultiplier tubes (PMT) located behind selective filters. The scatter PMT has a neutral density filter and the fluorescence PMT has a LP510 filter. PMTs are operated at 600 V with the gain set at $1.0 \times$ for scatter and $4.0 \times$ for fluorescence. Approximately 10^6 cells are passed through the sorter at a rate at which greater than 90% of the cells can be sorted, which requires less than 45 min.

The most highly fluorescent transformants are chosen from the population of cells labeled with *mnn4* Fab′, and the least fluorescent transformants from the population of cells labeled with *mnn2* Fab′. The collected cells are resuspended in 0.5 ml of YM medium plus 0.003% adenine and 0.002% histidine and spread on YM plates supplemented with 0.002% adenine and 0.008% histidine for isolation of individual clones.

Isolation of mnn Mutants by Binding to Anion Exchanger[8,28]

The following procedure is applied to the *S. cerevisiae mnn1 mnn9* haploid strain to obtain mutants with a reduced level of phosphorylation, but it can also be used with the *mnn1* or wild-type X2180 strains. In a *mnn1 mnn9* culture, most of the cells are clumped together, but the newly budded daughter cells exist as small single cells that can be isolated by allowing a log-phase culture suspended in 0.1 M potassium phosphate, pH 7.0, to stand for 10 min, during which most of the clumps settle out. The cells remaining in suspension in a culture grown to a density of 2 A_{600} units are then used in the following steps. To a tube with a Teflon-lined screw-cap, add 50 μl of ethylmethane sulfonate followed by 1 ml of the above partially settled cell suspension. Tightly cap the tube and shake it vigorously on a wrist-action shaker for 30 min. Transfer 0.2 ml of the cell suspension to each of four sterile tubes containing 6 ml of sterile 5% sodium thiosulfate and mix well. Collect the cells by centrifugation and wash them twice with sterile potassium phosphate buffer, pH 7.0. Resuspend the cells in 2 ml of YEPD containing 0.5 M KCl, an osmotic stabilizer for the fragile *mnn9* mutant, and grow the cultures with shaking at 30° for 48 hr. This mutagenesis treatment gives 80–90% survival.

[28] L. Hernandez, L. Ballou, and C. E. Ballou, unpublished results (1989).

Prehydrated QAE-Sephadex beads (acetate form) (40–120 mesh, Sigma Q-25-120, St. Louis, MO) are washed with distilled water, soaked in 95% ethanol for 30 min to sterilize the beads, and then are washed twice with sterile water. The beads are suspended in 25 mM acetic acid, pH 3.0, and 1 ml of the suspension is mixed with a sample of cultured mutagenized cells that have been washed twice with sterile water. (Treatment at pH 3.0 facilitates selection for phosphate-deficient cells but it reduces survival by about 70% for *mnn1 mnn9* cells.) The mixture is shaken for 5 min, the beads are allowed to settle for 5 min, and 1 ml of the turbid supernatant is transferred to a sterile tube for a second treatment with the QAE beads. The success of the QAE-Sephadex treatment is monitored by observing the beads under a microscope to see how many cells are bound. After a third treatment, the cells that remain unbound are removed, sonicated to disrupt clumps, and spread on YEPD + KCl plates to give about 100 colonies per plate. The plates are incubated at 30° for 48 hr or longer, when colonies are picked, grown up on fresh plates, and tested for an altered dye-binding phenotype.

Alcian Blue Dye-Binding Assay[23]

This cationic dye will bind to cells and organelles that possess a negative charge, and it distinguishes between yeast cells with different amounts of phosphate in the wall. A yeast culture is grown to stationary phase in liquid medium or on an agar plate, and an amount of cells that will pellet to just cover the bottom of a 10 × 75 mm disposable test tube is transferred to the tube and washed once with 0.5 ml of 0.9% NaCl. Care is used to avoid transferring any agar, which binds alcian blue strongly. To the cell pellet is added 0.5 ml of 0.1% alcian blue (ICN lot 38411-A) in 0.02 N HCl. The dye solution should be free of particles that will sediment at 5000 rpm, but it should not be filtered. Vortex the tube well, let it stand at room temperature for 15 min, and then centrifuge at 2000 rpm. Remove the solution by decantation, touching the rim of the tube to a paper towel to remove the last bit of liquid, and wash the pellet once with about 1 ml of 0.02 N HCl which is also decanted. Then resuspend the cells in 1 ml of 0.02 N HCl and centrifuge the tube a third time, but retain the supernatant above the undisturbed cell pellet. Align the tubes to be compared in a wire test tube rack and estimate the depth of the blue color of the pellet on a scale from 0 (for the *mnn4* or *mnn6* strains) to 5 (for the *mnn1* strain) by holding the rack up and looking at the bottoms of the tubes. Cells grown in presence of an osmotic stabilizer, such as 0.5 M KCl or 10% (w/v) sorbitol, show enhanced dye binding. (Tubes used for this assay are difficult to wash free of the dye, but they are suitable for reuse in the assay.)

Isolation of Mutants of mnn7 Class and Construction of Multiple
   mnn Mutants[16,19]

The *mnn7* class of mutant is derived from mutagenesis of the *mnn2-1* haploid strain with ethylmethane sulfonate, and the isolates are identified by a reduced ability to react with rabbit antiserum directed against the unbranched $\alpha 1 \rightarrow 6$-linked mannose backbone of the latter strain. Mutants of this class (*mnn7, mnn8, mnn9,* and *mnn10*) all show an increased ability to react with $\alpha 1 \rightarrow 3$-mannosyl-specific antiserum owing to exposure of the core after removal of all or part of the outer chain, and they gain the ability to bind alcian blue dye owing to an increased content or accessibility of mannosylphosphate groups in the mannoprotein. All of the strains retain the *mnn2* defect, although the effect of this mutation is expressed only in the *mnn7, mnn8,* and *mnn10* strains, which still contain short sections of the outer chain that become branched in the absence of the *mnn2* lesion.

Strains containing the *mnn1* mutation (defective in $\alpha 1 \rightarrow 3$-mannosyltransferase) are constructed by crossing a strain of the genotype *mnn1 mnn2* with, for example, *mnn2 mnn9* and selecting for recombinant haploids from the sporulated diploid that are clumpy (*mnn9* phenotype) and fail to agglutinate with antiserum against $\alpha 1 \rightarrow 3$-linked mannose (*mnn1* phenotype). Such a recombinant should have the genotype *mnn1 mnn2 mnn9,* and this is confirmed by backcrossing it to the *mnn2* strain and recovering the expected double mutants (*mnn1 mnn2* and *mnn2 mnn9*) from the sporulated diploid. In a similar fashion, a cross between the *mnn1 mnn2 mnn7* and *mnn1 mnn2 mnn9* strains gives a diploid that, following sporulation and dissection, yields some nonparental ditypes with two clumpy *(mnn1 mnn2 mnn7 mnn9)* and two nonclumpy *(mnn1 mnn2)* clones. The genotypes of such quadruple *mnn* mutants are checked by backcrosses to the *mnn1 mnn2* strain, with nonparental ditype recombinants showing four clumpy clones owing to the independent segregation of the *mnn7* and *mnn9* lesions.

Some difficulty is faced in genetic analysis of the *mnn7* class of mutant owing to the clumpiness of the cells, which interferes with dissection and mass mating procedures. Selection of zygotes by micromanipulation is the preferred technique because matings between newly budded nonclumpy cells can be sorted out from the aggregates of incompletely separated cells with defective walls and septums.

Since the "wild-type" strain in these studies has the *mnn1 mnn2* genotype, complementation between members of the *mnn7* class in the diplophase leads to the *mnn1 mnn2* phenotype. This phenotype is recognized by an enhanced reactivity of whole cells with an $\alpha 1 \rightarrow 6$-mannosyl-specific antiserum, by a reduced binding of alcian blue dye, and by a reduced rate of migration on gel electrophoresis of the external invertase, all consequences of the increased size of the N-linked polymannose chains.

Characterization of Glycosylation Defects by Native Gel Electrophoresis of External Invertase[29]

The rate of migration of invertase on a native gel depends on the number and size of the N-linked oligosaccharide chains (Fig. 2). Invertase from *S. cerevisiae* contains 14 potential glycosylation sites,[30] and the secreted enzyme usually possess 9–11 carbohydrate chains. On a native gel at room temperature, wild-type external invertase migrates as a diffuse band, while *mnn9* invertase migrates in a relatively sharp band as a dimer. At lower temperatures (5°), bands for the tetramer, hexamer, and octamer are also seen.[31,32] In crude extracts, the invertase bands are revealed by an activity stain, whereas pure invertase preparations can be visualized by the periodate–Schiff or silver stains, and show multiple bands when run on a denaturing gel, owing to heterogeneity in glycosylation. Gel electrophoresis of pure *mnn9* invertase under denaturing conditions reveals 3–4 bands that represent homologs with 9–11 oligosaccharide chains. Following exhaustive digestion with endoglucosaminidase H in presence of detergent, these are converted to a single band that migrates slightly slower than the internal nonglycosylated invertase owing to the presence of 9–11 N-acetylglucosamine units.

The presence of the *gls1* defect (glucosidase I)[33] in the *mnn1 mnn9* background increases the size of each oligosaccharide by 3 hexoses and the carbohydrate content of each invertase subunit by about 30 glucose units, and causes the dimer to migrate distinctly slower than the *mnn1 mnn9* invertase.[34] On the other hand, the *dpg1* or *alg5* mutations (dolichol-P-glucose synthetase) lead to underglycosylation and cause the invertase to migrate faster on a native gel than the *mnn1 mnn9* form and as a diffuse band.[29] On a denaturing gel, the *mnn1 mnn9 dpg1* invertase shows 4 bands

[29] L. Ballou, P. Gopal, B. Krummel, M. Tammi, and C. E. Ballou, *Proc. Natl. Acad. Sci. U.S.A.* **83**, 3081 (1986).

[30] R. Taussig and M. Carlson, *Nucleic Acids Res.* **11**, 1943 (1983).

[31] P. C. Esmon, B. E. Esmon, I. E. Schauer, A. Taylor, and R. Schekman, *J. Biol. Chem.* **262**, 4387 (1987).

[32] M. Tammi, L. Ballou, A. Taylor, and C. E. Ballou, *J. Biol. Chem.* **262**, 4395 (1987).

[33] B. Saunier, R. D. Kilker, Jr., J. S. Tkacz, A. Quaroni, and A. Herscovics, *J. Biol. Chem.* **257**, 14155 (1982).

[34] P.-K. Tsai, L. Ballou, B. Esmon, R. Schekman, and C. E. Ballou, *Proc. Natl. Acad. Sci. U.S.A.* **81**, 6340 (1984).

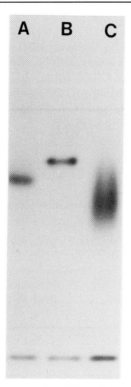

FIG. 2. Native gel electrophoresis and activity staining of invertases from *mnn* mutants. Whole cell extracts of (A) *mnn1 mnn9*, (B) *mnn1 mnn9 gls1*, and (C) *mnn1 mnn9 gls1 dpg1* cultures were electrophoresed as described in the text and stained for invertase activity. The lower band in all three extracts represents internal nonglycosylated invertase. The glycosylated invertases in lanes A and B migrate as dimers with 10–12 oligosaccharides of 10 mannoses and 2 *N*-acetylglucosamines per polypeptide subunit, and the slower migration of B results from the additional 3 glucoses that remain on each oligosaccharide owing to the glucosidase I defect. The invertase in lane C has oligosaccharides of the size present on the invertase of lane A, but it migrates faster because the polypeptides are underglycosylated owing to the less efficient transfer of the unglucosylated lipid-linked precursor.[29] The diffuse band might also reflect a reduced stability of the dimer.[32]

with 4–7 carbohydrate chains when the gel is treated with the silver stain. With experience, one can discriminate between other mutants such as *mnn1 mnn2, mnn1 mnn2 mnn8*, and *mnn1 mnn2 mnn10*, but the electrophoretic patterns are not always reproducible. Complementation between *mnn8, mnn9*, and *mnn10* in the *mnn1 mnn2* background to give the *mnn1 mnn2* phenotype is easily followed by the invertase patterns, and segregation of defects such as *alg5* and *dpg1* in the *mnn9* background can be scored for tetrad analysis by the invertase native gel patterns.

Preparation of Invertase Extract[35]

Cells, from a 5-ml mid-log phase (24-hr) culture on YEPD + 0.5 M KCl, are collected, washed twice with sterile saline + 0.5 M KCl, and resuspended in YEP + 0.5 M KCl containing 0.05% glucose to induce invertase for 2–4 hr on a shaker at 24°. (If the culture is grown until all glucose is used, the invertase will be induced spontaneously. KCl can be omitted if the cells do not tend to lyse.)

Cells are centrifuged, washed with 5 ml of cold 20 mM sodium azide and then with TBP buffer [5.52 g of diethylbarbituric acid and 1 g of Tris base per liter of water, pH 7.0; to 100 ml, add 1 ml of stock PMSF (0.174 g of phenylmethanesulfonyl fluoride in 10 ml of absolute ethanol) just before use]. Resuspend cells in TBP buffer (20 μl) in a 1.5-ml Eppendorf centrifuge tube, add clean dry glass beads (0.45 mm) up to the meniscus, and vortex in 30-sec pulses to break the cells, keeping the tubes on ice between treatments. The number of vortexings varies with strain, 2 times for *mnn9* and 5 times for wild-type cells. Finally, add 50 μl of TBP buffer containing 15% glycerol and 0.01% bromphenol blue, centrifuge, and take 5–20 μl of the supernatant solution to load on the gel.

Gel Electrophoresis of Invertase Extract[36]

> Stock acrylamide–bis solution: Acrylamide (30 g) and N,N'-methyl-enebisacrylamide (0.8 g) are dissolved in 100 ml of water, the solution is filtered through a 1.2 μm Millipore filter and stored at 4° in the dark
> Stock Tris buffer: Dissolve 9.69 g of Trisma base in 50 ml of water, titrate to pH 7.3 with HCl, and dilute to 100 ml with water
> Ammonium persulfate: Dissolve 0.1 g in 1.0 ml of water and keep on ice. Prepare freshly before use
> TEMED
> Tris–borate buffer: Dissolve 1 g of Tris base and 5.02 g of boric acid in 1 liter of water

In a 50-ml suction flask, mix 7.0 ml of acrylamide–bis solution, 3.0 ml of Tris buffer, and 19.9 ml of water. Degas under vacuum for 5 min, then add 0.1 ml of ammonium persulfate and 10 μl of TEMED, and swirl the solution gently to mix without forming bubbles. Cast into the gel mold (0.75 mm × 10 cm × 15 cm) without a stacking gel and add a comb on the top. Gels may be stored in the cold overnight before use. After adding samples, the gel is run for 3 hr at a constant 10 mA current with a Tris–borate buffer, pH 7.5.

[35] T. C. Huffaker and P. W. Robbins, *J. Biol. Chem.* **257,** 3203 (1984).
[36] J. Meyer and P. Matile, *Biochem. Physiol. Pflanz.* **166,** 377 (1964).

Activity Stain for Invertase[37]

> Sucrose solution (0.1 M pure sucrose in 0.1 M sodium acetate, pH 5.1): Dissolve 34.23 g of sucrose and 13.61 g of sodium acetate trihydrate in water, adjust to pH 5.1 with glacial acetic acid, and dilute to 1 liter
>
> Chromophore solution: Dissolve 50 mg of 2,3,5-triphenyltetrazolium chloride (TTC) in 50 ml of 0.5 M NaOH. Prepare freshly for each use
>
> Acetic acid, 10%: Dissolve 100 ml of glacial acetic acid in water to make 1 liter

Add the gel to 50 ml of cold (4°) sucrose solution and allow the substrate to diffuse into the gel for 10 min at 4°. Transfer the gel to a 37° sucrose solution (50 ml) and incubate it at that temperature for 10 min to hydrolyze the substrate. Pour off the sucrose solution and rinse the gel twice with water, then transfer it to a Pyrex dish containing 50 ml of TTC solution and heat the solution to boiling on a hotplate *in the hood* until the color develops satisfactorily. Finally, wash the gel with water, put it in 10% acetic acid, and then dry it for permanent storage.

Comments

When invertase extracts are run at room temperature, the internal nonglycosylated invertase dimer travels near the gel front and the external wild-type invertase runs as a diffuse band on the upper third of the gel. Invertase from the *mnn1 mnn9* strain runs as a fairly sharp band near the middle of the gel and can be resolved from the *mnn1 mnn9 gls1* invertase, which contains 3 glucose/oligosaccharide chain, but it is not well resolved from *mnn1 mnn2 mnn10* invertase. Invertase from the 4AL strain *(mnn1 mnn9 gls1 dpg1)* runs as a diffuse band ahead of the *mnn1 mnn9* invertase owing to underglycosylation. Invertases run on gels at 4° show bands for higher oligomers with 4, 6, and 8 subunits, the appearance of the bands depending on the number and size of the oligosaccharide chains.

Isolation of Yeast Mannoprotein

Extraction of Mannoprotein[38]

The wet cell paste (about 100 g), from a 10-liter culture of *S. cerevisiae mnn1 mnn9* cells grown on YEPD + 0.5 M KCl, is washed once with 0.9% NaCl and once with distilled water. The washed paste is suspended in

---

[37] O. Gabriel and S. F. Wang, *Anal. Biochem.* **27**, 545 (1969).
[38] S. Peat, W. J. Whelan, and T. E. Edwards, *J. Chem. Soc.*, 29 (1961).

50 ml of 0.02 M citrate buffer, pH 7.0, and the mixture is autoclaved at 125° for 90 min. The gelatinous solid is removed from the cooled mixture by centrifugation, the supernatant solution is saved, and the solid is extracted a second time with 75 ml of the same buffer. The extracts are combined and poured into 3 volumes of stirring methanol to precipitate the mannoprotein, and the mixture is stirred overnight at 4°. The precipitate is collected by centrifugation, redissolved in water, dialyzed against 2 changes of water, and lyophilized. The yield is about 1 g of crude mannoprotein, while the wild-type strain gives about 4 g owing to a higher carbohydrate content.

Cetavlon Precipitation[39]

One gram of crude mannoprotein is dissolved in 12.5 ml of water, 1 g of Cetavlon (Sigma; hexadecyltrimethylammonium bromide) dissolved in 12.5 ml of hot water is added, and the mixture is left at room temperature for 24 hr. Any precipitate is removed by centrifugation, and 12.5 ml of 2% boric acid is added to the supernatant solution which is then adjusted to pH 8.8 and left overnight at room temperature. The mannoprotein–borate–Cetavlon precipitate that forms is collected by centrifugation, washed once with 0.5% sodium borate, pH 8.8, and then dissolved in 12.5 ml of 2% acetic acid, which dissociates the complex. The free mannoprotein is precipitated by adding 3 volumes of ethanol, the precipitate is collected by centrifugation, washed once with 2% acetic acid in ethanol and once with ethanol, and then dissolved in water, dialyzed against water, and lyophilized. About 0.3 g of mannoprotein is obtained from the *mnn1 mnn9* strain and 2 g from the wild type.

Ion-Exchange Purification of Mannoprotein[40]

Mannoprotein, 2 g, is applied to a DEAE-Sephacel column (3 × 18 cm) in 20 mM Tris-HCl buffer, pH 7.5, and eluted with the same buffer in 11-ml fractions. A peak of unbound carbohydrate is eluted during the first 70 fractions, after which a 1-liter linear gradient of 0–0.5 M NaCl in the same buffer is applied, which elutes a second carbohydrate-containing peak. Some protein is associated with both beaks, but most is eluted at higher salt concentrations. The material in the second carbohydrate peak is combined and used for isolation of N-linked oligosaccharides.

---

[39] K. O. Lloyd, *Biochemistry* **9**, 3446 (1970).
[40] J. Frevert and C. E. Ballou, *Biochemistry* **24**, 753 (1985).

Isolation of N-Linked Oligosaccharides[11]

To 250 mg of mannoprotein in 1 ml of 50 mM citrate buffer, pH 5.6, containing 0.02% sodium azide, is added 0.01 unit of endoglucosaminidase H,[41] and the solution is incubated under toluene at 37° for 24 hr. A second addition of enzyme (0.005 units) is made and the incubation continued for 24 hr, followed by a third addition and incubation, when the reaction is stopped by heating the tube in boiling water for 5 min. The cooled solution is clarified by centrifugation, evaporated under vacuum to a small volume, and applied to a BioGel P-10 (−400 mesh) column (2 × 190 cm) equilibrated with 0.1 M ammonium acetate, pH 7.0. Elution with the same solvent separates the digest into an excluded peak of mannoprotein and an included peak of acidic and neutral oligosaccharides. The latter peak is collected and applied to a BioGel P-4 (−400 mesh) column (2 × 190 cm) equilibrated in water, and the column is eluted with water to yield an excluded peak of acidic (phosphorylated) oligosaccharides and an included peak(s) of neutral oligosaccharides. The neutral oligosaccharide fraction can be resolved into homologs by rechromatography on a BioGel P-4 column and the acidic oligosaccharide fraction can be separated into mono- and diphosphate isomers by ion-exchange chromatography. The latter are mostly in diester form and the small amount of phosphate in monoester form probably results from chemical hydrolysis during the high-temperature extraction of the mannoprotein.

Characterization of *mnn* Mutants by Nuclear Magnetic Resonance[42,43]

The N-linked oligosaccharides of yeast mannoproteins are based on the same Man$_8$GlcNAc$_2$-Asn structure found in all eukaryotes, and the H$_1$ chemical shifts of the mannose units provide intrinsic probes for the linkages present in this oligosaccharide and its processed derivatives. The comparison is more conveniently done on mutants in the *mnn1 mnn2* background because this eliminates the complexity of branching in the outer chain and substitution of the core with terminal $\alpha 1 \rightarrow 3$-linked mannose.

The empirical bases for making assignments are summarized in Fig. 3, which shows the H$_1$ chemical shifts for mannoses in the most common linkages found in yeast mannoproteins as well as the effects of substitution

[41] A. L. Tarentino, R. B. Trimble, and F. Maley, this series, Vol. 40, p. 574.

[42] R. E. Cohen and C. E. Ballou, *Biochemistry* **19**, 4345 (1980).

[43] J. F. G. Vliegenthart, L. Dorland, and H. van Halbeek, *Adv. Carbohydr. Chem. Biochem.* **41**, 209 (1983).

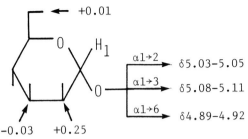

Fig. 3. Linkage and sequence effects on the H_1 chemical shifts of mannose in mannooligosaccharides. The chemical shifts of all H_1 signals are downfield from the outer ring protons and are found at $\delta 4.7 - 5.4$ because this hydrogen is attached to a carbon that is linked to two oxygens, but the exact chemical shift is determined by the position and configuration of the linkage to the next sugar and by the position of substitution by another sugar. A positive sign indicates a downfield shift and a negative sign an upfield shift. Thus, an $\alpha 1 \rightarrow 2$-linked mannose unit substituted at position 2 gives an H_1 signal at $\delta 5.29$.

at different positions around the ring.[42] Thus, a terminal $\alpha 1 \rightarrow 2$ mannose gives an H_1 signal at about $\delta 5.04$, and substitution at position 2 shifts this resonance downfield to $\delta 5.29$. Some interesting long-range effects are noted for the endoglucosaminidase H-released oligosaccharides owing to interaction through space with the N-acetylglucosamine anomers (Table I), and such effects are useful indicators of oligosaccharide conformation.[44]

Bulk mannoprotein is extracted from the cells with hot citrate buffer and precipitated from the extract with methanol. The precipitated mannoprotein is purified by ion-exchange chromatography and then digested with endoglucosaminidase H to release the N-linked oligosaccharides. The oligosaccharide fraction is separated by gel filtration in water and further fractionated by gel filtration, HPLC, or ion-exchange chromatography. ^{1}H-NMR spectra are obtained in D_2O at 40° on a 500 MHz instrument, with 0.5 μmol of sample sufficing for a one-dimensional spectrum, although useful spectra can be obtained with 0.03 μmol.

Spectra for three oligosaccharides are shown in Fig. 4, these being the intact oligosaccharides accumulated in the *mnn1 mnn2 mnn9* and *mnn1 mnn2 mnn10* mutants and the core fragment $Man_9GlcNAc$ obtained by endo-$\alpha 1 \rightarrow 6$-mannanase digestion of either the *mnn1 mnn2* or *mnn1 mnn2 mnn10* oligosaccharide. All three oligosaccharides have in common the signals at $\delta 4.70$, 4.77, 4.87, 5.04, $5.09/5.11$, 5.13, 5.24, 5.29, and 5.33, whereas the latter two have an additional signal at about $\delta 4.90$. The assignments of these signals are given in Table I.

[44] J. P. Carver and A. A. Grey, *Biochemistry* **20**, 6607 (1981).

TABLE I
¹H-NMR CHEMICAL SHIFTS FOR *S. cerevisiae* N-LINKED CORE OLIGOSACCHARIDES

Sugar residue and structure					δ (ppm)[a]					
E	D	C	B	A	E	D	C	B	Aα	Aβ
							5.085[b]			
	αM ↓2	αM ↓3				5.035	5.110			
	αM →6	αM →6 (↑3)	βM →4	αβGNAc		5.125	4.868	4.770	5.235	4.710
	αM →6	αM (↑2)				4.925	5.335			
		αM (↑2)					5.289			
			αM					5.045		
							5.086[b]			
	αM ↓2	αM ↓3				5.037	5.111			
	αM →6	αM →6 (↑3)	βM →4	αβGNAc		5.124	4.869	4.769	5.236	4.712
	αM →6 (↑2)	αM (↑2)				5.124	5.325			
	αM	αM (↑2)				5.045[b] / 5.055	5.289			
			αM					5.045		
							5.085[b]			
	αM ↓2	αM ↓3				5.037	5.113			
	αM →6	αM →6 (↑3)	βM →4	αβGNAc		5.126	4.868	4.771	5.238	4.712
αM →6 (↑2)	αMx →6	αM →6 (↑2)	αM		5.126	4.896[c]	4.908	5.335		
αM		αM (↑2)			5.037		5.286			
			αM					5.046		

[a] Referenced to acetone, δ2.217, at 40°. All signals are present in approximately unit ratios except that the *N*-acetylglucosamine anomers have different chemical shifts.

[b] Long-range effects of the two *N*-acetylglucosamine anomers split the H_1 signals.

[c] The intensity of this signal provides an indication of the length of the outer chain, which in the *mnn1 mnn2 mnn10* mannoprotein averages 10, with a range of 3–16.

Glycosidases for Analysis of Yeast Mannoprotein Structure

Several enzymes are particularly useful for this purpose. β-*N*-Acetylglucosaminidase H will remove all N-linked yeast oligosaccharides and its preparation has been reported elsewhere in this series.[41] Useful mannosidases include a nonspecific α-mannosidase from a bacterium that can grow

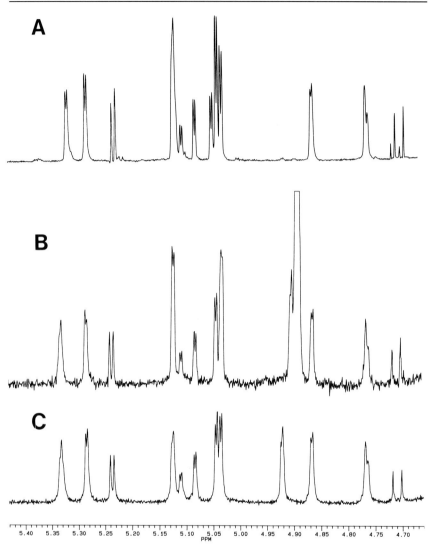

FIG. 4. ^{1}H-NMR spectra of yeast N-linked oligosaccharides. Spectra of (A) Man$_{10}$GlcNAc from Endoglucosaminidase H digestion of *mnn1 mnn9* or *mnn1 mnn2 mnn9* mannoprotein, (B) Man$_{22}$GlcNAc from Endoglucosaminidase H digestion of *mnn1 mnn2 mnn10* mannoprotein, and (C) Man$_9$GlcNAc obtained by endo-$\alpha1 \rightarrow 6$-mannanase digestion of *mnn1 mnn2* or *mnn1 mnn2 mnn10* oligosaccharide. The different resonances are assigned in Table I.

on wild-type yeast mannoprotein,[45] an endo-α1 → 6-mannanase from a bacterium that can grow on *S. cerevisiae mnn1 mnn2* mannoprotein,[46] and the fungal exo-α1 → 2-mannosidase.[47] Methods for purification of the latter three enzymes are given here. In contrast to the commonly used jackbean mannosidase,[48] all of these enzymes show good activity against the mannoproteins and large high-mannose oligosaccharides.

General Assay for Mannosidases That Act on Yeast Mannoproteins

The following procedure is modified for each of the enzymes described below by substitution of the appropriate mannoprotein substrate, buffer, and incubation time. Because the reducing sugar method[49] involves development of the color in a hot alkaline solution, the mannoprotein must first be subjected to β-elimination to reduce the blank reaction owing to the release of O-linked carbohydrate. Mannoprotein is heated in 0.5 N NaOH at 100° for 10 min to cause β-elimination, and the solution is neutralized with HCl, dialyzed overnight against water, and adjusted to 4% carbohydrate assayed by the phenol–sulfuric acid method.[50]

The enzyme assay contains 10 μl of the mannoprotein solution, 10 μl of buffer, and 10 μl of enzyme solution, and the mixture is incubated at the temperature and for the time indicated, when the release of reducing sugar is determined by the following procedure. One unit of enzyme liberates 1 μmol of reducing sugar equivalent in 30 min.

Just before use, mix together equal volumes of 0.05 M Na$_2$SO$_3$, 0.01 M CaCl$_2$, 0.02 M sodium citrate, and 0.3 M NaOH, and add 152 mg of p-hydroxybenzoic acid hydrazide (Sigma) to 20 ml of the solution. Add 2.0 ml of this reagent to the enzyme incubation from above, heat the tube in boiling water for exactly 10 min, and immediately cool it to room temperature by immersion in a cold-water bath. The absorbance is read at 420 nm, with 10 μg of glucose giving a reading of 0.3.

Purification of Endo-α1 → 6-mannanase[46]

The organism used in this preparation is a strain of *Bacillus circulans* (ATCC 29101) that was isolated from a soil enrichment culture on *S*.

[45] G. H. Jones and C. E. Ballou, *J. Biol. Chem.* **244**, 1043 (1969).
[46] T. Nakajima, S. K. Maitra, and C. E. Ballou, *J. Biol. Chem.* **251**, 174 (1976).
[47] E. Ichishima, M. Arai, Y. Shigematsu, H. Kumagai, and R. Sumida-Tanaka, *Biochim. Biophys. Acta* **658**, 45 (1981).
[48] Y.-T. Li and S.-C. Li, this series, Vol. 28, p. 702.
[49] J. Lever, *Biochem. Med.* **7**, 274 (1973).
[50] M. Dubois, K. A. Gilles, J. K. Hamilton, P. A. Rebers, and F. Smith, *Anal. Chem.* **28**, 350 (1956).

cerevisiae mnn2 mannoprotein. It is maintained on agar slants prepared from the same medium used for purification of the mannanase.

Endomannanase Assay. The assay mixture contains *mnn2* mannoprotein, 0.1 M citrate phosphate buffer, pH 6.0, containing 0.3 mM CaCl$_2$, and enzyme solution. After incubation at 50° for 30 min, the increase in reducing sugar is determined.

Purification. A liquid medium composed of 1 g of *mnn2* mannoprotein, 500 mg of (NH$_4$)$_2$SO$_4$, 400 mg of MgSO$_4$·7H$_2$O, 20 mg of FeSO$_4$·7H$_2$O, 60 mg of CaCl$_2$·2H$_2$O, 7.4 g of K$_2$HPO$_4$, 2.32 g of KH$_2$PO$_4$, and 500 mg of yeast extract in 1 liter of water is distributed in five 3-liter Fernbach flasks, each containing 1 liter of the medium, autoclaved, and inoculated with 10 ml of an 18-hr culture of the *B. circulans* strain. The flasks are incubated on a reciprocal shaker at 30° for 24 hr, at which time the mannanase activity reaches a plateau. All subsequent steps are carried out at 4°. The cells are removed by centrifugation (10,000 g for 30 min) and the supernatant is concentrated to 700 ml by dialysis against dry polyethylene glycol 8000 (Sigma P2139). The concentrate is dialyzed against distilled water for 24 hr and then against 0.05 M potassium phosphate, pH 7.0, containing 10^{-4} M CaCl$_2$ (buffer A) for 24 hr with several changes.

The concentrated and dialyzed cultural filtrate is adjusted with solid ammonium sulfate to 30% of saturation at 4°, the mixture is stirred for 30 min and allowed to stand for 1 hr when it is centrifuged at 15,000 g for 20 min. The precipitate is discarded and the supernatant is adjusted to 60% of saturation, stirred for 20 min, allowed to stand for 2 hr and again centrifuged. The precipitate is dissolved in 30 ml of buffer A and dialyzed against the same buffer for 24 hr.

The dialyzed ammonium sulfate fraction is applied to a DEAE-Sephacel column (2.5 × 10 cm) equilibrated with 0.05 M potassium phosphate, pH 7.0, and the column is eluted with a linear gradient of 0–0.6 M NaCl in buffer A. The mannanase activity is displaced at about 0.35 M NaCl, after most of the protein is eluted, and the active fractions are combined, concentrated as above with polyethylene glycol 8,000, and then dialyzed against buffer A.

The DEAE-Sephacel mannanase fraction is applied to a hydroxylapatite column (2.5 × 15 cm) equilibrated with 0.05 M potassium phosphate, pH 7.0, and the column is eluted with a linear gradient from 0.05 to 0.3 M potassium phosphate, pH 7.0. The fractions that contain mannanase activity, at a salt concentration of about 0.17 M, again come after most of the protein and are combined, dialyzed against 1 mM potassium phosphate, and lyophilized.

The lyophilized enzyme from the previous step is dissolved in a small amount of 0.1 M potassium phosphate, pH 7.0, and applied to a Sephadex G-200 column (2 × 80 cm) equilibrated in the same buffer, and the column is eluted with the same buffer. The active fractions are combined and lyophilized, and then dissolved in 10 mM potassium phosphate, pH 7.0, and dialyzed against buffer A.

Properties of Endomannanase. The enzyme preparation shows a single protein band on a denaturing gel (131 kDa) and a single band of protein and activity on a native gel. A broad pH optimum, pH 5.5 to 7, is observed, the enzyme remains fully active at 50° for many hours at pH 5.5, and it requires Ca^{2+} for full activity. The enzyme has an endolytic activity and degrades unbranched $\alpha1 \rightarrow 6$-mannan chains with the initial formation of large oligomannosides that are eventually reduced to a mixture of mannose and mannobiose. The N-acetylglucosamine-containing core oligosaccharide fragment retains one or two unsubstituted $\alpha1 \rightarrow 6$-linked mannose units at the nonreducing end, and any section of the outer chain that is branched contains two or three such mannose units at the reducing end. The K_m for large oligosaccharides is about 5 μM, but it increases to 1 mM for the mannotriose, whereas the maximal rate of hydrolysis increases up to the hexasaccharide and then reaches a plateau.

Purification of Exo-$\alpha1 \rightarrow 2$-mannosidase[47]

The organism used in this preparation is a strain of *Aspergillus phoenicis* (ATCC 14332, formerly *A. satoi*) and it is maintained on Czapek solution agar (Difco, Detroit, MI). After incubation on petri plates for 1 week at 30°, a heavy white filamentous growth covered with black spores is obtained that is suitable for inoculation.

Exo-$\alpha1 \rightarrow 2$-mannosidase Assay. The assay mixture contains *S. cerevisiae mnn1*mannoprotein, 0.1 M sodium acetate, pH 5, and enzyme solution. After incubation at 37° for 30 min, the release of reducing sugar is determined.

Purification. Two 3-liter Fernbach flasks, each containing 300 g of wheat bran from a local supermarket and 125 ml of water, are autoclaved for 30 min, cooled, and inoculated by adding to each flask half the agar, cut into small pieces, from a 1-week-old petri plate culture. The flasks are kept at 30° for 6 days and shaken once a day to break up the filamentous mass. To extract the mannosidase, 600 ml of 0.01 M sodium acetate, pH 5.0, is added to each flask which is allowed to stand at 4° for 6 hr. The mixture is then filtered overnight at 4° through a layer of Celite on a 24-cm Büchner funnel, with a rubber dam over the top to press out the liquid. The filtrate is adjusted to pH 5.0 with NaOH to give a volume of about 1300 ml.

Solid ammonium sulfate, 515 g, is added with stirring to the culture filtrate and, after 20 min, the precipitate is removed by centrifugation at 10,000 g for 20 min. To the supernatant is added 225 g of ammonium sulfate and the precipitate, which contains the enzyme activity, is collected by centrifugation. The precipitate is dissolved in 100 ml of 10 mM sodium acetate, pH 5, at 4° and the mannosidase is precipitated by adding 200 ml of cold (4°) ethanol and collected by centrifugation. The protein is dissolved in 10 ml of 10 mM sodium acetate, pH 5, containing 0.2 M NaCl to give 15 ml of solution. The yield of mannosidase is about 330 nkat. At this stage, the preparation contains an active cellulase, so dialysis in ordinary tubing is not feasible, but the cellulase activity is removed in the next gel filtration step.

Half of the enzyme solution (7.5 ml) is applied to an Ultrogel AcA 44 column (4 × 95 cm) equilibrated with 10 mM sodium acetate, pH 5, containing 0.2 M NaCl and 0.02% sodium azide, and the column is eluted with the same buffer in 10-ml fractions at a rate of 20 ml/hr. The fractions are assayed for mannosidase activity and for protein, and the peak of enzyme activity usually appears around fraction 65 to 70. The active fractions are combined from two column runs and dialyzed against 10 mM sodium acetate, pH 5. Side tubes of low activity should be avoided because the cellulase activity is eluted near the mannosidase peak.

The dialyzed enzyme solution from the gel filtration (200 ml) is applied at a rate of 40 ml/hr to a DEAE-Sephacel column (2.7 × 24 cm) equilibrated in 10 mM sodium acetate, pH 5, and the column is washed overnight with the same buffer at a rate of 20 ml/hr. The column is eluted with an 800-ml gradient of 0–0.5 M NaCl in the acetate buffer at a rate of 30 ml/hr, and 10-ml fractions are collected. The mannosidase appears in the first protein peak at a conductivity of 2–3 mmho. The active fractions are combined and dialyzed against 10 mM sodium acetate, pH 5, to give a total of 190 nkat of activity and 20 mg of protein. The specific activity, 9.5 nkat/mg, is similar to the 11.5 nkat/mg reported.[47]

Properties of Exomannosidase. The enzyme preparation described here migrates on a denaturing gel as a diffuse band of about 60 kDa but gives a size of 51 kDa by gel filtration.[47] The pH optimum is 5.0, and the enzyme is stable at 45° for at least 10 min. When frozen at pH 5, the enzyme remains active indefinitely. The K_m for *S. cerevisiae* mannoprotein is 0.45 mM and activity is shown only against α1 → 2-linked mannose.

Purification of a Nonspecific Mannoprotein α-Mannosidase[45]

The organism used in this preparation is a strain of *Oerskovia* sp. (ATCC 33522, formerly *Arthrobacter* GJM-1) that was isolated from a soil enrichment culture on mannoprotein from commercial bakers' yeast (Red Star Co., Oakland, CA). It is maintained on YEPD slants.

α-Mannosidase Assay. The assay mixture contains bakers' yeast mannoprotein, 0.1 M potassium phosphate, pH 6.8, containing 0.3 mM $CaCl_2$ and enzyme solution. After incubation at 37° for 15 min, the release of reducing sugar is determined.

Purification. A liquid medium composed of 1 g of crude bakers' yeast or *S. cerevisiae* X2180 mannoprotein, 500 mg of $(NH_4)_2SO_4$, 200 mg of $MgSO_4 \cdot 7H_2O$, 10 mg of $FeSO_4 \cdot H_2O$, 50 mg of $CaCl_2 \cdot 2H_2O$, 7.54 g of K_2HPO_4, 2.32 g of KH_2PO_4, and 500 mg of yeast extract in 1 liter of water is distributed in five 3-liter Fernbach flasks, each containing 1 liter of the medium, autoclaved, and inoculated with 10 ml of a 24-hr culture of the *Oerskovia* sp. strain. The flasks are incubated on a reciprocal shaker at 30° for 36 hr, at which time the mannanase activity reaches a plateau. All subsequent steps are carried out at 4°. The cells are removed by centrifugation (10,000 g for 30 min) and the supernatant is concentrated by dialysis against dry polyethylene glycol 8000 (Sigma P2139). The concentrate is dialyzed against 10 mM potassium phosphate, pH 6.8, containing 10^{-4} M $CaCl_2$ (buffer A) for 24 hr with several changes, and then it is lyophilized.

The protein precipitate is dissolved at a concentration of 2.5 mg/ml in 0.1 M potassium phosphate, pH 6.8, containing 10^{-4} M Ca^{2+} (buffer B), and solid ammonium sulfate is added to 85% of saturation at 4° (610 g/ liter). After standing overnight at 4°, the solution is centrifuged at 10,000 g for 30 min, the precipitate is collected, dissolved in 100 ml of 25 mM Tris-HCl, pH 7.8, containing 10^{-4} M $CaCl_2$ (buffer C), and dialyzed for 48 hr against the same buffer with two changes.

The dialyzed solution is applied to a DEAE-Sephacel column (1.8 × 12 cm) equilibrated in buffer C, and the column is washed with the same buffer at a rate of 30 ml/hr while collecting 10-ml fractions. The column is then eluted with a gradient of 0–0.6 M KCl (500 ml) in buffer C, and the fractions are assayed for protein and α-mannosidase activity, the latter appearing at about 0.33 M KCl. The enzyme-containing fractions are combined, dialyzed against buffer A, and lyophilized.

The lyophilized enzyme is dissolved in 5 ml of buffer B, containing 200 mM NaCl and 0.02% sodium azide, and centrifuged to remove a small amount of insoluble material. The solution is applied to an Ultrogel AcA 44 column (4 × 95 cm), equilibrated in the same buffer, and the column is eluted with this buffer at a rate of 20 ml/hr in 10-ml fractions. The major peak of enzyme activity is centered at fraction 60, with a minor peak of activity at fraction 48. The major peak of α-mannosidase activity is combined to give about 100 ml with an activity of 25 U/ml. The solution is dispensed in 1.5-ml portions and stored frozen.

Properties of Mannosidase. The enzyme is active from pH 5 to 9, with an optimum between 6.5 and 7. Maximal activity is observed in potassium phosphate buffers containing 10^{-2} M Ca^{2+}, and the enzyme is stable for at least 5 min at 50°, is able to survive repeated freezing and thawing, and can be stored indefinitely at $-10°$. The enzyme is an exomannosidase with activity against $\alpha1 \rightarrow 2$, $\alpha1 \rightarrow 3$, and $\alpha1 \rightarrow 6$ linkages in mannooligosaccharides and intact mannoprotein, but it hydrolyzes p-nitrophenyl-α-D-mannoside only very slowly. The K_m varies from 0.5 to 2 mM with di- to hexasaccharides. The enzyme acts on intact mannoprotein to give a product in which the highly branched N-linked chains are converted to long unbranched $\alpha1 \rightarrow 6$-polymannose units. The digestion of small oligosaccharides is sometimes incomplete.[11]

Methods for Selective Chemical Degradation of Mannoproteins[51]

The O-linked oligosaccharides of mannoproteins can be released by alkali-catalyzed β-elimination and the N-linked oligosaccharides by hydrazinolysis. Acetolysis is a selective method for cleaving the $\alpha1 \rightarrow 6$ linkages present in the N-linked oligosaccharides, while mild acid hydrolysis releases sugars in glycosyl linkage from the phosphodiester units and leaves monoesterified phosphate groups on the mannoprotein.

β-Elimination

Mannoprotein is dissolved in 0.1 N NaOH with or without 1 M NaBH$_4$ and kept at room temperature or at 37° for several days. The progress of the reaction is monitored by following the increase in absorbance at 220 nm due to the formation of dehydrolanine and dehydroaminobutyric acid, or the amount and composition of released oligosaccharides by gel filtration on BioGel P-2. In presence of NaBH$_4$, the oligosaccharides and the dehydroamino acids are reduced, which stabilizes both. In absence of NaBH$_4$, the reducing end mannose of each oligosaccharide is isomerized in part to glucose, a fact easily confirmed by the ^{1}H-NMR spectra. The release of O-linked oligosaccharides may not be complete for a week or more, and the composition of the oligosaccharides released with time changes, which suggests that the oligosaccharide structure or the peptide sequence can modulate the rate of elimination. During the reaction, the molecular size of the mannoprotein decreases by an extent that indicates peptide cleavage in addition to release of carbohydrate.

---

[51] C. E. Ballou, *Adv. Microbiol. Physiol.* **14**, 93 (1976).

Hydrazinolysis[52]

Mannoprotein, recovered from the β-elimination reaction, is mixed with an equal weight of hydrazine sulfate and dried at reduced pressure over P_2O_5 for several hours in an open screw-capped tube. Redistilled anhydrous hydrazine is added to the tube, at 100 times the weight of the mannoprotein; the tube is flushed with dry nitrogen, capped tightly, and heated at 105° for 10 hr. The tube is cooled, placed in a dry ice–propanol bath to freeze the hydrazine, and then opened and placed in a desiccator, over concentrated H_2SO_4 and P_2O_5, which is evacuated with an oil pump to sublime the hydrazine. The product is then dissolved in a small amount of water and fractionated by gel filtration on a BioGel P-4 or P-10 column, as appropriate, while the fractions are monitored by the phenol–sulfuric acid procedure for carbohydrate.[50] The recovered oligosaccharides contain a de-N-acetylated chitobiose unit at the reducing end that can be re-N-acetylated with saturated $NaHCO_3$ and acetic anhydride at room temperature.

Acetolysis[53]

Intact mannoprotein (1 g) is dissolved in a mixture of 6 ml of glacial acetic acid and 6 ml of acetic anhydride, and 0.6 ml of 98% sulfuric acid is added, while the temperature is kept below 25°. After 5 days at room temperature, the darkened mixture is centrifuged, the supernatant is poured into 200 ml of ice-water, and the solution pH is adjusted to 5.5 with NaOH. The solid acetylated oligosaccharides are collected by filtration, dissolved in dichloromethane, and the solution is decolorized with charcoal, washed with 1 M sodium bicarbonate until neutral, dried over anhydrous sodium sulfate, and evaporated to dryness. About 2 g of acetylated oligomannosides is obtained, which can be deacetylated in dry methanol by addition of sodium methoxide, during which the oligosaccharides precipitate. After deacetylation, water is added to dissolve the oligosaccharides, the solution is treated with Dowex 50 (H) to remove cations and is then evaporated to dryness under vacuum. The resulting oligosaccharides are fractionated by gel filtration on a BioGel P-4 column (−400 mesh) (2 × 190 cm) by elution with water. Extensive polymorphism is noted in the spectrum of oligosaccharides obtained by mannoprotein acetolysis, not only in the size but also in the linkages and in the sugar composition.

Oligosaccharides, obtained by hydrazinolysis or by endoglucosaminidase H digestion of mannoprotein (10 μg to 1 mg), are acetylated in 0.2 ml

[52] C. L. Reading, E. E. Penhoet, and C. E. Ballou, *J. Biol. Chem.* **253,** 5600 (1978).
[53] L. Rosenfeld and C. E. Ballou, *Carbohydr. Res.* **32,** 287 (1974).

of a 2:1 (v/v) mixture of trifluoroacetic acid–glacial acetic acid at room temperature for 30 min. The sample dissolves within 5 min and, at the end of the reaction time, the solution is evaporated below 40° to dryness. The acetylated product is dissolved in 0.5 ml of a 10:10:1 (v/v) mixture of acetic acid–acetic anhydride–concentrated sulfuric acid and kept at 40° for 90 min. Dichloromethane (2 ml) and water (2 ml) are added to the tube, the contents are mixed well, and the tube is centrifuged briefly to separate the two layers. The water layer is removed with a Pasteur pipet and the water wash of the dichloromethane layer is repeated several times until the wash is neutral. The organic layer is evaporated to dryness on a rotary evaporator and the acetylated oligosaccharide product is analyzed by FAB mass spectrometry or deacetylated in methanolic sodium methoxide and fractionated by gel filtration or HPLC.

Mild Acid Hydrolysis

Mannoprotein or whole yeast cells, when heated at 100° in 10 mM HCl for 30 min, release the sugar attached in glycosyl linkage to the phosphate units esterified to position 6 of mannose units in the N-linked polymannose chains. Bakers' yeast or *S. cerevisiae* X2180 mannoproteins yield mainly $\alpha 1 \rightarrow 3$-mannobiose; the *mnn1* mutant and some *S. cerevisiae* wild-type strains give free mannose; while the yeast *Hansenula polymorpha* releases glucose under these conditions.

Mannosyltransferase Reactions

Yeast microsomes provide a rich source of enzyme activities concerned with mannoprotein glycosylation, and several of these activities are affected in the *mnn* mutants.[6,14] The following sections deal with reactions involved in outer-chain biosynthesis.

Preparation of Solubilized Mannosyltransferase Activity[18]

Saccharomyces cerevisiae cells are grown to midlogarithmic phase in 3-liter Fernbach flasks and are washed twice with cold 1% KCl by centrifugation. The cell paste is suspended in 0.1 M Tris-HCl buffer, pH 7.2, containing 10 mM MnCl$_2$, 10 mM dithiothreitol, 1 mM phenylmethylsulfonyl fluoride, and 5% glycerol (homogenizing buffer), and glass beads are added to this suspension (1:2:2, w/v/w). The cells are broken in a Braun homogenizer at 2–4° with three 1-min pulses. Unbroken cells and beads are removed by centrifugation at 5000 rpm for 10 min, and the mixed membrane fraction is isolated from the supernatant liquid by centrifugation at 100,000 g for 90 min. The isolated membrane pellet is suspended in the homogenizing buffer to a protein concentration of 10 mg/ml, Triton X-100 is added to a final concentration of 2% (w/v), and the mixture is stirred at 4° for 18 hr. Solubilized membrane proteins are recovered by a second centrifugation at 100,000 g for 90 min. The preparation retains mannosyltransferase activity for at least 1 month when stored at 4°.

Mannosyltransferase Assay

In a total volume of 50 μl, the incubation mixture contains 50 mM Tris HCl, pH 7.2, 10 mM MnCl$_2$, 2 mM exogenous acceptor, 0.6 mM GDP-[^{3}H]mannose (50,000 cpm), and solubilized enzyme preparation (25 μg of protein). The mixture is incubated at 20–23° for 1 hr and the reaction is stopped by pouring the solution onto a Dowex AG-1 (acetate) column (0.5 ml in a Pasteur pipet), after which the neutral product is eluted with water. Radioactivity is counted in a portion of effluent with Scint A. Control reactions lack added acceptor.

TABLE II
MANNOSYLTRANSFERASE ACTIVITIES

Acceptor[a]	[^{3}H]Mannose incorporation (nmol)	Linkage	
		$\alpha 1 \rightarrow 2$	$\alpha 1 \rightarrow 6$
αM $\rightarrow^2$ $\alpha\beta$M	3.04	0.94	0.06
αM $\rightarrow^6$ αM $\rightarrow^6$ $\alpha\beta$M	2.55	0.60	0.40
(structure 1)	3.35	0.72	0.28
(structure 2)	0.29	ND	ND

(structure 1):
```
αM      αM
↓2      ↓3
αM →⁶ αM →⁶ βM →⁴ αβGNAc
              ↑³
        αM →⁶ αM
              ↑²
              αM
              ↑²
              αM
```

(structure 2):
```
αM      αM
↓2      ↓3
αM →⁶ αM →⁶ βM →⁴ αβGNAc
              ↑³
        αM →⁶ αM
       ↑²     ↑²
        αM    αM
              ↑²
              αM
```

[a] With mixed membranes from the *mnn1 mnn2* mutant. Similar results are obtained with membranes from the *mnn1 mnn9* mutant.

Acceptors and Product Analysis

The mixed membrane preparation contains activities for forming $\alpha 1 \rightarrow 2$, $\alpha 1 \rightarrow 3$, and $\alpha 1 \rightarrow 6$ linkages. By choice of an appropriate acceptor, the assay is made partially selective for each such transferase.[14]

$\alpha 1 \rightarrow 2$-Mannosyltransferase. Methyl α-D-mannopyranoside is a good acceptor for mannose transfer from GDP-mannose and the sole product is methyl-2-*O*-α-D-mannopyranosyl-α-D-mannopyranoside. Because the mannoprotein contains several structurally different $\alpha 1 \rightarrow 2$-linked mannoses, however, several transferases of this specificity may be present. More specific acceptors are available in oligosaccharides, such as $\alpha 1 \rightarrow 2$-mannobiose and $\alpha 1 \rightarrow 6$-mannotriose, and in the core oligosaccharide Man$_9$GlcNAc (Table II). Products formed in the $\alpha 1 \rightarrow 2$-transferase reaction are characteristically stable to partial acetolysis and release labeled mannose when digested with the exo-$\alpha 1 \rightarrow 2$-mannosidase.

$\alpha 1 \rightarrow 3$-Mannosyltransferase. Reduced $\alpha 1 \rightarrow 2$-mannotriose, αMan $\rightarrow^2$ αMan $\rightarrow^2$ rMan, is a good acceptor with membranes from wild-type cells but a poor acceptor with membranes from the *mnn1* mutant, which makes mannoproteins lacking terminal $\alpha 1 \rightarrow 3$-linked mannose. The product of the transferase reaction is stable to acetolysis, which is consistent with the structure αMan $\rightarrow^3$ αMan $\rightarrow^2$ αMan $\rightarrow^2$ rMan.

$\alpha 1 \rightarrow 6$-Mannosyltransferase. This enzyme is active with acceptors that possess a terminal unsubstituted $\alpha 1 \rightarrow 6$-linked mannose, and $\alpha 1 \rightarrow 6$-mannotriose is a good acceptor. This trisaccharide is also an acceptor for the $\alpha 1 \rightarrow 2$-transferase, however, so the product must be analyzed by acetolysis or exo-$\alpha 1 \rightarrow 2$-mannosidase digestion to assess the relative amount of each reaction.

[37] Molecular and Genetic Approach to Enhancing Protein Secretion

By VANESSA CHISHOLM, CHRISTINA Y. CHEN, NANCY J. SIMPSON, and RONALD A. HITZEMAN

Introduction

Although of enormous expression potential, *Saccharomyces cerevisiae* as a secretion system has yielded only moderate levels of heterologous proteins in secreted form.[1-4] The desirability of such products has raised the need for possible ways to enhance this process. Although there is a plethora of information on protein trafficking in yeast, it provides only clues to rate-limiting steps in this pathway.[5,6] Studies toward defining the parameters which appear to affect the efficiency of secretion of a particular protein, i.e., promoter strength, gene dosage, plasmid maintenance, nature of the signal sequence, and host mutations, all reflect the complexity inherent in the system.[7,8] These studies are further complicated by factors specific for different genes.

The most useful approaches to increasing the secreted levels of heterologous proteins in yeast have been screens for host mutations. These methods have exploited visual markers for the detection of a small number of strains within a large population which appear to secrete the product at higher levels. For example, the amount of calf prochymosin secreted by yeast colonies could be easily compared by observing the size and intensity of opaque regions of clotted milk generated in an overlay assay.[7,9] Another screening method, DNA–toluidine blue, was used by Wood and Brazzell to detect higher levels of secretion of *Staphylococcus* nuclease from a

[1] A. J. Brake, J. P. Merryweather, D. G. Coit, U. A. Heberlein, F. R. Masiarz, G. T. Mullenbach, M. S. Urdea, P. Valenzuela, and P. J. Barr, *Proc. Natl. Acad. Sci. U.S.A.* **81**, 4642 (1984).

[2] A. Singh, J. M. Lugoroy, W. J. Kohr, and L. J. Perry, *NAR* **12**, 8927 (1984).

[3] G. A. Bitter, K. K. Chen, A. R. Banks, and P. H. Lai, *Proc. Natl. Acad. Sci. U.S.A.* **81**, 5330 (1984).

[4] R. A. Hitzeman, C. Y. Chen, D. J. Dowbenko, M. E. Renz, C. Lui, R. Pai, N. J. Simpson, W. J. Kohr, A. Singh, V. Chisholm, R. Hamilton, and C. N. Chang, this volume [35].

[5] R. Schekman, *Annu. Rev. Cell Biol.* **1**, 115 (1985).

[6] R. Schekman and P. Novick, *in* "The Molecular Biology of the Yeast *Saccharomyces*" (J. N. Strathern, E. W. Jones, and J. R. Broach, eds.), Vol. 2, p. 361. Cold Spring Harbor Laboratory, Cold Spring Harbor, New York, 1982.

[7] R. A. Smith, M. J. Duncan, and D. T. Moir, *Science* **229**, 1219 (1985).

[8] J. F. Ernst, *DNA* **5**, 483 (1986).

[9] R. A. Smith and T. Gill, *J. Cell. Biochem. Suppl.* **9C**, 157 (1985).

mutagenized culture of yeast cells via the generation of a characteristic pink halo.[10] Recently, yeast mutants secreting increased amounts of mouse amylase were identified by the halos they generated on starch plates overlaid with iodine.[11] The above systems have identified a number of host mutations which increase the apparent secreted levels of some proteins, in one case up to 400-fold after several rounds of selection.[10] However, the pleiotropic nature of these mutant strains for other secreted proteins, as well as their direct effect on the actual process of secretion, is still under scrutiny.

In this article, we describe the approaches we have taken for generating strains of *S. cerevisiae* which have increased levels of secretion for the heterologous protein human serum albumin (HSA)[12] and for fusions of HSA with the yeast homologous invertase protein. Unlike a screening method, our approach relies on the potential for selecting directly for strains with increased levels of secretion for heterologous protein products via drug resistance.

Strain and Selection Methodology

Yeast Strains and Growth Conditions

The yeast strains originally used in the development of this selection scheme, particularly strains Ab101 *(MAT a, trpl leu2)* and Ab107 *(MAT α, trpl)*, were provided by Robert Hamilton and Anne Lucas (both of the Genentech Fermentation Research and Process Development Department). Strain Ab107 was derived from Ab101 via a selection which resulted in the phenotypic reversion of the *LEU2* locus. All of the strains listed in Table 1 are either direct derivatives of Ab107 or the results of subsequent rounds of selection. The DBY strains were provided by David Botstein. Strains DBY2440 *(MAT α, ura3-52 his4 ade2 suc2Δ9)* and DBY2441 *(MAT a, ura3-52 lys2 suc2Δ9)* contain complete deletions of the *SUC2* locus and lacked all of the invertase structural genes *(SUC1, SUC3 – SUC7).*

Saccharomyces cerevisiae strains were maintained on either YPD medium containing (w/v) 1% Bacto-yeast extract, 2% Bacto-peptone, and 2% glucose or minimal selective medium for plasmid maintenance using Difco

[10] J. S. Wood and C. Brazzell, in "Biological Research on Industrial Yeasts" (G. G. Stewart, I. Russell, R. D. Klein, and R. R. Hiebsch, eds.), Vol. 3, p. 105. CRC Press, Boca Raton, Florida, 1987.

[11] A. Sakai, Y. Shimizu, and F. Hishinuma, *Genetics* **119**, 499 (1988).

[12] R. M. Lawn, J. Adelman, S. C. Bock, A. E. Franke, C. M. Houck, R. C. Najarian, P. H. Seeburg, and K. L. Wion, *NAR* **9**, 6103 (1981).

TABLE I
OVERVIEW OF MUTANT STRAINS SELECTED USING DESIGN I

Round of selection	Strain	ng/ml/OD HSA									
		p335[a]				p206[b]		p335			
		Supernatant		Pellet		Supernatant	Pellet	Supernatant[c]		Supernatant[d]	
		−	+	−	+	+	+	4 hr	15 hr	24 hr	41 hr
Parent	Ab101	3.5	4.8	320	1440	4.3	1200	ND	ND[e]	ND	ND
Parent	Ab107	1.6	3.0	180	480	ND	ND	<5	<5	ND	ND
I	30-4	4.2	4.7	320	3040	2.4	480	11.4	2.1	4.6	6.3
I	R45	36.0	11.6	720	2480	18.2	1640	2.4	7.4	ND	ND
II	V7	>27	38.5	600	3600	>24	1600	ND	ND	ND	ND
III	V7-74	>27	33.6	640	3320	>24	1720	ND	ND	3.1	66

Round of selection	Strain	Drug resistance[f]	
		Growth on hygromycin (1000 μg/ml)	Growth on G418 (500 μg/ml)
Parent	Ab107	−	−
Parent	Ab107-3b	+	−
I	R45	+	+
I	30-4	−	−
I	30-4(8)	+	−
II	V7	+	+
III	V7-74	+	+

[a] Strains were grown in YNB + CAA medium both uninduced (−) and induced (+) with 0.1 mM copper sulfate. Dilutions of 1/500 from saturated cultures were used as inocula and assay points were taken after 48 hr and processed as described in the text. > or < indicates that HSA levels were above or below the linear range of the immunoassay standard curve.

[b] Strains, in this case, were transformed with plasmid p206 (see Fig. 1). Cells were grown and assay points processed as indicated above.

[c] Shake flask cultures were inoculated to an OD A_{660} of 0.5 and induced with the addition of 0.1 mM copper sulfate. Assay points were taken at the indicated times with OD values ranging from 2.0 to 3.6 at 4 hr and 7.1 to 10.0 at 15 hr.

[d] In this experiment, cell densities at 24 hr ranged from 5.5 to 6.0 and at 41 hr from 6.6 to 10.4 OD values.

[e] ND, Not determined.

[f] Resistance levels were determined for strains growing on media containing the indicated drug concentrations. Strains Ab107-3b and 30-4(8) are the integrative transformants of strains Ab107 and 30-4 expressing the phosphotransferase gene. + and − indicate the presence and absence of growth.

(Detroit, MI) yeast nitrogen base (YNB) without amino acids prepared according to label directions. The latter was supplemented with 20 mg/liter uracil, lysine, histidine, adenine, or 50 mg/liter tryptophan where necessary. Casamino acids (CAA) were added at a concentration of 0.2% (w/v) when rich, completely supplemented medium was indicated. For plates, all media were solidified with 3% Bacto-agar.[13] Strain manipulations were performed at 30°. Growth in liquid culture was monitored by the change in absorbance at 660 nm.

Escherichia coli Strains and Growth Conditions

Escherichia coli K12 strain 294 (*endA thi⁻ hsr⁻ hsm⁺*)[14] was cultured using Luria broth, or for plasmid maintenance, Luria broth supplemented with 50 μg/ml carbenicillin.[15] *Escherichia coli* strain JM101,[16] used in M13 mutagenesis procedures, was maintained on 2YT medium.[15]

Genetic Methods

Conventional genetic procedures, i.e., replica plating, tetrad analysis, and ethylmethane sulfonate (EMS) mutagenesis, were performed according to established methods as described.[13] Strains were normally mutagenized in 3% EMS to 40–60% viability.

Transformations

The transformation of *E. coli* was performed as described by Mandel and Higa.[17] Yeast strains were transformed according to the spheroplasting method.[18] Integration transformations were done by the method described by Rothstein.[19]

Yeast Drug and Sucrose Selection Methods

The following solid media were used in the generation of yeast mutants with increased resistance to the aminoglycoside hygromycin B (Calbio-

---

[13] F. Sherman, G. R. Fink, and J. B. Hicks, "Laboratory Course Manual for Methods in Yeast Genetics." Cold Spring Harbor Laboratory, Cold Spring Harbor, New York, 1986.

[14] K. Backman, M. Ptashne, and W. Gilbert, *Proc. Natl. Acad. Sci. U.S.A.* **73**, 4174 (1976).

[15] J. H. Miller, "Experiments in Molecular Genetics." Cold Spring Harbor Laboratory. Cold Spring Harbor, New York, 1972.

[16] J. Messing, this series, Vol. 101, p. 20.

[17] M. Mandel and A. Higa, *J. Mol. Biol.* **53**, 159 (1970).

[18] A. Hinnen, J. B. Hicks, and G. R. Fink, *Proc. Natl. Acad. Sci. U.S.A.* **75**, 1929 (1978).

[19] R. Rothstein, this, series, Vol. 101, p. 202.

chem, San Diego, CA) and/or to the drug Geneticin (G418 from Sigma, St. Louis, MO). Both drugs were dissolved in water, filter-sterilized, and added after medium had been autoclaved. Hygromycin was routinely stored at a concentration of 100 mg/ml and Geneticin at 50 mg/ml. For parent strains AB107-3b and 30-4(8), minimal YNB media with hygromycin concentrations of 3000 to 6000 μg/ml were used to establish maximal resistance levels. For parent strain DBY2441, YPD media with hygromycin concentrations of 50–350 μg/ml and G418 concentrations of 50–400 μg/ml were used to establish maximal resistance levels. In all cases, resistances were evaluated by either spotting 10^6 cells/10 μl directly or by replica-plating masters to a series of plates containing increasing drug concentrations and incubating them for 3–7 days at 30°.

Medium which selected for the ability to catabolize sucrose was made using, per liter, 6.7 g YNB without amino acids, 20 g sucrose, 30 g Bacto-agar, and 20 mg antimycin A (Sigma) dissolved in 100% ethanol.[20] Copper sulfate at a final concentration of 0.1 mM was added where induction of the Cu-metallothionein promoter was indicated. Cells were replica-plated to these plates and growth determined over a period of 8–12 days at 30°.

Other types of media used for testing the phenotypes of mutants were as described by McCusker et al.[21]

Preparation of Samples for HSA Determination

Cells were grown under the variety of conditions described in Table I. One-milliliter samples were collected at the given times and/or optical densities by centrifugation (5 min at 5000 g). Supernatant fractions were diluted directly into buffer [20 mM phosphate buffer, 0.14 M NaCl, 0.1% (w/v) gelatin, 0.05% (w/v) P20, 0.01% (w/v) thimerosal, pH 7.4] and stored at 0° until assay. Cell pellet samples were suspended in 250 μl of 20 mM phosphate buffer + 0.5% (w/v) Triton and vortexed on a Scientific Instruments Vortex-Genie 2 at room temperature at highest speed for 10 min in the presence of 0.5 volume of sterile glass beads (0.45–0.05 μm). These cell extract fractions were then centrifuged (5 min at 5000 g) and diluted into the same buffer as described for the supernatant. HSA cross-reacting material was measured by either radioimmunoassay or ELISA performed by the Genentech Immunoassay Group.

[20] V. A. Bankaitis, L. M. Johnson, and S. D. Emr, Proc. Natl. Acad. Sci. U.S.A. **83**, 9075 (1986).
[21] J. H. McCusker, D. S. Perlin, and J. E. Haber, Mol. Cell. Biol. **7**, 4082 (1987).

DNA Methodology

DNA Preparations

Standard molecular procedures were used for the construction, analysis, amplification, and restriction of plasmids from *E. coli*.[22] Restriction endonucleases and T4 DNA ligase were purchased from either BRL (Gaithersburg, MD) or New England Biolabs (Beverly, MA) and used as recommended by the manufacturers. Where DNA sequence analysis was necessary, the chain termination method using recombinant phage M13mp8 or direct plasmid priming were employed as described.[23] Standard M13 oligonucleotide-directed mutagenesis procedures were also employed.[24]

Construction of HSA – Hygromycin Phosphotransferase Fusion Cassette

For the initial experiments performed in the Ab107 background, plasmid constructions are outlined in Fig. 1. All of the HSA expression plasmids used in this work have been described in detail elsewhere.[4] For the construction of the fusion, *Xba*I and *Eco*RI sites were introduced into the amino terminus of the *E. coli* hygromycin B phosphotransferase gene[25] just 5' to the ATG by M13 mutagenesis. The use of this linker sequence allowed for the in-frame fusion of the HSA gene at an internal *Xba*I site to the *E. coli* gene. This *Xba*I site is located at position 1142 bp downstream from the initiation codon of HSA.

Construction of PreproHSA – Invertase Fusion Cassettes

For work in the DBY strain background, a fusion gene was employed linking the yeast invertase gene,[26] devoid of its signal sequence up to the second ATG, to the carboxy terminus of HSA (see Fig. 2). The level of invertase expression was modulated via the insertion of a single TAA (plasmid p336) or double ochre codons TAATAA (plasmid p337) within the linking sequence between the two genes. Limited translational readthrough, observed with both of the stop codon constructions, allowed sucrose catabolism to be used as a selective marker for expression of the fusion. Using a vector with a 2-μm origin of replication, no growth on

[22] T. Maniatis, E. F. Fritsch, and J. Sambrook, "Molecular Cloning: A Laboratory Manual." Cold Spring Harbor Laboratory, Cold Spring Harbor, New York, 1982.
[23] F. Sanger, S. Nicklen, and A. R. Coulson, *Proc. Natl. Acad. Sci. U.S.A.* **74**, 5463 (1977).
[24] M. J. Zoller and M. Smith, this series, Vol. 101, p. 468.
[25] L. Gritz and J. Davies, *Gene* **25**, 179 (1983).
[26] M. Carlson and D. Botstein, *Cell* **28**, 145 (1982).

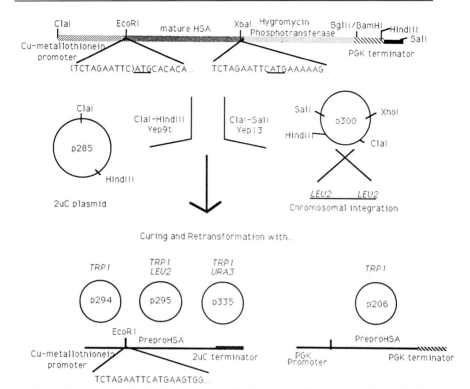

FIG. 1. Structures of the HSA–hygromycin B phosphotransferase gene fusion and other HSA gene expression plasmids used in this work.

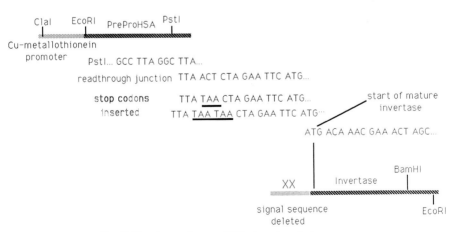

FIG. 2. Structures of preproHSA–invertase fusion genes.

sucrose could be observed for longer than 14 days of 30° for the double codon and 8 days for the single. Direct analysis of total cellular invertase activity for these constructions showed extremely low but measurable levels of activity. Using these fusions, we can now detect cells which will grow on sucrose, a process limited to secreted invertase.

Selection Design I

Phosphotransferase Fusion

Our original selection scheme was designed to detect strains with increased expression levels for HSA. It was not designed to target secretion mutations specifically. The system consisted of a HSA–hygromycin B phosphotransferase gene fusion transcriptionally regulated by the yeast Cu-metallothionein gene promoter (see Fig. 1). This promoter has been characterized in detail and shown to the inducible with copper.[27,28] Hygromycin B, an aminoglycosidic antibiotic, inhibits protein translation in S. cerevisiae and stops growth. The bacterial phosphotransferase gene, which inactivates hygromycin B, has been shown to be functional in yeast.[25] It provided a dominant selection for increased expression of the fusion gene via the monitoring of drug resistance levels. The natural signal and presequences of the HSA gene were also deleted. By using the mature HSA gene construction we expected to limit the selection to increases in the intracellular expression of the fusion protein. Two different fusion protein levels were employed in the selection by placing the cassette either on a 2-μm vector with selectable yeast and bacterial markers, (e.g., p 285) or integrated into the genome using the method defined by Rothstein.[19] Both types of transformants were then mutagenized and mutants selected for further analysis based on hygromycin B resistance levels. Strains showing stable increases in resistance levels in comparison to their respective parent strains, as in the case of 30-4 generated with p285, were cured of the selection cassette and retransformed with plasmid p294, p295, p335, or p206. These plasmids (see Fig. 1) carry the HSA gene, with its natural signal and prosequences intact. The homologous HSA signal has been shown to work efficiently in yeast.[4] All except plasmid p206 transcribe the HSA gene from the Cu-metallothionein promoter; p206 was constructed using the 3-phosphoglycerate kinase (PGK) promoter from yeast.[29] HSA

---

[27] T. Etcheverry, W. Forrester, and R. Hitzeman, Bio/Technology **4**, 726 (1986).

[28] T. Etcheverry, this volume [26].

[29] R. A. Hitzeman, C. Y. Chen, F. E. Hagie, E. J. Patzer, C.-C. Liu, D. A. Estell, J. V. Miller, A. Yaffe, D. G. Kleid, A. D. Levinson, and H. Oppermann, NAR **11**, 2745 (1983).

levels could be directly assayed from these plasmids using RIA or ELISA techniques. Mutant strains selected in the integrated background, such as R45, were cotransformed with one of these plasmids. This step immediately detected any strain capable of affecting HSA expression in trans. A further round of selection in strain 30-4 was performed using the integrated derivative 30-4(8). Strain V7 is a product of this selection. Mutant V7-74 was derived from a further round of selection using V7 as the parent strain.

Mutant Characterization

Both cell-associated and supernatant levels of HSA were determined for mutant strains derived from several rounds of selection (see Table I). A number of strains isolated showed increased levels of HSA in the supernatant. This was an unexpected phenotype considering the design of our construct. For strains V7, R45, and V7-74, the enhanced secretion phenotype was most obvious in late log/early stationary cells grown in test tubes at 30° in a rotating spinner. Under both induced and uninduced conditions these strains showed significant increases in the levels of HSA in the supernatant. Replacement of the copper-inducible promoter with the less-efficient *PGK* promoter in the HSA expression construct showed that the enhanced supernatant levels did not appear to depend on the promoter used (Table I, footnote *b*).

Increases in the levels of HSA found in the supernatant were more apparent for strain 30-4 at lower cell densities. For example, in comparison with strain R45, strain 30-4 showed an almost 5-fold increase in secreted HSA in early logarithmic phase cells (Table I, footnote *c*). In all cases, what is apparent is that the levels of HSA found associated with the supernatant have been increased in all of these strains. The levels of HSA present in the periplasm, i.e., secreted protein still associated with the pellet fraction, have not been determined.

Possible Relationship to Drug Resistance

Increases in hygromycin resistance levels of strains has been recently described for alleles of the *HYG2* locus, the yeast plasma membrane ATPase.[21] Many of the alleles of *HYG2* were also shown to be coresistant to G418. In our case, it was particularly interesting to note that many of the strains with the more remarkable supernatant levels were now much more resistant to G418 than their parent strains. The presence of the bacterial phosphotransferase gene in yeast was shown not to mediate this phenotype (Table I). These observations suggested that the increases in hygromycin resistance levels selected for in some of our strains may not be due to increases in the efficiency of expression of the bacterial gene, but to direct selection for hygromycin-resistant strains. Other characteristics of the *hyg2* mutants, i.e., pH and osmotic sensitivities, were also observed in many of the above mutants (data not shown). It became apparent that a direct correlation between the actual selection for drug resistance and the phenotypic increase in supernatant HSA levels may exist. Having been selected in highly mutagenized backgrounds, genetic analysis of these strains to correlate these two phenotypes directly was extremely difficult. Complicated segregation patterns existed even in first-round isolates for both drug resistances and supernatant levels of HSA. Levels of both hygromycin and G418 resistances in all the strains used in these selections including many sister spores and cross-out strains varied enormously. These patterns suggested that the highest levels of supernatant HSA attained were likely the result of multiple mutations.

In view of the above observations, a method was developed to determine whether these types of mutations could be selected directly via drug resistance. Given below is an outline of the method used to identify strains with potentially increased levels of secretion using a preproHSA–invertase fusion construction. This fusion was used to screen for increases in secreted levels of invertase activity within a selected population of hygromycin B- and G418-resistant strains.

Selection Design II: Procedures for Strain DBY2441

Maximal drug resistance levels for strain DBY2441 were established as described in methods. Hygromycin- and G418-resistant strains were generated spontaneously by spreading late-log cultures of cells transformed with p336 and p337 at concentrations of 10^8 cells per plate and allowing them to grow at 30° for 3–4 days. Resistant cell lines were selected from plates at concentrations of 300 and 400 μg/ml of either drug. Preliminary analysis of strains capable of some growth on sucrose is presented in Table II. Of interest is the fact that from 5 to 8% of the total drug-resistant strains isolated were of this type. The number of complementation groups involved in this phenotype is presently being determined. Analysis of both intracellular HSA levels and those within the supernatant fractions of cell cultures indicated similar but not so dramatic increases in supernatant levels of protein as compared with the highly selected background. Greater enhancement in secretion levels appears more in early logarithmically growing cultures than at later stages of cell growth. Curing and retransformation of these strains with plasmid p335, a *URA3* derivative of p294, established both the trans-acting nature of these mutations as well as the secretion phenotype (data not shown).

TABLE II
OVERVIEW OF MUTANT STRAINS SELECTED USING DESIGN II

Strain[a]	G418[b] (400 μg/ml)	Hygromycin[c] (250 μg/ml)	Sucrose[d]	ng/ml/OD HSA (p337[e])			
				18 hr		24 hr	
				Super	CPA	Super	CPA
G60	+	−	+/=	17.6	560	1.0	1440
G62	+	−	+/=	8.2	500	1.2	1580
G71	+	+/p	+/=	5.6	840	2.0	2140
G83	+	−	+/=	12.4	1040	1.0	2320
H110	+	+/−	+/=	ND		ND	
H111	+	+	+/=	5.2	520	0.6	1780
H29	−/p	+/−	+	4.6	720	1.3	2500
DBY2441	+	−/p	−	1.9	560	1.6	1280

[a] The initials G (G418) and H (hygromycin) signify the original drug resistance to which the strains were isolated. All mutants were first-round, spontaneously selected.

[b] Growth characteristics were determined after 7–8 days at 30°. Good growth is indicated by +, approximately half the confluency of good growth is designated by +/−, and a very weak growth by +/=. No growth after 8 days is indicated by −. /p signifies papillation to a lesser (−/p) or greater (+/p) extent on a no-growth background.

[c] Growth characteristics in the presence of hygromycin were determined in a manner equivalent to footnote b.

[d] Growth on sucrose was determined after 10 days.

[e] YNB + CAA shake flask cultures were inoculated at a 1/2000 dilution with fresh overnight cultures and grown in the presence of 0.1 mM copper sulphate. Samples for immunoassay were taken at the indicated times and processed as described in the text. OD values ranged from 0.9 to 1.7 at 18 hr and 4.0 and 5.4 at 24 hr. Both supernatant (Super) and cell pellet-associated (CPA) material were assayed. ND, Not determined.

Concluding Remarks

The procedure outlined above can be applied to any strain and any secretion construction, without the need of the invertase fusion, if an assay for the protein in question is available. The frequency of the enhanced secretion phenotype was high (5–8%). In this range one could, among drug-resistant mutants, obtain several potential secretion mutants by analyzing only a few hundred colonies. Since the selections can be done spontaneously (without mutagenesis), strain backgrounds would remain relatively congenic, allowing one to retain the beneficial properties of the parent strain. Of equal benefit is the specificity of this technique. If the enhancement phenotype is dictated by a particular complementation group or groups, the nature of the selection allows one to identify the specific alleles or genes that work the best with their individual construction. This selection technique is easy to implement and has been shown to enhance the secretion for three different strain backgrounds and for two secretion constructions.

Hygromycin and G418 resistances have been directly associated with mutations in a membrane protein.[21] The secretion effects generated in strains via these drug-selection methods may also provide a possible means of dissecting the role of the plasma membrane in protein secretion. Experiments are presently underway to determine the nature of these mutations.

Section V

Expression in Mammalian Cells
Article 38

A. Vectors
Articles 39 and 40

B. Transfection Methods
Article 41

C. Markers for Selection and Amplification
Article 42

D. Growth of Cell Lines
Article 43

E. Posttranslation Processing, Modification, and Secretion
Article 44

[38] Expression of Heterologous Genes
in Mammalian Cells

By ARTHUR D. LEVINSON

The prospect that mammalian cells could be manipulated in a manner to permit the high-level expression of heterologous genes was something that a few years ago was considered seriously by few investigators. Major and widely publicized successes had after all been achieved with *Escherichia coli,* to the point where success in this endeavor was considered unremarkable. What was not appreciated, however, particularly by the uninitiated, was the legion of failed attempts that went unreported. The charitable attributed their failures to bad luck while others questioned the integrity of their vectors or the competence of their compatriots. As we now know, success in expressing foreign genes at high levels of bacteria is far from assured, for any of the varied reasons pointed out in this volume. Partly as a consequence of the inability to express certain genes in bacteria, and partly for their own pressing scientific reasons, many investigators turned their attention to the engineering of mammalian cells. Because of complexity of the issues involved, we have sought here to highlight those with the most practical, as opposed to theoretical, significance; this information, we hope, will enable those unskilled in the art to approach the task of expressing foreign genes in mammalian cells in a straightforward manner and with a good understanding of the merits and pitfalls of the various approaches. This volume includes efforts by many of the pioneers in the field of mammalian expression; these contributions offer a perspective that should in its own right be enlightening.

There are two general methods of introducing genetic information into mammalian cells; those mediated by virus infection and those mediated by direct DNA transfer (see [39] in this volume). One advantage of the former relates to its efficiency; it is not unusual to achieve close to 100% targeting of cells by exploiting the ability of the virus to infect cells efficiently. In addition, great strides have been made of late in exploiting the host range specificity of viral coat proteins and promoter elements to target genes with great specificity to particular cell types for *in vivo* work. A disadvantage of this approach, and one that limits its utility, lies in the relative difficulty of its execution; it simply requires much more effort to engineer a virus to express a foreign gene in a manner that does not compromise its function than it does to administer DNA directly. As such, most attempts to express genes in mammalian cells rely on the construction of DNA vectors and their subsequent application by any of a variety of methods. The wide range of methods, each with its own advantages and disadvantages, is addressed in [47] in this volume. The reader will be introduced to virtually all currently employed methods, including direct application by microinjection; uptake facilitated by chemicals such as DEAE-dextran, calcium phosphate, Polybrene, and lipids; and uptake facilitated by physical methods, including electroporation and microprojection. Since the vast majority of expression efforts utilizing mammalian cells as hosts involve the incorporation of DNA by direct transfer, the reader will find most attention given to this direction.

Since the mere uptake of DNA does not ensure expression, much of what we presently understand about control signals that operate to regulate gene expression in mammalian cells is discussed in [40]. There has been an enormous increase in our understanding of how the expression of eukaryotic genes is controlled, and several critical *cis*-acting elements that mediate this control are reviewed; these include sequences that comprise promoters, enhancers, splice sites, and polyadenylation sites. Article [40] should be given serious attention by those attempting such manipulations for the first time; not only does it highlight the danger of inadequately considering the details of vector assembly, but it should leave the reader with an appreciation for how much we do not yet understand.

Many of the practicalities of expression efforts are presented in [42] and [44]. There is a straightforward description that instructs the reader in the ways of isolating stable cell lines that permanently express a foreign gene by linkage to a selectable marker that integrates into the host genome. There is also an extensive presentation of methods that are available to amplify the selectable marker (along with the linked gene of interest); properly, much of the attention is focused on the dihydrofolate reductase (DHFR) system. The steps in the biosynthesis of mammalian proteins are reviewed in [44], with an emphasis on methods of detecting the expression of introduced genes at the protein level. Detailed protocols are provided on how to radiolabel cells, prepare cell extracts, analyze the targeted protein by immunoprecipitation and Western blotting, and run SDS–polyacrylamide gels. There are also protocols for characterizing various posttranslational modifications that affect certain proteins, particularly those that are secreted.

Methods of optimizing the cell culture environment for the production of recombinant proteins are presented in [43]. Techniques of cell cloning by limiting dilution, growth in agar, and the fluorescence-activated cell sorter (FACS) and the relative merits of growing cells in roller bottles versus suspension cultures are reviewed, and guidelines on media optimization to achieve maximum levels of recombinant product are provided.

Mammalian cells are playing a role of increasing importance as hosts for the expression of foreign genes. Recent years have witnessed an explosion of interest in this area with the development of better methods of gene transfer and the identification of increasingly powerful promoters. Indeed, one may legitimately ask whether the pendulum has not swung too far in this direction, to the point where people may inappropriately dismiss the potential of alternative host systems (see [2] in this volume). Seen as a whole, the decision as to which host to utilize should be dictated not primarily by technical considerations — the engineering of all three host systems discussed herein has become too routine for that — but by considerations of a scientific nature. We have accordingly sought to communicate a sense of the advantages and disadvantages of using mammalian cells as hosts with respect to this standard. The protocols presented in Section V of this volume should be sufficiently detailed to execute a successful expression strategy.

[39] Vectors Used for Expression in Mammalian Cells

By Randal J. Kaufman

Introduction of Genetic Material into Cells

The techniques of gene isolation, modification, and transfer into appropriate host cells have provided a means to understand protein structure and function by enabling the production of large amounts of proteins that previously could be isolated in only minute quantities, and by allowing the generation of proteins with specific, designed alterations. The advantages of using mammalian cells as a host for the expression of a gene obtained from a higher eukaryote stem from experience that the signals for synthesis, processing, and secretion of these proteins are usually properly and efficiently recognized in mammalian cells. There are many purposes for the use of mammalian cells for expression of foreign genes: (1) To confirm that genes isolated by a variety of different approaches can direct the synthesis of a desired protein; (2) to evaluate the effect of specific mutations introduced into a gene; (3) to isolate genes directly based on screening or selecting recipient cells for the production of a particular protein; (4) to produce large amounts of proteins that are normally available in only limited quantity; and (5) to analyze the physiological consequences of expression of specific proteins in mammalian cells in order to study biological regulatory controls. Two major criteria in evaluating what type of expression vector to use for mammalian cells are (1) the method desired for introducing the foreign gene into the mammalian cell, and (2) the control elements to be utilized for efficient mRNA expression and protein synthesis. This article first discusses the advantages, utilities, and disadvantages of various host–vector systems available for transfer and expression or heterologous genes in mammalian cells. Then the genetic control elements for efficient expression are discussed. Finally, approaches to monitor efficient vector function are described.

Host–Vector Systems for Heterologous Gene Expression

Two general types of methods for transfer of genetic material into mammalian cells are those that are mediated by virus infection and those that are mediated by direct DNA transfer. Many viruses which infect mammalian cells have evolved mechanisms to usurp the protein synthesis machinery of the host to produce their viral proteins. The ability to engineer the genetic material of these viruses has made it possible to insert desired coding regions under the control of the viral expression elements and to produce infectious virus particles to obtain high levels of foreign gene expression. Viral-mediated gene transfer provides a convenient efficient means to introduce foreign DNA into the majority of the recipient cells. In addition, for many viruses, viral replication yields multiple copies of template DNA which can serve to amplify transcription of the foreign gene. Since some viruses can infect a wide range of cell types derived from different species, viral-mediated gene transfer may allow the convenient introduction of foreign genes into a variety of different cells. This article addresses those viral vector systems which have demonstrated success, which are used more frequently, or which have significant potential. For a more detailed review of the different eukaryotic viral vectors, see Ref. 1.

Viral Vector Systems

Papovaviruses. Papovaviruses are small, nonenveloped DNA-containing viruses of which SV40 and polyoma are the best studied.[2] SV40 virus replicates in simian cells and polyoma virus replicates in murine cells. The SV40 viral genome of approximately 5 kilobases (kb) is divided into an early region which encodes the transformation antigens, large T and small t antigens, and the late region which encodes the viral capsid proteins VP1, VP2, and VP3. After viral infection the early genes are expressed. After 12 hr, viral DNA replication is initiated and, at 36–48 hr, optimal expression of the late genes occurs. Newly replicated viral genomes are then assembled into progeny virus. About 72 hr postinfection, the cells detach and die.

---

[1] N. Muzyczka (ed.), *Curr. Top. Microbiol. Immunol.* in press (1990).
[2] J. Tooze (ed.), "DNA Tumor Viruses." Cold Spring Harbor Laboratory, Cold Spring Harbor, New York, 1981.

Since SV40 replicates to a higher copy number than polyoma virus, SV40 has been more widely utilized for successful heterologous gene expression. Foreign DNA can be inserted into either the early or late regions of SV40 to generate defective viral genomes. Replicating virus stocks can be prepared by providing a helper virus to complement the deficient region. Alternatively, it is possible to prepare helper-free recombinant viruses by propagation of the virus on appropriate cell lines which express the essential missing viral function. For example, African green monkey kidney CV-1 cells transformed with an origin-defective SV40 mutant virus (thus COS cells), express the SV40 large T antigen.[3] Large T antigen is required in trans to replicate DNA containing the SV40 origin of replication. COS cells can efficiently support the episomal replication of bacterial plasmids containing the SV40 origin of replication. This ability to replicate and amplify transfected DNA to greater than 100,000 copies/cell has provided a very efficient means of obtaining high-level transient expression of DNA directly introduced into these cells. Eventually the transfected cells die, presumably due to the very high level of DNA replication. The transient expression observed after plasmid DNA transfection of COS cells is probably the most efficient and convenient means to obtain expression of foreign genes (see below). Since COS cells express T antigen, they can also complement recombinant SV40 viruses which contain early region replacements.

Most success with SV40 recombinant viruses has been obtained by gene replacements into the late region of SV40. High expression levels (greater than 10^8 molecules/cell) have been obtained by insertion of cDNA into the 5' untranslated region of the SV40 late mRNAs and deleting the genes for VP1, VP2, and VP3.[4] Two vectors for this purpose are pSVL-HA[4] in which genes can be introduced between the HpaII and BamHI sites, and pspLT5,[5] in which genes can be introduced between the XhoI and BamHI sites. In order to complement such SV40 late gene replacements, cells are transfected with an equal amount of DNA from an SV40 mutant in the early region such as dl1055[1] or DNA from another SV40 recombinant containing an early gene replacement such as pSV40-rIns-pBR322.[6] For complementation, SV40 early region replacements are recommended since they minimize the potential for recombination to generate wild-type viruses.

[3] Y. Gluzman, *Cell* **23**, 175 (1981).
[4] M.-J. Gething and J. Sambrook, *Nature (London)* **292**, 620 (1981).
[5] X. Zhu, G. M. Veldman, A. Cowie, A. Carr, B. Schaffhausen, and R. Kamen, *J. Virol.* **51**, 170 (1984).
[6] M. Horowitz, C. L. Cepko, and P. A. Sharp, *in* "Eukaryotic Viral Vectors" (Y. Gluzman, ed.), Cold Spring Harbor Laboratory, Cold Spring Harbor, New York, p. 47. 1982.

The preparation of virus stocks has been described.[4,7] The SV40 recombinant DNA is prepared as bacterial plasmid DNA. The plasmid sequences are removed by restriction digestion and the DNA is then ligated at low concentrations (1 μg/ml) to generate closed circular DNA molecules. The ligated DNA is mixed with the complementing virus DNA prepared in a similar manner and transfected into monkey CV-1 cells (obtained from ATCC) using the DEAE-dextran transfection protocol as described below for COS-1 cells. Since the percent of cells that acquire DNA is usually less than 10%, it is necessary to prepare a primary virus stock and passage this once in order to obtain a high-titer virus stock that can be used subsequently to infect all the recipient cells. To prepare a high-titer stock, transfected cells are harvested by scraping into the growth medium at 5 days posttransfection. The cells in the suspension are lysed by quick freeze–thawing 3 times (from dry ice–ethanol to 37° water bath). One-half milliliter of the stock may then be used to infect a new 100-mm dish of CV-1 cells. Five days postinfection, the cells are again harvested and lysed as described above in order to obtain the high-titer virus stock. Twenty-four hours prior to infection, CV-1 cells are subcultured 1 : 4. For infection, the growth medium is removed and 0.3 to 0.5 ml of the virus stock is applied to the cells for 1 hr at 37°. The cells are then rinsed and fed with fresh medium. Generally, the cells or conditioned medium are harvested at 48 hr postinfection.

Although high expression levels (1 – 10 μg/10^6 cells) make SV40 recombinant viruses attractive, multiple disadvantages have limited their use. First, the range of permissible cell types is limited to monkey cells. Second, viral infection results in cell lysis after 3 to 4 days, necessitating only transient studies and limiting large-scale processes to a batch harvest procedure. Third, the requirements for viral packaging restrict the size of the foreign DNA which can be inserted. Generally, SV40 can accommodate up to 2.5 kb of foreign DNA inserted into either the early or the late region. Fourth, DNA rearrangements frequently occur during replication of these viruses. Finally, not all genes introduced into SV40 viruses behave the same. Since the reason for the variability is not understood, it is not possible to predict the probability of success. SV40 recombinant viruses may prove more useful for the expression of potentially toxic proteins.

Vaccinia Virus. Vaccinia virus is a member of the poxvirus family.[8] Vaccinia replicates in the cytoplasm of mammalian or avian cells. A linear double-stranded DNA genome of 185 kb is packaged into the virus core. Vaccinia encodes its own transcription and RNA processing system within

[7] C. Doyle, J. F. Sambrook, and M.-J. Gething, *J. Cell Biol.* **100,** 704 (1985).
[8] B. Moss, *in* "Virology" (B. N. Fields and D. M. Knipe, eds.), p. 2079, 2nd Ed. Raven, New York, 1990.

the viral particle. On infection, about 100 genes are expressed early and host protein synthesis is shut off. After DNA replication at 6 hr postinfection, about 100 late genes are expressed. There is no evidence of splicing for any vaccinia virus transcript. Virus is formed about 6 hr after infection and continues for about 2 days.

Because of the large size of the vaccinia genome, DNA must be inserted by homologous recombination. Plasmid vectors have been constructed in which foreign DNA can be inserted under control of either a vaccinia early or late promoter and flanked by DNA from a nonessential region of the vaccinia genome. After transfection of cells that have been infected with vaccinia virus, the foreign DNA recombines into the viral genome. Most commonly, insertion is directed into the thymidine kinase gene such that the TK⁻ phenotype can be used for selection of recombinants.[9] Alternatively, inclusion of the β-galactosidase gene within the expression plasmid permits a selection for recombinant plaques by screening with an appropriate color indicator.[10] Recombinant viruses may also be screened by DNA hybridization or antibody binding.

A vaccinia virus/bacteriophage T7 promoter vector system utilizes the efficient and selective T7 bacteriophage RNA polymerase to express mRNA in a mammalian cell. A recombinant vaccinia virus (vTF7-3) is used to express the T7 RNA polymerase.[11] Foreign DNA is cloned between two fragments of T7 DNA containing the ϕ10 promoter and the Tϕ termination sequences into a plasmid vector (pTF7-5) containing cloning sites and flanking vaccinia virus thymidine kinase gene sequences for homologous recombination. Transfection of the plasmid DNA containing the desired coding region insert into cells infected with the T7 polymerase vaccinia recombinant virus yields high-level expression of the foreign gene from the T7 promoter. As an alternative to transfection, the T7 promoter and desired gene can be introduced into another vaccinia virus and higher expression obtained by coinfection of the two recombinant viruses, one encoding the RNA polymerase and the other encoding the desired gene under control of the T7 promoter. With this coinfection approach, it is possible to introduce the desired gene into 100% of the recipient cells. Coinfection can yield very high levels of mRNA derived from the T7 promoter (about 30% of the total cell RNA) at 24–48 hr postinfection.[12] However, only 5–10% of the RNA is properly capped and the RNA is

[9] D. Panicali and E. Paoletti, *Proc. Natl. Acad. Sci. U.S.A.* **79**, 4927 (1982).

[10] S. K. Chakrabarti, K. Brechling, and B. Moss, *Mol. Cell. Biol.* **5**, 3403 (1985).

[11] R. R. Fuerst, E. G. Niles, F. W. Studier, and B. Moss, *Proc. Natl. Acad. Sci. U.S.A.* **83**, 8122 (1986).

[12] T. R. Fuerst, P. L. Earl, and B. Moss, *Mol. Cell. Biol.* **7**, 2538 (1987).

poorly translated. Current research is directed to identify means to increase the translation of the T7-derived transcript.

The expression level obtained with this system when successful is approximately 3% of the total protein by using either a strong late promoter of vaccinia or the vaccinia/T7 hybrid system. The major use for the vaccinia virus expression system is to deliver antigenic determinants for vaccination purposes. Progress in further attenuating vaccinia virus should lead the way to testing the efficacy of recombinant vaccinia viruses as vaccines.

Viral Vectors Maintained in Episomal State. BOVINE PAPILLOMA-VIRUS. Bovine papillomavirus (BPV) is a small, circular DNA containing virus that morphologically transforms rodent cells *in vitro.* Since BPV cannot generate infectious virus in tissue culture, it is necessary to transduce foreign DNA contained within BPV vectors by DNA-mediated transfection. Vectors containing the entire BPV genome or a 69% (5.5 kb) subgenomic transforming fragment encompassing the early region are, in some cases, stable as multicopy (20 to 100 copies/cell), extrachromosomal elements in murine NIH/3T3 and C127 fibroblast cells selected for morphological transformation.[13,14] In other cases, the vector sequences are maintained as multiple copies integrated into the host chromosome in a head-to-tail tandem array. The multicopy nature of the DNA in BPV-transformed cells is partly responsible for the high-level expression of foreign genes contained in BPV-based vectors. Derivatives of BPV contain selectable markers encoding, for example, neomycin resistance[15] or metallothionein for heavy metal ion resistance,[16,17] that obviate relying on morphological transformation to obtain cells harboring exogenous DNA (for review, see Ref. 18). One vector series, pBMT2X, contains both the human and mouse metallothionein (MT) genes as dominant selectable and amplifiable genetic markers for introduction of the BPV recombinant into cells.[19] Cells harboring the MT gene overexpress metallothionein, which binds excess cadmium and protects the cells against metal toxicity (up to 100 μM). Growth in increasing concentrations of cadmium can select for

[13] K. Zinn, P. Mellon, M. Ptashne, and T. Maniatis, *Proc. Natl. Acad. Sci. U.S.A.* **79,** 4897 (1982).

[14] D. DiMaio, V. Corbi, E. Sibley, and T. Maniatis, *Mol. Cell. Biol.* **4,** 340 (1984).

[15] M.-F. Law, D. R. Lowy, I. Dvoretsky, and P. Howley, *Proc. Natl. Acad. Sci. U.S.A.* **78,** 2727 (1981).

[16] M. Karin, G. Cathala, and M. C. Nguyen-Huu, *Proc. Natl. Acad. Sci. U.S.A.* **82,** 689 (1983).

[17] G. N. Pavlakis and D. H. Hamer, *Proc. Natl. Acad. Sci. U.S.A.* **80,** 397 (1983).

[18] D. DiMaio, *in* "The Papillomaviruses," p. 293. Plenum, New York, 1987.

[19] C. M. Wright, B. K. Felber, H. Paskalis, and G. N. Pavlakis, *Science* **232,** 988 (1986).

cells containing an increased copy number of BPV genomes.[20] pBMT 2X has a unique *Xho*I cloning site into the 5' untranslated region of the mouse metallothionein I gene.

Although there has been considerable success with BPV expression systems, the biology of BPV mRNA expression and the regulation of its plasmid maintenance and copy number are poorly understood. In addition, BPV vectors are prone to rearrangement. Because of this, results have been variable with different DNA inserts. As more features of the biology of BPV become understood, BPV may become a more useful system for expression of a wide variety of heterologous genes. Given the desirability of an episomal vector for mammalian cells and that BPV has significant disadvantages as described above, alternative vector systems have been explored. One potential system at this time is based on Epstein-Barr virus (EBV).

EPSTEIN-BARR VIRUS. EBV is a herpes virus which has a large DNA double-stranded genome of 135 kb. EBV infects human B lymphocytes *in vivo* and *in vitro*. The virus normally replicates as a plasmid in infected cells. The origin of replication of the plasmid has been identified *(oriP)* and plasmids containing *oriP* can be stably maintained in the presence of the Epstein-Barr nuclear antigen (EBNA) in a variety of human, monkey, and dog cell lines, but not in rodent cell lines.[21,22] The copy number ranges from one to several hundred per cell. The mechanism which controls plasmid copy number is not known. In the lytic cycle, the genome is amplified 500- to 1000-fold by the action of the viral protein BZLF-1 at a second origin of replication *(oriLyt)*.[23] Vectors are being constructed which contain both origins for the ability to maintain plasmids within cells and then to amplify as desired by regulated expression of BZLF-1.[23]

A cDNA mammalian cell expression cloning vector has been constructed based on the EBV vector system for efficient transformation and expression in human cells. For this approach, cDNAs are directly cloned into an EBV expression plasmid EBO-pcD-X.[24] This plasmid contains the

[20] G. N. Pavlakis, B. K. Felber, C. M. Wright, J. Papamatheakis, and A. Tse, *in* "Current Communications in Molecular Biology: Gene Transfer Vectors for Mammalian Cells" (J. H. Miller and M. P. Calos, eds.). Cold Spring Harbor Laboratory, Cold Spring Harbor, New York, 1987.

[21] J. L. Yates, N. Warren, and B. Sugden, *Nature (London)* **313**, 811 (1985).

[22] J. L. Yates, N. Warren, D. Reisman, and B. Sugden, *Proc. Natl. Acad. Sci. U.S.A.* **81**, 3806 (1984).

[23] W. Hammerschmidt, B. Sugden, *in* "Current Communications in Molecular Biology, Viral Vectors" (Y. Gluzman and S. H. Hughes, eds.). Cold Spring Harbor Laboratory, Cold Spring Harbor, New York, 1988.

[24] R. F. Margolskee, P. Kavatha, and P. Berg, *Mol. Cell. Biol.* **8**, 2837 (1988).

cDNA expression vector pcD-cDNA[25] within an EBV replication vector. DNA from the library is introduced by electroporation into recipient cell lines. Several million transformants are selected for hygromycin resistance encoded by the vector. The transformants are screened for expression of the desired gene product. Cells expressing the product are isolated and EBO-pcD episomes are rescued by the Hirt DNA extraction procedure. The isolated DNA is used to transform competent *Escherichia coli* to ampicillin resistance. The EBO-pcD plasmids are recovered from the bacterial colonies and are screened for expression of the desired product after reintroduction into mammalian cells.

Retroviral Vectors. Retroviruses are able to transfer their genetic information at high efficiency into eukaryotic cells. Because these enveloped RNA-containing viruses undergo a DNA stage in their replication cycle, they can be genetically manipulated to replace their own genes with exogenous genes. Since the DNA intermediate of the retrovirus can integrate into the host chromosome, these viruses can be utilized as vectors for gene transfer and insertion into the host chromosome. The development of cell lines which produce the proteins required for virus production, the reverse transcriptase (POL), the group-specific antigen (GAG), and the envelope (ENV) proteins enables the production of helper-free replication-defective recombinant viruses. Improvements in viral packaging lines have included modifications to increase the limited host range[26,27] and to avoid recombination leading to helper virus production.[28] Presently, the most useful cell line for recombinant retrovirus production is the PA317 amphotropic packaging cell line.[28] Two vectors which have demonstrated high titers and efficient expression[29] are the LNC vector, which expresses a selectable neomycin resistance gene and utilizes the CMV promoter for heterologous gene expression, and the LSN-2 vector, which utilizes the long terminal repeat for expression of foreign genes and the SV40 promoter for expression of neomycin resistance.

Significant advantages of retroviral vectors include their abilities: (1) to transduce genes into a variety of cell types and into a variety of species; (2) to produce stable cell lines as a result of retrovirus integration into the host chromosome; (3) to introduce DNA into nearly 100% of the host cells, due to the high infectivity and ability to produce high-titer virus stocks; and (4) to introduce foreign genes into animals. Expression of genes introduced by retroviral infection is usually more efficient than expression from genes

[25] H. Okayama and P. Berg, *Mol. Cell. Biol.* **3**, 280 (1983).
[26] R. D. Cone and R. C. Mulligan, *Proc. Natl. Acad. Sci. U.S.A.* **81**, 6349 (1984).
[27] A. D. Miller, M. F. Law, and I. M. Verma, *Mol. Cell. Biol.* **5**, 431 (1985).
[28] A. D. Miller and C. Buttimore, *Mol. Cell. Biol.* **6**, 2895 (1986).
[29] W. R. A. Osborne and A. D. Miller, *Proc. Natl. Acad. Sci. U.S.A.* **85**, 6851 (1988).

introduced by DNA transfection.[30] This may result from the selective retroviral integration into transcriptionally active loci or, alternatively, retroviral integration may result in transcriptional activation. However, in general, protein expression from retroviral-based vectors has been low due to problems with RNA splicing and mRNA translation. Since it is unknown which viral DNA sequences are essential for efficient expression in retroviral-based vectors, and since the insertion of different DNA sequences may impair propagation or expression of the recombinant retrovirus, success with these vectors has been variable. Furthermore, packaging constraints limit the size of the inserted segment to approximately 6–7 kb.

Direct DNA Transfer into Cells

With most of the methods of DNA transfer, 5–50% of the cells in the population acquire DNA and express it transiently over a period of several days to several weeks. Eventually, the DNA is lost from the population. This transient expression of transfected DNA is conveniently used instead of the more laborious procedure of isolating and characterizing stably transfected cell lines. Transient expression experiments obviate the effects of integration sites on expression and the possibility of selecting cells which harbor mutations in the transfected DNA. Transient expression offers a convenient means by which to compare different vectors and identify that an expression plasmid is functional before establishing stable cell lines. The efficiency of expression from transient transfection is dependent on the number of cells which take up the transfected DNA, the gene copy number, and the expression level per gene. The cell line most frequently used for transient expression is derived from African green monkey kidney cells by transformation with an origin-defective mutant of simian virus 40 (SV40).[3,31] These COS cells express high levels of the SV40 large tumor (T) antigen, which is required to initiate viral DNA replication. The T antigen-mediated replication can amplify the plasmid copy number to greater than 10,000 per cell. This large copy number results in high expression levels from the transfected DNA.

Vectors for Expression of cDNA Genes

Vector Components

A large variety of transient expression vectors have been described recently. In many cases, it is difficult to evaluate results from different vectors used in different laboratories. Most expression vectors are designed

[30] L. H. S. Hwang and E. Gilboa, *J. Virol.* **50,** 417 (1984).
[31] P. Mellon, V. Parker, Y. Gluzman, and T. Maniatis, *Cell* **27,** 279 (1981).

to accommodate cDNA rather than genomic clones. cDNA clones are more convenient to manipulate due to their smaller size. Today, most useful vectors contain multiple elements which include (1) an SV40 origin of replication for amplification to high copy number in COS monkey cells, (2) an efficient promoter element for transcription initiation, (3) mRNA processing signals which include mRNA cleavage and polyadenylation sequences and frequently also intervening sequences, (4) polylinkers that contain multiple endonuclease restriction sites for insertion of foreign DNA, and (5) selectable markers that can be used to select cells that have stably integrated the plasmid DNA.

The level of protein expression from heterologous genes introduced into mammalian cells depends on multiple factors, including DNA copy number, efficiency of transcription, mRNA processing, transport, stability, and translational efficiency, and protein processing, secretion, and stability. The rate-limiting step for high-level expression may be different for different genes.

Constitutive Promoters. Two important identified sequence elements that control transcription initiation are the promoter and the enhancer. The best-characterized promoter systems are those derived from viruses (SV40 and adenovirus), primarily due to the availability and analysis of mutations generated by natural variation of the virus (for a review, see Ref. 32). The promoter is generally composed of the TATAA box (often at approximately -30 bp with respect to the mRNA initiation site as $+1$),[33] which probably functions by designating the start site for RNA polymerase II transcription, and the CAAT box (at approximately -80 bp from the mRNA initiation site),[34] where factors bind to facilitate transcription initiation. Most vectors for mammalian cells contain promoter elements from efficient transcription elements such as the SV40 early promoter, the Rous sarcoma virus promoter, the adenovirus major late promoter, and the human cytomegalovirus (CMV) immediate early promoter.

Transcriptional enhancers encompass a variety of core sequences that act to increase transcription from a promoter in an orientation- and distance-independent manner.[35] The transcriptional enhancer appears to be a primary regulator of transcriptional activity. Some enhancers located in viruses, for example, polyoma[36] or Moloney murine sarcoma virus,[37] show

[32] S. McKnight and R. Tjian, *Cell* **46,** 795 (1986).
[33] R. Breathnach and P. Chambon, *Annu. Rev. Biochem.* **50,** 349 (1981).
[34] C. Benoist, K. O'Hare, R. Breathnach, and P. Chambon, *Nucleic Acids Res.* **8,** 127 (1980).
[35] E. Serfling, M. Jasin, and W. Schaffner, *Trends Genet.* **1,** 224 (1985).
[36] J. deVilliers, L. Olson, J. Banerji, and W. Schaffner, *Cold Spring Harbor Symp. Quant. Biol.* **47,** 911 (1983).
[37] L. A. Laimins, G. Khoury, C. Gorman, B. Howard, and P. Gruss, *Proc. Natl. Acad. Sci. U.S.A.* **79,** 6453 (1982).

a host cell preference and, thus, contribute to the host range of the virus. Others, like the SV40 enhancer,[38] the Rous sarcoma virus (RSV) long terminal repeat (LTR),[39] and the human CMV enhancer,[40] are very active in a wide variety of cell types from many species. Enhancers with strict cell-type specificity have been observed in many cellular genes which exhibit tissue specificity of expression, for example, the immunoglobulin genes[41-43] and the insulin gene.[44] Enhancers are likely the primary cis-acting determinants in tissue specificity of gene transcription. The addition of a strong enhancer can increase transcriptional activity by 10- to 100-fold. Thus, most expression vectors include a strong enhancer, frequently derived from SV40, RSV, or CMV.

Inducible Promoters. In order to express a protein which is potentially cytotoxic, it is advisable to use an inducible expression system which is regulated by an external stimulus. Sequences required for induced transcription from a number of promoters have been identified and function as enhancer elements. In selecting an inducible vector system for a particular gene, it is important to ensure that the inducing stimulus does not interfere with properties under study. It is also important to know what fold induction is desirable and the maximal achievable expression level. In many cases, the fold induction may be large but the maximal level of expression is low compared to a constitutive promoter. Alternatively, the fold induction may be low but the maximal expression level very high. Several of the inducible promoters which have demonstrated utility for expression of heterologous genes are described below.

INTERFERON β PROMOTER. The interferon β gene is highly inducible in fibroblasts by virus infection or by the presence of double-stranded RNA [poly(rI)–poly(rC)] (for review, see Ref. 45). The sequences responsible for its induction are found between bases from -77 to -36 from the start site for mRNA transcription. This system has been used to obtain a 200-fold induction of interferon β expression in Chinese hamster ovary cells after cointroduction and coamplification of interferon β genes with a dihydrofolate reductase (DHFR) gene.[46] These experiments demonstrated that the

---

[38] G. Neuhaus, G. Neuhaus-Uri, P. Gruss, and H.-G. Schweiger, *EMBO J.* **3**, 2169 (1984).

[39] C. M. Gorman, G. D. Merlino, M. C. Willingham, I. Pastan, and B. H. Howard, *Proc. Natl. Acad. Sci. U.S.A.* **79**, 6777 (1982).

[40] M. Boshart, F. Weber, J. Gerhard, K. Dorsch-Hasler, B. Fleckenstein, and W. Schaffner, *Cell* **41**, 521 (1985).

[41] J. Banerji, L. Olson, and W. Schaffner, *Cell* **33**, 729 (1983).

[42] S. D. Gillies, S. L. Morrison, V. T. Oi, and S. Tonegawa, *Cell* **33**, 717 (1983).

[43] C. Queen and D. Baltimore, *Cell* **33**, 741 (1983).

[44] H. Ohlsson, and T. Edlund, *Cell* **45**, 35 (1986).

[45] P. Lengyel, *Annu. Rev. Biochem.* **51**, 251 (1986).

[46] F. McCormick, M. Trahey, M. Innis, B. Dieckmann, and G. Ringold, *Mol. Cell. Biol.* **4**, 166 (1984).

normal inducibility of the interferon β gene is maintained in cells containing 25-fold amplified copies of the interferon gene. One potential problem with this induction system is that induction of murine fibroblasts results in a response in only 5–10% of the cells in the population.[47] These same studies demonstrate poor activation of the interferon β gene in human HeLa cells, but that pretreatment with interferon results in high inducibility. Thus, an interferon-inducible factor appears to be required for activation of the interferon β gene in the nonresponsive cells.

HEAT-SHOCK PROMOTER. Heat-shock genes are transcriptionally activated when cells are exposed to hyperthermia or a variety of other stresses.[48] The molecular basis for this gene activation has been extensively studied using the *Drosophila* heat-shock protein 70 gene (*hsp*70) promoter, which is efficiently expressed and tightly regulated in *Drosophila* and mammalian cells. The *Drosophila hsp*70 promoter contains a 15-bp sequence upstream from the TATA box which is responsible for activation of the promoter in response to heat and which can activate a heterologous thymidine kinase *(TK)* gene when placed upstream from the *TK* TATA box.[49] This sequence is the binding site for a specific heat-shock transcription factor that is activated during heat shock.[50,51]

This inducible system has been used to obtain high level of expression of the c-*myc* protein in CHO cells after coamplification of a *hsp*70 promoter–c-*myc* gene fusion with DHFR.[52] Highly amplified cell lines were obtained that contain 2000 copies of the introduced c-*myc* fusion gene, and undetectable levels of c-*myc* mRNA. Incubation of these cells at 43° resulted in at least a 100-fold induction of c-*myc* mRNA. Translation only occurred when the cells were returned to 37°. After 3–4 hr, the c-*myc* protein levels reached approximately 1 mg/10⁹ cells. These results demonstrate that, even at a level of 2000 copies, the *hsp*70 promoter retains inducibility. This suggests that any negative or positive acting factor involved in *hsp*70 promoter activity is not limiting. One potential problem with this system is that heat shock is detrimental to the cells and may severely affect the protein secretion machinery of some cell lines.[53]

METALLOTHIONEIN PROMOTER. Metallothioneins are small cysteine-rich proteins which play an important role in detoxification of heavy

[47] T. Enoch, K. Zinn, and T. Maniatis, *Mol. Cell. Biol.* **6**, 801 (1986).
[48] H. R. B. Pelham, *Trends Genet.* **1**, 31 (1985).
[49] H. R. B. Pelham and M. Bienz, *EMBO J.* **1**, 1473 (1982).
[50] C. Wu, *Nature (London)* **311**, 81 (1984).
[51] C. S. Parker and J. Topol, *Cell* **37**, 273 (1984).
[52] F. M. Wurm, K. A. Gwinn, and R. E. Kingston, *Proc. Natl. Acad. Sci. U.S.A.* **83**, 5414 (1986).
[53] M. J. Schlesinger, *J. Cell Biol.* **103**, 321 (1986).

metals and in heavy metal hemostasis.[54] Metallothionein *(MT)* gene transcription is induced by the presence of heavy bivalent metal ions, such as cadmium and zinc,[55] by glucocorticoid hormones,[56] and by interferon.[57] There is a cellular factor(s) that, in the presence of cadmium, interacts with the heavy metal-responsive element to induce transcription.[58] The mouse MT-I promoter region contains a metal regulatory element between bases −59 and −46, with respect to the start site for mRNA transcription as +1, which may act as an inducible enhancer.[59,60] Insertion of multiple copies of this sequence can confer zinc "inducibility" on a heterologous promoter.[61] The mouse MT-I promoter has been used successfully in bovine papilloma-based vector systems.[17,19] However, the basal level of expression is generally high, and the induction ratio is poor, generally never greater than 5- to 10-fold. In addition, high levels of cadmium are cytotoxic. Zinc may be used as a less toxic inducing agent.

GLUCOCORTICOID INDUCTION. Glucocorticoids and other steroid hormones mediate physiological responses in target cells via a specific, high-affinity interaction with cytoplasmic receptors. Hormone binding to receptors results in a conformational change, thereby allowing the hormone–receptor complex to bind specific sites on DNA.[62] Hormone–receptor complex binding to DNA results in transcriptional activation of the hormone-responsive gene. The best studied glucocorticoid-inducible promoter is the mouse mammary tumor virus (MMTV) LTR. The glucocorticoid-responsive element (GRE) behaves as an enhancer element and has been localized between −100 and −200 of the MMTV LTR. The purified glucocorticoid receptor protein binds to the GRE sequence *in vitro*[63,64] and has been cloned and extensively characterized.[65-67]

[54] P. E. Hunziker and J. H. R. Kaegi, *Top. Mol. Struct. Biol.* **7,** 5712 (1981).
[55] D. M. Durnam and R. D. Palmiter, *J. Biol. Chem.* **256,** 5712 (1981).
[56] L. J. Hager and R. D. Palmiter, *Nature (London)* **291,** 340 (1981).
[57] R. L. Friedman, S. P. Manly, M. McMohon, I. M. Kerr, and G. R. Stark, *Cell* **38,** 745 (1984).
[58] C. Sequin, B. K. Felber, A. D. Carter, and D. H. Hamer, *Nature (London)* **312,** 781 (1984).
[59] E. Serfling, A. Lubbe, K. Korsh-Hasler, and W. Schaffner, *EMBO J.* **4,** 3851 (1985).
[60] M. Karin, A. Haslinger, H. Holtgreve, G. Cathala, E. Slater, and J. D. Baxter, *Cell* **36,** 371 (1984).
[61] P. F. Searle, G. W. Stuart, and R. D. Palmiter, *Mol. Cell. Biol.* **5,** 1480 (1985).
[62] G. M. Ringold, *Annu. Rev. Pharmacol. Toxicol.* **25,** 529 (1985).
[63] F. Payvar, D. DeFranco, G. L. Firestone, B. Edgar, O. Wrange, S. Okret, J. A. Gustaffson, and K. R. Yamamoto, *Cell* **35,** 381 (1983).
[64] H. M. Jantzen, U. Strahle, B. Gloss, F. Stewart, W. Schmid, M. Boshart, R. Miksicek, and G. Schuts, *Cell* **49,** 29 (1987).
[65] V. Giguere, S. M. Hollenberg, M. G. Rosenfeld, and R. M. Evans, *Cell* **46,** 645 (1986).
[66] P. J. Godowski, S. Rusconi, R. Miesfeld, and K. R. Yamamoto, *Nature (London)* **325,** 365 (1987).
[67] S. Rusconi and K. R. Yamamoto, *EMBO J.* **6,** 1309 (1987).

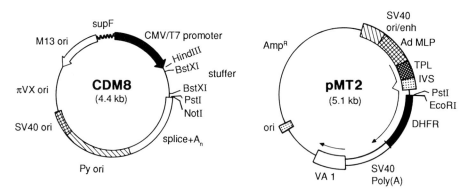

FIG. 1. Efficient cloning–expression vectors CDM8 [B. Seed and A. Aruffo, *Proc. Natl. Acad. Sci. U.S.A.* **84**, 3365 (1987)] and pMT2 [R. J. Kaufman, M. V. Davies, V. Pathak, and J. W. B. Hershey, *Mol. Cell. Biol.* **9**, 946 (1989)] for mammalian cells.

A conditional glucocorticoid-inducible expression vector has been developed for high-level inducible expression in Chinese hamster ovary cells.[68] This expression system requires high-level expression of the glucocorticoid receptor protein in the host cell and multiple copies of the glucocorticoid responsive element within the expression vector. To accomplish this, the rat glucocorticoid receptor was expressed at high level in DHFR-deficient Chinese hamster ovary cells (called GRA cells.)[68] When multiple copies of the glucocorticoid-responsive element from MMTV are inserted into the expression vector (pMT2; see Fig. 1) to yield pMG18PC, addition of dexamethasone yields a 7-fold induction of expression over the high constitutive level (Fig. 2). When the constitutive SV40 enhancer is deleted from pMG18PC to yield pMG18oS, basal expression is eliminated and maximal expression upon addition of dexamethasone is retained (Fig. 2). This expression system should be of general utility for studying gene regulation and for expressing heterologous genes in a regulatable fashion.

As the genes for other transactivators are cloned and characterized for the structural features required to activate gene transcription, it will be possible to engineer novel potent transactivators to elicit induction of specific promoters. This has recently been demonstrated by mixing DNA binding and transcriptional activation domains from yeast, bacteria, and mammalian cells to elicit transactivation of specific genes.[69–71]

[68] D. I. Israel and R. J. Kaufman, *Nucl. Acids Res.* **17**, 4589 (1989).
[69] N. Webster, J.-R. Jin, S. Green, M. Hollis, and P. Chambon, *Cell* **52**, 169 (1988).
[70] K. Lech, K. Anderson, and R. Brent, *Cell* **52**, 179 (1988).
[71] H. Kakidani and M. Ptashne, *Cell* **52**, 161 (1988).

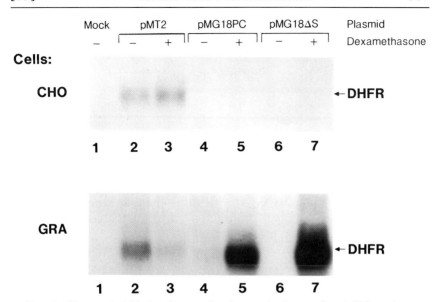

FIG. 2. Glucocorticoid-induced expression in transiently transfected Chinese hamster ovary cells. Plasmid DNA was transfected by the DEAE-dextran protocol into the indicated cell lines. At 24 hr posttransfection, dexamethasone was applied to one of two parallel plates (+ or − as indicated). Total RNA was isolated 48 hr posttransfection and analyzed for dihydrofolate reductase (DHFR) RNA by Northern blot hybridization to a DHFR-specific probe. The glucocorticoid expression vectors pMG18PC and pMG18ΔS indicated are similar to pMT2 except for the insertion of 18 glucocorticoid-responsive elements. pMG18ΔS has deleted the SV40 enhancer present in pMT2. The transfected cell lines are the original Chinese hamster ovary dihydrofolate reductase-deficient cells (CHO) [L. A. Chasin and G. Urlaub, *Proc. Natl. Acad. Sci. U.S.A.* **77,** 4216 (1980)] and the same cells (GRA) which express high levels of the rat glucocorticoid receptor. For details, see Ref. 68.

RNA Processing. SPLICING. Most higher eukaryotic genes contain introns which are processed from the precursor RNA in the nucleus to generate the mature mRNA. The sequences and factors responsible for RNA splicing have recently been reviewed.[72] Although many genes do not require introns for mRNA formation when introduced into mammalian cells, there are several examples of genes that have strong requirements for the presence of an intron.[73,74] In general, it is probably better to include introns in cDNA expression vectors for mammalian cells. For example, the intervening sequences from SV40 small t antigen mRNA or hybrid intron composed from adenovirus and immunoglobulin sequences are frequently used (see Fig. 2). In contrast to expression in mammalian cell lines, there is

[72] P. A. Sharp, *Science* **235,** 766 (1987).
[73] A. R. Buchman and P. Berg, *Mol. Cell. Biol.* **8,** 4395 (1988).
[74] B. Gruss and G. Khoury, *Nature (London)* **286,** 634 (1980).

a profound requirement for the presence of introns for expression of genes introduced into transgenic animals.[75]

TRANSCRIPTION TERMINATION AND POLYADENYLATION. Transcription termination, 3' end cleavage, and polyadenylation of precursor mRNAs are essential steps for the biogenesis of mRNA (for review, see Ref. 76). In higher eukaryotes, RNA polymerase II transcribes across the polyadenylation site(s) and terminates downstream of the mature 3' termini for the mRNA. To date, the DNA sequences which function as transcription termination signals have only been mapped to various restriction fragments at the 3' end of several genes.[77-79] Structural features within some of these regions have been compared to identify a sequence which represents a conserved element within the termination region of a number of different genes.[80] Further studies are required to identify whether this sequence has any functional significance.

The possibility that an upstream promoter may occlude transcription from a downstream promoter, due to polymerase transcribing through the downstream promoter,[81] suggests that insertion of transcription termination signals upstream from the second transcription unit in an expression vector may potentiate expression from that promoter by preventing transcription from extending through the upstream sequences. Indeed, this has been directly demonstrated by the insertion of a histone H2A gene termination region or a mouse β-globin gene termination region between two tandem α-globin transcription units to elicit an approximately 7-fold increase in expression from the downstream transcription unit.[82] Transcription termination signals would also provide an approach to minimize transcription from the opposite strand of DNA, which can suppress gene expression through the formation of antisense mRNA.[83,84]

The 3' end of eukaryotic mRNA is formed by polyadenylation, which involves cleavage of the precursor mRNA at a specific site and then polymerization of about 200 adenylate residues, poly(A), to the newly

[75] R. Brinster, J. Allen, R. Behringer, R. Gelinas, and R. Palmiter, *Proc. Natl. Acad. Sci. U.S.A.* **85**, 836 (1988).
[76] M. L. Birnstiel, M. Susslinger, and K. Strub, *Cell* **41**, 349 (1985).
[77] E. G. Frayne, E. J. Leys, G. F. Crouse, A. G. Hook, and R. E. Kellems, *Mol. Cell. Biol.* **4**, 2921 (1984).
[78] E. Falck-Pederson, J. Logan, T. Shenk, and J. E. Darnell, Jr., *Cell* **40**, 897 (1985).
[79] C. Birchmeier, D. Schumperli, G. Sconzo, and M. L. Birnstiel, *Proc. Natl. Acad. Sci. U.S.A.* **81**, 1057 (1984).
[80] E. G. Frayne and R. E. Kellems, *Nucleic Acids Res.* **14**, 4113 (1986).
[81] B. R. Cullen, P. T. Lomedico, and G. Ju, *Nature (London)* **307**, 241 (1984).
[82] N. J. Proudfoot, *Nature (London)* **322**, 562 (1986).
[83] J. G. Izant and H. Weintraub, *Science* **229**, 345 (1985).
[84] D. A. Melton, *Proc. Natl. Acad. Sci. U.S.A.* **82**, 144 (1985).

generated 3' end.[85] Removal of the polyadenylation site decreases expression up to 10-fold.[86] Two sequences important for polyadenylation have been identified. The first is a highly conserved hexanucleotide, AAUAAA, that is present 11–30 nucleotides upstream of most polyadenylation sites, and which forms the recognition sequence for the cleavage and polyadenylation reaction. Deletions or point mutations in this sequence prevent the appearance of properly polyadenylated mRNA *in vivo*.[87-89] There is also a requirement for a sequence downstream of the poly(A) site for efficient cleavage and polyadenylation.[90-92] A loose consensus sequence, for this potential second element, has been identified as either a U-rich or a (G + U)-rich tract.[93] However, removal of this sequence, in some instances, has no effect on the efficiency of polyadenylation[94] but may influence the position of 3' processing.[95] This sequence appears to be required for formation of a precleavage complex.[96] Efficient signals for polyadenylation from SV40 early transcription unit, the hepatitis B surface antigen transcription unit, and the mouse β-globin gene are commonly used.

Translation of mRNAs. The primary event limiting translation of mRNAs is at the initiation step (for a review, see Ref. 97). Polypeptide chain synthesis is initiated when a ternary complex of eukaryotic initiation factor 2, GTP, and initiator Met-tRNA binds a 40S ribosomal subunit. This 40S ribosomal subunit complex binds to the 5' end of the mRNA and migrates in the 3' direction until it encounters an AUG triplet which can efficiently serve as the initiator codon. Subsequently, a 60S ribosomal subunit joins to form an 80S initiation complex. Several approaches to develop expression vectors optimized for protein synthesis initiation are discussed below.

[85] C. L. Moore and P. A. Sharp, *Cell* **41,** 845 (1985).
[86] R. J. Kaufman and P. A. Sharp, *Mol. Cell. Biol.* **2,** 1304 (1982).
[87] C. Montell, E. F. Fisher, M. H. Caruthers, and A. J. Berk, *Nature (London)* **305,** 600 (1983).
[88] D. R. Higgs, S. E. Y. Goodbourn, J. Lamb, J. B. Clegg, D. J. Weatherall, and N. J. Proudfoot, *Nature (London)* **306,** 398 (1983).
[89] M. Fitzgerald and T. Schenk, *Cell* **24,** 251 (1981).
[90] M. Wickens and P. Stephenson, *Science* **226,** 1045 (1984).
[91] M. A. McDevitt, M. J. Imperiale, H. Ali, and J. R. Nevins, *Cell* **37,** 993 (1984).
[92] M. Sasofsky, S. Connelly, J. L. Manley, and J. C. Alwine, *Mol. Cell. Biol.* **5,** 2713 (1985).
[93] A. Gil and N. Proudfoot, *Cell* **49,** 399 (1987).
[94] D. Danner and P. Leder, *Proc. Natl. Acad. Sci. U.S.A.* **82,** 8658 (1985).
[95] P. J. Mason, J. A. Elkington, L. M. Malgorazata, M. B. Jones, and J. G. Williams, *Cell* **46,** 263 (1986).
[96] D. Zarkower and M. Wickens, *J. Biol. Chem.* **263,** 5780 (1988).
[97] V. Pain, *Biochem. J.* **235,** 625 (1986).

mRNA AND CONSENSUS SEQUENCE REQUIREMENTS. The sequence requirements for most efficient mRNA translation initiation are the following:[98] 5′CCᴬ₉CCATGG − 3′. Of most importance is the purine in the − 3 position and then the G in the +4 position. Mutations to alter this consensus sequence around the initiator methionine codon may reduce translation as much as 10-fold.

In addition to the consensus sequence requirements, there are other features that influence mRNA translation. The presence of upstream initiation codons is detrimental to initiation at the downstream site, particularly if they are not followed by in-frame termination codons before the start site for translation initiation of the desired coding region.[99,100] In addition, the presence of significant secondary structure (− 50 kcal/mol) near the AUG on the mRNA can be detrimental to translation.[101-103] Most vectors are designed to contain 5′ untranslated sequences derived from the particular promoter element used for transcription initiation.

TRANSLATIONAL EFFICIENCY MEDIATED BY eIF-2. Translation of mRNA derived from plasmid DNA after transient transfection of COS-1 cells may be inefficient. This deficiency can be corrected by the introduction of the adenovirus virus-associated (VA) RNA gene into the COS-1 cell expression vector.[104] VA RNA is a small RNA synthesized by RNA polymerase III which has been proposed to block the double-stranded RNA-activated (DAI) protein kinase.[105,106] It has been proposed that transcription from both strands of the episomal DNA template in transfected COS-1 cells can generate double-stranded RNA, leading to activation of DAI kinase and translation arrest. VA RNA stimulates translation by preventing the phosphorylation of the α subunit of eIF-2 α.[104,107,108] A similar stimulation in translation results from expression of serine to alanine mutant of eIF-2, which renders it phosphorylation resistant.[109] To date, there has been no demonstration of an effect of VA RNA on the

---

[98] M. Kozak, *Cell* **44**, 283 (1986).

[99] C.-C. Liu, C. C. Simonsen, and A. D. Levinson, *Nature (London)* **309**, 82 (1984).

[100] M. Kozak, *Nucleic Acids Res.* **12**, 3873 (1984).

[101] J. Pelletier and N. Sonenberg, *Cell* **40**, 515 (1985).

[102] M. Kozak, *Proc. Natl. Acad. Sci. U.S.A.* **83**, 2850 (1986).

[103] J. Pelletier and N. Sonenberg, *Biochem. Cell Biol.* **65**, 576 (1987).

[104] R. J. Kaufman, *Proc. Natl. Acad. Sci. U.S.A.* **82**, 689 (1985).

[105] J. Kitajewski, R. J. Schneider, B. Safer, S. M. Munemitsu, C. E. Samuel, B. Thimmappaya, and T. Schenk, *Cell* **45**, 195 (1986).

[106] R. P. O'Malley, T. M. Mariano, J. Siekierka, and M. B. Mathews, *Cell* **44**, 391 (1986).

[107] C. Svensson and G. Akusjarvi, *EMBO J.* **4**, 957 (1985).

[108] R. J. Kaufman and P. Murtha, *Mol. Cell. Biol.* **7**, 1568 (1987).

[109] R. J. Kaufman, M. V. Davies, V. Pathak, and J. W. B. Hershey, *Mol. Cell. Biol.* **9**, 946 (1989).

translation of mRNA derived from transcription units integrated in the host chromosome.

POLYCISTRONIC EXPRESSION VECTORS. It is now known that polycistronic mRNAs are translated in mammalian cells.[110] One mechanism which has been postulated is that translation termination occurs and ribosome scanning continues with reinitiation at downstream AUGs.[111] However, there are at least some viral mRNAs of the picornavirus family that contain specific sequences to promote internal ribosome binding and translation initiation.[112,113] The utilization of polycistronic mRNA expression vectors containing the desired gene in the 5' open reading frame and a selectable gene in the 3' position has been used to select cells that express high levels of the gene product in the 5' position.[114] Since the translation of the selectable gene in the 3' end is inefficient, polycistronic expression vectors exhibit preferential translation of the 5' desired gene and cells require high levels of polycistronic mRNA to survive selection.

Preparation of cDNA Inserts. The most common approach to express a foreign gene relies on insertion of a cDNA copy encoding the desired protein into an appropriate expression plasmid. Generally, cDNA copies are used, since they are easier to manipulate *in vitro* due to their smaller size. In addition, the appropriate coding region can be inserted behind an efficient promoter which also includes all expression elements to ensure efficient expression. In preparation of a cDNA insert for introduction into any particular vector, a couple of suggestions may improve expression. First, if the cDNA were obtained by addition of homopolymeric tails into the 5' noncoding region, they should be removed. Removal of GC tails may improve expression up to 10-fold.[115] Second, if the AUG does not conform to the consensus GCCAUGG, it is recommended that the context of the initiation codon should be changed so that it does conform. Third, in general and when convenient, it is better to remove as much as possible of the 5' and 3' untranslated regions of the cDNA. This is particularly true if the cDNA contains AUG initiation codons which are upstream of the authentic AUG or if the 5' untranslated region contains extensive secondary structure.[99-103] In addition, sequences within the 3' end of the mRNA may reduce mRNA stability.[116]

[110] D. Peabody and P. Berg, *Mol. Cell. Biol.* **6**, 2695 (1986).
[111] M. Kozak, *Cell* **47**, 481 (1986).
[112] J. Pelletier and N. Sonenberg, *Nature (London)* **334**, 320 (1988).
[113] S. K. Jang, M. V. Davies, R. J. Kaufman, and E. Wimmer, *J. Virol.* **63**, 1651 (1989).
[114] R. J. Kaufman, P. Murtha, and M. Davies, *EMBO J.* **6**, 187 (1987).
[115] C. C. Simonsen *et al., ICN-UCLA Symp. Mol. Cell. Biol.* **25**, 1 (1981).
[116] G. Shaw and R. Kamen, *Cell* **46**, 659 (1986).

Efficient Vectors for Expression of cDNAs in Mammalian Cells

Two vectors which have successfully been used in different laboratories are shown in Fig. 1. CDM8[117] contains a human cytomegalovirus (CMV) promoter for efficient transcription in a variety of cell types *in vivo* and a promoter from T7 bacteriophage DNA-dependent RNA polymerase for generation of transcripts *in vitro*. It contains SV40 and polyoma origins of replication for replication in COS monkey cells and polyoma virus-trans-formed mouse fibroblasts. Transcription is driven by the human CMV promoter and the vector contains an SV40 intervening sequence and polyadenylation signal. CDM8 contains a π VX origin of replication for propagation in *E. coli* and a suppressor tRNA gene (supF) for selection in *E. coli*. These elements permit the plasmid to be conveniently isolated from mammalian cells through isolation of episomal DNA by the Hirt procedure[118] and used to transform *E. coli* strain MC1061 (p3) to tetracy-cline and ampicillin resistance by suppression of amber mutations in the tetracycline resistance gene and β-lactamase genes carried on the p3 plas-mid. An alternative method to isolate the plasmid DNA from mammalian cells involves cloning DNA into amber mutant strains of bacteriophage λ and selecting phage by complementation mediated by *supF*. CDM8 also contains an M13 origin of replication which facilitates oligonucleotide-di-rected mutagenesis using single-stranded DNA templates, and it has two *Bst*NI restriction endonuclease sites for cloning cDNAs into the vector at high efficiency in the proper orientation by the use of oligonucleotide adapter sequences.

pMT2[109] contains plasmid sequences from puc18 which allow for prop-agation and selection for ampicillin resistance in *E. coli*. It contains the SV40 origin of replication and enhancer element. It utilizes the adenovirus major late promoter for transcription initiation. Contained within the 5' end of the mRNA is the tripartite leader from adenovirus late mRNA and a small intervening sequence. There are several cloning sites, including *Eco*RI and *Pst*I, for insertion of foreign DNA. In the 3' end of the tran-script there is a cleavage and polyadenylation signal from the SV40 early region. In addition, this plasmid contains the adenovirus *VAI* gene, which is a polymerase III transcription unit which encodes a small RNA that potentiates translation of the plasmid-derived mRNA in transfected COS monkey cells. This vector also contains a DHFR coding region in the 3' end of the transcript. Expression of DHFR from resulting bicistronic mRNAs can be used to select directly for DHFR expression in Chinese

[117] B. Seed and A. Aruffo, *Proc. Natl. Acad. Sci. U.S.A.* **84**, 3365 (1987).
[118] B. J. Hirt, *J. Mal. Biol.* **26**, 365 (1967).

hamster ovary cells which are DHFR deficient.[114,119] In addition, it is possible to select directly for amplification of the copy number by selection for resistance to increasing concentrations of methotrexate, an inhibitor of DHFR.

Transient Expression of Plasmid DNA in COS-1 Cells

The most widely used and convenient system for expression of a foreign gene is by introduction of DNA into COS-1 cells and then monitoring expression over the next 48–72 hr. Although there are many vectors suitable for this purpose, a very efficient vector is pMT2 (see Fig. 1). The following is a protocol that can be used to obtain efficient expression by this approach.

DEAE-Dextran-Mediated Transfection of COS-1 Cells[120]

Stock Solutions

(10×) DEAE-Dextran (Pharmacia, MW 500,000): 2.5 mg/ml in Dulbecco's modified essential (DME) medium and stored at 4°

(10×) 1 M Tris-HCl, pH 7.3: stored at 4°

(1000×) Chloroquine (Sigma): 0.1 M stored at $-20°$ in dark

(1×) 10% DMSO reagent (1 liter): 137 mM NaCl (8 g), 5 mM KCl (0.37 g), 0.7 mM NH$_2$HPO$_4$ (0.01 g), 6 mM D-glucose (1.08 g), 21 mM HEPES (5 g). Then add 900 ml H$_2$O to take pH to 7.1 and filter through a 0.2 μm filter. Finally, add 100 ml 100% DMSO

Cells. COS-1 cells are grown in DME medium supplemented with 10% heat-inactivated fetal calf serum. They are usually subcultured twice per week at a 1:4 or 1:6 split ratio.

Growth Medium. Dulbecco's modified essential medium with 2 mM glutamine, 100 U/ml streptomycin, 100 μg/ml penicillin, and 10% heat-inactivated fetal calf serum.

Transfection

1. Subculture cells 1:6 into 100 mM tissue culture plates at 12–24 hr before transfection.

2. Aspirate medium and wash 2 times with 7 ml each of serum-free DME. Note that the cells will die if the DEAE-dextran and the serum contact the cells at the same time.

3. Feed the cells the DNA–medium mix (4 ml per 100-mm culture

---

[119] L. A. Chasin and G. Urlaub, *Proc. Natl. Acad. Sci. U.S.A.* **77**, 4216 (1980).
[120] J. H. McCutchan and J. S. Pagano, *JNCI, J. Natl. Cancer Inst.* **41**, 351 (1968).

dish containing 8 μg of DNA) which is prepared as follows: (a) Add DNA [generally prepared in sterile Tris-HCl (10 mM, pH 8.0) and EDTA (1 mM)] to 0.4 ml of Tris-HCl (final DNA concentration should be 2 μg/ml in the medium). Mix well. (b) Add 0.4 ml of 10× DEAE-dextran to DNA–Tris-HCl. (c) Add 3.2 volumes of DME that contains 2 mM glutamine, 100 U/ml streptomycin, 100 μg/ml penicillin. Mix well.

    4. Incubate for 8–10 hr at 37°.

    5. Rinse once with 7 ml of serum-free DNA.

    6. Add 2 ml of 10% DMSO reagent to the dish. Leave for 2–3 min at room temperature. Aspirate.

    7. Add 5 ml/plate of DME + 10% fetal calf serum and 0.1 mM chloroquine for 2.5 hr.

    8. Remove chloroquine and rinse once with serum-free medium and add 10 ml of DME growth medium per plate.

    9. After 30 hr, aspirate medium and feed 10 ml of fresh growth medium (optional).

    10. After 48–72 hr, harvest medium and/or cells as desired.

Notes. This technique results in significant cell death. In the optimal experiment, approximately 25% of the cells will die. Of the remaining cells, 20% usually acquire and express the DNA. The toxicity is most evident when the transfected cells are less than 30% confluent at the start of the transfection. It is recommended to use CsCl-banded DNA for this procedure. However, miniplasmid DNA preparations may be used but yield significantly lower levels of expression.

Experimental Approach for Expression of Foreign Genes in Mammalian Cells

    Although many approaches may be taken to express a particular gene, the choice is very much dependent on the requirements for the questions that are to be answered. Table I lists the advantages and expression levels obtained with a variety of different expression systems. The most convenient system and the one which can yield the most rapid results is transient expression in COS cells. Success using the COS system will also provide incentive for any more laborious approach by ensuring that the cDNA carried in the expression plasmid can direct synthesis of the desired gene product.

    To monitor efficient expression in COS cells, it is necessary to include a positive control in order to compare expression of the desired gene with a gene that is known to be expressed well. This comparison rules out potential problems that may arise from the transfection. It will ensure that the transfection efficiency is adequate and that the COS cells are appropriate.

TABLE I
EXPRESSION LEVELS AND UTILITIES FOR DIFFERENT MAMMALIAN CELL EXPRESSION SYSTEMS

Cell line	Mode of DNA transfer	Optimal expression level (μg/ml)	Primary utility
Monkey cells			
CV-1	SV40 virus infection	1–10	Expression of wild-type and mutant proteins
COS	Transient DEAE-dextran DNA transfection	1	Cloning by expression in mammalian cells; rapid characterization of cDNA clones; expression of mutant proteins
CV-1	Transient DEAE-dextran DNA transfection	0.05	
Murine-fibroblasts			
C127	BPV stable transformant	1–5	High-level constitutive expression
3T3	Retrovirus infection	0.1–0.5	Gene transfer into animals; expression in different cell types
CHO-DHFR⁻	Stable DHFR + transformant	0.01–0.05	
	Amplified MTXʳ	10	High-level constitutive expression
Primate/rodent	Vaccinia virus	1	Vaccines
	EBV vector	NA	Cloning by expression

ᵃNA, Not available.

In addition, this comparison may also provide insight as to why a particular gene may not be efficiently expressed.

As an example of this comparison, Fig. 3 shows the results of transfected COS cells labeled with [³⁵S]methionine for 20 min at 48 hr posttransfection and then extracted by lysis in RIPA[121] buffer and electrophoresis on SDS–polyacrylamide gels. Cells transfected with the control DHFR expression plasmid pMT2 show a major product at 20 kDa which represents DHFR synthesis. When a heterologous coding region for human adenosine deaminase (ADA) is inserted into the cloning site to the 5' end of the DHFR-coding region, the transfected cells show a major band which represents ADA expression at approximately 44 kDa. The efficiency of ADA expression is roughly similar to that obtained for DHFR. Thus, we conclude that the expression plasmid is working appropriately for COS cells. Expression of an efficiently expressed secreted protein is shown in

[121] A. J. Dorner and R. J. Kaufman, this volume [44].

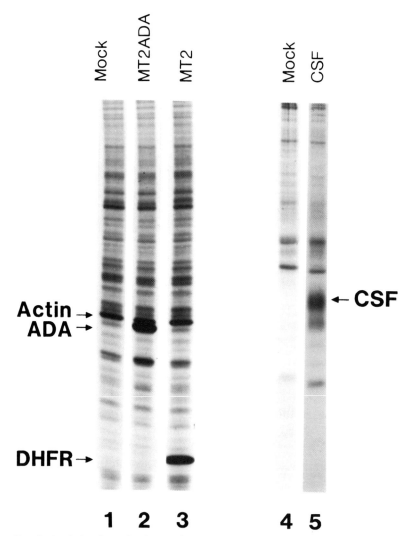

FIG. 3. Analysis of transiently transfected COS-1 monkey cells. The pMT2 expression plasmids encoding DHFR (MT2), human adenosine deaminase (ADA, MT2ADA), and a colony-stimulating factor (CSF) were transfected into COS-1 monkey kidney cells. At 48 hr posttransfection cells were pulse-labeled with [^{35}S]methionine for 20 min (cell extracts) or 3 hr (conditioned medium), and samples of the total cell extracts or conditioned medium were prepared for SDS–polyacrylamide gel electrophoresis. The results show the expression obtained for intracellular (ADA and DHFR) and secreted (CSF) proteins.

lane 5. In this case, transfected cells are labeled for 4 hr in order to permit secretion of the expressed protein and the labeled conditioned medium taken directly for analysis by SDS–polyacrylamide gel electrophoresis. In the labeled conditioned medium, a heterogeneous band is detected which represents the secreted colony-stimulating factor. It represents a major percent of the labeled secreted protein (approximately 1 μg/ml by antigen level). It migrates as a heterogeneous band due to heterogeneity in glycosylation. After treatment with N-glycanase, the band collapses to a tighter band which is more easily detected.[121]

In general, highest levels of expression in COS cells can be obtained with nontoxic intracellular proteins (greater than 1 μg/10^6 cells). Secreted proteins generally yield 0.3–1 μg/ml in the conditioned medium. Lower expression levels are usually obtained with membrane-associated proteins, possibly due to the lack of membrane surface area for which they can be deposited.

If the expression of the heterologus gene cannot be detected compared to the positive control, one must ensure that the vector was properly assembled. Then it is necessary to ensure that the mRNA was properly expressed. This is conveniently done by preparing RNA and analyzing it by Northern blot hybridization. When using the pMT2 vector, it is possible to use a DHFR probe and compare the level of DHFR mRNA obtained from pMT2-transfected cells to that from cells transfected with the same vector containing the insert, since they should both have a 3'-DHFR sequence provided from the pMT2 vector. If the mRNA is of the expected size and of the appropriate amount, then one must consider the possibility that the coding region may be interrupted so not to derive an intact translation product. If the RNA is not detected, then it is recommended that a different expression vector or system is tried, since it is always possible that some unforeseen reason may result in improper transcription or mRNA processing. However, examples of this actually occurring are very infrequent.

Acknowledgments

I gratefully thank Andrew Dorner for critically reading this manuscript, Michelle Wright for assistance in its preparation, and David Israel, Monique Davies, and Patricia Murtha for providing the data illustrated.

[40] Assembly of Enhancers, Promoters, and Splice Signals to Control Expression of Transferred Genes

By MICHAEL KRIEGLER

Long before the advent of molecular cloning in the early 1970s, a major objective of developmental and molecular biologists worldwide had been to uncover and describe the mechanisms that serve to regulate the expression of genes in both unicellular and multicellular organisms. The notion that we might develop the capacity to regulate the expression of genes of choice deliberately in both cells and animals was merely a fantasy. Ironically, since that time, we have achieved that fantasy and have developed remarkable facility at deliberately regulating the expression of foreign genes introduced into cells and animals, but we are just beginning to appreciate how the expression of resident genes are regulated in nature. This article presents a compilation of the data that have arisen as a result of the experimental manipulation of regulatory sequences of eukaryotic genes. Contained herein is a comparative analysis of the structure and biology of viral and cellular cis-acting elements that serve to regulate gene expression, including promoter and enhancers as well as splice signals. It is written to assist the experimentalist in the design of regulated recombinant eukaryotic cistrons for use in the construction of recombinant plasmids, recombinant viruses, transgenic mice, as well as in cells and animals that contain interesting recombinant molecules introduced into their genomes at the appropriate site through homologous recombination. In most cases, specific sequence information is not included. For that, the reader is encouraged to examine the cited reference.

First promoters are discussed, then enhancers are described, and the effects of splicing on gene expression are examined. Lastly, suggestions are made for practical application of these findings and the mechanisms by which these elements exert their influences are discussed.

What Are Promoters and Enhancers?

In our attempts to unravel eukaryotic gene expression, a collection of cis-acting elements and trans-acting factors that function through complex interplay to affect and regulate gene expression has been identified. The genetic dissection, through recombinant DNA techniques, of promoter elements, initially viral, led to a description of the molecular structure of eukaryotic promoters and to the discovery of a new class of cis-acting elements that appeared to modulate gene expression, elements that have come to be called enhancers. These elements appear to form the structural framework for DNA – protein interactions that may result in the formation of higher order chromatin structures that may, in turn, serve to regulate transcription of subject genes from RNA polymerase II promoters. The identification of enhancer elements has led to the isolation and characterization of trans-acting factors that either bind to or in some other way affect the promoters or enhancers with which they interact. For a compilation of transcription-regulating proteins, see Wingender.[1]

What has become clear from site-directed mutagenesis of eukaryotic expression elements is that both enhancers and promoters play pivotal roles in the regulation of gene expression. The elegant early studies of the herpes simplex virus thymidine kinase (HSVTK) promoter and the human β-globin promoter demonstrate that the structure of RNA polymerase II promoter elements is relatively set. One critical feature of all promoter elements is that they all contain mRNA cap sites, the point at which the mRNA transcript actually begins. Yet, another frequently occurring homology, the so-called TATA box, is centered around position − 25, just upstream of the mRNA cap site. The TATA box functions to position accurately the start of transcription. Further upstream, the relative position of promoter sequence homologies becomes more variable. They can, as is the case of the CCAAT homology in the HSVTK and human β-globin promoters, actually reside in different positions on different strands around − 70 to − 80 nucleotides upstream from the mRNA cap site. Another structural component of many housekeeping genes is multiple copies of GC-rich regions upstream of the mRNA cap. There appears to be an absolute requirement for these sequences in eukaryotic promoters and, as such, they must be included in all engineered recombinant cistrons.

With few exceptions, the most critical variable in the design of chimeric expression cistrons is the selection of an enhancer element or elements for inclusion in the recombinant molecule. First identified in the genomes of SV40 and murine retroviruses, enhancers are the most peculiar of all known expression elements. As mentioned above, little is known about their mechanism of action. The key properties that make an expression element an enhancer include (1) they are relatively large elements and may contain repeated sequences that can function independently. (2) They may act over considerable distances, up to several thousand base pairs. (3) They may function in either orientation. (4) They may function in a position-independent manner and can be within or downstream of the transcribed region, but can only function in cis. If several promoters lie nearby, the enhancer may preferentially act on the closest. (5) They may function in a cell-type- or tissue-specific manner.

[1] E. Wingender, *Nucleic Acids Res.* **16**, 1879 (1988).

Behavior of Wild-Type Enhancers and Promoters; Enhancer/
Promoter Chimeras

The analysis of native enhancer/promoter assemblies was greatly facilitated through the analysis of chimeric enhancer/promoter assemblies. Tissue culture transfection experiments as well as the generation and analysis
of transgenic mice containing engineered cistrons have revealed much
about the primary structure of enhancers and something about the molecular mechanism by which they operate.

Enhancers have been identified in association with both viral and
cellular genes. To minimize confusion, I will treat these two classes separately.

Viral Enhancers

The first enhancing DNA sequence identified was of viral origin and is
resident in the genome of SV40 around the origin of replication.[2,3]

The SV40 enhancer is composed of three functional units, A, B, and C,
each of which cooperates with the others or with duplicates of itself to
enhance transcription. When element C, containing the so-called enhancer
"core" consensus sequence, is inactivated by point mutations, revertants
with reduced enhancer function can be isolated, and they contain duplications of either one or both of the elements A and B.[4,5] In addition, each
element can act autonomously when present in multiple tandem copies.
Amplified copies of either the B or C element exhibit different cell-specific
activities.[6]

An analysis of the effects of the variation of position of the SV40
enhancer on the expression of multiple transcription units in a single
plasmid revealed two types of position effects. The first, promoter occlu-
sion, results in reduced transcription at a downstream promoter if transcription is initiated at a nearby upstream promoter. This effect does not
involve enhancer elements directly, even though the effect is most pronounced when the downstream promoter lacks an enhancer element. The
second effect stems from the ability of promoter sequences to reduce the
effect of a single enhancer element on other promoters in the same plasmid. This latter effect is mediated either by promoters adjacent to the

[2] J. Banjeri, S. Rusconi, and W. Schaffner, Cell 27, 299 (1981).
[3] P. Moreau, R. Hen, B. Wasylyk, R. Everett, M. P. Gaub, and P. Chambon, Nucleic Acids Res. 9, 6047 (1981).
[4] W. Herr and J. Clarke, Cell 45, 461 (1986).
[5] T. A. Firak and K. N. Subramanian, Mol. Cell. Biol. 6, 3667 (1986).
[6] B. Ondek, A. Sheppard, and W. Herr, EMBO J. 6, 1017 (1987).

enhancer element, or by promoters interposed between the enhancer element and the other promoters on the plasmid.[7] However, there is at least one enhancer element, the κ enhancer, cited below, that does not behave in this manner.[8]

Perhaps one of the more controversial observations regarding the SV40 enhancer is that enhancer-binding factors are required for the establishment but not the maintenance of enhancer-dependent transcriptional activation,[9] a conclusion that has been elegantly contested.[10] Such enhancer-binding proteins may be induced or activated, either directly or indirectly, in a human hepatoma cell line by the tumor promoter 12-O-tetradecanoylphorbol-13-acetate (TPA). In this cell line, TPA induces the transcriptional stimulatory activity of the SV40 enhancer.[11] Similarly, F9 embryonal carcinoma (EC) cells, when forced to differentiate, exhibit a marked increase in transcription from a transiently transfected genome whose transcription is driven from an SV40 promoter. Deletion of the enhancer from this plasmid removes the effect. Further, in both undifferentiated and differentiated F and EC cell types, the level of transcription was found to be limited by the availability and/or activity of cellular factors necessary for enhancer function.[12] Chimeras constructed between the SV40 enhancer and the virus response elements (VRE) of the interferon α promoter indicate that the two subsequences, Rep A and Rep B of the VRE, when individually inserted between the TATA box of the interferon α promoter and the SV40 enhancer serve to silence that promoter. Such promoter silencing is fully reversible after induction. Silencing is not observed when the intact VRE is placed between the TATA box and the SV40 enhancer.[13]

Thus, the SV40 enhancer element is a complex structure whose function is subject to some position effects, and whose cell-type-specific activation is dependent, in part, on the absence or presence of active cellular factors or proximal sequences. This scenario is repeated over and over again for viral and cellular enhancers, so the remainder of this discussion focuses on properties unique to each particular enhancer.

The polyoma virus enhancer, also found near the viral origin of replication, is similar in many ways to the SV40 enhancer. However, there are

[7] T. Kadesch and P. Berg, *Mol. Cell. Biol.* **6**, 2593 (1986).
[8] M. L. Atkinson and R. P. Perry, *Cell* **46**, 253 (1986).
[9] X. F. Wang and K. Calame, *Cell* **47**, 241 (1987).
[10] G. Schaffner, S. Schirm, B. Muller-Baden, F. Weber, and W. Schaffner, *J. Mol. Biol.* **201**, 81 (1988).
[11] R. J. Imbra and M. Karin, *Nature (London)* **323**, 555 (1986).
[12] M. J. Sleigh and T. J. Lockett, *EMBO J.* **4**, 3831 (1985).
[13] D. Kuhl, J. de al Fuenta, M. Chaturvedi, S. Parimoo, J. Ryals, F. Meyer, and C. Weissman, *Cell* **50**, 1057 (1987).

observations unique to this regulatory element. The polyoma virus enhancer is normally not active in undifferentiated F9 EC cells, but is active in these cells after differentiation. However, an enhancer point mutant has been isolated that is fully functional in both cell types. Mutant F441 carries a single base mutation at nucleotide 5233 of the viral genome. Enhancers of similar mutants that have been isolated have a duplicated segment of viral DNA encompassing nucleotide 5233. Furthermore, duplication of the point-mutated segment results in an even higher level of expression in the undifferentiated cell type. In addition, cotransfection of the F441 oligonucleotide, but not the wild-type sequence, inhibits the activity of the whole enhancer fragment of F441 attached to a reporter gene. Thus, it appears that the point mutation is a target for a cellular factor(s) which acts in a positive manner to increase the transcription of a gene in undifferentiated F9 EC cells.[14]

Studies have shown that both the SV40 enhancer and the wild-type polyoma enhancer can be repressed by adenovirus E1A gene products. Other work has shown that, while the wild-type polyoma enhancer cannot function in undifferentiated F9 EC cells, the SV40 enhancer, at least to a limited extent, can. On the basis of these observations, it has been postulated that undifferentiated F9 EC cells contain an E1A-like activity, and that this activity is responsible for the lack of polyoma virus enhancer activity in F9 EC cells. Hen et al. report that E1A gene products do not repress a point mutant of the polyoma virus enhancer that is active in undifferentiated F9 EC cells.[15] Their result is consistent with the notion that undifferentiated F9 EC cells contain a cellular repressor that blocks the polyoma virus enhancer and that this repressor has the same target sequence as the E1A proteins; however, the polyoma story becomes even more complex. The polyoma virus is normally not permissive for replication in most lymphoid lines. Deletion of the PvuII D fragment (spanning the F441 point mutation) facilitated replication in some T cells and mastocytoma cell lines; therefore, PvuII D can act as both a positive and a negative regulatory element. Substitution of the MuLV enhancer facilitates replication in 3T6 and B lymphoid cells. Substitution of the immunoglobulin heavy-chain enhancer facilitates replication in B lymphoid cells but not 3T6 cells or mastocytomas. Curiously, substitution with the SV40 enhancer does not allow for replication in any cell line tested.[16]

[14] M. Satake, K. Furukawa, and Y. Ito, J. Virol. 62, 970 (1988).
[15] R. Hen, E. Borrelli, C. Fromental, P. Sassone-Corsi, and D. Chambon, Nature (London) 321, 249 (1986).
[16] B. A. Campbell and L. P. Villarreal, Mol. Cell. Biol. 8, 1993 (1988).

Retroviruses carry potent transcriptional enhancers as well.[17-19] I will consider the enhancer elements of the murine leukemia and sarcoma viruses simultaneously because it makes for a more interesting story when one compares their activities. Not surprisingly, the type of tissue in which these retroviruses induce disease often, although not always, reflects the cell-type specificity of the enhancer elements resident in *both* long terminal repeats (LTRs) of the integrated provirus. The enhancers in these viruses are quite similar to one another, so mix-and-match experiments serve to reveal the nucleotide differences that determine tissue-type expression specificity. The transcriptional enhancers of Moloney murine leukemia virus (MuLV) and Moloney murine sarcoma virus (MuSV) exhibit different cell-type expression specificities from that of Friend murine leukemia virus. While the three enhancers are approximately equally active in erythroid cells, the Moloney MuSV and MuLV enhancers are 20- to 40-fold more active than the Friend MuLV enhancer in T lymphoid cells. There appears to be an element, repeated several times within the enhancer, which modulates the activity of the enhancer in T cells without affecting it in erythroid cells, and thus appears to be one of the determinants of the tissue specificity of the enhancer.[20,21]

Enhancer elements within the Moloney MuSV, and the nonleukemogenic (AKV) and T cell leukemogenic (SL3-3) murine retroviruses also exhibit strong cell-type preference in transcriptional activity. These elements are additionally regulated by the glucocorticoid dexamethasone. Mapping studies in combination with DNase I footprinting experiments define the presence of glucocorticoid regulatory elements at the promoter proximal ends of each enhancer repeat. These elements behave like inducible enhancers. Their regulatory activity is independent of position and orientation when they are linked in cis to a heterologous promoter. The sequences required for dexamethasone regulation for both AKV and SL3-3 include a 17-nucleotide consensus sequence termed the glucocorticoid response element (GRE), which is located at the promoter proximal ends of each enhancer repeat. Although the GREs are identical for these enhancers, the sequences surrounding these elements differ.[22-24]

[17] M. Kriegler and M. Botchan, *in* "Eukaryotic Viral Vectors" (Y. Gluzman, ed.), p. 11. Cold Spring Harbor Laboratory, Cold Spring Harbor, New York, 1982.
[18] M. Kriegler and M. Botchan, *Mol. Cell. Biol.* **3,** 325 (1983).
[19] B. Levinson, G. Khoury, G. VanDeWoude, and P. Gruss, *Nature (London)* **295,** 79 (1982).
[20] H. J. Thiesen, Z. Bosze, L. Henry, and P. Charnay, *J. Virol.* **62,** 614 (1988).
[21] Z. Bosze, H. J. Thiesen, and P. Charnay, *EMBO J.* **5,** 1615 (1986).
[22] D. Celander and W. A. Haseltine, *J. Virol.* **61,** 269 (1987).
[23] R. Miksicek, A. Heber, W. Schmid, U. Danesch, G. Posseckert, M. Beato, and G. Schutz, *Cell* **46,** 203 (1986).
[24] D. Celander, B. L. Hsu, and W. A. Haseltine, *J. Virol.* **62,** 1314 (1988).

Our experiments with both infected and transfected retroviral genomes have led us to some interesting observations regarding promoter and enhancer strength in specific cell types. First, we observed that infected genomes, be they of DNA or RNA viruses, express their gene products at substantially higher levels than do the same transfected genomes present in the same copy number. Second, comparison of the synthetic capability of wild-type SV40 virus and an engineered murine retrovirus (MuLV) expressing SV40 T antigen revealed that, in mouse fibroblasts, an engineered SV40 retrovirus could produce approximately 10× the amount of T antigen as its wild-type SV40 equivalent. This observation seems to hold true for a variety of structural genes that we have analyzed.[17,18,25-29]

The papillomaviruses, both human and bovine, induce tissue-specific diseases in their hosts and have complex enhancer architecture. Cells transformed by these viruses maintain the viral DNA as nuclear plasmids. Human papillomavirus type 18 contains three different enhancer domains, two of which are inducible and one of which is constitutive. The inducible enhancers are responsive to papillomavirus-encoded trans-acting factors while the constitutive enhancer requires only cellular factors for activity. The inducible enhancer IE2 is located proximal to the E6 cap site and responds to the E2 trans-activator. Inducible enhancer IE6 is located 500 base pairs upstream of the E6 cap and responds to the E6 gene product. The third enhancer, which is constitutive, lies between the two inducible enhancers. Each enhancer functions independently of the other and may function at different stages of the viral life cycle.[30] Human papillomavirus 16 contains a keratinocyte-dependent enhancer in sequences 51 to the P97 coding region. The E2 trans-activator of HPV16 as well as that of the related bovine papillomavirus further enhances HPV16 transcription. Similarly, the "Short E2" (SE2) gene products of both viruses repress E2 activation and suppress activation in keratinocytes.[31] Furthermore, Gloss et al.[32] report that the region upstream of P97 in HPV16 functions as a GRE and shares sequence homology to the consensus GRE. Elements of

[25] M. Kriegler, C. Perez, and M. Botchan, in "Gene Expression" (D. Hamer and M. Rosenberg, eds.), p. 107. Alan R. Liss, New York, 1983.

[26] M. Kriegler, C. Perez, C. Hardy, and M. Botchan, Cell 38, 483 (1984).

[27] M. Kriegler, C. Perez, K. DeFay, I. Albert, and S. D. Liu, Cell 53, 45 (1988).

[28] M. Kriegler, C. Perez, D. DeFay, I. Albert, and S. D. Liu, in "Cytotoxic and Growth Inhibitory Peptides," p. 203. Alan R. Liss, New York, 1989.

[29] K. Choi, C. Chen, M. Kriegler, and I. B. Roninson, Cell 53, 519 (1988).

[30] D. Gius, S. Grossman, M. A. Bedell, and L. A. Laimins, J. Virol. 62, 665 (1988).

[31] T. P. Cripe, T. H. Haugen, J. P. Turk, F. Tabatabai, P. G. Schmid III, M. Durst, L. Gissmann, A. Roman, and L. P. Turek, EMBO J. 6, 3745 (1987).

[32] B. Gloss, H. U. Bernard, K. Seedorf, and G. Klock, EMBO J. 6, 3735 (1987).

the enhancer behavior described above are shared by HPV types 1, 66, 7, and 11 as well as cottontail rabbit papillomavirus.[33]

Another potentially tissue-specific viral enhancer has been described in hepatitis B virus. This enhancer is located 3' to the hepatitis B surface antigen coding sequences but is contained within the mature viral transcripts.[34] Vannice and Levinson[35] have reported that the HBV enhancer can dramatically increase expression levels of genes controlled by the SV40 enhancer/promoter, but only when the enhancer is located within the transcribed region of the gene. Further, this effect appears to be orientation dependent, a violation of enhancer rules. Lastly, these authors report that when this enhancer is located either 5' to the promoter or in the 3' untranslated region of a recombinant cistron, its enhancing spectrum is not cell type specific. This finding is in striking contradiction to the report of Jameel and Siddiqui,[36] who report that the enhancer displays strict host and tissue specificity in that it is functional only in human liver cells. This specificity requires trans-acting factors, potentially those described by Shaul and Ben-Levy.[37] Additional studies support the contention that the X gene product of hepatitis B virus can trans-activate the HBV enhancer.[38]

The human immunodeficiency viruses (HIV) exhibit strikingly complex regulatory strategies including virally encoded trans-activating proteins (tat) and their cis-acting element congeners (tar). Most surprising is that the cis-acting elements (tar) with which these factors interact are located downstream of the cap site and do not function in an orientation-independent manner and do not, on their own, activate promoters lacking their own enhancers. They are not categorized formally as enhancers and therefore are not included in this discussion. Nevertheless, HIV does carry an enhancer, upstream of the SP1 binding sites within the *U3* region between nucleotides -137 and -17. The enhancer is nonspecific in that it can be replaced by either the RSV or SV40 enhancer.[39] Viral enhancers have been identified in adenovirus,[40] Epstein-Barr virus,[41] gibbon ape leukemia virus,[42] and *Herpesvirus saimiri*[43] as well.

---

[33] H. Hirochika, T. R. Browker, and L. T. Chow, *J. Virol.* **61,** 2599 (1987).
[34] G. A. Bulla and A. Siddiqui, *J. Virol.* **62,** 1437 (1988).
[35] J. L. Vannice and A. D. Levinson, *J. Virol.* **62,** 1305 (1988).
[36] S. Jameel and A. Siddiqui, *Mol. Cell. Biol.* **6,** 710 (1986).
[37] Y. Shaul and R. Ben-Levy, *EMBO J.* **6,** 1913 (1987).
[38] D. F. Spandau and C. H. Lee, *J. Virol.* **62,** 234 (1988).
[39] C. A. Rosen, J. G. Sodroski, and W. A. Haseltine, *Cell* **41,** 813 (1988).
[40] P. Hearing and T. Shenk, *Cell* **45,** 229 (1986).
[41] D. Reisman and B. Sugden, *Mol. Cell. Biol.* **6,** 3838 (1986).
[42] N. Holbrook, A. Gulino, and F. Ruscetti, *Virology* **157,** 211 (1987).
[43] S. Schirm, F. Weber, W. Schaffner, and B. Fleckstein, *EMBO J.* **4,** 2669 (1985).

Cellular Enhancers

The initial demonstration of enhancers of cellular genes has led to the identification of a great number of cellular enhancers. For the purposes of these discussions they will be divided into two groups, immune function genes and other genes. The immune function genes include those encoding the immunoglobulin heavy and light chains, the histocompatibility genes, the T cell receptor, and the interferon β genes. The other group includes humans β-actin, muscle creatine kinase, prealbumin, elastase I, metallothionein, collagenase, α-fetoprotein, γ-globin and β-globin, c-*fos*, c-*ras*, and insulin.

A substantial amount of time has been spent on developing an understanding of how expression of immune function genes is regulated. These studies have led to the identification of cellular enhancers and the subsequent identification of the trans-acting factors with which they interact, in particular, those of the immunoglobulin heavy-chain and light-chain genes. Simultaneous publication by two groups of the existence of a tissue-specific enhancer in the immunoglobulin heavy-chain genes indicated that the enhancer resided in an intron, downstream of the joining region of the heavy-chain genes.[44,45] Concurrently, an enhancer was identified in the major intron of the immunoglobulin κ (light-chain) genes.[46,47] These enhancers are interesting for at least three reasons: (1) They are located within introns, (2) they appear to be lymphoid specific, and (3) enhancer activity may be only transiently required in the case of the κ gene.[48] The heavy-chain enhancer is considerably stronger than the κ light-chain enhancer.[47] The light-chain enhancer has been further mapped by Queen and Baltimore.[49] The tissue specificity of the heavy-chain enhancer appears to be determined in part by two distinct negative regulatory elements in and around the heavy-chain enhancer, such that, in nonlymphoid cells, activation by ubiquitous transcription factor is repressed.[50] It has been shown that removal of this repressive sequence allows the heavy-chain enhancer to function in fibroblasts, cells in which the wild-type enhancer does not normally function.[51] In fact, in the heavy-chain enhancer, three different domains serve to alter the expression of the immunoglobulin heavy-chain gene in a tissue-specific manner. They are the promoter, an enhancer, and

---

[44] J. Banerji, L. Olson, and W. Schaffner, *Cell* **35**, 729 (1983).
[45] S. D. Gilles, S. L. Morris, V. T. Oi, and S. Tonegawa, *Cell* **33**, 717 (1983).
[46] C. Queen and D. Baltimore, *Cell* **35**, 741 (1983).
[47] D. Picard and W. Schaffner, *Nature (London)* **307**, 83 (1984).
[48] M. L. Atkinson and R. P. Perry, *Cell* **48**, 121 (1987).
[49] C. Queen and D. Baltimore, *Cell* **35**, 741 (1983).
[50] J. L. Imler, C. Lemaire, C. Wasvlyk, and B. Waslyk, *Mol. Cell. Biol.* **7**, 2558 (1987).
[51] J. Weinberger, P. S. Jat, and P. A. Shart, *Mol. Cell. Biol.* **8**, 988 (1988).

a third element.[52] Polymerization of the entire enhancer results in an increase in overall transcription. Polymerization of defined enhancer DNA segments and subsequent analysis revealed two functional domains within the enhancer. In fact, greater than wild-type activity can be obtained through polymerization of these subenhancers. One domain contains three regions thought to be involved in protein binding (E1, E2, E3), the other domain contains the fourth E Motif (E4) and the conserved oligonucleotide ATTTGCAT.[53] Each element is necessary but insufficient to direct high-level gene expression. Although there exists an abundance of evidence suggesting that enhancer binding proteins exist and are important for the immunoglobulin enhancer to manifest its activity, there is some evidence to suggest that the immunoglobulin enhancer is not merely an entry site for RNA polymerase or a transcription complex. Atkinson and Perry[8] dramatically demonstrated that tandem immunoglobulin promoters are equally active in the presence of the κ enhancer even when the promoters are located 7 kb from the enhancer. Transgenic mice that bear a recombinant cistron containing the c-myc oncogene driven by a immunoglobulin κ enhancer develop a fatal lymphoma within a few months of birth. The observed lymphomas are of B cell origin and represent both immature and mature B cells.[54] As such, the immunoglobulin enhancers achieve their dramatic cell-type specificity through a combination of both positive and negative, so-called "silencing," enhancer elements.

   An enhancer of the α chain of the human T cell receptor has been localized to a 1.1-kb BamHI–HindIII fragment located 5′ to the first exon of the C α gene. This enhancer's activity is specific for lymphoid cells of both B and T cell origin.[55]

   The HLA DQ α and DQ β genes have also been shown to possess enhancer elements. Two regions have been demonstrated in DQ α and one in DQ β.[56]

   The interferon β gene contains a virus-inducible enhancer between nucleotides −65 and −109 relative to the cap site. This region is composed of a series of tandemly repeated 6-bp segments. Synthetic oligomers of the repeats also function as virus-inducible enhancers.[57] Deletion analysis within the enhancer region indicates that this enhancer is under negative control. Deletion of sequences from the 3′ end of the enhancer leads to a

[52] R. Grosschedl and D. Baltimore, Cell 41, 885 (1985).

[53] M. Kiledjian, L. K. Su, and T. Kadesch, Mol. Cell. Biol. 8, 145 (1988).

[54] J. M. Adams, A. W. Harris, C. A. Pinker, L. M. Corcoran, W. S. Alexander, S. Cory, R. D. Palmiter, and R. L. Brinster, Nature (London) 318, 533 (1985).

[55] S. Luria, G. Gross, M. Horowitz, and D. Givol, EMBO J. 6, 3307 (1987).

[56] K. M. Sullivan and B. M. Peterlin, Mol. Cell. Biol. 7, 3315 (1987).

[57] T. Fukita, H. Shibuya, H. Hotta, K. Yamanishi, and T. Taniguchi, Cell 49, 357 (1987).

dramatic increase in the basal level of interferon β mRNA and a corresponding decrease in the induction ratio. The remaining 5' region of the enhancer serves as a strong constitutive expression element. Thus, in the case of interferon β, depression of a constitutive transcription element appears to play a key role in the regulation of the expression of this gene.[58]

The second set of cellular genes is more diverse and includes genes encoding structural proteins, cellular enzymes, and secretory proteins.

β-Actin is one of the most abundant proteins in eukaryotic cells and is expressed in a variety of cells. Kawamoto *et al.* constructed a series of deletion mutants in an attempt to localize the β-actin enhancer.[59] They found that the first of five introns contains enhancer activity. Their results indicate that enhancer activity resides between positions $+770$ and $+793$ in the β-actin gene and contains a so-called enhancer "core." The β-actin promoter–enhancer assembly functions well in recombinant cistrons and when driving the expression of the *neo*^{r} gene. In transfection experiments, β-actin *neo*^{r} constructions consistently yield substantially greater numbers of stable transformants than do equivalent constructions driven by either the MuLV LTR or the SV40 promoter (M. Kriegler and C. Perez, unpublished observations, 1986).

Muscle creatine kinase (MCK) is induced to high levels during skeletal muscle differentiation. The MCK gene contains multiple upstream regulatory elements, including one element, located between 1.0 and 1.256 kb upstream of the transcription start site that has the properties of a transcriptional enhancer. It is not clear, however, if this enhancer confers cell-type expression specificity on the MCK gene. It appears that, even in the absence of the enhancer, low-level expression from the MCK promoter retains differentiation specificity.[60]

Cell-specific expression of the transthyretin (prealbumin) gene has been localized to two upstream elements. One of these elements, located between 1.96 and 1.86 kb 5' to the mRNA cap site, behaves as an enhancer element. This element appears to be cell specific because it can stimulate the β-globin promoter in HEP G2 cells but not in HeLa cells.[61]

The elastase I gene promoter and enhancer have been analyzed and it is clear that the elastase enhancer, located between positions -205 and -73, exhibits tissue-type expression specificity for pancreatic acinar cells. The elastase enhancer can activate both the metallothionein and human growth

[58] S. Goodburn, H. Burtein, and T. Maniatis, *Cell* **45**, 601 (1986).
[59] T. Kawamoto, K. Makino, H. Niw, H. Sugiyama, S. Kimura, M. Anemura, A. Nakata,and T. Kakunaga, *Mol. Cell. Biol.* **8**, 267 (1988).
[60] J. B. Jaynes, J. E. Johnson, J. N. Buskin, C. L. Gartside, and S. D. Hauschka, *Mol. Cell. Biol.* **8**, 62 (1988).
[61] R. H. Costa, E. Lai, D. R. Grayson, and J. E. Darnell, *Mol. Cell. Biol.* **8**, 81 (1988).

hormone (hGH) promoters. Combinations of immunoglobulin and elastase enhancers with a heterologous promoter result in expression in all of the tissues predicted by the sum of each enhancer acting alone. Thus, these enhancers act independently of each other and, unlike the immunoglobulin genes, appear to lack silencing activity in cell types in which they normally do not function.[62]

The human metallothionein IIA gene contains two enhancer elements whose activity is induced by a variety of heavy metal ions. Deletion of the metal-responsive elements of the enhancers has no effect on the basal activity of the enhancer but prevents further induction by heavy metal ions. Replacement of the basal level enhancer element with a DNA linker served to inactivate the enhancer both before and after induction, leading to the conclusion that the metal-responsive elements act as positive regulators of metallothionein enhancer activity.[63]

TPA-induced expression of the human collagenase gene is mediated by an enhancer element located 5′ to the mRNA start site. The 32-bp enhancer is located between positions -73 and -42 relative to the transcription start site. When inserted next to the herpes simplex virus thymidine kinase promoter, the new chimeric enhancer–promoter combination can be induced by TPA.[64] Transgenic mice were utilized to identify an enhancer element in the albumin gene responsible for efficient, liver-specific gene expression. This enhancer does function in an orientation-independent manner but does not function well with a heterologous promoter — a phenomenon similar to some other enhancer–promoter pairs as well.[65]

The regulatory region of the α-fetoprotein gene contains at least three distinct enhancer elements. Each enhancer directs gene expression in the appropriate tissues, including the visceral endoderm of the yolk sac, the fetal liver, and the gastrointestinal tract. The DNA sequence elements responsible for directing the activation, repression, and reinduction of transcription are contained within, or 5′ to, the gene itself.[66] These enhancers have been fine structure mapped to regions 200 to 300 bp in length that are 2.5, 5.0, and 6.5 kb upstream of the transcription start site.[67]

Another embryonic gene product, human γ-globin, carries an enhancer

[62] D. M. Ornitz, R. E. Hammer, B. L. Davison, R. L. Brinster, and R. D. Palmiter, *Mol. Cell. Biol.* 7, 3466 (1987).

[63] M. Karin, A. Haslinger, A. Heguy, T. Dietlin, and T. Cooke, *Mol. Cell. Biol.* 7, 606 (1987).

[64] P. Angel, I. Bauman, B. Stein, H. Delius, H. J. Rahmsdorf, and P. Herrlich, *Mol. Cell. Biol.* 7, 2256 (1987).

[65] C. A. Pinkert, D. M. Ornitz, R. L. Brinster, and R. D. Palmiter, *Genes Dev.* 1, 268 (1987).

[66] R. E. Hammer, R. Krumlauf, S. A. Camper, R. L. Brinster, and S. M. Tilghman, *Science* 235, 53 (1987).

[67] R. Godbout, R. S. Ingram, and S. M. Tilghman, *Mol. Cell. Biol.* 8, 1169 (1988).

at the 3' end of the gene approximately 400 bp downstream from the polyadenylation site. This enhancer is less than 750 bp in length and behaves in a nontissue-specific manner in tissue culture cells.[68] However, in transgenic mice, the G γ-globin gene is active only in embryonic erythroid cells, while the β-globin gene is expressed only in fetal and adult erythroid cells. There exists an enhancer at the 3' end of the β-globin gene that, when attached to the G γ-globin gene in either orientation, activates transcription of the otherwise silent G γ-globin gene in fetal liver. This enhancer is within the region 600–900 bp 3' to the β-globin polyadenylation signal and obviously contributes to the stage-specific expression of the β-globin gene.[69]

A serum-responsive enhancer element has been found associated with the c-*fos* oncogene. The c-*fos* transcriptional enhancer is located between −332 and −276 bp relative to the mRNA cap site and does not appear to be tissue specific in nature.[70,71] The function of this so-called serum response element (SRE) cannot be replaced by either the SV40 or the Moloney MuLV enhancers.

An enhancer that lies 3' to the human c-HA-*rasI* gene has been identified as well. This enhancer is a repetitive sequence element, variants of which have been associated with malignant disease. This element can function in a somewhat position- and orientation-independent manner.[72]

Two regulatory elements, one of which functions as an enhancer, have been identified 5' to the rat insulin gene. The enhancer activity is spread over a region (−103 to −333). Internal portions of the enhancer elements (−159 to −249), which showed only low activity, stimulated full activity when duplicated. The activity of the insulin enhancer appears to be confined to insulin-producing cells.[73]

This is but a partial listing of the viral and cellular enhancers identified to date. New elements are described monthly, and the identification of their trans-acting enhancer binding protein congeners follows shortly thereafter. What should be apparent from this listing is that enhancers are highly varied and function in a variety of ways. The identification of an ever-increasing variety of enhancer elements places in the hand of the experimentalist the ability to designate the cell or tissue type in which an experimental gene will be expressed.

[68] D. M. Bodine and T. J. Ley, *EMBO J.* **6**, 2997 (1987).
[69] M. Trudel and F. Constantini, *Genes Dev.* **1**, 954 (1987).
[70] R. Treisman, *Cell* **42**, 889 (1985).
[71] J. Deschamps, F. Meijlink, and I. M. Verma, *Science* **230**, 1174 (1985).
[72] J. B. Cohen, M. V. Walter, and A. D. Levinson, *J. Cell. Physiol.* **5**, 75 (1987).
[73] T. Edlund, M. D. Walker, P. J. Barr, and W. J. Rutter, *Science* **230**, 912 (1985).

Are Introns Required for Efficient Gene Expression?

Yes and No. One of the least ballyhooed but most incredible observations about factors affecting the efficacy of recombinant cistrons is experimental evidence that indicates that the requirement for a functional intron in a recombinant cistron is promoter dependent.

In transfection assays of recombinant cistrons, an intron is required when transcription is driven by an immunoglobulin μ promoter–enhancer combination, although the requirement is not specific for a particular intron. No intron is required for efficient expression from either the cytomegalovirus or heat-shock promoters[74] nor from murine retroviral. We have in fact found that the presence of splice donors and acceptors in recombinant retroviruses can substantially reduce the yield of both intact retroviral genomes as well as correct translational templates.[27,28] β-Globin gene expression driven by the SV40 promoter appears to require sequence information from intervening sequences as well; however, this intervening sequence information cannot support the stable production of β-globin mRNA when placed downstream of the polyadenylation site. Further, sequences derived from the transcribed herpes simplex virus thymidine kinase gene can support substantial intron-independent expression, a greater than 100-fold increase, of a β-globin cDNA whose expression is driven by the SV40 early promoter.[75]

The ramifications of intron inclusion appear vitally important in determining the level of accumulation of mRNAs in exogenous cistrons introduced into transgenic mice. Brinster et al. examined the behavior of enhancer–promoter–structural gene chimeras by fusing either the mouse metallothionein I or the rat elastase I enhancer–promoter to the rat growth hormone gene.[76] The mouse metallothionein I–growth hormone chimera was assayed in fetal liver. The rat elastase I–growth hormone chimera was assayed in the pancreas. The behavior of a mouse metallothionein gene marked with a synthetic oligonucleotide and the human β-globin gene, also introduced into transgenic mice, was examined in parallel. In each case, there was 10- to 100-fold more mRNA produced from the intron-containing constructs. Most surprising perhaps is the observation that both the intron minus and intron plus mouse metallothionein constructs function equally well when transfected into tissue culture cells. This inexplicable observation strongly suggests that introns play a role in facilitating tran-

[74] M. S. Nueberger and G. T. Williams, Nucleic Acids Res. 16, 6713 (1988).
[75] A. R. Buchman and P. Berg, Mol. Cell. Biol. 8, 4395 (1988).
[76] R. L. Brinster, J. M. Allen, R. R. Behringer, R. E. Gelinas, and R. D. Palmiter, Proc. Natl. Acad. Sci. U.S.A. 85, 836 (1988).

scription of microinjected genes and that this effect may be manifest only when those genes have been placed in a developmental setting.

Related Phenomena and Practical Applications

These data make one point very clear. Despite our initial prejudice to the contrary, enhancers, promoters, and even splice signals do not function as fully independent, interchangeable elements. In many cases, appropriate temporal and tissue-specific expression is dependent on the assembly of an appropriate expression ensemble: enhancer with promoter and promoter with splice signal. As such, attention to these details in the construction of a recombinant cistron is critical in achieving the desired biochemical function.

The initial discovery of enhancers was motivated, in part, by the search for eukaryotic origins of replication as cis-acting elements that could enhance the transforming efficiency of a marker gene. In fact, numerous studies link enhancer function with DNA replication. Several recent studies of the effects of DNA replication on transcription and presumably enhancer function indicate that such effects are cell specific.[77-79] Such effects should not be ignored by the researcher.

Other factors may also affect the function of a recombinant cistron introduced into cells and animals. For instance, there exists a body of evidence that the pattern of DNA methylation may affect developmental regulation of enhancer function within retroviral genomes integrated into the germ line of mice[80] as well as the developmental regulation of the immunoglobulin κ gene in tissue culture.[81] In addition, the site of integration of the recombinant cistron into the genome of the host cell can dramatically alter the "expressibility" of that cistron. This phenomenon is referred to as position effect. Recent experimental evidence indicates that such random position effects can be overcome by exploiting a methodology that facilitates targeted integration to the recombinant molecule into the host genome through homologous recombination. In such experiments the neo^r gene, driven by the SV40 promoter, was homologously recombined into the chromosomal β-globin locus. Postintegration, the foreign cistron was subject to the same transcriptional inducers as the natural β-globin allele.[82] Implementation of this technology should enable the

[77] D. S. Grass, D. Read, E. D. Lewis, and J. L. Manley, *Genes Dev.* **1**, 1065 (1987).

[78] T. Enver, A. C. Brewer, and R. K. Patient, *Mol. Cell. Biol.* **8**, 1301 (1988).

[79] B. A. Campbell and L. P. Villarreal, *Mol. Cell. Biol.* **6**, 2068 (1986).

[80] D. Jahner and R. Jaenisch, *Mol. Cell. Biol.* **5**, 2212 (1985).

[81] D. E. Kelley, B. A. Pollok, M. L. Atchison, and R. P. Perry, *Mol. Cell. Biol.* **8**, 930 (1988).

[82] A. K. Nandi, R. S. Roginski, R. G. Gregg, O. Smithies, and A. I. Skoultchi, *Proc. Natl. Acad. Sci. U.S.A.* **85**, 3845 (1988).

experimentalist to retain authentic biological regulation of a recombinant cistron postintegration.

What is the molecular basis of the enhancer effect? We know that enhancers dramatically affect transcription even when they are placed at great distances from their cognate promoters. We also know that, in many cases, enhancer multimers function more efficiently than the same enhancer when present only as a monomer.[10] Enhancer binding proteins have been isolated and we know, through the conduct of cis/trans tests, that these sequence-specific enhancer binding proteins are required for efficient enhancer function. We can assume that a multimeric enhancer binds more enhancer binding proteins than does the corresponding monomer. The simplest explanation for the action at a distance enhancer effect is that the enhancer binding protein–DNA complex directly interacts with the transcription complex, including RNA polymerase and other promoter binding proteins, at the promoter. The simplest mechanism by which this could be accomplished is to bend the DNA into a loop structure in a manner such that the enhancer binding complex is brought into direct physical contact with the appropriate promoter binding complex. Such an interaction might allow for the formation of a more stable, more easily regulatable multimolecular enhancer–promoter binding complex. As for the molecular mechanism of promoter–intron cross talk, that remains a mystery.

[41] Methods for Introducing DNA into Mammalian Cells

By WAYNE A. KEOWN, COLIN R. CAMPBELL, and RAJU S. KUCHERLAPATI

The ability to introduce defined genetic information into mammalian cells by various methods has revolutionized various aspects of the study of gene structure–function relationships. These methods can be classified as direct or indirect methods. Direct methods involve the introduction of genetic material (usually DNA) into the nucleus of a somatic cell, or the male pronucleus of a fertilized egg, by microinjection. Indirect methods involve the active or passive uptake of the genetic information by the cell which is to be transfected. The indirect methods most commonly used include delivery of genetic information by viral vectors, formation of complexes between DNA and chemical agents followed by active cellular uptake, or physical methods of introduction of nucleic acids into cells. The use of viral vectors, particularly that of viral RNA vectors, has become quite popular and has been dealt with extensively in recent articles.[1-3] The discussion of procedures in this article is restricted to direct methods and to physical or chemical-mediated indirect methods for gene transfer. Chemical-mediated procedures discussed herein include formation of complexes of DNA with calcium phosphate, DEAE-dextran, Polybrene, or natural or synthetic cationic lipids. Physical methods include electric field-mediated DNA transfer (electroporation) and a microprojectile method in which DNA packaged into a "projectile" is "fired" into cells. Detailed protocols are described for some of the more popular procedures. The relative advantages of other procedures are also discussed. Specific protocols for these less widely used methodologies may be obtained from the references cited for each procedure.

Calcium Phosphate Coprecipitation

One of the most commonly used methods of introducing DNA into mammalian cells is to coprecipitate the DNA with calcium phosphate and present the mixture to cells. The technique was originally used to increase the infectivity of adenoviral DNA,[3] and was made popular by Wigler and colleagues[4] and Maitland and McDougall.[5] Since the methods were originally described a number of permutations on the basic theme have been described. All of these procedures involve the mixing of purified DNA with buffers containing phosphate and calcium chloride which results in the formation of a very fine precipitate, and the presentation of this mixture to cells in culture. A typical protocol for cells which grow attached to a substratum follows:

1. On day 1, seed $2-3 \times 10^4$ cells/cm^2 in normal growth medium. Allow the cells to attach. At the time of transfection the cells should be 80–90% confluent.

2. On day 2, prepare the DNA–calcium phosphate coprecipitate:

DNA	x μl
TE	$440 - x$ μl
2× HBS	500 μl
2 M CaCl$_2$	60 μl

TE: 10 mM Tris, 1 mM EDTA pH 8.0

[1] H. Temin, in "Gene Transfer" (R. Kucherlapati, ed.), p. 149. Plenum, New York, 1986.
[2] V. R. Baichwal and B. Sugden, in "Gene Transfer" (R. Kucherlapati, ed.), p. 117. Plenum, New York, 1986.
[3] F. L. Graham and A. J. Van Der Eb, Virology 52, 456 (1973).
[4] M. Wigler, S. Silverstein, L. S. Lee, A. Pellicer, V. C. Cheng, and R. Axel, Cell 11, 223 (1977).
[5] N. J. Maitland and J. K. McDougall, Cell 11, 233 (1977).

2X HBS: Hanks' balanced salts, 1.4 mM Na$_2$HPO$_4$, 10 mM KCl, 12 mM glucose, 275 mM NaCl, and 40 mM HEPES, pH 6.95
2 M CaCl$_2$: calcium chloride in 10 mM HEPES, pH 5.8
Mix and allow to stand at room temperature for 30 min.

3. Remove medium from cells and replace with fresh medium (60-mm dish: 3.5 ml, 100-mm dish: 9 ml).

4. Mix the precipitate gently by shaking or pipetting and add directly to the medium in dishes containing cells (0.5 ml/60-mm dish, 1 ml/100-mm dish).

5. Incubate the cells at 37° for 4 hr.

6. Remove the medium containing the precipitate. Add 2 ml of 10-20% (w/v) dimethyl sulfoxide (DMSO) in 1 × HBS. After 2 min, add 4 ml of serum-free medium to each dish. Aspirate the mixture, wash the cell layer twice with serum-free medium, and add medium (4 ml/60-mm dish, 10 ml/100-mm dish). Incubate overnight at 37°.

7. Trypsinize the cells and split the contents of each plate into 3-4 fresh plates. If selection is applied for stable transfectants, selective medium can be used at this time or a day later.

Notes. (1) The quality of the cells is important. Use rapidly dividing cultures for experiments. (2) The pH of the 2× HBS used to prepare the precipitate is critical. (3) The quality of the precipitate that is optimal for transfection can be assessed visually. It should be a hazy white suspension. If DNA precipitates to the bottom of the tube, repurify the DNA by protease digestion and phenol extraction before preparing another precipitate. (4) The time of incubation of the cells with the precipitate can be varied. Four hours is usually the minimum time. (5) Treatment of cells with DMSO or glycerol facilitates uptake of the DNA. The concentrations of these chemicals (usually 10-20%) and the time of treatment that are optimal vary from cell line to cell line. Use a concentration that is sublethal to cells. The time for which the cells are exposed to the chemical is crucial. If they are allowed to stay too long, cell viability may be dramatically reduced. If the cells are overly sensitive to these chemicals, they can be left with the DNA precipitate for 8-24 hr followed by a serum-free medium wash, omitting the chemical shock.

Using a variety of different cell types, we have obtained transfection efficiencies of up to 10^{-3}.

A modified calcium phosphate transfection protocol which yielded as much as 10-15% or more stable tranfectants has been described by Chen and Okayama.[6] They indicate that the following factors played a role in the increased transfection efficiencies:

[6] C. Chen and H. Okayama, *Mol. Cell. Biol.* **7**, 2745 (1987).

1. Temperature and CO_2 levels: The precipitate is allowed to form as the cells are incubated at $35°$ in $2-4\%$ CO_2 for $15-24$ hr.

2. DNA concentration: A much finer precipitate is observed when the concentration of DNA is at $20-30$ $\mu g/ml$ than at other DNA concentrations.

3. DNA form: Circular DNA is far more active in this protocol than is linear DNA.

4. Nature of vector: A promoter–enhancer system which permits high levels of expression of the gene is found to be an efficient vector system.

No chemical shock is employed in this protocol.

DEAE-Dextran-Mediated DNA Transfection

DEAE-dextran, which was initially used to increase viral infectivity of cells, was modified for transfection of DNA into mammalian cells. This method is ideally suited for experiments in which transfection is used for transient biological assays. The procedure is simple and efficient, but has not been shown to result in stable transfection.[7,8] As with calcium phosphate coprecipitation procedures, many variations on the procedure have evolved. The basic protocol used in our laboratory is a variation of the procedures of Lopata et al.[9] and Sussman and Milman.[10] It is described below.

1. On day 1, seed cells to be tranfected into tissue culture plates at a cell density of $13,000/cm^2$, or at a density such that the monolayer will be about 90% confluent by the next day.

2. On day 2, to a sterile tube add DNA (usually $5-50$ μg), DEAE-dextran (MW = 500,000) from a sterile 100 mg/ml stock, and serum-free medium and mix. The final concentration of DEAE-dextran should be 200 $\mu g/ml$ and the final concentration of DNA should be $0.5-6.5$ $\mu g/ml$.

3. Remove medium from cells and wash once with serum-free medium. Add DNA–DEAE-dextran mixture directly onto the cell monolayer (50 μl for each cm^2 of monolayer surface). Incubate at $37°$ for 4 hr.

4. Remove DNA–DEAE-dextran mixture from dishes and wash once with serum-free medium. Shock with 10% DMSO in $1\times$ HBS for 2 min. Wash with an equal volume of serum-free media three times. Add normal growth medium and incubate at $37°$.

5. Assay $48-72$ hr after transfection.

[7] J. H. McCutchan and J. S. Pagano, *J. Natl. Cancer Inst.* **41**, 351 (1968).

[8] G. Milman and M. Herzberg, *Somatic Cell Genet.* **7**, 161 (1981).

[9] M. A. Lopata, D. W. Cleveland, and B. Sollner-Webb, *Nucleic Acids Res.* **12**, 5707 (1984).

[10] D. J. Sussman and G. Milman, *Mol. Cell. Biol.* **4**, 1641 (1984).

Notes. (1) Because of the rather small volume of DNA–DEAE-dextran used, the plates should be incubated in a humidified atmosphere. (2) As with calcium phosphate transfections, the sensitivity of cells to DMSO or glycerol must be determined empirically. (3) This method has been reported to yield transfection frequencies as high as 80%. (4) This method is not useful to isolate stable transfectants. (5) DNA introduced into cells by this method and by calcium phosphate coprecipitation is subject to rather high rates of mutation.[11–13]

Electric Field-Mediated DNA Transfection (Electroporation)

Electric field-mediated DNA transfection, commonly called electroporation, is rapidly becoming the method of choice for many transfection applications. Advantages presented by this procedure include ease of operation, reproducibility of conditions, applicability to cells which grow attached or in suspension, utility for both stable and transient transfection procedures, and the capacity to control the copy number of transfected DNA molecules.

When membranes are subjected to an electric field of sufficiently high voltage, regions of the membrane undergo a reversible breakdown, resulting in the formation of pores large enough to permit the passage of macromolecules. Unlike calcium phosphate coprecipitation, in which DNA entering the cell is taken up into phagocytic vesicles,[14] electroporated DNA remains free in the cytosol and nucleoplasm.[15] Perhaps because of this difference, lower mutation frequencies are often observed for electroporated DNA than for DNA introduced into cells by calcium phosphate coprecipitation.[16]

Many cell types which are resistant to transfection by other procedures are readily transfected by electroporation, including lymphocytes,[17] hematopoietic stem cells,[18] and rat hepatoma cells.[19] By altering the parameters

[11] M. P. Calos, J. S. Lebkowski, and M. R. Botchan, *Proc. Natl. Acad. Sci. U.S.A.* **80**, 3015 (1983).
[12] A. Razzaque, H. Mizusawa, and M. M. Seidman, *Proc. Natl. Acad. Sci. U.S.A.* **80**, 3010 (1983).
[13] C. R. Ashman and R. L. Davidson, *Somatic Cell Mol. Genet.* **11**, 499 (1985).
[14] F. L. Graham and A. Van der Eb, *Virology* **52**, 456 (1973).
[15] W. Bertling, K. Hunger-Bertling, and M. J. Cline, *J. Biochem. Biophys. Methods* **14**, 223 (1987).
[16] N. R. Drinkwater and D. K. Klinedinst, *Proc. Natl. Acad. Sci. U.S.A.* **83**, 3402 (1986).
[17] H. Potter, L. Weir, and P. Leder, *Proc. Natl. Acad. Sci. U.S.A.* **81**, 7161 (1984).
[18] F. Toneguzzo and A. Keating, *Proc. Natl. Acad. Sci. U.S.A.* **83**, 3496 (1986).
[19] C. Sureau, J.-L. Romet-Lemonne, J. I. Mullins, and M. Essex, *Cell* **47**, 37 (1986).

for specific experiments, it is also possible to introduce as little as one or a few copies of the tranfected DNA by electroporation.[20]

Exponentially growing cells are harvested and pelleted, then washed once with electroporation buffer (140 mM NaCl, 20 mM HEPES at pH 7.15, 750 μM Na$_2$HPO$_4$) and resuspended in electroporation buffer at a cell concentration of $2-20 \times 10^6$ cells/ml. DNA is added to the cell suspension, which is then incubated on ice for 10 min. The cell and DNA suspension is then subjected to an electric field and returned to ice for 10 min prior to plating in nonselective medium. Selection (or transient assay) can be carried out 48 hr later. The strength of the electric field appropriate for each cell type must be determined empirically. For most cell types we have tested, optimal transfection occurs at a field strength which results in a cell death of 50% or more with the conditions described. This generally corresponds to a capacitance of about 1 mF, a pulse time of 100 msec, and a potential of 100–400 V. Many units are commercially available which allow the experimeter to adjust the different parameters as desired.

Notes. Several investigators have examined the effects of various parameters on levels of stable transfection obtained by this procedure.[20-22] Results from these experiments indicate that a variety of conditions yield successful electroporation. Refer to these articles for specific conditions.

Polybrene-Mediated DNA Transfection

Polybrene-mediated DNA transfection, first described as a method for efficiently transfecting chick embryo fibroblasts with cloned Rous sarcoma virus DNA,[23] has more recently been optimized for stable transfection of CHO cells with DNA from a variety of sources.[24] The major advantages of this procedure over calcium phosphate coprecipitation are a 15-fold better stable transfection frequency in CHO cells and elimination of a requirement for carrier DNA to maximize transfection efficiency. Although the procedure also yields stable transfectants in other cell types, such as HeLa and L cells, it has not been shown to improve transfection frequencies in these cell types over those obtained with calcium phosphate methods. It is not yet clear whether the procedure can be optimized in cell lines other than CHO to yield improved transfection frequencies.

[20] S. S. Boggs, R. G. Gregg, N. Borenstein, and O. Smithies, *Exp. Hematol.* **14,** 988 (1986).
[21] G. Chu, H. Hayakawa, and P. Berg, *Nucleic Acids Res.* **15,** 1311 (1987).
[22] R. Tur-Kaspa, L. Teicher, B. J. Levine, A. I. Skoultchi, and D. A. Shafritz, *Mol. Cell. Biol* **6,** 716 (1986).
[23] S. Kawai and M. Nishizawa, *Mol. Cell. Biol.* **4,** 1172 (1984).
[24] W. G. Chaney, D. R. Howard, J. W. Pollard, S. Sallustio, and P. Stanley, *Somatic Cell Mol. Genet.* **12,** 237 (1986).

In the protocol described by Chaney *et al.*,[24] 5×10^5 CHO cells are plated in alpha medium containing 10% (v/v) fetal calf serum (FCS) and 3 μg/ml Polybrene in 100-mm tissue culture dishes and incubated at 37° overnight. The medium is removed and replaced with 3 ml of alpha medium containing DNA and 10 μg/ml Polybrene. Plates are gently rocked every 1.5 hr. After 12 hr, the Polybrene–DNA mixture is removed and the cells are shocked with DMSO (30% in FCS) for 4 min, washed once with alpha medium, and fed with fresh alpha medium containing 10% FCS. Selection is initiated 24–48 hr later. Using this procedure with the shuttle vector pSV2neo, Chaney observed a transfection frequency of 36,000 G418-resistant colonies/μg plasmid DNA for 5×10^5 cells plated. The system also worked well for transfection of an amplified cosmid library DNA and total genomic DNA. High transfection efficiencies were obtained with 10–30 ng DNA/dish, and the addition of carrier DNA did not result in an increase in transfection efficiency.

Lipid-Mediated Transfection

A number of techniques based on the use of lipids have been developed. The best established of these methods utilizes unilamellar phospholipid vesicles (liposomes) which have been loaded with DNA.[25–27] This approach relies on the fusion of the DNA-containing vesicles with the plasma membrane of the recipient cells. The DNA appears to traverse the cytoplasm and subsequently enter the nucleus.[28] This technique has been shown to be effective in both transient and stable expression, and has been used with both adherent and suspension cell types. A major drawback to this approach is the complexity of liposome preparation techniques. While the transfection efficiency of the liposome-mediated methods is not generally greater than that obtained by the other methods, such as calcium phosphate coprecipitation, it may nevertheless be useful for certain cell types which are difficult to transfect with other techniques. The ability to produce liposomes of roughly physiological composition reduces toxic effects observed with other protocols. A particularly exciting aspect of this method is the ability to perform *in vivo* transfections. Nicolau and colleagues[29] have demonstrated transient expression of a preproinsulin gene

[25] R. Fraley, S. Subramani, P. Berg, and D. Papahadjopoulous, *J. Biol. Chem.* **255**, 10431 (1980).
[26] T. K. Wong, C. Nicolau, and P. H. Hofschneider, *Gene* **10**, 87 (1980).
[27] M. Schaefer-Ridder, Y. Wang, and P. H. Hofschneider, *Science* **215**, 166 (1982).
[28] T. Itani, H. Arigg, N. Yamaguchi, T. Tadakuma, and T. Yasuda, *Gene* **56**, 267 (1987).
[29] C. Nicolau, A Lepapo, P. Soriano, F. Fargette, and M. F. Juhel, *Proc. Natl. Acad. Sci. U.S.A.* **80**, 1068 (1983).

in the liver and spleen of rats injected with DNA-loaded liposomes. Three different methods which have been successfully used for the production of DNA-loaded unilamellar vesicles (liposomes) are (1) the ether-infusion method,[30] (2) reversed-phase evaporation method,[31] and (3) phosphatidylserine calcium-induced fusion method.[32]

Lipofection

A novel technique which also relies on the use of DNA-containing liposomes to introduce genetic material into cultured cells has been described. This protocol, referred to as lipofection, utilizes a synthetic cationic lipid, N-[1-(2, 3-dioleyloxy)propyl]-N,N,N-trimethylammonium chloride (DOTMA). This molecule spontaneously associates with DNA and, on sonication, forms unilamellar vesicles in which 100% of the DNA is trapped.[33] Felgner et al.[33] described high levels of transient expression of introduced DNA in three different cell lines. Transfection efficiencies were 6–10 times greater than those obtained with DEAE-dextran, and stable transfection efficiencies which are 6–80 times greater than those obtained with calcium phosphate precipitation have been reported. The major advantage of this technique over the other liposome-mediated transfection techniques is the apparently greater transfection efficiency as well as the quantitative incorporation of DNA into liposomes (the previously described techniques incorporated only approximately 10% of DNA into the liposomes). However, the potential toxicity of the DOTMA may reduce the general usefulness of this method.

Red Blood Cell-Mediated Transfection

The extremely plastic structure of the erythrocyte and the ability to remove its cytoplasmic contents and reseal the plasma membrane have led a number of investigators to trap different macromolecules within the so-called hemoglobin free "ghost." Combining these "ghosts" with target cells and a fusogen such as polyethylene glycol has permitted the introduc-

---

[30] D. Deamer and A. D. Bangham, *Biochim. Biophys. Acta* **443**, 629 (1976).
[31] F. Szoka, Jr., and D. Papahadjopoulos, *Proc. Natl. Acad. Sci. U.S.A.* **75**, 4194 (1978).
[32] D. Papahadjopoulos, W. J. Vail, K. Jacobson, and G. Poste, *Biochim. Biophys. Acta* **394**, 483 (1975).
[33] D. L. Felgner, T. R. Gadek, M. Holm, R. Roman, H. W. Chan, M. Wenz, J. P. Northrop, G. M. Ringold, and M. Danielsen, *Proc. Natl. Acad. Sci. U.S.A.* **84**, 7413 (1987).

tion of a variety of macromolecules into mammalian cells.[34-36] Both transient and stable expression of introduced DNA have been achieved by this method.[37-39] The complexities of obtaining and preparing erythrocyte ghosts and the variability of DNA entrapment obtained with this procedure have precluded its widespread use.

DNA Microinjection

As the name implies, this technique involves the direct microinjection of DNA into the nucleus of recipient cells. Unlike the other methods described in this article, microinjection does not expose the DNA to the cytoplasm or organelles within it. This is considered beneficial since it has been suggested that DNA sustains considerable damage during the transit from the cell exterior to the nucleus.[40]

Diacumakos and colleagues[41] first described microinjection of a variety of substances into human cultured cells. Capecchi[42] described the use of this approach to transform thymidine kinase-deficient (TK^-) mouse cells with the herpes simplex virus TK gene.

A particularly useful application of DNA microinjection is in the production of transgenic mice.[43] Microinjection into mouse embryos has largely been replaced by the method developed by Gordon et al.,[44] in which the DNA is injected into the male pronucleus of the fertilized mouse egg. DNA sequences introduced in this manner are detectable and expressable in some of the progeny animals.

Although the method requires use of sophisticated equipment and technical expertise, it is very useful in cases where high efficiencies of

[34] M. Furasawa, T. Nishimura, M. Yamaizumi, and Y. Okada, *Nature (London)* **249**, 449 (1974).

[35] K. Kaltoft, J. Zeuthen, F. Engbaek, P. W. Piper, and J. E. Celis, *Proc. Natl. Acad. Sci. U.S.A.* **73**, 2793 (1976).

[36] C. Boogard and G. H. Dixon, *Exp. Cell Res.* **143**, 175 (1983).

[37] F. C. Wiberg, P. Sunnerhagen, K. Kaltoft, J. Zeuthen, and G. Bjursell, *Nucleic Acids Res.* **11**, 7287 (1983).

[38] F. C. Wiberg, P. Sunnerhagen, and G. Bjursell, *Mol. Cell. Biol.* **6**, 653 (1986).

[39] F. C. Wiberg, P. Sunnerhagen, and G. Bjursell, *Exp. Cell Res.* **173**, 218 (1987).

[40] C. T. Wake, T. Gudewicz, T. Porter, A. White, and J. A. Wilson, *Mol. Cell. Biol.* **4**, 387 (1984).

[41] E. G. Diacumakos, S. Holland, and P. Pecora, *Proc. Natl. Acad. Sci. U.S.A.* **65**, 911 (1970).

[42] M. R. Capecchi, *Cell* **22**, 479 (1980).

[43] M. L. De Pamphilis, S. A. Herman, E. Martinez-Salas, C. E. Chalifour, D. O. Wirak, D. Y. Cupo, and M. Miranda, *BioTechniques* **6**, 662 (1988).

[44] J. W. Gordon, G. A. Scangos, D. J. Plotkin, J. A. Barbosa, and F. H. Ruddle, *Proc. Natl. Acad. Sci. U.S.A.* **77**, 7380 (1980).

transfection as well as control of the copy number of integrated sequences are required. Stable integration frequencies as high as 1 in 5 injected cells have been reported. In addition, high transfection efficiencies can be obtained even when as few as 5 copies of DNA are introduced per cell.[42] It has been reported that calcium phosphate-mediated transfection introduces an average of 1000 kb of DNA per recipient cell.[45] Though DNA microinjection has been used by a limited number of investigators, the advent of automated systems which permit injection of a large number of cells promises a wider use.

Laser Method

A sophisticated technology for the introduction of DNA into cultured cells has been described by Kurata et al.[46] In this method, the DNA to be introduced into cells is dissolved in the culture medium surrounding the cells. Uptake of DNA by the cells is mediated by the introduction of minute holes in the cell membrane by brief pulses with a finely focused laser. Cells treated in such a manner repair the holes in their membrane within a fraction of a second, but nevertheless succeed in taking up DNA. Stable transfection efficiencies depend on the concentration of DNA in the culture medium and have been observed to be as high as 0.6%. The experimental apparatus is at least as complex as that required for microinjection, but allows for the more efficient treatment of a larger number of cells.

Microprojectile-Mediated Gene Transfer

A novel method for gene transfer into maize cells has been reported[47] which utilizes a ballistic approach for introducing DNA into cells. In this procedure, DNA is adsorbed to microscopic tungsten particles in the presence of calcium chloride and spermidine. The DNA-adsorbed tungsten particles are then placed on the front surface of a cylindrical polyethylene macroprojectile, which is placed in the barrel of a particle gun device[48] which is fired by a blank gunpowder charge. The barrel of the gun is directed toward suspension cells of maize on the bottom of a standard tissue culture dish. When the gun is fired, the macroprojectile moves down the barrel where it is impeded by a stopping plate. The microprojectiles

[45] M. Perucho, D. Hanahan, and M. Wigler, Cell 22, 309 (1980).
[46] S. Kurata, M. Tsukakoshi, T. Kasuya, and Y. Ikawa, Exp. Cell Res. 162, 372 (1986).
[47] T. M. Klein, M. Fromm, A. Weissinger, D. Tomes, S. Schaff, M. Sletten, and J. C. Sanford, Proc. Natl. Acad. Sci. U.S.A. 85, 4305 (1988).
[48] J. C. Sanford, T. M. Klein, E. D. Wolf, and N. Allen, Part. Sci. Technol. 5, 27 (1987).

continue past the stopping plate, penetrating the cells. When such a system was used for transfecting maize cells with an expressable chloramphenicol acetyltransferase (CAT) gene, CAT activity up to 200-fold over background was observed 24–96 hr after bombardment. Repeated bombardments of the same cell samples resulted in a proportionate increase in the observed levels of CAT activity. The major advantages of this system for gene transfer into maize cells are elimination of the need to generate protoplasts prior to transfection and elimination of the subsequent difficulty of generating whole plants from transfected protoplasts.

The applicability of this method to mammalian cells has not been determined. Factors believed to be important to the success of the method include the size of the microprojectile, the velocity at which the microprojectiles are delivered, and the atmospheric pressure under which the bombardment takes place.

[42] Selection and Coamplification of Heterologous Genes in Mammalian Cells

By RANDAL J. KAUFMAN

Introduction

In the early 1950s, investigators established cell culture systems to study the mechanism by which cancer cells become resistant to a variety of chemotherapeutic agents, such as methotrexate (MTX). The initial investigations led to the observation that stepwise selection for growth of cultured animal cells in progressively increasing concentrations of MTX results in cells with increased levels of the target enzyme, dihydrofolate reductase (DHFR), as a consequence of a proportional increase in the DHFR gene copy number.[1] Subsequently, it has been observed that gene amplification is ubiquitous in nature and many, if not all, genes become amplified at some frequency (approximately $1/10^4$, although this number can vary extensively) (for recent review, see Ref. 2). With appropriate selection conditions, where the growth of cells harboring amplification of a particular gene is favored, a population of cells that contain the amplified gene will outgrow the general population. In the absence of drug selection, the amplified gene is most frequently lost.

Since the degree of gene amplification, in most cases, is proportional to

[1] F. W. Alt, R. E. Kellems, J. R. Bertino, and R. T. Schimke, *J. Biol. Chem.* **253**, 1357 (1978).
[2] R. T. Schimke, *J. Biol. Chem.* **263**, 5989 (1988).

the level of gene expression, it offers a convenient means to increase expression of any particular gene. Although the copy number of a wide variety of genes can be amplified as a consequence of applying appropriate selective pressures, for many genes, direct selection methods are not available. In these cases, it is possible to introduce the gene of interest with a selectable and amplifiable marker gene into the cell and subsequently select for amplification of the marker gene to generate cells that have coamplified the desired gene. DHFR is the most widely used amplifiable marker gene for this purpose. This article reviews the characteristics of gene amplification and describes the methods and available selectable genetic markers for cotransfection and coamplification to obtain high-level expression of heterologous genes in mammalian cells.

Characteristics of Gene Amplification

Several general features are characteristic of gene amplification in cultured mammalian cells: (1) amplification is usually obtained after stepwise selection for increasing resistance, (2) the amplified DNA displays variable stability, and (3) the size and structure of the amplified unit are variable and these characteristics may change with time. These characteristics suggest that multiple mechanisms are likely responsible for gene amplification.

Stepwise Selection

Gene amplification generally occurs as a result of stepwise selection for resistance to increasing concentrations of the selective agent. When larger selection steps are employed, mechanisms other than gene amplification are observed. For example, single-step high-level resistance to MTX may result in cells with an altered, MTX-resistant DHFR enzyme[3,4] or in cells that have altered MTX transport properties.[5,6] Since gene amplification requires DNA replication and cell division, optimal amplification occurs when cells are subject to severe, but not absolute growth-limiting conditions. In some cases, resistant colonies may appear but fail to continue growing, whereas in other cases, only after prolonged selection does the growth rate eventually return to that of the original parental line. These observations suggest that many events may occur from the time a cell first becomes resistant until it is established as a drug-resistant clonal line.

---

[3] W. F. Flintoff, S. V. Davidson, and L. Siminovitch, *Somatic Cell Genet.* **2,** 245 (1976).

[4] D. A. Haber, S. M. Beverley, and R. T. Schimke, *J. Biol. Chem.* **256,** 9501 (1981).

[5] F. M. Sirotnik, S. Kurita, and D. J. Hutchison, *Cancer Res.* **28,** 75 (1968).

[6] Y. Assaraf and R. T. Schimke, *Proc. Natl. Acad. Sci. U.S.A.* **84,** 7154 (1987).

A number of approaches have been taken to increase the frequency of gene amplification. Perturbation of DNA synthesis by treatment with hydroxyurea,[7] aphidicolin,[7] UV γ irradiation,[8] hypoxia,[9] carcinogens,[10] and arsenate[11] can enhance the frequency of specific gene amplification 10- to 100-fold. Agents which promote cell growth, such as phorbol esters or insulin, can also increase the frequency of gene amplification.[12] Interestingly, cells resistant to one selective agent as a result of gene amplification exhibit increased frequencies for amplification of other genes.[13] This "amplificator" phenotype may be one particular property of cells selected for stepwise increasing levels of drug resistance or it may reflect the finding that disruption of DNA replication results in an enhancement of gene amplification events in general.[14]

Variable Stability of Amplified Genes

When drug-resistant cells are propagated in the absence of the selective agent, the amplified genes may be maintained or lost. The amplified genes in newly selected resistant cell lines are unstable. As cells are propagated for increasing periods of time in the presence of the selection agent, the amplified genes exhibit increased stability.[15,16] The degree of stability of the amplified genes in the absence of drug selection may correlate with the localization of the amplified genes and the complexity of the associated karyotypic alterations. Amplified genes localized to double minute chromosomes are usually lost upon propagation in the absence of selection.[16] Double minute chromosomes are small paired extrachromosomal elements which lack centromeric function and thus, segregate randomly at mitosis. Most evidence suggests that double minute chromosomes are circular DNA structures.[17] In contrast, stably amplified genes are usually integrated into the chromosome and frequently are associated with expanded chromosomal regions termed homogeneously staining regions

[7] C. A. Hoy, G. C. Rice, M. Kovacs, and R. T. Schimke, *J. Biol. Chem.* **262,** 11927 (1987).
[8] T. D. Tlsty, P. C. Brown, and R. T. Schimke, *Mol. Cell. Biol.* **4,** 1050 (1984).
[9] G. C. Rice, C. Hoy, and R. T. Schimke, *Proc. Natl. Acad. Sci. U.S.A.* **83,** 5978 (1986).
[10] S. Lavi, *Proc. Natl. Acad. Aci. U.S.A.* **78,** 6144 (1981).
[11] T. C. Lee, P. W. Lamb, N. Tanaka, T. N. Gilmer, and J. C. Barrett, *Science* **241,** 79 (1988).
[12] A. Varshavsky, *Cell* **25,** 561 (1981).
[13] E. Giulotto, C. Knights, and G. R. Stark, *Cell* **48,** 837 (1987).
[14] R. N. Johnston, J. Feder, A. B. Hill, S. W. Sherwood, and R. T. Schimke, *Mol. Cell. Biol.* **6,** 3373 (1986).
[15] R. J. Kaufman and R. T. Schimke, *Mol. Cell. Biol.* **1,** 1069 (1981).
[16] R. J. Kaufman, P. C. Brown, and R. T. Schimke, *Proc. Natl. Acad. Sci. U.S.A.* **76,** 5669 (1979).
[17] B. Hamkalo, P. J. Farnham, R. Johnston, and R. T. Schimke, *Proc. Natl. Acad. Sci. U.S.A.* **82,** 1126 (1985).

(HSRs).[18] This term reflects the lack of banding in these regions after Giemsa–trypsin banding procedures. Both amplified endogenous[19] and heterologous[20] DHFR genes in MTX-resistant mouse fibroblasts are usually associated with double minute chromosomes. In contrast, amplified endogenous and transfected heterologous DHFR genes in MTX-resistant Chinese hamster ovary (CHO) cells are frequently associated with expanded chromosomal regions.[18,21] In a few cases, the amplified transfected DNA may be present with little cytological perturbation.[21] This may reflect a smaller size of the amplified unit within the chromosome. CHO cells selected in the presence of MTX for amplification of heterologous DHFR genes have generally proved to be very stable upon propagation in the absence of MTX.[18] Instability associated with chromosomal amplified genes may correlate with more complex chromosomal alterations. For example, cells selected for amplified genes may become tetraploid, a characteristic possibly associated with increased instability. In other cases, transfected and amplified DNA may be observed in HSRs associated with extremely large or dicentric chromosomes. These chromosomes are prone to breakage events which may be responsible for instability.[21-24]

Variable Size and Structure of Amplified Unit

The size of the amplified unit estimated by cytogenetic analysis is highly variable and may range from less than 100 kilobase pairs to greater than 500 kilobase pairs.[18,25] In several cases, the amplified DNA consists of inverted duplications.[25-27] During propagation of cells in culture, the amplified DNA may undergo changes which involve (1) extrachromosomal circles becoming larger through tandem duplications,[28,29] (2) the positions

[18] J. H. Nunberg, R. J. Kaufman, R. T. Schimke, G. Urlaub, and L. A. Chasin, *Proc. Natl. Acad. Sci. U.S.A.* **75**, 5553 (1978).

[19] P. C. Brown, S. M. Beverly, and R. T. Schimke, *Mol. Cell. Biol.* **1**, 1077 (1981).

[20] M. J. Murray, R. J. Kaufman, S. A. Latt, and R. A. Weinberg, *Mol. Cell. Biol.* **3**, 32 (1983).

[21] R. J. Kaufman, P. A. Sharp, and S. A. Latt, *Mol. Cell. Biol.* **3**, 699 (1983).

[22] D. M. Robins, R. Axel, and A. S. Henderson, *J. Mol. Appl. Genet.* **1**, 191 (1981).

[23] B. Fendrock, M. Destremps, R. J. Kaufman, and S. A. Latt, *Histochemistry* **84**, 121 (1986).

[24] J. L. Andrulis and L. Siminovits, *in* "Gene Amplification" p. 75. (R. T. Schimke, ed.). Cold Spring Harbor Laboratory, Cold Spring Harbor, New York, 1982.

[25] J. E. Looney and J. L. Hamlin, *Mol. Cell. Biol.* **7**, 569 (1987).

[26] M. Ford and M. Fried, *Cell* **45**, 425 (1986).

[27] J. C. Ruiz and G. M. Wahl, *Mol. Cell. Biol.* **8**, 4302 (1988).

[28] S. M. Carroll, M. L. DeRose, P. Gaudray, and C. M. Moor, *Mol. Cell. Biol.* **8**, 1525 (1988).

[29] N. A. Federspeil, S. M. Beverley, J. W. Schilling, and R. T. Schimke, *J. Biol. Chem.* **259**, 9127 (1984).

of novel recombination breakpoints between amplified units may change,[29,30] or (3) the propagation of mutations from one amplified unit to other units within an amplified array, possibly by gene conversion.[31]

The mechanism of amplification of a transfected CAD gene upon selection for PALA resistance has been extensively studied.[27,28,32] The integrated CAD sequence is first deleted and then undergoes replication as an extrachromosomal element. The extrachromosomal intermediate may increase in size by forming multiples of itself to eventually create double minute chromosomes. The extrachromosomal sequence may eventually integrate to form a homogeneously staining chromosomal region.

Amplification of Cotransfected DNA

With most methods of DNA transfer, 5–50% of the cells in the population acquire DNA and express it transiently for several days to several weeks. However, the DNA is eventually lost from the cell population unless some selection procedure is used to iosolate cells that have stably integrated the foreign DNA into their genome. The limiting event for obtaining stable DNA transfer is the frequency of DNA integration and not the frequency of DNA uptake into the cell. Different cell lines as well as different transfection methods yield dramatically different efficiencies with respect to the frequency of stable integration, as well as to the amount of foreign DNA incorporated. The ability to select for incorporation of one gene by selection for a second cotransfected gene has been termed cotransformation.[33] In cotransformation, separate DNA molecules become ligated together inside the cell and subsequently cointegrate as a unit via nonhomologous recombination into the host chromosome. When separate DNA molecules are sequentially introduced into recipient cells, the molecules do not become linked and are not cointegrated into the same chromosomal position. If DNA molecules are linked together within the chromosome, selection for amplification of one of the molecules usually results in coamplification of the linked DNA molecule. Different cell lines and DNA transfection methods exhibit different potentials for cotransformation. For example, the frequency of cotransformation in CHO cells is lower than that observed in mouse L cells. In this case, the difference may be attributa-

[30] E. Giulotto, I. Saito, and G. R. Stark, *EMBO J.* **5**, 2115 (1986).

[31] J. M. Roberts and R. Axel, *Cell* **29**, 109 (1982).

[32] S. Carroll, P. Gaudray, M. DeRose, J. Emery, J. Meinkoth, E. Nakkim, M. Subler, D. Von Hoff, and G. Whal, *Mol. Cell. Biol.* **7**, 1740 (1987).

[33] M. Wigler, S. Silverstein, L. S. Lee, A. Pellicer, Y. Cheng, and R. Axel, *Cell* **11**, 223 (1977).

ble to the lesser amount of DNA incorporated into CHO cells compared to mouse L cells. Cotransformation by CaPO$_4$ DNA-mediated transfection is very efficient,[33] whereas cotransformation by protoplast fusion of separate bacteria harboring two independent plasmids is very rare. In situations where cotransformation is inefficient, it is better to use plasmid vectors which express both the product gene and the selection gene. Vectors useful for this purpose are described below and in Fig. 3.

In order to coamplify a particular DNA sequence, it is important that the DNA sequence or its products not interfere with amplification or be toxic to the cell. Since a minor subset of cells from the total population survives during each step of the selection process, it is possible to select for mutations in the coamplified gene which permit cell viability. Thus, the protein obtained after selection for expression at high level may be different from that encoded by the original transfected gene. For example, selection for coamplification of DHFR and SV40 small t antigen in CHO cells resulted in highly MTX-resistant cells which expressed SV40 small t antigen at 15% of the total protein synthesis.[34] However, the t antigen expressed at high level migrated at a molecular weight of 2000 less than the wild-type t antigen from the same cells at a lower expression level. Although the basis for this change was not characterized, it demonstrates the potential difficulty in assuring the fidelity of proteins produced after selection for gene amplification.

Amplifiable Genetic Markers

Resistance to cytotoxic drugs is the characteristic most widely used to select for stable transformants. Drug resistance can be recessive or dominant. Genes conferring recessive drug resistance require a particular host which is deficient in the activity which is being selected. Genes conferring dominant drug resistance can be used independent of the host. Many of the recessive genetic selectable markers are involved in the salvage pathway for purine and pyrimidine biosynthesis. When *de novo* biosynthesis of purines or pyrimidines is inhibited, the cell can utilize purine and pyrimidine salvage pathways, providing that the enzymes (i.e., thymidine kinase, hypoxanthine-guanine phosphoribosyltransferase, adenine phosphoribosyltransferase, or adenosine kinase) necessary for conversion of the nucleoside precursors to the corresponding nucleotides are present (see Fig. 1). Since these salvage enzymes are not required for cell growth when *de novo* purine or pyrimidine biosynthesis is functional, cells deficient for a particular salvage pathway enzyme are viable under normal growth conditions. How-

[34] R. J. Kaufman and P. A. Sharp, *J. Mol. Biol.* **159**, 601 (1982).

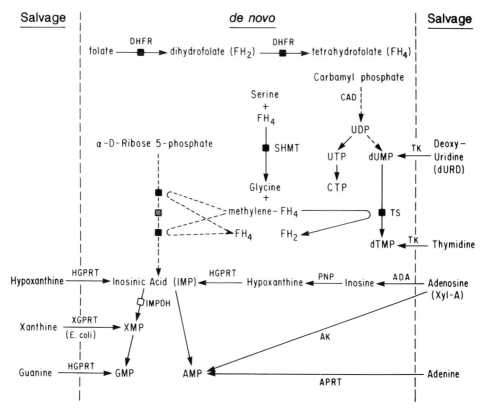

FIG. 1. *De novo* and salvage biosynthetic pathways for purines and pyrimidines involving available selectable markers. *De novo* enzymes: DHFR, dihydrofolate reductase; CAD, carbamoyl-P synthase, aspartate transcarbamoylase, and dihydroorotase (only the latter two activities are indicated); SHMT, serine hydroxymethyltransferase; TS, thymidylate synthase; IMPDH, inosine-monophosphate dehydrogenase. Salvage enzymes: TK, thymidine kinase; ADA, adenosine deaminase; PNP, purine nucleoside phosphorylase; AK, adenosine kinase; APRT, adenine phosphoribosyltransferase; HGPRT, hypoxanthine-guanine phosphoribosyltransferase; XGPRT, *E. coli* xanthine-guanine phosphoribosyltransferase. Solid arrows indicate single reactions. Dashed arrows indicate multiple reactions. Solid squares indicate reactions that are inhibited by the folate analogs methotrexate and aminopterin. The hatched square indicates the principal reaction inhibited by azaserine. The open square indicates the reaction inhibited by mycophenolic acid. [Reprinted with permission from R. J. Kaufman, *in* "Genetic Engineering" (J. K. Setlow, ed.), Vol. 9, p. 155. Plenum, New York, 1988.]

ever, addition of drugs which inhibit the *de novo* biosynthesis of purines or pyrimidines results in death of deficient cells because the salvage pathway becomes essential to provide purines and pyrimidines. Cells which acquire the capability to express the deficient activity via gene transfer can be selected for growth under these conditions. This has become a common selection technique and is the basis for the thymidine kinase (TK), adenine phosphoribosyltransferase (APRT), and hypoxanthine-guanine phosphoribosyltransferase (HGPRT) selection.

Many selectable recessive genetic markers can be used in a dominant fashion if there is a suitable drug to titrate the activity of the marker gene product and if there is a suitable expression vector to produce significantly more gene product than is present in the host cell to be transfected. Table I lists genes for which selection schemes for gene amplification are available. Most, if not all, the genes listed in Table I could be used as selectable markers for coamplification. In addition to these genes, a large number of genes can become amplified upon transformation and tumor progression.[35] However, there is no direct means for which to select for their amplification. The most useful dominant and amplifiable genetic marker selection systems for coamplification of heterologous genes are described below.

Dihydrofolate Reductase–Methotrexate Resistance

Dihydrofolate reductase (DHFR) catalyzes the conversion of folate to tetrahydrofolate (FH4). FH4 is required for the biosynthesis of glycine from serine, for the biosynthesis of thymidine monophosphate from deoxyuridine monophosphate, and for purine biosynthesis (see Fig. 1). MTX is a folic acid analog which binds and inhibits DHFR, leading to cell death. When cells are selected for growth in sequentially increasing concentrations of MTX, the surviving population contains increased levels of DHFR which result from an amplification of the DHFR gene.[1,2] Highly MTX-resistant cells may contain several thousand copies of the DHFR gene and express several thousandfold elevated levels of DHFR. MTX resistance selection for DHFR gene amplification is the most convenient and widely used system for coamplification to obtain high-level expression of heterologous genes.

The convenience of the DHFR selection system relies on the availability of CHO cells that are deficient in DHFR.[36] The origin and genetics of CHO cells have been reviewed.[37] DHFR-deficient CHO cells were isolated after ethylmethane sulfonate and UV γ irradiation induced mutagenesis and selection for growth in high-specific-activity [³H]deoxyuridine.[36] Cells containing functional DHFR convert [³H]deoxyuridine to [³H]thymidylate, which is incorporated into DNA. This incorporation into DNA is lethal due to disruption of the DNA by radioactive damage. DHFR-deficient cells survive this selection procedure. The DHFR-deficient cells re-

[35] K. Alitalok, *Med. Biol.* **62**, 304 (1984).
[36] G. Urlaub and L. A. Chasin, *Proc. Natl. Acad. Sci. U.S.A.* **77**, 4216 (1980).
[37] M. M. Gottesman (ed.), "Molecular Cell Genetics." Wiley, New York, 1985.

TABLE I
GENE AMPLIFICATION IN DRUG-RESISTANT MAMMALIAN CELLS

Selection	Gene	Ref.[a]
Methotrexate	Dihydrofolate reductase	(1)
Cadmium	Metallothionein	(2)
PALA	CAD	(3)
Xyl-A-or adenosine and 2'-deoxycoformycin	Adenosine deaminase	(4)
Adenine, azaserine, and coformycin	Adenylate deaminase	(5)
6-Azauridine, pyrazofuran	UMP Synthetase	(6)
Mycophenolic acid	IMP 5'-dehydrogenase	(7)
Mycophenolic acid with limiting xanthine	Xanthine-guanine phosphoribosyltransferase	(8)
Hypoxanthine, aminopterin, and thymidine (HAT)	Mutant HGPRTase or mutant thymidine kinase	(9, 10)
5-Fluorodeoxyuridine	Thymidylate synthetase	(11)
Multiple drugs[b]	P-glycoprotein 170	(12)
Aphidicolin[c]	Ribonucleotide reductase	(13)
Methionine sulfoximine	Glutamine synthetase	(14)
β-Aspartyl hydroxamate or Albizziin	Asparagine synthetase	(15)
Canavanine	Arginosuccinate synthetase	(16)
α-Difluoromethylornithine	Ornithine decarboxylase	(17)
Compactin	HMG-CoA reductase	(18)
Tunicamycin[c]	N-Acetylglucosaminyl transferase	(19)
Borrelidin[c]	Threonyl-tRNA synthetase	(20)
Ouabain	Na$^+$, K$^+$-ATPase	(21)

[a] Key to references: (1) F. W. Alt, R. E. Kellems, J. R. Bertino, and R. T. Schimke, *J. Biol. Chem.* **253**, 1357 (1978); (2) L. R. Beach and R. D. Palmiter, *Proc. Natl. Acad. Sci. U.S.A.* **78**, 210 (1981); (3) G. M. Wahl, R. A. Padgett, and G. R. Stark, *J. Biol. Chem.* **254**, 8679 (1979); (4) C. Y. Yeung, D. E. Ingolia, C. Bobonis, B. S. Dunbar, M. E. Riser, J. J. Siciliano, and R. E. Kellems, *J. Biol. Chem.* **258**, 8338 (1983); (5) M. Debatisse, O. Hyrien, E. Petit-Koskas, B. R. de Saint-Vincent, and G. Buttin, *Mol. Cell. Biol.* **6**, 1776 (1986); (6) J. J. Kanalas and D. P. Suttle, *J. Biol. Chem.* **259**, 1848 (1984); (7) E. Huberman, C. K. McKeown, and J. Friedman, *Proc. Natl. Acad. Sci. U.S.A.* **78**, 3151 (1981); (8) A. B. Chapman, M. A. Costello, R. Lee, and G. M. Ringold, *Mol. Cell. Biol.* **3**, 1421 (1983); (9) J. Brennand, A. C. Chinault, D. S. Konecki, D. W. Melton, and C. T. Caskey, *Proc. Natl. Acad. Sci. U.S.A.* **79**, 1950 (1982); (10) J. M. Roberts and R. Axel, *Cell* **29**, 109 (1982); (11) C. Rossana, L. G. Rao, and L. F. Johnson, *Mol. Cell. Biol.* **2**, 1118 (1982); (12) J.R. Riordan, K. Deuchars, N. Kartner, N. Alon, J. Trent, and V. Ling, *Nature (London)* **316**, 817 (1985); (13) C. L. K. Sabourin, P. F. Bates, L. Glatzer, C.-C. Chang, J. E. Trosko, and J. A. Boezi, *Somatic Cell Genet.* **7**, 255 (1981); (14) A. P. Young and G. M. Ringold, *J. Biol. Chem.* **258**, 11260 (1983); (15) I. L. Andrulis, C. Duff, S. Evans-Blackler, R. Worton, and L. Siminovitch, *Mol. Cell. Biol.* **3**, 391 (1983); (16) T. S. Su, C. S. G. O. Bock, W. E. O'Brien, and A. L. Beaudet, *J. Biol. Chem.* **256**, 11826 (1981); (17) L. McConlogue and P. Coffino, *J. Biol. Chem.* **258**, 12083 (1983); (18) K. L. Luskey, J. R. Faust, D. J. Chin, M. S. Brown, and J. L. Goldstein, *J. Biol. Chem.* **258**, 8462 (1983); (19) B. A. Crisculolo and S. S. Krag, *J. Cell Biol.* **94**, 586 (1982); (20) J. S. Gantt, C. A. Bennett, and S. M. Arfin, *Proc. Natl. Acad. Sci. U.S.A.* **78**, 5367 (1981); (21) P. G. Pauw, M. D. Johnson, P. Moore, M. Morgan, R. M. Fineman, T. Kalka, and J. F. Ash, *Mol. Cell. Biol.* **6**, 1164 (1986).

[b] Adriamycin, vincristine, colchicine, actinomycin D, puromycin, cytocholasin B, emetine, maytansine, Bakers' antifolate.

[c] Examples of increased enzyme activity likely resulting from gene amplification.

quire the addition of thymidine, glycine, and hypoxanthine to the growth medium, and do not grow in the absence of added nucleosides unless they acquire a functional DHFR gene. Most frequently used cell clones for heterologous expression are DUKX-B11 (DXB-11) (isolated from the proline auxotroph CHO-K1) and DG44 [isolated from the CHO (Toronto) cell line]. These cell lines require the presence of a purine source (either hypoxanthine or adenosine and deoxyadenosine), thymidine, and glycine to complement the DHFR deficiency. These lines can be obtained from Dr. Lawrence Chasin, Columbia University, NY.

Coamplification of heterologous genes with DHFR in DHFR-deficient CHO cells can yield cell lines that express high levels of a protein from heterologous genes. The advantages of CHO cells for the expression of heterologous genes include (1) the amplified genes are integrated into the host chromosome and as a result are stably maintained even in the absence of continued drug selection, (2) a variety of proteins have been properly expressed and secreted at high levels in CHO cells, (3) CHO cells adapt well to growth in the absence of serum, (4) CHO cells can grow either attached or in suspension, and (5) CHO cells have been scaled up to greater than 5000 liters. Among the genes expressed in this manner are human tissue plasminogen activator,[38] human interferon γ,[39,40] human interferon β,[41] human factor IX,[42] human factor VIII,[43] the herpes simplex glycoprotein D,[44] and the α and β subunits of bovine luteinizing hormone.[45]

Transformant Selection and Coamplification of DHFR in DHFR-Deficient CHO Cells. GROWTH OF DHFR-DEFICIENT CHO CELLS. Medium: Ham's nutrient mixture F12 was specially designed for growth of CHO cells. Alternatively, alpha medium may be used in place of Ham's F12. Both media can be ordered specifically without nucleosides from Gibco Inc. (Grand Island, NY). For complementation of the DHFR deficiency, the following supplements need to be added if they are not present in the growth medium: hypoxanthine (10 μg/ml) or adenosine (10 μg/ml) and deoxyadenosine (10 μg/ml), thymidine (10 μg/ml), glycine (50 μg/ml). In

[38] R. J. Kaufman, L. C. Wasley, A. T. Spiliotes, S. D. Gossels, S. A. Latt, G. R. Larsen, and R. M. Kay, *Mol. Cell. Biol.* **5**, 1730 (1985).
[39] J. Haynes and C. Weissman, *Nucleic Acids Res.* **11**, 687 (1983).
[40] S. J. Schahill, R. Devos, J. V. Heyden, and W. Fiers, *Proc. Natl. Acad. Sci. U.S.A.* **80**, 4654 (1983).
[41] F. McCormack, M. Trahey, M. Innis, B. Dieckmann, and G. Ringold, *Mol. Cell. Biol.* **4**, 166 (1984).
[42] R. J. Kaufman, L. C. Wasley, B. C. Furie, B. Furie, and C. Shoemaker, *J. Biol. Chem.* **261**, 9622 (1986).
[43] R. J. Kaufman, L. C. Wasley, and A. J. Dorner, *J. Biol. Chem.* **263**, 6362 (1988).
[44] P. W. Berman, D. Dowbenko, C. C. Simonsen, and L. A. Laskey, *Science* **222**, 524 (1983).
[45] D. M. Kaetzel, J. K. Browne, F. Wondisford, T. M. Nett, A. R. Thomason, and J. H. Nilson, *Proc. Natl. Acad. Sci. U.S.A.* **82**, 7280 (1985).

addition, the medium is supplemented with 10% (v/v) fetal calf serum, streptomycin sulfate (100 μg/ml), and penicillin (100 units/ml). The serum is heat-inactivated by treatment for 30 min at 56° in order to inactivate complement factors and potential virus and mycoplasma contaminants.

Cells are grown in 5% CO_2-air at 37°. Cultures are never incubated for more than 24 hr past the achievement of confluency, or else they become detached and/or sick.

The cells are typically subcultured at 1:20 twice per week. This is accomplished by detaching the cells from the plate by treatment with 0.05% (w/v) trypsin (1:250) (Gibco) in phosphate-buffered saline (PBS). Trypsinization is usually for 5 min at room temperature, but the exact time varies and the cells should be checked microscopically for rounding up. If overtrypsinized, the cells will clump. Monolayers are first rinsed with a few milliliters of trypsin solution before the final trypsin application. Trypsinization is stopped by removal of trypsin and cells are resuspended in serum-containing medium.

DHFR SELECTION CONDITIONS. Selection for DHFR-positive transformants is performed by subculturing cells 1:15 at 48 hr posttransfection into selective medium which lacks nucleosides. DHFR-selective medium may be either Ham's F12 minus thymidine and hypoxanthine or alpha medium minus nucleosides (obtained from Gibco). Either growth medium is supplemented with 10% dialyzed heat-inactivated fetal bovine serum and penicillin and streptomycin sulfate. The serum is dialyzed against PBS to remove low-molecular-weight nutrients such as nucleosides and amino acids which may act to bypass the selective conditions. Dialysis is performed in large dialysis 40-mm tubing (1 liter fetal bovine serum) versus 4 liters of PBS containing penicillin (10–20 unit/ml) and streptomycin (10–20 μg/ml) for 12–15 hr with one change of PBS during this time. The serum is then sterilized by filtration through a 0.2-μm nitrocellulose filter. Thus, the initial selection in medium minus nucleosides requires low levels of DHFR expression. Macroscopic colonies appear after 10 days and can be isolated after 14 days. If no colonies are observed the cells should be left to grow for another week. If DHFR is poorly expressed, cell growth is slower and colonies take longer to appear.

CONDITIONS FOR DHFR COAMPLIFICATION. The transformed DHFR-positive cells are expanded in DHFR-selective medium and subjected to increasing MTX selection. The initial selection for MTX resistance is accomplished by subculturing approximately 2×10^5 cells into a 10-cm tissue culture dish in DHFR-selective medium with either 0.005 or 0.02 μM MTX. The concentration is dependent on the efficiency of DHFR expression obtained from the transfected plasmid and should be determined experimentally. This MTX resistance selection should result in death of greater than 90% of the cells. At the appropriate concentration of

MTX, between 10 and several hundred colonies may appear within 2 weeks of selection. Since MTX-mediated cell death is slow (generally taking 3–5 days) and requires cell growth, it is important to ensure the cells under selection do not reach stationary phase of growth. If cells reach confluency within 4–5 days after growth in MTX selection, it is advised to subculture the cells 1 : 15 in the same concentration of MTX as well as to increase the concentration severalfold. The colonies which appear on individual dishes can be pooled and propagated for approximately 10 cell doublings in the same concentration of MTX. After the cells resume rapid growth, the concentration of MTX should be increased 2- to 5-fold and cells selected as described above. This process can be repeated until final levels of MTX resistance reach 1–10 μM MTX. Selection for resistance above this level does not usually yield significantly higher degrees of amplification. MTX (+) is obtained from Sigma (St Louis, MO).

DHFR Coamplification in Cells That Are Not DHFR-Deficient. Although DHFR is most frequently used as a recessive selectable marker in the DHFR-deficient CHO cells, several adaptations have made DHFR a useful dominant selectable and amplifiable genetic marker for cells that are not DHFR-deficient. These approaches rely on growth conditions that preferentially select for expression and amplification of the transfected DHFR gene over amplification of the endogenous DHFR gene. Although the appropriate concentration of MTX for selection varies widely between different cell lines, typical MTX concentrations for the initial selection step range from 0.1 to 0.5 μM.

COTRANSFECTION WITH ANOTHER DOMINANT SELECTABLE MARKER.[46] Cotransfection of both the gene of interest and the DHFR amplification marker with another dominant selectable marker [for example, pSV$_2$Neo encoding resistance to geneticin[47] (G418, Gibco)] can be used to isolate initial transformants. This selection for cotransformants yields a population of cells which have incorporated the heterologous DNA. This allows subsequent selection for amplification of the heterologous DNA, which usually occurs at frequencies above the frequency of amplification of endogenous DHFR genes. If one were to transfect and select directly for DHFR expression and amplification in a single step, the frequency of transfection (approximately 10^{-4}) would have to be multiplied by the frequency for amplification (around 10^{-3} to 10^{-4}) and would be less than the frequency of spontaneous amplification of the endogenous DHFR gene.

For this procedure, the transfected cells are first selected for resistance

---

[46] S. K. Kim and B. J. Wold, *Cell* **42**, 129 (1985).
[47] P. Southern and P. Berg, *J. Mol. Appl. Genet.* **1**, 327 (1982).

to the dominant selectable marker. In the case of pSV$_2$Neo, the colonies that appear after 2 weeks of selection for growth in 0.5 – 1 mg/ml G418 can be pooled and selected in moderate concentrations of MTX (usually around 0.3 μM) for amplification of the transfected DNA. Most frequently, the foreign DHFR gene is preferentially amplified over the endogenous DHFR gene.[46] For this approach, the vector pRK1-4 shown in Fig. 3 is a useful expression vector since it contains two transcription units: one which has a cloning site for insertion of a foreign coding region and which encodes a wild-type murine DHFR. After insertion of the desired coding sequences into this vector, the plasmid DNA can be cotransfected with a dominant selectable marker such as pSV$_2$Neo.

USE OF AN EFFICIENT DHFR EXPRESSION PLASMID. Another approach relies on use of a strong promoter for transcription of the foreign DHFR gene to permit sufficiently high levels of DHFR expression from the transfected DHFR gene to allow selection with MTX at concentrations above the natural resistance level of the cell.[20] Upon growth in the appropriate concentration of MTX, the transfected cells with amplified foreign DHFR genes can be selected over MTX-resistant cells which are a result of amplification of the endogenous DHFR gene.

USE OF MTX-RESISTANT DHFR EXPRESSION PLASMIDS.[48] Mutations within the DHFR coding region can render enzymes less sensitive to inhibition by MTX.[4] A murine DHFR cDNA which contains a single amino acid change from a leucine to an arginine at amino acid residue 22 encodes an altered enzyme which has a 270-fold elevated K_i relative to the wild-type enzyme.[4,49] It is possible to select directly for the presence of this enzyme by growth in moderate levels of MTX (0.5 – 1.0 μM).[49] In addition, the transfected DNA can be amplified by further selection in increasing concentrations of MTX. However, the large change in K_i makes it difficult to obtain high degrees of amplification with the mutant DHFR gene since lower levels of the enzyme are able to confer MTX resistance.

An alternative approach to MTX-resistant DHFR amplification vectors relies on the isolation of trimethoprim (a folic acid analog)-resistant DHFRs from microbial sources. In E. coli there are two plasmid-derived DHFRs that confer resistance to trimethoprim.[48] The type 1 DHFR is contained within the transposon Tn7 as part of the multidrug resistance episome R483. In contrast to the type I DHFR which is resistant to intermediate levels of trimethoprim, the type II enzyme is extremely resist-

[48] C. C. Simonsen, in "Molecular Genetics of Mammalian Cells" (G. M. Malacinski, ed.), p. 99. Macmillan, New York, 1986.
[49] C. C. Simonsen and A. D. Levinson, Proc. Natl. Acad. Sci. U.S.A. 80, 2495 (1983).

ant to trimethoprim. The genes for the type I[50] and the type II[51] enzymes can function as dominant MTX resistance selection markers in mammalian cells. However, since these DHFRs are extremely resistant to MTX, their amplification is not possible by MTX selection.

Monitoring Stability of Amplified DHFR Genes. The most direct method to monitor stability of the amplified genes is to remove the selective agent from the growth medium and monitor the gene copy number by Southern blot hybridization analysis as the cells propagate in the absence of selection. However, this is somewhat cumbersome and time consuming. The stability may be predicted by karyotypic analysis of the resistant cells. The presence of double minute chromosomes, tetraploidy, or dicentric or extremely aberrant chromosomes is usually indicative of instability.[16,18,21,23]

Alternatively, the stability of amplified DHFR genes can be predicted by analysis of the histogram of the DHFR content per cell for the population. This is performed by staining cells with a fluorescent analog of MTX (MTX-F) (from Molecular Probes, Eugene, OR) and analysis using a fluorescence-activated cell sorter.[52] MTX-F quantitatively stains cells for DHFR. The pattern of the histogram of fluorescence/cell can be used to predict stability. By this procedure, there is negligible staining of DHFR-deficient cells (Fig. 2a). An unstable MTX-resistant population exhibits a heterogeneous skewed distribution with many cells exhibiting low levels of DHFR (Fig. 2b). A stable MTX-resistant population exhibits a uniform population of highly fluorescent cells (Fig. 2c). For MTX-F staining, the cells are propagated for 2–3 days in the absence of MTX since MTX will effectively compete with MTX-F for DHFR binding. Then the staining of cells for DHFR is performed by addition of 10–30 μM MTX-F to the culture medium and incubation at 37° for at least 12 hr prior to analysis. After incubation, the cells are rinsed several times with medium, harvested by trypsinization, centrifuged at low speed, and resuspended in tissue culture medium for analysis of fluorescence intensity with a fluorescence-activated cell sorter.

Carbamoyl-Phosphate Synthase–Aspartate Transcarbamoylase–
   Dihydroorotase

Carbamoyl-phosphate synthase–aspartate transcarbamoylase–dihydroarotase (CAD) is a multifunctional protein that catalyzes the first three steps in UMP biosynthesis. The complex includes carbamoyl-phosphate

[50] C. C. Simonsen, E. Y. Chen, and A. D. Levinson, *J. Bacteriol.* **155,** 1001 (1983).
[51] K. O'Hare, C. Benoist, and R. Breathnach, *Proc. Natl. Acad. Sci. U.S.A.* **78,** 1527 (1981).
[52] R. J. Kaufman, J. R. Bertino, and R. T. Schimke, *J. Biol. Chem.* **253,** 5852 (1978).

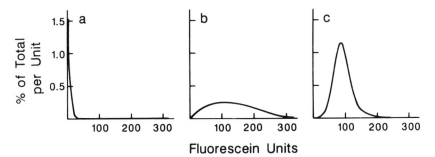

Fig. 2. MTX-F fluorescence of stable and unstable MTX-resistant populations. The distribution of DHFR content per cell determined by MTX-F fluorescence intensity per cell is shown for the original CHO DHFR-deficient cells (a), for a MTX-resistant cell line with unstable amplified DHFR genes (b), and for a MTX-resistant cell line with stable amplified DHFR genes (c).

synthase, aspartate transcarbamoylase, and dihydroorotase. PALA (*N*-phosphonoacetyl-L-aspartic acid) is a transition state analog inhibitor of the aspartate transcarbamoylase which can kill cells as a result of UMP depletion. Growth in sequentially increasing concentrations of PALA selects for amplification of the CAD gene.[53] In practice, the use of this selection system has been limited to CAD-deficient CHO cells.[54] The full-length cDNA for the mammalian gene has not been isolated intact, due to its large size. A cloned copy of the genomic CAD gene has been introduced into CAD-deficient CHO mutants (Urd-A)[55] by selection for growth in the absence of uridine as described below.[56] Individual transformants selected for growth in increasing concentrations of PALA exhibit amplification of the introduced CAD gene.[54] Alternatively, the *E. coli* aspartate transcarbamoylase gene *(PyrB)* carried in a eukaryotic expression plasmid can be selected and amplified up to 4-fold by PALA selection in aspartate transcarbamoylase-deficient CHO cells[57] as described below.

Conditions for Selection and Coamplification with CAD. A vector encoding the *E. coli pyrB* gene product (the catalytic subunit of aspartate transcarbamoylase) and the Tn5 phosphotransferase (geneticin, G418, resistance marker) in plasmid pSPN[57] can be used to transfer and amplify in CHO cells deficient in CAD (*Urd⁻A*, clone D20 from David Patterson, University of Colorado). *Urd⁻A* cells are maintained in Ham's F12 me-

[53] T. D. Kemp, E. A. Swryrd, M. Bruist, and G. R. Stark, *Cell* **9**, 541 (1976).
[54] G. M. Wahl, B. R. de Saint Vincent, and M. L. DeRose, *Nature (London)* **307**, 516 (1984).
[55] D. Patterson and D. V. Carnright, *Somatic Cell Genet.* **3**, 483 (1977).
[56] B. R. de Saint Vincent, S. Delbruck, W. Eckhaert, J. Meinkoth, L. Vitto, and G. Wahl, *Cell* **27**, 267 (1981).
[57] J. C. Ruiz and G. M. Wahl, *Mol. Cell. Biol.* **6**, 3050 (1986).

dium containing 10% (v/v) fetal calf serum and supplemented with 60 μM uridine.[58] At 48 hr posttransfection, cells are selected for CAD expression by plating 10^5 cells per 10-cm tissue culture dish in medium lacking uridine and containing 10% dialyzed fetal calf serum. Selection for amplification is performed by selection cells in 100 μM PALA (obtained from the National Institutes of Health) and then increasing the concentration to 250 μM and 1 mM as the cells become sequentially resistant to each concentration. It should be possible to select and amplify this plasmid in cells that are not CAD-deficient by initial selection for transformants that are G418 resistant and then increasing the selection for CAD by growth in uridine-free medium with increasing concentrations of PALA.

Adenosine Deaminase

Adenosine deaminase (ADA) is a useful dominant selectable and highly amplifiable genetic marker for mammalian cells.[59] Although ADA is present in virtually all mammalian cells, it is not an essential enzyme for cell growth. However, under certain conditions, cells require ADA.[60-63] Since ADA catalyzes the irreversible conversion of cytotoxic adenine nucleosides to their respective nontoxic inosine analogs, cells propagated in the presence of cytotoxic concentrations of adenosine or cytotoxic adenosine analogs such as 9-β-D-xylofuranosyl adenine (Xyl-A) require ADA to detoxify the cytotoxic agent for survival. Once functional, ADA is required for cell growth, then 2'-deoxycoformycin (dCF), a tight-binding transition-state analog inhibitor of ADA can be used to select for amplification of the ADA gene.[59] As a result of one selection protocol, ADA was overproduced 11,400-fold and represented 75% of the soluble protein synthesis of the cell.[64] ADA can function as a dominant selectable and amplifiable genetic marker[59] because most cells synthesize minute quantities of ADA. Thus, it is possible to select for introduction of a heterologous ADA gene carried in an efficient expression vector. One vector, pMT3SV$_2$ADA, useful for ADA coamplification is shown in Fig. 3. It contains a transcription unit to direct

[58] J. N. Davidson, D. V. Carnright, and D. Patterson, *Somatic Cell Genet.* **5**, 175 (1979).
[59] R. J. Kaufman, P. Murtha, D. E. Ingolia, C.-Y. Yeung, and R. E. Kellems, *Proc. Natl. Acad. Sci. U.S.A.* **83**, 3136 (1986).
[60] C.-Y. Yeung, D. E. Ingolia, C. Bobonis, B. S. Dunbar, M. E. Riser, J. J. Siciliano, and R. E. Kellems, *J. Biol. Chem.* **258**, 8338 (1983).
[61] G.-Y. Yeung, M. E. Riser, R. E. Kellems, and M. J. Siciliano, *J. Biol. Chem.* **258**, 8330 (1983).
[62] C. Fernandez-Mejia, M. Debatisse, and G. Buttin, *J. Cell. Physiol.* **120**, 321 (1984).
[63] P. A. Hoffee, S. W. Hunt III, and J. Chiang, *Somatic Cell Genet.* **8**, 465 (1982).
[64] D. E. Ingolia, C.-Y. Yeung, I. F. Orengo, M. L. Harrison, E. G. Frayne, F. B. Rudolph, and R. E. Kellems, *J. Biol. Chem.* **260**, 13261 (1985).

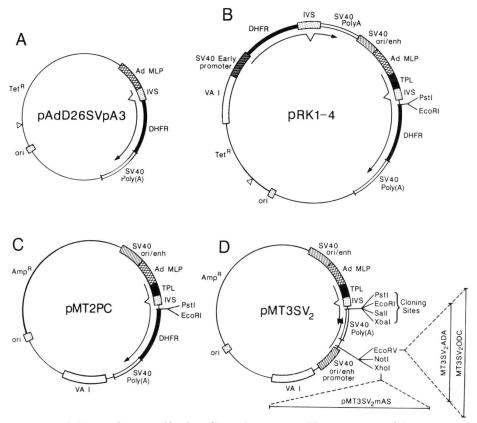

FIG. 3. Vectors for coamplification of heterologous genes. The components of the vectors described are as follows: *Tet^R* and *Amp^R*, tetracycline and ampicillin resistance genes for selection in *E. coli;* ori, bacterial origin of replication from pBR322 (A, B) or from pUC18 (C, D) for replication in *E. coli;* Ad MLP, adenovirus major late promoter for transcription initiation; IVS, intervening sequence composed of 5' splice site from the adenovirus first leader of late mRNA and a 3' splice site from an immunoglobulin gene for mRNA processing; TPL, a cDNA copy of the majority (2-$\frac{1}{3}$) of the adenovirus tripartite leader for efficient translation; SV40 ori/enh, origin of replication and enhancer element from SV40 for replication in COS monkey cells and for transcriptional activation; SV40 promoter, the early promoter of SV40 for transcription initiation; SV40 poly(A), polyadenylation signal from SV40 for transcription termination and mRNA processing; VA I, adenovirus VA I gene product to enhance translation. Selection markers: DHFR, dihydrofolate reductase; mAS, human asparagine synthetase; ADA, human adenosine deaminase; ODC, ornithine decarboxylase. The unique sites for cloning are indicated. (A) Enhancerless DHFR expression vector for cotransfection and coamplification. (B) DHFR selection vector with cloning sites for insertion for foreign coding region (*Pst*I, *Eco*RI) into an efficient transcription unit. (C) Selection vector for expression of DHFR from a polycistronic transcript. (D) General vector with alternate amplifiable selection markers and several restriction sites for insertion of a foreign coding region into an efficient transcription unit.

expression of a coding region inserted into a unique cloning site and a separate transcription unit for ADA expression.

Conditions for Selection and Coamplification with ADA. Two selection schemes have been described which can be used for selection and amplification of ADA as a dominant selectable genetic marker.

XYL-A. The adenosine analog Xyl-A is cytotoxic due to its incorporation into DNA via conversion to the purine nucleotide catalyzed by adenosine kinase. Cells may become resistant to Xyl-A by action of ADA to inactivate the adenosine analog or by deletion of adenosine kinase. A selection scheme that allows for convenient transfer of ADA genes into a variety of cells involves the use of low concentrations of Xyl-A (4 μM) in the presence of 2′dCF (0.01–0.03 μM) to inhibit endogenous ADA. This Xyl-A selection is very cytotoxic and cells usually die within 48 hr. The Xyl-A selection is the most convenient scheme for introduction of heterologous ADA genes into a variety of cells. However, amplification by selection for increased resistance to higher concentrations of dCF is variable in different cell lines. The variability is most likely due to the frequency of deletion of adenosine kinase.[59,60] For this reason, the 11-AAU selection procedure was devised to allow for amplification of heterologous genes since growth in 11-AAU requires functional adenosine kinase.

11-AAU. The 11-AAU (adenosine, alanosine, uridine) selection protocol was modified from the selection protocol for adenosine kinase[65] by increasing the adenosine concentration 11-fold to 1.1 mM. In the presence of 1.1 mM adenosine, ADA is required to alleviate adenosine toxicity. In this selection, alanosine (50 μM) blocks *de novo* AMP synthesis and thus cells require adenosine kinase to convert adenosine to AMP. Since adenosine depletes phosphoribosylpyrophosphate (PRPP), resulting in the inhibition of endogenous pyrimidine synthesis, the medium is supplemented with uridine (240 μg/ml) to allow RNA synthesis to occur.[66]

For selection, 2 days after transfection, cells are subcultured 1:15 into medium containing 11-AAU with 0.03 or 0.1 μM dCF. Cells from one plate can be subcultured 1:15 into 11-AAU medium, with some plates containing 0.03 μM and the others containing 0.1 μM dCF. Cell death upon growth in the 11-AAU selection is slower than with the Xyl-A selection. Colonies appear at 10–14 days and are then pooled and grown in sequentially 3-fold increasing concentrations of dCF.

The results of selection for ADA are variable since different cell lines have differing capacities to grow in 11-AAU medium, in the absence of dCF. If required, it may be necessary to select first for a variant of the host

[65] T. S. Chan, R. P. Creagan, and M. P. Reardon, *Somatic Cell Genet.* **4**, 1 (1978).
[66] H. Green and T. Chan, *Science* **182**, 836 (1973).

cell line that is capable of growing in 11-AAU. The basis of this initial selection for growth in 11-AAU is unknown. For selection in CHO cells, it is not necessary to preselect for growth in 11-AAU. Once a cell line is obtained that can grow in 11-AAU, then heterologous ADA genes can be transfected and selected for by growth in 11-AAU with addition of dCF $(0.03-0.1 \ \mu M)$.

Since fetal calf serum contains significant, variable levels of bovine ADA, it is important to prepare selection medium fresh prior to use. For convenience, the selection medium may be made and stored without serum. Ten percent serum can be added just prior to use. For appropriate selection, it is advisable to feed the cells frequently (every 4–5 days) with fresh selection medium. This is especially true when using the Xyl-A selection procedure. As an alternative to fetal calf serum, horse serum, which has significantly lower levels of ADA, should be used if it can support the growth of the host cells. Xyl-A and alanosine can be obtained from the Drug Synthesis and Chemistry Branch, Division of Cancer Treatment, National Cancer Institute. 2'-Deoxycoformycin can be obtained from Sigma (St. Louis, MO).

Multidrug Resistance

Cross-resistance to a variety of lipophilic cytotoxic agents such as adriamycin, colchicine, vincristine, and actinomycin D frequently results from overproduction of a 170-kDa plasma membrane glycoprotein (P-glycoprotein) due to gene amplification.[67] This membrane protein is involved in transport of these drugs out of the cell.[68-70] Some multiple drug-resistant (mdr) cell lines are significantly more resistant to the drug used in their selection than to other drugs.[71] The preferential resistance found for colchicine transport can be accounted for by the appearance of mutations within the P-glycoprotein structural gene that preferentially confer resistance to colchicine.[72] Thus, expression of different cDNA clones derived from cells resistant to different drugs may confer preferential resistance to the drug for which the original cells were initially selected. The cloned murine mdr and human mdr1 cDNAs can function as dominant selectable

[67] J. R. Riordan, K. Deuchars, N. Kartner, N. Alon, J. Trent, and V. Ling, *Nature (London)* **316,** 817 (1985).

[68] G. F.-L. Ames, *Cell* **47,** 323 (1986).

[69] P. Gros, J. Croop, and D. Housman, *Cell* **47,** 371 (1986).

[70] G.-J. Chen, J. E. Chin, K. Ueda, D. P. Clark, I. Pastan, M. M. Gottesman, and I. B. Roninson, *Cell* **47,** 381 (1986).

[71] K. W. Scott, J. L. Biedler, and P. W. Melera, *Science* **232,** 751 (1986).

[72] K. Choi, C. Chen, M. Driegler, and I. B. Roninson, *Cell* **53,** 519 (1988).

and amplifiable genetic markers.[73,74] The potential degree of gene amplification using mdr may be limited since cells exhibiting very high levels of drug resistance contain only 50 copies of the gp170 gene.[73] The human mdr1 gene has been used to coamplify major excreted protein (MEP) in NIH 3T3 cells.[74] In this case, there was approximately 10-fold amplification as colchicine resistance increased from 80 ng/ml to 1 μg/ml of colchicine.

Conditions for Selection and Coamplification with mdr. At 48 hr posttransfection, cells are subcultured at $2-4 \times 10^5$ cells/10-cm dish in growth medium containing 60 ng/ml of colchicine. Clones or pools of transformants are then grown in 80 ng/ml colchicine and are selected in 2-fold increments of colchicine. At each step, the cells should grow for 5 – 10 days before being plated into the next higher step. The particular concentration of colchicine used may vary depending on the expression vector and the particular cDNA encoding mdr. It has been possible to select for mdr transformants with as little as 7 ng/ml colchicine.[74]

An alternate to colchicine resistance selection for mdr is to select for adriamycin resistance. At 0.1 μg/ml of adriamycin, the murine mdr cDNA carried in an efficient expression vector was capable of transmitting resistance to transfected cells.[73]

Ornithine Decarboxylase

Ornithine decarboxylase (ODC) is the first enzyme in the synthesis of polyamines and is an essential enzyme for cell growth.[75] An expression plasmid encoding the ODC cDNA can complement the growth of ODC-deficient CHO cells[76] in medium lacking putrescine.[77] In these cells, it is possible to select for greater than 700-fold amplification of the ODC gene by growth in increasing concentrations of the suicide-substrate inhibitor difluoromethylornithine (DFMO). By introduction of the ODC cDNA into an efficient expression vector it is possible to select directly for ODC in wild-type cells by growth in moderate concentrations of DFMO.[77] Selection in increasing concentrations of DFMO has resulted in a 300-fold amplification of the transfected gene in wild-type cells.

Conditions for Selection and Coamplification of ODC. At 48 hr posttransfection, cells are subcultured at 10^5/10-cm dish into medium contain-

[73] P. Gros, Y. B. Neriah, J. M. Croop, and D. E. Housman, *Nature (London)* **323,** 728 (1986).
[74] S. E. Krane, B. R. Troen, S. Gal, K. Veda, I. Pastan, and M. M. Gottesman, *Mol. Cell. Biol.* **8,** 3316 (1988).
[75] C. W. Tabor and H. Tabor, *Annu. Rev. Biochem.* **53,** 749 (1984).
[76] C. Steglich, A. Grens, and I. E. Scheffler, *Somatic Cell. Mol. Genet.* **11,** 11 (1985).
[77] T.-R. Chiang and L. McConlogue, *Mol. Cell. Biol.* **8,** 764 (1988).

ing DFMO at 160 μM. The plates are refed every 5 days until a resistant population emerges. Multiple colonies are pooled and selected sequentially for resistance to 600 μM, 1 mM, 3 mM, 9 mM, and 15 mM DFMO. DFMO may be obtained from Merrell Dow Research Institute (Cincinnati, OH).

Asparagine Synthetase

Asparagine synthetase (AS) is a housekeeping enzyme responsible for the biosynthesis of asparagine from glutamine and aspartic acid. Selection for expression of heterologous AS in cells is possible by selection for growth in asparagine-free medium with addition of an appropriate concentration of an inhibitor of AS. Two AS inhibitors which can be used are albizziin, which is a glutamine analog, and β-aspartyl hydroxamate (β-AHA), an aspartic acid analog.[78] The use of AS as a dominant selection and amplification marker relies on its efficient expression to select its amplification over that of the endogenous AS gene.

The *E. coli* AS gene is more convenient to use as a dominant selectable and amplifiable marker. Since the *E. coli* enzyme uses ammonia instead of glutamine as a nitrogen source, it is resistant to albizziin. Mammalian AS uses glutamine as a nitrogen source and so is inhibited by albizziin. Dominant selection for the bacterial AS gene is possible by growth in asparagine-free medium containing albizziin to inhibit the endogenous mammalian AS. Subsequently, the transfected DNA can be amplified up to 100-fold by growth in increasing concentrations of β-AHA.[79]

Cotransformation and Coamplification with Bacterial AS. For cotransfection and coamplification with AS, plasmid DNA encoding the desired gene product is cotransfected with the bacterial AS expression plasmid pSV$_2$-AS[79] [pSV$_2$-AS can be obtained from Clifford Stanners (McGill University, Quebec, Canada)]. Forty-eight hours after transfection, cells are rinsed with selective medium and subcultured at 2×10^5/10-cm dish in selective medium. Selective medium consists of α-MEM lacking asparagine, with 30 μg/ml of glutamine, 10% dialyzed fetal calf serum, and 2 mM albizziin (purchased from Sigma). The exact concentration of albizziin required for death of untransfected cells should be determined for each cell type.

For amplification, transfected clones are grown in asparagine-free medium with 30 μg/ml of glutamine and containing 0.1 mM β-AHA (β-AHA can be obtained from Sigma). Resistant cells are cultured in medium containing increasing concentrations of β-AHA starting at 0.2 mM and

[78] I. Andrulis, J. Chen, and P. Ray, *Mol. Cell. Biol.* **7**, 2435 (1987).
[79] M. Cartier, M. Chang, and C. Stanners, *Mol. Cell. Biol.* **7**, 1623 (1987).

increasing in successive steps of 0.2 mM up to 1.5 mM. From this concentration up to 5 mM β-AHA, the steps are 1 mM each.

Cotransformation and Coamplification with Mammalian AS. Generally better success for amplification is obtained by using a mammalian AS expression vector and albizziin as the selection agent. One vector, pMT3SV$_2$mAS (Fig. 3D), contains the human AS gene[78] under transcriptional control of the SV40 early promoter. Selection for this vector in transfected cells is performed by growth in asparagine-free medium containing 2 mM glutamine and 2 mM albizziin. Resistant transformants can be amplified by increasing albizziin stepwise up to 50 mM. Further selection can be accomplished by reducing 10-fold both the concentration of glutamine (to 0.2 mM) and of albizziin. Then selection for further increases in albizziin resistance can be performed.

Other Strategies for Amplification of Transfected DNA

In addition to the drug selections described above, transformants can be selected on the basis of expression of surface antigens [for example, the T cell antigen leu-2[80] and the transferrin receptor[81]] by cell sorting using a fluorescence-activated cell sorter and specific antibodies. Sequential selection for positive fluorescence yields cells that have amplified the gene for the surface antigen.[82] Other fluorescently labeled compounds that can be used to select cells include a fluorescent conjugate of methotrexate to select for elevated DHFR levels,[52] or aryl hydrocarbons to select for cells exhibiting elevated levels of aryl hydrocarbon hydroxylase.[83] It is also possible to isolate living cells by staining of living cells for β-galactosidase activity that is introduced by transfer of specific chimeric genes encoding β-galactosidase.[84]

Other Dominant Selectable Markers

The most commonly used dominant selectable marker confers resistance to neomycin, kanamycin, and similar compounds in bacterial cells. It is encoded by the bacterial transposon Tn5 aminoglycoside phosphotransferase gene *(Neo^r)*. G418 (obtained from Gibco) is a compound related to neomycin, which can act in mammalian cells to block translation. The *Neo^r* gene introduced into a mammalian expression vector (pSV$_2$neo)[47]

[80] P. Kavathas and L. A. Herzenberg, *Proc. Natl. Acad. Sci. U.S.A.* **80**, 524 (1983).
[81] A. McClelland, L. C. Kuhn, and F. H. Ruddle, *Cell* **39**, 267 (1984).
[82] P. Kavathas and L. A. Herzenberg, *Nature (London)* **306**, 385 (1983).
[83] A. G. Miller and J. P. Whitlock, Jr., *Mol. Cell. Biol.* **2**, 625 (1982).
[84] G. P. Nolan, S. Fiering, J. F. Nicholas, and L. A. Herzenberg, *Proc. Natl. Acad. Sci. U.S.A.* **85**, 2603 (1988).

can confer G418 resistance to mammalian cells. For selection, cells are grown in 0.5–1.0 mg/ml G418. Similarly, the hygromycin B phosphotransferase gene from *E. coli,* when introduced into mammalian cells in an appropriate expression vector (pHyg 141-31, obtained from B. Sugden, University of Wisconsin, Madison, WI), can confer resistance to the antibiotic hygromycin.[85] Hygromycin can be obtained from Calbiochem (San Diego, CA) and is generally used at 100–300 μg/ml. The biological half-life of hygromycin in solution at 4° is approximately 10 days. G418 and hygromycin resistance selection schemes are independent, and both resistance genes can be selected for in cells transfected with both genes. However, neither of these selection markers has been demonstrated to be amplifiable.

Strategic Considerations

In designing an approach to obtain efficient expression of a heterologous gene through coamplification, several considerations should be taken into account: (1) the method of DNA transfer, (2) the vector for coamplification, and (3) the protocol for selection and cloning of the cells.

Methods of DNA Transfer

Since a detailed description of the different methods for transfer of DNA into mammalian cells is beyond the scope of this article,[37] discussion will be limited to how the different methods may influence the selection protocol for gene amplification. The most common method for gene transfer is to add DNA directly to the cells in the form of a $CaPO_4$ coprecipitate. Transfection frequencies may vary widely between cell lines to as high as 1 transformant per 10^4 transfected cells.[33] Via this approach DNA enters the cell though an endocytic vesicle. As a consequence, the DNA frequently becomes rearranged, possibly as a result of passage through cellular compartments of low pH or containing endonucleases. In contrast, other methods directly introduce DNA into the cytoplasm of the cell. One of these approaches relies on the polyethylene glycol (PEG)-induced fusion of bacterial protoplasts with mammalian cells.[86] This method can be very efficient (10^{-2}–10^{-4}/cell) and can be used for cells that are difficult to transfect by the $CaPO_4$ procedure. Protoplast fusion frequently yields multiple copies of the plasmid DNA tandemly integrated into the host chromosome in a head-to-tail array.[56] Since it is not likely that two different plasmids harbored in two different bacteria will integrate into the same cell, protoplast fusion is not useful for cotransfection of independent

[85] J. L. Yates, N. Warren, and B. Sugden, *Nature (London)* **313,** 812 (1985).
[86] W. Schaffner, *Proc. Natl. Acad. Sci. U.S.A.* **77,** 2163 (1980).

plasmids, in contrast to $CaPO_4$ cotransfection. Thus, protoplast fusion necessitates that the selection and amplification marker be on the same plasmid as the gene of interest. Similarly to protoplast fusion, electroporation can efficiently introduce DNA directly into the cytoplasm of recipient cells.[87,88] Electroporation does not require the selection gene and gene of interest to be on the same plasmid. However, there is less experience with this method to compare it directly with protoplast fusion and $CaPO_4$-mediated cotransfection for propensity to generate rearrangements.

Selection Vectors for Coamplification

There are two classes of vectors for coamplification: (1) those in which the product gene transcription unit is on a separate plasmid than the selection gene transcription unit, and (2) those in which the product and selection gene transcription units are contained within the same vector. The efficiency of cotransfection of separate plasmids depends on the ability of the two DNAs to be ligated within the cell. It is also possible to linearize and ligate the selection and product transcription unit plasmids before transfection. One advantage with using separate plasmids is that the ratio of the product and selection gene can be varied in favor of the product gene to ensure a greater copy number relative to the selection gene copy number. One approach to select for the cells which have incorporated both the selection gene and the product gene is to use a defective expression vector for the selection gene, for example, one that has deleted an enhancer element. In this case, cotransfection with the product gene plasmid containing an enhancer element yields cells which efficiently express the selection gene only when its transcription unit has become linked to the product gene. By this approach, the enhancer-deficient DHFR expression plasmid pAdD26SVpA3[38] (Fig. 3A) is useful for cotransformation of CHO DHFR-deficient cells and coamplification by MTX resistance selection.

Two types of vectors that have both the selection gene and the product gene within the same plasmid have been described. The first contains one transcription unit encoding the selection marker and another transcription unit containing a unique cloning site for insertion of heterologous coding regions. Different vectors described (Fig. 3B,D) have been constructed to contain DHFR, ADA, mdr, and AS. These vectors are useful for protoplast fusion and ensure that cells transformed with the selection marker will likely also incorporate the product gene.

Another approach relies on the ability of ribosomes to translate internal

---

[87] H. Potter, L. Weir, and P. Leder, *Proc. Natl. Acad. Sci. U.S.A.* **81,** 7161 (1984).
[88] K. Shigekawa and W. J. Dower, *BioTechniques* **6,** 742 (1988).

cistrons within polycistronic mRNAs. The "scanning" hypothesis for translation initiation states that ribosomes bind to the 5' end of capped mRNAs and migrate in the 3' direction until they encounter an AUG in an appropriate context that can serve as an initiator codon.[89] However, there is a low level of internal initiation of translation at internal AUG codons. This inefficient translation at internal AUG codons may yield sufficient quantities of gene product to allow selection for transcription units which harbor selection genes within the 3' end of the transcript.[90] The vector pMT2PC (Fig. 3C) contains a DHFR coding sequence within its 3' end which can be used for DHFR selection in CHO DHFR-deficient cells. Since one proposed mechanism of this internal initiation involves translation termination, continued scanning, and reinitiation, it is important that no AUG codons are present between the termination codon for the 5' open reading frame and the DHFR coding region. The optimal number of bases between the two coding regions is estimated to be approximately 100 in order to optimize DHFR translation.[91] Cell lines which are selected for heterologous gene expression by DHFR expression by this approach generally express 10-fold greater levels of the heterologous gene which is inserted into the cloning site at low levels of MTX selection. However, at high levels of MTX resistance selection, the 5' open reading frame frequently becomes deleted to allow more efficient DHFR expression. Thus, while higher expression levels may be obtained earlier in selection, the maximal expression obtained with vectors encoding polycistronic transcription units is not effectively greater than that obtained by other approaches.

Protocols for Selection for Coamplification

Two strategies for transformant selection and coamplification are as follows. In scheme A (Fig. 4), individual clonal tranformants are isolated and subsequently independently grown in increasing concentrations of the selection agent for gene amplification. In scheme B (Fig. 4), a pool of transformants is collected and grown in increasing concentrations of the selection agent for gene amplification. The final resistant pool is cloned by limited dilution plating at the end of the selection process. The advantage of scheme B is that it requires less effort to amplify larger numbers of individual transformants within the pool. Cells selected are those that become resistant to increasing MTX because the heterologous DNA has integrated into a chromosomal position that is efficiently expressed and amplified. Approximately 30% of the selected pools exhibit coamplifica-

---

[89] M. Kozak, *Cell* 22, 7 (1980).
[90] R. J. Kaufman, P. Murtha, and M. V. Davies, *EMBO J.* 6, 187 (1987).
[91] M. Kozak, *Mol. Cell. Biol.* 7, 3438 (1987).

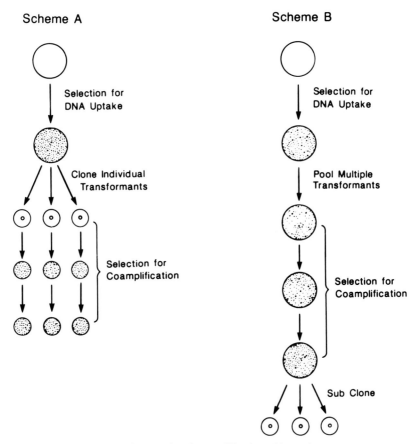

FIG. 4. Strategies for selection for amplification of heterologous genes.

tion of the desired gene with the selection gene. After several rounds of increasing MTX selection, the final cell line obtained is usually composed of cells derived from one or a very few transformants. In contrast, when high-producing individual clonal transformants are selected for amplification as depicted in scheme A (Fig. 4), it is infrequent that the cotransfected gene is coamplified with the selection gene.[38] One disadvantage of scheme B (Fig. 4) is the potential to lose good producing cell lines. It is possible to avoid this by cloning after the first step in the selection for coamplification and subsequently selecting the cloned lines for increased coamplification. The success of the particular strategy depends on the type of expression vector used and the method of DNA transfer. For this reason, the selection protocols provided below are examples that have proved most successful for a particular vector system and mode of DNA transfer.

Cotransformation by CaPO₄ Cotransfection with DHFR

1. DNA from an enhancer containing vector encoding the product gene (25 μg) is coprecipitated with pAdD26SVpA3 (2.5 μg) in 0.3 M sodium acetate, pH 4.5 (300 μl) and 2.5 volumes of ethanol at $-20°$ for 20 min.

2. DNA is pelleted by centrifugation for 5 min in an Eppendorf centrifuge at 4°. The supernatant is removed. In the tissue culture hood (the operator must wear gloves), the tubes are decanted and turned over to drain. The DNA pellets are air dried in the hood. The tubes should have no residual ethanol before continuing.

3. DNA is resuspended in 2 × HEBSS (0.25 ml) (2 × HEBSS: HEPES, 10 g/liter; NaCl, 16 g/liter; KCl, 0.74 g/liter; $Na_2HPO_4 \cdot H_2O$, 0.25 g/liter; dextrose, 2 g/liter; pH 7.05). The appropriate pH of the HEBSS is crucial and varies depending on the cell line. Typically, buffers of varying pH should be prepared and tested for each cell line for optimal transfection efficiency. The 2 × HBSS should be stored at $-70°$. The solution may be vortexed lightly to ensure the DNA is in solution.

4. To the 2 × HEBSS–DNA suspension, an equal volume of 2 × $CaCl_2$ (2 × $CaCl_2$, 0.25 M) is added slowly while bubbling. The 0.25 M $CaCl_2$ can be stored at $-20°$. The $CaPO_4$–DNA precipitate is briefly vortexed and allowed to rest for 20 min at room temperature. It is important that the precipitate be very fine and it should appear opalescent.

5. Medium is removed from CHO DHFR-deficient cells which were subcultured at 5 × 10⁵/10-cm dish 24 hr prior to transfection. The precipitate is then applied and the cells are incubated for 30 min at room temperature.

6. Then 4.5 ml of alpha medium with 10% fetal calf serum is added without removing the DNA precipitate and the cells are incubated at 37° for 4–5 hr.

7. The medium is then removed and 2 ml of alpha medium containing 10% glycerol is added for 3 min at room temperature.

8. The glycerol is removed and cells rinsed and fed with alpha medium containing 10% fetal calf serum, and 10 μg/ml each of thymidine, adenosine, and deoxyadenosine. Penicillin (100 U/ml) and streptomycin (100 μg/ml) are also included.

9. Two days later, cells are subcultured 1 : 15 into alpha medium.

10. After 10–14 days, approximately 10–100 colonies should appear per dish. Pools of 10–100 colonies are made and are selected for resistance to increasing concentrations of MTX. Selection of 6 pools for amplification will most likely ensure success in coamplification of the heterologous product gene.

For a control for the transfection and selection, it is good to include cells that do not receive the DHFR expression plasmid. Under appropriate selection, no colonies should appear in cells that did not receive a DHFR plasmid.

Coamplification with Vectors Containing Transcription Units for Both Selection Gene and Product Gene. The foreign DNA coding region is inserted into a cloning site within the appropriate expression vector, as shown in Fig. 3B,D. The DNA can be introduced into cells by electroporation after linearization by restriction endonuclease digestion of naked DNA or by protoplast fusion. Below is a protocol for protoplast fusion.

PREPARATION OF PROTOPLASTS

1. *Escherichia coli* HB101 or DH5 harboring the expression plasmid and resistant to either ampicillin or tetracycline are used for protoplast fusion. A fresh overnight culture of bacteria is inoculated (100 μl) into 50 ml of L broth containing $8-10$ μg/ml tetracycline or 50 μg/ml ampicillin and incubated at $37°$ on a rapid shaker for approximately $5-6$ hr (depending on the strain), to an $A_{600\ nm}$ of 0.6. Bacterial growth should be carefully monitored during the growth period so the 0.6 A_{600nm} is not surpassed.

2. Chloramphenicol is added to 250 μg/ml (from a 125 mg/ml stock) and the culture incubated at $37°$ for an additional $12-16$ hr in order to amplify the plasmid copy number. At the end of this incubation, the cultures can rest on ice for several hours if necessary until ready to continue. It is advisable at this point to isolate plasmid DNA from a small portion of the bacteria by a miniplasmid preparation procedure[92] and analyze it by restriction endonuclease digestion and gel electrophoresis. This analysis will ensure that no plasmid rearrangements have occurred during propagation of the cells to prepare protoplasts.

3. Centrifuge 50 ml at 3000 rpm for 10 min at $4°$. From this point it is important to keep the bacteria on ice-water.

4. Resuspend the pellet in 2.5 ml of chilled 20% sucrose in 50 mM Tris-HCl, pH 8.0.

5. Add 0.5 ml of a freshly prepared solution of lysozyme (5 mg/ml in 0.25 M Tris-HCl, pH 8.0), and swirl the tube very gently to mix.

6. Incubate on ice for 5 min.

7. Add 1 ml of 0.25 M EDTA, pH 8.0.

8. *Slowly* add 1 ml of 0.05 M Tris-HCl, pH 8.0, one drop at a time.

9. Incubate at $37°$ in a water bath for 15 min.

10. Carefully dilute the protoplast solution with 20 ml of prewarmed

[92] T. Maniatis, E. F. Fritsch, and J. Sambrook, "Molecular Cloning: A Laboratory Manual," Cold Spring Harbor Laboratory, Cold Spring Harbor, New York, 1982.

DME medium containing 10% sucrose and 10 mM MgCl$_2$. Let the solution drip down the side of the tube and invert the tube slowly and gently to mix. It is important to be gentle at this point in order to not disrupt the protoplasts.

11. Hold at room temperature for 15 min (the protoplasts can remain in the hood for up to 2 hr, and at 4° for up to 2–3 days). Before doing the fusion, invert the tube several times, gently. At this time the solution will be slightly viscous and consists of protoplasts which are bacteria with their cell walls stripped. Removal of the cell wall now permits fusion of the cell membrane with that of the mammalian cell, an event that is potentiated by polyethylene glycol treatment.

PROTOPLAST FUSION

1. One day before fusion, subculture cells into 6-well plates (Costar Cluster 6 #3506; 24-mm diameter/well). Inoculate only 3 wells/plate at 1.5 × 10^5 cells/ml, 5 ml/well in complete medium. Incubate overnight until attached. It is better to seed the cells earlier in the day in order to give them maximum time for attachment.

2. Prepare polyethylene glycol (PEG) solution: Solid PEG (Baker, Phillipsburg, NJ; U218-9, average MW 950–1050) is melted by microwaving in a 50-ml conical test tube for 1–2 min until liquid. The PEG is then diluted with an equal volume of prewarmed Dulbecco's minimal essential medium (DMEM) (50/50, v/v) and is kept at room temperature.

3. Rinse the cells three times with serum-free medium.

4. Add 1.5 ml of protoplast suspension per well.

5. Centrifuge for 5 min at 1500 rpm (only two dishes at a time) in a table-top centrifuge with appropriate adapters (IEC Centra-7).

6. Aspirate by tilting the wells gently.

7. Add 1.5 ml of PEG/well and leave on for 1 min (for CHO cells). The PEG time will vary for different cell types. The appropriate time must be determined accordingly by testing for cell death and efficiency of transformation. The time may vary from 15 sec to 2 min.

8. The cells are rinsed six times with serum-free medium. Do not tip the plate when aspirating PEG from the wells. Dilute the wells with 5 ml of the medium and then aspirate it gently from one angle of the well, aspirating with one hand while adding medium in the well with the other hand.

9. Add 5 ml of complete medium containing 100 µg/ml kanamycin (Flow laboratories, McLean, VA; #16-720-48; 5000 µg/ml = 50×). This will kill any bacteria that have escaped conversion to protoplasts.

10. Incubate for 48–72 hr at 37°, then subculture cells into 5 × 10cm plates/well into selective medium.

11. After colonies appear, pools of 25–300 transformants/pool are prepared and selected for increasing resistance to selective agent for amplification.

Coamplification Using Polycistronic mRNA Expression Vectors. Polycistronic mRNAs can be translated in mammalian cells; however, the efficiency of translation of the internal cistron is variable and inefficient.[89-91] Insertion of a coding region upstream from DHFR within the expression vector pMT2PC (Fig. 3C) yields a bicistronic transcript in which DHFR is poorly translated. However, selection for the low level of DHFR expression from this vector is possible by growth selection of DHFR-deficient CHO cells in the absence of nucleosides. Since there is a strong selection for deletion of the upstream open reading frame when selecting for expression of the 3' open reading frame in polycistronic vectors, it is important to take precautions to minimize the frequency of deletions as well as the potential for selection cells harboring deletions within the 5' open reading frame to outgrow the population. For this reason, the DNA should be introduced by an efficient means and one which does not frequently result in rearrangements, such as protoplast fusion or electroporation. In addition, the strategy for transformant amplification should follow that of Fig. 4 scheme A, in which independent clonal transformants are isolated and subsequently grown in increasing selection for amplification. If one were to pool the transformants as in scheme B, the likelihood is greater that a cell harboring a deletion will outgrow. In addition, the increasing steps of selection for gene amplification should be small to encourage amplification and also to minimize deletion. Although this approach has potential difficulties, the advantages are that it may render higher degrees of expression at early stages of the amplification process.

Acknowledgments

I gratefully thank Patricia Murtha, Kimberly Marquette, Louise Wasley, Monique Davies, Debra Pittman, Maryann Krane, David Israel, and Andrew Dorner for their experiments which were important in formulating many of the ideas presented. I also thank Andrew Dorner for critically reading this article and Michelle Wright for assistance in its preparation.

[43] Optimizing Cell and Culture Environment for Production of Recombinant Proteins

By JENNIE P. MATHER

Introduction

Cloning a gene is only the first step in the study of genetic regulation and gene function. Further studies usually require that the gene be expressed *in vitro,* frequently in a mammalian cell culture system. The cell culture and expression systems chosen will have a significant impact on the amount and, frequently, the biochemical characteristics of the protein produced from the cloned gene. While the exact parameters necessary for successful expression of a given protein in mammalian cells cannot yet be predicted with great accuracy, several factors have been shown to contribute to optimizing expression. This article outlines some of the critical elements in producing acceptable expression levels in minimum time, especially where the techniques for handling cell lines expressing recombinant proteins might differ from normal cell culture practice.

Choice of Cell Line, Plasmid, and Transfection Method

The choice of cell line, plasmid construction, and transfection method are interdependent and beyond the scope of this article. However, in all cases, cells which are cultured in conditions which lead to optimum growth will yield higher transfection frequencies and be easier to select, clone, and grow in order to obtain optimal expression of the desired recombinant protein. The choice of a cell line which can be grown in serum-free medium may later prove to be a significant advantage in detecting the recombinant protein at low levels of expression where the only available assays are not specific for the protein of interest or do not work in the presence of serum. The use of serum-free medium may also provide an advantage in purifying secreted proteins from the medium.

Different cell types expressing the same protein and the same cell type expressing different proteins should be treated as separate cell lines, and each one will frequently have distinctly different properties. However, the optimal medium and growth supplements will be most closely related in cells derived from the same parent cell line. This is important in selecting the conditions for comparing recombinant proteins produced from two separate transfection events (e.g., expressed in two different cell lines). It is

METHODS IN ENZYMOLOGY, VOL. 185

frequently more meaningful to compare two lines each grown in conditions optimal for that line, rather than in identical conditions.

It should be remembered that substances such as methotrexate or antibiotics, which are used to select for successfully transfected cells or to amplify genes, are toxic to cells and frequently mutagenic. In addition, the recombinant product being expressed may be detrimental to cell growth and viability at the high concentrations desired. Therefore, cells transfected with recombinant genes are frequently required to grow in adverse conditions. For this reason, it is best to optimize the culture conditions to the fullest extent possible, first for the parent line to be used for transfection and then for the selected recombinant line.

Cloning

The purpose of cloning is to ensure that all of the cells in the culture are descended from a single cell, i.e., are genetically identical. This prevents the rapid and unpredictable changes in culture phenotype which may occur in mixed cell populations when conditions change to favor one cell type over another (e.g., a nonproducer cell over one producing a recombinant product). Perhaps more importantly, it allows the separation of high producer from low or nonproducing cells after the initial transfection and selection events. It should be emphasized, however, that there can be considerable change with time even in cloned populations. These changes can be genetic, and therefore irreversible, or only a phenotypic response to changing or marginal culture conditions which can be controlled or reversed. There are two methods of cloning which will be discussed here: (1) cloning by limiting dilution; and (2) cloning with the use of cloning rings.

Cloning in soft agar is also frequently used, but works best with highly transformed cells and is somewhat more cumbersome than the other two methods described. Clones can also be picked from plates using sterile cotton swabs. The cells are then transferred by dipping the swab in fresh medium in another plate. This method is rapid but is more likely to result in the cross-contamination of clones than the two methods mentioned above unless the colonies are very widely spaced on the plate from which the clones are picked. Cloning by limiting dilution can be used with suspension or attached cells and should be used with suspension cells or with cells which are very mobile when attached.

Conditioning of the medium is sometimes necessary to get good growth at the very low cell densities used in cloning. This is especially true when cloning amplified recombinant cells which may be growing under adverse conditions or in low-serum or serum-free conditions. To condition medium, the parent line used for transfection should be grown to high density, the medium changed, and this "conditioned" medium collected 24–48 hr later, before the medium components have been exhausted. This medium is then filtered through a 0.1 μm filter to remove any cells which may be floating in the medium and then it is resterilized. The conditioned medium mixed 1:1 (v/v) with fresh medium is used when plating the cells for cloning. Cells should be diluted so that approximately half of the wells of a 96-well multiwell plate contain cells when 50 μl/well of medium is added (10 cells/ml if the cells have a 100% plating efficiency).

The cloning ring method allows one to inspect visually the colonies to be selected more easily and thus to select cells with desired, or different, morphologies. It can also be easier to use with cells with very low plating efficiencies. Cloning rings are usually used when cloning between amplification steps, as the cells are frequently already growing in widely dispersed colonies at the end of the selection. The plates should be examined microscopically and selected colonies marked on the plate. After placement of the cloning rings they can be checked visually to ensure that only one colony has been included. The following points refer to both methods.

1. When plating cells for cloning it is essential that they be in a dispersed single-cell suspension. Visually inspect to see that most cells in the suspension are separate before plating.

2. Check colonies visually early during growth and mark wells or colonies which arise from a single cell. If two cells are in a well, or a colony starts from two cells it should be discarded.

3. When using any of the above methods, the cloning should be repeated 2–3 times in succession to ensure that one has a true clone.

4. To obtain the maximum plating efficiency when cloning, the medium used should be the normal maintenance medium without methotrexate or other selective agent during the low density growth prior to cloning and during the first subsequent passage. Afterward the selective agent can be returned to the medium. This is less stressful on the cells and allows a higher plating efficiency. An exception to this may be cloning from plates in which only a few colonies have survived after amplification or selection.

5. When clones are picked, they should be passaged first into one well of a 24-well dish, then to 35-mm and 100-mm plates. This maintains a relatively high cell density throughout the growth of the clone. Conditioned medium may also be used in the first passage after cloning, if necessary.

Pools

For some applications, such as transient expression and expression of stably integrated genes in some cell types, cloning may not be necessary. This saves a good deal of time initially. However, the pool may not be as stable to prolonged culture as a clone would be. The pool is also likely to be a mixture of cells with widely differing expression levels, including cells not producing the desired recombinant protein, and therefore will not produce at as high a level as a clone selected for high production. Adequate numbers of vials of the pool should be frozen down as soon as possible so that new cells can be thawed periodically (every few weeks).

Fluorescence-Activated Cell Sorting

If there is a way of fluorescence tagging living cells a fluorescence-activated cell sorter (FACS) may be used to select for, and clone, cells producing high levels of the desired recombinant protein. Conditioned medium, as described above for cloning by limiting dilution, may improve cloning efficiency. The FACS may also be used to assess the homogeneity of the population and the need for cloning from the original pool. If the selected pool is heterogeneous for expression of the recombinant protein, it is best to obtain a clonal population. It should be emphasized that FACS sorting is an excellent method for obtaining cells which produce high amounts of cell-associated protein (e.g., cytoplasmic, membrane-bound), but cellular levels of secreted proteins may not correlate well with specific productivity.

Amplification

If large quantities of the recombinant protein are desired for purification and further studies, amplification of the transfected gene may be accomplished using one of several amplifiable systems, such as dihydrofolate reductase/methotrexate,[1] ornithine decarboxylase/dihydrofluoromethylornithine,[2] or glutamine synthase/methionine sulfoxamine.[3]

These systems frequently require the use of selective media which are deficient in some essential nutrients and selective agents which are toxic to

---

[1] R. Kaufman and P. A. Sharp, *J. Mol. Biol.* **159,** 601 (1982).

[2] T. R. Chang and L. McConologue, *Mol. Cell. Biol.* **8,** 764 (1988).

[3] C. R. Bebbington and C. C. G. Hentschel, *in* "DNA Cloning, Volume III: A Practical Approach" (D. M. Glover, ed.), p. 163. IRL Press, Oxford, 1987.

normal cells. In these instances, it is again especially important to use cell culture media that are optimal for the growth of the cell used to express the protein to be amplified. Amplification of the desired gene will frequently result in poor growth performance of the resulting cell population. At some point this poor growth may offset the gains of the increased specific productivity of the cells. This may be counteracted to some extent by selecting for cells with good growth characteristics, as well as high specific productivity when picking clones. In its simplest form, this would mean assaying for protein production after a short (24 hr) and a more extended (e.g., 1 week) period in culture. The initial value would primarily reflect the amount of product produced per cell, while the extended production of protein will be a reflection of both specific productivity and viable cell days in culture (which in turn reflects both growth rate and viability), as well as other parameters such as protein stability.

Choice of Culture Method

The choice of culture method will depend on the characteristics of the recombinant protein, the type of studies to be carried out in the transfection system, the amount of recombinant protein desired, and the properties of the recombinant cell line obtained. Thinking through these factors carefully initially can save a good deal of time later in obtaining a workable experimental or production system.

If the purpose of the transfection is to demonstrate that the transfected gene can be expressed and/or to study the genetic regulation of that expression, very little of the actual protein will be required and any one of a number of established cell lines can be used for expression. The choice might then be dictated by the efficiency of transfection and expression or the characteristics of the cell line itself (e.g., use of a liver cell line to study regulation of expression of a secreted protein).

If large amounts of protein will be required for future studies, then some type of culture system should be used which can be scaled up to produce large amounts of cells or culture fluid. Roller bottle production is the most straightforward, but is labor intensive and limited in the degree to which it can be scaled up. The next choice might be growth in spinners, or fermentors. This requires a cell line which naturally grows in suspension or has been adapted to do so (see below). The fermentor systems have been scaled up to thousands of liters.[4]

[4] W. R. Arathoon and J. R. Birch, *Science* 232, 1390 (1986).

Suspension Adaptation

The purpose of suspension adaptation is to obtain a cell line which will grow as single cells unattached to a substrate. These cells are then capable of being easily scaled up for production in spinners or fermentors using existing technologies. This is usually the fastest and least expensive method of obtaining relatively large amounts of recombinant protein for further study.

There are several approaches to obtaining this end: (1) alter the medium so that the ability of cells to attach is eliminated or diminished, (2) select for cells which will not attach in the standard medium conditions, or (3) select for cells which will grow in suspension in the standard media when the surfaces available have been treated to prevent attachment but will attach to standard tissue-culture-treated surfaces in standard serum-containing medium.

The first approach has the disadvantage that the media devised to promote cell detachment (generally with much reduced magnesium and calcium, e.g., Joklik's medium) are frequently suboptimal for supporting high titers of desired proteins. The second approach is adequate for production cell lines but is more difficult than the third and the resulting cell lines less flexible in use. The third approach is generally (but by no means always) reasonably rapid, and results in a line that can be grown in an attached state for further manipulation such as cloning or transfection.

The approach outlined below is designed to suspension adapt cells in this third sense with as little alteration in other cell properties as possible. The one exception to this rule is that we have in several cases chosen to suspension adapt in a reduced-serum, hormone-supplemented medium in order to obtain a line which will grow continuously in these conditions. In at least one case, this strategy also improved our ability to suspension adapt the cells and obtain a stable phenotype after transfection. There are other ways to suspension adapt cells but we have been most successful with this one.

In cases where maintaining selective pressure on the cells during suspension adaptation is desired, this may be done, but it may make the adaptation to suspension more lengthy or more difficult. Cultures can most easily be adapted to suspension and then the selective pressure reintroduced after the cells are growing well in suspension, or the cells recloned and clones selected which are high producers and grow well in suspension. After the suspension adaptation described below, cells may be grown as attached cells for cloning, and cloned populations grown up and reintroduced into suspension with little or no difficulty.

Protocol

Medium. The medium used should be the medium selected for the optimal growth of the desired cell line (see below). Supplement the medium with 1–5% (v/v) fetal bovine serum and insulin (5 μg/ml) or the ITS supplement (which can be obtained from Collaborative Research, Bedford, MA). Determine the minimum amount of serum necessary for optimal growth in the medium to be used by obtaining a serum dose–response curve in the presence and absence of the insulin or ITS supplement. Further supplementation with a polyol such as F-68[5] at 0.1% (w/v) and an organic buffer such as HEPES (10–20 mM) will help protect the cells from mechanical damage and provide additional buffering capacity during the suspension growth. The presence of polyols can be critical for success in suspension adapting some cell lines (see Fig. 1).

Spinners. Use well siliconized spinners. Solutions for silicone coating glassware can be obtained from Dow Chemical (Dowcoat). We prefer using 250-ml spinners with 50- to 100-ml volumes of culture medium. Spinners are run at a rate sufficient to keep cells and small aggregates in suspension (50–100 rpm) while minimizing damage to the cells. Set up two spinners in parallel, subculturing on different days. If the cells accumulate around the spinner shaft and on the sides of the spinner at the medium surface the spinner is not properly siliconized. If this occurs, cells tend to form clumps which break off and allow other cells to attach to them, thus reducing the likelihood that they will grow as single cells.

Culture. Trypsinize starting culture and set up the spinners at 2–5 × 10^5 cells/ml. Check cells daily. Sterile sodium bicarbonate solution may be added if the pH drops below 6.8. After the first day or so, the caps on the spinners should be loosened to allow for increased oxygen exchange. On day 3 or 4, the cells should be counted and passaged. Initially, cells should be centrifuged and fresh medium added to bring the cell number back to between 2 and 5 × 10^5/ml. As the cells start growing to densities over 10^6/ml they may be passaged by dilution of the suspended cells with fresh medium. The cell density should be such as to allow at least a 1 : 5 split. If cells clump excessively, large clumps should be allowed to settle before subculture and not passaged.

After 1–2 months, the cells should be capable of logrithmic growth in suspension and reach densities of > 10^6 cells/ml when inoculated at densities of 5–10 × 10^4 cells/ml. Cell viability should remain at > 90% throughout the growth period. At this point, the cells are termed "suspension

[5] A. Mizrahi, *J. Clin. Microbiol.* **2,** 11 (1975).

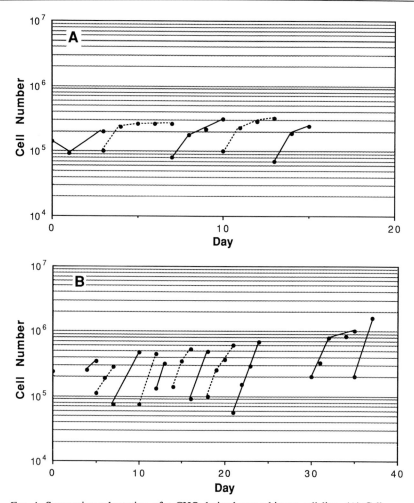

FIG. 1. Suspension adaptation of a CHO-derived recombinant cell line. (A) Cells were grown in spinners in medium which did not contain a polyol. The cells never grew to high densities. (B) The same cells were grown in medium with the addition of 0.1% F-68. The cells grew well by the second week and were fully adapted to suspension growth by 6 weeks in culture.

adapted" even though they will still attach to tissue culture plastic in serum-containing medium and may exhibit some clumping in suspension under some conditions.

In Fig. 1 the cell counts and passage times (arrows) are shown for the course of adapting a CHO recombinant clone. The cells were considered suspension adapted after 5 weeks. The amount of time required for this degree of adaptation may vary widely from one cell line to another or from one clone to another. Where suspension growth is essential, it is wise to start adapting several clones and eliminate those that do not adapt well.

Growth to High Cell Densities

High-density growth may be desired to obtain high titers of secreted recombinant proteins. If this is desirable, cells should be screened for the ability to grow to high densities during the cloning stage. The extended screen described above will tend to identify cells which can grow to high densities. Selection of a medium designed for high-density cell growth (e.g., Dulbecco's modified Eagle's medium or F12/DME) rather than the use of medium designed for clonal growth (e.g., Ham's nutrient mixtures) is essential.

Selection of Optimal Medium

The optimization of culture media, especially for serum-free or low-serum growth, has been carefully outlined elsewhere.[6-8] The optimization of each individual component of the medium is the only way to ensure that one is using the best medium for a given cell line. However, it is relatively easy, and much quicker, to screen a number of commercially available media as a good first approximation. For best results, these media formulations should be purchased as dry powdered mixtures, prepared in the laboratory using highly purified water, and stored for no more than 1 month at 6°. Serum, F-68, and HEPES may be added at the time of medium preparation but other additions, such as insulin and other hormones, should be prepared as separate stocks and added at the time of use. These media should be screened in the presence of the minimal amount of serum required, as optimizing the nutrient mixture frequently results in a decreased serum and/or hormone requirement. Media designed for the growth of cells frequently used for expression of recombinant proteins such as CHO (e.g., Hana Biologicals, Alameda, CA, CHO medium) in low-serum levels (e.g., 0.1–2%) are becoming more widely available.

[6] R. G. Ham, in "Methods for Preparation of Media, Supplements, and Substrata for Serum-Free Animal Cell Culture" (D. W. Barnes, D. A. Sirbasku, and G. H. Sato, eds.), p. 3. Alan R. Liss, New York, 1984.
[7] C. Waymouth, in "Methods for Preparation of Media, Supplements, and Substrata for Serum-Free Animal Cell Culture" (D. W. Barnes, D. A. Sirbasku, and G. H. Sato, eds.), p. 23. Alan R. Liss, New York, 1984.
[8] J. P. Mather and M. Tsao, in "Large Scale Mammalian Cell Culture Technology" (A. R. Lubiniecki, ed.), in press. Marcel Decker, New York, 1990.

Serum-Free Medium

A great deal has been written about the development of serum-free media. The use of serum-free media is advantageous in instances where experimental control over the culture environment is required. Thus, if one wishes to study the hormonal regulation of gene expression, it is best to do these studies in a defined, serum-free, hormone-supplemented medium where all of the culture variables can be studied independently. Serum-free or low-serum media can also provide a significant advantage when a secreted protein is to be purified from the culture medium. Elimination of the serum results in the desired protein constituting a significant percentage of the protein in the conditioned medium.

The most straightforward way to achieve this end is to transfect the gene of choice into a cell line for which a defined medium is already available. A number of these culture systems have been described in detail in the last decade.[9-11] This same medium should then be sufficient to grow the recombinant cell line, particularly if clones picked for study are tested for production of the protein in the defined medium. Any clones unable to grow in this medium can be eliminated initially. Persistent inability to grow transfected cells in the medium that the parent line grew in probably reflects an inhibitory action of the recombinant protein on the cells or medium components.

Selection for Desired Characteristics

The use of mammalian cells to express recombinant proteins, especially for the large-scale production of these proteins, has tremendous potential for engineering the cells as hosts in ways that go beyond transfecting the gene coding for the desired product. The methods employed can be divided into four categories: (1) selection of cells with the desired characteristic, (2) introducing the desired characteristic through mutation, (3) fusion of two cells each of which exhibits a desired trait followed by selection for these traits, or (4) introduction of new genes whose expression results in the desired phenotype. The most critical step for the success of any of these approaches is setting up the conditions which will eventually be used to select for the desired phenotype. Examples of these three approaches are

[9] G. H. Sato and R. Ross, "Hormones and Cell Culture," Books A and B. Cold Spring Harbor Laboratory, Cold Spring Harbor, New York, 1979.
[10] D. W. Barnes, D. A. Sirbasku, and G. H. Sato, "Cell Culture Methods for Molecular and Cell Biology," Vols. 1–4. Alan R. Liss, New York, 1984.
[11] G. H. Sato, A. B. Pardee, and D. A. Sirbasku, "Growth of Cells in Hormonally Defined Media," Books A and B. Cold Spring Harbor Laboratory, Cold Spring Harbor, New York, 1982.

the suspension adaptation of cells as outlined above, the selection of a *dhfr⁻* mutant for use in gene amplification, and the creation of hybridomas producing monoclonals of desired specificities. Each of these techniques is extremely powerful and not yet used to its full potential. The future of recombinant gene expression should see the optimization of the characteristics of the cells themselves as well as the media and cell culture systems.

[44] Analysis of Synthesis, Processing, and Secretion of Proteins Expressed in Mammalian Cells

By ANDREW J. DORNER and RANDAL J. KAUFMAN

Introduction

The secretion of biologically active protein from mammalian cells is the final step in a complex pathway of posttranslational modifications performed in the endoplasmic reticulum (ER) and Golgi complex (GC). Many of the steps of the secretory pathway in mammalian cells have been reviewed (see Fig. 1).[1,2] Proteins destined for the exocytic pathway are first cotranslationally translocated into the lumen of the ER. During translocation, an amino-terminal leader peptide is, in most cases, proteolytically removed and a high-mannose oligosaccharide core is enzymatically transferred to asparagine residues located in the sequence Asn-X-Ser/Thr, where X can be any amino acid except proline. In the ER the initial steps of carbohydrate processing occur. Terminal glucose residues are rapidly removed by glucosidases I and II and at least one α-1,2-linked mannose residue is removed by an α-1,2-mannosidase in the ER. Acylation with long-chain fatty acids may also occur in the ER, usually by addition of palmitate or myristate to cysteine or serine residues. Protein is transported to the GC where further modifications occur. Transit out of the ER has been identified as a potential rate-limiting step in secretion.[3]

During traversal of the GC a series of reactions separated spatially and temporally involve the removal of mannose residues by mannosidases I and II and addition of *N*-acetylglucosamine, fucose, galactose, and sialic acid residues by specific transferases to modify high-mannose carbohydrate to complex forms (Fig. 1). In addition to N-linked glycosylation, proteins can also have carbohydrate attached to serine and threonine residues. The initial step in this O-linked glycosylation is the direct transfer of *N*-acetyl-

[1] R. Kornfeld and S. Kornfeld, *Annu. Rev. Biochem.* **54,** 631 (1985).
[2] A. M. Tartakoff, *Int. Rev. Cytol.* **85,** 221 (1983).
[3] H. F. Lodish, N. Kong, M. Snider and G. J. Strous, *Nature (London)* **304,** 80 (1983).

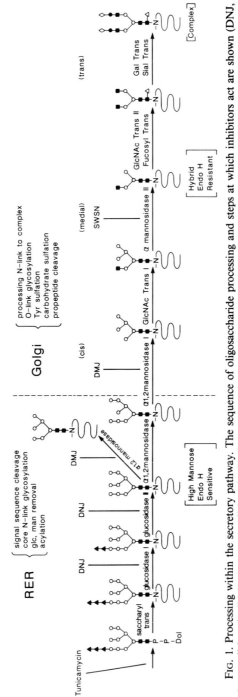

FIG. 1. Processing within the secretory pathway. The sequence of oligosaccharide processing and steps at which inhibitors act are shown (DNJ, deoxynojirimycin; DMJ, deoxymannojirimycin; SWSN, swainsonine). Also indicated are the compartmentalization of the GC (cis, medial, trans) and the point of acquisition of Endo H resistance. Sugars: ▲, glucose; ○, mannose; ■, N-acetylglucosamine (GlcNAc); △, fucose; ●, galactose (Gal); ◇, sialic acid (Sial). trans, Transferase. (Adapted from Kornfeld and Kornfeld[1] and S. Libresco and Genzyme.)

galactosamine to the protein.[4] A consensus amino acid sequence for O-linked glycosylation has not been identified. The GC is the site of proteolytic cleavage of precursor proteins to their mature forms[5] and of the addition of sulfate to tyrosine residues and carbohydrate.[6]

The modification of a protein as it transits the ER and GC is dependent on the conformation it presents to the processing enzymes. Each modification may further alter the conformation to make regions of the protein more or less accessible to successive processing enzymes. Thus, the major determinant of the final form of a protein is its primary structure and the conformation presented through successive processing steps. Alteration of the conformation of a protein by incorporation of amino acid analogs or inhibition of an initial processing step such as N-linked glycosylation can change later processing steps or reduce secretion competency.[7,8] Our experience in the expression of heterologous proteins in mammalian cells is that each protein will present its own set of characteristics and potential problems for efficient secretion in a biologically active form, and that the protein produced by heterologous expression in a mammalian cell usually carries modifications very similar to protein derived from the natural source. However, it has been observed that variation in the extent of modification may occur for the same protein expressed in different cell types, perhaps as a result of differences in the level or activity of processing enzymes.[9,10] This article presents various techniques for examining the glycosylation state, proteolytic processing, sulfation, and acylation of proteins synthesized in mammalian cells.

Radiolabeling Protein and Preparation of Cell Lysates

The most commonly used procedure to radiolabel protein is the incorporation of [^{35}S]methionine or [^{35}S]cysteine, both of which are essential amino acids. Since these amino acids must be supplied in the medium, the intracellular pool can first be reduced by incubation of cells in methionine-free or cysteine-free medium before addition of radioactive amino acids. This allows high incorporation of radioactive amino acid into protein. The efficiency of incorporation is dependent on several factors. First,

[4] R. Kornfeld and S. Kornfeld, in "The Biochemistry of Glycoproteins and Proteoglycans" (W. Lennarz ed.), p. 34. Plenum, New York, 1980.
[5] J. S. Bond and P. E. Butler, *Annu. Rev. Biochem.* **56**, 333 (1987).
[6] P. A. Baeuerle and W. B. Huttner, *J. Cell. Biol.* **105**, 2655 (1987).
[7] M. Green, *J. Biol. Chem.* **257**, 9039 (1982).
[8] A. J. Dorner, D. G. Bole, and R. J. Kaufman, *J. Cell Biol.* **105**, 2665 (1987).
[9] P. Hsieh, M. R. Rosner, and P. W. Robbins, *J. Biol. Chem.* **258**, 2548 (1983).
[10] D. B. Williams and W. J. Lennarz, *J. Biol. Chem.* **259**, 5195 (1984).

the protein of interest must contain Met or Cys residues. Second, the cells should be used at logarithmic stage of growth when they are metabolically active since the intensity of incorporation depends on the rate of protein synthesis. Third, the amount of radiolabeled protein detected in the cell extract or secreted into the conditioned medium will depend on its half-life and rate of secretion.

In many cases, initial labeling experiments will be done in a transient expression system such as COS-1 cells containing SV40-driven vectors[11] to check the functioning of the expression vector prior to establishment of stable cell lines. Transiently transfected monolayer cultures of approximately 2×10^6 cells in 100-mm dishes are usually radiolabeled 48–60 hr posttransfection. For stable cell lines such as those derived from Chinese hamster ovary (CHO) cells, approximately 2×10^6 to 10^7 cells can be labeled in a 100-mm plate depending on cell type and confluency. In this case, the cells are usually split out 48 hr and fluid changed 24 hr prior to the experiment to ensure a population of growing cells.

1. Remove medium from cell monolayers by aspiration.

2. Wash once with prewarmed medium minus Met or Cys. Add 2 ml of medium minus Met or Cys. Incubate cells for 10 min at 37°. Methionine-free medium can be obtained from Flow Laboratory (McLean, VA) as Eagle's minimum essential medium (MEM) (modified) with Earle's salts and without methionine or glutamine and should be supplemented with 1 mM glutamine. Cysteine-free medium can be prepared with the MEM Select-Amine kit available from Gibco Laboratories (Grand Island, NY).

3. Remove medium. Add 1 ml of fresh medium minus Met or Cys supplemented with 100–500 μCi/ml of ^{35}S-labeled amino acid and incubate at 37°. For suspension cell cultures, cells should be centrifuged, washed in prewarmed medium minus Met or Cys, and recentrifuged. Cells are then resuspended in this same medium at 10^7 cells/ml and incubated for 10 min at 37°. ^{35}S-Labeled amino acid then can be added at 100–500 μCi/ml and the incubation continued at 37°. Suspension cells should be gently agitated if labeling time is greater than 30 min.

4. If serum is required for cell viability, dialyzed serum can be included in the labeling medium. Dialysis removes low-molecular-weight molecules such as nucleosides and amino acids from the serum. For dialysis, 1 liter of heat-inactivated serum is dialyzed against 8 liters of phosphate-buffered saline (PBS) for 16 hr at 4° with one change of PBS after 8 hr.

5. The length of time for pulse labeling is dependent on the purpose of the experiment. Short labeling periods of 5–15 min serve to identify the

[11] R. J. Kaufman, *Proc. Natl. Acad. Sci. U.S.A.* **82,** 689 (1985).

primary translation product and assess the expression level of the protein of interest. Times greater than 1 hr may require that the labeling medium be supplemented with complete medium or the addition of unlabeled methionine or cysteine to a concentration of 1 mM, since depletion of the high-specific-activity radiolabeled amino acids can occur.

6. Following a pulse with radiolabeled amino acids, a chase with medium containing excess unlabeled amino acid can be performed to follow the intracellular processing and fate of the radiolabeled protein synthesized during the pulse. By varying the length of chase time, the rate of appearance or secretion of the mature form can be determined. Remove the labeling medium and dispose of as radioactive waste. Rinse the cells once with prewarmed complete medium. Add 2 ml of prewarmed complete medium containing serum and 0.5% (v/v) aprotinin (Sigma, St. Louis, MO) and incubate at 37°. Aprotinin or other protease inhibitors is added to the chase medium if degradation of secreted protein is a problem. The length of chase can vary from 5 min to 24 hr and is dependent on the half-life and secretion rate of the protein.

7. Remove the medium into tubes on ice and save the medium for analysis of the secreted protein. To prevent proteolytic degradation after the conditioned medium has been collected, it is advisable to add more protease inhibitors to the harvested medium, such as 1 mg/ml soybean trypsin inhibitor (Sigma) and 1 mM phenylmethanesulfonyl fluoride (PMSF; Sigma), and freeze immediately.

8. Wash the cell monolayer 3–4 times with ice-cold PBS and keep the plates on ice. Make sure that the PBS is completely removed before lysing the cells.

9. Cell lysis can be accomplished with a variety of ionic and nonionic detergent-containing buffers. Since in many cases the proteins of interest will be found within the secretory pathway or associated with membranes, two lysis buffers which give good solubilization of cells will be described below. Following cell lysis, proteins become increasingly susceptible to degradation by released intracellular proteases. Thus, it is important to add protease inhibitors to lysis buffers and keep the cell extracts on ice.

Radioimmune Precipitation Buffer

Radioimmune precipitation (RIPA) buffer consists of 150 mM NaCl, 20 mM Tris, pH 7.5, 0.1% (w/v) SDS, 1% (w/v) sodium deoxycholate, 1% (v/v) Triton X-100, 1 mM PMSF, and 1 mg/ml soybean trypsin inhibitor. EDTA (0.5 mM) can also be included to reduce nonspecific immunoprecipitation. Since pH and ionic strength can have an important influence on the extent of solubilization and subsequent recovery of protein, several variations of RIPA buffer can be made by increasing the NaCl to 0.5 M or changing the pH. One characteristic of lysis with RIPA buffer is the formation of a viscous cell extract, probably due to disruption of nuclei and release of DNA. This viscous suspension can be sheared by 10 passages through a 23-gauge syringe needle and debris pelleted by centrifugation for 5 min in a microfuge at 4°.

NP-40 Lysis Buffer

NP-40 lysis buffer consists of 150 mM NaCl, 50 mM Tris, pH 7.5, 0.05% SDS, 1% (v/v) NP-40, 1 mM PMSF, and 1 mg/ml soybean trypsin inhibitor.

Lysis with Nonidet-P40 (NP-40) lysis buffer does not produce a viscous suspension compared to RIPA buffer, perhaps because it does not lyse nuclei. Both buffers provide similar recovery of protein from the secretory pathway. NP-40 lysis buffer is the buffer of choice when the affinity of the antibody is unknown, since the level of ionic detergent is lower compared to RIPA buffer and will allow the binding of lower affinity antibodies.

10. For a 100-mm plate, 1 ml of cold lysis buffer containing protease inhibitors is added and the cells scraped off the plate with a rubber policeman. The cell extract is transferred to a 1.5-ml microfuge tube and kept on ice for 10 min. Debris is pelleted by centrifugation in a microfuge for 5 min at 4° and the cleared cell extract can be transferred to a fresh tube for storage at −20°.

11. Cell extracts can be prepared as described above following radiolabeling with a variety of ³H-labeled sugars such as mannose, galactose, or galactosamine. Usually, the labeling medium is glucose-free medium such as Dulbecco's modified Eagle's medium (DMEM) without glucose (GIBCO; special order) which has been supplemented with 100 μg/ml glucose. Cells are washed twice with this low-glucose medium and then 1 ml of medium containing radioactive sugar at concentrations of 0.25 – 1 mCi/ml is added. Labeling times can range from 30 min to overnight.

Immunoprecipitation

An essential regeant for the analysis of heterologous expression is an antibody which can recognize different forms of the protein of interest. In some cases, it may be possible to use a mixture of monoclonal antibodies which recognizes a number of different epitopes or a polyclonal antiserum. It is important to note that it is not possible to predict the epitopes which will be exposed on a protein following detergent extraction of cells, so that a wide specificity of the antibody is desirable in the initial analysis. It is therefore preferable to use a polyclonal antibody which has been raised against denatured protein.

1. Trichloroacetic acid (TCA) precipitable radioactivity is a measure of the amount of radiolabeled amino acid which has been incorporated into protein since protein is insoluble in TCA while free amino acid is soluble. A 2- to 5-μl aliquot of cell lysate is spotted onto a filter of Whatman 3MM paper and air dried. The filter is then washed on ice three times with 10% TCA, once with 100% ethanol, air dried, and the radioactivity left on the filter measured by liquid scintillation counting. A 1-ml extract of 2×10^6 cells which had been labeled for 30 min with [^{35}S]methionine can produce 5×10^5 to 10^6 TCA-precipitable counts per 5-μl aliquot. A normal immunoprecipitation should use approximately $10-20 \times 10^6$ counts which will correspond to $50-100$ μl of extract.

2. The dilution of antibody used for efficient precipitation will depend on the titer and avidity of the antibody preparation and must be determined empirically by doing a series of immunoprecipitations with constant volume of cell extract and varying the amount of antibody. Before adding antibody to the cell extract, it is advisable to supplement the antibody with protease inhibitors at the same concentration as used in the cell extract. To remove any precipitated material or debris in the antibody preparation it is advisable to centrifuge the antibody for 5 min in a microfuge at 4° prior to use.

3. An aliquot of cell extract or conditioned medium is added to a 1.5-ml microfuge tube on ice along with antibody and the tube is placed at 4° on a rotating platform such as hematology/chemistry mixer model 346 (Fisher Scientific, Fairlawn, NJ). Incubation time with the primary antibody can range from 2 hr to overnight. If degradation or instability of the protein of interest is a problem, then the shorter time should be used. Cell extracts and conditioned medium may be subjected to a preclearing step prior to immunoprecipitation to reduce nonspecific precipitation. This is accomplished by adding 100 μl of protein A–Sepharose (described below) to 1 ml of extract or medium and incubating for 1 hr at 4° in the presence of protease inhibitors. The protein A–Sepharose is pelleted by centrifugation for 3 min at 4° and the supernatant removed for immunoprecipitation.

4. After the primary antibody has been bound, a precipitating agent which can either be another antibody directed against the primary antibody or protein A bound to Sepharose beads must be added. The precipitating agent must be added in excess to ensure a quantitative precipitation of the immune complexes. For rabbit antibodies, protein A from *Staphylococcus aureus* bound to Sepharose beads can be used. For mouse monoclonal antibodies, protein A will not adsorb well and so a secondary antibody, usually rabbit anti-mouse immunoglobulin, is used. Secondary antibodies against immunoglobulins from different species can be obtained from Organon Teknika (BCA/Cappel Products, West Chester, PA). The binding

capability of protein A for various immunoglobulin classes of different species is shown in Table I. Protein A–Sepharose beads are prepared by resuspending 0.1 g of protein A–Sepharose CL-4B (Pharmacia, Piscataway, NJ) in 20 ml of TENN buffer (50 mM Tris, pH 7.5, 5 mM EDTA, 0.15 M NaCl, and 0.5% NP-40) and allowing the beads to settle. The buffer is removed and the procedure repeated. Two milliliters of TENN buffer is added to the final pellet to give a final volume of 2.5 ml and a final protein concentration of 40 mg/ml. Fifty microliters of 40 mg/ml protein A–Sepharose beads (for rabbit antibodies) or 50 μl of 1 mg/ml rabbit anti-mouse immunoglobulin (for mouse antibodies) is added to the immunoprecipitation reaction and incubated for 1 hr or more at 4°. Immune complexes are pelleted by centrifugation in a microfuge for 2 min at 4°. Pellets are washed twice with PBS containing 1% Triton X-100 and once each with PBS containing 0.5% Triton X-100 and PBS containing 0.05% Triton X-100. Other wash protocols are possible, such as one wash with RIPA buffer containing 0.5 M NaCl and two washes with RIPA buffer containing 0.15 M NaCl. Washes are done by adding 0.5 ml of wash buffer to the pellet, vortexing briefly, and spinning for 2 min in the microfuge at 4°. In most cases, the final pellet is resuspended in polyacrylamide gel electrophoresis (PAGE) sample buffer and can be stored at −20°. 1× PAGE sample buffer consists of 62.5 mM Tris, pH 6.8, 2% SDS, 5% glycerol, 0.7 M 2-mercaptoethanol, 0.025% bromphenol blue (BPB), and is usually made up as a 2× stock as follows: 2.5 ml of 0.5 M Tris, pH 6.8, 4 ml of 10% SDS, 2 ml of 100% glycerol, 1 ml of 2-mercaptoethanol, and 0.5 ml of 1% BPB.

SDS–Polyacrylamide Gel Electrophoresis

By far the most commonly used PAGE system is based on the method of Laemmli.[12] The formulation and resolution of 15 × 17 cm 1.5-mm-thick slab gels [BRL (Gaithersburg, MD) vertical gel system model V16] are given below (see Fig. 2).

	Running Gel		
	6%: 250–45 kDa	10%: 150–20 kDa	15%: 90–5 kDa
30% acryl/0.8% bis (ml)	8	13	20
1.5 M Tris, pH 8.8 (ml)	10	10	10
10% SDS (ml)	0.4	0.4	0.4
H$_2$O (ml)	22	17	10
30% APS (ml)	0.1	0.1	0.1
TEMED (ml)	0.05	0.05	0.05

[12] U. K. Laemmli, *Nature (London)* **227,** 680 (1970).

TABLE I

BINDING OF PROTEIN A TO IMMUNOGLOBULIN CLASSES OF DIFFERENT SPECIES[a]

Species	Immunoglobulin class	Notes
Rabbit	IgG	Binds quantitatively to all serum IgG
Mouse	$IgG_{2a,2b,3}$	Binds to IgG_1 at elevated pH (8.1)
Goat	IgG_2	Binds to IgG_1 at elevated pH (9.1)
Dog	IgG	Binds qunatitatively to all serum IgG and most serum IgM
Guinea pig	$IgG_{1,2}$	
Rat	$IgG_{1,2c}$	
Sheep	IgG_2	
Pig	$IgG_{1,2}$	
Cow	IgG_2	
Syrian hamster	$IgG_{1,2}$	
Man	$IgG_{1,2,4}$	Binds to some IgM

[a]Compiled from R. Lindmark, K. Thoren-Tolling, and J. Sjoquist, *J. Immunol. Methods* **62**, 1(1983). The various classes of immunoglobulin from different species which bind well to protein A are listed.

The running gel is the lower gel and is poured first. For this gel size, approximately 30 ml of the above mix should be poured to 4 cm below the top of the smaller glass plate to leave room for the stacking gel. To ensure an even surface the gel should be overlayed with a small amount of *n*-butanol. After polymerization the butanol is removed, the upper stacking gel is poured, and the lane comb inserted. The same stacking gel is used for all three running gel concentrations:

30% acryl/0.8% bis	2.7 ml
0.5 *M* Tris, pH 6.8	5 ml
10% SDS	0.6 ml
H_2O	7.7 ml
30% APS	0.05 ml
TEMED	0.04 ml

Gels can be run overnight at 30–50 V or in 6–7 hr at 100–150 V until the BPB dye runs off the gel. For the shorter running times, it is preferable to run the sample through the stacking gel at a lower rate to ensure good resolution. This usually corresponds to 90 V and 30 mA, and the voltage is increased after the samples have entered the running gel.

Samples are resuspended in the PAGE sample buffer described above and denatured by heating to 90–100° for 3 min. Running buffer can be made up as a 10× stock as follows: 120 g of Tris base, 575 g of glycine, 20 g of SDS in a final volume of 4 liters. This will have a pH of 8.3 and should be diluted to 1× prior to use. Following electrophoresis, gels are

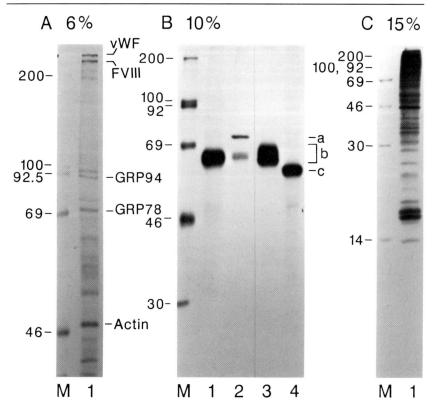

FIG. 2. Resolution of different percentage polyacrylamide gels. (A) Total cell extract of [³⁵S]methionine-labeled Chinese hamster ovary (CHO) cells expressing heterologous proteins (6% gel). Molecular weight markers are indicated, as are various proteins: FVIII, 230K factor VIII; vWF, 260K von Willebrand factor; GRP78, glucose-regulated protein of 78K; GRP94, glucose-regulated protein of 94K; actin, 46K. (B) Immunoprecipitation of [³⁵S]methionine-labeled cell extracts and conditioned medium of CHO cells expressing tissue plasminogen activator (tPA) (10% gel). Molecular weight markers are indicated as well as (a) GRP78, (b) glycosylated tPA, heterogeneous doublet of 68K arising from variability in glycosylation site usage, and (c) unglycosylated tPA, 60K. (C) (15% gel). Total cell extract of [³⁵S]methionine labeled pre-B cells (15% gel). Molecular weight markers are indicated.

processed by treating with Fix solution, such as 10% TCA, 10% acetic acid, 30% methanol, for 30 min to overnight. After fixing, gels should be treated with a fluorographic agent according to the supplier's recommendations,[13] dried, and exposed to film such as Kodak X-OMAT AR at −70°. This fluorographic treatment is essential for ³H-labeled material since autora-

[13] G. Perng, R. D. Rulli, D. L. Wilson, and G. W. Perry, *Anal. Biochem.* **173**, 387 (1988).

diographic sensitivity is increased > 500-fold, and useful for ^{35}S- and ^{14}C-labeled material since sensitivity is increased 10-fold.[14]

Analysis of N-Linked Glycosylation

Many secretory and membrane-associated proteins are glycoproteins which have oligosaccharide attached to asparagine residues. The addition and processing of this N-linked carbohydrate can be crucial for proper folding and secretion of proteins.[15] The carbohydrate composition of a protein can have dramatic effects on biological function, half-life, and antigenicity.[16] Thus, an analysis of the N-linked glycosylation of a heterologous protein is important. Aside from a complete carbohydrate analysis of a protein, which is beyond the scope of this article, several glycosidase digestions can be performed to give an indication of the amount of complex carbohydrate compared to high-mannose carbohydrate that is present on the secreted form of the protein. The acquisition of Endo H-resistant complex carbohydrate can also be used to monitor the movement of intracellular protein from the ER to the GC. Finally, several inhibitors of the addition or processing of N-linked glycosylation can be used on cells to examine the role of N-linked glycosylation in secretion and biological function.

Glycosidases

Two glycosidases of different specificities are most widely used. Endo-β-N-acetylglucosaminidase (Endo H) cleaves high-mannose and certain hybrid-type carbohydrate at the GlcNAcβ1 – 4GlcNAc linkage to leave a single GlcNAc residue attached to the asparagine.[17] The modification to a complex structure which confers resistance to Endo H digestion occurs in the medial GC following the action of GlcNAc transferase I and mannosidase II.[18] Peptide-N^4-(N-acetyl-β-glucosaminyl)asparagine amidase (N-gly) cleaves all N-linked carbohydrate regardless of complexity by hydrolyzing the asparginyl–oligosaccharide bond and leaving no sugar residue on the protein.[19] Thus, a comparison of the mobility of a protein after treatment with endoglycosidase H or N-glycanase will indicate the amount of high mannose compared to complex carbohydrate which has been added to the

[14] W. M. Bonner and R. A. Laskey *Eur. J. Biochem.* **46**, 83 (1974).
[15] R. Kornfeld and S. Kornfeld, *Annu. Rev. Biochem.* **54**, 631 (1985).
[16] T. W. Rademacher, R. B. Parekh, and R. A. Dwek, *Annu. Rev. Biochem.* **57**, 785 (1988).
[17] A. L. Tarentino and F. Maley, *J. Biol. Chem.* **249**, 811 (1974).
[18] M. G. Farquhar, *Annu. Rev. Cell Biol.* **1**, 447 (1985).
[19] A. L. Tarentino, C. M. Gomez, and T. H. Plummer, *Biochemistry*, **24**, 4665 (1985).

protein and the contribution that carbohydrate makes to the molecular weight observed by PAGE. In many cases, the protein secreted from cells may appear heterogeneous in size. Digestion with glycosidases can determine if this heterogeneity is due to N-linked glycosylation or some other modification. If a natural source of the protein is available, then a comparison of the utilization of the N-linked sites can be done to ascertain whether they are being processed in the heterologous system in the same way as the natural source.

The isolation of protein for digestion can be accomplished by immunoprecipitation of cell extracts or conditioned medium. Protein can also be concentrated without immunoprecipitation by precipitation with acetone. Acetone precipitation is accomplished by adding 8 volumes of 80% acetone to cell extract or conditioned medium and placing it on dry ice until frozen. Usually this can be done in a 1.5-ml microfuge tube and the protein precipitate recovered by centrifugation for 4 min in a microfuge. The supernatant is removed and the pellet briefly dried. The pellet is resuspended in 60 μl of 50 mM Tris, pH 6.8, 0.5% SDS, 0.1 M 2-mercaptoethanol and heated to 90–100° for 5 min. The completeness of digestion is dependent on the accessibility of the oligosaccharide to the enzyme and so denaturation is essential. After denaturation the sample is divided into three 20-μl aliquots to which the following additions are made.

Additions	Final concentration
Control	
20 μl of 0.15 M sodium acetate, pH 6.0	75 mM
0.5 μl of 0.1 M PMSF	1.25 mM
Endo H	
20 μl of 0.1U/ml Endo H in 0.15 M sodium acetate, pH 6.0	50 mU/ml Endo H 75 mM
0.5 μl of 0.1 M PMSF	1.25 mM
Ngly	
5 μl of H$_2$O	
8 μl 1 M sodium phosphate, pH 8.3	0.2 M
5 μl of 14% NP-40	1.75%
1.5 μl of 250 U/ml N-gly	9.4 U/ml
0.5 μl of 0.1 M PMSF	1.25 mM

For Endo H digestion, the concentration of SDS must be reduced by dilution to 0.3% or less. For N-gly digestion, NP-40 in a 7-fold excess compared to SDS must be added prior to addition of the enzyme to prevent inactivation of N-gly. The digestion is usually allowed to incubate overnight at 37°, although shorter times for complete digestion can be determined. The exact amounts of enzyme required for digestion will depend on the concentration of total protein in the reaction and it is best to refer to the supplier's recommendations. Although the amount of radiolabeled protein is probably low, an immunoprecipitate will contain a significant amount of immunoglobulin which must be taken into account. The above conditions are suitable for immunoprecipitated material from 150 to 200 μl of cell extract or conditioned medium or 3 μg of a purified glycoprotein. After incubation the reactions can be stopped by the addition of an equal volume of 2× sample buffer and analyzed by SDS–PAGE.

Inhibitors of Carbohydrate Addition or Processing

The use of inhibitors acting at specific steps in the glycosylation pathway allows the synthesis of protein which is either unglycosylated or has altered processing of its oligosaccharides (Fig. 1). This allows one to determine the role of N-linked glycosylation in the processing and secretion of the protein and in its biological properties if the protein is secreted. Tunicamycin is an antibiotic which blocks the formation of the dolichol phosphate-linked oligosaccharide donor and so prevents N-linked glycosylation.[20] Deoxynojirimycin (DNJ) is a glucose analog which inhibits ER glucosidases I and II, blocking the removal of the three glucose residues from the oligosaccharide core unit.[21] Removal of these glucose residues is necessary for further processing of the oligosaccharide, and so proteins synthesized in the presence of DNJ have an increased proportion of high-mannose carbohydrate. Deoxymannojirimycin (DMJ) is a mannose analog which blocks several α-mannosidases, preventing the processing of high-mannose oligosaccharides to the complex forms.[22] Two different α-1,2-mannosidases are inhibited by DMJ: GC mannosidase I, which is responsible for conversion of Man_8 oligosaccharide to Man_6 and an ER mannosidase of similar specificity which appears to act on some proteins which fail to reach the GC.[23] Another ER mannosidase which removes only a single mannose residue from proteins before they move to the GC is not inhibited by DMJ. Swainsonine (SWSN) is an inhibitor of GC mannosidase II which removes terminal α-1,3- and α1,6-linked mannose residues.[24] Swainsonine treatment prevents the formation of complex carbohydrate structures and results in the accumulation of Man_5-containing oligosaccharides. The inhibitors are added to the normal medium of the cells. The following concentrations have been reported efficacious but it is

[20] A. Takasuki, K. Kohno, and G. Tamura, *Agric. Biol. Chem.* **39**, 2089 (1975).
[21] B. Saunier, R. D. Kilker, J. S. Tkacz, A. Quaroni, and A. Herscovics, *J. Biol. Chem.* **257**, 14155 (1982).
[22] U. Fuhrmann, E. Bause, G. Legler, and H. Ploegh, *Nature (London)* **307**, 755 (1984).
[23] J. Bischoff, L. Liscum, and R. Kornfeld, *J. Biol. Chem.* **261**, 4766 (1986).
[24] H. P. Broquist, *Annu. Rev. Nutr.* **5**, 391 (1985).

advisable to determine an effective inhibitor concentration for each cell type: tunicamycin (Sigma), 10 μg/ml; DNJ (Genzyme, Boston, MA), 2–5 mM; DMJ (Genzyme), 1 mM; SWSN (Sigma), 0.1 mM.

1. Cells are washed once with medium and then fresh medium containing inhibitor is added. Cells should be incubated for 1–2 hr with the inhibitor before beginning the radiolabeling procedure.

2. The same pulse–chase procedure is used for inhibitor-treated cells as normal cells. Inhibitors should be present in the medium throughout the labeling and chase periods. Cell extracts and conditioned medium can be immunoprecipitated to determine if secretion has been impaired.

3. Secreted radiolabeled protein can be examined for the presence of high-mannose carbohydrate by Endo H digestion to determine the extent of inhibition of processing since DNJ, DMJ, and SWSN treatment will result in increased amounts of Endo H-sensitive high-mannose carbohydrate on proteins. Conditioned medium from treated cells can also be harvested and analyzed for biological activity.

Analysis of O-linked Glycosylation

O-Linked oligosaccharides are linked to the hydroxyl group of serine or threonine residues through an O-glycosidic bond to N-acetylgalactosamine (CalNAc).[25] Galactose, fucose, and sialic acid are commonly found attached to the core GalNAc. GalNAc is not found on N-linked carbohydrate. Since GalNAc is the linkage sugar for O-linked carbohydrate GalNAc is required for proper O-linked glycosylation, but its absence should have no effect on N-linked glycosylation. The addition of O-linked carbohydrate occurs in the GC concomitant with complex N-linked carbohydrate processing.[26,27] Analysis of O-linked glycosylation has been hampered by the lack of specific inhibitors such as are available for N-linked glycosylation and a dearth of glycosidases which cleave O-linked carbohydrate.

The best system now available to study the role of O-linked glycosylation on protein secretion and function is a mutant CHO cell line which has a reversible defect in O-linked glycosylation. This line, 1d1D, is deficient in UDP-galactose and UDP-N-acetylgalactosamine 4-epimerase and so cannot synthesize Gal or GalNAc under normal culture conditions.[28] Without

[25] R. Kornfeld and S. Kornfeld, in "The Biochemistry of Glycoproteins and Proteoglycans" (W. Lennarz, ed.), p. 34. Plenum, New York, 1980.

[26] C. Abeijon, and C. B. Hirschberg $J.$ $Biol.$ $Chem.$ **262**, 4153 (1987).

[27] R. D. Cummings, S. Kornfeld, W. J. Schneider, K. K. Hobgood, H. Tolleshaug, M. S. Brown, and J. L. Goldstein, $J.$ $Biol.$ $Chem.$ **258**, 15261 (1983).

[28] D. M. Kingsley, K. F. Kozarsky, L. Hobbie, and M. Krieger, $Cell$ **44**, 749 (1986).

these sugars normal N-linked and O-linked glycosylation cannot occur. The addition of Gal to the medium allows normal N-linked glycosylation to take place but, since GalNAc is the linkage sugar used only in O-linked glycosylation, proteins will not have O-linked carbohydrate. Addition of both Gal and GalNAc permits normal N- and O-linked glycosylation. The 1d1D cell line can be used as a host for expression of a heterologous protein.[29] Cells grown in medium supplemented with both Gal and Gal-NAC at 0.1 mM produce normally glycosylated protein. When supplemented with 0.1 mM Gal alone, these cells produce proteins deficient in O-linked carbohydrate.

The only enzyme available to remove O-linked carbohydrate from protein is endo-α-N-acetyl-D-galactosaminidase (O-glycanase) (Genzyme; Boehringer Mannheim Biochemicals, Indianapolis, IN). The enzyme cleaves the Galβ1,3galNAc disaccharide unit linked to serine or threonine.[30] Substituents on the Gal or GalNAc will inhibit O-glycanase cleavage. For many structures the substituent will be sialic acid, which can be removed by digestion with neuraminidase (Genzyme). Immunoprecipitated protein is thus sequentially digested with neuraminidase and O-glycanase.

1. Protein is immunoprecipitated or acetone precipitated from cell extracts or conditioned medium and the final pellet is resuspended in 40 μl of 0.5% SDS, 0.1 M 2-mercaptoethanol. Protein is denatured by heating to 90–100° for 5 min. The following additions are made.

Additions	Final concentration
10 μl of 14% NP-40	1.75%
2 μl of 0.4 M calcium acetate	10 mM
4 μl of 0.4 M Tris–maleate, pH 6.0	20 mM
22 μl of H$_2$O	
1 μl of 0.1 M PMSF	1 mM
Neuraminidase	1 U/ml

The galactosidase inhibitor D-galactolactone (Genzyme) can also be added to the reaction at a concentration of 10 mM to prevent removal of galactose residues from N-linked glycosylation by any contaminating galactosidases. These conditions are designed for enzymes purchased from Genzyme. The requirement for calcium in the reaction is dependent on the source of neuraminidase, and exact digestion conditions should be referred

[29] M. M. Matzuk, M. Krieger, C. L. Corless, and I. Boime, *Proc. Natl. Acad. Sci. U.S.A.* **84**, 6354 (1987).
[30] J. Umemoto, V. P. Bhavanandan, and E. A. Davidson, *J. Biol. Chem.* **252**, 8609 (1977).

to in the supplier's recommendations. Tris-HCl should not be used as a buffer since chloride ions inhibit O-glycanase. The addition of NP-40 in a 6- to 7-fold excess over SDS is necessary to retain enzyme activity.

2. Neuraminidase treatment should be for 1 hr at 37°. After this incubation, the reaction is divided into two 40-μl aliquots and O-glycanase added to a final concentration of 100–200 U/ml to one aliquot. The incubation should continue overnight at 37°. The reaction can be terminated by addition of an equal volume of 2× PAGE sample buffer and analyzed by SDS–PAGE.

Reduction in molecular weight or a decrease in the heterogeneity of a protein species following neuraminidase and O-glycanase treatment compared to neuraminidase alone indicates the presence of O-linked carbohydrate. Neuraminidase treatment alone is also useful to examine the extent of sialation, since the decrease in molecular weight following treatment is indicative of the amount of sialic acid which has been added to the protein. Unsialated or incompletely sialated protein will have terminal Gal residues exposed which can result in *in vivo* clearance by the liver and thus reduced half-life.[31]

Analysis of Tyrosine Sulfation

The addition of sulfate to tyrosine as an O^4-sulfate ester is a common posttranslational modification of secretory proteins. The sulfation reaction occurs in the trans GC and may be one of the last modifications before secretion.[32] Sulfated tyrosines are usually located within acidic regions of the protein. All sites which have been characterized have aspartic and glutamic acid residues near the sulfated tyrosine and an acidic amino acid is usually found immediately N-terminal to it in the -1 position.[33] At least three acidic residues are found between -5 and $+5$ around the tyrosine. No known tyrosine sulfation sites are adjacent to N-linked glycosylation sites or within seven residues of a cysteine.

Cells can be labeled with [^{35}S]sulfate to incorporate sulfate at both tyrosine residues and carbohydrate. These two sites can be differentiated by labeling cells in the presence of an inhibitor of N-linked glycosylation or treating [^{35}S]sulfate-labeled protein with N-glycanase to remove carbohydrate.

1. For radiolabeling with [^{35}S]sulfate similar cell densities of 0.2–1 × 10^7 per 10-cm dish as described above for [^{35}S]methionine labeling can be

[31] G. Ashwell and A. G. Morell, *Adv. Enzymol.* **41,** 99 (1974).
[32] P. A. Baeurle and W. B. Huttner, *J. Cell Biol.* **105,** 2655 (1987).
[33] W. B. Huttner, this series Vol. 107, p. 200.

used. [^{35}S]Sulfuric acid can be obtained from New England Nuclear (Boston, MA) (NEX-042). The labeling medium employed should be sulfate free and can be obtained from Hazelton Research Products (Lenexa, KS). In addition, sulfate-free medium with reduced levels of methionine and cysteine to 2% of normal concentration may be prepared from a kit available from Hazelton. This lower level of methionine and cysteine appears to increase the efficiency of sulfate incorporation. However, it must be determined that the reduced amino acid level is not having an adverse effect on protein synthesis, particularly if the labeling period is long. This can be checked by monitoring incorporation of [^{35}S]methionine into proteins in a separate experiment. To reduce unlabeled sulfate levels further, dialyzed serum should be used when required for cell viability.

2. Cells are washed once with medium and then 2 ml of sulfate-free medium is added and the cells are incubated for 30 min at 37°. This medium is removed and 3 ml of fresh sulfate-free medium containing 0.5–1 mCi [^{35}S]sulfuric acid and dialyzed fetal calf serum is added. Cells can be labeled overnight in this medium.

3. Cell extracts and conditioned medium are harvested and treated as described for ^{35}S-amino acid-labeled proteins. The tyrosine sulfate ester bond is acid labile, making it important to avoid low pH in lysis and gel conditions. The lysis buffers and gel buffers outlined in this article are safe to use but it is advisable to avoid the use of TCA. After a 24-hr labeling period, approximately 20–30% of the radioactive sulfate is incorporated into protein.

4. Since sulfate can also be added to carbohydrate it is necessary to differentiate between tyrosine and carbohydrate sulfation. One way is to perform the sulfate label in the presence of tunicamycin to block N-linked glycosylation. This has its drawbacks, since some proteins are not secreted efficiently in the absence of appropriate glycosylation and it has been reported that inhibition of glycosylation can result in different posttranslational modifications than normally occur.[34] Alternatively, immunoprecipitated protein can be treated with *N*-glycanase to remove N-linked carbohydrate before PAGE analysis. This treatment should leave only sulfate linked to tyrosine on the protein.

Fatty Acid Acylation

The addition of long-chain fatty acid residues such as myristate and palmitate to proteins along the secretory pathway has been described. A number of enveloped virus glycoproteins and membrane-associated recep-

[34] P. A. Baeurle and W. B. Huttner, *EMBO J.* **3**, 2209 (1984).

tors have been reported to contain covalently bound fatty acids.[35] It is possible that the presence of acyl groups may facilitate intermolecular interactions with membranes by providing a hydrophobic group on the region of the protein. The fatty acylation of proteins appears to be an early event in the secretory pathway occurring in a late ER compartment. It has been demonstrated that a hybrid glycoprotein which fails to be transported from the ER to the GC becomes acylated in the ER.[36] The acyltransferase has also been localized to the ER by subcellular fractionation experiments.[37]

1. Cells can be labeled with [³H]myristic acid or [³H]palmitic acid. The cells should be labeled in serum-free medium or dialyzed serum which has been delipidated by extraction with butanol: diisopropyl ether should be used.[38] The labeled fatty acids are dissolved in dimethyl sulfoxide and added to the medium at a concentration of 0.2–0.5 mCi/ml. Similar cell numbers and media volumes can be used as described for radioactive amino acid labeling. A pulse as short as 20 sec has been used to obtain incorporation of palmitate into a viral glycoprotein. In general, a metabolic labeling of 4 hr is more commonly used.

2. Cell extracts can be prepared and immunoprecipitations performed as described above and the acylated proteins resolved by SDS–PAGE.

3. The nature of the covalent bond can be examined by treatment of the gels with either hydroxylamine or alkali. Material is run on several gels in parallel. After fixation gels are treated for 16 hr at room temperature with 1 M hydroxylamine, pH 7.0, or 1 M Tris, pH 10. As a control, gels are treated with 1 M Tris, pH 7.0. Hydroxylamine or alkali lability of the acylation is indicative of a thioester or O-ester linkage of the fatty acid to the protein. Resistance to this treatment is characteristic of an amide linkage. If the linkage is labile, the radioactivity associated with the acylated protein will disappear from the gel following treatment compared to the control gel as monitored by autoradiography. Palmitate is primarily bound by ester linkages and is labile while myristate is attached by amide bonds and is resistant.[39]

4. Significant interconversion of myristate to palmitate has been observed during the 4-hr labeling period and this possibility must be kept in mind when interpreting labeling results.[39] Shorter labeling periods such as 90 min can be employed to circumvent this possibility.

[35] D. A. Towler and J. I. Gordon, *Annu. Rev. Biochem.* **57,** 69 (1988).
[36] L. J. Rizzolo and R. Kornfeld, *J. Biol. Chem.* **263,** 9520 (1988).
[37] M. Berger and F. G. Schmidt, *FEBS Lett.* **187,** 289 (1985).
[38] B. E. Cham and B. R. Knowles, *J. Lipid Res.* **17,** 176 (1976).
[39] E. N. Olson, D. A. Towler, and L. Glaser, *J. Biol. Chem.* **260,** 3784 (1985).

Western Immunoblot Analysis

An alternative method to immunoprecipitation for the analysis of protein is the Western immunoblot technique.[40,41] This method involves the electrophoretic transfer of protein from SDS–polyacrylamide gels to nitrocellulose filter. Antibody is then incubated with the filter and binds to the protein of interest. Detection of antibody complexes on the filter is accomplished by incubation with ^{125}I-labeled protein A or ^{125}I-labeled secondary antibody which binds to the primary antibody. The advantage of this technique is that large numbers of cell lines can be analyzed simply by harvesting conditioned medium or cell extracts and resolving them by PAGE. In this way, the level of secretion of different cell lines can be detected and compared.

1. Resolve the conditioned medium or total cell extracts by PAGE.

2. For electrophoretic transfer, a Trans-Blot apparatus (BioRad Laboratories, Richmond, CA) and nitrocellulose sheets with pore size of $0.45 \mu m$ (BA85) or $0.1 \mu m$ (PH79) are used (Schleicher and Schuell, Keene, NH) depending on the molecular weight of the protein of interest. For proteins below 20K in molecular weight it is advisable to use the smaller pore size. Cut the nitrocellulose and two pieces of Whatman 3MM filter paper to the size of the gel and wet with transfer buffer (25 mM Tris, pH 8.3, 192 mM glycine, 20% methanol). Transfer buffer can be made as follows: 3.03 g of Tris, 14.4 g of glycine, 200 ml of methanol in a volume of 1 liter. If made correctly, the pH should be 8.3

3. The blot sandwich is constructed in the following order: (a) Place a piece of Whatman 3MM on top of the Scotch-Brite pad supplied with the blotting apparatus. (b) Place the polyacrylamide gel on the Whatman paper. (c) Place the nitocellulose sheet on top of the gel. (d) Place a piece of Whatman 3MM on the nitrocellulose sheet. Care must be taken to eliminate any air pockets within the sandwich, particularly between the gel and the nitrocellulose.

e) Place the other Scotch-Brite pad on top and insert this blot sandwich in the transfer cell containing transfer buffer with the nitrocellulose sheet nearest the positive electrode.

4. Electrotransfer can be accomplished at 30 V (120 mA) for 16 hr at room temperature. We have transferred protein of greater than 200 kDa with this procedure.

5. The following buffers are prepared:

Low-salt blocking: 20 mM Tris, pH 7.6, 0.15 M NaCl, 0.05% Triton X-100, 0.5% casein (sodium salt)

[40] V. C. W. Tsang, J. M. Peralta, and A. R. Simons, this series Vol. 92, p. 377.
[41] W. N. Burnette, *Anal. Biochem.* **112,** 195 (1981).

Low-salt wash: 20 mM Tris, pH 7.6, 0.15 M NaCl, 0.05% Triton X-100

High-salt wash: 20 mM Tris, pH 7.6, 1 M NaCl, 0.4% N-laurylsarcosine

6. Following transfer, disassemble the blot sandwich and remove the nitrocellulose from the gel. Wash two times in low-salt blocking buffer. Washes are done in a shallow container for 15 min each at room temperature with slight agitation. Nitrocellulose should be completely immersed and free to move with agitation.

7. The primary antibody is diluted into 10 ml of low-salt blocking buffer. The antibody concentration used will depend on its titer and must be determined. It is essential to use antibody which recognizes denatured protein and will be bound by protein A. If the antibody is not bound by protein A, it is possible to use radioiodinated secondary antibody against the primary antibody. Not all antibodies work well in a Western immunoblot and it is best to use a polyclonal rabbit antiserum, although we have had success with goat- and burro-derived antiserum.

8. Incubation with the primary antibody is done for 1 hr at room temperature in a sealed plastic bag with slight agitation.

9. Following incubation, wash four times for 15 min each in low-salt wash buffer.

10. The blot is then incubated in 10 ml of low-salt blocking buffer containing 2 μCi/ml ^{125}I-labeled protein A (IM144; Amersham, Arlington Heights, IL) for 1–2 hr at room temperature with slight agitation.

11. Remove protein A mixture to radioactive waste. Wash four times for 15 min each with high-salt wash buffer at room temperature with agitation. Allow nitrocellulose to air dry and expose to Kodak X-OMAT film at room temperature.

Acknowledgments

We wish to thank our colleagues L. Wasley, P. Murtha, K. Kerns, D. Pittman, M. Davies, and C. Wood for contributing methods and discussion, F. Lindon for artwork, and E. Fritsch and E. Alderman for critical reading of the manuscript.

Section VI

Mutagenesis

[45] Saturation Mutagenesis Using Mixed Oligonucleotides and M13 Templates Containing Uracil

By BRUCE H. HORWITZ and DANIEL DiMAIO

The development of site-directed mutagenesis techniques has greatly enhanced our ability to study biological phenomena on the molecular level. It is now possible to change predetermined nucleotides at virtually any position on a DNA molecule. Sometimes, rather than producing one specific mutant, it is desirable to produce a set of mutants containing nucleotide substitutions within a relatively localized region. Such an approach is termed saturation mutagenesis and it can be a valuable tool for studying and modifying gene expression. The uses of saturation mutagenesis include the definition of essential nucleotides in a cis-acting regulatory region and determining the effect of random amino acid substitutions on the function of a protein. A variety of saturation mutagenesis procedures have been described. These procedures generally entail subcloning a small mutagenized DNA fragment into an appropriate vector. To achieve this goal, several techniques have been developed that increase the efficiency and target specificity of chemical mutagenesis.[1,2] One approach developed by Myers *et al.*[1,3] relies on a physical technique to separate small segments of chemically mutagenized DNA from wild-type DNA, and usually requires the use of a specialized cloning vector. Cassette methods have also been described for saturation mutagenesis that take advantage of the advances made in the production of synthetic DNA.[4-7] In these methods, a population of oligonucleotides is chemically synthesized so that each oligonucleotide in the population contains, on average, a single base substitution compared to the wild-type sequence. Such a "doped" oligonucleotide is then directly cloned into the appropriate vector. A drawback of the methods discussed above is that the ends of the mutagenized target sequence must be compatible with unique restriction endonuclease cleavage sites present at the proper positions in the vector. Moreover, we have occasionally had difficulty cloning short oligonucleotides.

---

[1] R. M. Myers, L. S. Lerman, and T. Maniatis, *Science* 229, 242 (1985).

[2] R. Pine and P. C. Huang, this series, Vol. 154, p. 415.

[3] R. M. Myers, S. G. Fischer, T. Maniatis, and L. S. Lerman, *Nucleic Acids Res.* 13, 3111 (1985).

[4] M. D. Matteucci and H. L. Heyneker, *Nucleic Acids Res.* 11, 3113 (1983).

[5] K. M. Derbyshire, J. J. Salvo, and N. D. F. Grindley, *Gene* 46, 145 (1986).

[6] C. A. Hutchinson, S. K. Nordeen, K. Vogt, and M. H. Edgell, *Proc. Natl. Acad. Sci. U.S.A.* 83, 710 (1986).

[7] D. E. Hill, A. R. Oliphant, and K. Struhl, this series, Vol. 155, p. 558.

This article describes a rapid and simple technique for saturation mutagenesis that combines several modifications of the oligonucleotide-directed mutagenesis method developed by Zoller and Smith to generate specific nucleotide substitutions.[8,9] In the standard method, an oligonucleotide containing a specific mismatch is annealed to the target region cloned into single-stranded phage DNA, forming a partial heteroduplex. The oligonucleotide is then used to prime second strand synthesis *in vitro* to produce double-stranded DNA that is used to transfect *Escherichia coli.* Progeny phage that have incorporated the mutation can be identified by a number of methods, including differential hybridization, restriction site analysis, or DNA sequencing. This procedure has been modified by Kunkel *et al.*[10,11] to increase the efficiency of mutant production. The single-stranded template is prepared from an *E. coli* host harboring mutations in the *dut* and *ung* genes. The absence of the enzymes dUTPase and uracil N-glycosylase results in the production of uracil-substituted phage DNA. After priming of the template by a mismatched oligonucleotide and conversion to the double-stranded form, the DNA is transfected into an *ung*[+] host, which imposes a strong biological selection against the uracil-containing wild-type strand. Progeny phage are therefore much more likely to be derived from the strand synthesized *in vitro,* which contains the mutagenic primer.

The method for oligonucleotide-directed mutagenesis, as it is described above, is not a practical approach for saturation mutagenesis since each mutation requires an individual mutagenic primer. We have adapted this method by replacing the mutagenic primer containing a specific mismatch with a mixture of "doped" mutagenic oligonucleotides produced as described by Derbyshire *et al.*[5] (Fig. 1). Thus, a population of mutant phage is produced which is, in essence, a library of mutant sequences. The mutants are produced with high enough efficiency to be identified by DNA sequencing. A similar approach has been independently described by others.[12-16] In this article, the rationale for this method and a detailed protocol are presented. The saturation mutagenesis of a 16-amino acid

[8] M. J. Zoller and M. Smith, this series, Vol. 100, p. 468.
[9] B. H. Horwitz, A. L. Burkhardt, R. Schlegel, and D. DiMaio, *Mol. Cell. Biol.* **8**, 4071 (1988).
[10] T. A. Kunkel, *Proc. Natl. Acad. Sci. U.S.A.* **82**, 488 (1985).
[11] T. A. Kunkel, J. D. Roberts, and R. A. Zakour, this series, Vol. 154, p. 367.
[12] R. A. Ach and A. M. Weiner, *Mol. Cell. Biol.* **7**, 2070 (1987).
[13] A. Peterson and B. Seed, *Nature (London)* **392**, 842 (1987).
[14] S. S. Ner, D. B. Goodin, and M. Smith, *DNA* **7**, 127 (1988).
[15] P. Hubner, S. Iida, and W. Arber, *Gene* **73**, 319 (1988).
[16] D. D. Loeb, C. A. Hutchison III, M. H. Edgell, W. G. Farmerie, and R. Swanstrom, *J. Virol.* **63**, 111 (1989).

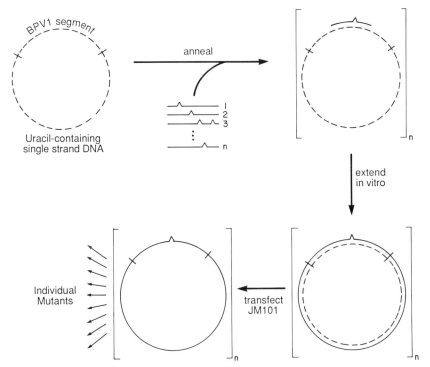

Fig. 1. Mutagenesis procedure. Uracil-containing single-stranded phage DNA was prepared in the appropriate *E. coli* host. A mixture of mutagenic oligonucleotide primers was hybridized to the phage DNA and extended *in vitro* using the Klenow fragment of DNA polymerase and the four dNTPs. Extension products were transfected into *E. coli* strain JM101, which selectively degrades the uracil-containing wild-type strand, and individual phage plaques were isolated and analyzed by DNA sequencing to identify mutations. Dashed lines indicate uracil-containing DNA. Solid lines indicate DNA not substituted with uracil. Indented lines indicate a base not complementary to the wild-type base present in the template DNA. Populations of intermediates that differ only as a consequence of the different mutagenic primers are surrounded by brackets. (Reproduced from Ref. 9.)

region of the bovine papillomavirus type 1 (BPV1) ORF E5 transforming protein is used as an example.

Experimental Rationale

Selection of Target for Mutagenesis

Any region that is contained on a piece of DNA that can be cloned into an M13 phage is a potential target. The target size cannot exceed the maximum length of a synthetic oligonucleotide (currently approximately

100 bases). To minimize aberrant hybridization by the primer, the target should not share a high level of similarity with other sequences present on the molecule.

Production of the Mutagenic Oligonucleotide

To produce a mutagenic primer, an oligonucleotide complementary to the single-stranded target region is designed. However, the pure nucleotide derivatives used in the synthesis are contaminated with a mixture containing all four nucleotides. This leads to the occasional substitution of the wild-type nucleotide at any one position with one of the three other nucleotides. To calculate the proper level of contamination, one must consider the length of the oligonucleotide and the average number of mutations desired per molecule. The distribution of mutations as a function of the level of contamination can be calculated from Eq. (1)[5]

$$P(x) = \frac{N!}{x! \, (N - x)!} \, C_m{}^x \, (1 - C_m)^{N-x} \qquad (1)$$

where $P(x)$ is the probability of x mutations per oligonucleotide, N is the number of bases in the oligonucleotide, and C_m is the total fractional concentration of contaminating nucleotides. The maximum number of single base substitutions is produced when C_m equals $1/N$, the reciprocal of the length of the oligonucleotide. At this level of contamination, approximately 37% of the oligonucleotides in the mixture are predicted to contain single base substitutions, while the remainder would contain no substitutions ($\sim 35\%$) or multiple substitutions. Using similar procedures, it is possible to synthesize mutagenic oligonucleotides containing randomly distributed single base deletions.[17]

Production of Template for Mutagenesis

To produce the appropriate template for mutagenesis, a DNA fragment containing the target sequence is cloned into an M13-based vector. It is not necessary for the target region to be directly flanked by unique restriction sites. Phage from an individual plaque are used to reinfect fresh host cells to produce a high-titer phage stock. This stock is then passed through two rounds of infection of an *E. coli dut⁻ ung⁻* host.[10] There are two *E. coli dut⁻ ung⁻* hosts that are generally used for mutagenesis. *Escherichia coli* strain RZ1032[8] contains an amber suppressor and is therefore appropriate for M13 phages containing amber mutations such as M13mp7 and mp8. Strain CJ236[10] lacks a suppressor but contains an F′ selectable with chlor-

---

[17] S. S. Ner, T. C. Atkinson, and M. Smith, *Nucleic Acids Res.* **17**, 4015 (1989).

amphenicol, which helps to maintain infectability.[11,18] It is therefore a better host for M13mp18 and 19. Single-stranded phage DNA is then isolated[5] from a large volume of phage-containing supernatant since the efficiency of phage production is relatively low in a dut^- ung^- host. We have found that the single-stranded template is the most critical and variable factor in the mutagenesis protocol. Small amounts of double-stranded DNA can reduce the overall efficiency of mutagenesis, as can contaminating RNA species, which may serve as nonspecific primers or as inhibitors of the *in vitro* reactions.

Mutagenesis

The population of phosphorylated mutagenic oligonucleotides is hybridized to uracil-containing template DNA at a molar ratio of between 2 and 10 to 1 (primer to template). Ratios exceeding this are not suggested since an excess of primer can cause hybridization to nontarget sequences, and contaminants in the oligonucleotide preparations can inhibit subsequent transformation. It is also possible that an excess of primer during the annealing reaction could favor hybridization of oligonucleotides that have the wild-type sequence, thus lowering the efficiency of mutagenesis. The primer is elongated to produce double-stranded DNA using either the Klenow fragment of DNA polymerase I or T4 DNA polymerase, and the second strand is covalently closed by the action of T4 DNA ligase.[8,11] Single-stranded phage DNA, double-stranded nicked circular DNA, and covalently closed circular DNA are all well resolved by electrophoresis on agarose gels containing ethidium bromide, allowing the success of the enzymatic procedures to be easily monitored. A fraction of the elongation reaction is transfected into a wild-type *E. coli* host without further purification. Generally, elongation reactions are performed in the presence and absence of the mutagenic primer to assess the amount of endogenous priming in a template preparation. Reactions done in the presence of primer should yield at least 5-fold more plaques than in its absence. However, in our experience, we have readily isolated mutants in situations where this difference was only 2-fold.

Identification and Subcloning of Mutants

Mutations are identified in individual phage isolates by DNA sequencing. Once mutant phage are identified, a DNA fragment containing the target sequence can be subcloned, if necessary, into the appropriate vector for biological assay. To subclone we prefer the method of primed synthe-

---

[18] C. Joyce, personal communication, 1987.

sis[19] in which the M13 universal primer is used in conjunction with DNA polymerase to convert the insert to double-stranded DNA. The fragment can then be excised with the appropriate restriction enzymes and cloned. Further purification to remove vector sequences is not necessary since M13 and its standard derivatives will not give rise to antibiotic-resistant bacterial colonies. In practice, this method of subcloning has been rapid, allowing us to subclone a large number of mutations easily. The fragment subcloned can be considerably larger than the target region, but we subclone the smallest fragment possible to reduce the chances of carrying over spurious mutations induced during the mutagenesis procedure. Finally, as with any mutagenesis procedure, we perform genetic mapping experiments to ensure that the detected mutation is actually responsible for the observed phenotype.

Materials and Methods

The procedures we use in our laboratory are described below. Detailed protocols for many of the individual steps have been published elsewhere, and undoubtedly could be used for this application.[8,10,11,20]

Materials

All enzymes were purchased from New England Biolabs (Beverly, MA) with the exception of SalI, which was from International Biotechnologies, Inc. (New Haven, CT). All reagents are from standard suppliers.

Bacterial Strains

JM101: Δlacpro, supE, thi, F' traD36, proAB, laqI^q ZΔM15
RZ1032: HfrKL16 PO/45 [lysA(61–62)], dut1, ung1, thi1, relA1, Zbd-279::Tn10, SupE44

Synthesis and Phosphorylation of Mutagenic Primer

To produce a 48-nucleotide mutagenic primer specific for the 3' end of BPV1 ORF E5, the sequence complementary to BPV1 nucleotides 3963–4010 was programmed into an Applied Biosystems 380A automated DNA synthesizer. The total volume of each nucleotide substrate that was necessary to complete the synthesis was calculated. Because we wished to generate the maximum number of single base substitutions, we contaminated each nucleotide pool at a fractional level of 1/48, or 2.08%, with the

---

[19] G. Winter, S. Fields, and G. Ratti, *Nucleic Acids Res.* **9**, 6907 (1981).
[20] J. Messing, this series, Vol. 101, p. 20.

improper nucleotides by adding the appropriate volume of a commercially available mixture containing an equimolar amount of all four nucleotides. To account for the fact that the mix contained the wild-type nucleotide in addition to the incorrect ones, we adjusted the volume of the mixture added by 4/3. Therefore, the final concentration of the nucleotide mixture was 2.08% × 4/3 or 2.77%. After synthesis, the oligonucleotide was deprotected and purified by preparative electrophoresis on denaturing polyacrylamide gels, followed by desalting on a Sep-Pak C_{18} column (Waters Associates, Milford, MA).

Fifty picomoles (0.8 μg) of the primer was phosphorylated in 100 mM Tris, pH 8.0, 5 mM dithiothreitol (DTT), 100 μM ATP, 10 mM MgCl$_2$ in a total volume of 30 μl. Incubation with 4.5 U of polynucleotide kinase was for 45 min at 37°, followed by 10 min at 65° to inactive the enzyme.[8] No further purification was required.

Production of Phage Containing the Target Sequence

The *Eco*RI to *Bam*HI fragment of BPV1 DNA was cloned into the polylinker of M13mp8.[21] After transfection into JM101 cells,[20] an individual plaque was transferred into 0.5 ml of Luria broth (LB), vortexed, and incubated at 37° for 15 min with shaking to liberate the phage. The residual bacteria were then inactivated at 65° for 10 min. The titer of this stock was > 10^8 plaque forming units/ml. To produce a high-titer phage stock, 0.25 ml of the primary phage stock was used to infect 10 ml of exponentially growing JM101 cells which were aerated by vigorous shaking. After 8–24 hr the bacteria were removed by centrifugation and the supernatant heat inactivated. The titer of phage in this supernatant was > 10^{11}/ml. Use of phage stocks with titers lower than this is discouraged.

Growing Phage in dut⁻ ung⁻ Host

Bacterial strain RZ1032 was routinely propagated at 30° on LB plates containing tetracycline. One milliliter of LB was inoculated with bacteria from a confluent area of the plate and grown to saturation at 37°. (A confluent area of the plate was used to avoid the possibility that an individual colony had lost its F′ and would therefore be uninfectable. This is not necessary with strain CJ236.) The saturated culture was diluted 1 : 100 in 10 ml of LB and grown for approximately 2 hr at 37° until log phase was reached (as judged by inspection). Uridine was then added to a final concentration of 0.25 μg/ml, and the culture was inoculated with 0.2 ml of the high-titer phage stock. After 8 hr of further incubation with shaking,

[21] D. DiMaio, *J. Virol.* **57**, 475 (1986).

the supernatant was harvested and heat inactivated. It is difficult to determine the phage titer of the supernatant at this stage because the phage do not form distinct plaques on RZ1032 cells. Therefore, we usually spot 10 μl of supernatant containing the phage on a freshly plated mixture of LB top agar and RZ1032 cells, and then inspect the lawn for clearing after incubation for several hours. This phage-containing supernatant (1.2 ml) was then used to infect a fresh 25 ml logarithmically growing culture of RZ1032 cells supplemented with uridine. After 8 hr the supernatant was harvested and single-stranded uracil-substituted DNA was prepared from it as described below.

Isolating Uracil-Substituted Single-Stranded DNA from Phage

Twelve 1.5-ml aliquots of the culture from above were spun in microfuge tubes for 10 min to remove bacteria. Then 1.2 ml of supernatant from the top of each tube was removed with a Pasteur pipet, placed in a new tube, and the centrifugation was repeated. One milliliter from each sample was removed and placed in a fresh tube. To precipitate the phage contained in the clear supernatant, 150 μl of 20% polyethylene glycol (PEG) 8000 (Sigma, St Louis, MO), 2.5 M NaCl was added to each sample. After incubation for 2 hr at room temperature, the tubes were spun for 5 min, which yielded a small white pellet at the bottom of each tube. The supernatant was discarded and the tubes inverted for at least 1 hr to drain. The pellets were then each resuspended in 50 μl of 10 mM Tris, pH 8.0, 1 mM EDTA (TE) and pooled into two tubes. To isolate DNA from the phage particles, an equal volume of buffer (100 mM Tris, pH 8.0, 1 mM EDTA)-saturated phenol was added to the phage, mixed vigorously, and incubated at 37° for 15 min. After centrifugation for 30 sec, the aqueous phase was removed and extracted once with chloroform:isoamyl alcohol (24 : 1, v/v). The phage DNA was precipitated with 0.1 volume 3 M sodium acetate, pH 5.2, and 3 volumes ethanol at −70° for 30 min. The DNA was pelleted by centrifugation for 10 min, the supernatant was removed, and the sides of the tubes were washed with 70% cold ethanol (we usually do not see a pellet at this stage). Each pool was then dissolved in 25 μl of TE. Agarose gel electrophoresis of 2 μl of the preparation yielded a band easily visible by UV illumination after ethidium bromide staining.

Mutagenesis

Two microliters of template, 2 μl of kinased mutagenic primer (diluted 1 : 10 in TE), 2 μl of TM (100 mM Tris, pH 7.4, 100 mM MgCl$_2$), and 14 μl of H$_2$O were combined in a microfuge tube. A parallel reaction was set up without the primer as a control for endogenous priming. To anneal

TABLE I
DISTRIBUTION OF SUBSTITUTIONS PER MOLECULE

| Substitutions per molecule | Percentage | |
	Predicted[a]	Observed
0	68	68
1	18.5	18
2	9.5	12
3 and above	4	2

[a] To calculate these percentages, it was assumed that 50% of the phage would have substitutions distributed as calculated using Eq. (1), and that an additional 50% would be wild type. This correction is based on the observation that the mutation frequency using this procedure with oligonucleotides containing specific mismatches is approximately 50%.[10,25]

the primer, tubes were heated to 55° in a water bath for 5 min and then cooled on the bench top to room temperature. After 5 min, the tubes were centrifuged briefly to remove condensation from the sides. Elongation was initiated by adding 12 μl of chase solution (250 μM of all four deoxyribonucleotide triphosphates in 10 mM Tris, pH 8.0, 1 mM EDTA), 4 μl of TM, 2 μl of 10 mM ATP, 0.4 U of Klenow polymerase, and 2 μl of T4 DNA ligase (6000 Weiss units per milliliter). Incubation was then carried out for 10 min on ice, followed by 10 min at room temperature, and finally 2 hr at 37°.

At the completion of the incubation period, 2 μl of each reaction was compared to single-stranded phage DNA and RF double-stranded DNA by electrophoresis on a 1% agarose gel containing 2 μg/ml ethidium bromide. The reaction with primer resulted in almost complete conversion of the single-stranded template to double-stranded forms, whereas a significant amount of single-stranded template remained in the reaction without the primer.

Competent JM101 cells were transfected with 2 μl of each reaction mix. The reaction with primer yielded several hundred plaques, while the reaction without yielded less than half this number. Individual plaques were picked and, after amplification by a round of infection in JM101 cells, single-stranded DNA was isolated. The region of interest was sequenced by standard chain termination methods.[22] To facilitate identification of mu-

[22] F. Sanger, S. Nicklen, and A. R. Coulson, *Proc. Natl. Acad. Sci. U.S.A.* **74**, 5463 (1977).

TABLE II
SUMMARY OF NUCLEOTIDE SUBSTITUTIONS

From[a]	To			
	C	T	A	G
C (11)	—	5	1	4
T (7)	0	—	4	5
A (19)	2	4	—	1
G (11)	1	3	18	—

[a] Substitutions reflect changes occurring in the DNA strand with the same polarity as the oligonucleotide. Numbers in parentheses indicate the total number of positions occupied by that particular nucleotide in the wild-type target sequence.

tant sequences, the reactions terminating in A for all the mutants were electrophoresed side by side, as were the reactions terminating in G, C, and T.[23] Mutant sequences were rapidly identified by visual inspection of the autoradiograms, since the positions of all bands within a group of reactions were identical except at the positions where mutations occurred.

Subcloning of Mutant Sequences

A small double-stranded DNA fragment containing the mutation was generated by primer extension and subcloned back into a plasmid containing the BPV1 genome. Ten microliters of the mutant single-stranded template used for the sequencing reaction was annealed with 2 μl of the M13 universal primer (2.5 ng/μl) following addition of 1.5 μl of TM. After cooling, 2 μl of chase solution and 0.5 U of Klenow polymerase were added, and incubation was carried out at room temperature for 20 min, followed by inactivation of the enzyme at 75° for 10 min. The reaction volume was increased to 50 μl after addition of 5 μl of the appropriate 10× restriction buffer and several units of BstXI and SalI were added. After 2 hr at 37° the reaction was extracted with an equal volume of phenol, the aqueous phase extracted with ether, and the DNA was precipitated with sodium acetate and ethanol. The fragment was dissolved in 10 μl of TE and cloned without further purification into the appropriate sites in pBPV142-6, which contains the full-length BPV1 genome.[24] DNA isolated

[23] J. D. Parvin, A. Moscona, W. T. Pan, J. M. Leider, and P. Palese, J. Virol. 59, 377 (1986).
[24] N. Sarver, J. C. Byrne, and P. M. Howley, Proc. Natl. Acad. Sci. U.S.A. 79, 7147 (1982).

```
                              M
                              M
                              M              S
                              M              S    S              S
                              M              S    S              S
                              M         M    S    S              S
                              M         M    M    M    M    M    S
M    S    S         M    M    M    M    M    M    M    M
M    M    S    N    M    M    M    M    N    M    M    M    M    S
leu-val-try-trp-asp-his-phe-glu-cys-ser-cys-thr-gly-leu-pro-phe
```

FIG. 2. Distribution of substitution mutations isolated at each amino acid position within the target region. The sequence of the 16-residue target region is shown, and each mutation isolated is indicated by the letter M, N, or S (missense, nonsense, and silent mutations, respectively) above the position where it occurs.

from transformed bacteria was prepared by isopycnic centrifugation in cesium chloride. We have found that sequenase T7 DNA polymerase (U. S. Biochemicals, Cleveland, OH) is occasionally more effective than Klenow polymerase for primed synthesis and subcloning of large inserts.

Results

After mutagenesis with the 48-nucleotide mutagenic primer, 108 independent phage were isolated and sequenced to detect mutations. Thirty four of these, or 31%, contained nucleotide substitutions compared to the wild-type sequence, while 14, or 13%, contained small insertions or deletions. The insertions and deletions often occur on the same molecule as a point mutation, suggesting that they were caused by mispairing of a mutant primer. In the 200- to 300-nucleotide segment containing the target region that was routinely sequenced, one frameshift mutation and no substitution mutations were detected outside of the target region. Within the group of mutants that contained nucleotide substitutions there were 19 with single base substitutions, 13 with double substitutions, and one each containing triple and quadruple substitutions. The distribution of substitution mutations per molecule is close to the theoretical prediction, as shown in Table I.[25] A summary of the substitutions observed is shown in Table II. All possible base substitutions were isolated with the exception of T to C, although we have isolated this particular substitution in other experiments using this method, albeit infrequently.[9] There also is an overrepresentation of G to A transitions, and this has also been observed in other experiments. The reasons for these biases are unclear, but they may reflect differences in coupling efficiencies during synthesis of the mutagenic oligonucleotide or differences in the way certain mismatched nucleotide pairs are repaired during propagation of heteroduplex molecules in bacteria. These biases can

[25] B. H. Horwitz and D. DiMaio, unpublished observation, 1988.

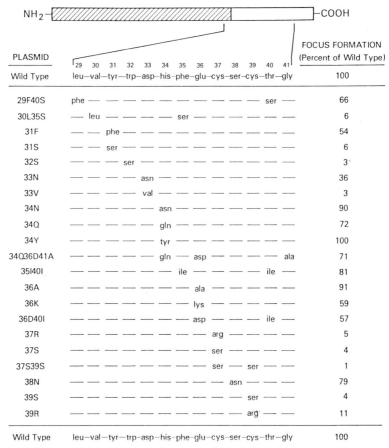

FIG. 3. Focus-forming efficiency of mutants with substitutions in the carboxyl-terminal one-third of the BPV1 E5 protein. The top bar represents the E5 protein. The open area is the carboxyl-terminal portion of the protein. The expanded sequence shows the amino acids targeted for mutagenesis in this region. The leftmost column indicates the name of mutant plasmids. The central section shows the predicted amino acid substitution(s) encoded by the corresponding mutants. The right-hand column indicates the percentage of wild-type focus-forming activity of each mutant plasmid. (Reproduced from Ref. 9.)

be eliminated by incorporation of additional *in vitro* manipulations into the mutagenesis scheme.[15]

The goal of this experiment was to produce point mutations in the carboxyl-terminal region of the BPV1 E5 protein and assay their effects on focus-forming activity. Figure 2 shows the distribution of missense, nonsense, and silent mutations isolated in each codon throughout the target region. Nucleotide substitutions were produced within all targeted codons with the exception of the carboxyl-terminal 2 amino acids. These positions correspond to the 5' end of the mutagenic primer, and it is possible that primers with mutations in their 5' end are displaced by Klenow polymerase on completion of second strand synthesis. It has been reported that T4 DNA polymerase has less strand displacement activity than the Klenow polymerase and may therefore be of use in eliminating this problem.[11]

After sequencing all 108 phage, there were still several mutations that we desired to analyze but had failed to detect. Since in any saturation mutagenesis procedure the chances of isolating any one specific mutation are quite low, we constructed these additional mutations by standard site-directed techniques. Figure 3 shows all the mutants with missense mutations in this region that have been analyzed in detail, and their relative performance in the C127 cell focus-formation assay. As is shown, we have succeeded in isolating and characterizing a diverse set of mutations at almost all of the targeted residues, and the resulting mutants show a wide range of biological phenotypes.

In conclusion, the method of saturation mutagenesis described here has few of the drawbacks of other methods. It is generally applicable since there is no need for the target region to be closely flanked by unique restriction sites. In addition, there is no requirement for specialized vectors, physical enrichment procedures, or difficult enzymatic reactions. Because only a single chemical synthesis is required, it is far less expensive and time consuming than engineering multiple individual mutations with standard oligonucleotide-directed mutagenesis. This method provides site specificity with target sizes up to about 100 nucleotides, relatively high efficiency of mutant isolation, a large proportion of single nucleotide substitutions, and a mutant population of high complexity. Under favorable circumstances, it is possible to use these approaches to generate missense mutations at every amino acid residue of a short protein.[26]

Acknowledgments

We would like to thank Susan J. Baserga, Keith M. Derbyshire, Cathy Joyce, and other members of the DiMaio and Grindley laboratories for introducing us to many of the techniques described in this article, and for useful discussions.

[26] D. D. Loeb, R. Swanstrom, L. Everitt, M. Manchester, S. E. Stamper, and C. A. Hutchison III, *Nature (London)* **340**, 397 (1989).

Author Index

Numbers in parentheses are footnote reference numbers and indicate that an author's work is referred to although the name is not cited in the text.

Evans, R. M., 499
Evans-Blackler, S., 545
Everett, R., 514
Everitt, L., 611
Eytan, E., 390

F

Fahnestock, S. R., 145, 146(17), 217, 222
Falck-Pederson, E., 502
Falco, S. C., 238(20), 241(20), 243(20), 275, 306
Falkenburg, P. E., 390
Falkow, S., 20, 173, 174, 423
Fandl, J. P., 171, 172
Fangman, W. L., 235(24), 271(163), 275, 279
Fara, J. W., 401
Farabaugh, P. J., 66
Fargette, F., 533
Farmerie, W. G., 600
Farnham, P. J., 539
Farquhar, M. G., 587
Farr, A., 380
Fasiolo, F., 423
Fath, M. J., 171
Faure, T., 30, 36(64)
Faust, J. R., 545
Fayet, O., 127
Feder, J., 539
Federoff, H. J., 245(59), 276
Federoff, H., 246(45), 276
Federspeil, N. A., 541
Feigner, D. L., 534
Feinstein, S. I., 118, 119(9)
Felber, B. K., 492, 493, 499(19)
Feldman, J., 235(32), 275
Feldman, L. T., 162
Felmlee, T., 173
Fendrock, B., 540, 550(23)
Fenimore, C. M., 265(133), 266(133), 278, 408, 418(2)
Fennie, C. W., 431, 437(47), 439(47)
Ferber, S., 389, 395
Ferguson, J., 247(60), 276
Fernandez-Meljia, C., 552
Ferrari, E., 200, 222, 225, 227
Ferrari, F. A., 225
Ferretti, J. J., 180

Ferretti, L., 30
Fester, R. J., 192
Fettes, I., 131
Fiechter, A., 33
Field, C., 409
Fields, S., 604
Fiering, S., 558
Fiers, W., 29, 95, 546
Fieschko, J. C., 241(175), 279, 415, 417(44)
Figursky, D., 19
Fiil, N. P., 319
Fiil, N., 171
Fikes, J. D., 171
Filpula, D., 217, 220(9)
Fincham, J. R. S., 400
Findeli, A., 266(109), 267(109), 278
Fineman, R. M., 545
Fink, G. R., 251(140, 141), 263(17), 271(18), 275, 278, 280, 282, 285, 288, 292(1), 294, 335, 347, 355, 418, 474
Fink, G., 314
Finkelstein, D., 273
Finkelstein, R. A., 174, 175
Finley, D., 389, 392(9), 396, 401, 402(23)
Finnie, M. D. A., 400
Firak, T. A., 514
Firestone, G. L., 499
Fischer, S. G., 599
Fischhoff, D. A., 244(61), 276
Fisher, E. F., 503
Fisher, K. E., 217, 222
Fisher, P. A., 236(173), 279
Fitzgerald, M., 503
Fjellstedt, T. A., 282
Flamm, E., 119
Flashner, Y., 49
Fleckenstein, B., 497
Fleckstein, B., 519
Flessel, M. C., 266(54), 276, 412
Flinta, C., 398, 405(5)
Flintoff, W. F., 538
Flores, N., 15, 20(1), 21(1), 22(1), 23(1), 25(1), 26(1), 31
Fogel, S., 320, 425
Fonseca, R., 175
Ford, M., 540
Forrester, W., 322, 426, 429(28), 478
Forstrom, J., 319
Foster, T. J., 21, 22(12)
Foster, T., 145

T

Subject Index

A

11-AAU, for selection and amplification of adenosine deaminase genetic marker, 554–555

N-Acetylgalactosamine, transfer to protein, 577–579

β-N-Acetylglucosaminidase H, 459

N-Acetyl-DL-phenylalanine β-naphthyl ester, overlay test for S. cerevisiae protease B, 374–376

N$^\alpha$-Acetyltransferase, 399, 402–405
actions of, distribution of protein N termini resulting from, 404–405

Actin, D. discoideum, with unblocked N-terminus, synthesis in cell-free system, 406

β-Actin gene, cellular enhancer, 522

Adenosine deaminase
amplification of transfected genes using, 545, 552–555
mammalian-cell expression system for, 509–511

Adenovirus
E1A gene products, suppression of SV40 enhancer and polyoma enhancer, 516
as transcriptional enhancer, 519

Affinity handles
gene fusion systems using, 131, 140–143
protein A as, 145–146

Ala^{64}-subtilisin, site-specific cleavage of fusion proteins, 139–140

Aminopeptidase. See Saccharomyces cerevisiae, aminopeptidase I

Amino terminus. See also Cotranslational amino-terminal processing
effect on function and stability, 399–403
fusion, in gene fusion systems, 133–135
of protein from B. subtilis, analysis of, by radiolabeling, 221–222
of proteins from E. coli expression system, 13
structure

manipulation, strategies for, 403–407
selection, 404–406

Ampicillin, in E. coli expression system, 22, 65–66, 80–82

Amplificator phenotype, 539

Amylase, mouse, secretion from yeast, overlay assay, 472

α-Amylase, wheat, expression in yeast, 329

Angiotensin, production, as fusion protein in E. coli, 131

Antibiotic resistance genes
rationale for use of, 20–21
in T7 expression systems, 65–69

Antibiotics, in E. coli expression system, 21
mode of action, 22–23
mode of bacterial resistance, 22–23
working concentrations, 22–23

Antibody production
expression system for preparation of proteins for, 4
using protein A fusions, 147–148

Antisense promoters, in E. coli T7 expression system, 88

Anti-Shine–Dalgarno sequence, 104–105

Antitermination control
pause sites in, 45
in plasmid vectors for E. coli expression system, 17, 26
trans-acting elements, 17, 26

α$_1$-Antitrypsin, expression in yeast, 341–342, 348–349, 439

APE. See N-Acetyl-DL-phenylalanine β-naphthyl ester

Asoaragine synthase system, for amplification of transfected genes, 545, 557–558

B

Bacillus subtilis
expression systems, 5–6, 199–201
advantages of, 199
dual-plasmid repressible system for, 208–213

transformed cells, storage of, 35, 78
trp attenuator, 44
trp promoter, 25
  in direct expression of proteins, 54–60
  functional features, 54
  induction, 57–59
  induction ratio, 54–55
Exo-$\alpha1\rightarrow2$-mannosidase
  assay, 463
  fungal, 461
  properties, 464
  purification, 463–464
Export, definition of, for *E. coli* expression
  system, 167
Expression plasmid vectors. *See* Plasmid
  vectors
Expression system, selection of, and purpose
  of expressed protein, 4

F

Fab proteins, mouse–human, secretion in
  E. coli, 184–186
Factor Xa, site-specific cleavages of fusion
  proteins, 139, 406
Fatty acid acylation, analysis of, 593–594
F9 EC cells
  polyoma virus enhancer in, 516
  SV40 virus enhancer in, 515
α-Fetoprotein gene, enhancer elements, 523
Fingers, of mammalian proteins, 129
Flag peptide, gene fusion system used to
  facilitate purification of, 138
Fluorescence-activated cell sorting, 570
  of mammalian cells, amplification of
  transfected DNA by, 558
Formic acid, site-specific cleavages of fusion
  proteins, 139–140
Fusion proteins, 15, 17, 31, 59, 129. *See*
  also Gene fusions
  advantages of, 161
  affinity purification of, 136–137,
  141–143
  antibodies raised against, 143
  degradation, protection against, 128
  expression strategies, 3, 5
  for immunization and diagnostics, 143
  production, in yeast, 431–433
  site-specific cleavage of, 136–140, 149,
  405

solubility, 131–132
stability, 120
in yeast expression systems, to minimize
  ubiquitin-dependent degradation, 395

G

G418. *See* Geneticin
β-Galactosidase
  assays, 356–357
  expression, in yeasts transformed with
  temperature-regulated plasmids,
  362–366
  induction, in *B. subtilis*, with IPTG,
  226–228
  purification of, gene fusion system used to
  facilitate, 138, 140–141, 143
gal operon, expression, control of, 48
Gene, amplified
  stability of, 539–540
  variable size and structure of, 540–541
Gene amplification, 537–538. *See also*
  Mammalian cells, amplification of
  transfected gene
  characteristics of, 538–541
  in drug-resistant mammalian cells,
  544–545
  and gene expression, 537–538
  by stepwise selection, 538–539
Gene expression
  control of, gene-specific mechanisms,
  48–51
  global control mechanisms, 51–54
  in heterologous cells. *See also specific*
  expression system
  systems for, 3–7
  types of protein expressed, 3–4
  modulation by genome position, 53–54,
  57
Gene fusions, 129–143. *See also* Fusion
  proteins
  cII, 131, 141
  examples of systems, 140–142
  for expression of recombinant proteins,
  129–130
  hybrid, construction of, for *B. subtilis*
  secretion system, 217
  product, localization of, 132–133
  protein A. *See* Protein A